Modern Manufacturing Processes

Modern Manufacturing Processes

Edited by
Muammer Koç and Tuğrul Özel

WILEY

This edition first published 2020

Registered Office
John Wiley & Sons, Inc., 111 River Street, Hoboken, NJ 07030, USA

Editorial Office
111 River Street, Hoboken, NJ 07030, USA

For details of our global editorial offices, customer services, and more information about Wiley products visit us at www.wiley.com.

Wiley also publishes its books in a variety of electronic formats and by print-on-demand. Some content that appears in standard print versions of this book may not be available in other formats.

Library of Congress Cataloging-in-Publication Data
Names: Koç, Muammer, 1968- editor. | Özel, Tuğrul, 1967- editor.
Title: Modern manufacturing processes / edited by Muammer Koç, Tuğrul Özel.
Description: Hoboken, NJ, USA : John Wiley & Sons, [2020] | Includes bibliographical references and index. |
Identifiers: LCCN 2019000636 (print) | LCCN 2019001369 (ebook) | ISBN 9781119120674 (Adobe PDF) | ISBN 9781119120391 (ePub) | ISBN 9781118071922 (hardcover)
Subjects: LCSH: Manufacturing processes.
Classification: LCC TS183 (ebook) | LCC TS183 .M616 2019 (print) | DDC 670–dc23
LC record available at https://lccn.loc.gov/2019000636

Cover Design: Wiley
Cover Image: © Baloncici/Getty Images

Set in 10/12pt WarnockPro by SPi Global, Chennai, India

Printed in United States of America

V10012698_080219

Contents

Foreword *xvii*
List of Contributors *xix*

Part I Advanced Forming Processes 1

1 **Advances in Stamping** *3*
Ilyas Kacar and Fahrettin Ozturk
1.1 Introduction *3*
1.1.1 Blankholder and Drawbeads *3*
1.1.2 The Drawbead Restraining Force (DBRF) *6*
1.1.3 The Use of Active Drawbeads *7*
1.1.4 Control Techniques of Active Drawbeads *8*
1.1.5 Variable Blankholder Force (VBHF) *10*
1.1.6 Flexibility *11*
1.1.7 Future Challenges *12*
References *13*

2 **Hydroforming** *15*
C Hartl
2.1 Introduction *15*
2.2 Fundamentals *16*
2.2.1 Classification of Hydroforming Processes *16*
2.2.1.1 Tube Hydroforming *17*
2.2.1.2 Sheet Hydroforming *17*
2.2.1.3 Integrated Operations in Hydroforming Processes *19*
2.2.2 Process Parameters *23*
2.2.2.1 Tube Hydroforming *24*
2.2.2.2 Sheet Hydroforming *26*
2.2.3 Forming Limits *29*
2.2.3.1 Necking and Bursting *29*
2.2.3.2 Buckling and Wrinkling *30*
2.2.4 Tribology *31*
2.3 Process Development and Design *33*
2.3.1 Part Design *33*
2.3.2 Semifinished Products *35*
2.4 Hydroforming Systems *37*
2.4.1 Forming Tools *37*

2.4.2 Machine Systems 38
2.5 Concluding Remarks 39
References 40

3 Incremental Sheet Forming 47
Rogelio Pérez-Santiago, Isabel Bagudanch, and Maria Luisa Garcia-Romeu
3.1 Incremental Sheet Forming: General Overview 47
3.2 ISF Variants 49
3.2.1 Single Point Incremental Forming (SPIF) 49
3.2.2 Two Point Incremental Forming (TPIF) 49
3.2.3 Other ISF Variants 50
3.2.3.1 Warm ISF 50
3.2.3.2 Stretch Forming with ISF 50
3.2.3.3 Roboforming 51
3.3 Process Cycle 51
3.4 Materials 52
3.5 Formability in ISF 52
3.5.1 FLD 52
3.5.2 ISF Deformation Mechanisms 53
3.5.3 Forces 54
3.6 ISF Process Parameters 55
3.7 Accuracy 55
3.8 Simulation 57
3.9 Future Trends in ISF 58
3.9.1 Application Fields 58
3.9.2 Materials 58
3.9.3 Sustainability 59
3.10 Case Study 59
3.11 Concluding Remarks 59
References 60

4 Powder Forming 65
Rahmi Ünal
4.1 Introduction 65
4.2 Reasons for Using PM Route 67
4.3 Powder Production 69
4.3.1 Mechanical Production Techniques 71
4.3.2 Atomization 72
4.3.3 Chemical Fabrication Techniques 72
4.3.4 Precipitation from Solution 72
4.4 Consolidation Techniques 73
4.4.1 Cold Powder Compaction 74
4.4.2 Hot Pressing 79
4.4.3 Powder Forging 79
4.5 Sintering 79
4.6 Powder Injection Molding (PIM) 82
4.7 Summary and Future Work 84
References 85

5 **Injection Molding at Multiscales** *89*
Danyang Zhao, Minjie Wang, and Donggang Yao
5.1 Introduction *89*
5.2 Overview of Injection Molding *91*
5.2.1 Overview of the Injection Molding Process *91*
5.2.1.1 Injection Molding Machine and Cycle *91*
5.2.1.2 Basic Structure of an Injection Mold *92*
5.2.2 Developments of Injection Molding *93*
5.2.2.1 Materials *93*
5.2.2.2 Equipment *94*
5.2.2.3 Processing *95*
5.2.2.4 Mold Manufacturing Technology *103*
5.3 Injection Molding of Precision Parts *105*
5.3.1 Precision Parts *105*
5.3.2 Injection Compression Molding *106*
5.3.2.1 Overview of Injection Compression Molding *106*
5.3.2.2 Mold Base for Injection Compression Molding *107*
5.3.3 Processing Parameters *107*
5.4 Injection Molding of Thin Wall Parts *109*
5.4.1 Scalable Filling *109*
5.4.2 Rapid Thermal Response Molding *110*
5.4.3 Effect of Processing Parameters *112*
5.5 Injection Molding of Microstructured Parts *116*
5.5.1 Materials *116*
5.5.2 Mold Manufacturing Technologies *116*
5.5.2.1 Mechanical Micromachining *117*
5.5.2.2 Electroforming *118*
5.5.2.3 Electron Beam Lithography (EBL) *119*
5.5.2.4 Deep Reactive-ion Etching (DRIE) *119*
5.5.2.5 Combination of Mechanical Micromachining and LIGA *120*
5.5.3 Molding Machine and Mold Design *121*
5.5.4 Process Parameters *121*
5.5.4.1 Effect of Process Parameters *121*
5.5.4.2 Design of Experiments Approach *122*
5.5.4.3 In-process Monitoring of Process Parameters *122*
5.5.4.4 Multiple Quality Optimization of Process Parameters *122*
5.6 Injection Molding of Microparts *124*
5.6.1 CO_2-assisted Plasticization *125*
5.6.2 X-melt Process *126*
5.6.3 Process Study on Micropart Injection Molding *126*
5.7 Simulation of Injection Molding *127*
5.7.1 Standard Injection Molding Simulation *128*
5.7.1.1 3D Flow Model *128*
5.7.1.2 Simulation Example *129*
5.7.2 Challenges for Simulating Miniature Injection Molding Processes *129*
5.7.3 Simulation of Miniature Injection Molding *130*
5.8 Summary and Outlook *131*
References *132*

6 Manufacturing Techniques of Bulk Metallic Glasses *137*
Mustafa Bakkal, Umut Karagüzel, and Ali T. Kuzu
6.1 Introduction *137*
6.2 Mechanical Properties and Usage of Bulk Metallic Glasses *139*
6.3 Rapid Quenching Methods *140*
6.4 Water-Quenching Method *141*
6.5 Arc Melting Drop/Suction Casting Method *142*
6.6 High-Pressure Die Casting Method *143*
6.7 Copper Mold Casting Method *144*
6.8 Cap Casting Method *144*
6.9 Centrifugal Casting Method *145*
6.10 Metal Foaming Method *146*
6.11 Concluding Remarks *147*
References *147*

7 Micromanufacturing *149*
Ömer N. Cora and Muammer Koç
7.1 Introduction *149*
7.1.1 Micromanufacturing Market Projections *149*
7.1.2 Trends in Micromanufacturing *150*
7.2 Classification of Micromanufacturing Processes *150*
7.2.1 Classification Based on Manufacturing Approach (Top-down vs. Bottom-up) *150*
7.2.2 Classification Based on Lithography Technique Involvement *151*
7.2.3 Brinksmeier's Classification *152*
7.2.4 Classification Based on the Effect of Process *152*
7.3 Micromanufacturing Processes *154*
7.3.1 Mechanical Processes *155*
7.3.1.1 Micromachining Processes *155*
7.3.1.2 Micromilling *155*
7.3.1.3 Microdrilling *156*
7.3.1.4 Microturning *156*
7.3.1.5 Microgrinding *157*
7.3.1.6 Micropolishing *158*
7.3.1.7 Micro Abrasive Waterjet Machining (μ-AWM) *159*
7.3.2 Microforming Processes *161*
7.3.2.1 Microforging *161*
7.3.2.2 Micro Extrusion *162*
7.3.2.3 Progressive Microforming *163*
7.3.2.4 Micro Deep-drawing *163*
7.3.2.5 Micro Stamping *164*
7.3.2.6 Micro Hydroforming *165*
7.3.2.7 Micro Injection Molding *166*
7.3.2.8 Powder Compaction *167*
7.3.3 Electrical/Chemical Micromanufacturing Processes *168*
7.3.3.1 Micro Wire Electro Discharge Machining (Micro WEDM, μ-WEDM) *168*
7.3.3.2 Electro-chemical Micromachining (ECM) *169*
7.3.4 Electron Beam-Based Micromanufacturing Processes *170*
7.3.4.1 Electron Beam Drilling *172*
7.3.4.2 Electron Beam Welding *172*
7.3.5 Laser-assisted Microforming *173*

7.3.5.1 Laser Ablation *175*
7.3.5.2 Laser Cutting *175*
7.3.5.3 Laser Micro Welding *176*
7.3.5.4 Laser Microwelding of Polymers *177*
7.3.5.5 Laser Drilling *178*
7.3.5.6 Laser Etching *178*
7.3.5.7 Other Laser-Assisted Microforming Processes *178*
References *179*

Part II Thermal and Energy-assisted Manufacturing Processes 185

8 Warm Stamping *187*
Fahrettin Ozturk, Serkan Toros, and Ilyas Kacar
8.1 What Is Stamping? *187*
8.2 Benefits and Usage Areas of Warm Stamping *187*
8.3 Warm Stamping and Recent Developments *188*
8.4 Effects of Temperature on Strain Hardening for Warm Stamping *194*
8.5 Interrelation of Temperature and Strain Rate *196*
8.6 Effect of Temperature and Deformation on Elasticity Modulus *198*
8.7 Effect of Temperature on Springback *201*
8.8 Effect of Temperature on Forming Limit Diagrams (FLD) *204*
8.9 Analyze Techniques on Formability at Warm Stamping *205*
8.10 The Effects of Lubrication *215*
8.11 Future Directions *215*
References *216*

9 Warm Hydroforming *219*
Muammer Koç, Ömer N. Cora, Hüseyin S. Halkacı, and Mevlüt Türköz
9.1 Introduction *219*
9.2 Warm Sheet Hydroforming *220*
9.2.1 Implementation of Warm Sheet Hydroforming *220*
9.2.2 Parameters in Warm Sheet Hydroforming Process *223*
9.2.3 Studies on Warm Sheet Hydroforming *228*
9.2.4 Studies on Material Characterization *229*
9.2.5 Studies Conducted on the Optimization of the Parameters *229*
9.2.6 Warm Sheet Hydroforming with Die *229*
9.3 Warm Hydromechanical Deep Drawing *230*
9.3.1 Current/Future Trends in Warm Sheet Hydroforming *230*
9.4 Warm Tube Hydroforming *231*
9.4.1 Implementation of Warm tube Hydroforming Process *231*
9.4.2 Parameters in Warm Tube Hydroforming Process *232*
9.4.3 Studies on Warm Tube Hydroforming *232*
References *237*

10 Hot Stamping *239*
Fahrettin Ozturk, Ilyas Kacar, and Muammer Koç
10.1 Introduction *239*
10.2 Process Description and Motivation *240*
10.3 Why Hot Stamping? *241*

10.4 Automotive Parts by Hot Stamping and Potentials *241*
10.5 Advantages and Disadvantages *243*
10.6 Process Description and Methods *245*
10.6.1 Hot Stamping Process Types *245*
10.6.2 Materials for Hot Stamping *247*
10.6.3 Coating and Lubrication for Hot Stamping *248*
10.6.4 Tools for Hot Stamping *251*
10.6.5 Heating for Hot Stamping *251*
10.6.6 Conventional Chamber Furnace (Radiation) *252*
10.6.7 Conduction Heating *252*
10.6.8 Induction Heating *253*
10.7 Cooling for Hot Stamping *254*
10.8 Process Control *255*
10.9 Modeling and Analysis *255*
10.10 Design and Optimization in Hot Stamping *256*
10.11 FEA in Hot Stamping *257*
10.12 Research and Development Trends and Needs *258*
References *262*

11 High-Speed Forming (Electromagnetic, Electrohydraulic, and Explosive Forming) *265*
Brad Kinsey and Yannis Korkolis
11.1 Introduction *265*
11.1.1 Advantages *266*
11.1.2 Disadvantages *267*
11.1.3 Process Variations Mentioned Briefly *267*
11.2 Electromagnetic Forming and Magnetic Pulsed Welding *267*
11.2.1 Methodology/Basics *267*
11.2.2 Generation of the Current Pulse *270*
11.2.3 Coil (and Field Shaper Where Applicable) *271*
11.2.4 Optional Female Forming Die *273*
11.2.5 Typical Applications and Equipment *273*
11.2.6 Other Applications *274*
11.3 Electrohydraulic Forming *274*
11.3.1 Methodology/Basics *274*
11.3.2 Generating the Pressure Pulse and the Exploding Wire Technique *277*
11.3.3 The Pressure Chamber *278*
11.3.4 The Wave Shaper *278*
11.3.5 Typical Applications and Equipment *279*
11.4 Explosive Forming *279*
11.4.1 Methodology/Basics *279*
11.4.2 Setup and Equipment *281*
11.4.3 Typical Applications *282*
11.5 Emerging Technologies *282*
11.6 Metrology and Measurements *284*
11.7 Material Characterization *286*
11.8 Modeling of High-Speed Forming Processes *288*
11.9 Summary and Future Work *291*
References *292*

Part III Advanced Material Removal Processes 295

12 High-Speed Machining *297*
Elisa Vázquez and Guillem Quintana
12.1 High-Speed Machining Overview *297*
12.2 High-Speed Machining Processes and Capabilities *298*
12.3 Machine Tools for High-Speed Machining *298*
12.4 Tools for High-Speed Machining *300*
12.4.1 Cutting Tool Materials *300*
12.4.2 Coating Application and Materials *302*
12.4.3 Tool Microgeometry *302*
12.4.4 Tool Wear and Failure *304*
12.5 High-Speed Machining Applications and Future Trends *305*
References *306*

13 Hard Machining *309*
Durul Ulutan and Tuğrul Özel
13.1 Introduction *309*
13.1.1 Hard Machining Operations *309*
13.1.2 Characteristics of Hard Machining *310*
13.2 Mechanics of Hard Machining *312*
13.3 Cutting Tools *313*
13.3.1 Cutting Tool Materials *313*
13.3.2 Coating Application and Materials *315*
13.3.3 Tool Microgeometry *315*
13.3.4 Tool Wear and Failure *316*
13.4 Surface Quality and Integrity *316*
13.4.1 Surface Defects *317*
13.4.2 Microstructural Alterations *317*
13.4.3 Microhardness *318*
13.4.4 Surface Roughness *318*
13.4.5 Residual Stresses *319*
13.5 Summary and Conclusions *320*
References *320*

14 Advances in Material Modeling for Manufacturing Analysis and Simulation (Deformation and Cutting Processes) *323*
Elisabetta Ceretti, Claudio Giardini, and Antonio Fiorentino
14.1 Introduction on Material Characterization and Modeling *323*
14.2 Material Models and Applications *324*
14.2.1 Johnson–Cook Model *325*
14.2.2 Oxley Model *325*
14.2.3 Maekawa Model *325*
14.2.4 El-Magd Model *326*
14.3 Failure Models *327*
14.3.1 Prediction of Ductile Fracture *327*
14.3.2 Ductile Fracture Criteria *328*
14.3.2.1 Effective Strain *328*

14.3.2.2 Cockroft and Latham *328*
14.3.2.3 McClintock *329*
14.3.2.4 Other Criteria *330*
14.3.3 Critical Damage Value *331*
14.4 Modeling of Contact, Friction, and Wear *331*
14.4.1 Coulomb Friction Model *332*
14.4.2 Schley Friction Model or Shear Model *332*
14.4.3 Parameters Influencing Friction *334*
14.4.3.1 Process Parameters *334*
14.4.3.2 Workpiece and Tool Materials *334*
14.4.3.3 Lubricants *334*
14.4.4 Lubricant *334*
14.4.5 Tool Wear *335*
14.4.5.1 Abrasive and Adhesive Tool Wear Model *336*
14.4.5.2 Diffusion Tool Wear Model *337*
14.4.6 Applications of Material Modeling in 2D and 3D Simulations *337*
14.4.6.1 Orthogonal Cutting *337*
14.4.6.2 Blanking *338*
14.4.6.3 Forward Extrusion *339*
14.4.6.4 Die Wear in 2D Stamping FEM Model *340*
14.4.6.5 Micro Processes and Simulation *341*
References *347*

15 Advanced Grinding *351*
Taghi Tawakoli and Amir Daneshi
15.1 Introduction *351*
15.2 Grinding Wheels *351*
15.2.1 Conventional Wheels *352*
15.2.2 Superabrasive Wheels *353*
15.3 Bond Materials *353*
15.4 Grinding Wheel Conditioning *354*
15.4.1 Profiling, Sharpening, and Cleaning of Grinding Wheels *355*
15.4.2 Grinding Wheel Structuring *361*
15.5 Grinding Force and Energy *363*
15.6 Thermal Damages in Grinding *363*
15.7 Environmentally Friendly Grinding *364*
15.8 High-efficiency Deep Grinding (HEDG) *367*
15.9 Ultrasonic-Assisted Grinding (UAG) *367*
15.10 Ultrasonic-Assisted Dressing *371*
References *373*

16 Electro-Discharge Machining (EDM) *377*
Muhammad P. Jahan
16.1 Introduction *377*
16.2 Principle of the EDM Process *378*
16.2.1 Physics *378*
16.2.2 Material Removal Mechanism *378*
16.3 EDM System Components *379*

16.3.1 Pulse Generators or Discharging Unit *380*
16.3.1.1 RC-Type Pulse Generator *380*
16.3.1.2 Transistor-Type Pulse Generator *382*
16.3.2 Servo Control System or Gap Control Unit *382*
16.3.3 Dielectric Circulation System or Flushing Unit *383*
16.4 Analysis of the Pulses Used in the EDM Process *383*
16.5 Brief Overview of the EDM Parameters *384*
16.6 EDM Variants: Working Principles and Application Examples *385*
16.6.1 Die-sinking EDM *385*
16.6.2 Wire EDM *386*
16.6.3 EDM Milling *388*
16.6.4 EDM Drilling *388*
16.6.5 Rotary Disc Electrode Electrical Discharge Machining (RDE-EDM) *389*
16.6.6 Dry or Near-Dry EDM Milling *390*
16.6.7 Planetary EDM *391*
16.6.8 Reverse EDM *392*
16.7 Examples of Research Advances in EDM and Micro-EDM *393*
16.7.1 Abrasive Electro-Discharge Grinding (AEDG) *394*
16.7.2 Abrasive Wire EDM (AWEDM) *394*
16.7.3 Powder-Mixed EDM and Micro-EDM *396*
16.7.4 Vibration-Assisted EDM and Micro-EDM *396*
16.7.5 EDM/Micro-EDM and ECM-Combined Process *397*
16.7.6 Magnetic Field–Assisted EDM/Micro-EDM *399*
16.7.7 Laser-Assisted EDM and Micro-EDM *400*
16.8 Research Focus Toward Micro- and Nano-EDM *402*
16.9 Summary *403*
References *404*

17 MicroMilling Operations *411*
Simon S. Park, Martin B.G. Jun, and Gerardo García
17.1 Introduction *411*
17.2 Machine Tools for Micromilling *413*
17.2.1 Spindles *414*
17.2.2 Drives and Guides *415*
17.2.3 Ultraprecision Machine Tools *415*
17.2.4 Coolant and Lubricant Delivery *416*
17.2.5 Micro-end Mills *418*
17.2.6 Measurements and Monitoring *419*
17.2.7 Burr Removal *420*
17.3 Micromilling Forces *420*
17.3.1 Minimum Uncut Chip Thickness (MUCT) *421*
17.3.2 Elastic Recovery *422*
17.3.3 Mechanistic Micromilling Force Modeling *424*
17.3.3.1 Shearing Dominant Regime *424*
17.3.3.2 Ploughing Dominant Regime *424*
17.4 Tool Tip Dynamics *427*
17.5 Summary *430*
References *431*

18 Laser Machining *427*
Dani Teixidor, Inés Ferrer, Luis Criales, and Tuğrul Özel
18.1 Introduction *435*
18.2 Laser–Material Interaction *437*
18.2.1 Laser Ablation *437*
18.3 Laser Processing of Materials *438*
18.3.1 Laser Processing of Metals And Alloys *439*
18.3.2 Laser Processing of Polymers *439*
18.3.2.1 Laser Processing of Composites *440*
18.3.3 Laser Processing of Ceramics *440*
18.3.3.1 CVD Diamond *441*
18.3.3.2 Silicon *441*
18.3.3.3 Glass *441*
18.4 Laser-Processing Parameters *442*
18.4.1 Pulse Duration *442*
18.4.2 Pulse Repetition Rate *443*
18.4.3 Wavelength *443*
18.4.4 Beam Quality *443*
18.4.5 Laser Power *444*
18.4.5.1 Pulse Energy *444*
18.4.5.2 Fluence *445*
18.4.5.3 Peak Power *445*
18.4.6 Pulse Overlap *445*
18.5 Laser Drilling *445*
18.5.1 Laser Drilling Without Relative Motion Between Laser Spot and Workpiece *446*
18.5.1.1 Single-Pulse Drilling *446*
18.5.1.2 Percussion Drilling *446*
18.5.2 Laser Drilling With Relative Motion Between Laser Spot and Workpiece *447*
18.5.2.1 Trepanning Drilling *447*
18.5.2.2 Helical Drilling *448*
18.6 Laser Cutting *448*
18.6.1 Melt Cutting *448*
18.6.2 Laser Ablation Cutting *449*
18.7 Laser Milling *450*
18.7.1 Single-Shot Laser Milling *451*
18.7.2 Single-Pass Laser Milling (Laser Scribing) *451*
18.7.3 MultiPass Laser Milling (3D Laser Milling) *451*
18.8 Concluding Remarks *452*
References *453*

19 Laser-assisted Machining Operations *459*
Eneko Ukar, Ivan Tabernero, Silvia Martínez, Aitzol Lamikiz, and Asier Fernández
19.1 Introduction *459*
19.2 Heat-assisted Processes *460*
19.2.1 Plasma-assisted Machining *460*
19.2.2 Laser-assisted Machining (LAM) *462*
19.2.3 Direct Laser Machining *466*
19.3 Analysis of LAM Processes *470*
19.3.1 Energy Source *471*

19.4 Laser-assisted Applications *474*
19.5 Conclusions *477*
References *478*

20 Selective Laser Sintering *481*
Jordi Delgado, Lidia Serenó, Karla Monroy, and Joaquim Ciurana
20.1 General Overview *481*
20.1.1 Selective Laser Sintering Process *481*
20.1.2 Lasers *483*
20.2 Mechanisms *483*
20.2.1 Indirect Selective Laser Sintering *483*
20.2.2 Direct Selective Laser Sintering *484*
20.2.2.1 Solid-State Sintering (SSS) *484*
20.2.2.2 Liquid-phase Sintering (LPS) *484*
20.2.2.3 Selective Laser Melting (SLM) *485*
20.3 Process Parameters *486*
20.3.1 Geometric Parameters *487*
20.3.1.1 Hatch Spacing *487*
20.3.1.2 Scanning Strategy *487*
20.3.1.3 Layer Thickness *488*
20.3.2 Laser Parameters *488*
20.3.2.1 Scanning Speed *489*
20.3.2.2 Laser Power *489*
20.3.2.3 Beam Size *489*
20.3.3 Material Properties *489*
20.4 Materials *490*
20.4.1 SLS of Polymers *491*
20.4.2 SLS of Reinforced Polymers *491*
20.4.3 SLS of Metals and Ceramics *492*
20.4.3.1 Stainless Steel *492*
20.4.3.2 Aluminum *492*
20.4.3.3 Cobalt Chrome *493*
20.4.3.4 Nickel Chrome *493*
20.4.3.5 Steel *493*
20.4.3.6 Titanium *493*
20.5 Capabilities and Limitations *494*
References *496*

Index *501*

Foreword

It is our pleasure to share this book on Modern Manufacturing Processes with prospective readers of various background and purposes. The book project has taken longer than we expected due to various unexpected reasons and issues, but with the strong will and continuous support of the contributing colleagues, friends, and authors, we have finalized it.

We believe that this book will bring a new face and approach to the knowledge-body, learning and teaching of the manufacturing science, technology, and art. It contains 20 distinct chapters focusing on various, quite broad, and novel topics and methods of manufacturing processes ranging from unconventional deformation process technologies to innovative material-removal processes and from tooling-based to tool-less rapid manufacturing technologies. It differs from traditional manufacturing books mainly in that it only focuses on nontraditional, emerging, and innovative manufacturing process technologies, some of which have been used only recently in mass production, and some are still viable choices for small volume and high-end production cases.

We would like to take this opportunity to present our gratitude and thanks to all contributing authors and their co-workers for their commitment and continuous updates on the content of their chapters. We also acknowledge the support of our publishing staff members for their facilitating and encouragements.

June 2019 *Muammer Koç and Tuğrul Özel*

List of Contributors

Isabel Bagudanch
Departament d'Enginyeria Mecànica i de la Construcció Industrial
EPS – Universitat de Girona
c/M. Aurèlia Capmany
Girona 17003, Spain

Mustafa Bakkal
Faculty of Mechanical Engineering
Istanbul Technical University
Istanbul, Turkey

Elisabetta Ceretti
Department of Mechanical and Industrial Engineering
University of Brescia
Via Branze 38, 25123 Brescia, Italy

Joaquim Ciurana
Department of Mechanical Engineering and Industrial Construction, Product Process and Production Engineering Research Group
School of Engineering, University of Girona
Girona, Spain

Ömer N. Cora
Department of Mechanical Engineering
Karadeniz Technical University
Trabzon, Turkey

Luis Criales
Manufacturing and Automation Research Laboratory, Department of Industrial and Systems Engineering
School of Engineering, Rutgers University
Piscataway, NJ 08854, USA

Amir Daneshi
Institute of Grinding and Precision Technology (KSF)
Hochschule Furtwangen University
Furtwangen im Schwarzwald, Germany

Jordi Delgado
Department of Mechanical Engineering and Industrial Construction, Product Process and Production Engineering Research Group
School of Engineering, University of Girona
Girona, Spain

Asier Fernández
Department of Mechanical Engineering
University of Basque Country UPV/EHU
Bilbao, Spain

Inés Ferrer
Department of Mechanical Engineering and Industrial Construction
University of Girona
Girona, Spain

Antonio Fiorentino
Department of Mechanical and Industrial Engineering
University of Brescia
Via Branze 38, 25123 Brescia, Italy

Gerardo García
Center for Innovation in Design and Technology
Tecnológico de Monterrey
Monterrey, México

Maria Luisa Garcia-Romeu
Departament d'Enginyeria Mecànica i de la Construcció Industrial
EPS – Universitat de Girona
c/M. Aurèlia Capmany
Girona 17003, Spain

Claudio Giardini
Department of Engineering
University of Bergamo
Via Pasubio 7b
24044 Dalmine (Bg), Italy

Hüseyin S. Halkacı
Department of Mechanical Engineering
Konya Technical University
Konya, Turkey

MK-C Hartl
Koeln University
Cologne, Germany

Muhammad P. Jahan
Department of Mechanical and Manufacturing Engineering
Miami University
Oxford, OH 45056, USA

Martin B.G. Jun
School of Mechanical Engineering
Purdue University
585 Purdue Mall, West Lafayette
IN 47907, USA

Ilyas Kacar
Department of Mechatronics Engineering
Nigde Omer Halisdemir University
Nigde, Turkey

Umut Karagüzel
Mechanical Engineering Department
Isik University
Istanbul, Turkey

Brad Kinsey

Muammer Koç
College of Science and Engineering, Divison of Sustainable Development
Hamad Bin Khalifa University
Doha, Qatar

Yannis Korkolis

Ali T. Kuzu
Yeditepe University
Istanbul, Turkey

Aitzol Lamikiz
Department of Mechanical Engineering
University of Basque Country UPV/EHU
Bilbao, Spain

Silvia Martínez
Department of Mechanical Engineering
University of Basque Country UPV/EHU
Bilbao, Spain

Karla Monroy
Department of Mechanical Engineering and Industrial Construction, Product Process and Production Engineering Research Group
School of Engineering, University of Girona
Girona, Spain

Tuğrul Özel
Manufacturing and Automation Research Laboratory, Department of Industrial and Systems Engineering
School of Engineering, Rutgers University
Piscataway, NJ 08854, USA

Fahrettin Ozturk
Department of Mechanical Engineering
Ankara Yıldırım Beyazıt University
Ankara, Turkey

and

Strategy and Technology Management
Turkish Aerospace Industries Inc.
Ankara, Turkey

Simon S. Park
Department of Mechanical and Manufacturing Engineering
University of Calgary
Calgary, Alberta, Canada

Rogelio Pérez-Santiago
Department of Industrial and Mechanical Engineering
Universidad de las Americas Puebla
Cholula, Mexico

Guillem Quintana
ASCAMM Technology Centre
Barcelona, Spain

Lidia Serenó
Department of Mechanical Engineering and Industrial Construction, Product Process and Production Engineering Research Group
School of Engineering, University of Girona
Girona, Spain

Ivan Tabernero
Department of Mechanical Engineering
University of Basque Country UPV/EHU
Bilbao, Spain

Taghi Tawakoli
Institute of Grinding and Precision Technology (KSF)
Hochschule Furtwangen University
Furtwangen im Schwarzwald, Germany

Dani Teixidor
Department of Mechanical Engineering and Industrial Construction
University of Girona
Girona, Spain

Serkan Toros
Department of Mechanical Engineering
Nigde Ömer Halisdemir University
Nigde, Turkey

Mevlüt Türköz
Department of Mechanical Engineering
Konya Technical University
Konya, Turkey

Eneko Ukar
Department of Mechanical Engineering
University of Basque Country UPV/EHU
Bilbao, Spain

Durul Ulutan
Manufacturing and Automation Research Laboratory, Department of Industrial and Systems Engineering
School of Engineering, Rutgers University
Piscataway, NJ 08854, USA

Rahmi Ünal
Mechanical Engineering Department
Gazi University
Ankara, Turkey

Elisa Vázquez
Department of Mechanical Engineering and Civil Construction
Universitat de Girona
Girona, Spain

Minjie Wang
School of Mechanical Engineering
Dalian University of Technology
Dalian, PR China

Donggang Yao
School of Materials Science and Engineering
Georgia Institute of Technology
Atlanta, GA, USA

Danyang Zhao
School of Mechanical Engineering
Dalian University of Technology
Dalian, PR China

and

School of Materials Science and Engineering
Georgia Institute of Technology
Atlanta, GA, USA

Part I

Advanced Forming Processes

1

Advances in Stamping

Ilyas Kacar[1] and Fahrettin Ozturk[2,3]

[1] *Department of Mechatronics Engineering, Nigde Omer Halisdemir University, Nigde, Turkey*
[2] *Department of Mechanical Engineering, Ankara Yıldırım Beyazıt University, Ankara, Turkey*
[3] *Strategy and Technology Management, Turkish Aerospace Industries Inc., Ankara, Turkey*

1.1 Introduction

Stamping is the general name of sheet metal pressing to produce sheet metal parts from sheet metal blanks. Millions of sheet metal parts are produced by stamping operations. Complex shaped parts can easily be produced at low cost. Increased production demands, competitive markets, and government regulations force the sheet metal companies to do research and make innovations in order to produces high quality products at low cost. The main research areas are failure prediction, design and control of a variable blankholder force (VBHF), active drawbeads, and the segmented die design to provide the controllability and flexibility of stamping processes.

The stamping processes are widely used in the automotive, aerospace, and appliance industries due to its high volume productivity, low manufacturing costs, and high strength to weight ratios of their final products. Controllability and flexibility are most important variables for manufacturing techniques. Perfection in stamped parts is important to avoid assembly problems. Tearing, wrinkling, and springback are considered as major problems to affect the quality of the final part. There have been lots of studies on how parameters affect the material flow and elastic-plastic deformation.

In a typical sheet metal stamping operation, energy exerted on the press machine's punch is transferred to sheet metal through a set of tooling for plastic deformation on the sheet metal. Blankholders and drawbeads restrain the sheet metal flow into the die cavity as seen in Figure 1.1. All new developments about sheet metal forming are focused on the control of material flow. Blank holders, active drawbeads, flexible dies, and VBHF techniques are used to provide regulated material flow during deformation by controlling material flow locally.

In this chapter, first of all, whole process and its parameters are explained briefly. Advanced manufacturing techniques on three-dimensional sheet metal stampings such as blank holders, active drawbeads, flexible dies and punches with multi point force, and VBHF techniques are explained in detail.

1.1.1 Blankholder and Drawbeads

Three main tools in the sheet metal stamping process are the punch, the die, and the blank-holder (or binder). A workpiece from sheet metal (or blank) is clamped between the die and

Modern Manufacturing Processes, First Edition. Edited by Muammer Koç and Tuğrul Özel.

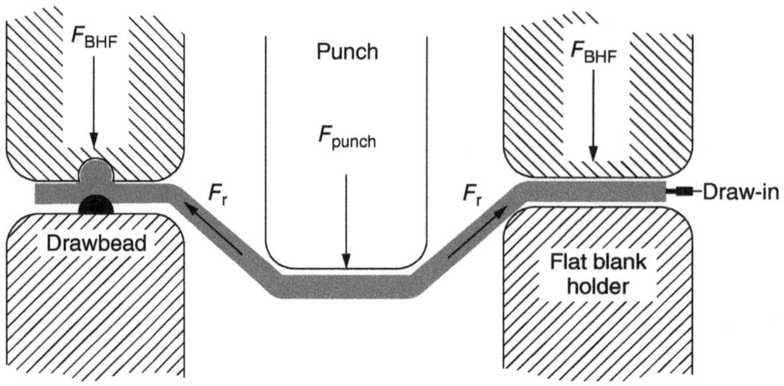

Figure 1.1 Schematic of a stamping process.

blankholder then is deformed by the punch. In sheet metal stamping, the quality of the final product depends on the control of the metal flow significantly. Blankholders are aimed to produce blankholder force (BHF) (or binder hold down force [BHDF]) to control material flow. BHF is the main factor affecting the occurrence of failure modes such as wrinkling and tearing in sheet metal parts. While the insufficient BHF leads to wrinkling, the excessive BHF causes tearing and eventually splitting of the sheet blank. The BHF should be sufficient to eliminate or reduce the mentioned defects on stamped parts. Drawbeads are mounted on blankholder for much more constrain to control material flow, satisfy the desired force, and prevent wrinkling. They play an important role in stamping operations as a main actor to control the sheet metal, flowing into the die cavity, in the forming of nonsymmetric, complex or irregular shaped parts especially large automotive panels. The cover panel, the trunk lid or roof of an automotive body commonly consist of complex 3D surfaces. When these surfaces are being deformed, defects such as wrinkles, fractures, and surface distortion may be encountered owing to nonuniform material flow. The BHF and drawbeads are used to eliminate these defects. The restraining force level should increase in the forming of nonsymmetric or complex parts so that metal flows slowly and uniformly. A drawbead is often used to change the restraining force regionally and to control uncertain material flow that may lead to failures such as wrinkling, tearing, fractures, surface distortion, and springback. These failures stem from excessive or insufficient restraining of blank.

Drawbeads typically cause localized tensile stresses along the blank periphery. A drawbead behaves as a boundary condition technically at applied regions on sheet blank by constraining sheet metal's flow and tightly holding the sheet partially and forcing metal to flow around itself. These constrains generate local restrain force from friction between sheet and tooling interface. The regulation of the whole deformation process can be ensured by these types of local manipulations, especially during a production run.

A drawbead consists of two components: a semicircular, rectangular or other shaped bead on the one side of the blankholder surface, and a matching groove on the opposite side of the blankholder surface as seen in Figure 1.2. The configurations of drawbeads depend on the stamping requirements. The drawbead operates in two steps: clamping (or locking) and drawing (or pulling) steps.

- In the clamping, blankholder starts to touch and hold the blank up to a desired drawbead penetration value.
- In the drawing phase, pulling is begun. Bending, sliding, and re-bending happen on the blank cyclically when the blank passes over the shoulders of the groove and around the drawbead

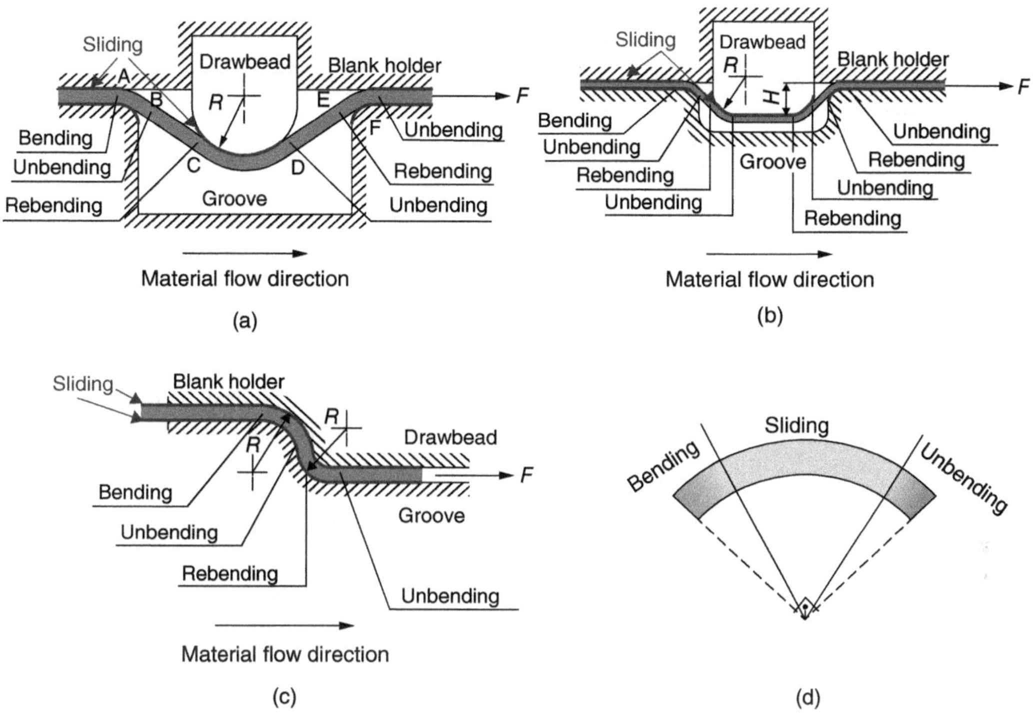

Figure 1.2 Deformation sequences for three drawbead geometries (a) dome ($R = H$); (b) rectangular ($R \neq H$); and (c) edge ($R = 90°$). (d) Deformations on a unit sheet stripe element.

as seen in Figure 1.2d. In order to perform a sheet metal strip pulling around a drawbead, the applied pulling force should overcome the **drawbead restraining force (DBRF)** caused by bending-re-bending sequences and frictional force. In some applications a layer of urethane material is inserted into groove to increase friction and constrain. A drawbead can provide high restraining force at relatively low punch pressure.

For dome shaped bead, there are six bending and re-bending deformations over the grooves as in Figure 1.2a, eight times for a rectangular shaped bead as in panel (b), and four times for an edge deformation as in panel (c). It is clear that the restraining force obtained would be higher in panel (b) than in panel (a). Although the total number of deformation is four for an edge formation, the maximum restraining force is obtained in that case due to the fact that full penetration is obtained while the bending angle is 90° that the effective bending radius is equal to the tool radius.

Besides these advantages, in some cases, drawbeads may cause deformation damages. Wrinkling and fracture are two important defects in drawing process and both determine forming limits. Figure 1.3 illustrates these defects. The BHF has a significant effect for preventing these failure types. In the low BHF case, sheet material has the risk of wrinkling, while the excessive BHF may lead to fracture. Wrinkling stems from excessive compressive stresses normal to thickness (insufficient tensile stresses) and causes the sheet blank to buckle locally. It may be seen on the flange or in the side walls of the finished part. Splitting (or tearing) is caused by excessive tensile stresses.

Drawbeads may lead to work hardening of the sheet metal before stamping. Work hardening can lead to crack occurrences just at the beginning of the stamping.

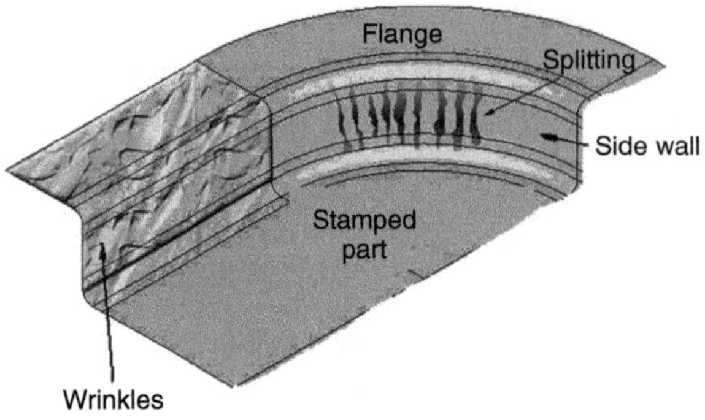

Figure 1.3 Some defects on stamping deformed without drawbead.

Bead geometry may cause defects. During bending on sheet blank, the drawbead with smaller corner radii like a rectangular bead carries out higher strains than the bigger radius like a semi-circular bead. Excessive strains can cause much more surface damage on the blank and the tool wear on the bead and groove shoulders.

Stamped part's geometry itself may cause defects. A large restraining force must be applied for a shallow geometry. In the forming of nonsymmetric or complex parts, the metal flow will be higher. In case of higher restraining force requirements, it is practical but not convenient that blank-holder pressure is increased generally. Higher BHF leads to excessive wear on the tool surfaces or galling in the blank or splitting in the sheet, so, higher BHF is not desired. The design and estimation of drawbead forces are still often based on a trial and error procedure [1].

In some cases, such as complex shaped stampings or shallow parts, the required BHF may exceed the tonnage capacity of the press machine. Therefore, the usage of drawbeads is inevitable. By means of a drawbead, the BHF is reduced and better part quality can be obtained. Drawbead reduces BHF requirements, and minimize the blank size. When stamping equipment includes drawbeads, the drawbead trajectory in addition to the BHF must be considered in the forming operation. The DBRF can be also controlled by movement of active drawbead in a very effective mean.

1.1.2 The Drawbead Restraining Force (DBRF)

When BHF is applied to the blank, the restraining force occurs in the planar form on the contact surface between tool and sheet metal. **The blankholder creates the restraining force by friction between the blank and tools.** The frictional resisting strength should overcome the strength from material's strain hardening in consequence of stretching and bending/re-bending sequences through the drawbead radii. To obtain uniform deformation on the whole blank, the restraining force should be applied locally for regulation of the material flow. Knowing or prediction of the DBRF and BHF is a necessary item for design and implementation of the drawbead into forming processes.

The DBRF is exerted by both deformation resistance and frictional resistance. Bending and re-bending leads to deformation resistance. Sliding of the blank leads to frictional resistance. The amount of the sheet metal movement into the die cavity depends on the frictional force. In Figure 1.2a sliding between points A and B generates a friction force. At point B, the metal is straightened (unbended) and slides at the radius of the groove shoulder. This sequence of bending, sliding, and unbending accounts for restraining force from the first groove shoulder. The same sequence occurs in the arcs, CD and EF. The entire restraining force of the drawbead is due to the work done in the three bending and unbending sequences plus the friction force between the bead and the sheet.

The DBRF mainly depends on some geometric parameters such as the drawbead radius R, drawbead penetration H, and bending angle as seen in Figure 1.2. While bending angle is getting increased, a bigger DBRF is obtained in the case of $R = H$. Also while H is getting increased, a bigger DBRF is obtained. And while R is getting decreased, a bigger DBRF is obtained. Influence of drawbead geometry on sheet metal forming has been analyzed in the bending, unbending, and reverse bending stages by modeling of blank holding conditions and friction by most of the researchers.

The DBRF's magnitude depends on the size and shape of the workpiece. The force magnitude for shallow geometries such as a car roof must be larger relatively. In the literature, there have been lots of efforts on the study of the DBRF.

1.1.3 The Use of Active Drawbeads

The size and shape of the drawbeads have important roles on restraining force. Fixed drawbeads are immovable components on the binder. Different shaped and sized drawbeads have been analyzed to present their relationship with the BHF and DBRF. Locations and trajectories of the drawbeads, size and shape of them, gap and friction conditions, and cross section of the blank are most important parameters to adjust the DBRF. In the active drawbeads, the drawbead penetration is controlled **as a function of location and time (or punch stroke) in the stamping control process.** These parameters including the effect of BHF on the sheet metal part quality are investigated and suggestions have been presented for controlling the BHF as a function of punch travel during the forming process [2] and results of the simulation on the variation of the parameters such as material type, bead penetration, and friction conditions and analytical studies of drawbeads.

Making drawbead's penetration adjustable was not practical in early applications. But nowadays, real-time drawbead penetration adjustment (*active drawbead*) is common practice. With active drawbead, the amount of drawbead penetration into the material can be adjusted by a hydraulic actuator quickly for a suitable DBRF as seen in Figure 1.4.

There may be more than one drawbead in process depending on the shape and dimension requirements of the part. Active drawbeads are used for the production of both large and complex shaped panels to be able to gain both setup time and process parameters' reduction. In addition to the drawbead penetration, the BHF can also be adjustable while the sheet strip is being drawn.

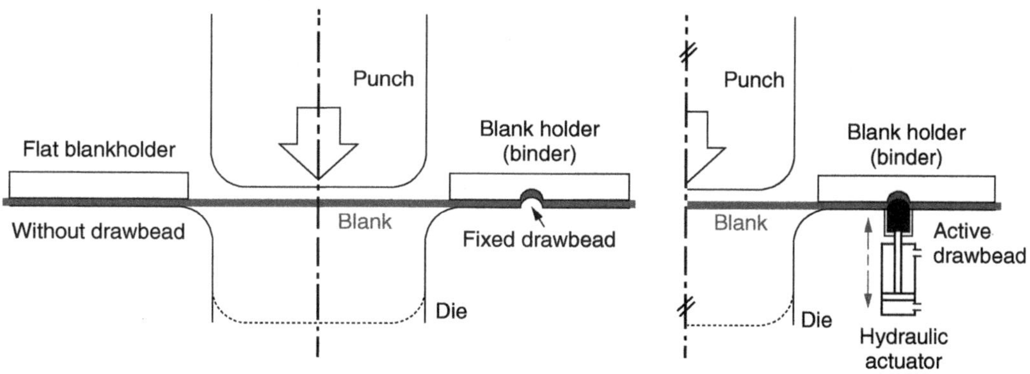

Figure 1.4 Drawbead types.

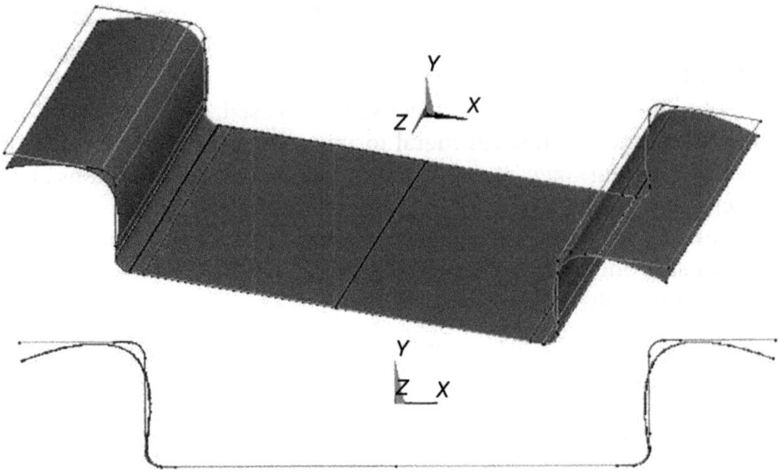

Figure 1.5 Sidewall curl occurrences.

Adjustable penetration and better formability can be obtained by blankholders with active or fixed heightened drawbeads rather than flat blankholders. For example, maximal drawing depth was obtained by using active drawbeads in the stamping process of on Al 6111-T4. The optimal drawbead trajectory depends on the BHF selection. The formability of aluminum has been improved by restraining force's control with the penetration adjusting of active drawbeads. Stamping process gains enhanced flexibility due to the fact that active drawbeads can penetrate to variable depths even if during deformation.

In stamping, another defect is sidewall curls as in Figure 1.5. Reducing or possibly eliminating the sidewall curl depends on achieving to the adequate strain levels by using active drawbeads. In the absence of the active drawbead, compressive stresses arise in the side wall regions on the blank. While the penetration into material in the flange is getting increased by active drawbeads, material flow into the die cavity is restricted near the end of the stroke. This restriction makes the region in the sidewall stretched and stretching makes the stress distribution in the side wall balanced.

1.1.4 Control Techniques of Active Drawbeads

Stampings without any defects and reduction or eliminating time requirements for tool change are the purpose of automatic control on stamping. The control system used in the press machines with fixed drawbead or flat blankholder is called "open loop control" system. This type of systems has not got measurement device to provide feedback the current process variables such as, force, displacement, etc.

With the development of the computer-controlled servo hydraulic press machines, it has been possible to control the punch *displacement* or *force* to be applied on punch by actuators using differential signals between set values and feedback even if during process. These systems are called "closed-loop control" as in Figure 1.6. Controlled system is the blankholder with active drawbeads. BHF, drawbead penetration, punch speed, and punch displacement are process variables. The best values of these parameters have been determined through various optimization techniques. In the closed-loop systems, response time of the system must be determined and optimized by designing parameters of the controller. So controller lets punch and drawbead apply the set value of force or travel during the process.

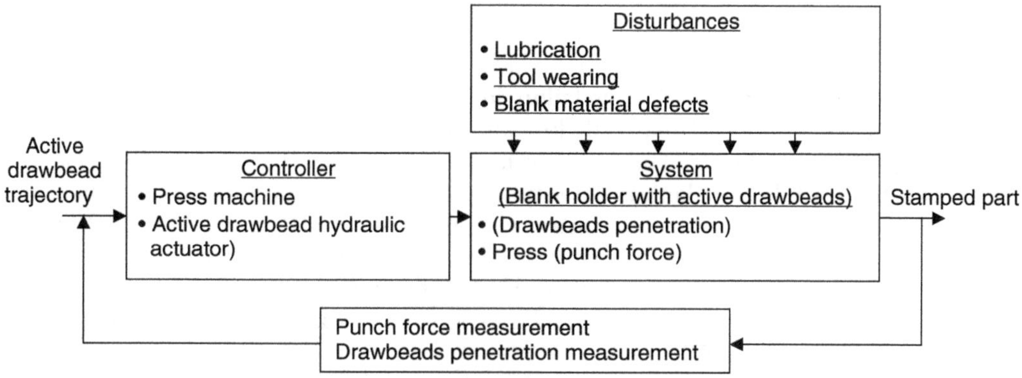

Figure 1.6 Active drawbead penetration control of sheet metal forming.

Feedback devices send the current process value to the controller. The controller compares set value from user demand with measured value from feedback device. In the *force* control case, critical points like excessive tensile forces are sensed by loadcells. The trajectory of the drawbeads is adjusted again to eliminate the difference between desired and current drawing forces. If the difference between desired and actual punch forces is zero, it means that desired values for bead penetration are achieved and process is controlled just on set. These control strategies can be applied successfully and quickly thanks to the hydraulic actuators. By means of these actuators, researchers have developed different controller types such as linear controllers (proportional [P] and proportional-integral [PI]) on an electro-hydraulic die-cushion driver. In these types of systems, a target trajectory determined by press control system is applied onto hydraulic actuator of the active drawbead. Once an optimal path is obtained as in Figure 1.7, this pattern can be used easily whenever necessary. So time and engineering requirements for designing and optimization are reduced or eliminated.

Nowadays, the forming press machines equipped with a servo hydraulic control system are used to apply force or punch kinematics in any desired function in open or closed-loop control types. Servo presses control the BHF, punch displacement, punch velocity, and each die segment separately (if segmented) during stamping process. These types of servo-driven press machines provide accurate speed and position control. These features contribute the flexibilities to the manufacturing process and equipment.

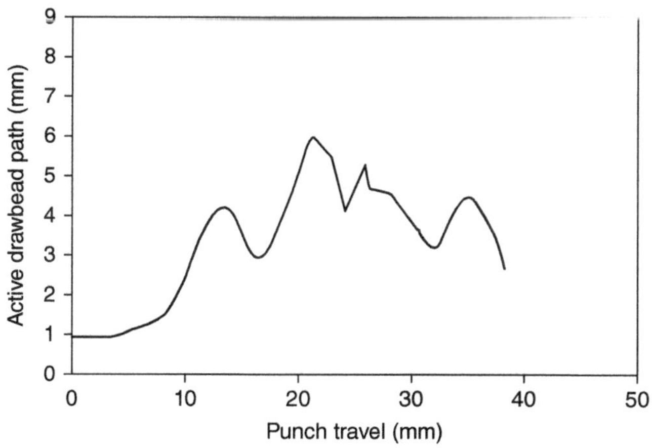

Figure 1.7 An example path for an active drawbead.

1.1.5 Variable Blankholder Force (VBHF)

Improvements on defects are based on the control of material flow during stamping. Before any failure does not occur, the BHF at the failure location can be reduced/raised by the VBHF. This strategy allows more local material flow and thus the crack occurrence is eliminated similarly as in active drawbeads. The essential mission of blankholders is to apply local and real-time constrain, while the blank is deforming. Control can be performed with the time-independent techniques. The real-time control system follows predetermined BHF patterns in a loop. Desired BHF is a function of punch travel. In the case of constant punch velocity, time can be evaluated as an independent variable instead of the punch displacement.

The VBHF technology provides mainly flexibility to processes and makes BHF profile varied in the axisymmetric forming. When the VBHF is applied instead of a constant BHF, it is seen that product quality is increased remarkably. Thanks to its high capability on local material flow control and reduced scrap rates, VBHF technology offers a powerful solution for stamping applications.

At the beginning time of the VBHF applications, the pressure distribution on the blankholder could be varied on the whole contact surface only. Nowadays, the segmented blankholders driven by individual actuators have been developed to satisfy different BHF at the same time. In this technique, the blankholder is divided into multiple segments, so height is adjustable by die cushion pins. The segmented blankholders provide more flexibility to deformation process for both simple and irregular shapes.

The key to implementing VBHF technology is to have a system that can be commanded to provide local BHF in very precise steps during the press stroke. An actuator array consisting of two type actuator units is aligned in hexagonal shape and cross sections can be seen in Figure 1.8 and a press machine mechanism equipped multipoint VBHF control is as in Figure 1.9. To generate the desired BHF in each cylinder separately during the stroke, an advanced control system is needed. Those press machines are able to adapt to multiple stroke speeds.

Radial and tangential stresses act on the flange during the forming process. If these stresses are not balanced, the blank is stamped irregularly. Especially in the stamping of irregular shaped parts, manufacturers had to use larger sized blanks to compensate these unbalanced stress distributions. This caused much more scrapes as blanking skeletons. These problems are eliminated thanks to BHF control technology.

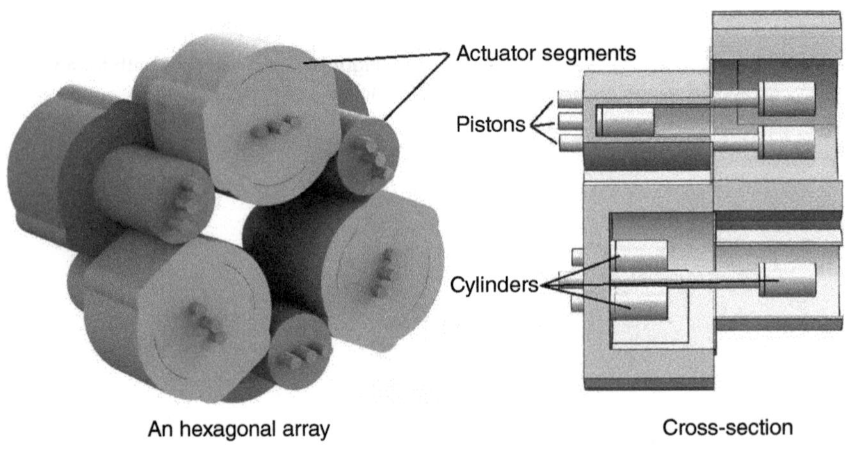

Figure 1.8 Variable force binder: actuator segments and their hexagonal array.

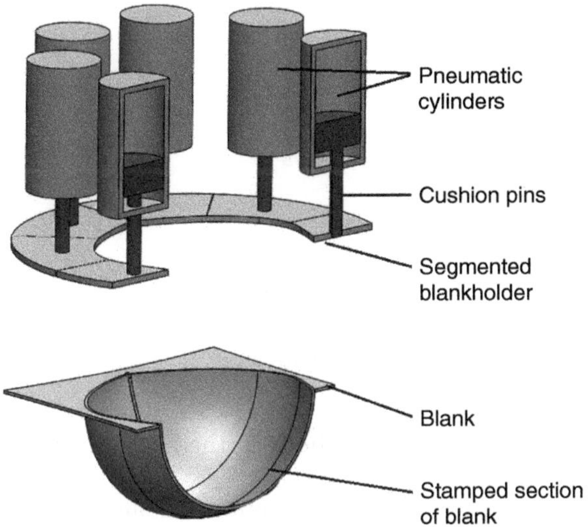

Figure 1.9 Variable binder force control using height-adjustable die cushion pins.

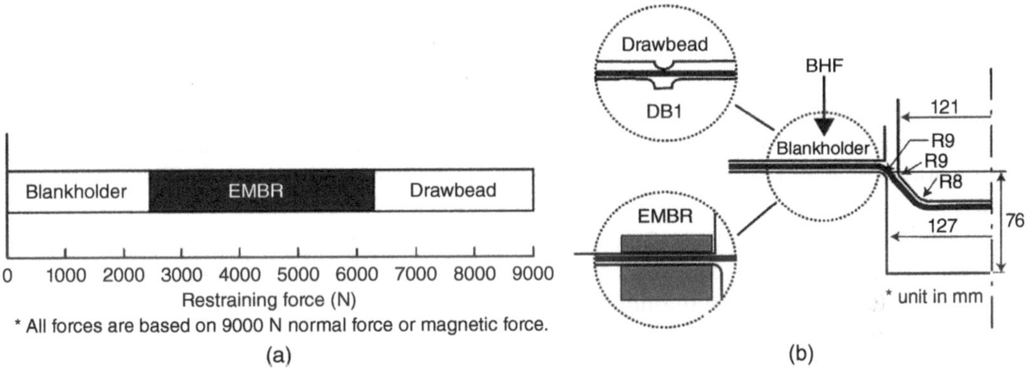

Figure 1.10 (a) Restraining force comparison. (b) EMBR assembly in the die. *Source:* Seo 2008 [4]. Reproduced with permission of Elsevier.

Change in the material flow locally around the blank regulates the stress distribution as equal as possible during deformation. In the press machines equipped with VBHF, BHFs can be tuned during process. Measuring the thickness variation throughout stamped part is one of the most common methods to compare. In ideal condition, large deviation of the thickness variation is not desired. Wang et al. [3] obtained reduction on thickness variations up to 70% by means of VBHF. The necking is postponed effectively with VBHF.

Another active blank holder force control is based on electromagnetic holding and used efficiently. Comparison between a drawbead and an electromagnetic blank restraining (EMBR) design is seen in Figure 1.10.

1.1.6 Flexibility

Press tools are important elements in stamping operation. Sheet metal forming is a process that depends on a combination of the punch, blankholder, and dies. These tools play important roles on process. Tool changing during process is undesirable due to the time consumption. It

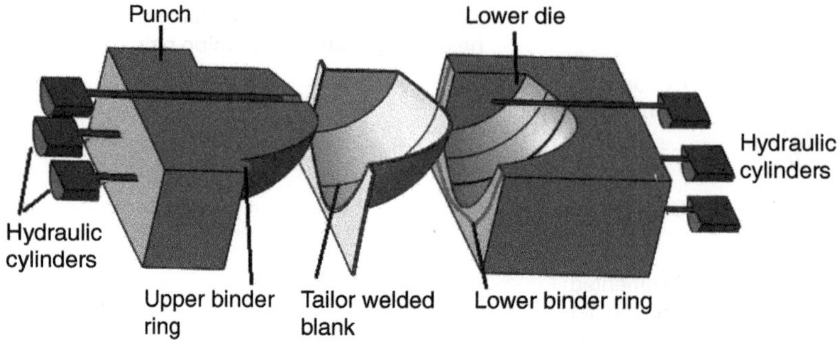

Figure 1.11 Schematic representation of segmented die with local adaptive controllers.

should be eliminated through enhanced controllability and flexibility in the next generation stamping tooling.

One of the methods to improve both product and process is to put more flexibility to them. The flexibility makes controllability easier by increasing degree of freedom (DOF) that give the relative movement to tool or workpiece as in Figure 1.11. Segmented dies have one die composed from lots of subdies (die segments). They provide flexibility, weight, and cost reduction by eliminating additional die designs. While conventional presses have tool movement with one DOF, flexible press machines have path-velocity controlled tool movements with more DOFs. Tool segmentation is not new concept in press working. It has been used in a lot of applications such as ultrasonic machining, image processing, sheet metal stamping, and medical applications. It is more practice in the complex-shaped, three-dimensional sheet metal part stamping.

Joints especially in automotive industry have to be more rigid and strength against to impact. Lap joints are not effective due to corrosion from interface. Spot welding requires flange preparation before welding. So manufacturers are forced to use tailor-welded blanks (TWBs) before their stamping operations to satisfy market demands such as improved dimensional accuracy, corrosion resistance, and increased stiffness.

Die or blank holder segmentation mainly provide homogeny thickness variation through the whole workpiece. Multipoint forming (MPF) technology is also an advanced technology on punch that provides flexibility on forming of three-dimensional sheet metal parts. A MPF process of sheet metal is illustrated in Figure 1.12. Punch is in the form of a series of mini punches in 3D matrix form. Every different part shape can be adopted by height adjustment of every mini punch in the punch matrix.

1.1.7 Future Challenges

Nowadays, tendency on the application of various methods is based on the control of material flow during the forming process. The main target is to produce the defect-free part quickly at low cost. Blankholders, active drawbeads, flexible dies, multipoint force punches, and VBHF techniques are advanced technologies on forming of three-dimensional sheet metal parts. It is expected that they will be the next generation stamping issues. Applications especially in the automotive and aerospace have drawn attention. Wide range of the future usage is expected by means of stamping with a servo hydraulic control system and closed-loop control types.

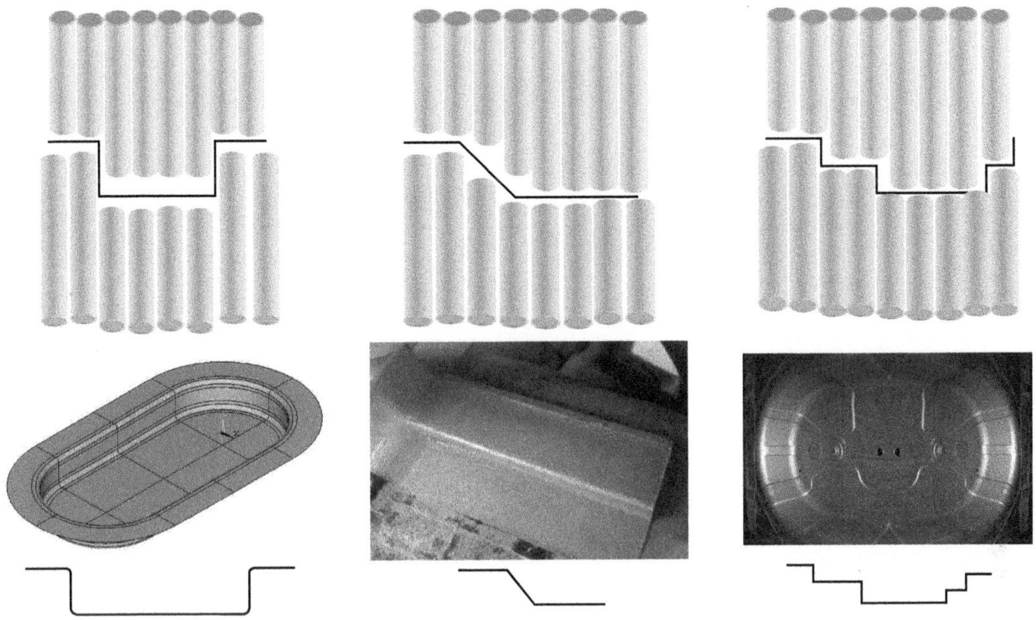

Figure 1.12 Schematic illustration of multi point forming process of sheet metal.

References

1 Tufekci, S., Wang, C., Kinzel, G., and Altan, T. (1994). Estimation and Control of Drawbead Forces in Sheet Metal Forming. SAE Technical Paper 940941.

2 Ahmetoglu, M., Coremans, A., Kinzel, G., and Altan, T. (1993). Improving Drawability by Using Variable Blankholder Force and Pressure in Deep Drawing of Round and Non-Symmetric Parts. SAE Technical Paper 930287.

3 Wang, W.R., Chen, G.L., and Lin, Z.Q. (2010). Application of new VBHF optimization strategy to improve formability of automobile panels with aluminum alloy sheet. *Transactions of Nonferrous Metals Society of China* 20 (3): 471–477.

4 Seo, Y.R. (2008). Electromagnetic blank restrainer in sheet metal forming processes. *International Journal of Mechanical Sciences* 50 (4): 743–751.

2

Hydroforming

C Hartl

Koeln University, Cologne, Germany

2.1 Introduction

Hydroforming is applied in the industrial production of thin-walled, complex-shaped metal components, ranging from small to large series. The automotive industry is one of major user of this technology today. Furthermore, examples of applications are to be found in aerospace industry, sanitary, and piping products manufacture.

The term "hydroforming" comprises forming processes that use pressurized liquid or gaseous media to plastically form an initial blank into a defined shape. For the example of tube hydroforming, as shown in Figure 2.1, the medium pressure is applied to the inner of the tubular workpiece and causes its expansion in direction to the surrounding die cavity. Tubular semifinished products, like straight or bent tubes or profiles, as well as single or multiple sheets are of importance in practice as initial blanks are subject to investigations, as summarized, for example in [1–7]. The use of these different types of semifinished products led to the general distinction which is made between "tube hydroforming" and "sheet hydroforming" [8].

Hydroforming can offer technical and economical advantages for the manufacture of thin-walled complex components in comparison with conventional manufacturing techniques like deep drawing or casting. This applies in particular to the production of light weight components for the automotive industry, as for example presented in [9–11]. The technology enables the manufacturing of complex-shaped geometries with a reduced number of process steps, combined with a reduction of components and joints. Typical automotive applications, produced in series today, are chassis components, engine cradles, exhaust system parts, body components, and components for power transmission.

Against the background, to improve competitiveness of hydroforming, various developments were made in the past few years, dealing with measures to reduce initial investment costs, increasing production rate and material utilization, consolidation of multiple parts in one workpiece, integration of further manufacturing processes, and eliminating drawbacks such as excessive thinning of formed component wall. This resulted in a technical progress of hydroforming technology with new and improved potential, which is part of the here-presented content, besides an insight in the fundamentals of hydroforming.

Modern Manufacturing Processes, First Edition. Edited by Muammer Koç and Tuğrul Özel.

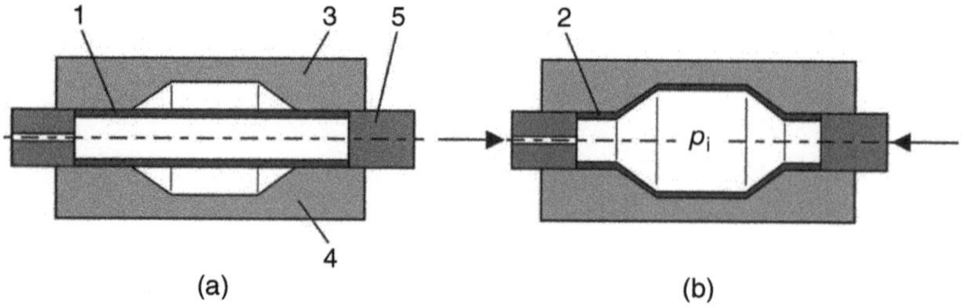

Figure 2.1 Principle of tube hydroforming. (a) Start of process and (b) expansion. 1, Initial tubular blank; 2, hydroformed component; 3, top die; 4, bottom die; 5, sealing punch; p_i, internal pressure.

2.2 Fundamentals

2.2.1 Classification of Hydroforming Processes

According to the current and past developments in hydroforming, various distinctions can be drawn between hydroforming processes that are used by industry or that are subject of investigations respectively. This concerns the type of blanks that are formed, whether these are tubular or sheet metal blanks, the applied forming temperature like cold or warm forming, and the combination of the hydroforming process with additional, integrated operations such as mechanical forming, joining or piercing by media pressurization. Figure 2.2 summarizes

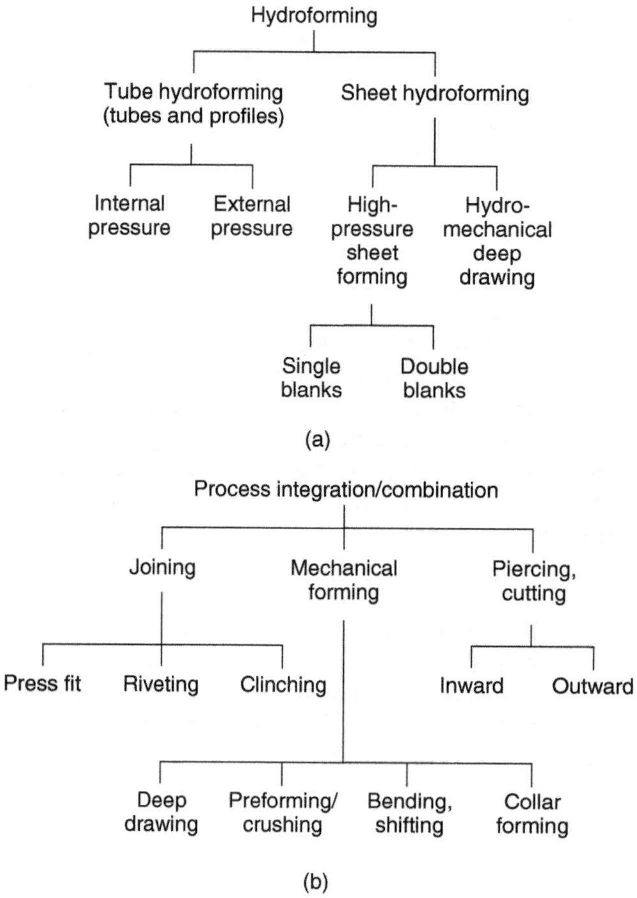

Figure 2.2 Classification of hydroforming processes. (a) Hydroforming process types and (b) processes integrated in or combined with hydroforming.

distinguishing characteristics of hydroforming processes. However, it does not consider the level of dissemination of the individual process types in industrial application.

2.2.1.1 Tube Hydroforming

Concerning the different process types, hydroforming of tubes with internal pressurization at room temperature is the one which is most commonly applied in industrial production. Forming tools for tube hydroforming consists in general of die halves that contain the forming die cavity, and sealing punches to seal the tube ends against the inner pressure, as depicted in Figure 2.3. Depending on the component geometry, which is to be produced, additional tool elements can be involved in the process. As an example, the hydroforming of T-shaped components, as shown in Figure 2.3b, or Y-shaped, requires an additional counter punch that acts on the end of the expanded protrusion and is displaced by the pressurized workpiece during forming. The complexity of industrially produced components requires in most cases that additional preceding operations be considered with the hydroforming process itself. These operations can involve, for example bending and preforming (crushing, flattening) of the tubular blank to ensure that it is capable of insertion into the die or to obtain an optimized material distribution [12]. Figure 2.4 shows examples of components produced by such tube hydroforming processes.

The variant of tube hydroforming with external pressure was investigated by [13] in detail. This process requires a rigid tool element to define the inner shape of the formed component as shown in Figure 2.5. In comparison to the hydroforming with internal pressurization, this process enables to improve the quality of the inner dimensions of the formed tubular component [13].

The application of superimposed external pressurization applied to tube hydroforming with internal pressure forming was presented by [14]. Objective of these investigations was to improve the ductility of tubular material in hydroforming at room temperature by increasing the hydrostatic pressure. For copper tubes a tripling of feasible expansion could be demonstrated [14].

2.2.1.2 Sheet Hydroforming

Sheet hydroforming can be divided into high-pressure sheet forming and hydromechanical deep drawing. In high-pressure sheet forming, the sheet blank is formed in a die cavity by pressurization, as shown in Figure 2.6. Major advantages in comparison to forming with rigid tools consist in the more equal distribution of plastic strains within the workpiece and with this, in an increased possible forming depth of the manufactured components [15, 16]. Similar to

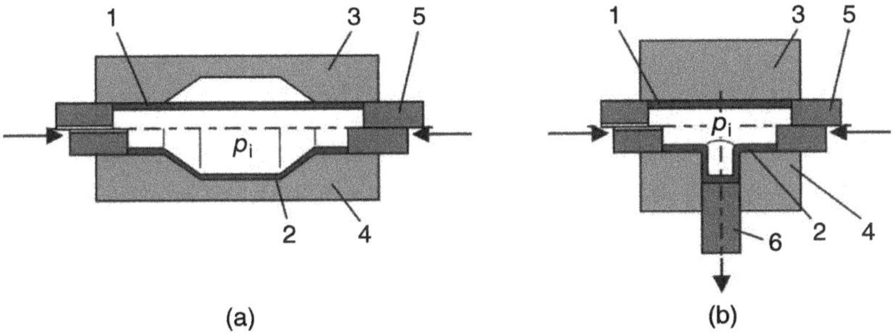

Figure 2.3 Principles of tube hydroforming processes with internal pressurization. (a) Expansion of longitudinal parts and (b) expansion of T-shaped component. 1, Initial tube; 2, hydroformed component; 3, top die; 4, bottom die; 5, sealing punch; 6, counter punch.

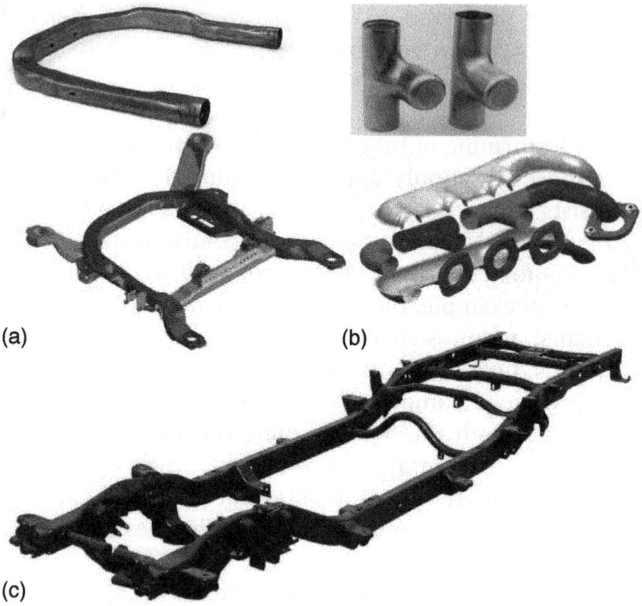

Figure 2.4 Examples of components manufactured by tube hydroforming. (a) Engine cradle, (b) exhaust system components and (c) light truck frame with hydroformed side and cross members. *Source:* Hartl 2005 [3]. Reproduced with permission of Elsevier.

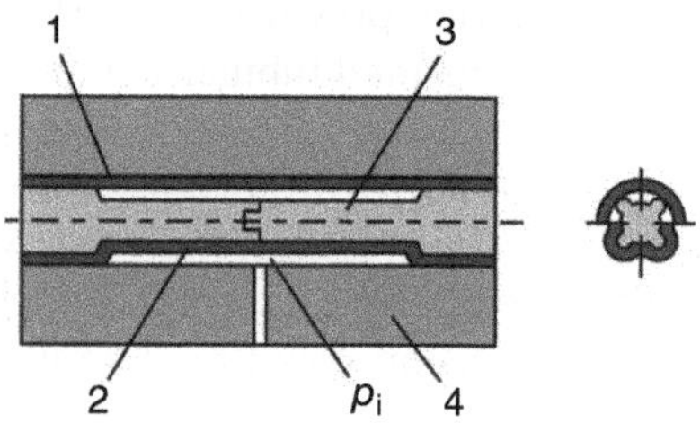

Figure 2.5 Principle of tube hydroforming with external pressurization. 1, Initial tube; 2, formed component; 3, mandrel; 4, pressure chamber.

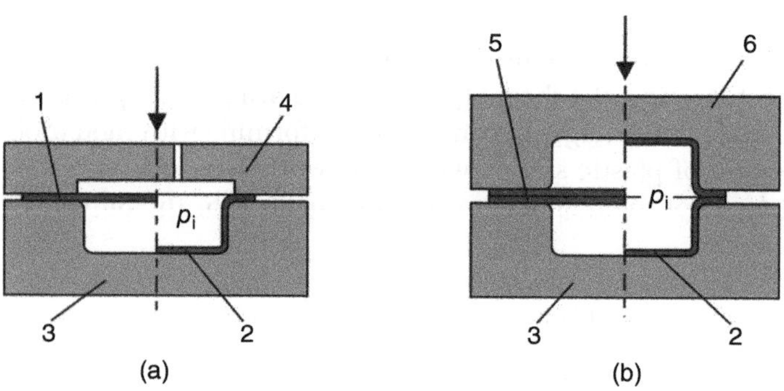

Figure 2.6 Principles of high-pressure sheet forming processes. (a) Single blank hydroforming, (b) forming of double blanks. 1, Initial sheet metal blank; 2, formed component; 3, bottom die with draw ring and die cavity; 4, top die with blank holder; 5, double blanks; 6, top die with draw ring and die cavity.

high-pressure sheet forming are processes that use an additional rubber diaphragm that is in contact with the sheet metal blank and that is pressurized with a liquid media to form the workpiece, as for example presented by [17]. Besides the forming of single blanks, high-pressure forming enables also the expansion of double blanks as depicted in Figure 2.6b. This type of process requires additional tool elements to apply the pressurized medium between the two blanks. Suitable docking systems are discussed, for example in [18]. Conducted investigations considered blank pairs that are welded before, as well as after the hydroforming. Nonwelded blanks allow a relative movement between the both blanks during the forming operation,

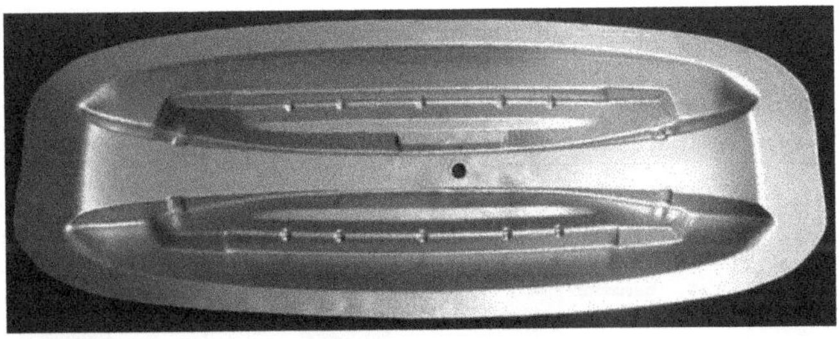

Figure 2.7 Example of a high-pressure sheet formed body component. *Source:* Neugebauer 2007 [20]. Reproduced with permission of Springer.

which is of advantage for forming of components with different depths of the upper and lower shell [18]. For hydroforming of sheet pairs with different forming characteristics, for example different materials or blank thicknesses, a variant is proposed and investigated in [19], which applies an additional outer fluid pressure to the blank that requires less pressure. An example of a component formed by high-pressure sheet forming is depicted in Figure 2.7.

Hydromechanical deep drawing is similar to conventional deep drawing of sheet metal, except the application of a counterpressure that presses the formed blank against the drawing punch during the forming operation, as shown in Figure 2.8. In comparison to conventional deep drawing, this pressure-assisted process offers the advantages of drawing complex shapes within one step, with higher drawing ratio and improved dimensional quality [15, 21]. Different process variants of hydromechanical deep drawing have been developed in the past:

- Processes that differ in techniques to seal the formed blank against the pressurized medium, as documented, for example in [21],
- A process variant that applies a hydraulic prebulging of the blank against the direction of punch movement at process start, to obtain a strain hardening of the blank that is useful to increase strength of flat components [15, 22],
- Variants that are working without a drawing ring and using a uniform pressurization of the overall blank by applying the medium pressure either directly to the workpiece [23] or via multiple pressurized rubber membranes [24].

2.2.1.3 Integrated Operations in Hydroforming Processes

According to Figure 2.2b, hydroforming can involve integrated processes that can be divided into mechanical forming, joining, and piercing/cutting operations. Concerning mechanical forming operations, deep drawing, bending, flattening, stamping, and collar forming are to be found in published research work and industrial applications, respectively. Deep drawing can

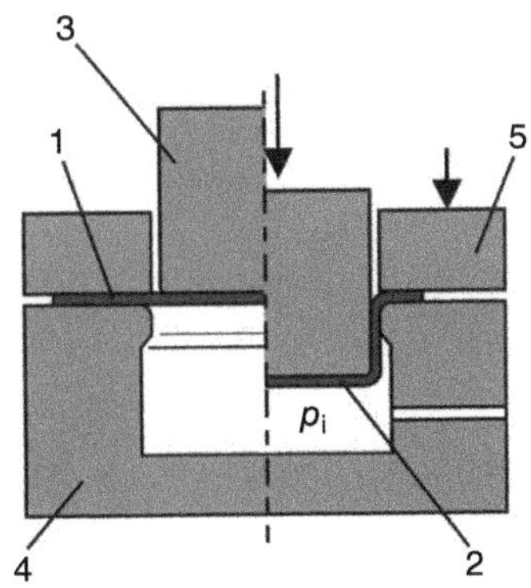

Figure 2.8 Principle of hydromechanical deep drawing. 1, Initial blank; 2, formed component; 3, punch; 4, drawing die with pressure chamber; 5, blank holder.

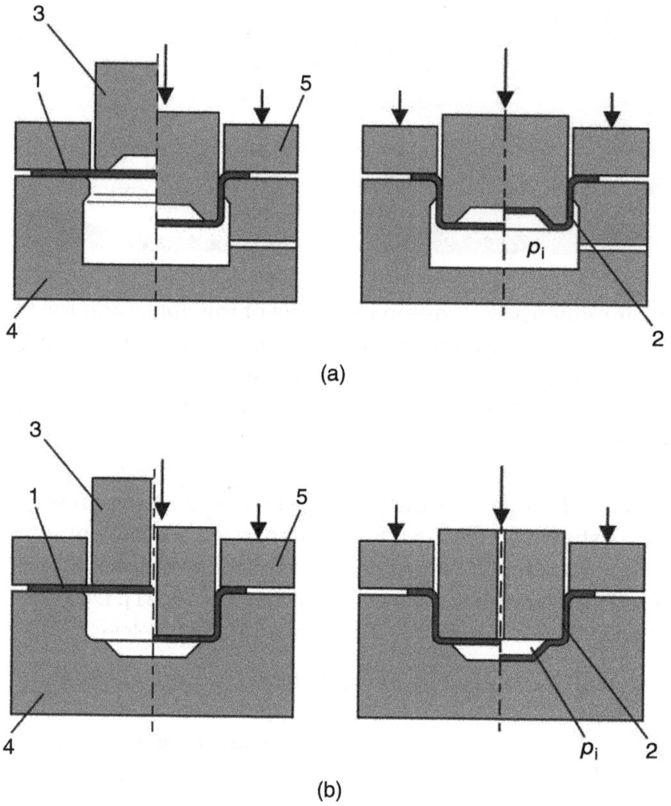

Figure 2.9 Combination of deep drawing and subsequent hydroforming according to [2]. (a) Hydroforming toward punch and (b) hydroforming in direction of drawing die. 1, Initial blank; 2, formed component; 3, punch; 4, drawing die; 5, blank holder.

be applied as a process combination in sheet hydroforming as well as in tube hydroforming applications. Figure 2.9 shows the principle of a deep drawing operation of single blanks with a subsequent hydroforming as described in [2] and investigated, for example by [25, 26]. The combination of deep drawing double blanks with subsequent hydroforming was investigated by [27] in detail.

Deep drawing is also applicable in combination with tube hydroforming, as described in [20] for the example of a reverse-drawing of a hydroformed section (Figure 2.10). This process sequence includes the hydroforming of a T-piece in a first step, the mechanical preforming without internal pressure by inserting the part in a second hydroform tool in a second step, and the hydroforming with reverse drawing against an internal pressure with a third and fourth steps, respectively.

In tube hydroforming, bending is in general applied to the tubular blanks as a mechanical forming operation prior to the hydroforming operation, when components with bent main axis

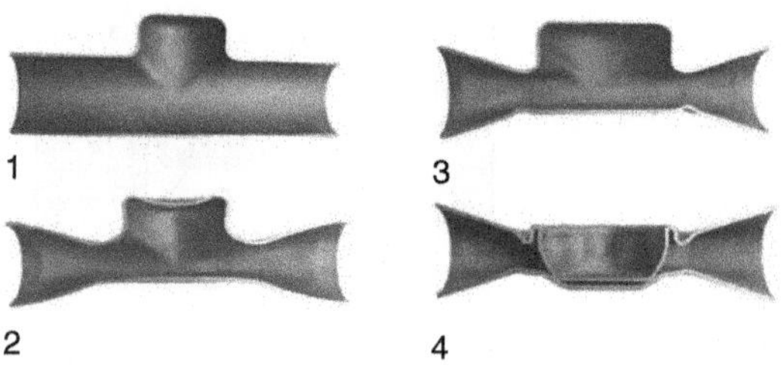

Figure 2.10 Process sequence of tube processes with mechanical deep drawing operations. *Source:* Neugebauer 2007 [20]. Reproduced with permission of Springer.

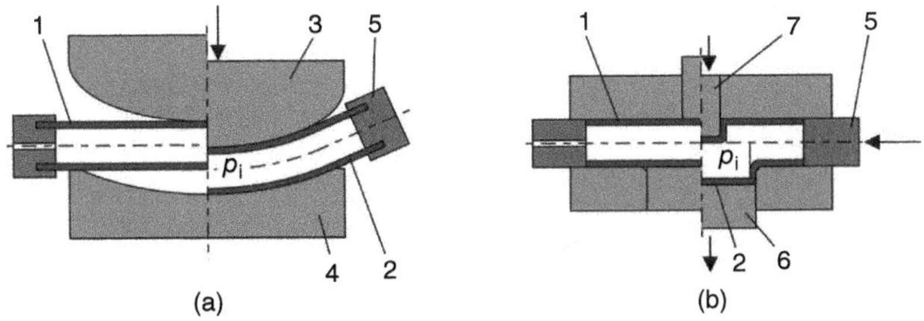

Figure 2.11 Integrated processes with modification of workpiece longitudinal axis. (a) Principle of bending with internal pressure and (b) principle of shifting/displacement. 1, Initial tubular blank; 2, formed component; 3, top die; 4, bottom die; 5, sealing punch; 6/7, displacing tool slides.

have to be formed. The integration of bending in a hydroforming process with inner pressure support of the blank during the bending may be advantageous for thin-walled tubular blanks with comparatively large diameter, as demonstrated in [28]. Those blanks tend to excessive wrinkling and buckling when bent with conventional bending operations. Figure 2.11a shows the principle of the process. Similar to integrated bending operations of tubular blanks, the deflection of the workpiece longitudinal axis can be performed also by shifting a section of the pressurized part, as already shown by [29] for aluminum profiles (Figure 2.11b and Figure 2.12).

In some cases, integrated flattening (crushing) of tubular blanks can be performed by closing the hydroforming tools, as shown in Figure 2.13. However, this requires a suitable design of the tool joint face in consideration of the intermediate shape of the formed blank to avoid a clamping of workpiece material within the closing joint face. The movement of closing the hydroforming tool can be used in sheet metal forming processes to conduct bending and stamping operations, as shown schematically in Figure 2.14. A suitable design of piercing punches within the hydroforming tool enables the collar forming within the hydroforming process itself (Figure 2.15). Those elements can be used for datum holes or threads. Information about attainable shapes is to be found in [3, 30].

Besides additional forming operations, as previously described, also joining processes can be integrated into the hydroforming process to enhance profitability. Press fits can be obtained by expanding a tube within a surrounding component, where the elastic residual stresses of the expanded components ensure their frictional connection (Figure 2.16). Joining of cams on

Figure 2.12 Hydroformed aluminum prototype part with displaced axis.

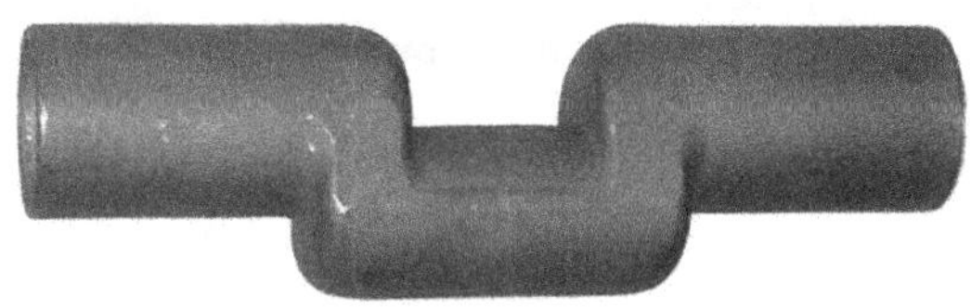

Figure 2.13 Flattening of tubular blanks in hydroforming. 1, Cross section of initial tube; 2, preformed shape; 3, hydroformed cross section; 4, top die; 5, bottom die.

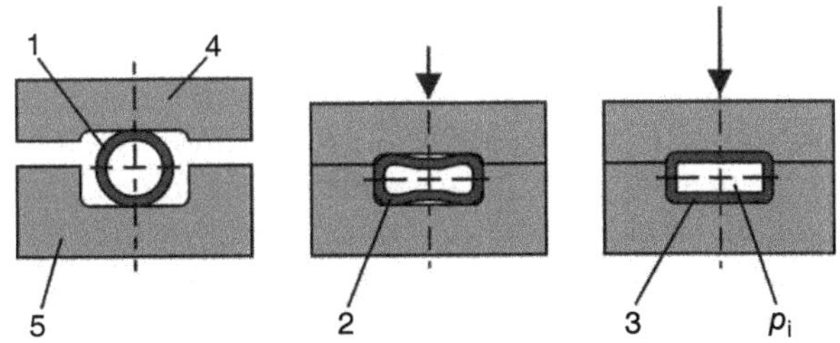

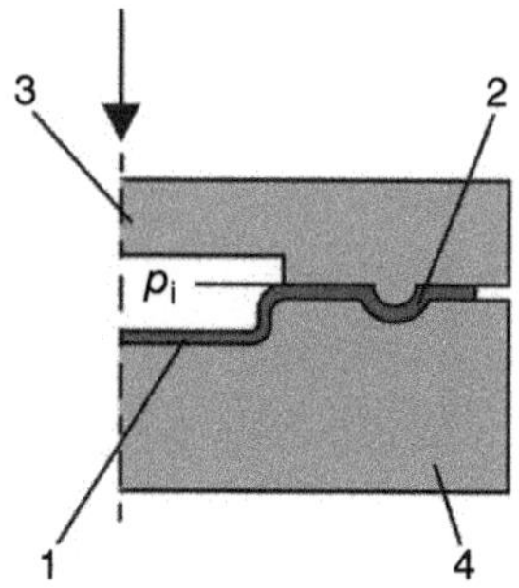

Figure 2.14 Examples of integrated mechanical forming in sheet hydroforming tools. 1, Hydroformed blank; 2, integrated preforming; 3, top die; 4, bottom die.

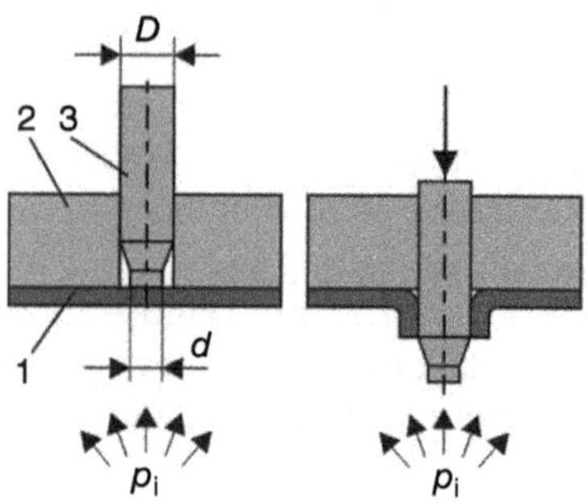

(a)

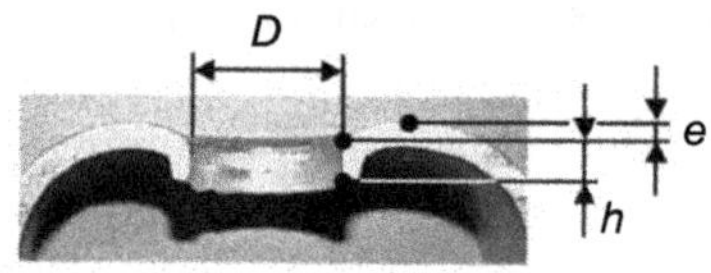

D/d	h (mm)	e (mm)	Transition from D to d
1.4	5.0	1.7	
1.5	5.4	1.8	Conical
1.6	5.5	1.9	
1.6	5.9	1.2	Traktrix

(b)

Figure 2.15 Integrated collar forming in hydroforming processes. (a) Principle and (b) experimental results (tube made from unalloyed steel, wall thickness t_0 3 mm, p_i 1800 bar, D 16 mm). *Source*: After Hartl 2005 [3]. Reproduced with permission of Elsevier. 1, Cross section of expanded workpiece wall; 2, forming die; 3, piercing punch.

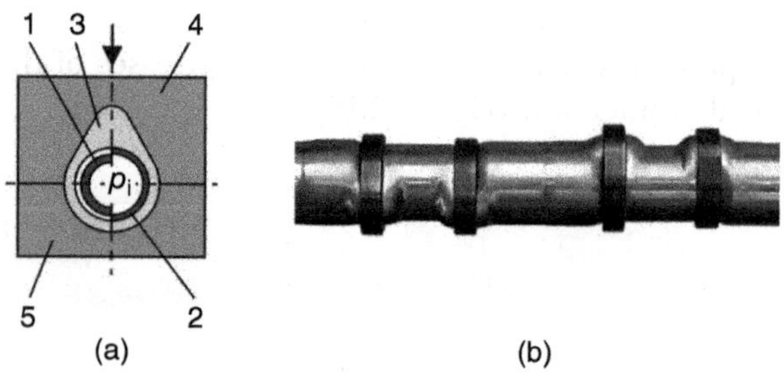

Figure 2.16 Joining of components by hydroforming for the example of camshafts. (a) Principle and (b) hydrojoined camshaft. *Source*: After Neugebauer 2005 [20]. Reproduced with permission of Springer. 1, Initial tubular blank; 2, expanded tube; 3, cam; 4, top die; 5, bottom die.

camshafts as well as the joining of tubes into the plates of heat exchangers are a typical applications of this technique [20, 31]. This press fit can be combined with a form-fitting shape of the mating face [31] to improve the joint strength. Furthermore, the internal pressurization of the tube can be applied to the overall workpiece or locally by inner mandrels, as presented in [32] for the example of camshaft manufacture. Also, it is possible to conduct the expansion without a die that supports the expanded components, as for example investigated in [33]. Results of joining T-shaped connections of tube-with-tube are presented in [34]. The feasibility of joining of self-piercing riveting, clinching and fastening of screws in combination with hydroforming processes was shown by [20, 35], Figure 2.17.

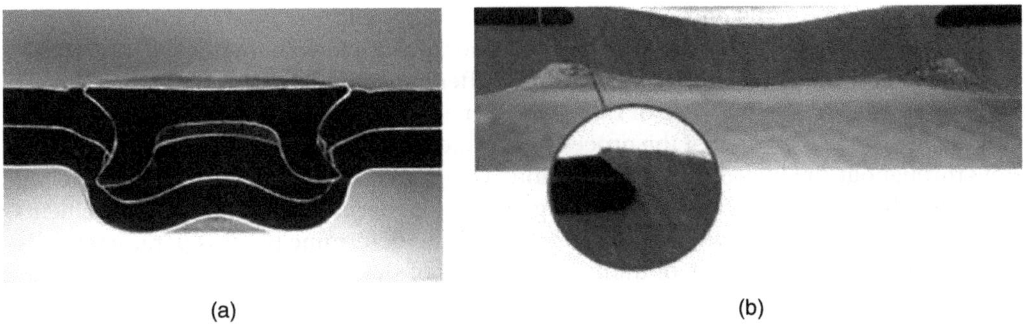

Figure 2.17 Different joining techniques in hydroforming processes. *Source*: Neugebauer 2005 and 2008 [20, 35]. Reproduced with permission of Springer. (a) Self-piercing riveting with wobble movement, steel sheets made of ZStE180BH, thickness 1.2 mm, and ZStE260BH, thickness 1.5 mm, internal pressure 1800 bar. (b) Hydro-clinching with active punch movement, blank materials DC04 1.0 mm, and AlMg0.7Si thickness 4.0 mm; point diameter 17.5 mm; internal pressure 1600 bar.

Piercing and cutting, as an integrated operation in hydroforming processes, can be applied to tube as well as to sheet hydroforming. It is common today in series production of tube hydroforming. A distinction can be drawn between inward and outward piercing as shown in Figure 2.18. Inward piercing requires a driven piercing punch that forces the cut material into the pressurized tube or sheet, respectively. For outward piercing, the pressurized medium cuts the slug into a cutting ring within the die cavity. The two methods provide different advantages and disadvantages concerning the slug removal after forming [36]. The application of a sequenced inward–outward piercing has shown a considerable reduction in burr formation [37]. Besides piercing of holes and openings, also cutting of sections at tubular hydroformed components, as shown in Figure 2.18 in principle, and of T-shaped parts are feasible [38].

2.2.2 Process Parameters

To obtain the required forming result, the control of forming loads should be appropriate to maintain plastic yielding of the hydroformed blank material with avoiding instabilities such as wrinkling, buckling, necking, or bursting. Maximum load parameters are of interest for machine and tool design, whereas the layout of a most suitable time-dependent path of load change is of importance for the reliable, faultless production process. The amounts of forming loads and courses of appropriate load paths are dependent on the material forming behavior,

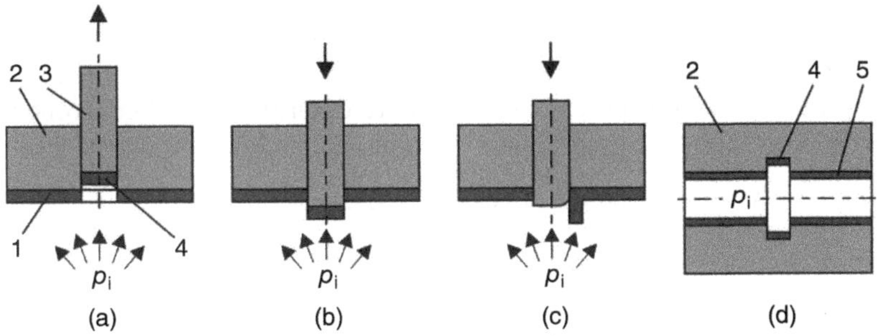

Figure 2.18 Piercing and cutting in hydroforming processes. (a) Outward piercing, (b) inward piercing, (c) inward piercing with attached slug, and (d) cutting of sections. 1, Cross section of expanded workpiece wall; 2, forming die; 3, punch; 4, slug; 5, tubular workpiece.

blank dimensions, the shape and complexity of the component that is to be formed, and on lubrication conditions in the contact area between workpiece and forming tool.

The individual types of hydroforming processes differ in the kind of loads that are to be considered additionally to the pressure of the forming medium. However, all types have in common that the selected load path of the process parameters is to be kept within a certain process window throughout the hydroforming operation. This window is limited by the possible failures as well as by limits of forces of the forming machine or permissible loads of tool elements. The following summary of methods and fundamentals, to determine process parameters, distinguishes between hydroforming of tubular blanks and of sheet metal blanks.

2.2.2.1 Tube Hydroforming

Several methods and strategies were developed and investigated in the past to determine suitable load paths for hydroforming processes, predominately for tube hydroforming. At an early stage of the application of hydroforming, modeling of tube hydroforming processes was conducted by the use of the Membrane Theory of Shells and the Levi and von Mises correlations, for example presented in [39–42], or by applying the Theory of Shells and the Continuum Theory of Plasticity, as described in [43]. Further analytical models, partially also considering failure determination, were than developed and presented, as for example by [44–48]. Nowadays, the finite element analysis (FEA) provides an economic way to support load path determination, as demonstrated for example with the work of [49–51]. Based on the use of FEA, several optimization strategies were proposed and investigated as for example hybrid-constrained multiobjective genetic algorithm [52] or the application of Taguchi Method [53] as well as fuzzy control technique [54]. In [55], such methods for load path determination were divided into fast methods, optimization methods, and adaptive simulation approaches, and compared for as an example of a specific hydroforming workpiece. In optimization methods, design sensitivities are used to establish optimized solutions in an iterative way. Adaptive methods apply continuously monitoring for failures and update the process parameters accordingly.

In tube hydroforming processes, the internal pressure p_i is applied to the inside of the tubular part to achieve its expansion, and the axial force F_a acts on the component ends to achieve its sealing as well as a material transport in direction of the die cavity, Figure 2.19. The component is formed under the simultaneously controlled action of p_i and F_a. The stress state, which is resulting from these loads, has to support the plastic yielding of the workpiece material, to enable its expansion. At the same time, instabilities such as necking, bursting, wrinkling, or buckling are to be avoided by the suitable definition of the individual amounts and combination of these loads during the forming process. According to [43], a range of feasible parameter variations can be specified, which is limited by the yield start of the workpiece, the occurrence of instabilities, and the minimum required force to seal the workpiece, as depicted in Figure 2.20.

Figure 2.21 shows the sequence of a hydroforming process, including sealing of the tubular part, the forming with axial feeding, internal pressurization, and the calibration with the subsequent relief of loads. Commonly, these process parameters are determined and controlled over the course of time of the forming process, as shown in Figure 2.21. However, it was dem-

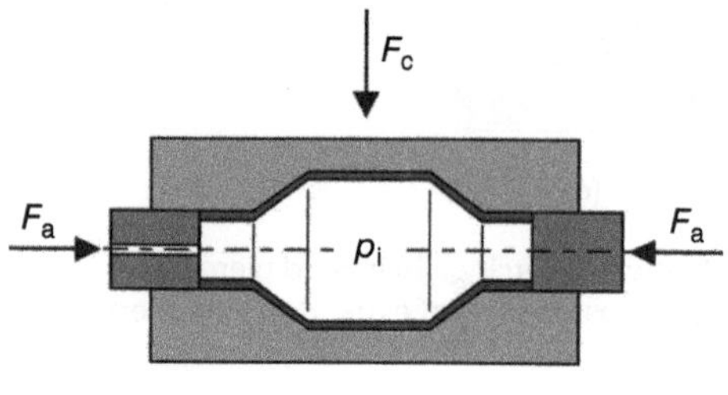

Figure 2.19 Loads in tube hydroforming.

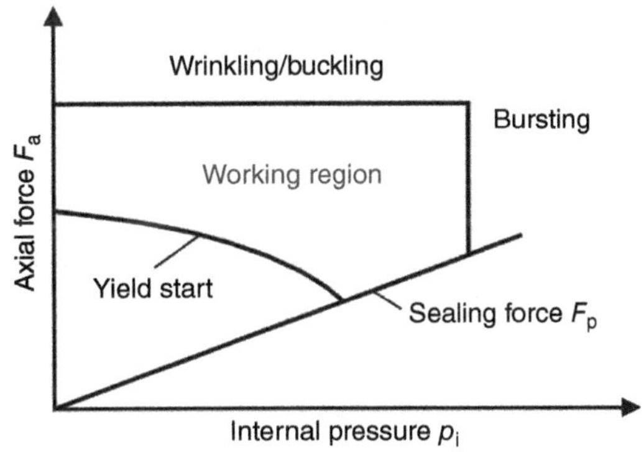

Figure 2.20 Range of process controls for tube hydroforming. *Source:* From Klaas 1987 [43].

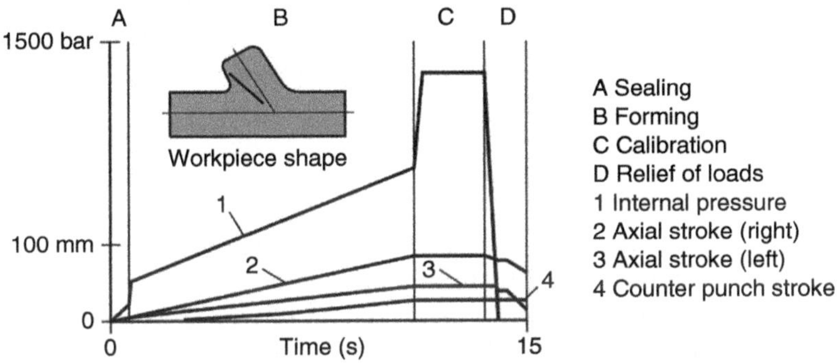

Figure 2.21 Process control for hydroforming of a tubular component. *Source:* From Hartl 2008 [48] and Jirathearanat et al. 2004 [51].

onstrated in [56] that with the control of the volume flow of the pressurizing medium, instead of control of the internal pressure, an enhancement in process stability can be achieved.

As shown by [43] for the state of free expansion of a cylindrical tube, the maximum applicable internal pressure p_b at the moment of bursting can be determined with

$$p_b = \sigma_{UTS}\, 2t_0 / (d_0 - t_0) \tag{2.1}$$

Here, σ_{UTS} is the tensile strength of the material, t_0 the initial wall thickness, and d_0 the initial diameter of the tube. The amount of p_b should not be exceeded within a hydroforming process while the tube is not in contact with the surrounding die cavity. Only when large areas of the expanded tube wall are aligned to the die cavity, the internal pressure p_i can be increased higher than p_b. Various correlations were developed, concerning the determination of the maximum internal pressure p_c at the end of the forming process, the calibration, when the component wall has to be formed into the corner radii of the die cavity. Examples are to be found in [46, 57]. In [58], an empirically deduced equation was presented, which have shown to be suitable for a first estimation to determine the maximum necessary internal pressure

$$p_c = 1.2\sigma_{UTS}\, t_0 / r_c \tag{2.2}$$

with the smallest corner radius r_c of the die cavity that is to be formed via the inner pressurization of the part.

According to the different types of hydroforming processes, various correlations were developed to determine a suitable amount of the axial force F_a which is acting on the end section of the hydroformed component, applied by the sealing punch. Examples are to be found in [43, 46, 48]. In general, the axial force F_a can be made up of three individual forces [48]

$$F_a = F_z + F_s + F_f \tag{2.3}$$

The force F_z is the component initiated in the tube wall and maintaining the plastic yielding of the tube wall together with the action of the internal pressure. For the hydroforming of components with predominantly free expansion, the correlation

$$F_z = \alpha\,\sigma_{UTS}\,t_0\,\pi(d_0 - t_0) \tag{2.4}$$

has shown to be applicable with $\alpha = 1.2, \ldots, 2.0$. When plastic yielding is predominantly induced by the axial loads, e.g. hydroforming of T-shaped components, the use of equations derived from continuum theory of plasticity is to be recommended, as for example summarized in [48].

The component F_s results from the reaction force of the internal pressure on the front punch face and can be determined with

$$F_s = 0.25\,p_i\,\pi\left(d_0 - 2t_0\right)^2 \tag{2.5}$$

For the case that the axial force F_a is less than F_s, leakage between the hydroformed tube ends and the punches will occur, and with this a premature termination of the process.

The frictional force F_f must be overcome when an axial feeding of workpiece material is applied, as the tube ends are in contact with the hydroforming tool throughout the forming process. When Coulomb's frictional behavior is taken as a basis, the friction force component is given as

$$F_f = \mu\,\sigma_N\,d_0\,\pi l_f \tag{2.6}$$

with the friction coefficient μ between tube and surrounding hydroforming tool, the contact pressure σ_N in this area, and the length l_f of the tube section moved along the tool under friction. For thin-walled tubes, it can be assumed as a first estimation that $\sigma_N \approx p_i$. A detailed correlation to determine σ_N as a function of tube dimensions, yield stress, and internal pressure is to be found in [48].

To ensure that the hydroforming tool remains closed throughout the hydroforming operation, the closing force F_c has to be applied. The minimum required force, can be determined by

$$F_c \geq p_i\,A_p \tag{2.7}$$

with the projected surface A_p of the hydroformed component shape, perpendicular to the tool closing direction.

2.2.2.2 Sheet Hydroforming

For the high-pressure sheet forming, as shown in Figure 2.22, analytical models to determine process parameters were developed and presented for example by [18, 59, 60]. However, predominantly numerical methods with the aid of FEA are used today for investigations into process parameters and suitable process controls of sheet hydroforming operations. This concerns investigations into optimization of blank holder forces to reduce elastic spring back of

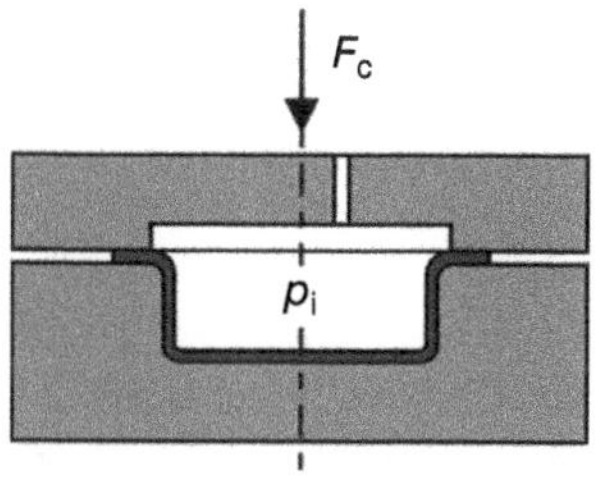

Figure 2.22 Loads in high-pressure sheet forming.

the formed component [61] or to improve drawing depth and wall thickness distribution with active-elastic tools [62].

To form the component in high-pressure sheet forming, the initial blank is pressurized by the pressure p_i which is acting on the blank on the opposite side to the forming die cavity, or in the case of double blanks, acting between the two blanks. The pressurization causes the expansion of the individual blank into the die cavity. Additionally to the forming load p_i, a blank holder force F_n is applied to the flange of the workpiece, which in general is a component of the closing force F_c. On the one hand, this force component enables the sealing of the pressurized medium volume, and on the other hand, it allows to control the material flow from the flange into the die cavity. Analogous to tube hydroforming, the stress state within the formed blank, resulting from the combination of these loads, has to enable the plastic yielding of the material without the occurrence of instabilities like necking or bursting. Also for high-pressure sheet forming, a range of feasible parameter variations can be specified according to [18], limited by instabilities, the minimum required force to seal the workpiece and the maximum applicable force provided by the forming machine (Figure 2.23).

Similar to tube hydroforming, also the high-pressure sheet forming process consists of an initial process stage, where the required amount of blank material flows into the die cavity, and a final process stage where the component shape is calibrated with increased internal pressure [62]. Figure 2.24 shows schematically a typical process control of the loads p_i and F_c for high-pressure sheet forming.

An upper bound of the forming pressure p_i is given by instabilities through an excessive strain state. The resulting bursting pressure can be estimated according to [18, 63] with

$$p_b = \sigma_{UTS} \, 2t_0 \, / \, \rho_p \tag{2.8}$$

where σ_{UTS} is the tensile strength of the material, t_0 the wall thickness of the initial sheet metal blank, and ρ_p is the radius of curvature of the expanded spherical cap at the moment of burst-

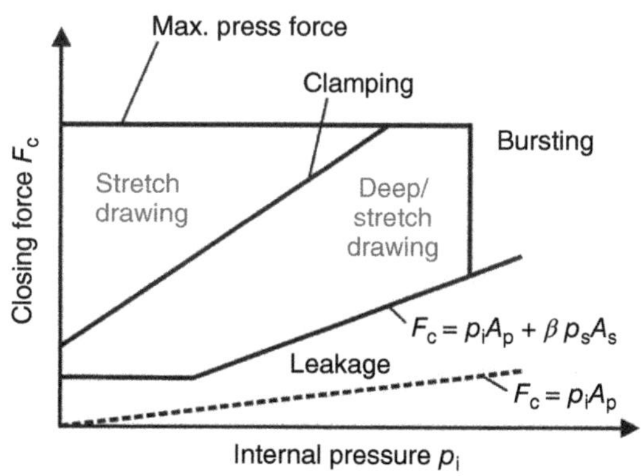

Figure 2.23 Range of process controls for high-pressure sheet forming. *Source:* After Hein and Vollertsen 1999 [18]. Reproduced with permission of Elsevier.

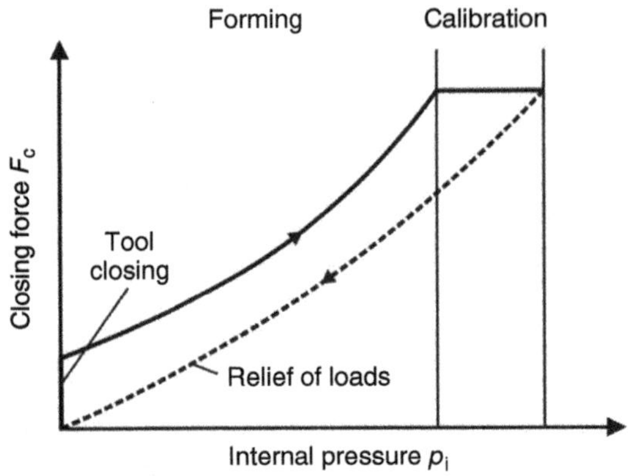

Figure 2.24 Schematic representation of process control for high-pressure sheet forming. [62]

ing. It has to be taken into consideration that this pressure p_b, in contrast to the bursting pressure for tube hydroforming according to Eq. (2.1), depends not only on the workpiece geometry but also on the chosen load path for F_c and p_i [18, 63]. Equation (2.2) can be used to estimate the maximum pressure p_c for calibration of the component into the corner radii of the die cavity.

The closing force F_c in high-pressure sheet hydroforming should be controlled in a way to ensure a defined draw in of material from the flange into the die cavity. For the amount of this force, the sealing limit provides a lower bound, and the maximum press force an upper bound (Figure 2.23). The contact pressure at the flange, induced by F_c, must be kept higher than a minimum pressure p_s in order to prevent an opening of contact area between flange and tool, where for increasing values of forming pressure p_i, the pressure p_s tends toward p_i [18]. According to [18], the sealing limit can be determined with

$$F_c > p_i A_p + \beta p_s A_s \tag{2.9}$$

where A_p is the projected area of the expanded shell within the die cavity and A_s the projected area of the flange. The parameter β is a safety factor with $\beta > 1$.

In hydromechanical deep drawing, the blank is drawn by a punch with the force F_p through a draw die, while in general, the flange is loaded with the blank holder force F_n, and a medium pressure p_i is pressing the formed blank against the punch during the forming operation (Figure 2.25). Analytic correlations to determine the process parameters for the hydromechanical deep drawing of sheet metal blanks, derived on the basis of fundamentals of conventional deep drawing, are presented for example in [2, 15]. According to [2], the higher the pressure p_i is during the drawing process, the higher is the feasible drawn component depth. An upper limit of p_i consists, besides applicable machine loads, in the bursting pressure p_b that causes the

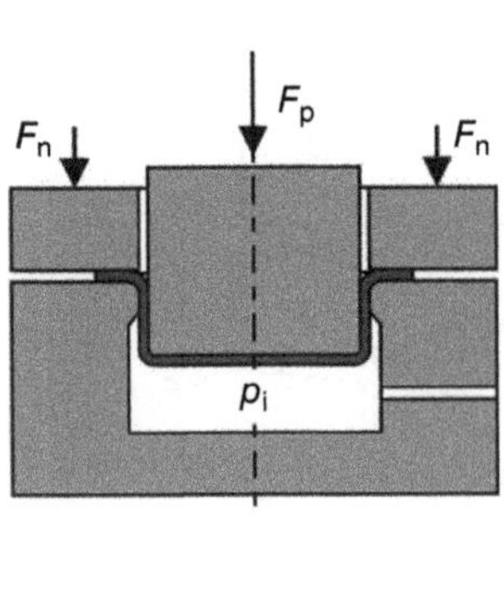

Figure 2.25 Loads in hydromechanical deep drawing.

bursting of the unsupported area of the blank between punch and draw die [2]. So that the correlation

$$p_i < p_b \tag{2.10}$$

applies for the maximum of the applicable pressure p_i in hydromechanical deep drawing.

2.2.3 Forming Limits

The attainable component geometries with hydroforming processes are limited by the occurrence of failures at the formed workpiece that can consist in necking with subsequent bursting, local or extensive wrinkling, and buckling. Figure 2.26 shows schematically typical locations of these failures at hydroformed geometries.

2.2.3.1 Necking and Bursting

Necking of tubular blanks as well as of sheet metal blanks is a result of an exceeded formability of the workpiece material. It can occur during the free expansion of the blank, as long as it is unsupported by the forming tool, and it can appear at corner radii of the workpiece during the calibration, when the predominant part of material is already in contact with the surrounding die cavity. A distinction is to be drawn between diffuse necking, characterized in general by a maximum point of a force or a pressure, and subsequent localized necking, which involves strain localization through the thickness. When the forming pressure is increased after necking has occurred, bursting of these areas will be the consequence.

The occurrence of the instability of necking with subsequent bursting is predominantly influenced by the formability of the blank material. Further on, forming operations preceded to the hydroforming process can reduce the remaining formability, like bending [48] or the roll forming of sheet metal blanks to tubular products [64]. In the free-forming stage of the hydroforming process, the acting stress state plays an essential role concerning the appearance of necking. For tube hydroforming, the application of axial compression stresses by the axial force F_a enables to reduce the risk of necking and with this an increase in expansion of workpiece perimeter, e.g. [29, 43, 46]. In high-pressure sheet forming processes, the stress

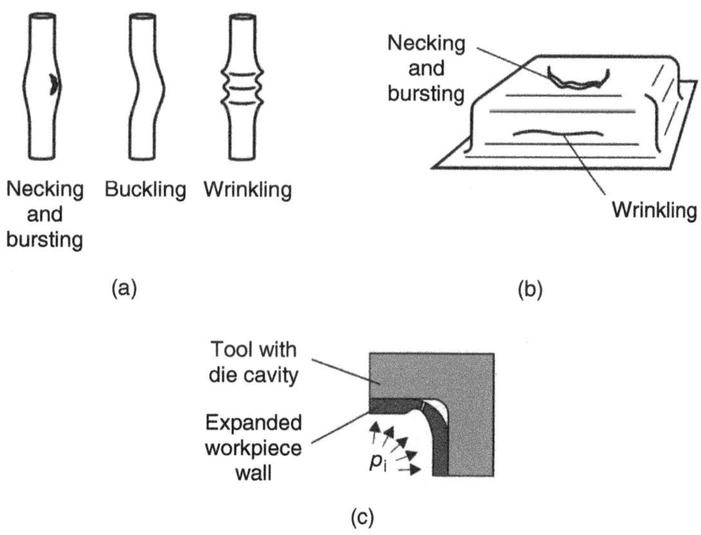

Figure 2.26 Possible failure modes in hydroforming, schematically. (a) Tube hydroforming, (b) high pressure sheet hydroforming, and (c) necking and bursting at calibration stage.

state is determined by the draw in of blank material into the die cavity, controlled via the amount of the closing force F_c, and the friction conditions within the contact area of flange and tool. In general, a reduction in material draw in by a high amount of F_c increases the risk of necking and bursting [20]. Suitable measures to counteract these instabilities at this early stage of the forming process consist in an adapted control of the closing force, as investigated in [18, 61, 62]. Necking and bursting in the calibration stage of the process is influenced by the friction between blank material and forming tool as depicted in Figure 2.26c. High friction forces avoid a material flow from those regions of the blank that are already in contact with the tool in direction to the free expansion regions where the radii have to be formed. This results in a necking of the free expansion region in the case that the formability of the blank material is exhausted. Similar to conventional deep drawing processes, necking and bursting in high-pressure sheet forming can appear also for nonoptimized shapes of the initial blank or at small radii where material is drawn in [20].

Currently, a large spectrum of different criteria exists to determine and predict fracture in forming processes. Concerning hydroforming, the analysis with forming limit diagrams (FLD) and according forming limit curves (FLC) is a common tool to investigate the feasibility of components in practice, as documented in [65]. Furthermore, the application of ductile fracture criteria in conjunction with FEA has shown to be suitable for failure prediction in hydroforming processes as presented for example by [66], and also damage-based criteria were investigated, e.g. [67]. Examples of closed solution for tube hydroforming are to be found for example in [40, 46, 68, 69].

2.2.3.2 Buckling and Wrinkling

In tube hydroforming processes, buckling as well as wrinkling results predominantly from excessive axial loads. These failures limit the amount of axial stress that can be applied to counteract the decrease in wall thickness during the expansion process. For axisymmetric components, wrinkling can occur during the course of their free expansion, whereas the wrinkling of T-pieces and Y-shaped components can appear despite the fact that the workpiece wall is in contact with the surrounding tool [48]. Through this, the transport of material into the forming die cavity is limited and necking with subsequent bursting can result. Buckling is observed predominantly for long, unsupported tubular blanks with thick walls, whereas wrinkling has to be considered for thin walled tubular workpieces [46]. For high-pressure sheet forming, wrinkling can occur when in the first forming stage, more than the required amount of material is drawn into the die cavity, and irreversible wrinkles are formed during the calibration stage. In contrast to conventional deep drawing processes, wrinkling within the corner of the flange at high-pressure sheet forming does not appear in general, due to the comparatively high blank holder forces [20].

First investigations concerning the determination of buckling in tube hydroforming were presented by [70]. The authors developed a criterion, based on buckling with plastic material behavior, to determine load paths for hydroforming of rotationally symmetrical workpieces with maximum applicable compressive stress. The axial load that resulted in buckling was taken as a limit for the maximum applied load, with the objective to reduce wall thinning. Reference [46] presented a simplified model to determine the minimum internal pressure to avoid wrinkling in tube hydroforming. In [68], closed-formed expressions for the critical axial compressive stress were derived to predict the onset of buckling, axisymmetric wrinkling and asymmetric wrinkling in tube hydroforming by treating these failure modes as an elastoplastic bifurcation problem.

2.2.4 Tribology

Tribological conditions at the interface of the formed blank and the forming tool influence decisively the achievable forming result in hydroforming processes. In general, for tube hydroforming as well as for sheet hydroforming a mixed-layer lubrication governs the tribological situation. That implies that, despite an intermediate layer of lubricant, also local contact between blank and tool occurs, in conjunction with an increase in friction at these areas. Furthermore, the resulting tribological conditions are characterized by time-dependent and locally varying interface pressures, sliding velocities, alteration of workpiece surface, and new surface generation.

In tube hydroforming processes, three different friction zones can be defined [71, 72], as depicted in Figure 2.27 for the example of hydroforming a rotational symmetrical component: The guiding zone, where the workpiece and the tool are in full contact during the process, with comparatively high interface pressures and sliding velocities due to the movement of the axial punches, and wall thickening of workpiece with surface contraction; the transition zone with reduced sliding velocity and start of surface enlargement; and the expansion zone with minimum value of sliding velocity and maximum of surface enlargement.

For high-pressure sheet forming, four different friction zones can be identified [73], as shown in Figure 2.28. The friction zone at the flange where compression stresses due to the blank holder force and the radial compression of this area are acting; the friction zone at the transition between flange and wall, where the material is subjected to bending and unbending stresses; the friction zone at the wall of the formed component, with tensile loading in axial and circumferential direction and compression in thickness direction; and the friction zone at the base of the workpiece, where the material undergoes bending stress mode.

According to [73], adequate lubricants for hydroforming processes should have the following attributes:

- Reduction of friction at the tool-workpiece interface to reduce required forming loads and to enable a more homogeneous material forming,
- Separation of surfaces and ability to respond to new surface generation to prevent metal-to-metal contact,
- Adaptability to varied conditions during forming, concerning varying interface pressures, sliding velocities, material deformation modes, and temperatures,

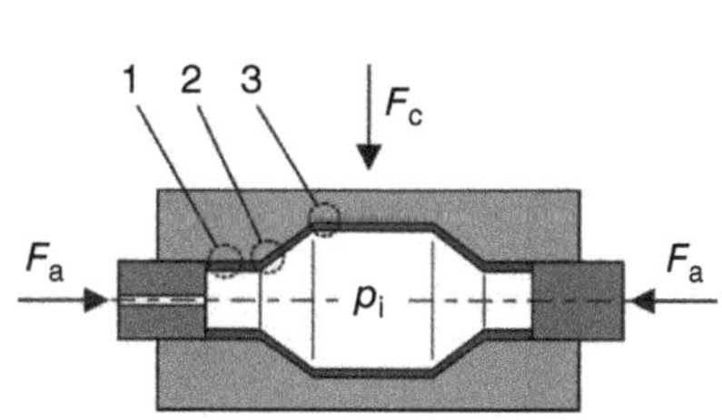

Figure 2.27 Friction zones in tube hydroforming, according to [71, 72]. 1, Guiding zone; 2, transition zone; 3, expansion zone.

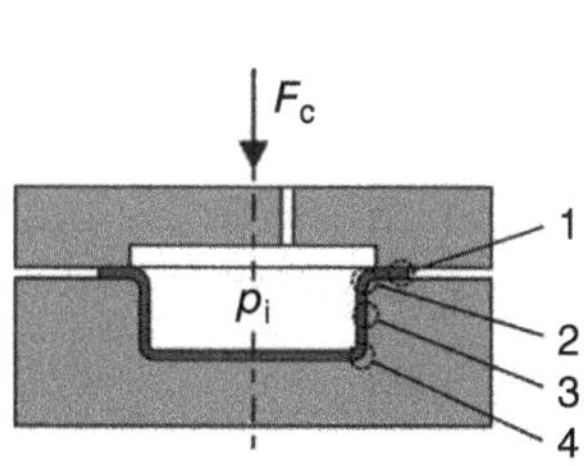

Figure 2.28 Friction zones in high-pressure sheet hydroforming. *Source:* From Ngaile 2008 [73].

- Compatibility with materials during forming and with subsequent processes like welding, painting, or annealing,
- Maintaining an acceptable surface finish on the hydroformed component without metal-pick-up and with reduced tool wear,
- Stability against extreme temperatures, oxidation or bacteriological attack, without corroding effect to the hydroformed component or tool, and a chemical reactivity that is solely beneficial to reducing friction,
- Easy to apply and to remove, without environmental or health impairments,
- Cost effectiveness of the lubricant and its application.

A wide spectrum of different lubricants for hydroforming processes has been developed in the past few years, predominantly based on lubricant formulation for sheet metal forming [73]. The lubricants used today in hydroforming production can be classified in oil-based lubricants, aqueous emulsions and aqueous suspensions, as well as solid lubricants [20]. For the evaluation of lubricants and tribological conditions in tube hydroforming processes, simulative test methods were developed that consider the situation at the guiding zone and expansion zone, e.g. documented in [72–76]. For example, the difference in maximum wall thinning in hydroforming of pear-shaped specimens can be used as a criteria to rank lubricants, as shown in Figure 2.29 [75].

Furthermore, surface coating of the forming tools is a measure to optimize the tribological conditions in the interface of forming tool and workpiece [3]. In particular, coatings are used for the guiding zone of tools for tube hydroforming. The aim is to reduce wear in this area which can be caused due to the higher relative movement between workpiece and tool. Analogous to textured sheet metal blanks for deep drawing processes, also textured surfaces of tubular blanks for hydroforming processes influence the friction during forming. Investigations, described in [77], into the frictional behavior of textures at tubes made by electrical discharge machining, knurling, and sand blasting, for example, have shown that the latter exhibit the lowest friction coefficient. This study also has shown that an increase in interface pressure can result in either lower or higher friction, depending on the surface texture conditions.

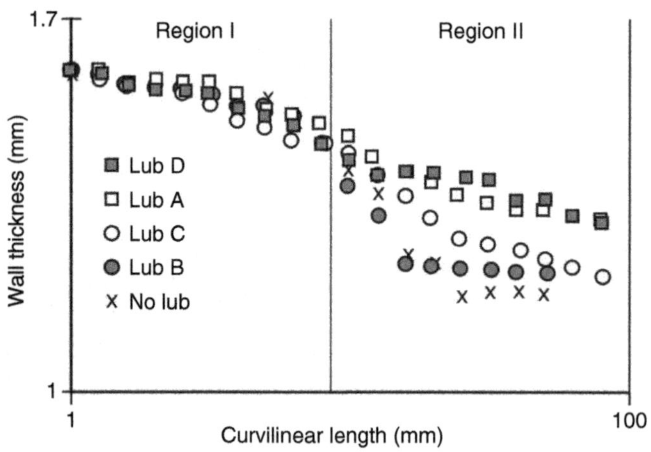

Figure 2.29 Effect of different lubricants on wall thickness distribution. *Source:* After Ngaile et al. 2004 [75]. Reproduced with permission of Elsevier.

2.3 Process Development and Design

The development of a hydroforming part production generally starts with a given component geometry, definition of number of parts produced per year, details of necessary geometrical accuracy, requirements to component strength, and cost specifications. Based on the selection of the semifinished product, its material, as well as the definition of additionally preceding, integrated or following process steps, the hydroforming process as such is defined. This includes the adequate design of component shape and tool, selection of lubricant, the determination of process loads, and the definition of suitable machine size. Typically, feasibility studies with the aid of FEA and experimental prototyping support the design of part and process.

2.3.1 Part Design

Figure 2.30 shows a possible classification of hydroformed components. Whether tubular or sheet blanks are to be recommended as semifinished products for the manufacture depends on the type of cross sections, whether these are closed or opened, the amount of difference in perimeters along the axis, whether this is significant or can be neglected and the shape of the main geometry, whether this is straight or bent [20]. Furthermore, the complexity of the cross sections, together with the decision about the position of the part in the hydroforming tool, determines whether preforming operations are necessary, for example bending and flattening of tubes or deep drawing of sheets.

Important topics in hydroform part design are the definition of parting line, the applied preceding operations, and the amount of expansion during forming. Concerning the parting line, the following recommendations apply to the hydroforming of tubes and sheet materials, respectively:

- Removal of the component has to be feasible without undercuts, to avoid the requirement of additional tool slides,
- Flat die cavities should result from the design of parting line (Figure 2.31a), as deep cavities increase the stresses in tool elements under forming loads (Figure 2.31b) and increase the necessary material deformation,
- For tube hydroforming, the closing of the tool has to be possible without clamping of tube material in the area of the parting line (Figure 2.31c),
- For sheet hydroforming, a bending of the initial blank with closing the tool has to be feasible without wrinkling of the sheet material,
- Largely symmetrical die cavity cross sections should result from the parting line to avoid lateral forces (Figure 2.31d).

Preforming operations can consume a significant amount of formability of the blank material [48, 78, 79]. Thus, it has to be taken into account that the formability of a preformed component

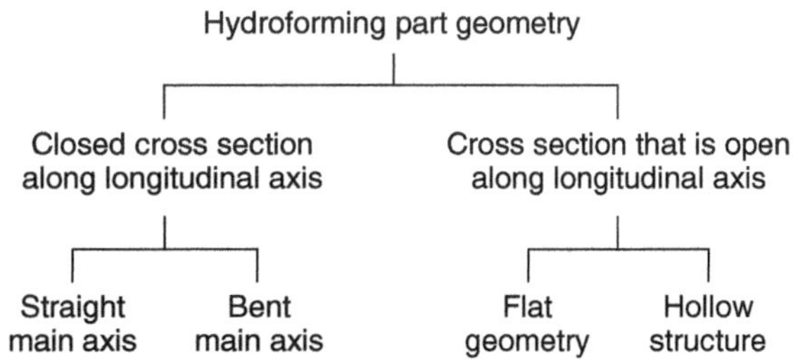

Figure 2.30 Classification of hydroformed components. [20]

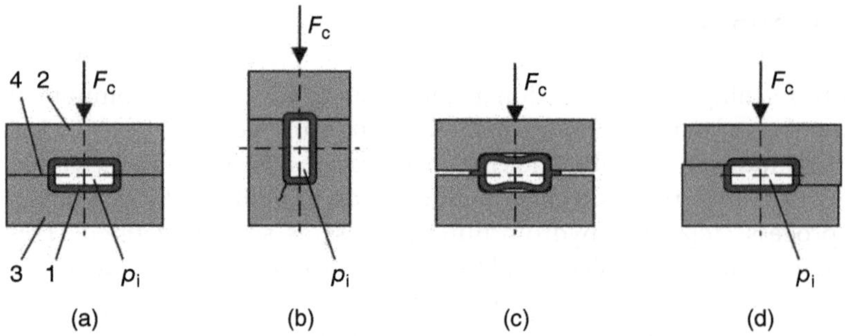

Figure 2.31 Design of tool parting line. (a) Symmetrical die halves with flat die cavities, (b) deep die cavity with risk of reduced tool lifetime, (c) clamping of workpiece material, (d) lateral forces and vertical shifting of die halves due to unsymmetrical parting line. 1, Cross section of workpiece; 2, top die; 3, bottom die; 4, tool parting line.

is restricted in the hydroforming process. This applies particularly for bending of tubes as a preforming operation. The wall thickness of bent tubes decreases at the outer radius and increases at the inner radius. Depending on the ratio of tube diameter to bending radius, an expansion in the following hydroforming operation can cause an immediate local necking and bursting at the outer radius of the workpiece. Furthermore, elastic spring back of preformed blanks has to be taken into account when designing the individual steps and the corresponding tooling [79].

The attainable expansion or drawing depth, respectively, in hydroforming processes is restricted by forming limits, as described before. In tube hydroforming, the applicable amount of axial material feeding, to provide axial stresses in the forming zone, increases the feasible expansion. However, it is limited by the occurrence of wrinkling or buckling of the tubular workpiece. In sheet hydroforming, the control of blank holder force is the decisive parameter to obtain improved drawing depths. Besides some specific design recommendations, as documented for example in [20, 78–82], the following general design rules can be defined for the development of hydroforming component shapes:

- Sharp corners and edges are to be avoided as these require increasing forming pressures and impede the material flow.
- It should be aimed at an evenly distributed strain to reduce large differences in strength as well as to reduce influences of elastic spring back (the higher the amount of plastic strain, the smaller is the spring back and the higher is the formed component accuracy [20]); in particularly large flat component areas should be avoided.
- For tube hydroforming components, large expansions should be close to the tube ends where axial feeding can be provided (an expansion of tube perimeter of about 10% is feasible without influence of axial compressive stresses in straight areas of the component, while about 30–60% expansion can be attained with axial feeding).
- Practical experience in tube hydroforming has shown that ratios of tube diameter to tube wall thickness d_0/t_0 from 20 up to 45 are advantageous for processes with axial feeding; higher values of this ratio increase the risk of wrinkling and buckling, and requires the reduction of the free and unsupported tube length below a length of $2\,d_0$, which is recommended not to be exceeded in general [20].

In general, the effort to achieve similar accuracies for hydroformed components as for conventional sheet metal parts or bulk metal formed parts, is comparatively high. Various parameters influence the accuracy and the scattering of dimensions, such as the used material, process chain and process parameters, tool design and the aimed component shape. In particular, the comparatively high-forming loads result in according elastic deformation of the tools, which influences the produced dimensions, superimposed by the elastic spring back of the formed component itself. Only for hydromechanical deep drawing a crucial improvement in accuracy compared with conventional deep drawing was documented [21]. Influences of the process chain on the produced component accuracy were investigated by [83]. Investigations into possibilities to improve the accuracy in high-pressure sheet forming by controlled blank holder force are presented in [61].

2.3.2 Semifinished Products

In general, all metal materials that commonly are used in cold forming operations, and providing a sufficient formability, are suitable for semifinished products in hydroforming processes. Advantageous for a good formability are a fine-grained structure of the material, large values of uniform elongation until necking and elongation at fracture, as well as a high strain-hardening coefficient. A high value of work-hardening of the formed material improves the strength of the produced component. However, this increases also the necessary medium pressure for the final calibration of the hydroformed part.

The suitability of a semifinished product for hydroformed component is also determined by its manufacturing process [64] and preceding forming operation that might have been performed, for example bending or preforming. In principle, annealing processes can be applied to regain material formability used up by any preceding forming operations. However, as additional costs are arising from this, it is commonly avoided in industrial mass production of hydroformed components. Furthermore, the application of annealing processes can cause deterioration in dimension qualities of bent or preformed tubes, resulting in unstable conditions during hydroforming.

In most cases, the used semifinished products show an anisotropic forming behavior. This means that the material parameters differ in longitudinal direction of the blank and across to this direction. For sheet metal blanks, the anisotropy results from the manufacturing process by rolling. It is recommended to consider this behavior in sheet hydroforming processes by positioning the blank in the forming tool with the direction of the highest applicable material strains in the direction of the highest required strains for the component forming [20]. Tubular products, made from sheet metal blanks, consequently, show an anisotropic behavior as well. Furthermore, anisotropy caused by the manufacturing process can apply to extruded profiles. According to [42], an increasing anisotropic parameter in longitudinal direction of a tubular blank provides a decrease in thickness reduction and an increase in feasible expansion diameter for tube hydroforming. Based on experimental results into anisotropy of tubular material, it is recommended in [84] to select tubes with high anisotropic parameters to reduce change in wall thickness in tube-bending processes. Investigations into bulge tests of dual-phase steel tubes, presented by [64], have shown the influence of anisotropic behavior when expanding discontinuously formed and laser-welded tubes that were manufactured with the longitudinal axis parallel to the rolling direction of the initial sheet as well as perpendicular to it. In the latter case, the tubes showed a higher amount of feasible expansion. The importance to consider the influence of anisotropy in simulations of tubular hydroforming of seamless aluminum tubes made of a 6260-T4 aluminum alloy was demonstrated by [85, 86], based on hydroforming experiments.

Today, predominantly ductile low carbon steels and aluminum alloys were applied in hydroforming processes [20, 87]. Tube hydroforming of stainless steel can be found in piping applications, and in mass production of automotive exhaust system components [3, 87]. Several applications used within the body in white structures of cars are documented in [9]. However, concerning steel materials, there is a trend toward the use of alloys with increased strength, where weight reduction in automotive vehicle design is the driving force for this development. Examples of investigations into hydroforming of corresponding tubes made from steel alloys with higher strength are to be found in [88–90]. A wide spectrum of aluminum alloys is applied in industrial hydroforming production or has been investigated, respectively, as summarized for example in [6, 20, 87]. Work hardening aluminum 5000 alloys are used when the priority is for a high amount of formability and corrosion resistance, whereas precipitation-hardening aluminum 6000 alloys are applied in cases where components made from tubular profiles with higher strengths are required.

Against the background to meet further demands for weight reduction of hydroformed automotive components, research facilities, hydroforming technology suppliers, and users developed innovative semifinished products for tubular as well as for sheet metal blanks with tailored properties. Such alternative manufacturing methods for the initial blanks comprise laser welded blanks, as well as flexible rolled blanks or patchwork blanks, as depicted in Figure 2.32. Those products provide an optimized distribution of wall thickness and/or material strength. Investigations into high-pressure sheet forming of laser-welded tailored and patchwork blanks, presented by [91], showed that the blank layout and weld line geometry are decisive factors to prevent premature material failures. This applies in comparable form to hydroforming of tubular laser-welded semifinished products also, as analyzed with the aid of FE-simulations for example by [92, 93]. Flexible rolled blanks are sheet metal blanks where thickness distribution in rolling direction is modified by adjusting the roll gap during the rolling process. According semifinished products are available for sheet metal blanks as well as for tubes [94]. Hydroformed prototypes for automotive applications manufactured with high-pressure sheet-hydroforming and tube hydroforming are presented in [94, 95], respectively.

For tubular components with significant varieties in perimeters along the main axis, which are not feasible to be produced in a single hydroforming step from a cylindrical tube, initial tubular blanks with an adjusted shape may be a solution. Examples are conical tubes [96] or customized geometries [97] from individually shaped and laser-welded sheet metal blanks, like shown in Figure 2.33.

To tailor properties of initial blank or formed component made from aluminum alloy for hydroforming, local heat treatment showed to be a feasible strategy. In [98], experimental results of hydroformed tubes made from aluminum alloy 6063-T4 are presented, where a local modification of the formed components to T6-condition could be conducted. Simulation results of a local heat treatment to soften certain areas of extruded aluminum profiles, as initial blanks, are described in [99].

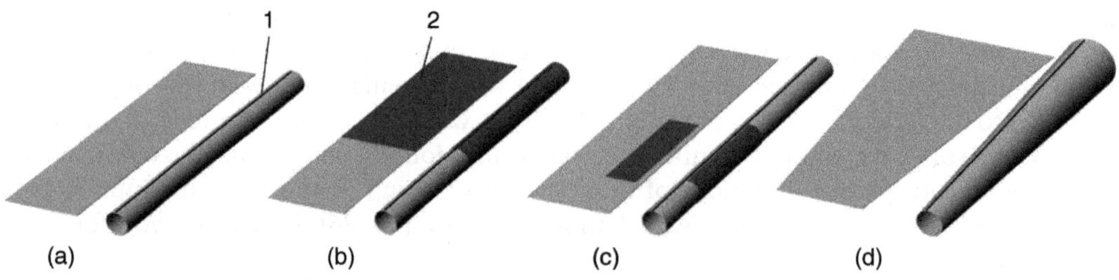

Figure 2.32 Examples of tailored blanks. (a) One-piece cylindrical tube, (b) tailored blank and tailored tube, (c) patchwork blank and patchwork tube, and (d) one-piece conical tube. 1, Welding seam; 2, modified thickness and/or properties. *Source:* After Hartl 2004 [87]. Reproduced with permission of Elsevier.

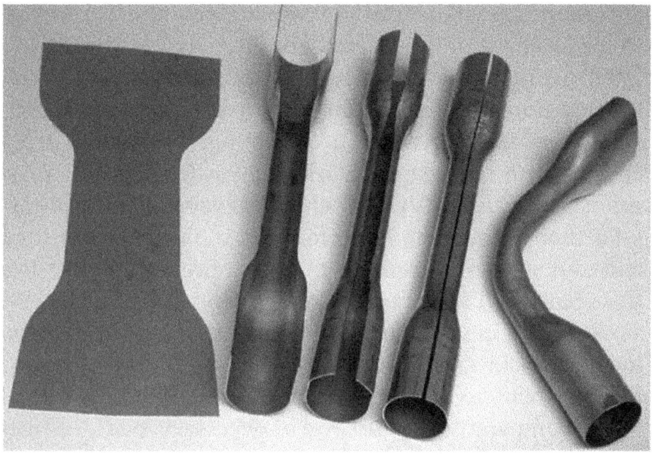

Figure 2.33 Customized blank and tubular product with varying cross sections. *Source:* From weil technology 2005 [97].

2.4 Hydroforming Systems

Besides the applied process parameters and quality of the used blanks, the produced quality of hydroformed parts is decisively determined by the appropriate design of the hydroforming tool and machine. Furthermore, important production parameters such as cycle time, production reliability, equipment availability, and production costs depend on their design.

2.4.1 Forming Tools

Figure 2.34 represents a schematic drawing and an example of hydroforming tooling for mass production. The tool inserts contain the die cavity and are in general made of heat-treated tool steel. Adjusting plates, covering the contact surfaces between the tool inserts and the basic tool block, are used to adjust the correct position of the inserts to each other. The basic tool blocks

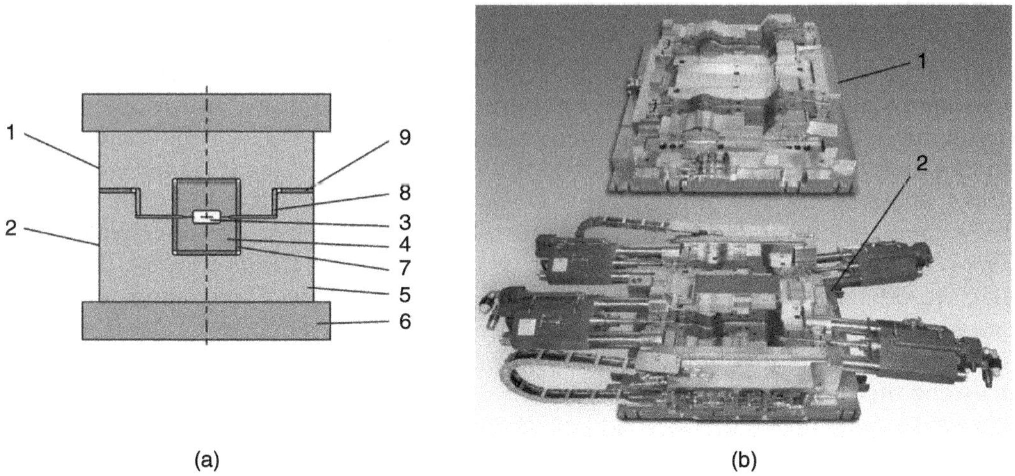

(a) (b)

Figure 2.34 Hydroforming tools. (a) Schematic drawing of the cross section of a tool for hydroforming of tubular components and (b) example of a hydroforming tool for manufacture of left- and right-hand part for engine cradles. *Source:* Siempelkamp Pressen Systeme. 1, Top die; 2, bottom die; 3, die cavity; 4, tool inserts; 5, basic tool blocks; 6, base plates; 7, matching plates; 8, guiding plates; 9, matching plates.

are commonly made from heat-treatable tool steel with a strength that is lesser in comparison to the strength of the inserts. For large hydroforming tools, also casted tool blocks were used [100]. Stiff guiding systems, mounted at the basic tool blocks, serve for the positioning of the top and bottom when closing the hydroforming tool. To reduce wear of the die cavity, corresponding surface treatments or coatings are applied in general to the cavity surface which is in contact with the formed workpiece, as for example plasma nitriding or coatings by physical vapor deposition (PVD)- or chemical vapor deposition (CVD)-technique [75, 101, 102].

Further elements in hydroforming tools consist in sealing punches, ejectors, piercing units, and tool slides. The sealing punches, predominantly driven by hydraulic cylinders, are made from hardened tool steel. Different sealing systems were developed and tested in the past, as summarized in [103] for hydroforming of tubular blanks or in [18] for hydroforming of sheet metal pairs. Ejectors are integrated to remove the formed part from the die cavity but offer also the possibility to locate the starting blank during loading [12].

The comparatively high level of necessary loads, resulting from internal pressure and closing force, is one important difference between hydroforming and other forming processes. These acting loads generate elastic deformation of the tool elements that crucially influences part quality and tool lifetime. General information about adequate hydroforming tool design are presented, for example, in [12, 20, 100, 102], suitable to consider the requirements for mass production and sufficient tool stiffness in tubular hydroforming. Design guidelines and details for sheet hydroforming tools are presented for example in [15, 20, 61, 62, 104] for high-pressure sheet forming, and in [2, 15, 105, 106] for hydromechanical deep drawing. Concerning the consideration and determination of reduction of lifetime of hydroforming tools caused by fatigue, strategies and experimental results are published in [107]. The presented analysis results showed that additionally carbides and nonmetallic inclusions within the forming tool elements, as well as marks at the cavity surface from manufacturing processes influences the lifetime [107].

2.4.2 Machine Systems

The predominant tasks of the hydroforming machine are to open and close the hydroforming tool, to supply and pressurize the forming medium, and to enable the controlled movement of tool elements with application of loads for the production process. In general, a water-oil emulsion is used as pressurizing medium. Several different press concepts have been developed within the past few years for tube and sheet hydroforming, aiming at further optimization concerning cycle time reduction by improved movement of the systems for loading and unloading the workpiece and regarding cost reduction by simplified press design. Examples and summaries of according machine concepts are to be found in [1, 3, 12, 20, 102, 108–110].

The systems basically differ in either the structure of the press frame or principle of press slide movement and locking mechanism for the press ram. Figure 2.35 shows the two principles that are predominately used in industry today, both based on the use of hydraulic actuators for the press slide movement. The one principle is based on traditional hydraulic presses with one or more large cylinders that move the press slide, carrying the top die and apply the closing force during the hydroforming process, Figure 2.35a. The second principle uses a hydraulic drive with low force but fast movement to close the hydroforming die, combined with a mechanical locking or stop of the slide during the forming process. One or more cylinders with short stroke are used then to apply the closing force. These short stroke cylinders can be integrated into the slide or the bed of the press.

Typical hydroforming presses for large tubular components like side members for automotive applications are applied with pressure intensifiers for maximum pressures between 2000

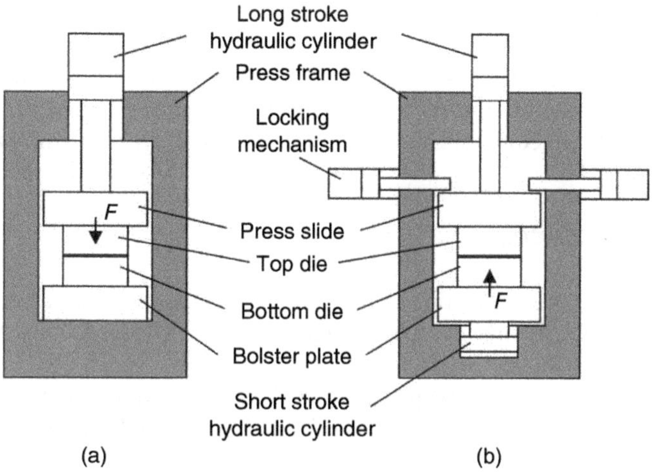

Figure 2.35 Typical principles of hydroforming presses, schematically. (a) One hydraulic drive for movement and closing of hydroforming tool and (b) fast hydraulic drive for closing the tool and short stroke cylinder for application of closing force. *Source:* Hartl 2005 [3]. Reproduced with permission of Elsevier.

Figure 2.36 Hydroforming production press (50 MN). *Source:* Schuler.

and 4000 bar, and enable closing forces in the range of 70 up to 80 MN. Figure 2.36 shows an example of a production machine for tube hydroforming. At present, presses up to 100 MN are in operation for high-pressure sheet forming.

2.5 Concluding Remarks

Research and development work carried out during the past few years by research facilities, suppliers of hydroforming equipment, and users has led to the status that hydroforming achieved today, which enables an economic mass production of components from tubes, profiles, and sheet metals. Against the background, to further raise the feasible component geometries with hydroforming processes, the use of evaluated temperatures is one important subject of current research and development work, as summarized in [1]. The use of oscillation, as a method to enhance material formability and improve tribological conditions, was investigated successfully in [111, 112]. Continuous optimizations in machine and tool technology, in simulation methods, as well as in semifinished product designs, are current measures which are contributing to improvements of hydroforming applications.

References

1 Koç, M. and Cora, O.N. (2008). Introduction and state of the art of hydroforming. In: *Hydroforming for Advanced Manufacturing* (ed. M. Koç), 1–29. Cambridge: Woodhead Publishing.

2 Siegert, K. and Wagner, S. (2008). Hydroforming sheet metal forming components. In: *Hydroforming for Advanced Manufacturing* (ed. M. Koç), 216–237. Cambridge: Woodhead Publishing.

3 Hartl, C. (2005). Research and advances in fundamentals and industrial application of hydroforming. *Journal of Materials Processing Technology* 167: 383–392.

4 Lang, L.H., Wang, Z.R., Kang, D.C. et al. (2004). Hydroforming highlights: sheet hydroforming and tube hydroforming. *Journal of Materials Processing Technology* 151: 167–177.

5 Vollertsen, F. (2001). State of the art and perspectives of hydroforming of tubes and sheets. *Journal of Materials Science and Technology* 17 (3): 321–324.

6 Koç, M. and Altan, T. (2001). An overall review of the tube hydroforming technology. *Journal of Materials Processing Technology* 108: 384–393.

7 Ahmetoglu, M. and Altan, T. (2000). Tube hydroforming: state-of-the-art and future trends. *Journal of Materials Processing Technology* 98: 25–33.

8 Schmoeckel, D., Hielscher, C., Huber, R., and Geiger, M. (1999). Metal forming of tubes and sheets with liquid and other flexible media. *CIRP Annals* 28 (2): 497–513.

9 Tolazzi, M. (2010). Hydroforming applications in automotive: a review. *International Journal of Material Forming* 3: 307–310.

10 Koganti, R., Balzer, J., and Hertell, K. (2007). Current trends of hydroforming process into automotive body structure and chassis applications. In: *ASME International Mechanical Engineering Congress and Exposition*, Seattle, Washington, USA (11–15 November), 33–38.

11 Wendt, A. and Delker, M. (2005). Hydroforming in the BMW Group in correlation with economic efficiency, innovation and process excellence. In: *International Conference Hydroforming*, Fellbach, Germany (25/26 October), 1–16.

12 Hartl, C. and Abbey, T. (1999). Product development of complex hydroformed parts and requirements regarding tool manufacture. In: *Proceedings of 6th ICTP, Advanced Technology of Plasticity*, Erlangen, Germany (19–24 September), vol. 2, 1183–1188.

13 Siegert, K. and Guel-López, P. (1999). Hydroforming of tubes with external high pressure. In: *International Conference Hydroforming*, Fellbach, Germany (12/13 October), 463–480.

14 Fuchs, F.J. (1965). Production metal forming with hydrostatic pressures. ASME paper 65 Prod. (17 April). pp. 34–40.

15 Siegert, K. and Loesch, B. (1999). Sheet metal hydroforming. In: *International Conference Hydroforming*, Fellbach, Germany (12/13 October), 221–248.

16 Kang, B.-S., Son, B.-M., and Kim, J. (2004). A comparative study of stamping and hydroforming processes for an automobile fuel tank using FEM. *International Journal of Machine and Tool Manufacture* 44: 87–94.

17 Bergkvist, M. (2008). Fluid cell pressing in the aerospace industry. In: *Hydroforming for Advanced Manufacturing* (ed. M. Koç), 315–334. Cambridge: Woodhead Publishing.

18 Hein, P. and Vollertsen, F. (1999). Hydroforming of sheet metal pairs. *Journal of Materials Processing Technology* 87: 154–164.

19 Geiger, M., Merklein, M., and Cojutti, M. (2009). Hydroforming of inhomogeneous sheet pairs with counterpressure. *Production Engineering* 3: 17–22.

20 Neugebauer, R. (2007). *Hydro-Umformung.* Berlin: Springer Verlag.

21 Finckenstein, E.V. and Brox, H. (1990). Sonder-Tiefziehverfahren. In: *Umformtechnik,* vol. 3 (ed. K. Lange), 449–468. Berlin: Springer Verlag.

22 Nakamura, K. (1987). Sheet metal forming with hydraulic counter pressure in Japan. *CIRP Annals* 36 (1): 191–194.

23 Lang, L., Danckert, J., and Nielsen, K.B. (2004). Study on hydromechanical deep drawing with uniform pressure onto the blank. *International Journal of Machine and Tool Manufacture* 44: 495–502.

24 Vollertsen, F. (2002). Process layout avoiding reverse drawing wrinkles in hydroforming of sheet metal. *CIRP Annals* 51 (1): 203–208.

25 Wagner, S., Jaeger, S., and Rudolf, S. (2005). Combination of deep drawing with subsequent hydroforming. In: *International ESAFORM Conference on Material Forming,* Cluj-Napoca, Romania (27–29 April), 471–474.

26 Geiger, M. and Lamprecht, K. (2006). Hydroforming of deep-drawn performs – a novel approach towards innovative sheet-metal components. In: , Annals of the German Academic Society for Production Engineering, vol. 13(2), 121–124.

27 Wagner, S. and Jaeger, S. (2005). Forming of double sheets by the process combination deep drawing with subsequent hydroforming. In: *International Conference Hydroforming,* Fellbach, Germany (25/26 October), 317–334.

28 Lovric, M. (2005). Press-bending of thin-walled tubes – increasing the productivity of internal high pressure forming processes. In: *International Conference Hydroforming,* Fellbach, Germany (25/26 October), 365–386.

29 Dohmann, F. and Hartl, C. (1994). Liquid-bulge-forming as a flexible production method. *Journal of Materials Processing Technology* 45: 377–382.

30 Ortmann, S. (2006). Innenhochdruckumformen-Kragenziehen. *wt Werkstatttechnik online* 96 (10): 758–760.

31 Dohmann, F. (1993). Innenhochdruckumformen. In: *Umformtechnik,* vol. 4 (ed. K. Lange), 252–270. Berlin: Springer Verlag.

32 Schroeder, M. (1999). Hydroforming of BIW structural parts and components. In: *International Conference Hydroforming,* Fellbach, Germany (25/26 October).

33 Marré, M., Gies, S., Maevus, F., and Tekkaya, A.E. (2010). Joining of lightweight frame structures by die-less hydroforming. *International Journal of Material Forming* 3 (1): 1031–1034.

34 Groche, P. and Tibari, K. (2006). Fundamentals of angular joining by means of hydroforming. *CIRP Annals* 55 (1): 259–262.

35 Neugebauer, R., Mauermann, R., and Gruetzner, R. (2008). Hydrojoining. *International Journal of Material Forming* (Suppl. 1): 1303–1306.

36 Smith, L.M. (2008). Hydroforming: hydro-piercing, and-cutting and welding. In: *Hydroforming for Advanced Manufacturing* (ed. M. Koç), 202–215. Cambridge: Woodhead Publishing.

37 Shiomi, M., Ueda, Y., and Oskada, K. (2006). Piercing of steel sheet by using hydrostatic pressure. *CIRP Annals* 55 (1): 255–258.

38 Bietke, D. (2005). Innenhochdruckabschneiden mit Schneidring von Hohlprofilen. Doctoral thesis. Otto-von-Guerike-Universitaet Magdeburg, Magdeburg, Germany.

39 Woo, D.M. (1973). Tube bulging under internal pressure and axial force. *Journal of Engineering Materials and Technology* (10): 219–223.

40 Sauer, W.J., Gotera, A., Robb, F., and Huang, P. (1978). Free bulge forming of tubes under internal pressure and axial compression. In: *Sixth North American Metalworking Research Conference*, Gainsville (16–19 April), 228–235.
41 Fuchizawa, S. (1984). Influence of strain hardening exponent on the deformation of thin-walled tubes of finite length subjected to hydrostatic internal pressure. *Advanced Technology of Plasticity* 1: 297–302.
42 Fuchizawa, S. (1987). Influence of plastic anisotropy on deformation of thin-walled tubes in bulge forming. *Advanced Technology of Plasticity* 2: 727–732.
43 Klaas, F. (1987). *Aufweitstauchen von Rohren durch Innenhochdruckumformen.* Düsseldorf: VDI.
44 Asnafi, N. (1999). Analytical modelling of tube hydroforming. *Thin-Walled Structures* 34: 295–330.
45 Xia, Z. (2001). Failure analysis of tubular hydroforming. *ASME Journal of Engineering Materials and Technology* 123: 423–429.
46 Koç, M. and Altan, T. (2002). Prediction of forming limits and parameters in the tube hydroforming process. *International Journal of Machine Tools and Manufacture* 42: 123–138.
47 Yang, C. and Ngaile, G. (2008). Analytical model for planar tube hydroforming: Prediction of formed shape, corner fill, wall thinning, and forming pressure. *International Journal of Mechanical Sciences* 50: 1263–1279.
48 Hartl, C. (2008). Deformation mechanism and fundamentals of hydroforming. In: *Hydroforming for advanced manufacturing* (ed. M. Koç), 52–76. Cambridge: Woodhead Publishing.
49 Koç, M., Allen, T., Jirathearanat, S., and Altan, T. (2000). The use of FEA and design of experiments to establish design guidelines for simple hydroformed parts. *International Journal of Machine Tools and Manufacture* 40: 2249–2266.
50 Aue-U-Lan, Y., Ngaile, G., and Altan, T. (2004). Optimizing tube hydroforming using process simulation and experimental verification. *Journal of Materials Processing Technology* (145): 137–143.
51 Jirathearanat, S., Hartl, C., and Altan, T. (2004). Hydroforming of Y-shapes-product and process design using FEA simulation and experiments. *Journal of Materials Processing Technology* 146: 124–129.
52 An, H., Green, D.E., and Johrendt, J. (2011). A hybrid-constrained MOGA and local search method to optimize load path for tube hydroforming. *International Journal of Advanced Manufacturing Technology* https://doi.org/10.1007/s00170-011-3648-0.
53 Li, B., Nye, T.J., and Metzger, D.R. (2007). Improving the reliability of the tube-hydroforming process by the Taguchi Method. *ASME Journal of Pressure Vessel Technology* 129 (2): 242–247.
54 Aydemir, A., de Vree, J.H.P., Brekelmanns, W.A.M. et al. (2005). An adaptive simulation approach designed for tube hydroforming processes. *Journal of Materials Processing Technology* 159: 303–310.
55 Jansson, M., Nilsson, L., and Simonsson, K. (2007). On process parameter estimation for the tube hydroforming process. *Journal of Material Processing Technology* 190: 1–11.
56 Groche, P., Steinheimer, R., and Schmoeckel, D. (2003). Process stability in the tube hydroforming process. *CIRP Annals* 52 (1): 229–232.
57 Birkert, A. (1997). Abschaetzung der Kalibrierdruecke beim Innenhochdruckumformen. *Blech Rohre Profile* 9: 102–107.
58 Braeutigam, M. and Rutsch, H. (1992). *Hydroformen – als Ausweg aus der Investitionsklemme*, VDI Berichte Nr. 946, 185–202. Duesseldorf: VDI.
59 Banabic, D. (1999). Closed-form solution for bulging through elliptical dies. In: *International Conference Sheet Metal*, Erlangen-Nuremberg, Germany (27/28 September), 623–628.

60 Geiger, M. (2005). Sheet hydroforming: analytical modeling and experimental verification of complex structures. *Production Engineering* 12 (2): 69–72.

61 Tekkaya, E. and Trompeter, M. (2010). Untersuchungen zum Rückfederungsverhalten wirkmedienbasiert umgeformter Blechteile. In: *30th EFB-Kolloquium Blechverarbeitung*, Bad Boll, Germany (2/3 March), 359–374.

62 Groche, P. and Metz, C. (2006). Investigation of active blank holder systems for high-pressure forming of metal sheets. *International Journal of Machine Tools and Manufacture* 46: 1271–1275.

63 Shang, H.M., Qin, S., and Tay, C.T. (1997). Hydroforming sheet metal with intermittent changes in the draw-in condition of the flange. *Journal of Materials Processing Technology* 63: 72–76.

64 Groche, P. and Breitenbach, G.v. (2005). Influence of tube manufacture processes on hydroforming. In: *International Conference Hydroforming*, Fellbach, Germany (25/26 October), 219–240.

65 Green, D.E. (2008). Formability analysis for tubular hydroformed parts. In: *Hydroforming for advanced manufacturing* (ed. M. Koç), 93–120. Cambridge: Woodhead Publishing.

66 Saboori, M., Gholipour, J., Chamliaud, H. et al. (2011). Prediction of burst pressure using a decoupled ductile fracture criterion for tube hydroforming of aerospace alloys. In: *International ESAFORM Conference on Material Forming*, Belfast, Northern Ireland (27–29 April), 301–306.

67 Butcher, C., Cheng, Z., Bardelcik, A., and Worswick, M. (2009). Damage-based finite-element modeling of tube hydroforming. *International Journal of Fracture* 155: 55–65.

68 Chu, E. and Xu, Y. (2004). Hydroforming of aluminum extrusion tubes for automotive applications. Part I: buckling, wrinkling and bursting analyses of aluminum tubes. *International Journal of Mechanical Sciences* 46: 263–283.

69 Chu, E. and Xu, Y. (2004). Hydroforming of aluminum extrusion tubes for automotive applications. Part II: process window diagram. *International Journal of Mechanical Sciences* 46: 285–297.

70 Dohmann, F., Boehm, A., and Dudziak, K.-U. (1993). The shaping of hollow shaft-shaped workpieces by liquid bulge forming. *Advanced Technology of Plasticity* 447–452.

71 Altan, T. and Ngaile, G. (2001). Practical methods for evaluation lubricants for tube hydroforming. *Hydroforming Journal* 8–12.

72 Prier, M. and Schmoeckel, D. (1999). *Tribology of Internal High Pressure Forming*, 379–390. Frankfurt: MAT-INFO Werkstoff-Informationsgesellschaft mbH.

73 Ngaile, G. (2008). Tribological aspects in hydroforming. In: *Hydroforming for Advanced Manufacturing* (ed. M. Koç), 144–178. Cambridge: Woodhead Publishing.

74 Ngaile, G., Jaeger, S., and Altan, T. (2004). Lubrications in tube hydroforming (THF) Part I: lubrication mechanism and development of model tests to evaluate lubricants and die coatings in the transition and expansion zones. *Journal of Materials Processing Technology* 146: 108–115.

75 Ngaile, G., Jaeger, S., and Altan, T. (2004). Lubricants in tube hydroforming (THF) Part II: performance evaluation of lubricants using LDH test and pear-shaped tube expansion test. *Journal of Materials Processing Technology* 146: 116–123.

76 Vollertsen, F. and Plancak, M. (2002). On possibilities for the determination of the coefficient of friction in hydroforming of tubes. *Journal of Materials Processing Technology* 125–126: 412–420.

77 Ngaile, G. (2006). Enhancing tribological conditions in tube hydroforming by using textured tubes. *Journal of Tribology* 128 (3): 674–676.

78 Strano, M. (2008). Design and modeling parts, process and tooling in tube hydroforming. In: *Hydroforming for Advanced Manufacturing* (ed. M. Koç), 121–142. Cambridge: Woodhead Publishing.

79 Khodayari, G. (2008). Preforming: tube rotary draw bending and per-flattening/crushing in hydroforming. In: *Hydroforming for Advanced Manufacturing* (ed. M. Koç), 181–200. Cambridge: Woodhead Publishing.

80 Dohmann, F. and Hartl, C. (1998). Hydroforming components for automotive applications. *The Fabricator* (February), pp. 30–38.

81 Klaas, F. (1997). Innovations in high-pressure hydroforming. In: *The 2nd International Conference on Innovations in Hydroforming Technology*, Columbus, OH, USA (15–17 September).

82 Anonymous (1996). *Handbuch der Umformtechnik.* Berlin: Springer Verlag.

83 Vollertsen, F. (2000). Accuracy in process chains using hydroforming. *Journal of Materials Processing Technology* 103: 424–433.

84 Tautenhahn, H. (1996). Anisotropie und Textur von Rohren. *Blech Rohre Profile* 43 (5): 234–239.

85 Korkolis, Y.P. and Kyriakides, S. (2011). Hydroforming of anisotropic aluminum tubes: Part I experiments. *International Journal of Mechanical Sciences* 53: 75–82.

86 Korkolis, Y.P. and Kyriakides, S. (2011). Hydroforming of anisotropic aluminum tubes: Part II analysis. *International Journal of Mechanical Sciences* 53: 83–90.

87 Hartl, C. (2008). Materials and their characterization for hydroforming. In: *Hydroforming for Advanced Manufacturing* (ed. M. Koç), 77–92. Cambridge: Woodhead Publishing.

88 Flehmig, T. and Schwarz, S. (2003). Hydroforming complex hollow sections. *Steel Grips* 1 (6): 408–412.

89 Peters, B.-M. and Saeuberlich, T. (2005). Economic production of high-strength tubes – a possibility to cost reduction of hydroforming processes. In: *International Conference Hydroforming*, Fellbach, Germany (25/26 October), 89–100.

90 Abedrabbo, N., Worswick, M., Mayer, R., and Riemsdjik, I.V. (2009). Optimization methods for the tube hydroforming process applied to advanced high-strength steels with experimental verification. *Journal of Material Processing Technology* 209: 110–123.

91 Tolazzi, M., Lamprecht, K., and Geiger, M. (2007). Hydroforming of laser welded tailored and patchwork blanks. In: *Proceedings of the LANE 2007*, Erlangen (25–28 September), 1289–1300.

92 Natal Jorge, R.M., Roque, A.P., Valente, R.A.F. et al. (2007). Study of hydroformed tailor-welded tubular parts with dissimilar thickness. *Journal of Material Processing Technology* 184: 363–371.

93 Liu, G., Yuan, S., and Chu, G. (2007). FEA on deformation behavior of tailor-welded tube in hydroforming. *Journal of Material Processing Technology* 187–188: 287–291.

94 Pohl, S. and Hauger, A. (2005). Tailor rolled tubes – weight and function optimized tubes for hydroforming. In: *International Conference Hydroforming*, Fellbach, Germany (25/26 October), 29–36.

95 Krahn, M., Kopp, R., Krux, P., and Kleiner, M. (2001). Umformen von flexibel gewalzten Blechen mittels Wirkmedien. In: *Kolloquium Wirkmedien-Blechumformung – Sheet Metal Hydroforming 2001"*, Dortmund, Germany (11 December), 137–148.

96 Flehmig, T. and Floeth, T. (2005). Innovative longitudinal member for automobiles – design, manufacture, mounting and costs situation. In: *International Conference Hydroforming*, Fellbach, Germany (25/26 October), 203–218.

97 Weil, W. (2005). From blank via forming to laser welded profile. In: *International Conference Hydroforming*, Fellbach, Germany (25/26 October), 275–286.

98 Hong, S.-T. and Lavender, C.A. (2007). Tailoring of mechanical properties of hydroformed aluminum tubes. *Journal of Material Processing Technology* 189: 477–482.

99 Vollertsen, F., Prange, T., and Sander, M. (1999). Hydroforming: needs, developments and perspectives. *Advanced Technology of Plasticity* 2: 1197–1210.

100 Zoephel, B. and Schiek, F. (2002). Werkzeugkonzepte zur Fertigung von großen IHU-Karosseriebauteilen. In: *The 3rd Chemnitz Car Body Colloquium Hydroforming for Car Body Components*, Chemnitz, Germany (25/26 September), 259–276.

101 Vollertsen, F. (2001). Challenges and chances of hydroforming of aluminum alloys. In: *Germa-Chinese Ultralight Symposium*, vol. 218, 71–79. DVS-Berichte.

102 Hartl, C., Luecke, H.-U., and Boehm, A. (2002). Produktivitätssteigerung beim Innenhochdruckumformen – Maßnahmen und Strategien. In: *The 3rd Chemnitz Car Body Colloquium Hydroforming for Car Body Components*, Chemnitz, Germany (25/26 September), 169–182.

103 Krei, M. (1999). State of the art of sealing techniques for hydroforming. In: *International Conference Hydroforming*, Fellbach, Germany (25/26 October), 441–462.

104 Neugebauer, R., Schnaepel, J., Muth, E., and Braeunlich, H. (2003). Hydroforming sheet metals as a tooling scheme for series production. In: *International Conference Hydroforming*, Fellbach, Germany (28/29 October), 371–388.

105 Aust, M. (1999). Hydromechanical deep-drawing of fuel tanks. In: *International Conference Hydroforming*, Fellbach, Germany (25/26 October), 309–324.

106 Khandeparkar, T. and Gehle, A. (2005). Equipment and die design optimisation for hydromechanical deep drawing. In: *International Conference Hydroforming*, Fellbach, Germany (25/26 October), 387–423.

107 Groche, P., Elsenheimer, D., Berger, C., and Kaiser, B. (2008). Durability-based design for endurance strength of hydroforming tools. *Material Science and Engineering Technology* 39 (9): 643–648.

108 Luecke, H.-U. (2003). Hydroforming manufacturing concepts. In: *International Conference Hydroforming*, Fellbach, Germany (28/29 October), 423–434.

109 Schnupp, M.K. (2003). Presses for hydromechanical drawing of panels for automobiles. In: *International Conference Hydroforming*, Fellbach, Germany (28/29 October), 409–412.

110 Siegert, K., Schwager, A., Rieger, R., and Haeussermann, M. (1999). New press concept for hydroforming. In: *International Conference Hydroforming*, Fellbach, Germany (25/26 October), 123–138.

111 Bunget, C. (2008). Mechanics of ultrasonic tube hydroforming. PhD thesis. North Carolina State University, Raleigh, North Carolina, USA.

112 Loh-Mousavi, M., Bakhshi-Jooybari, M., Mori, K.-I., and Hyashi, K. (2008). Improvement of formability in T-shape hydroforming of tubes by pulsation pressure. *Proceedings of the Institution of Mechanical Engineers, Part B, Journal of Engineering* 222 (9): 1139–1146.

3

Incremental Sheet Forming

Rogelio Pérez-Santiago[1], *Isabel Bagudanch*[2], *and Maria Luisa Garcia-Romeu*[2]

[1] *Department of Industrial and Mechanical Engineering, Universidad de las Americas Puebla, Cholula, Mexico*
[2] *Departament d'Enginyeria Mecànica i de la Construcció Industrial, EPS – Universitat de Girona, c/M. Aurèlia Capmany, Girona 17003, Spain*

3.1 Incremental Sheet Forming: General Overview

Deep drawing is one of the most common forming processes used in industrial applications due to its low cycle time and process maturity. In order to carry out this process, it is necessary to manufacture a tooling system (die, punch, binder, etc.) specific for each product. The tooling system has to be manufactured using high-resistant materials able to withstand the high forces generated in the process. Furthermore, high accuracy must be guaranteed to avoid collisions between parts of the tooling system and ensure the final part's dimensional accuracy. To reach these requirements, the materials used for the tooling system and the manufacturing processes to obtain it are very expensive. The cost of the process is only justified for high volume production. This effect can be observed in Figure 3.1.

In order to be able to produce highly customized products with a reasonable manufacturing cost, several innovative forming processes have been developed. *Incremental sheet forming* (ISF) is one of these new technologies, and it has gained importance in the last years, becoming the focus of interest for many researchers and institutions.

The first research work related with the ISF process was done by Manson, in 1978, as it has been reported in the historical review of Emmens et al. [1].

Although the first reference of the technology is dated more than three decades ago, it is not until 2004 that ISF process research attracted the attention of research groups of different universities around the world (Figure 3.2). The increasing effort in the development of the technology is aligned with the changing trends in manufacturing like the more frequent use of small production batches or one-of-a-kind products in order to meet the customer's requirements and obtain customized product features.

The functioning principle of ISF is simple: based on the geometry of the target component, a tool path planning software generates the trajectory to be followed by a CNC controlled machine. A forming tool attached to the driving machine, incrementally deforms the sheet material, which is fixed at its boundaries. Deformation occurs at successive layers below the initial sheet surface until the geometry of the part is generated. The incremental movement implies that the processing time is long (usually within hours, depending on the part size) compared with deep drawing that can produce a part in a few seconds.

Modern Manufacturing Processes, First Edition. Edited by Muammer Koç and Tuğrul Özel.

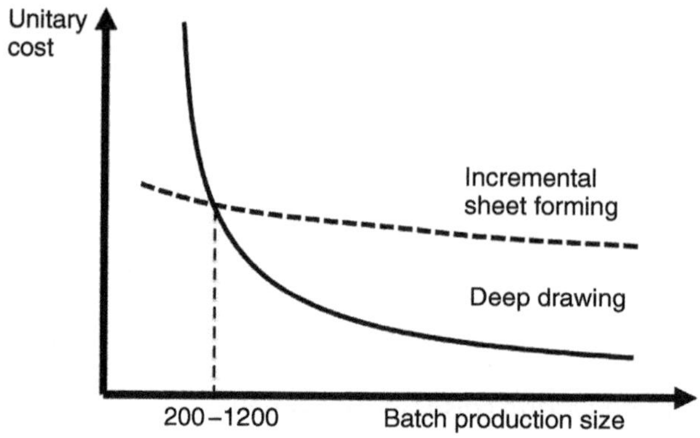

Figure 3.1 Graphical representation of the unitary cost as a function of the batch production size for the forming processes deep drawing and ISF.

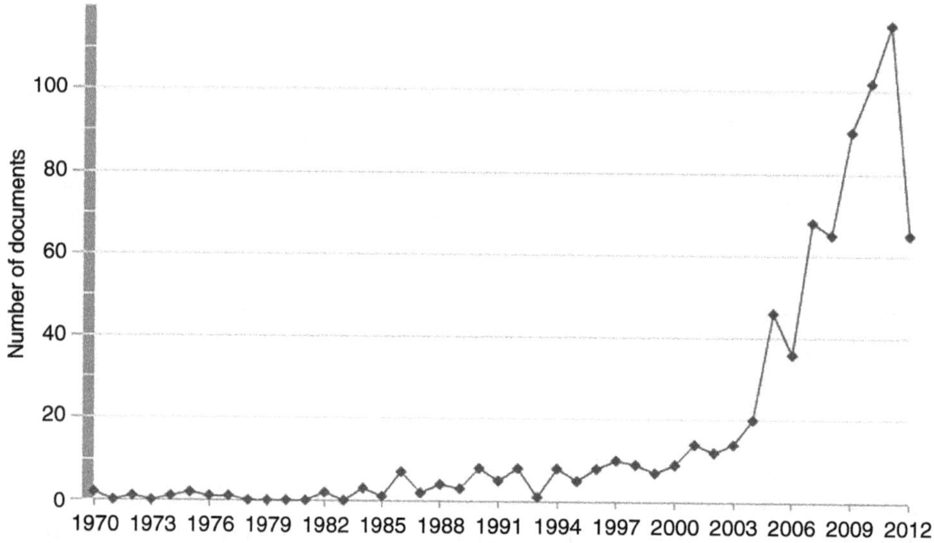

Figure 3.2 Evolution of the number of papers published in indexed journals from 1982 to June 2012 (information retrieved from Scopus data base).

The forming tool used in ISF process is not dedicated for a specific product but can be used in a wide range of parts with different geometries and characteristics. This feature becomes one major advantage of the process: its high flexibility. Another advantage can be derived from this fact: the cost related with the manufacturing of the punch is low, making the ISF technology economically viable for low batch production.

The ISF process can be conducted in any numerically controlled three-axis machine as they allow high feed rates, significant work volume, and stiffness. Indeed, one of the most frequent resources used for ISF is a CNC milling machine because it is available in most workshops, meaning that it is not necessary to make a large investment in machinery. Besides the forming tool, a clamping system is required to convert a milling machine into an ISF system, therefore is a relatively economic solution.

There is also machinery especially dedicated to the ISF technology. These machines are equipped with a movable clamping system, allowing the realization of different process variants; see Section 3.2 for more detail on the variants.

A third type of equipment that can be used to realize ISF is an industrial robot. The main limitations are the low stiffness and maximum force that it can withstand. The die can be substituted by another robot acting as a support tool.

In the following sections of this chapter, a deeper insight of the process is provided in terms of different variants of the technology; formability; deformation mechanisms; the most influencing process parameters, and future trends.

3.2 ISF Variants

In the present section, the most common variants of the ISF processes will be described in Sections 3.2.1 and 3.2.2, respectively: single point incremental forming (SPIF) and two point incremental forming (TPIF). Additionally, a review of other ISF variants developed to improve the process' performance in terms of accuracy or formability will be summarized in Section 3.2.3.

3.2.1 Single Point Incremental Forming (SPIF)

In terms of set-up, this is the simplest variant of the technology and the most deeply and widely studied in the academic field. There is only one point in which force is applied: the contact zone between the forming tool and the blank sheet (Figure 3.3). In this case neither partial nor total die is utilized; hence, the blank is only supported at the edges by the clamping system. Precisely due to the lack of die, the main drawback of this variant is the low accuracy. In order to compensate the geometrical deviation, one of the commonly used solutions is the modification of the toolpath to impose larger deformation on the sheet. Then, when the springback effect occurs, the sheet will be located in the desired point.

3.2.2 Two Point Incremental Forming (TPIF)

In this case, there are two points in which the force is applied, hence the name of this process variant. The first point, as in SPIF, is in the contact zone between the forming tool and the blank. The second point is at the blank–die interface. The die can be total or partial and the forming process can be negative (Figure 3.4b) or positive (Figure 3.4a). In the negative case, the tool is in contact with the inner side of the part, while the outer side of the part is contacted in the positive case. In the later situation, the complexity of the set-up increases because there is need to move the fixing device in perfect synchrony with the axial feed of the tool.

The die can be manufactured using materials that are cheap and easy to machine (wood, resins, etc.). Therefore, the costs associated with the die production are still not comparable

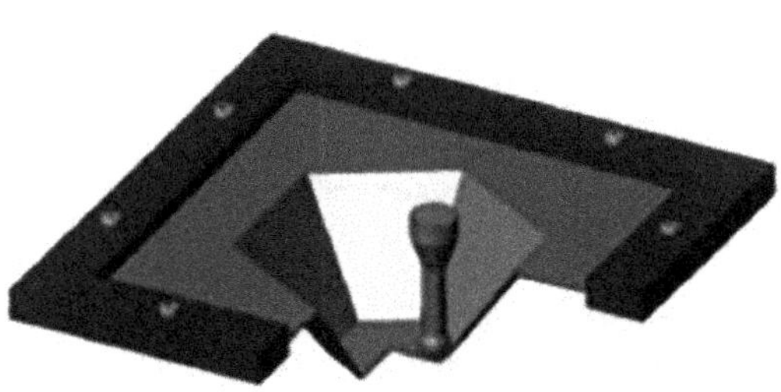

Figure 3.3 SPIF variant. *Source:* Allwood et al. 2010 [2]. Reproduced with permission of Elsevier.

Figure 3.4 TPIF process variant (a) positive die and (b) negative die. *Source:* Allwood et al. 2010 [2]. Reproduced with permission of Elsevier.

with the ones related to the production of dies used in deep drawing processes. Consequently, the TPIF technology is also viable to manufacture small batch products.

Although the complexity of the set-up is higher in TPIF than in SPIF the dimensional accuracy obtained in the former variant can be significantly improved [3].

3.2.3 Other ISF Variants

3.2.3.1 Warm ISF

This variant is the combination of the ISF technology (either in SPIF or TPIF variant) with a heating system. The most important advantage of using the heating system is that the formability of the sheet can be significantly increased. This is applied in materials that are very difficult to deform at room temperature like magnesium (e.g. AZ31) or titanium alloys (e.g. Ti–6Al–4V).

Several heating systems have been used: (i) local heating produced due to the friction between the tool rotating at a high spindle speed and the material [4]; (ii) heating device attached to the clamping system [5]; (iii) heating chamber covering the entire ISF process (forming tool, clamping system, and blank) in order to obtain a global warming [6], and (iv) localized heating system generated from electrical circuit [7].

3.2.3.2 Stretch Forming with ISF

In order to solve the principal drawbacks of the ISF technology (sheet thinning, accuracy, and process time), a new hybrid process that combines ISF with Stretch Forming has been proposed by Araghi et al. [8] (Figure 3.5).

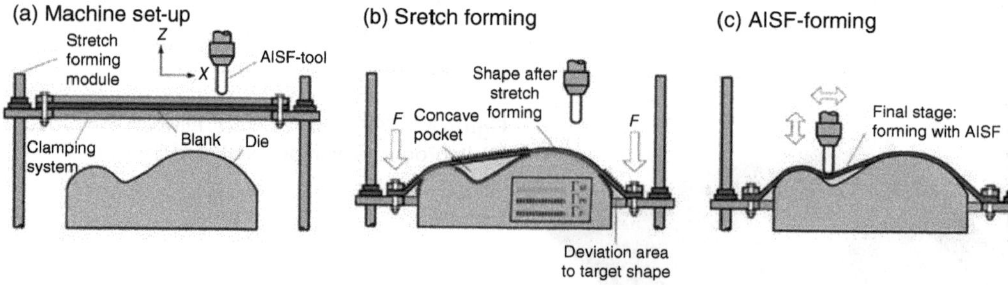

Figure 3.5 Operation principle of the hybrid process (a) experimental setup, (b) first stage: stretch forming, and (c) second stage: ISF. *Source:* Araghi et al. 2009 [8]. Reproduced with permission of Elsevier.

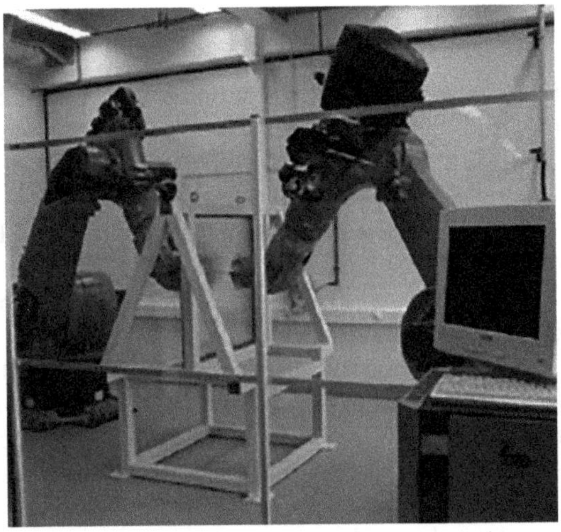

Figure 3.6 Experimental set-up of roboforming. *Source:* Meier et al. 2009 [9]. Reproduced with permission of Elsevier.

The first stage of the hybrid variant consists on obtaining a pre-form using Stretch Forming. In this first stage, an approximation to the final geometry can be reached. The slots, cavities, or deep corrugations are achieved in a subsequent ISF manufacturing stage.

3.2.3.3 Roboforming

Many manufacturing industries have in their installations industrial robots, therefore, some researchers are performing the ISF technology with industrial robots, a variant known as Roboforming. The main features of the robots are fast movements, low stiffness, and low capacity of withstanding high forces. The last two features might limit its use for the manufacturing of incremental formed parts.

With the use of robots it is possible to replace the partial or full die by another robot, which acts as a support tool of the sheet, as it can be seen in Figure 3.6.

3.3 Process Cycle

The basic ISF process cycle (SPIF) has four main stages: (i) the modeling of the component, (ii) the obtaining of the tool path, (iii) the production of the part, and (iv) secondary operations (cutting, measuring, polishing, etc.).

Regarding the first stage, a three dimensional (3D) model of the target part has to be defined; CAD tools are mainly used in this stage, in the case of customized products reverse engineering techniques can be applied for obtaining a 3D model. Regarding tool path generation, CAM systems can be used. Since there does not exist a specific CAM for ISF, machining CAMs are mostly used. For simple geometries, there is also the possibility of defining the contouring tool path by user-programmed routines. Once the tool path in the appropriate format is obtained, the part can be manufactured and finally analyzed and compared to the original design, in order to quantify geometric discrepancies.

With the intention of decreasing the number of iterations or adjusting tests, finite element method (FEM) models can be employed. An optimization process can be done to improve the process even before the physical manufacturing. The main objective of these optimizations is to minimize the geometric inaccuracies, and comprise mainly the modification of the tool-path and/or the use of a partial or full die (TPIF) as geometry complexity grows.

3.4 Materials

In the early days of the technology, only metallic materials were experimentally tested, therefore, the technology was also known as incremental sheet metal forming (ISMF). Nowadays, this nomenclature is currently not used due to the diversification of materials that can be employed in the process.

Commonly, the metallic materials used in ISF are soft aluminum alloys (AA1050) [8, 10, 11] or some steels: DC04 [12], AISI304 [13–15], etc. The main reason of using these materials is that they are relatively easy to deform at room temperature. They have been used very frequently and therefore their behavior and properties are widely known.

More recently, researchers are focused on other metal materials such as magnesium alloys [5, 6] or titanium ([7]). These materials cannot be successfully formed at room temperature, requiring the use of a heating system, either locally [4] or globally [6].

Afterwards, the process has also been applied to polymers. The combination of polymers and incremental deformation has as main advantage that can be done at room temperature, thereby reducing heat generation costs required in most processing techniques of plastics. The first polymer used in the ISF was PVC [16], using the SPIF variant.

3.5 Formability in ISF

3.5.1 FLD

In sheet forming processes, each point of the sheet might be subjected to a different combination of strain depending on the process conditions.

In order to predict "formability,"[1] the methods based on forming limit diagrams (FLD) graphically locate the strain state of different points of a part in a (Y-positive) Cartesian plane bounded by the major and minor principal strains axes, in combination with curves indicating the defect (necking or fracture) limits of the sheet material. The minor strain is represented in the horizontal axis and the major strain in the vertical one. The FLD is also called the Keeler–Goodwin diagram.

The FLD are characterized by three differentiated regions (Figure 3.7a):

- *Safe*: points within this zone will be successfully formed.
- *Critical*: there is a high probability of slight defects appearing (necking).
- *Failure*: high probability to obtain material cracks.

In the case of ISF, the FLD has a radically different trend in comparison of the typical FLD obtained for deep drawing (presented in Figure 3.7a): In ISF, the FLD is a straight line with a negative slope indicating higher deformation limits than those allowed under deep drawing conditions (see Figure 3.7b).

Thus, the asymmetric incremental sheet forming (AISF) process is characterized by increased material formability. Much higher strains than normally observed in deep drawing are observed in AISF, for instance strains can be well over 3. This is because in ISF there is the suppression of necking and fracture appears without the formation of instabilities. The stabilization mechanisms that can avoid necking are widely discussed in the comprehensive review done by Emmens and Van Den Boogaard [18].

1 Ability of a certain material to undergo plastic deformation without damage.

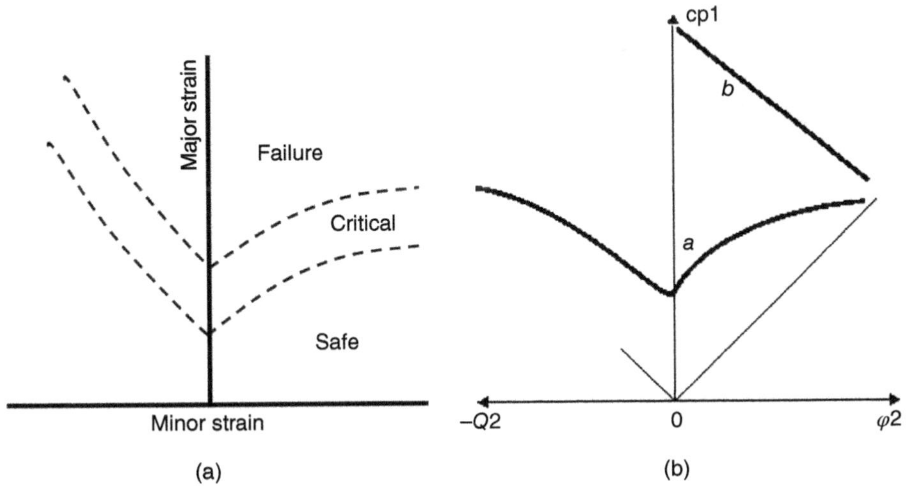

Figure 3.7 (a) FLD regions. (b) FLD in traditional sheet metal forming processes *-a-* and in incremental sheet forming *-b-*. *Source:* Kopac and Kampus 2005 [17]. Reproduced with permission of Elsevier.

Another difference in ISF can be found when the deformation pattern is considered. During the in-plane movement of the forming tool in a straight-line, the deformation that occurs is plane-strain stretching, while in the curved zones the deformation becomes biaxial stretching [19].

3.5.2 ISF Deformation Mechanisms

It has to be said that although extensive research in ISF over the last decade has been carried out, the deformation mechanism under the ISF process is not fully understood [20]. The common deformation mechanism assumed in the literature (as well in SPIF as in TPIF) has been a mechanism of pure shear through the thickness of the sheet and plane strain in the plane parallel to the undeformed sheet. According to [20], this mechanism has become associated to ISF process over the last 7–10 years but it has not been experimentally verified. The mechanism is based on shear spinning rotary forming process. In shear spinning, the part is formed over the mandrel by a shear deformation process in which the outside diameter remains constant and the wall thickness is therefore reduced. This shear straining and consequent thinning of the metal distinguish this process from the bending action in conventional spinning. The resulting thickness of the spun wall can readily be determined by the sine law.

As it was pointed out by Jackson and Allwood [20] "a real understanding of the deformation mechanism is important to allow accurate numerical models of the process to be developed for tool path design and process control and, to develop an understanding of the increased forming limits observed in ISF in comparison to pressing." In that sense the findings of their work based on an annealed copper (C101) plate formed to a truncated cone of wall angle of 30° are noteworthy: first, the deformation mechanisms of both SPIF and TPIF are significantly different to the idealized mechanism of shear spinning. In both SPIF and TPIF, the deformation is a combination of stretching and shear that increases on successive tool laps, with the greatest strain component being shear in the tool direction. Shear occurs perpendicular to the tool direction in both SPIF and TPIF, which is more significant in SPIF resulting in a piling up of the material at the center of the plate.

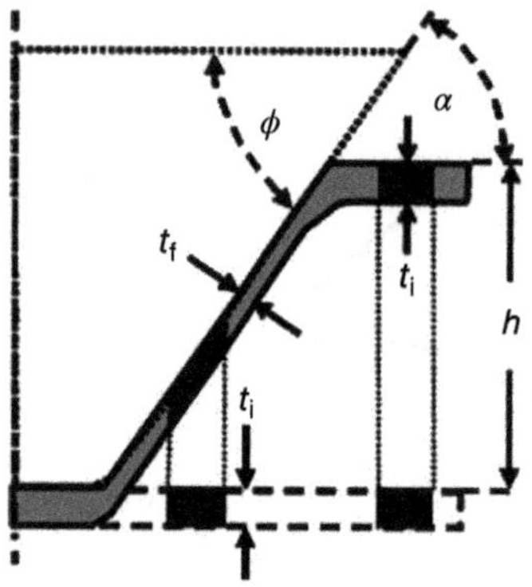

Figure 3.8 Sine law. *Source:* Jeswiet et al. 2005 [23]. Reproduced with permission of Elsevier.

Furthermore, based on experimental results, Jackson and Allwood [20] concluded that the deformation mechanism is intrinsically different for SPIF and TPIF. Therefore, the two processes should be distinguished in future discussions of their deformation mechanics.

Finally, despite the ideal shearing deformation assumed by the sine law does not correspond to the deformation mechanism of ISF, the formula allows to estimate sheet thinning during both processes, SPIF and TPIF, in cases when there is no radial displacement of material elements trough the thickness [8, 20–22] (Figure 3.8).

3.5.3 Forces

Forming forces estimation in ISF is especially important in the case of using machinery adapted for the process like milling centers and robots. It has been demonstrated that the predominant force in ISF is developed in the axial direction of the tool while this, in general, is not the case in milling. As a consequence, an accurate estimation of the maximum axial force developed during the forming process is required in order to ensure the safe utilization of the hardware. Another reason to study forming forces is their direct relationship with the structural integrity of the work piece.

Aiming for quick estimations of the maximum force, several researchers have proposed analytical models intended for uniform wall angle (UWA) geometries formed with different materials. The first ISF force analytical model was proposed by Iseki [24], who assumed plane-strain conditions, uniform sheet deformation, and neglected friction for its derivation. When designing and manufacturing the first dedicated ISF equipment, Allwood et al. [25] proposed a second analytical model to estimate the vertical force limits of its equipment. However, recent research of the writers [26] has confirmed that the model developed by Aerens et al. [27] provides accurate estimations of the steady-state axial forming force (F_{Zs}) of truncated UWA cones formed with the metallic alloys, for instance aluminum or stainless steel, generally utilized in ISF works. This analytical model, Eq. (3.1), is a phenomenological correlation originated from the results of an experimental campaign conducted on two aluminum alloys, AA3003 and AA5754, and three steel alloys DC01, AISI304, and 65Cr2.

$$F_{Zs} = 0.0716\, R_m\, t_0^{1.57}\, d_t^{0.41}\, \Delta h^{0.09}\, \alpha \cos\alpha \tag{3.1}$$

R_m is the tensile strength of the formed material aligned with the wall direction (MPa); t_0, initial sheet thickness (mm); d_t, tool diameter (mm); Δh, the scallop height (mm); and α initial wall angle (°).

3.6 ISF Process Parameters

Table 3.1 summarizes the main process parameters and specifications that should be considered in the ISF.

Regarding to the toolpath, the most common strategies are

- *Unidirectional*: the tool is always moved in the same direction (clockwise or counterclockwise) along a plane. Torsion appears to the final part. The profile of the geometry is followed in a discrete way.
- *Bidirectional*: in this case the tool movement changes its direction in each plane, avoiding the appearance of torsion. The profile of the geometry is followed in a discrete way.
- *Spiral*: basically is used in order to improve the surface finishing, because the profile of the geometry is followed in a continuous way, and there is no tool-indentation in the surface of the sheet. Torsion appears to the final part.

There are three different types of tools: *flat* (the less used in the literature), *ball end* (the ball has free rotation in order to have the minimum friction), and *hemispherical*, being the last one the most common.

Finally, another process parameter with strong influence in the accuracy, surface finishing, forming forces, and operational time is *step down*. This parameter determines the tool displacement along the different forming layers. It can be defined according to three strategies:

- *Constant depth step* (Δz): the axial movement of the tool is defined with a constant value. The surface finishing is better in steep walls.
- *Constant scallop height* (Δh): the separation of the different layers varies with the wall angle, allowing better surface finishing than in the previous case.
- *Constant angular increment* ($\Delta\theta$): the angular increment between two successive layers has to be constant. The main disadvantage is that for very steep walls the surface finishing is not acceptable.

3.7 Accuracy

One of the most important drawbacks of the ISF process is its poor dimensional accuracy compared with the requirements of the industrial applications [28] showed that most of the industrial applications where ISF could fit-in require geometric tolerances below 1 mm, which are unfeasible with the normal ISF process chain: CAD modeling → CAM processing → CNC manufacturing.

Specifically, for the SPIF variant, the main geometric error typologies were classified by Micari et al. [29], Figure 3.9) as follows:

- Sheet bending close to the blank's clamping region. The deviation in the outermost zone of the part is considerable. This deviation takes place in the first stages of the process because sheet is prone to bending rather than being locally deformed. In order to reduce this deviation, it is necessary to use a backing plate. However, this support is not able to eliminate the low accuracy in the central zones.
- Deviations in the wall caused by the elastic recovery (springback) of the sheet when the load is removed.
- The pillow effect in the bottom of the part, which is originated by the spread of elastic stresses of zones where tool has not exerted plastic deformation to other parts of the piece.

Table 3.1 Summary of the main process parameters in ISF.

<table>
<tr><td rowspan="3">Design</td><td>Geometry</td><td colspan="2">Simple
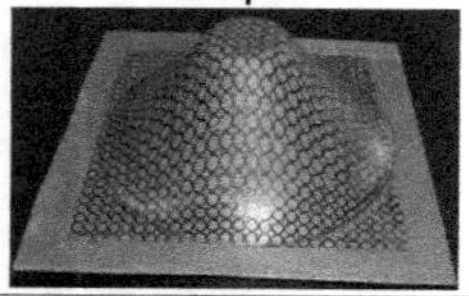</td><td colspan="2">Intermediate
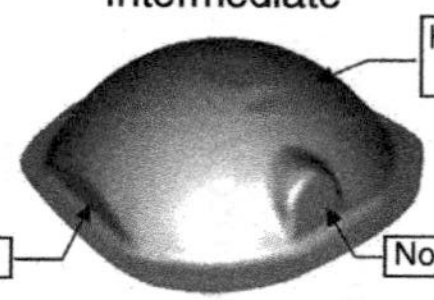
</td><td colspan="2">Complex
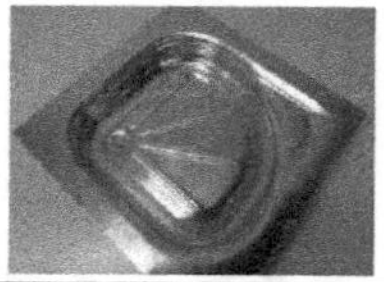</td></tr>
<tr><td>Material</td><td>Aluminium</td><td>Magnesium</td><td colspan="2">Steel</td><td>Titanium</td><td>Polymers</td></tr>
<tr><td>Sheet thickness</td><td colspan="3">Microforming (<0.1 mm)</td><td colspan="3">Macroforming (0.5–5 mm)</td></tr>
<tr><td rowspan="3">Tool</td><td>Tool path</td><td colspan="2">Unidirectional (clockwise or counterclockwise)</td><td colspan="2">Bidirectional</td><td colspan="2">Spiral</td></tr>
<tr><td>Type</td><td colspan="2">Flat
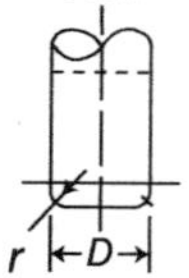
</td><td colspan="2">Hemispherical
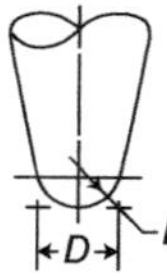
</td><td colspan="2">Ball end
</td></tr>
<tr><td>Dimensions</td><td colspan="3">Microforming (<1 mm)</td><td colspan="3">Macroforming (5–20 mm)</td></tr>
<tr><td rowspan="3">Process</td><td>Step down</td><td colspan="2">Δz constant
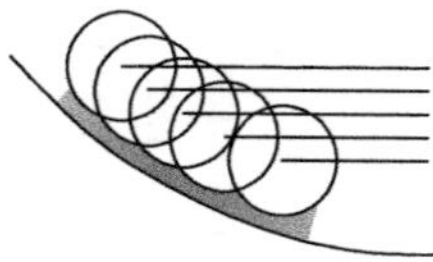
</td><td colspan="2">Δh constant
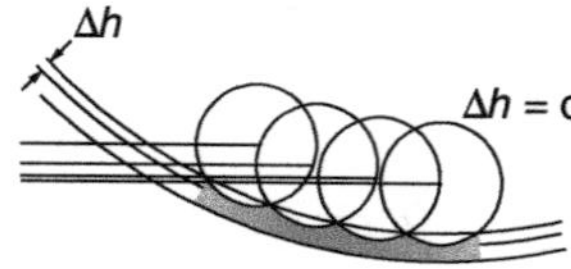
</td><td colspan="2">$\Delta\theta$ constant
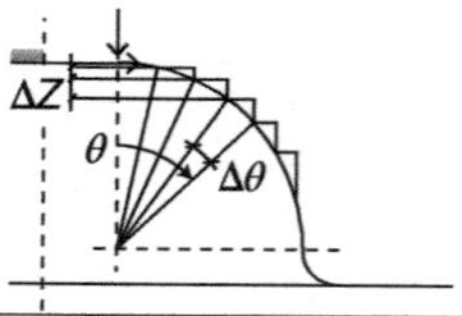
</td></tr>
<tr><td>Feed rate</td><td colspan="6">400–3000 mm/min</td></tr>
<tr><td>Spindle speed</td><td colspan="6">0–4000 rpm</td></tr>
</table>

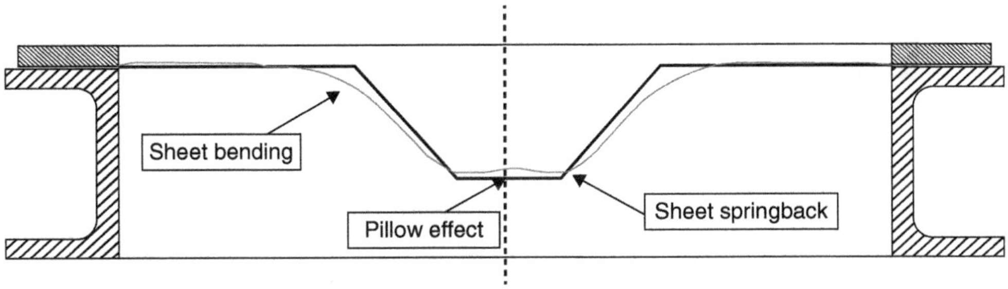

Figure 3.9 Geometrical error topologies observed during the SPIF of UWA parts. *Source:* Micari et al. 2007 [29]. Reproduced with permission of Elsevier.

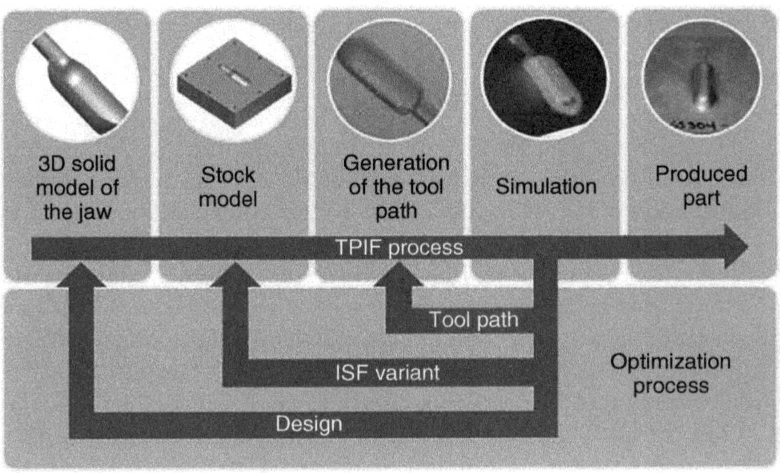

Figure 3.10 TPIF optimized process map.

One alternative for offsetting these undesired effects is the optimization of the tool path, as it was previously mentioned in Section 3.3. Hence, in ISF process, optimization cycles have to be carried out (Figure 3.10). These optimization procedures are often based on trial and error process and simulations until acceptable tolerances are achieved.

3.8 Simulation

According to Tekkaya [30], the first simulations of industrial sheet metal components appeared at the early 1980s. Taking into account that ISF is a relatively new process, it is not surprising to find the first papers regarding ISF simulations dated around 2000.

Since the deformation mechanics of ISF is still an open question, FEM results became a complementary source of information to experimental and theoretical analyses. One of the earliest ISF simulations was reported by Shim and Park [31]. The authors modeled the SPIF of a squared frustum formed with the fully annealed Aluminum Alloy AA1050-O (Figure 2.19). Bi-axial and plane-strain stretching, respectively, occurring at the corners and walls of the virtual part were confirmed by the experimental results.

As a follow-up of this work, Kim and Park [19] studied the influence of several parameters in the forming of a straight groove. Filice and Fratini [32] compared the strains obtained by FEM

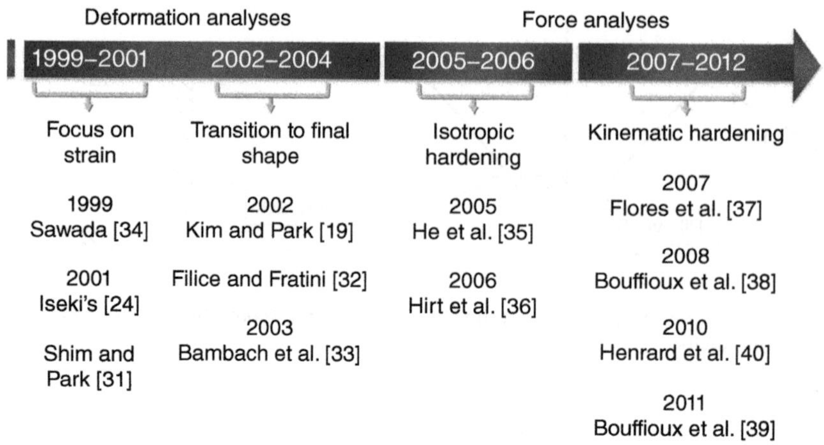

Figure 3.11 Historical timeline for ISF simulation.

simulation of a cross-shaped component with the FLD obtained by experiments. The previous two works also utilized the AA1050-O alloy. Then, Bambach et al. [33] reported a complete study about modeling and simulation of ISF. This publication is frequently cited because of its significant contributions.

Simulations with force orientation appeared short after their experimental counterparts. While different element formulations and time integration schemes have been utilized, the choice of a constitutive law seems to fit two separate stages: (i) up to 2007, most models utilized von Mises and isotropic exponential hardening laws which showed important discrepancies from the experimental measurements and (ii) afterward, mixed isotropic-kinematic hardening laws have been preferred. While the most recent publications suggest that this approach gives the highest accuracy, their limited scope on aluminum alloys and UWA geometries make results difficult to generalize on other scenarios. Figure 3.11 summarizes the historical timeline regarding the simulation of the ISF process.

3.9 Future Trends in ISF

3.9.1 Application Fields

In the automotive industry, the massive production of sheet metal parts is dominated by conventional stamping technologies. However, ISF could be applied in the early stages of product development for rapid prototyping of parts complying with the fit and form requirements of the final product.

In architecture and industrial construction, there is also a potential growth, because it could be interesting obtain metal parts of customized geometries. There are recent studies working with the forming of sandwich panels [41].

ISF has an important potential in the biotechnology field. It is possible to produce customized products: ankle prosthesis [42], cranial implant [43], palate implant [44], or a part for a knee implant [45].

3.9.2 Materials

In the last few years, some researchers have been focused on the use of polymeric materials in ISF. A deep analysis of the effect that the process parameters have on the final results is required.

Furthermore, it would be interesting to work with a wide range of polymers, including the biocompatible ones in order to be able to produce customized medical prosthesis.

3.9.3 Sustainability

Through the years, manufacturing processes have been analyzed under several points of view. One of the most recent is related to the world concern on environmental and sustainability issues. Duflou et al. [46] described results of two case studies performed at the Belgian K. University Leuven on air bending and laser cutting processes. Ingarao et al. [47] and Anghinelli et al. [48] have studied sustainability issues for metal forming processes and the environmental cost in ISF, respectively. A recent work has also been developed by the authors in the Spanish environment [49]. Dittrich et al. [50] demonstrated that for small batches production of (up to 300 parts), ISF is an environmentally friendlier process in comparison to conventional stamping and hydroforming processes.[2]

3.10 Case Study

This case study aimed at the examination and comparison of performance between the two main ISF variants, SPIF, and TPIF on a scaled prototype of a biopsy micro forceps (BMF) jaw [51]. Two materials were used, AA1050-H24 and SS304, both 0.5 mm thick. Experimental tests were carried out initially on SPIF set. Rupture of specimens and their low accuracy due to bending of the material lead to build TPIF set-up. The main results are summarized in Figure 3.12.

Using the SPIF variant, a sound component was manufactured with stainless steel (Figure 3.12d), whereas the aluminum specimen broke (Figure 3.12a). The trials on TPIF configuration using a single tool path were both failures (Figure 3.12b,e). After several adjustments to the tool-path in the new configuration, shifting to multi tool-path strategy in two phases in this case, led to successful BMFs manufactured by TPIF on AA1050-H24 and AISI304 (Figure 3.12c,f).

3.11 Concluding Remarks

The ISF process has been introduced, including its main process variants and technological parameters. As explained, the process flexibility and set-up simplicity allows any academic laboratory or industrial workshop already owning three-axis machining centers or robots to manufacture sheet metal parts with an affordable equipment investment.

Because of its long cycle time and low accuracy, ISF cannot substitute the conventional sheet metal processes. Nevertheless, application opportunities in the rapid prototyping of sheet metal products intended to be massively produced, low batch runs, and unique products in the architectural or biomedical field have been identified. To exemplify the former, a case study demonstrating the feasibility to use ISF for the rapid prototyping of sheet metal medical devices was presented.

Simplified models to estimate two important process outcomes, wall thinning, and process peak axial force have been exposed, as well as the progress obtained with simulations based on the FEM. Regarding the latter, even if long computational times are undesirable in any

2 Specifically, the elimination of die manufacturing means CO_2 savings around 1500 kg per die set. Nevertheless, the comparison in terms of system useful work potential (exergy) was unfavorable because of the electric energy consumed by the idle operation of the CNC machine and the amount of lubricant utilized for the process.

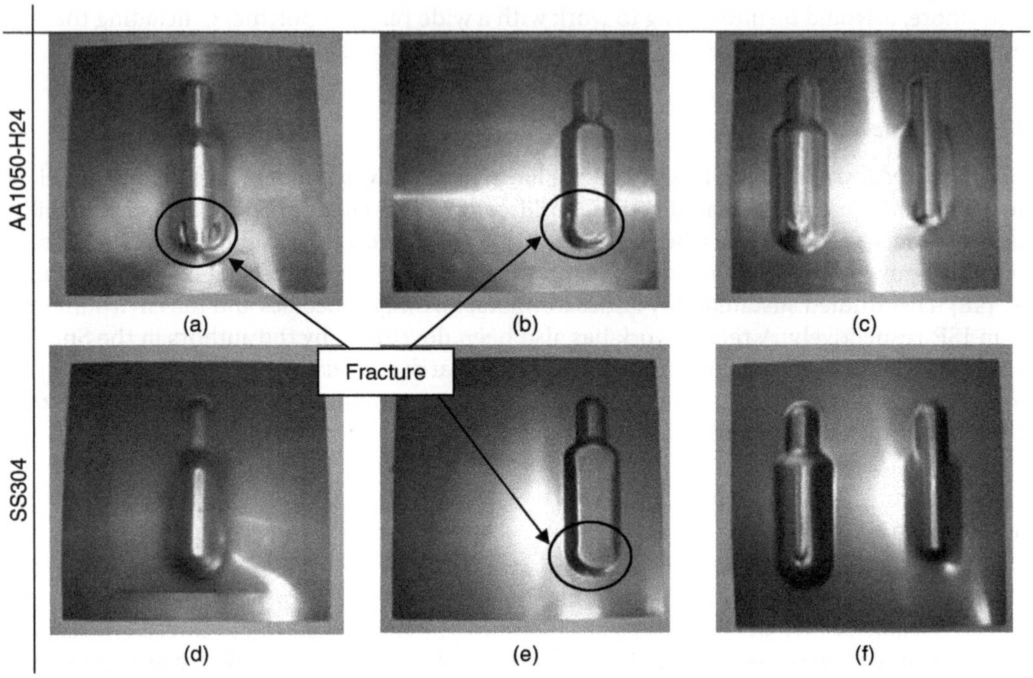

Figure 3.12 BMF jaws using the (a and d) SPIF, (b and e) TPIF (single tool-path), and (c and f) TPIF (multi tool-path) variants.

numerical model, this condition becomes a significant handicap for ISF since at this moment physical experiments require less effort and time than their virtual counterparts. As a consequence, despite providing valuable information about the deformation mechanics and the qualitative gains obtained with process parameters modification, FEM simulations are difficult to be implemented in the design process of an ISF-intended part.

Finally, the sustainability advantages of ISF make it a feasible solution for the future production of customized and low-quantity sheet metal products.

References

1 Emmens, W.C., Sebastiani, G., and van den Boogaard, A.H. (2010). The technology of incremental sheet forming – a brief review of the history. *Journal of Materials Processing Technology* 210 (8): 981–997.

2 Allwood, J.M., Braun, D., and Music, O. (2010). The effect of partially cut-out blanks on geometric accuracy in incremental sheet forming. *Journal of Materials Processing Technology* 210 (11): 1501–1510.

3 Attanasio, A., Ceretti, E., Giardini, C., and Mazzoni, L. (2008). Asymmetric two points incremental forming: improving surface quality and geometric accuracy by tool path optimization. *Journal of Materials Processing Technology* 197 (1–3): 59–67.

4 Park, J., Kim, J., Park, N., and Kim, Y. (2009). Study of forming limit for rotational incremental sheet forming of magnesium alloy sheet. *Metallurgical and Materials Transactions A* 41 (1): 97–105.

5 Ambrogio, G., Filice, L., and Manco, G. (2008). Warm incremental forming of magnesium alloy AZ31. *CIRP Annals – Manufacturing Technology* 57 (1): 257–260.

6 Zhang, Q., Guo, H., Xiao, F. et al. (2009). Influence of anisotropy of the magnesium alloy AZ31 sheets on warm negative incremental forming. *Journal of Materials Processing Technology* 209 (15–16): 5514–5520.

7 Fan, G., Gao, L., Hussain, G., and Wu, Z. (2008). Electric hot incremental forming: a novel technique. *International Journal of Machine Tools and Manufacture* 48 (15): 1688–1692.

8 Araghi, B.T., Manco, G.L., Bambach, M., and Hirt, G. (2009). Investigation into a new hybrid forming process: incremental sheet forming combined with stretch forming. *CIRP Annals – Manufacturing Technology* 58 (1): 225–228.

9 Meier, H., Buff, B., Laurischkat, R., and Smukala, V. (2009). Increasing the part accuracy in dieless robot-based incremental sheet metal forming. *CIRP Annals – Manufacturing Technology* 58 (1): 233–238.

10 Ambrogio, G., Costantino, I., Denapoli, L. et al. (2004). Influence of some relevant process parameters on the dimensional accuracy in incremental forming: a numerical and experimental investigation. *Journal of Materials Processing Technology* 153–154: 501–507.

11 Fratini, L., Ambrogio, G., Di Lorenzo, R. et al. (2004). Influence of mechanical properties of the sheet material on formability in single point incremental forming. *CIRP Annals – Manufacturing Technology* 53 (1): 207–210.

12 Bambach, M., Taleb Araghi, B., and Hirt, G. (2009). Strategies to improve the geometric accuracy in asymmetric single point incremental forming. *Production Engineering* 3 (2): 145–156.

13 Aerens, R., Eyckens, P., Bael, A., and Duflou, J.R. (2010). Force prediction for single point incremental forming deduced from experimental and FEM observations. *The International Journal of Advanced Manufacturing Technology* 46 (9–12): 969–982.

14 Ceretti, E. (2004). Experimental and simulative results in sheet incremental forming on CNC machines. *Journal of Materials Processing Technology* 152 (2): 176–184.

15 Perez-Santiago, R., Bagudanch, I., García-Romeu, M., and Hendrichs, N. (2012). Effect of strain hardening exponent in the incremental sheet forming force. *Steel Research International* 439–442.

16 Franzen, V., Kwiatkowski, L., Martins, P., and Tekkaya, A. (2009). Single point incremental forming of PVC. *Journal of Materials Processing Technology* 209 (1): 462–469.

17 Kopac, J. and Kampus, Z. (2005). Incremental sheet metal forming on CNC milling machine-tool. *Journal of Materials Processing Technology* 162–163: 622–628.

18 Emmens, W.C. and Van Den Boogaard, A.H. (2009). An overview of stabilizing deformation mechanisms in incremental sheet forming. *Journal of Materials Processing Technology* 209 (8): 3688–3695.

19 Kim, Y. and Park, J. (2002). Effect of process parameters on formability in incremental forming of sheet metal. *Journal of Materials Processing Technology* 130-131: 42–46.

20 Jackson, K. and Allwood, J. (2009). The mechanics of incremental sheet forming. *Journal of Materials Processing Technology* 209 (3): 1158–1174. https://doi.org/10.1016/j.jmatprotec.2008.03.025.

21 Ziran, X., Gao, L., Hussain, G., and Cui, Z. (2009). The performance of flat end and hemispherical end tools in single-point incremental forming. *The International Journal of Advanced Manufacturing Technology* 46 (9–12): 1113–1118.

22 Takano, H., Kitazawa, K., and Goto, T. (2008). Incremental forming of nonuniform sheet metal: possibility of cold recycling process of sheet metal waste. *International Journal of Machine Tools and Manufacture* 48: 477–482. https://doi.org/10.1016/j.ijmachtools.2007.10.009.

23 Jeswiet, J., Micari, F., Hirt, G. et al. (2005). Asymmetric single point incremental forming of sheet metal. *CIRP Annals – Manufacturing Technology* 54 (2): 88–114.

24 Iseki, H. (2001). An approximate deformation analysis and FEM analysis for the incremental bulging of sheet metal using a spherical roller. *Journal of Materials Processing Technology* 111 (1–3): 150–154.

25 Allwood, J.M., Houghton, N.E., and Jackson, K.P. (2005a). The design of an incremental sheet forming machine. *Advanced Materials Research* 6–8: 471–478.

26 Perez-Santiago, R. (2012). Forming force estimation in single point incremental forming of uniform and variable wall angle components. PhD thesis. Instituto Tecnológico Y De Estudios Superiores De Monterrey, Campus Monterrey, Mexico.

27 Aerens, R., Duflou, J., Eyckens, P., Van Bael, A. (2009). Advances in force modelling for SPIF. *International Journal of Material Forming* 2 (S1): 25–28.

28 Allwood, J.M., King, G.P.F., and Duflou, J. (2005b). A structured search for applications of the incremental sheet-forming process by product segmentation. *Proceedings of the Institution of Mechanical Engineers, Part B: Journal of Engineering Manufacture* 219 (2): 239–244.

29 Micari, F., Ambrogio, G., and Filice, L. (2007). Shape and dimensional accuracy in single point incremental forming: state of the art and future trends. *Journal of Materials Processing Technology* 191 (1–3): 390–395.

30 Tekkaya, A.E. (2000). State-of-the-art of simulation of sheet metal forming. *Journal of Materials Processing Technology* 103 (1): 14–22.

31 Shim, M.-S. and Park, J.-J. (2001). The formability of aluminum sheet in incremental forming. *Journal of Materials Processing Technology* 113 (1–3): 654–658.

32 Filice, L. and Fratini, L. (2002). Analysis of material formability in incremental forming. *CIRP Annals – Manufacturing Technology* 2: 3–6.

33 Bambach, M., Hirt, G., and Junk, S. (2003). Modelling and experimental evaluation of the incremental CNC sheet metal forming process. In: *7th International Conference on Computational Mechanics, COMPLAS 2003*, 1–16.

34 Sawada, T., Matsubara, S., Sakamoto, M., and Fukuhara, G. (1999). Deformation analysis for stretch forming of sheet metal with CNC machine tools. In: *Proceedings of the 6th ICTP*, vol. II, 1501–1504.

35 He, S., Van Bael, A., Van Houtte, P., Szekeres, A. et al. (2005). Finite Element Modeling of Incremental Forming of Aluminum Sheets. *Advanced Materials Research* 6–8: 525–532.

36 Hirt, G., Ames, J., and Bambach, M. (2006). Validation of FEA for asymmetric incremental sheet forming by on-line measurements of deformation and tool forces.154 Production Engineering. *Annals of the German Academic Society for Production Engineering* XIII (1): 39–44.

37 Flores, P., Duchêne, L., Van Bael, A. et al. (2007). Model identification and FE simulations: Effect of different yield loci and hardening laws in sheet forming. *International Journal of Plasticity* 23 (3): 420–449.

38 Bouffioux, C., Eyckens, P., Henrard, C. et al. (2008). Identification of material parameters to predict Single Point Incremental Forming forces. *Journal of Materials* 1147–1150.

39 Bouffioux, C., Lequesne, C., Vanhove, H. et al. (2011). Experimental and numerical study of an AlMgSc sheet formed by an incremental process. *Journal of Materials Processing Technology* 211 (11): 1684–1693.

40 Henrard, C., Bouffioux, C., Eyckens, P. et al. (2010). Forming forces in single point incremental forming: prediction by finite element simulations, validation and sensitivity. *Computational Mechanics* 47 (5): 573–590.

41 Jackson, K., Allwood, J., and Landert, M. (2007). Incremental forming of sandwich panels. *Journal of Materials Processing Technology* 204 (1–3): 290–303.

42 Ambrogio, G., Denapoli, L., Filice, L. et al. (2005). Application of incremental forming process for high customised medical product manufacturing. *Journal of Materials Processing Technology* 162–163: 156–162.
43 Duflou, J.R., Lauwers, B., Verbert, J. et al. (2005). Medical application of single point incremental forming: cranial plate manufacturing. *Virtual Modelling and Rapid Manufacturing – Advanced Research in Virtual and Rapid Prototyping* 161–166.
44 Tanaka, S., Nakamura, T., Hayakawa, K. et al. (2007). Residual stress in sheet metal parts made by incremental forming process. *AIP Conference Proceedings* 908: 775–780.
45 Oleksik, V., Pascu, A., Deac, C. et al. (2010). The influence of geometrical parameters on the incremental forming process for knee implants analyzed by numerical simulation. *AIP Conference Proceedings* 1252: 1208–1215.
46 Duflou, J.R., Clarke, R., Merklein, M. et al. (2011). Environmental performance of sheet metal working processes. *Key Engineering Materials* 473: 21–26.
47 Ingarao, G., Di Lorenzo, R., and Micari, F. (2011). Sustainability issues in sheet metal forming processes: an overview. *Journal of Cleaner Production* 19 (4): 337–347.
48 Anghinelli, O., Ambrogio, G., Di Lorenzo, R., and Ingarao, G. (2011). Environmental costs of single point incremental forming. *Steel Research International* 525–530.
49 Bagudanch, I., Garcia-Romeu, M.L., Ferrer, I., and Lupiañez, J. (2013). The effect of process parameters on the energy consumption in single point incremental forming. In: *Proceedings MESIC Conference.*
50 Dittrich, M.A., Gutowski, T.G., Cao, J. et al. (2012). Exergy analysis of incremental sheet forming. *Production Engineering Research and Development* 6 (2): 169–177.
51 Garcia-Romeu, M.L., Pérez-Santiago, R., Bagudanch, I., and Puigpinós, L. (2012). Fabrication of a biopsy micro-forceps prototype with two point incremental sheet forming. In: *1st International Conference on Design and Processes for Medical Devices, PROMED 2012*, 103–106.

4

Powder Forming

Rahmi Ünal

Mechanical Engineering Department, Gazi University, Ankara, Turkey

4.1 Introduction

Material forming processes based on powders, or more specifically, powder metallurgy (PM) process, permit the production of finished products with the minimum amount of steps, notably by limiting the number of machining stages, which normally result in lower production costs. Powder forming technologies start with powders – that is, collections of particles. A particle is defined as the smallest unit of a powder that are larger than smoke (0.01–1 μm), but smaller than sand (0.1–3 mm). Many useful powders are on the scale of diameter of a human hair (25–200 μm) [1].

PM is a process that involves converting powder into a solid object. PM processing encompasses an extensive range of ferrous and nonferrous alloy powders, ceramic powders, and mixes of metallic and ceramic powders (composite powders). PM process is a near-net shape manufacturing process that combines the features of shape-making technology for powder compaction with the development of final material and design properties (physical and mechanical) during subsequent densification or consolidation processes (e.g. sintering). It is critical to recognize this interrelationship at the outset of the design process because a subtle change in the manufacturing process can cause a significant change in material properties [2].

The primary commercial PM processes for the production of parts and components are cold pressing and sintering; hot pressing; direct powder consolidation by hot isostatic pressing (HIP); and the densification of a preform by forging. The PM techniques have the advantages of low energy cost and material loss for producing parts with complex geometries. For many PM parts, secondary operations are eliminated completely or reduced significantly, compared to conventional processes because PM results in close dimensional tolerances. Furthermore, these techniques are attractive for the development of advanced materials because of high flexibility [3].

The cold die powder compaction manufacturing process is a popular manufacturing route (Figure 4.1) for small and relatively simple geometrical components. This is due to its high material usage as the parts are produced to near-finished dimensions from the powder [4]. Sintering is an essential process for achieving necessary powder product properties because compact bodies manufactured via compaction only have insufficient cohesive force. Defects such as distortions or cracks can arise during sintering as the compact body undergoes dimensional changes [5, 6].

Modern Manufacturing Processes, First Edition. Edited by Muammer Koç and Tuğrul Özel.

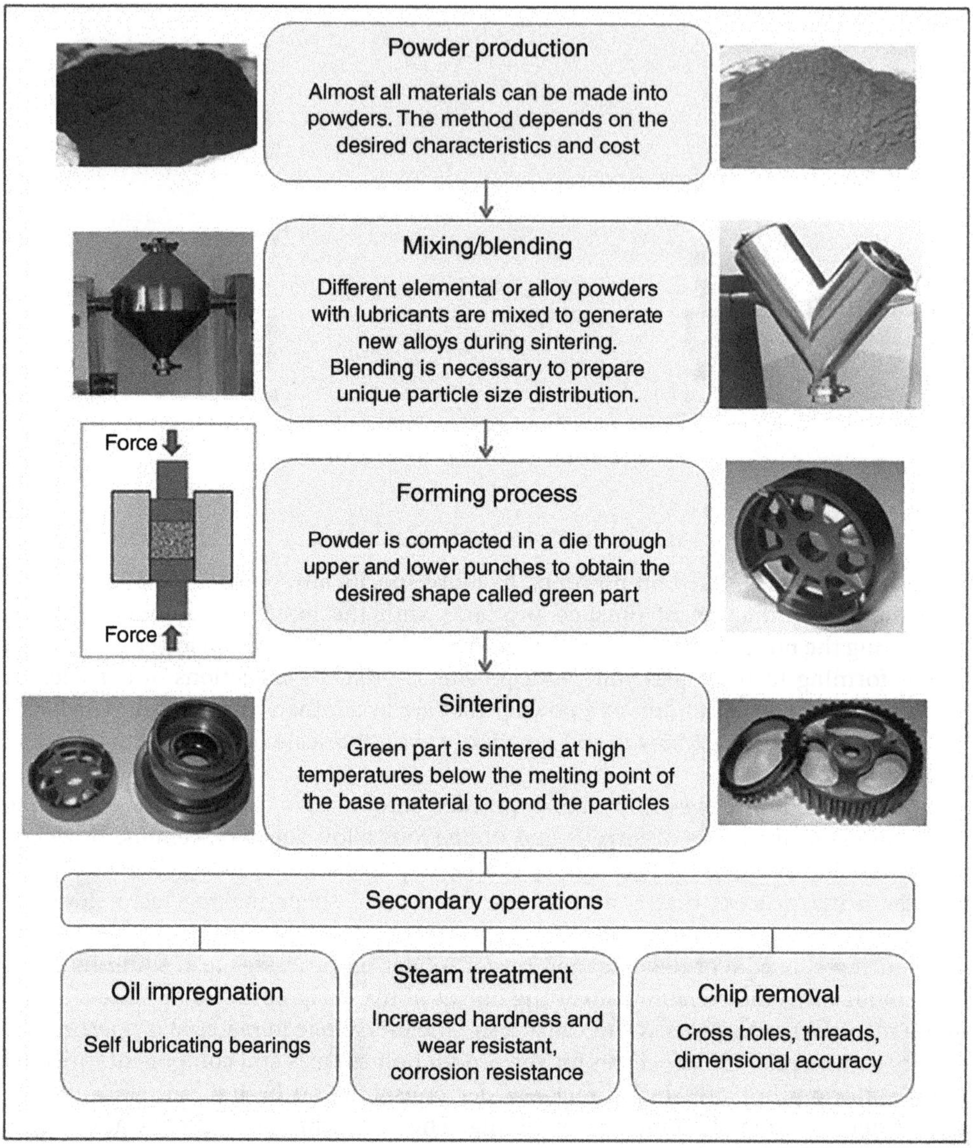

Figure 4.1 Basic PM process part manufacturing route.

These defects are a result of the sintering process and pre-existing structural factors. For example, after compaction a compact body will, in general, have a nonuniform density distribution and residual stress [6, 7], and such attributes will be reflected in the final product after sintering. The sintering temperature and pressure will also affect the dimensional accuracy and quality of the final product [8–10].

Powder metallurgy is used in various industrial sectors: automobile industry (motors, gear assemblies, and brake pads), abrasives (polishing and grinding wheels), manufacturing (cutting and drilling tools), electric and magnetic devices (magnets, soft magnetic cores), medical and dental (implants, prostheses, and amalgams), aerospace (motors, heat shields, and structural parts), welding (solder, electrodes), energy (electrodes, fuel cells), other (porous filters, sporting goods).

Ceramic components are formed by compacting a powder into the desired engineering shape. The void space is eliminated with a high-temperature heat treatment called densification. Because advanced ceramic powders such as silicon nitride and alumina usually lack the plastic properties of traditional clay-based powders when dispersed in water, shape forming is carried out by either pressing a dry powder mixed with a polymer binder or by injection molding the powder mixed into a polymer that imparts plasticity. The reliability of engineering components shaped by these commercial methods suffer from inclusions, present in the powder and retained during shape forming and densification, that concentrate an applied stress to severely degrade the component's strength [11].

4.2 Reasons for Using PM Route

Powder metallurgy, as the most rapidly growing metal-forming technology, is expected to feature heavily in the expanded usage. PM is a highly flexible and automated process that is environmentally friendly, with low relative energy consumption and high level of materials utilization. The PM process has the highest raw material utilization (>95%) and the lowest energy requirement per kilogram of finished part, comparing with other manufacturing processes. Thus, it is possible to fabricate high-quality parts to close tolerance at low cost. PM competes with casting processes, solid forming, and cutting techniques in a number of applications. The decision in favor of powder metallurgy is therefore based not only on technological criteria. In fact, the economic parameters are frequently the decisive factor. There are certain materials that can only be produced in a powder metallurgical operation, e.g. hard metals or alloys of metals with widely differing melting points.

PM offers many advantages compared to other manufacturing technologies. The advantages could be classified into three principal reasons for using a powder metallurgy product (Figure 4.2): (i) cost savings compared with alternative processes, (ii) unique properties attainable only by the PM route, (iii) necessity of PM for certain applications like hard and high temperature materials [1]. The cost-saving feature is one reason for the increasing use of powder metal for making a variety of parts. If the decision is for necessity or unique properties, they cover the cost saving also.

Product cost effectiveness is by far the predominant reason for choosing PM and is the main driver of the structural (or mechanical) parts sector. PM wins the cost competition on the basis of its lower energy consumption, higher material utilization, and reduced numbers of process steps, in comparison with other production technologies.

All of these factors, in turn, are dependent on powder metallurgy's ability to reduce, or even possibly eliminate entirely, the machining operations that would be applied in conventional manufacture. In order to eliminate machining operations, powder metallurgy relies on its abilities to form complex geometrical shapes directly and to hold close dimensional tolerance control in the sintered product. Powder metallurgy's cost effectiveness generally also requires that the particular product be made in large production quantities. If production quantity requirements are too low, there would be no opportunity to amortize the costs of the (long-lasting) forming tooling over a sufficient numbers of parts or to avoid the loss of significant fractions of potential production time in tool changeover/setting operations (http://www.ipmd.net/Introduction_to_powder_metallurgy/Why_Powder_Metallurgy).

T15 tool steel is strong and tough and is used to make the punches and dies that press metal powder into shapes. T15 is a tungsten type super high-speed steel containing high vanadium for excellent abrasion resistance and cobalt for good red hardness. If the foundry melts the tool steel, the alloying elements are all uniformly dissolved in the liquid steel. When the melt is

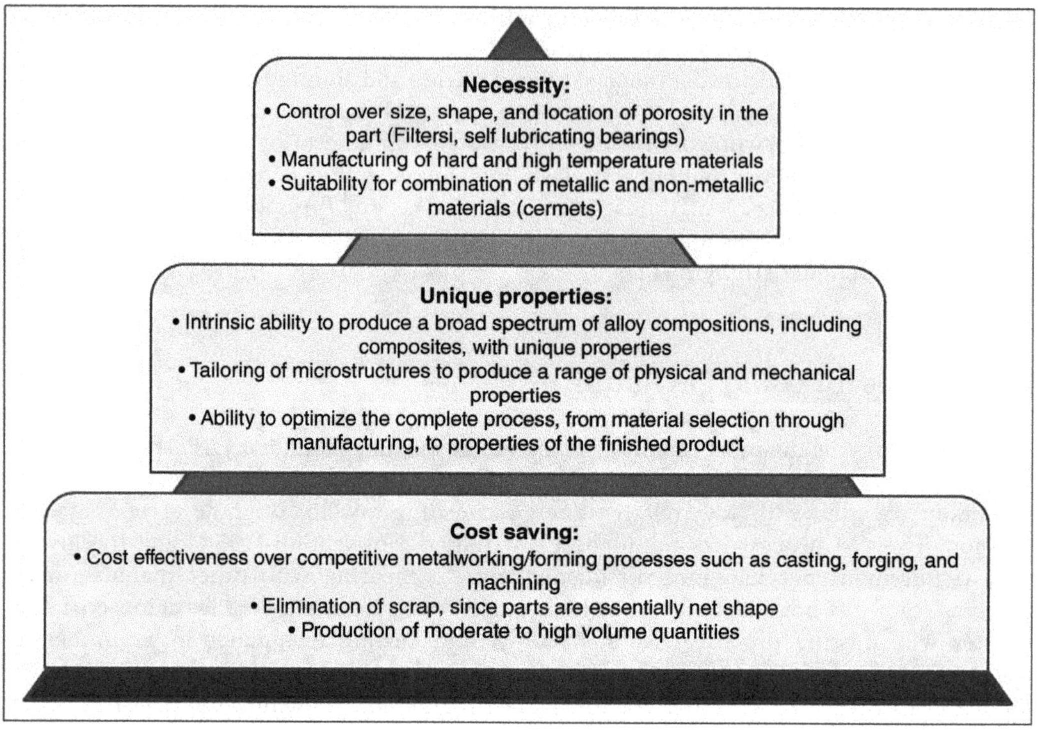

Figure 4.2 Reasons for using PM technology.

poured into a large casting, the elements don't stay uniformly distributed. During the slow cooling process, the chromium and the vanadium have time to react with carbon to form carbide inclusions (lumps). These carbides lower the strength and make the wrought tool steel hard to machine and grind. In PM, the liquid tool steel is gas atomized into small particles that cool rapidly. The carbides don't have time to grow and are very small compared to those in a regular casting. The tool steel particles are then joined together by heat and pressure to form large billets. The billets still have the small carbides that the powder had, and they are much stronger and easier to grind than the cast product. PM T15 has a bend fracture strength more than twice the conventional wrought or ingot based tool steel. The toughness is higher and grindability is three times better [12].

One of the earliest uses (1921) of metal powder was the making of porous bearings. Mixture of copper and tin powder was pressed. After heating to melt the tin so it would dissolve in the copper, the result was a porous bronze bearing. The advantage of over a solid cast bronze was that the porous could be filled with lubricating oil. Then, when the bearing run against a shaft and heated up, the oil oozed out and lubricated the shaft. When the rotation stopped, and the bearing cooled, the oil was sucked back into the bearing. So that the bearing requires no further lubrication during the whole life of the machine in which it is used. This feature accounts for the use of the term "self-lubricating bearings," and typically these are made of bronze. Porosity and oil retention are genuine technical advantages for PM. However, PM bearings are also less expensive than ball bearings or teflon lined bearings [12].

Rapid solidification processing (RSP) is a means of producing an entirely new range of alloys that have mechanical and corrosion properties that are far superior to those obtainable using conventional ingot metallurgy. As suggested by the name, RSP involves extreme rates of

Table 4.1 Key structure property relationships in RSP materials.

Structural features	Expected property improvements
Size refinement	Hall-Petch strengthening
	Improved fracture and impact toughness
	Enhanced super-plasticity
Extended solid solubility	Increased solid solution strengthening
	Increased precipitation and dispersion strengthening
	Minimized brittle equilibrium phase precipitation
Chemical homogeneity	Improved corrosion/oxidation resistance
	Better response to working
	Better response to heat treatments
Precipitation of nonequilibrium crystalline phases	Improved physical properties
	Improved mechanical properties
Formation amorphous phases	Improved physical properties
	Improved mechanical properties
	Improved corrosion resistance

cooling from the melt. Instead of cooling rates of 10^0–10^2 °C/s that occur during conventional casting processes, cooling rates of 10^4–10^{10} °C/s can be achieved during RSP. Such rapid cooling is achieved by atomization (10^4–10^5), melt spinning (10^6) or surface melting (>10^6). Atomization generates fine particles, <500 μm. Water or air is the usual atomizing fluids, but inert gases are used for high-performance applications. In melt spinning, the melt stream forms a single ribbon 10–100 μm thick by impingement onto a copper chill rotating at a surface speed of tens of meters per second. Ribbons vary in width from a 5 to 500 mm. The ribbon is then pulverized into platelets and consolidated using powder metallurgy techniques.

RSP has been applied to a wide range of alloys, and in limited cases to ceramics, to improve mechanical, electrochemical and physical properties for structural and nonstructural applications. In order to understand how we can exploit the opportunity for microstructural control and property enhancement, it is instructive to discuss what RSP does to the microstructure and hence, to the properties of materials in generic terms. Rapid solidification (RS) has five major types of effects on material structure, namely, size refinement, extended solid solubility, increased chemical homogeneity, evolution of nonequilibrium crystalline phases, and formation of amorphous phases. Table 4.1 summarizes what property improvements we might expect to see for each of the five effects of RSP. The first three effects are important mainly for improved mechanical properties and corrosion/oxidation resistance, whereas the other two effects lead to the improvement of both mechanical and nonmechanical properties [13].

4.3 Powder Production

Metal powder production techniques are used to manufacture a wide spectrum of metal powders designed to meet the requirements of a large variety of applications. Almost all materials can be made into powders [2]. Since a close correlation exits between the quality of the powder and the quality of the finished parts, processes for the production of powders from special materials are constantly being improved [14].

Suitable methods for powder production depend on required production rates, powder properties, and the physical and chemical properties [1, 15]. The main categories for fabricating powders are based on mechanical comminution, chemical reactions, electrolytic deposition, liquid atomization, and vapor condensation [1]. Chemical and electrolytic methods are useful for producing high-purity powders. Mechanical comminution (or milling) is the most widely used method of powder production of hard metals and ceramics. Secondary milling of oxide-reduced, atomized, or electrolytic powders is also a very common and economical practice. Atomization is perhaps the most versatile method that produces metal powders over a wide range of production rates and wide range of powders sizes from 10 to 1000 μm [2].

In principle, all existing processes used for particle manufacturing can be included in one of the following three distinct categories. The methods in the first group are based on the so-called "phase-breakdown" or "top-down" approach, which consists in reducing the size of the bulk metal to desired dimensions by supplying the energy required to increase the surface area of the system. Unfortunately, the processes are low in efficiency, so the energy required to form a powder is many times larger than the newly created surface energy. With low process efficiency the input energy can be much larger, often 30 times more than the newly created surface energy [1]. The size reduction can be achieved either by the mechanical breakdown or by dispersing a molten metal into small droplets that are subsequently solidified by rapid cooling (atomization). Although both these methods are routinely used to produce large quantities of metal powders for many applications, they are not capable of generating highly uniform metallic particles required in most high technology fields. The second group includes the "phase transformation" methods, which rely on the conversion of fine metal compounds particles into metal ones trough either thermal decomposition (thermolysis, pyrolysis) or chemical reduction. If the precursor particles are uniform in size and shape and the aggregation process is prevented during the entire transformation (by using aerosol or in bulk after particle encapsulation), high performance metallic particles with a broad range of well-defined properties can be obtained by these techniques. The third group of methods is based on the "phase buildup" or "bottom up" approach, meaning that the metallic particles are "constructed" starting from the buildings blocks, the metal atoms. The condensation process of the atoms leading to the formation of the metallic phase can take place either in gas phase (chemical or physical vapor deposition) or in a liquid phase (chemical precipitation). Since nucleation, particle growth, and the interactions between particles at all stages of the process can be much better controlled in the liquid phase, the chemical precipitation process is the most suitable for the generation of uniform metallic particles ranging in size from few nanometers to several microns [16].

The modern powder-related technologies highly rely on powders with tailored properties, such as microstructure and morphology [17]. The morphology (particle shape, mode, and size distribution) of the produced particles has a crucial role on pressing/filling behavior, sintering activity, shrinkage, and porosity of the final parts and products [18–20]. Metal powders exhibit a diversity of shapes ranging from spherical to acicular. Particle shape is an important property, since it influences the surface area of the powder, its permeability and flowability, and its density after compaction. The microstructure of the produced metal powders directly affects several material properties such as hardness and strength; microstructure is also important in fabricating specific structures, such as bulk metallic glasses (BMG) [21–23]. Chemical composition and purity also affect the compaction behavior of powders.

The production method and the processing conditions determine the characteristics of the obtained metal powders. Due to the importance of final product properties, different approaches for producing metal powders with better production rate and economy have been developing, such as mechanical alloying [24], twin fluid melt atomization [25, 26], centrifugal atomization [27, 28], electro-deposition [29], etc. These metal powder production processes differ in their production rate, structure, and morphology of the obtained materials and the economical aspects especially in energy consumption. However, the main goals that become prominent in high-tech applications are control and adjustment of particle size, structure, and morphology [17]. Among above mentioned methods, melt atomization is widely used in industrial application. The selection of one fabrication method over another depends on understanding the process, its economics, the resulting powder characteristics, and how those characteristics match with the intended application.

4.3.1 Mechanical Production Techniques

The principle of mechanical disintegration lies in the transfer of the energy of mechanical movement to the material to be crushed in the grinding unit [14]. Mechanical comminution is the most widely used method of powder production for hard metals and ceramic materials. There are four fundamental mechanical comminution process: impaction, attritioning, shearing, and compression [1]. The impact process shown in Figure 4.3 named as milling. A ball mill is a type of grinder used to grind materials into extremely fine powder. The most widely used objectives of milling process include particle size reduction of brittle metals and for producing metal flakes of ductile materials. Secondary milling of spongy cakes of oxide-reduced, atomized, or electrolytic powders is the most common milling process; hammer and rod mills are used for this type of milling [2].

During milling four types of forces act on particulate material: impact, attrition, shear, and compression. Impact is the instantaneous striking of one object by another. Attrition is the production of wear debris on particles created by the rubbing action between two bodies. Shearing fractures the material by a cutting process, such as occurs in machining. Compression is the slow application of compressive forces to a body (crushing or squeezing of a particulate material) [2].

Under optimized grinding conditions, metallurgical processes such as solid phase dispersion strengthening (mechanical alloying) can take place in addition to any dispersion mixing/agglomeration. Grinding can be processed in this way via intermediate products with a layered structure until particle agglomerates with particle sizes of 10 and 100 μm are reached. These agglomerates are made up of amorphous or nano-crystalline particles with individual crystal

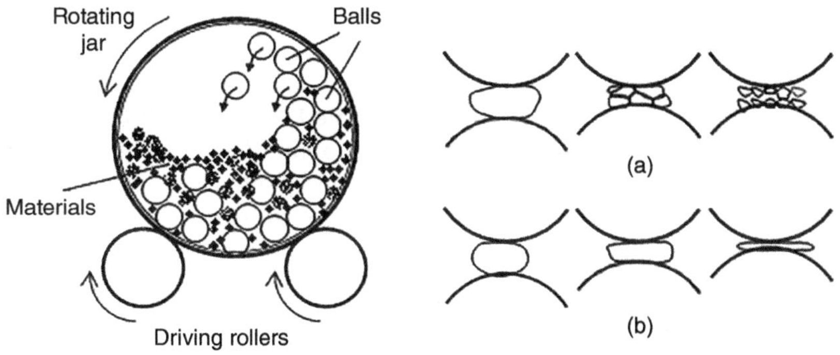

Figure 4.3 A view of the action in a jar mill and brittle (a)/ductile (b) behavior of single particles. *Source:* Adapted from Refs. [1] and [2].

sizes of about 10 nm. Ultrafine powders with individual particle sizes 100 nm cannot be obtained from grinding as singular particles of powder agglomerates, neither can they be separated from the grinding medium [14].

4.3.2 Atomization

Atomization can be defined as manufacturing powder metal by spraying liquid metal with water, air, or inert gas in an atomization chamber. The spray causes the metal to cool and solidify into particles. Atomization to produce powder provides the dual advantages of having the metal or alloy available in a readily dispensable and economical form as well as providing a beneficial and uniform microstructure in the final powder metallurgy product. Atomization is particularly useful for the production of alloys in powder form, since the constituents metals are fully alloyed in the molten state. Thus each powder particle has the same chemical composition [30]. Generally speaking, if a melt can be prepared and contained, it can be atomized, to a greater or lesser degree of success. Success can be quantified in terms of yield of the desired size fraction in the powder, purity which can affect chemical activity and thus the ease of subsequent consolidation and structure which, if it can be completely or partially retained in subsequent processing, determines the performance of the consolidated bulk material or coating. Generally, a fine and uniform microstructure is desirable in an engineering material, although there are notable examples where the opposite is the desired outcome. In gas atomization solidification rates vary, in practical terms, within the range 10^3–10^6 K/s and it might be supposed that, for a bulk material production process, this is as much as can reasonably be achieved in an attempt to generate a fine structure by virtue of high cooling rates without resorting to extreme atomization conditions [31].

4.3.3 Chemical Fabrication Techniques

There are several variants to chemical synthesis; powders are formed by solid, liquid, or vapor phase reactions. A classic form of metal powder fabrication is oxide reduction [1]. Solid-state reduction process has been for long the most widely used method for the production of iron powder. Selected ore is crushed, mixed with carbon, and passed through a continuous furnace where reducing reaction takes place leaving a cake of sponge iron which is then further treated by crushing, separation of nonmetallic material, and sieving to produce powder [30]. Iron ore is reduced in solid state at 800–1050 °C either by reducing gas or coal. The reducing gas is a mixture majority of hydrogen (H_2) and carbon monoxide (CO) that acts as reducing agent. Since no refining operation is involved, the purity of the powder is dependent on that of the raw materials. The irregular sponge like particles are soft, and readily compressible, and give compacts of good green strength.

Thermal decomposition process is generally used for the production of iron and nickel powders. The process starts with the metal carbonyls, for example $Fe(CO)_5$ or $Ni(CO)_4$. The carbonyls are liquids at normal temperature with a low boiling point. These are formed by reaction of metal and carbon-monoxide gas under pressure and high temperature. The carbonyls can be decomposed by heating the vapor at atmospheric pressure. Carbonyl iron powder is usually spherical in shape and very fine (10 μm), while the nickel powder is usually is quite irregular, porous, and fine [32].

4.3.4 Precipitation from Solution

Production of metal powders by hydrometallurgical processing is based on leaching an ore or ore concentrate, followed by precipitating the metal from leach solution [2]. A dissolved compound such as a nitrate, chloride, or sulfate can be chemically treated [1]. The most widely used commercial processes based on hydrometallurgy are copper cementation and the separation and precipitation of copper, nickel, and cobalt from salt solutions [2].

Tin oxide (SnO_2) is one of the representative materials for a semiconductor type of gas sensor. Also, it has been widely used as an electrode material, a transparent electrode, a glass melting electrode, and so on. As an n-type semiconductor, tin oxide shows very high sensitivity to many reducing gases such as H_2, CO, hydrocarbon, and alcohol. It is known that the sensitivity can be improved by increasing the sensor surface area so as to increase the number of oxygen sites on the surfaces in these reducing gas species. Thus, for the application of tin oxide to a sensing layer material of chemical gas sensors, a higher surface area of the tin oxide powder is desirable. The tin oxide used for gas sensors is usually prepared from precursor hydroxides precipitated by the addition of $NH_4{\cdot}OH$ to $SnCl_4$ aqueous solution. However, using this precipitation method, it is not easy to obtain high surface area powders because of the high degree of agglomeration and the irregular particle morphology of the tin oxide powders [33].

4.4 Consolidation Techniques

It has long been known that as the density of powder materials increases, the mechanical properties of that material also increase [34]. The properties of sintered parts depend on the density of the material as shown in Figure 4.4.

Powder forming technologies are based on three basic methods. Densification could be obtained during cold compaction, compressing at high temperature, or sintering as shown in Figure 4.5. The box in the figure indicates the densification step during the forming technologies.

Even though there are some exceptions, in the structural parts production by cold compaction a densification during sintering is not striven for, because systematic dimensional change during sintering impair the geometrical precision and increase the dimensional scatter. The dimensional reproducibility is for the majority of all applications utmost importance for economics of the technology. Therefore, the density required must essentially be achieved during

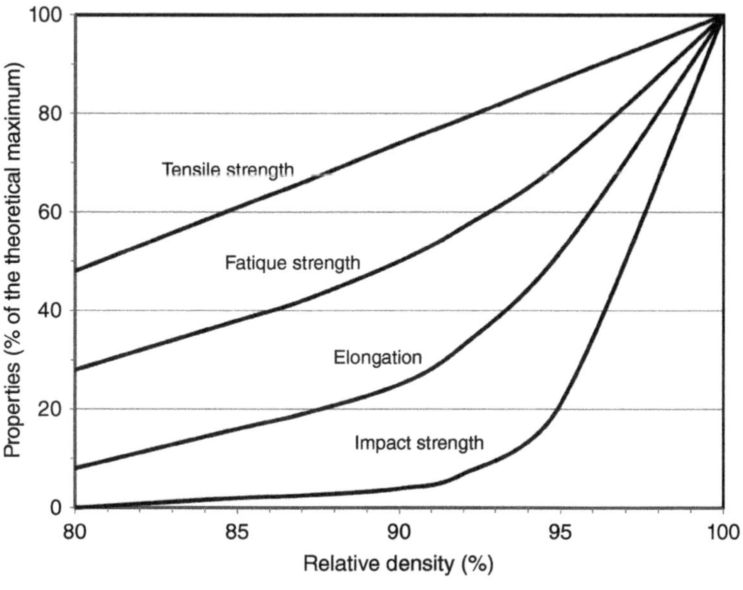

Figure 4.4 Density dependence of mechanical properties schematically. *Source:* From Yu 2005 [35].

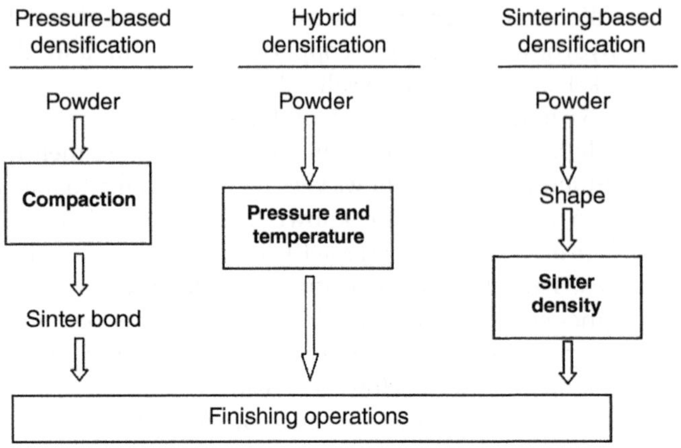

Figure 4.5 Powder forming technologies.

compaction; in subsequent processing steps only minor density increase are possible mainly by cold deformation of the bulk part in closed dies or by local densification of functional areas [36]. The pores, whether they are discrete or connected, in material, which is not fully dense, can be thought of as internal notches. Because higher density means fewer pores, there are fewer internal notches available to initiate failure. The conclusion: Increase density to improve mechanical properties [34].

In deciding whether a higher density powder product is needed, the specific application must be considered to determine exactly what properties are required. This, in turn, will fix the density that must be achieved for that application. As a general rule, higher performance applications require higher densities. Improved mechanical properties can be achieved by using hot pressing, HIP, powder extrusion, or powder forging processes that go along with fully dense state [34].

One major difference between the hot consolidation processes and press and sinter process is the geometry and purity of the powders that are routinely used in each. Press and sinter powder is usually irregularly shaped. Although this shape powder can be used for both hot pressing and powder forging, it is necessary to use spherical powder for HIP and powder extrusion. Spherical powders are usually produced by inert gas atomization, while the preferred production method of irregularly shaped powder is water atomization. Gas atomization is more expensive than water atomization, so the starting raw material, even of the same composition, is more costly for HIP and extrusion [34]. Another difference between gas and water atomized powder is the oxygen content. Water atomization encourages the pick-up of oxygen, while gas atomization can produce extremely low oxygen contents. It was reported that oxygen content in water atomized Stainless T316L was between 2000 and 2600 ppm. This same alloy when gas atomized will have between 80 and 150 ppm oxygen [2]. Along with higher density, lower oxygen content is a contributor to improved mechanical properties [34].

4.4.1 Cold Powder Compaction

A wide range of engineering components is manufactured using the powder route. Die compaction of powders followed by sintering is used in powder metallurgy, industrial ceramics, and the hard metal sectors. Pharmaceutical tablets and detergents are also made from powders using compaction. Thus, die compaction is a common operation across a range of industries. Powders are compacted under high pressure by various methods where the application of

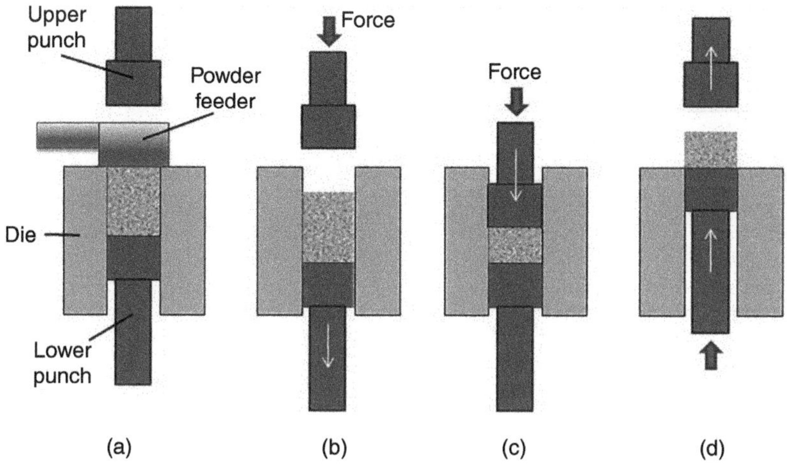

Figure 4.6 Cold die compaction of powder steps. (a) Die filling, (b) positioning, (c) compaction, and (d) ejection.

pressure packs the powders and reduces porosity. Many compaction methods are used, but the most prevalent method for PM parts production is uniaxial compaction in a rigid die. This method is cost effective with relatively straightforward tooling [2]. The process involves filling a powder formulation into a die, compression using rigid punches, and ejection from the die as shown in Figure 4.6. Post-compaction operations such as sizing, sintering, coating, etc., are usually also applied depending on the application [37].

The compacted powder is termed as green body, so the density after pressing is the green density and the strength after pressing is the green strength [1]. The compressed parts (also referred to as green parts) require a high density in order to achieve the required specifications in terms of strength and dimensional tolerances after sintering. The density distribution in green parts must be kept at a minimum in order to avoid distortions during sintering. Structural ceramics and hard metal components are also subject to similar requirements. Ceramic parts in the green state usually have low strength and can present handling difficulties. The control of tolerances is particularly important for hard metals that are expensive to machine after sintering [37]. Compacting pressures (Table 4.2) depend on the material type and desired final density of the applications [38].

The density required must essentially be achieved during compaction; in subsequent processing steps, only minor density increases are possible mainly by cold deformation of the bulk part in closed dies or by local densification of functional areas. This way the property profile is more or less determined with the powder compaction. Many properties obey a power law relationship with the density

$$(P / P_0) = (\rho / \rho_0)^m$$

where P is property, P_0 is property in the fully dense state, ρ is density, ρ_0 is density of the pore-free state, m is the property and pore morphology-dependent constant.

A few examples are given in Figure 4.7. There is virtually no material characteristic that is proportional to the density. For better mechanical strength values, it would be most interesting to compact to full density, as the fractional property increase is largest on approaching the pore-free state [36].

Table 4.2 Typical compacting pressures for various applications.

Application	Compaction pressures (MPa)
Porous metals and filters	40–75
Refractory metals and carbides	69–205
Porous bearings	145–355
Machine parts	265–695
High density iron and steel parts	695–1655

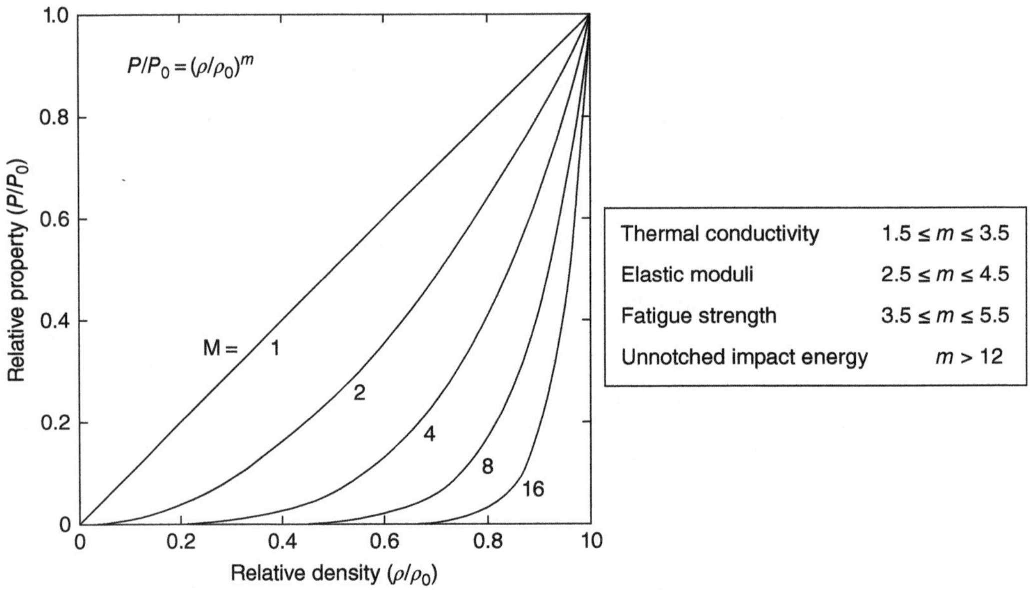

Figure 4.7 Density dependence of properties schematically. *Source:* From Beiss 2005 [36].

With increasing applied pressure, the density of the powder mass increases or porosity decreases [2]. The mathematical relation between the pressure and density that has been widely used was developed by Balshin:

$$\ln P = -A \cdot V + B \tag{4.1}$$

In which $\ln P$ is the natural logarithm of the applied pressure; V is the relative powder volume (the ratio of the volume of the powder compact and the volume of solid metal of the same mass); A and B are constants. This equation can not be valid at high pressure, because it would predict a relative powder volume smaller than one at these pressures. Also for zero pressure, the equation would predict an infinitely large relative powder volume [2]. The developed another relationship between pressure and density as follows:

$$P = \frac{1}{K}\left(\ln\frac{1}{1-D} + B\right) \tag{4.2}$$

where D is the relative density of the compact (ratio of green density to solid density): P is the applied pressure, and K and B are constants. The powders compacted in a rigid die [2].

When pressure is transmitted from only one punch, the compact is less uniform and the process termed single-action pressing [1]. Simple thin parts can be made on single action presses as shown in Figure 4.8a. Thin parts can be done this way because there is not enough die wall friction to substantially lower the transmitted pressure from the upper punch only. The die is stationary, and the part is ejected by a single lower punch [12]. When the pressure is transmitted from both bottom and upper punches (Figure 4.8b), the process is termed double-action pressing. A similar effect is possible using a floating die (Figure 4.8c), with the bottom punch held stationary and the die allowed to move while the upper punch applies pressure [1].

Die compaction gives rise to friction on the tool walls. Because of this, the forces inside the compact and the resulting internal density distribution is not uniform as shown in Figure 4.9. In simple single action compaction effective pressure and the resulting density are highest at the face in contact with the upper punch while the lowest pressure and density values occur at

Figure 4.8 PM part compaction press types: (a) Single action, (b) double action, and (c) floating die.

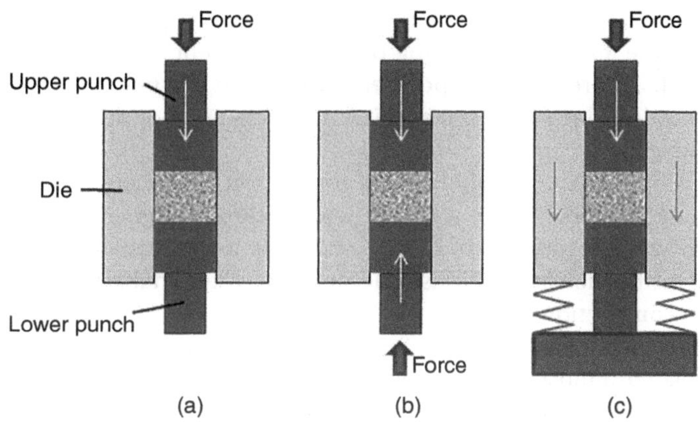

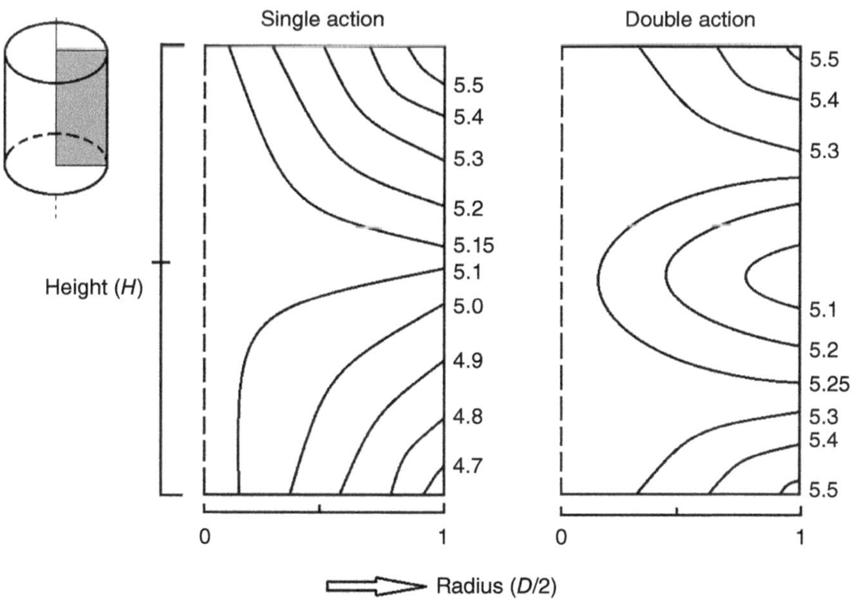

Figure 4.9 Constant density lines in cylinders of compacted copper powder. *Source:* From German 2005 [1].

the bottom edge of the compact [14]. The presence of green density gradients is a problem because the component will wrap in sintering. Hence, minimal green density variations ensure uniformity in both properties and dimensions after sintering [1].

Double action compaction allows the height of a compact that can practicably be produced to be doubled. Each side of so-called neutral axis is a mirror image of the other, with identical distribution of pressure and density. Thus using double acting compaction the height of the compact can be doubled while all other conditions remain unchanged. Where compacts have particularly complicated shapes, the distribution of density and pressure become increasingly nonuniform which can easily cause cracks to form during pressing. This is especially the case during the application of load for ejection of the compact from the die [14]. The compact experiences considerable friction with the tooling during ejection. The strength of the green compact is low, yet the ejection stresses are high. Cracks occur when the ejection stresses exceed the green strength of the compact [1]. Die wall lubrication or the addition of lubricants to the powder improve the density distribution and the conditions for ejection from the die [14].

The mechanical performances of the final materials closely depend on the quality of the green powder compacts. In particular, the compacted metal pieces must outlive ejection from the die and allow handling before subsequent processing, such as sintering or hot extrusion [39]. The research on powder compaction has been mainly focused on densification of pure powders due to the complexity in densification of mixed powders and it has been extended to mixed powders [40].

Powders are densified by the application of pressure, initially by the particles sliding past one another and then by particle deformation at higher pressures. As shown in Figure 4.10, the increase in density is initially rapid at low pressures, but the powder progressively resists densification as the pores collapse. It is evident that particle hardness is an important parameter in compaction [1].

Typical green densities for die-compacted parts are 75–85% of full density. Green density is related almost exponentially to the applied load. At low densities, a small increase in load causes a major increase in density, while at high-density levels, a large increase in applied load is required to get a small increase in density. The required compaction pressure to achieve a desired level of density is a function of the following [2]:

- The powder shape (i.e. sponge, flake, and spherical)
- Particle size and size distribution
- Powder chemistry (i.e. prealloyed, blended master alloy)
- Lubrication practice

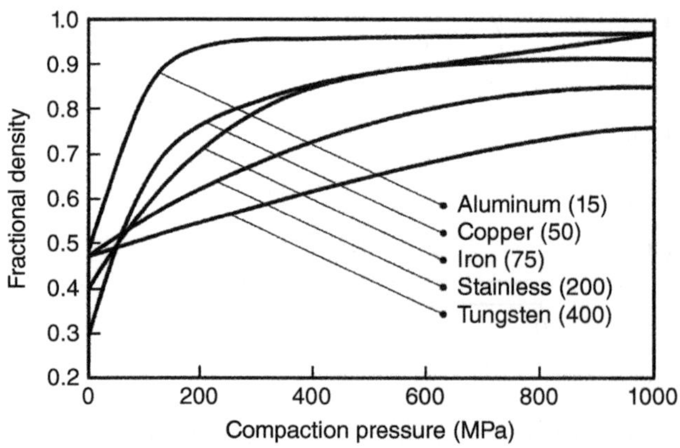

Figure 4.10 Compaction behavior of several metals, shoving how density increases with pressure. *Source:* From German 2005 [1].

For steel powders, commonly used pressures are in the 400 MPa range to achieve green densities from 80% to 85% of full density [2].

4.4.2 Hot Pressing

Hot pressing process is used to consolidate numerous high temperature materials, which require the simultaneous application of pressure and temperature. Hot pressing is commonly used for ceramic and other high temperature composites, sputtering targets, metal-bonded diamond products, beryllium powder products, and tungsten carbide products that by itself is the single biggest commercial application.

4.4.3 Powder Forging

Powder forging have been studied extensively and raising much interest in many parts of the industry as economic method of producing high-strength, high-ductility parts from metal powders [41]. As powder metallurgy (PM) method competes with other methods on the basis of cost that can be lower for high volume production of complicated components. The PM technology is conducive nearly any material that can be processed in powder form. This technology is sometimes the only manufacturing method used to produce parts using materials such as porous materials, composite materials, refractory materials, and special high duty alloys [42]. The vast application of ferrous powder metallurgy materials is in automotive and aerospace industries [43].

Sintered PM compacts are made by the process of compacting and sintering ferrous powder and nonmetal powder. A known limitation of this route is the large number of small voids left in components after sintering. Plastic deformation is a main way to improve the performance of sintered ferrous material and obtain the final product. In general, the preform produced by the conventional process will undergo a large degree of plastic deformation with enhanced level of densification [44, 45]. Though plastic deformation of powder preforms is similar to that of conventional fully dense material, the additional complications are because of substantial amount of void fractions. Because there is a large number of residual porosity in the sintered powder materials, plastic volume change of sintered compacts will result from the void reducing and closing during plastic deformation. During the elastic deformation of fully dense material, Poisson's ratio remains constant, and it is a property of the material; this ratio being 0.5 for all materials that conform to volume constancy. However, in the plastic deformation of sintered PM preforms, density changes occur, resulting in Poisson's ratio remaining less than 0.5 and tending to approach 0.5 only in the near vicinity of the theoretical density (Figure 4.11) [45–47].

4.5 Sintering

The ISO definition of sintering process is "The thermal treatment of a powder or compact at a temperature below the melting point of the main constituent, for the purpose of increasing its strength by bonding together of the particles." Sintering converts the green powder compact into a product with the desired size, shape, and quality. In defining a sintering cycle, the first consideration is how to include the desired level of sinter bonding. Along with particle bonding, sintering cycles have secondary objectives such as oxide reduction [1].

Sintering is a complex high temperature process that consolidates and strengthens loose or compacted particles into a more dense coherent body. During sintering powder particles form

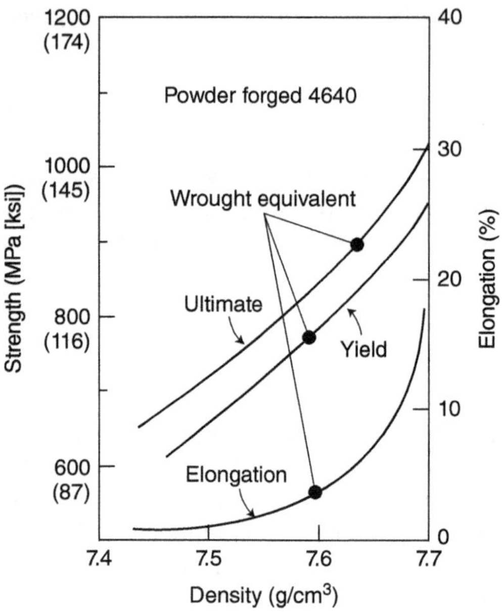

Figure 4.11 Density effect on the strength and ductility of a forged prealloyed 4640 steel. *Source:* From German 2005 [1].

coherent bonds and densify by pore shrinkage. During the sintering process a wide variety of physical, chemical and metallurgical phenomena occur within the mass of metal powder particles. These effects are influenced by the sintering conditions (time, temperature, and atmosphere), as well as the chemical composition of the powder mass [48]. Particle size, compact porosity, and powder type (mixed, prealloyed, and diffusion-alloyed) also influence sintering practices. Sintering time, temperature, and atmosphere are the most significant factors from a practical perspective, with temperature being the most important variable [2].

Sintering is generally performed at temperatures around $^2/_3$ to $^4/_5$ of the absolute melting point or solidus of the material for a single-component system. Table 4.3 lists typical sintering temperatures for various PM materials and ceramics. Multicomponent powder mixtures are generally sintered near the melting point of the constituent with the lowest melting temperature. Sintering times are typically 20–60 minutes under protective atmosphere. Widely used furnace atmospheres include endothermic gas, exothermic gas, dissociated ammonia (DA), hydrogen,

Table 4.3 Sintering temperatures for powder metal alloys and special ceramics.

Sintered material	Sintering temperatures (°C)
Aluminum alloys	590–620
Bronze	740–780
Brass	890–910
Iron, carbon steel	1120–1150
High alloyed stainless steels	1200–1280
Hard metals (cemented carbides)	1350–1450
Molybdenum and its alloys	1600–1700
Alumina	1400–1800
Zirconia (with different additives)	1400–1750

hydrogen–nitrogen mixtures, and vacuum. Of these endothermic gas is the most widely used, followed by dissociated ammonia. The main function of the atmosphere is to protect a part from oxidation or nitridation, as might occur when heating in air. Frequently, however, critical aspects of a sintering atmosphere also include reducing and carburizing power and capability for efficient removal of the lubricant. Some sintering furnaces contain so-called rapid burn-off zones for rapid and efficient removal of the lubricant. This usually entails an atmosphere with a higher oxidation potential, that is, a higher concentration of gases such as steam and carbon dioxide.

Inadequate sinter generally is indicated by low strength, low hardness, and improper dimensions. Causes of inadequate sintering are often related to the atmosphere, but several factors may be involved, such as sintering temperature too low, insufficient reducing agent, dew point too high in the hot zone, high O_2 content in the hot zone, incorrect green density, incorrect belt speed, or time at temperature. Corrective actions would include the following: (i) to measure and control dew point and O_2 content in the hot zone, (ii) to measure and increase H_2 content, (iii) check and correct belt speed and/or powder compositions [2].

The properties available cover a broad spectrum, even for a single composition because of differences in microstructure and density. Thus, composition alone is not sufficient basis for selection of a material for an application. A major differentiation in properties centers around porosity or density. For example Figure 4.12 plots strength, elongation, and impact toughness versus porosity for a sintered steel. All of these properties improve with density, so the highest properties come from full-density PM products [1].

The growth of PM is dependent on the ability of technological advancements to deliver products with higher densities than are currently available. Parts distribution by density has progressed and the anticipated use of higher density parts in the future. The density increases to date are mainly due to improvements in powder making and compaction processing. The future advances are likely to come mostly from our understanding of the sintering process [49].

High temperature sintering is arbitrarily defined as processing above 1150 °C, since this is the practical limit for metal wire belts that are used to convey product in continuous sintering furnaces. The higher temperature processing is, therefore, performed in ceramic belt, ceramic pusher, ceramic walking beam, or vacuum furnaces. Traditional sintering at 1120 °C yields entirely satisfactory powder metallurgy products for the vast majority of applications for iron based alloys, but sintering at higher temperatures -as high as 1370 °C promotes additional particle-to-particle bonding and more complete alloying because of the higher diffusion rates.

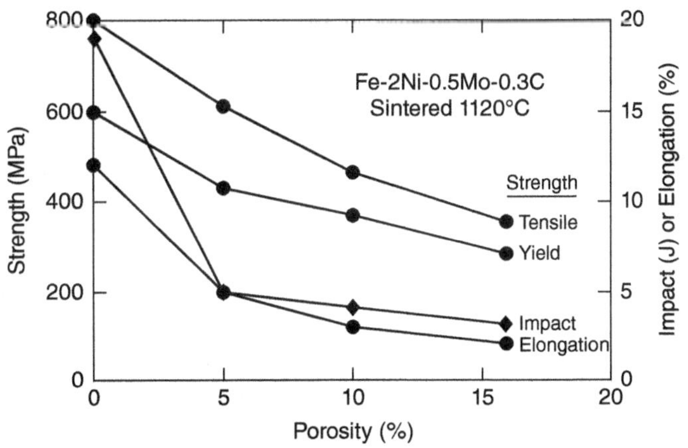

Figure 4.12 Property dependence on porosity. *Source:* From German 2005 [1].

Much of the PM industry uses mixtures of hydrogen and nitrogen as the protective atmosphere during sintering. While the hydrogen has the primary role of promoting reduction of oxides, the nitrogen can result in a few different outcomes. Nitrogen is not a true inert gas for processing steels since it does have some interaction with both ferrite and austenite phases. For structural applications, a small amount of nitrogen absorption is positive because it results in solid solution strengthening of the matrix. However, for soft magnetic and corrosion resistance applications, the gas can have a detrimental effect. In the case of stainless steels, the nitrogen readily combines with chromium, depleting it from the matrix and preventing it from providing the protective layer to the material that is so essential to corrosion resistance. In the 400 series stainless steels, there is the added complication that nitrogen dissolves in the austenite phase and, if the nitrogen content is sufficiently high, martensite formation can occur.

In soft magnetic materials, nitrogen can form nitrides in the grains and grain boundaries, which pin the magnetic domain walls and lower performance. To some degree, the deleterious effects of nitrogen can be minimized by sintering at higher temperatures since the solubility of the element in austenite decreases as the temperature is increased. Ironically, the maximum solubility of nitrogen in steels occurs near 1120 °C. Therefore, it is important to cool through this temperature range as quickly as possible in order to minimize the impact of nitrogen. High temperature sintering is essential when processing tool steels such as M2, D2, and T15 (http://mppd-11a01.buffalo.mediasauce.com/tech-insider/).

4.6 Powder Injection Molding (PIM)

Powder injection molding (PIM) is a new manufacturing technology for net-shape production of small, intricate, and precise metal or ceramic components in large quantities. This new technology uses the shaping advantage of injection molding but is applicable to metals and ceramics such as stainless steel, titanium, nickel, intermetallics, alumina, silicon carbide that are difficult to produce through conventional techniques like machining, casting, forging etc. It utilizes the advantage of shape complexity of plastic injection molding and the materials flexibility of powder metallurgy [34, 50]. PIM has five key features: low production costs, shape complexity, tight tolerances, applicability to several materials, and high final properties. Many successful applications rely on particular combinations of these attributes. Examples include orthodontic brackets for straightening teeth, porous filters for treating hot waste water, magnets for controlling computer disk drives, small gears for electrical hand tools and toothbrushes, cleats on sporting shoes, surgical tools such as scalpels, electrical connectors, handgun components, and microwave filters for high-frequency microelectronics [2, 34]. The evolution of the PIM technology has resulted in many variations, reflecting different combinations of powders, binders, molding techniques, debinding routes, and sintering practices. Metal injection molding, commonly known by its acronym MIM, is by far the most widely used PIM process.

PIM consists of four major steps (Figure 4.13): mixing of fine metal/ceramic powder (50–65 vol%) with binder (a blend of polymer, wax, surfactants, additives) to form feedstock, injection molding of the molten feedstock to a desired shape to form green compact, removal of the binder from the green compact called debinding and then sintering of the debound part to form strong metallic/ceramic component in much the same way as traditional die compacted parts. The binders can be removed by gradual heating up to 450 °C in air, in up to 60 hours. This decomposes the organic material and oxidizes the metal. The oxides hold the particles together. Alternatively, some of the binder can be dissolved in solvent. When this partially debindered part is slowly heated in atmosphere or vacuum the rest of the binder goes off. When things work right, the metal particles begin to sinter before all of the binder is gone,

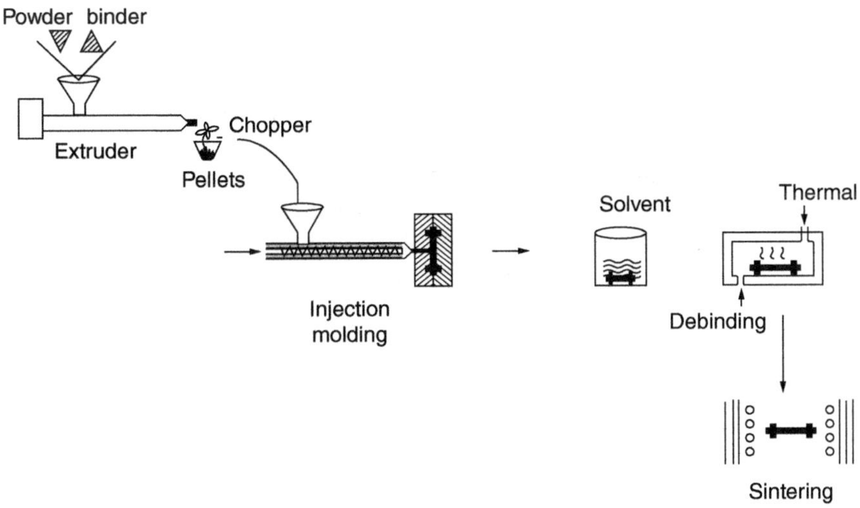

Figure 4.13 The sequence of steps involved in PIM, where a binder and powder are mixed to form a feedstock which is molded, debound, and sintered. *Source:* Ye et al. 2008 [51]. Reproduced with permission of Elsevier.

and this provides strength to hold the part together. After sintering densities of 95% or more are reached, and the mechanical properties are, for that reason, generally superior or equivalent to those of traditional PM parts [12, 30, 34, 52].

PIM competes with other shaping technologies, namely plastic injection molding, die and investment casting, machining, cold isostatic compaction, and slip casting. It overcomes the property limitations inherent to plastics, the shape limitations of traditional powder compaction, the cost of machining, the productivity limits of isostatic pressing and slip casting, and the defect and tolerance limitations of casting [34].

The role of the binder is to carry the powder to the intricate parts of the mold by reducing the viscosity of the metal/ceramic powder. Binder formulation plays the most important role as the success of PIM depends on it. The rheological properties of the feedstock, that is the powder/binder mix, are of major importance [50]. Since torque rheometers are routinely used in measuring viscosity during mixing, maximum torque values have been suggested. During molding, the shear rate usually ranges between 10^2 and $10^5\,s^{-1}$. In this range, the maximum useful viscosity for the mixture is 10^3 Pa s at the molding temperature [1]. If binder volume percent is more in the feedstock, there is a chance of separation of powder and binder during molding, which will cause uneven shrinkage during sintering, and shape retention will be difficult during debinding. On the other hand, less binder will increase the feedstock viscosity, which will cause improper filling of the mold. Binder formulation affects the debinding process as improper binder formulation causes shape distortion during debinding [50]. The viscosity at the molding temperature must be such that the mix flows smoothly into the die without any segregation, and the viscosity should be as constant as possible over a range of temperature. However, the mix must become rigid on cooling. These requirements dictate the properties of the binders used, and to some extent, granulometry of the powder. PIM has many advantages albeit with certain limitations, e.g. the defects that occur in the previous steps cannot be rectified in the subsequent steps rather the defects become enlarged in the next subsequent steps causing rejection of components [50].

Almost any metal that can be produced in a suitable powder form can be processed by MIM. The list of metals includes many common and several less common metals and their alloys – plain and low alloy steels, stainless steels, high speed steels, copper base alloys, nickel and cobalt base superalloys, titanium, intermetallics, magnetic alloys, refractory metals, and hard metals. Aluminum and magnesium are exceptions because the adherent oxide film that is always present on the surface of powder particles inhibits sintering. The most promising candidates from the economic point of view are the more expensive materials. [53].

PIM begins with the mixing of selected powders and binders. The particles are small to aid sintering, usually between 0.1 and 20 μm with spherical or near-spherical shapes. For example a 5 μm carbonyl iron powder is widely used in the PIM process, as −16 μm gas-atomized stainless steel powder. Most common engineering alloys are used, including various steels, tool steels, and stainless steels. Likewise ceramics, refractory metals, and cemented carbides are processed in a similar manner. Many types of powders can be used, but great differences exist in mixing and molding, especially if the particle shape is nonspherical [2].

Among all the stages of PIM, injection stage is the most important as many potential defects such as weld line, powder binder separation, jetting, short shot occur due to improper specification of the injection process parameters. Trial and error techniques are employed to optimize the process parameters in injection stage to get consistent quality product through this process. Nowadays, commercial computational fluid dynamics (CFD) simulation software's such as Moldflow, Cadmould, etc., meant for plastic injection molding, have been used to aid mold design and optimize the process parameters, especially in the injection stage of PIM.

The fabrication of titanium components using the metal PIM process (Ti MIM) turned a significant corner in recent years. In the middle of the 1990s, there was concern on meeting the quality standards needed for medical devices. By then, gun parts, hand tools, toys, golf clubs, and a variety of demonstration surgical hand tools were available, but the production of demanding medical devices was still pending. As the Ti-MIM field matured, the situation has shifted dramatically in recent years [54]. Applications include chisels, curettes, cutters, bone cutting forceps, various knives, skin punches, dissectors, osteotomes, reamers, scalpels, scissors, and various orthopedic instruments. From the materials perspective, a majority of these components are fabricated from stainless steels, and MIM stainless steels are now reasonably well established in the medical field. In the medical device area, there is a significant demand for components made from titanium and its alloys and to a lesser extent from cobalt-chromium-molybdenum-base alloys; this reflects the unique properties of these alloys [55].

The attractive features of the MIM process can be applied advantageously to the fabrication of metal matrix composite (MMC) or ceramic matrix composite (CMC) parts. Although many MMC or CMC offer unique properties that cannot be normally achieved by monolithic materials, their commercial use are often restricted by the high cost of materials and manufacturing. By applying the MIM process, the cost for commercial use of composite materials can be significantly reduced. In recent years, comprehensive work has been conducted to explore the potential of MIM for the fabrication of metal- and ceramic-based composites and components [56].

Interestingly for MIM, there is a progressive move toward smaller products. Several firms report typical components of 5 g, and the overall industry median was under 9 g, with a range from 0.002 g to over 100 g [57].

4.7 Summary and Future Work

Worldwide demand for lowering the cost of sintered part including automobile parts is rising. Thus, development of materials that achieve high strength with low cost elements has become necessary. Nanocrystalline materials possess many novel properties that are absent in bulk materials and, therefore, offer a wide range of potential applications. In recent years, nanoscale powders have proven to be of increasing interest due to their ability to produce nanocrystalline hardmetals with superior hardness and toughness. Currently, there is no scalable, domestic source for the production of nanosize tungsten, tungsten carbide, and other refractory metal powders [58].

Ceramic microsystems are gaining increasing significance in applications, such as healthcare, where properties such as biocompatibility and chemical resistance are desirable. Several fabrication techniques have been developed for producing ceramic micro-components, but a major limitation of such techniques is in their ability to produce truly three-dimensional component with relatively complex geometries. One approach to overcoming process limitations on component geometry is to use lost core techniques. For relatively large, "macro-scale," ceramic components, there has been increased interest in the idea of using lost cores as part of the PIM process, in order to be able to mold internal geometries that could not be produced with slides or cores in conventional molding [59].

References

1 German, R.M. (2005). *Powder Metallurgy & Particulate Materials.* Princeton: MPIF.

2 (1998). *ASM Handbook,* Powder Metal Technologies and Applications, vol. 7. Materials Park OH: ASM International.

3 Mori, K. (2006). Finite element simulation of powder forming and sintering. *Computer Methods in Applied Mechanics and Engineering* 195: 6737–6749.

4 Rolland, S.A., Mosbah, P., Gethin, D.T., and Lewis, R.W. (2012). Load dependency in the cold die powder compaction process. *Powder Technology* 221: 123–136.

5 German, R.M. (1996). *Sintering Theory and Practice.* Wiley.

6 Coube, O. and Riedel, H. (2000). Numerical simulation of metal powder die compaction with special consideration of cracking. *Powder Metallurgy* 43 (2): 123–131.

7 Briscoe, B.J. and Rough, S.L. (1998). The effects of wall friction in powder compaction. *Colloids and Surfaces A: Physicochemical and Engineering Aspects* 137 (1–3): 103–116.

8 Gillia, O. and Bouvard, D. (2000). Phenomenological analysis of densification kinetics during sintering: application to WC–Co mixture. *Materials Science and Engineering A* 279 (1–2): 185–191.

9 Munir, Z.A., Anselmi-Tamburini, U., and Ohyanagi, M. (2006). The effect of electric field and pressure on the synthesis and consolidation of materials: a review of the spark plasma sintering method. *Journal of Materials Science* 41 (3): 763–777.

10 Jeong, M.S., Yoo, J.H., Rhim, S.H., and Lee, S.K. (2012). A unified model for compaction and sintering behavior of powder processing. *Finite Elements in Analysis and Design* 53: 56–62.

11 Lange, F.F. (2001). Shape forming of ceramic powders by manipulating the interparticle pair potential. *Chemical Engineering Science* 56: 3011–3020.

12 Pease, L.F. and West, W.G. (2002). *Fundamentals of Powder Metallurgy.* Princeton: MPIF.

13 Ünal, R. (1999). Production, Consolidation and Investigation of PM Al–Mg–Sc Alloys. Ph.D dissertation. Technische Univesitaet Clausthal.

14 Schatt, W. and Wieters, K.-P. (1997). *Powder Metallurgy Processing and Materials.* Shrewsbury: EPMA, Liveseys Ltd.

15 Lawley, A. (1992). *Atomization: The Production of Metal Powders.* Princeton: MPIF.

16 Dan V Goia-p6-Nanotechnology.

17 Liu, H. (2000). *Science and Engineering of Droplets: Fundamentals and Applications.* Noyes Publications.

18 Abreu, C.R.A., Tavares, F.W., and Castier, M. (2003). Influence of particle shape on the packing and on the segregation of spherocylinders via Monte Carlo simulations. *Powder Technology* 134: 167–180.

19 Miyajima, T., Yamamoto, K.I., and Sugimoto, M. (2001). Effect of particle shape on packing properties during tapping. *Advanced Powder Technology* 12: 117–134.

20 Fayed, M.E. and Otten, L. (1997). *Handbook of Powder Science & Technology*, 2e. New York: Chapman & Hall.

21 Senkov, O.N., Srisukhumbowornchai, N., Ovecoglu, M.L., and Froes, F.H. (1998). Microstructural evolution of a nanocrystalline Ti–47Al–3Cr alloy on annealing at 1200 °C. *Scripta Materialia* 39: 691–698.

22 Kima, J.C., Ryu, H.J., Kim, J.S. et al. (2009). Synthesis and densification of Cu added Fe-based BMG composite powders by gas atomization and electrical explosion of wire. *Journal of Alloys and Compounds* 483: 28–31.

23 Lee, S.Y., Kim, T.S., Lee, J.K. et al. (2006). Effect of powder size on the consolidation of gas atomized $Cu_{54}Ni_6Zr_{22}Ti_{18}$ amorphous powders. *Intermetallics* 14: 1000–1004.

24 Suryanarayana, C., Ivanov, E., and Boldyrev, V.V. (2001). The science and technology of mechanical alloying. *Materials Science and Engineering A* 306: 151–158.

25 Metz, R., Machado, C., Houabes, M. et al. (2008). Nitrogen spray atomization of molten tin metal: powder morphology characteristics. *Journal of Materials Processing Technology* 195: 248–254.

26 Markus, S., Fritsching, U., and Bauckhage, K. (2002). Jet break up of liquid metal in twin fluid atomization. *Materials Science and Engineering A* 326: 122–133.

27 Li, H. and Deng, X. (2007). Prediction of powder particle size during centrifugal atomisation using a rotating disk. *Science and Technology of Advanced Materials* 8: 264–270.

28 Eslamian, M., Rak, J., and Ashgriz, N. (2008). Preparation of aluminum/silicon carbide metal matrix composites using centrifugal atomization. *Powder Technology* 184: 11–20.

29 Jovic, V.D., Jovic, B.M., and Pavlovic, M.G. (2006). Electrodeposition of Ni, Co and Ni–Co alloy powders. *Electrochimica Acta* 51: 5468–5477.

30 Dowson, G. (1998). *Booklet: "Powder Metallurgy: The Process and its Products".* Shrewsbury: EPMA.

31 Hopkins, W.G. (2002). Co-atomization of liquid melts. *Metal Powder Report* 57 (2), Pages 14, 16–17.

32 Upadhyaya, G.S. (2002). *Powder Metallurgy Technology.* Cambridge International Science Publishing Ltd.

33 Song, K.C. and Kang, Y. (2000). Preparation of high surface area tin oxide powders by a homogeneous precipitation method. *Materials Letters* 42: 283–289.

34 German, R.M. and Bose, A. (1997). *Injection Molding of Metals and Ceramics.* Princeton: MPIF.

35 Yu, Y. (2005). PM Training Course for Young Materials/Design Engineers, Course notes. EPMA, Aachen-Germany (3–11 September).

36 Beiss. (2005). PM Training Course for Young Materials/Design Engineers, Course notes. EPMA, Aachen-Germany (3–11 September).

37 Sinka, I.C. (2007). Modelling Powder Compaction, KONA No. 25, pp. 4–22.

38 Degarmo, E.P., Black, J.T., and Kohser, R.A. (1997). *Materials and Processes in Manufacturing,* 8e. Upper Saddle River, NJ: Prentice Hall.

39 Carles, V.B., Arnal, A., Poquillon, D., and Tailhades, P. (2008). Correlation between the morphology of cobalt oxalate precursors and the microstructure of metal cobalt powders and compacts. *Powder Technology* 185: 231–238.

40 Kim, K.T. and Cho, J.H. (2001). A densification model for mixed metal powder under cold compaction. *Int. J. Materials Sciences* 43: 2929–2946.

41 Poshal, G. and Ganesan, P. (2009). Neural network approach for the selection of processing parameters of aluminium–iron composite preforms during cold upsetting. *Journal of Engineering Manufacture* 224: 459–472.
42 Rosochowski, A., Beltrando, L., and Navarro, S. (1998). Modeling of density and dimensional changes in re-pressing/sizing of sintered components. *Journal of Materials Processing Technology* 80: 188–194.
43 Lindskog, P. (2003). Economy in car-making – powder metallurgy, Business briefing: global automotive manufacturing and technology. London, UK: Technology & services.
44 Hua, L., Qin, X., Mao, H., and Zhao, Y. (2006). Plastic deformation and yield criterion for compressible sintered powder materials. *Journal of Materials Processing Technology* 180: 174–178.
45 Liu, Y., Chen, L.F., Tang, H.P. et al. (2006). Design of powder metallurgy titanium alloys and composites. *Materials Science and Engineering A* 418: 25–35.
46 Narayanasamy, R. and Selvakumar, N. (2005). Deformation behaviour of cold upset forming of sintered Al–Fe composite preforms. *Journal of Engineering Materials and Technology* 127: 251–256.
47 Rajeshkannan, A., Pandey, K.S., Shanmugam, S., and Narayanasamy, R. (2008). Sintered Fe–0.8%C–1.0%Si–0.4%Cu P/M preform behaviour during cold upsetting. *Journal of Iron and Steel Research International* 15: 92–97.
48 Sanderow, H.I. (ed.) (1990). *High Temperature Sintering*. Princeton: MPIF.
49 Narasimhan, K.S. and Semel, F.C. Sintering of powder premixes, Hoeganaes Corp. NJ. A Brief Overview. Paper no. 2007-01-045.
50 Samanta, S.K., Chattopadhyay, H., and Godkhindi, M.M. (2011). Thermophysical characterization of binder and feedstock for single and multiphase flow of PIM 316L feedstock. *Journal of Materials Processing Technology* 211: 2114–2122.
51 Urtekin, L. (2008). Investigation of the effect of molding and sintering parameters on properties of powder injected molded steatite ceramics. Ph.D thesis. Gazi University.
52 Zauner, R. (2006). Micro powder injection molding. *Microelectronic Engineering* 83: 1442–1444.
53 Schlieper, G., Dowson, G., Williams, B., and Petzoldt, F. (2013). *Metal Injection Molding*. EPMA, Free Publications.
54 German, R.M. (2012). Infrastructure emergence for metal injection molded titanium medical devices. *International Journal of Powder Metallurgy* 48 (2): 33–38.
55 Bose, A. (2012). Powder injection molding for medical and dental markets. *International Journal of Powder Metallurgy* 48 (2): 4–9.
56 Ye, H., Liu, X.Y., and Hong, H. (2008). Fabrication of metal matrix composites by metal injection molding – a review. *Journal of Materials Processing Technology* 200: 12–24.
57 German, R.M. (2008). PIM breaks the $1 bn barrier. *Metal Powder Report* 63 (3): 8–10.
58 Sherman, A.J., Doud, B., and Hardy, D. (2012). Properties of nanocrystalline fluidized-bed-reduced tungsten, tungsten-carbide, and tungsten rare-earth alloy powders. *International Journal of Powder metallurgy* 47 (5): 41–47.
59 Attia, U.M. and Alcock, J.R. (2012). Fabrication of ceramic micro-scale hollow components by micro-powder injection moulding. *Journal of the European Ceramic Society* 32: 1199–1204.

5

Injection Molding at Multiscales

Danyang Zhao[1,2], Minjie Wang[1], and Donggang Yao[2]

[1] School of Mechanical Engineering, Dalian University of Technology, Dalian, PR China
[2] School of Materials Science and Engineering, Georgia Institute of Technology, Atlanta, GA, USA

5.1 Introduction

Since 1976, plastics have been the most widely used materials in the United States, surpassing steel, copper, and aluminum in combined volume. Among all polymer processing methods, injection molding accounts for more than one-third of all polymeric materials processed, and is widely used for mass producing discrete plastic parts of complex shape cost-effectively with high precision [1].

Injection molding is a manufacturing process for producing parts from both thermoplastic and thermosetting plastic materials. The molding material is fed into a heated barrel, mixed, and forced into a mold cavity where it cools and hardens to the configuration of the cavity. Some advantages of injection molding are high production rates, repeatable high tolerances, the ability to use a wide range of materials, low labor cost, minimal scrap losses, and little need to finish parts after molding. It is capable of producing a vast array of plastic products for a variety of applications including automotive, aerospace, medical engineering, consumer goods, plumbing, packaging, and construction [2].

In recent years, with the rapid development of and increasing demand for biomedical devices, high-quality optics and precision micro-electro mechanical systems (MEMS), the size of plastic products becomes smaller and smaller, and the accuracy of molded products becomes higher and higher. At the same time, the weight and dimension of plastic products emerge themselves at multiscales. In terms of differences in size, weight, and dimensional accuracy, these emerging plastic parts can be divided into four distinct groups:

1) *Precision parts*: The overall dimension of precision parts is the same of conventional parts, but the dimensional accuracy is in microns or submicrons. Therefore, the part can be large in size, but the required dimensional accuracy is extremely high. One example is a plastic optical lens with a high-accuracy surface contour, as shown in Figure 5.1a.
2) *Thin-wall parts*: Thin-wall parts are conventionally defined as parts that have a nominal wall thickness of 1 mm or less and a surface area of at least 50 cm^2 with a flow length/thickness ratio above 100, as shown in Figure 5.1b.
3) *Microstructured parts*: The weight of a microstructured part may still exceed 1 g, but the accuracy of microstructures (holes, slots, ribs, etc.) is in microns or even submicrons. Many plastic parts belong to this category, such as diffractive optical elements, grating

Modern Manufacturing Processes, First Edition. Edited by Muammer Koç and Tuğrul Özel.

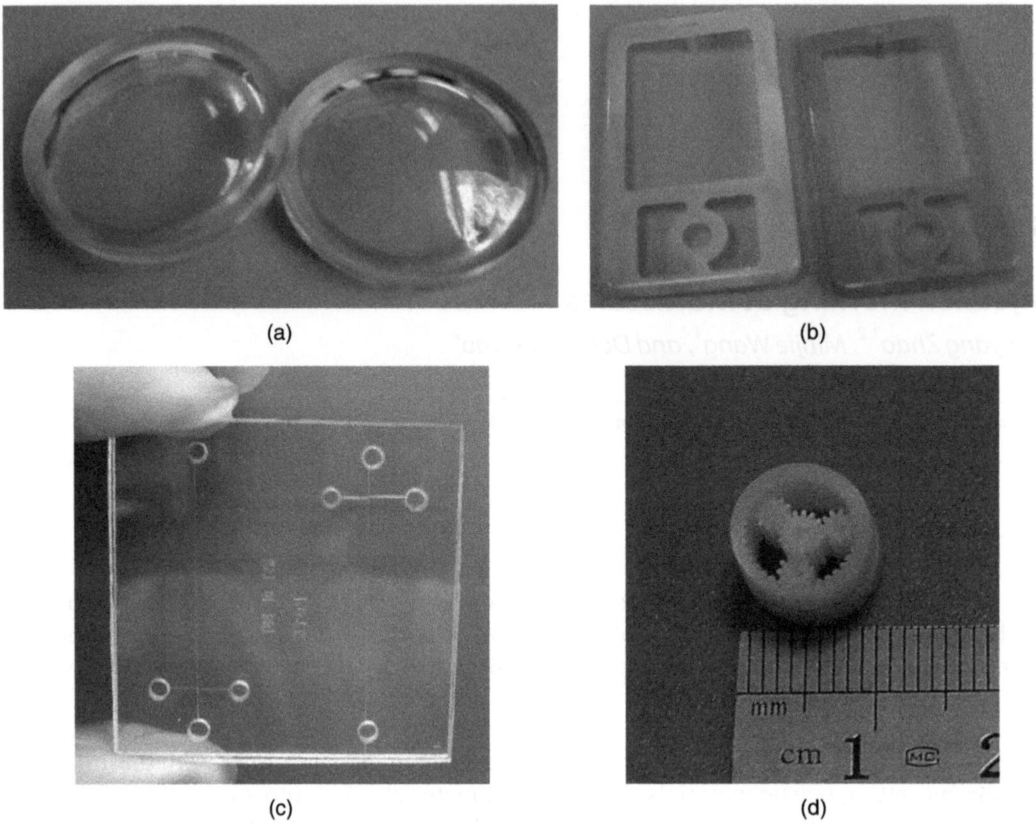

Figure 5.1 Injection molded parts: (a) plastic optical lenses, (b) mobile-phone shells, (c) microfluidic chips, and (d) microgears.

optical elements, multi-fiber connector, light guiding plates, DVD, microfluidic chips, etc. The microstructures of these parts typically sit on the surface of a relatively thick substrate that is injection molded simultaneously as the microstructures in a single shot. A microfluidic chip is shown as an example in Figure 5.1c.

4) *Microparts*: The weight of micropart is in milligrams, and the dimensional accuracy is in microns or submicrons. Some plastic parts belong to this category, such as micro gears, micro needles, micro lenses, hearing aid supports implanted in the ear. Micro gears are shown as examples in Figure 5.1d.

In the injection molding process, the polymer property, equipment, and processing have a significant effect on the part quality and the production efficiency. Generally speaking, a successful process of injection molding needs a lot of knowledge and experience. The general knowledge and experience of standard injection molding can be useful in all areas of injection molding, but rethinking is often needed when the size is changed at multi scales. As we know, when a system is reduced isomorphically in size, the changes in length, area, and volume ratios alter the relative influence of various physical effects that determine the overall operation of the process in unexpected ways. Consequently, some proven designs and processing strategies at one size scale might not work at a different size scale. For example, common polymers used by standard injection molding may encounter additional difficulty in filling microcavities. At the same time, a conventional mechanical machining technique suitable for macroscale can quickly lose its capability as size reduces. As a result, special micro fabrication technologies are needed in making micromolds, including UV-LIGA, micro-EDM (electric discharge machining), micro-milling, focused energy beam machining, etc. Likewise, special techniques such as mold rapid heating, ultrafast injection, dual screw injection, expansion injection, cavity vacuum, and others are often needed to overcome scaling caused processing difficulties at miniature size scales.

Therefore, injection molding at different size scales can be quite different. Different considerations of materials, tooling, and processing are often needed at different size scales and dimensional accuracies. The scaling-related question needing to be answered in injection molding is how the standard molding know-how for part and tool design, process setup, materials characterization, and modeling and simulation can be properly scaled and used for miniaturized parts [3]. Therefore, this chapter aims at providing a general perspective on different considerations on materials, equipment, and processing needed for successful molding at multi scales. Particular processing strategies and techniques for injection molding precision parts, thin wall parts, microstructured parts, and microparts are highlighted. Moreover, the standard simulation for injection molding is introduced, and its adaptations to miniature size scales are discussed.

5.2 Overview of Injection Molding

5.2.1 Overview of the Injection Molding Process

5.2.1.1 Injection Molding Machine and Cycle

Injection molding is the most important process used to manufacture net-shape plastic products. Today, more than one-third of all thermoplastic materials are injection molded and more than half of all polymer processing equipment is for injection molding [4]. The injection molding process is ideally suited to manufacture mass-produced parts of complex shapes requiring precise dimensions. The process goes back to 1872 when the Hyatt brothers patented their stuffing machine to inject cellulose into molds. However, today's injection molding machines are mainly related to the reciprocating screw injection molding machine patented in 1956. A modern injection molding machine with its most important elements is shown in Figure 5.2.

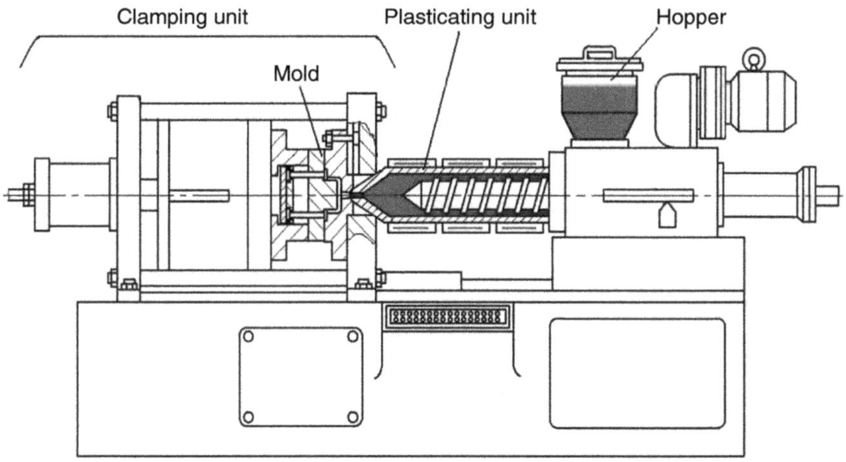

Figure 5.2 Schematic of an injection molding machine.

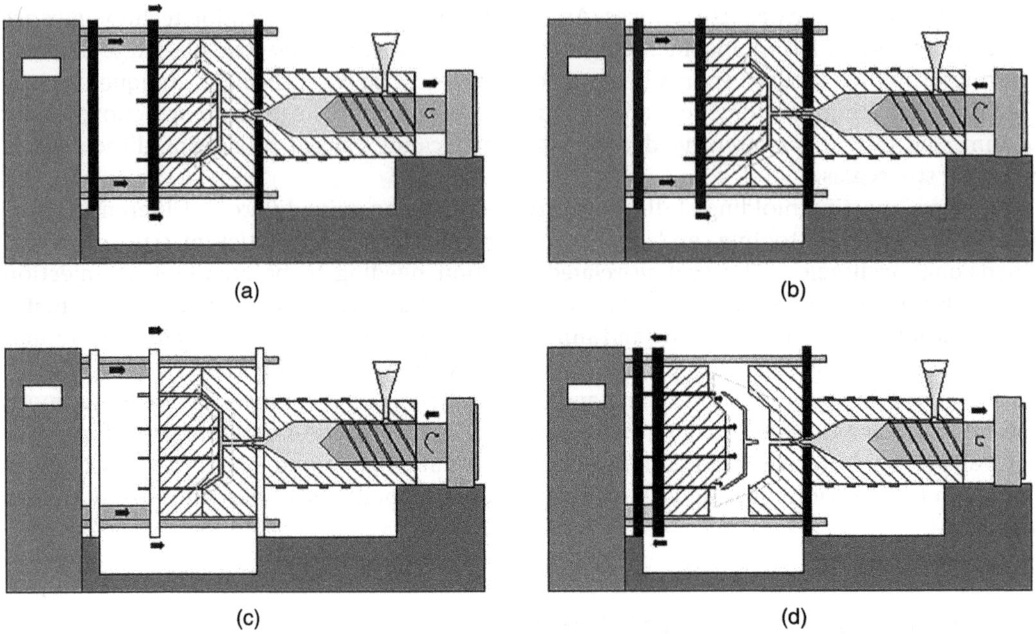

Figure 5.3 Schematic of an injection molding cycle: (a) plasticizing, (b) injecting, (c) cooling, and (d) ejecting.

The major components of the injection molding machine are the plasticating unit, the clamping unit, and the mold.

The sequence of events during the injection molding of a plastic part, as shown in Figure 5.3, is called the injection molding cycle. The cycle begins when the mold closes, followed by the injection of the polymer melts into the mold cavity. Once the cavity is filled, a holding pressure is maintained to compensate for material shrinkage. In the next step, the screw turns, feeding the next shot to the front of the screw. This causes the screw to retract as the next shot is prepared. Once the part is sufficiently cooled, the mold opens and the part is ejected.

5.2.1.2 Basic Structure of an Injection Mold

The injection molding process uses molds, typically made of steel or aluminum. An injection mold is generally custom-made consisting of a sprue and runner system, a gate or gates, a cooling system, and an ejection system. The basic structure of an injection mold is shown in Figure 5.4. The mold contains many components and can be split into two halves. Each half is attached inside the injection molding machine, and the rear half is allowed to slide so that the mold can be opened and closed along the mold parting line. The two main mold components are the mold core and the mold cavity. When the mold is closed, the space between the mold core and the mold cavity forms the part cavity. The mold core and mold cavity are each mounted to the corresponding mold base, which is then fixed to the platens inside the injection molding machine. The front half of the mold base includes a support plate, to which the mold cavity is attached, the sprue bushing, into which the material will flow from the nozzle, and a locating ring used to align the mold base with the nozzle. The rear half of the mold base includes an ejection system, to which the mold core is attached, and a support plate. The ejector pins push the solidified part out of the open mold cavity.

In order for the molten plastic to flow into the mold cavities, several channels are integrated into the mold design. First, the molten plastic enters the mold through the sprue. Additional

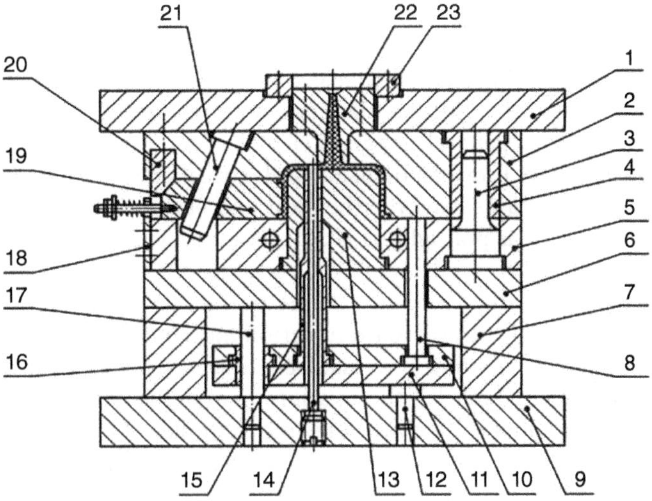

Figure 5.4 Basic structure of an injection mold. 1, Clamping plate of fixed half; 2, cavity plate; 3, shouldered guide pillar; 4, headed guide bush; 5, core-retainer plate; 6, support plate; 7, spacer; 8, return pin; 9, clamping plate of moving half; 10, ejector retainer plate; 11, ejector plate; 12, stop pin; 13, 14, core; 15, ejector sleeve; 16, ejector guide bush; 17, ejector guide pillar; 18, stop block; 19, side core-slide; 20, wedge block; 21, angle pin; 22, sprue bush; 23, locating ring.

channels, called runners, carry the molten plastic from the sprue to all of the cavities that must be filled. At the end of each runner, the molten plastic enters the cavity through a gate that directs the flow. Another type of channel that is built into the mold is cooling channels. These channels allow water to flow through the mold walls, adjacent to the cavity, and cool the molten plastic.

There are many other design issues that must be considered in the mold design. First, the mold must allow the molten plastic to flow easily into all of the cavities. Equally important is the removal of the solidified part from the mold, so a draft angle must be applied to the mold walls. The design of the mold must also accommodate any complex features on the part, such as undercuts or threads, which require additional mold pieces. Most of these devices slide into the part cavity through the side of the mold and are therefore known as slides, or side-actions. The most common type of side-action is a side-core that enables an external undercut to be molded. Other devices enter through the end of the mold along the parting direction, such as internal core lifters, which can form an internal undercut. To mold threads into the part, an unscrewing device is needed, which can rotate out of the mold after the threads have been formed.

5.2.2 Developments of Injection Molding

As the global trend in the plastics injection molding industry slants toward precision molding and miniaturization, there are growing demands for cost-effective manufacturing of precision or micro components with plastics. In the recent years, there have been consistent developments on materials, equipment, mold making, and processing to address the different needs for successful injection molding at different size scales [5].

5.2.2.1 Materials

There are many types of materials that may be used in the injection molding process. The selection of a material for creating injection molded parts is not solely based upon the desired characteristics of the final part. While each material has different properties that will affect the strength and function of the final part, these properties also dictate the parameters used in processing these materials. Each material requires a different set of processing parameters in the injection molding process, including the injection temperature, injection pressure, mold temperature, ejection temperature, and cycle time.

Moreover, thermoplastic materials represent a very large material class, which allows one to find a suitable polymer for nearly every application. There are polymers that are stable at temperatures as high as 250 °C (e.g. poly ether ether ketone [PEEK]) and others which resist aggressive chemicals such as alkaline solutions, acids, and solvents (e.g. perfluoroalkoxy [PFA]). Polymers, in general, are good electrical and thermal insulators, but when filled with appropriate powders they can be used as electrical conductors, heat sinks, and even magnets. Molded microstructures can be either soft and elastic such as polyoxymethylene (POM) or hard and brittle such as polysulfone (PSU). They are available from optically transparent materials such as cycloolefin copolymer (COC) and opaque ones such as polyamide (PA) filled with graphite. Polymers such as polyvinylidene fluoride (PVDF) even exhibit a piezo-electrical effect.

Compared with standard injection molding at macroscales, only some materials are suitable to injection molding at miniature size scales, because the size or accuracy is in microns or even submicrons. Table 5.1 shows a summary of the polymers commonly employed in precision and micromolding and points out some examples of practical use. Due to the small mass or volume of micro components, material costs are less important so that technical and high-performance polymers are often applied [6, 7].

5.2.2.2 Equipment

Besides the mold, the injection molding machine is another key equipment to control the quality of plastic products. The standard injection molding machine suitable for molding macroscale parts is not a dedicated injection molding machine at other size scales. As the size reduces, the quality of the molded part becomes more difficult to control. Therefore, for the injection molding machine to adapt to molding of miniature and precision parts modifications on the machine is often needed. Some specific modifications that have been proven to work for miniature molding processes include the following:

1) *High injection rate*: The injection process at smaller size scales requires a shorter time to prevent premature freezing of the polymer melt. As a result, higher injection speed is needed. The ram speed for traditional hydraulic-driven injection is typically lower than 200 mm/s, while injection speeds as high as 800 mm/s are often needed for successful thin wall molding.

Table 5.1 Polymer materials often used in injection molding at multiscales.

Polymer	Maximum aspect ratios	Minimum structural thicknesses (μm)	Application example
PMMA	20	20	Optical fiber connector
PC	7	350	Cell container
PA	10	50	Micro gear
POM	5	50	Filter with defined pore diameters
PSU	5	270	Housings for microfluidic devices
PEEK	5	270	Housings for micropumps
LCP	5	270	Microelectronic devices
HDPE	8	125	High aspect ratio squares
COC	0.02–2	0.1–0.9	Microfluidic patterns
PA 12-C	10	50	Housings for electrostatic microvalves

Source: From Piotter et al. 2002 [6] and Attia et al. 2009 [7].

Due to shear thinning of the polymer melt, the higher the melt viscosity, the lower the injection speed becomes, thus enabling filling into thinner walls and smaller geometries. In particular, for typical thin wall parts such as tiny memory chip holders and battery cases for mobile phones or digital cameras, a high-speed and high-acceleration capability is required to inject the polymer melt into the mold cavity within a small fraction of a second.

2) *Precise dosing of injection volume*: The weight of some microparts is at milligram level, so precision measurement and control of the injection volume is critical. The traditional single reciprocating screw is unsuitable for such a purpose. Some other metering mechanisms such as lobar stroke are needed to achieve micron-level accuracy.
3) *Rapid response capabilities*: The movement of the screw/plunger of the injection molding machine is quite small; thus the drive unit must have a rapid response capability for quickly ramping the injection pressure in an instant.

To meet the growing need for miniature molding, various types of special precision injection molding machines have been developed by some companies and research institutions in Europe, Japan, and the United States. Based on different types of plasticizing and injection unit involved, precision injection molding machines can be generally classified into four types: screw type, piston type, screw plunger hybrid, and other special forms.

1) *Screw type*: The plastication, metering, and injection of the polymer are fulfilled by a reciprocating screw. The structure of such a machine is simple, and it is easy to control the motion of the screw. However, due to the presence of a check valve at the front of the screw, accurate control of the injection volume is challenging. Representative models of such precision injection molding machines are BOY12A from Boy Co. in Germany, HM7-DENKEY from Nissei Co. in Japan, JMW-015S-5T from Juken Machine Works Co. in Japan, and EC5 from Toshiba Machine Co. in Japan.
2) *Piston type*: The plastication, metering, and injection of polymer are fulfilled by a single plunger or two plungers. These injection molding machines are usually small, but the plasticizing quality is also poor. Representative machine models are Babyplast6/10 from Cronoplast Company in Spain, Rabbit2/3 from MCP Group in the United Kingdom, and Sesame from Medical Murray, Inc., in the United States.
3) *Screw plunger hybrid*: The plasticizing and mixing of polymer are fulfilled by a screw, and the metering and injecting of polymer are done by a small diameter plunger. Representative machine models are TR18S3A from Sodick Co. in Japan, Au3E from Nissei Co. in Japan, Microsystem50 from Battenfeld IMT in Austria, and FX25 from IKV at University of Aachen in Germany.
4) *Other forms*: The ET-10/16-1 injection molding machine from Ettlinger Co. in Germany has a coaxially combined screw/plunger injection unit, and it can produce 0.1–5.0 g of plastic parts.

Many choices exist today for precision molding machines. In 2010 only, several new machines using the screw over plunger technology have been introduced to the industry. Table 5.2 lists several popular precision injection molding machines for micromolding applications [8].

5.2.2.3 Processing

Compared to standard injection molding for macroscale parts, it is more difficult to gain a high-quality part at miniature scales due to increased demand for dimensional accuracy and consistency. Some new processing technologies have therefore been developed to improve the quality of plastic parts at reduced sizes, including rapid thermal cycling, hot runner, mold vacuum, injection compression, in-mold assembly, and mold visualization, among others.

Table 5.2 Parameters of different precision injection molding machines.

Company	Battenfeld	Boy	Ferromatik Milacron	Nissei	Sodick	Sumitomo	Toshiba
Model	Microsystem	129-11	Babyplast	AU3	TR20EH	SE18S	EC5-0.1A
Tonnage	5.6	3.1	6.6	3	22	18	5.5
Injection capacity (oz)	0.04	0.21	0.13	0.11	0.16	0.21	0.24
Injection pressure (psi)	36 259	26 106	38 425		39 841	32 429	29 000
Screw diameter (in.)	0.55	0.55		0.55	0.6	0.55	0.55
Piston diameter (in.)	0.197	n/a	n/a	0.47	0.5	n/a	n/a
Max. daylight (in.)	11.8	11.8	5.51	8.7	13.8		9.8
Footprint (in.)	73 × 81 × 95	90 × 38 × 63	35 × 18 × 26	54 × 20 × 61	97 × 33.9 × 86.3	89 × 24 × 57	71 × 31 × 55

Source: From Bibber 2004 [8].

Rapid Thermal Cycling Technology A heated mold with temperature above the polymer-softening temperature is highly desired in precision injection molding. The elevated mold temperature reduces unwanted freezing during the injection stage, thus improving moldability and enhancing part quality. The resulting advantages include, but are not limited to, longer flow path, improved feature replication and surface transcription, reduced molecular orientations and residual stresses, smoother surfaces of composites, better control of crystallization, stronger weld lines, etc. However, the heated mold needs to be rapidly cooled during the cooling stage to maintain a short cycle time. Despite the large body of available literature, mold rapid heating and cooling does not represent a well-developed area of practice. Development of capable techniques for rapidly heating and cooling a mold with a relatively large mass is technically challenging because of the constraints set by the heat transfer process and the endurance limits set by the material properties [9].

The primary driver stems from the growing need of precision parts, optical parts, and parts with microfeatures in the electronics and biomedical industries. Without an elevated mold temperature during the filling stage, it is difficult to mold a thin and long part without short shots, a precision part with minimal residual stress and thus acceptable warpage and dimensional stability, an optical part with a low level of birefringence, and a microstructured part with high aspect ratio surface microfeatures. Different from the earlier work, most of the more recent work focused on selectively heating only the surface portion of the mold. This not only enhances the heating efficiency but also reduces the burden for cooling. Different terminologies for mold rapid heating and cooling have been coined in the literature; the authors intended to differentiate a particular technology from others. Examples are low thermal inertia molding, variotherm injection molding, momentary mold surface heating process, rapid thermal response (RTR) molding, rapid thermal processing, and dynamic mold surface temperature control. A number of innovative approaches for rapidly heating only the surface portion of the mold have been presented, including methods such as resistive heating of a thin conductive layer, convective heating using hot fluid, heating using condensing vapor, induction heating, high-frequency proximity heating, infrared heating, heating and cooling using a volume-controlled variable conductance heat pipe, heating and cooling using thermoelectric Peltier modules, passive heating by the incoming polymer, microwave heating, contact heating, flame heating, etc.

The overall design objective for a rapidly heatable and coolable mold is to design a mold with a low thermal mass, yet having the necessary mechanical stiffness, strength, and durability to endure the harsh environment of injection molding. This mold should also have means for rapid heat generation inside the mold or at the mold surface during heating and means for rapid heat removal during the cooling stage. Therefore, at least three constituent elements are needed for a rapidly heatable and coolable mold: (i) a stiff, strong, and durable mold with a low thermal mass; (ii) means for rapid heat generation in the mold surface portion; and (iii) means for rapid heat removal in the mold surface portion.

It should be noted that each constituent element represents a unique physical mechanism that is needed in the final mold design. For example, for the low thermal mass, the physical mechanism is self-explained by the name; that is, the product of mass and specific heat should be minimal for the mold material involved in heating and cooling. Based on an understanding of this physical mechanism, one can start to devise building blocks for realizing such a physical mechanism, for example using a mold material with low-density or low-thermal capacity, or both. Further understanding of this mechanism may result in other feasible mechanisms, for example adding insulation beneath the mold surface portion. Possible building blocks for all three constituent elements are listed in Table 5.3 [9].

Hot Runner Technology A hot runner system is an assembly of heated components in a plastic injection mold and is used as heated channels for the polymer melt to flow through before reaching the mold cavity. A hot runner system usually includes a heated manifold and a number of heated nozzles, as shown in Figure 5.5. The main task of the manifold is to distribute the plastic entering the mold to the various nozzles which then meter it precisely to the injection points in the cavities.

Hot runners are fairly complicated systems; they have to maintain the plastic material within them heated uniformly, while the rest of the injection mold is being cooled in order to solidify the product quickly. For this reason, they are usually assembled from components premanufactured by specialized companies. Two main types of hot runner systems are the externally heated and internally heated. In the externally heated type, molten plastic runs within a solid manifold and within the nozzles. In the internally heated (now obsolete), the plastic flows directly over slender heaters inside oversized runners. The outside boundaries of the runners normally solidify so that the plastic material flows only in proximity of the internal heaters or "torpedoes."

Compared with conventional cold runner systems, the required injection pressure with a hot runner system is considerably lower, and yet a more effective holding stage can be performed [10]. For injection molding at miniature size scales, the hot runner system is a useful technology to improve the moldability of the plastic melt and enhance the quality of molded parts.

Vacuum Venting Technology As the size reduces, it is more difficult to fill the mold cavity because of the air compression effect; most air initially occupying the runners is forced into the tiny cavity. Therefore, mold venting becomes an issue for injection molding at miniature size

Table 5.3 Constituent elements and building blocks for mold rapid thermal cycling.

Low thermal mass	Rapid heating	Rapid cooling
a. *Low density* b. *Low specific heat* c. *Small volume* d. *Combination of a, b, or c* e. *Porous material* Scaffolded molds Porous metal from gas foaming Porous metal from powder metallurgy f. *Insulated mass* Multilayer structure Gradient materials Orthotropic materials89	a. *Electrical resistive heating* With low-frequency current With direct current With alternating current With high-frequency current Induction heating Proximity heating b. *Dielectric heating (microwave)* c. *Heating with thermoelectric or the Peltier effect* d. *Convective heating* With internal media Hot oil Steam heating Hot gas With external media Hot gas Condensing vapor Flame e. *Radiation heating* Infrared radiation Focused beams f. *Contact heating* With a heated solid With the polymer melt–passive heating g. *Other methods, e.g. ultrasonic heating*	a. *Convective cooling* With internal media Water Cold gas including air b. *Cooling with heat pumps* Peltier effect Vortex tubes Condensers and evaporators

Source: Yao et al. 2008 [9]. Reproduced with permission of John Wiley & Sons.

scales. In the recent years, vacuum venting has been developed and evolved into a standard practice for molding miniature plastic parts.

Vacuum venting, a process where the cavity air is forcibly evacuated during mold filling, is an improved venting method that overcomes the limitations of conventional venting pins. A typical vacuum venting system consists of a vacuum pump, a vacuum reservoir, and a solenoid operated valve between the mold and the vacuum reservoir, which is interfaced with the molding machine controls. It is important that air flow restrictions between the mold cavity and the vacuum system be minimized to facilitate air movement. Ideally, a vacuum is drawn before the melt enters the cavity and must be maintained up until the last vent is sealed off. The solenoid valve opens at the start of (or slightly before) injection and closes once filling is complete. During the remainder of the molding cycle, the vacuum reservoir is evacuated so that it is ready for the injection phase of the next molding cycle begins. Vacuum levels in the vacuum reservoir must be maintained throughout the filling phase of the process, even during the initial rush of air when the solenoid valve is open, necessitating the use of a relatively large vacuum reservoir. Vacuum venting systems can be physically connected to the mold through wide parting line vents, or through sealed ejection boxes. Air infiltration from outside sources (e.g. through the parting line, ejector pin holes) must be minimized. This can be accomplished using o-rings and seals of various types.

For example, vacuum venting was applied to produce a small cell container by injection molding, as shown in Figure 5.6. Figure 5.7 shows venting slots on the mold insert. A photo of the vacuum venting system is shown in Figure 5.8. Figure 5.9 shows parts molded with vacuum

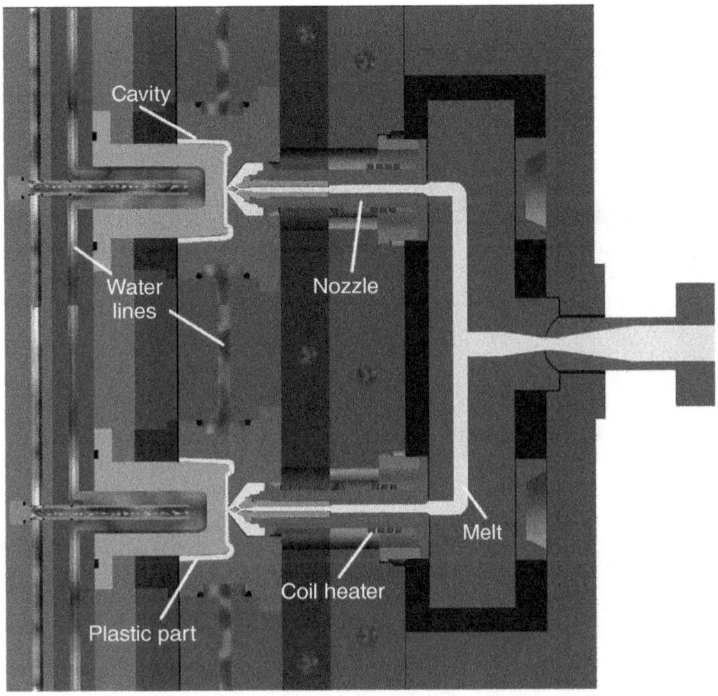

Figure 5.5 Cross-section showing hot runner system of plastic injection mold.

Figure 5.6 Cell container.

venting and without vacuum venting. This comparison indicates that vacuum venting has a significant effect on filling the micro-mold cavity.

In-mold Assembly Technology Polymer processing is an important technology in the field of microstructures and microsystems. By the use of polymers, microstructures can easily be replicated in medium- and large-scale production by micro injection molding. As components and functional structures become smaller, effective micro assembly is gaining more importance. The increasing complexity of assembly processes in microsystem technology requires new and advanced assembly processes.

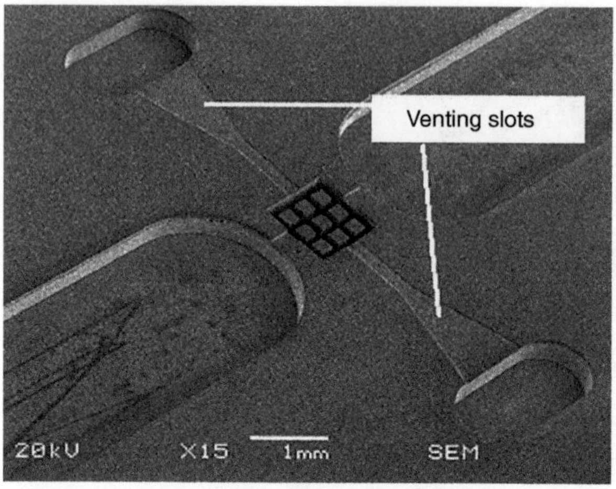

Figure 5.7 Venting slots on mold insert.

Figure 5.8 Photo of vacuum venting system.

In the area of polymer microsystems, micro injection molding can be used for the generation and direct assembly of hybrid microsystems. Using this process, one process step leads to compound part consisting of two thermoplastics or a thermoplastic and an insert part (e.g. metal, silicon, glass, or ceramic). The fact that only one step is necessary to generate a hybrid microsystem, the high reproducibility and the ability to highly automate the injection molding process justifies the existence of this process next to conventional joining processes such as soldering, bonding, or welding, especially in high series production. All variants of in-mold assembly have in common that the integration of the assembly into the generation of the actual part contains a high potential for rationalization by eliminating additional handling and joining steps [11].

Figure 5.10 shows the basic structure of in-mold assembly. The mold is designed as a split mold with horizontal jaws. The lower jaw includes a conical mold insert with the microstruc-

Figure 5.9 Cell container pictures: (a) molded without vacuum venting and (b) molded with vacuum venting.

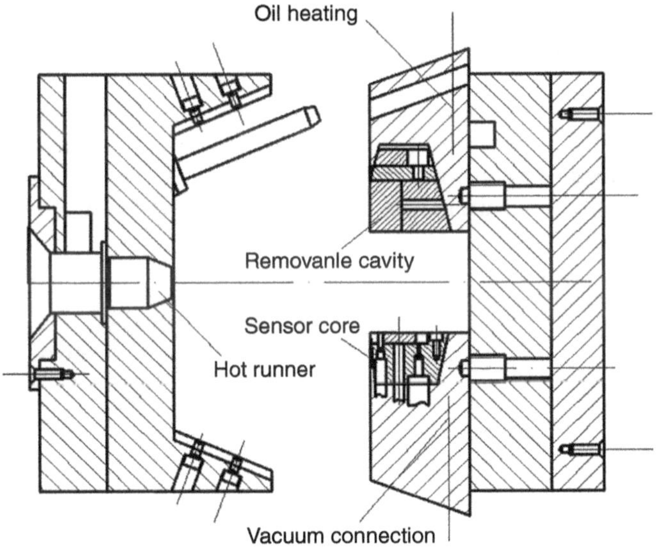

Figure 5.10 Basic structure of in-mold assembly.

tured cavity, which is taken out of the mold after each shot, and can therefore easily be replaced. By using this exchangeable cavity, the process steps including demolding, positioning of inlay parts, and heating of the cavity for variothermal processing can all be done at an external station.

The material combination used for in-mold assembly is chosen such that the materials are chemically incompatible (i.e. they have no tendency of adhering to each other during and after injection molding) [12]. Figure 5.11 shows a picture of micro-hinges assembly.

Mold Visualization Technology Injection molding is a widely used process for production of various plastic parts of simple to complex geometries. The process has various aspects of research interest particularly in the filling stage. Knowledge on the material behavior during flowing in the mold cavity can yield a better insight into the design issues and final product

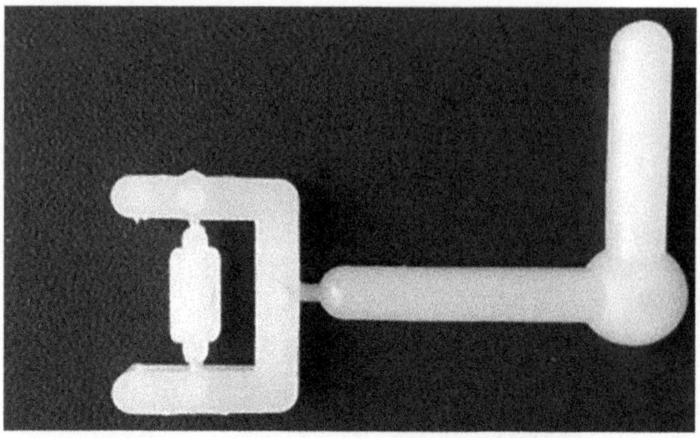

Figure 5.11 Picture of micro hinges assembly.

quality. However, the melt flow behavior during the filling stage of injection mold seems to be inside a gray box. In addition to sensing techniques used for measuring the pressure and temperature of polymer melt, flow visualization is another important technique for analyzing various phenomena during plastic injection molding. In particular, the understanding of resin behavior inside the cavity is an important theme, as it will also help clarify the causes of molding defects [13, 14].

The design and fabrication of a visualization mold is the key of flow visualization. Two main methods for flow visualization have been reported. Methods for visualizing the melt behavior inside the mold can be divided into two groups: static and dynamic. Static methods illustrate the pattern of velocity gradient at the surface of a solid boundary. Dynamic methods involve the observation of motion of tracer particles by using a high-speed video camera. Basically, there are two types of molds for dynamic visualization: light-reflection molds and light-transmission molds, as shown in Figure 5.12a,b.

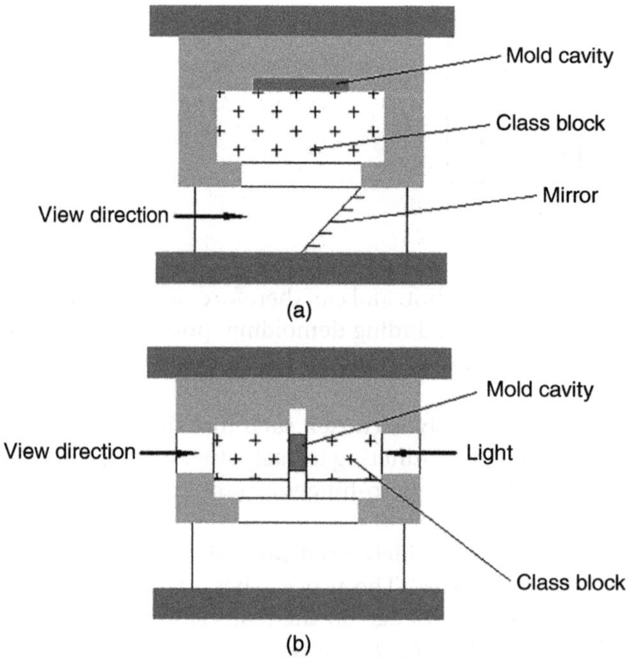

Figure 5.12 Two different types of visualization molds: (a) Reflective type of visualization and (b) transmissive type of visualization mold.

For a plastic part with microfeatures molded by injection molding, it is important to investigate the filling behavior of melt into the microcavity in a short time. The filling behavior of the melt into microfluidic chip was observed using a long distance microscope and an ultrahigh-speed video camera, as shown in Figure 5.13. With this visualization system, the effects of factors, such as injection rate, mold temperature, on the replication process during the filling stage were investigated. This visualization capability offers a unique opportunity to understand and improve the filling process in small cavities. Figure 5.14 shows melt flow fronts at the different time by visualization.

5.2.2.4 Mold Manufacturing Technology

It is well known that the configuration of mold insert is a major factor in affecting the dimension and geometry accuracy of molded parts. Therefore, how to make a mold insert suitable for

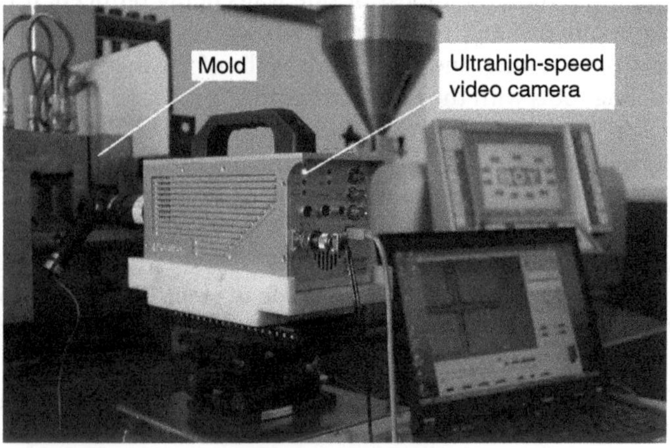

Figure 5.13 Experimental setup for visualization of melt flow in microfluidic chip.

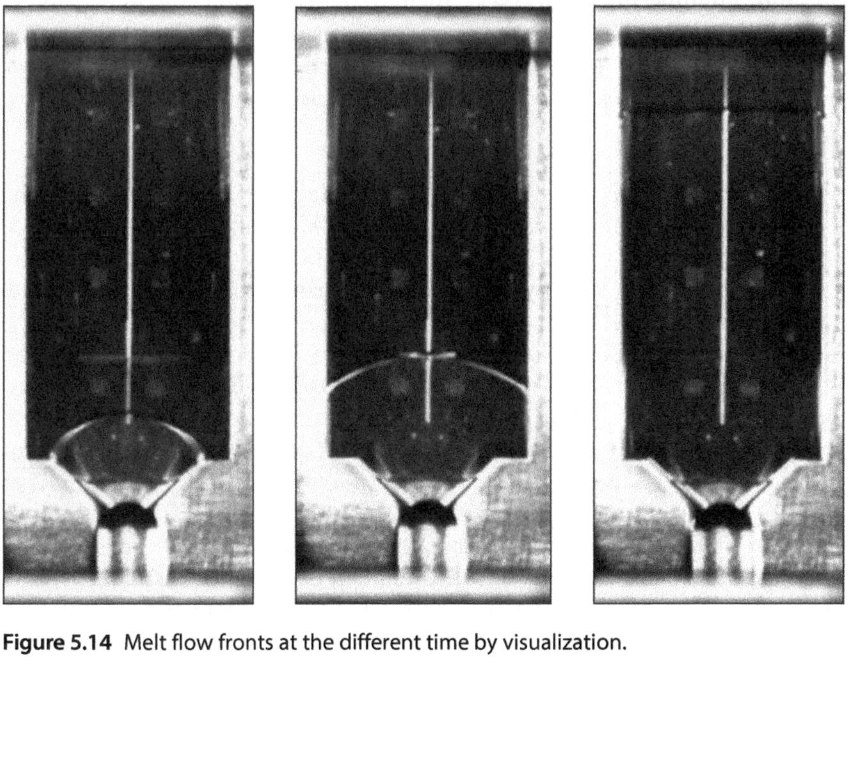

Figure 5.14 Melt flow fronts at the different time by visualization.

molding at miniature size scales is of a critical importance. By now, various methods have been developed and applied to mold insert fabrication for miniature injection molding. Among them, some are direct manufacturing methods, e.g. mechanical micromachining, laser structuring, and micro EDM, and some others are lithographic processes with X-rays or UV radiation combined with electroplating [15].

Mechanical Micromachining Mechanical micromachining is similar to traditional tooling technology. Techniques such as turning, drilling, or milling are employed for fabricating mold inserts. The smoothest side walls of microstructures are obtained with diamond tools, but these are not suitable for work in steel which is a favorite material for mold inserts. Moreover, the smallest diameter of diamond tools currently available is approximately 200 μm. When narrower grooves need to be fabricated on a mold insert or the mold insert needs to be made of tool steel, milling and drilling tools made of hard metal can be used, with the requirements regarding the smoothness of side walls being reduced. It is easy to fabricate mold inserts with three-dimensional (3D) microstructures even with curved surfaces by mechanical micromachining, and sloped side walls of the microstructures can be achieved by simply using milling tools with the required profile.

Micro EDM EDM has been traditionally used to make macroscopic parts. Today, micro EDM has also been developed to fabricate complex micro-mold inserts for injection molding. This is achieved by using wires as thin as 30 μm for wire erosion or by sinking erosion with electrodes produced by electroplating. In this way, microstructures from nickel or other metals can be transformed into mold inserts made of steel. However, a disadvantage of this technique is that the side walls of microstructures fabricated in this way are typically rough compared to milling.

Laser Micromachining Laser micromachining is a developing technology for high precision microfabrication, but has enormous potential in terms of aspect ratios and minimum structural dimensions. The use of lasers offers many distinct advantages over conventional micro-manufacturing and engineering processes, including processing flexibility, high resolution and precision, unlimited material coverage, multiprocess compatibility, noncontact processing, 2D and 3D structures, etc. This technology is of particular interest, as it allows processing of materials such as stainless steel or tungsten carbide that are difficult to be processed otherwise.

LIGA/UV-LIGA LIGA (a German acronym which stands for lithography, electrodeposition, and molding) is particularly suited for minute structures. A resist patterned with a microstructure is electroplated to build up a mold insert. Classical LIGA technology based on X-ray deep-etch lithography is characterized by extremely high structural heights, ultra-small lateral dimensions, and super-smooth side walls with a roughness of less than 50 nm. UV LIGA utilizes an inexpensive ultraviolet light source, like a mercury lamp, to expose a polymer photoresist, typically SU-8. UV-LIGA is a less complex and less expensive alternative to X-ray technology but is able to meet less demanding specifications. Both methods are based on the lithographic generation of nonconductive polymer microstructures that are filled by electroplating.

Etching Mold inserts from silicon or glass microstructures by wet etching or reactive ion etching can also be employed for micromolding. However, silicon is a very brittle material, and silicon wafers tend to break. Therefore, the wafer is bonded on top of a quartz glass to obtain a stable compound. If less than 10 molding steps are to be performed for prototyping purposes, mold inserts from silicon may be suitable. Silicon treated by wet etching is more favorable, because the resulting microstructure shows inclined side walls and a smoother surface than that from dry-etched silicon.

5.3 Injection Molding of Precision Parts

5.3.1 Precision Parts

The overall dimension of precision parts is the same as of conventional parts, but the dimensional accuracy is in microns or submicrons. Therefore, the parts can be large in size, but the accuracy is high. These plastic precision parts are widely involved in fields such as electronics, photonics, and medical plastics, and have become economical substitutions for more expensive metallic and ceramic parts. For example, a plastic optical lens with a high-accuracy surface contour and a super surface finish is one such precision part.

To achieve the high accuracy of the molded part, one has to first choose a suitable plastic material. Material properties are responsible for process capability and manufacturability and also dictate the shrinkage and distortion of the molded part and thus the tolerances. There are a wide variety of polymers available to precision molding, including COC, polymethylmethacrylate (PMMA), PC, polystyrene (PS), POM, PVC, PP, PET, PEEK, PA, and others. However, choices become more limited if precision optical components are concerned. Polymers suitable for such parts desire high optical transmission and low optical aberration and often also need to be coated with dielectric anti-reflex coatings.

Important material properties in precision molding are viscosity, melting temperature, glass transition temperature, water intake, and gas absorption. The latter two parameters are particularly important for thin film coating processes on polymers because such processes are commonly running with process temperatures above 80 °C and require a water and residual gas free atmosphere. Mold shrinkage is another important measure of a given material to accurately replicate fine product features and meet tight dimensional tolerances. Amorphous polymers typically exhibit lower shrinkage (0.3–0.8%) than semicrystalline polymers (1–3%). Material specific feature shrinkage is dependent on process parameters and typically exhibits some batch to batch variations. Hence, shrinkage prediction in injection molding is still a quite challenging problem and practical experience with material behavior is irreplaceable. To improve the shrinkage behavior inorganic fillers like TiO nanoparticles are commonly used. In general, dimensional stability improves with a higher glass transition temperature and a larger difference between the operation temperature and the glass transition temperature. For these reasons, COCs and cyclic olefin polymers (COPs) are more suitable materials for optical components due to their low water intake and their high thermal stability. Polycarbonate (PC) is not well suited for high precision applications because the relatively high water intake causes swelling which will spoil high tolerances immediately. For thin film processes, water vapor from the part needs to be suppressed [16].

Conventional transparent thermoplastic resins for optical modules processed by injection molding have poor heat-resistant characteristics and cannot withstand the reflow process. Therefore, optical modules produced using existing thermoplastic resins (COP, PMMA, PC, etc.) require a surface mount process other than the reflow process, for example, laser soldering or connector. Recently, electron beam irradiation cross-linking has been applied to molded parts for improving heat resistance [17]. A correcting aspheric lens for a 635-nm laser diode was fabricated by injection molding and was then irradiated with an electron beam.

5.3.2 Injection Compression Molding

5.3.2.1 Overview of Injection Compression Molding

The injection compression molding process was developed to meet the growing demands for high dimensional accuracy and production efficiency of precision plastic parts. As shown in Figure 5.15, injection compression molding is the process of injecting molten resin into a slightly open mold, then deforming the injected resin by clamping the two mold halves to the fully closed position, and after mold cooling finally opening the mold and ejecting the part. Therefore, injection compression molding is a combination of injection molding and compression molding, and it synergizes the advantages of both technologies. This process provides uniform distribution of the resin, yields extremely high accuracy of surface patterns, and renders excellent dimensional stability for the finished product.

Injection compression molding aside from reduced material shear and less orientation offers numerous advantages for enhancing quality of molded parts. It also permits reductions in injection pressure, clamping force, and cycle time. Added to this is often an improved holding pressure effect, which minimizes sink mark and warpage. Representative advantages of injection compression molding are summarized as follows:

(a) Compensates shrinkage by compressing the melt through the clamping movement;
(b) Results in uniformly distributed holding pressure;
(c) Reduces the holding time and shortens the cycle time;
(d) Permits overpacking of cavity;
(e) Causes less orientation and molecular alignment during injection;
(f) Eliminates sink marks in thick wall sections and at the end of the flow path;
(g) Reduces warpage susceptibility and improves long-time dimensional stability;
(h) Reduces stresses in mats or films in direct back injection.

Injection compression molding particularly excels in the production of optical lens, thin-walled parts, long-glass fiber reinforced components, and parts exposed to heat cycles. It is considered to be the optimum production process for these components among all available polymer molding processes.

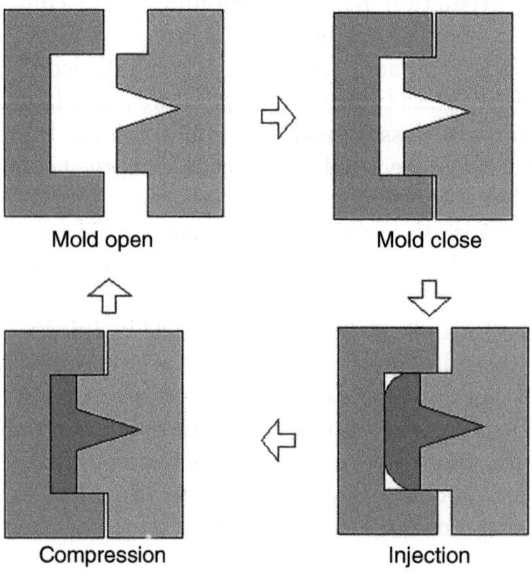

Figure 5.15 Cycle of injection compression molding.

5.3.2.2 Mold Base for Injection Compression Molding

For precision plastic parts (e.g. optical lenses), the alignment accuracy and the structure of mold base significantly affect the part quality. Recently, Michaeli et al. [18] proposed a special mold design that simultaneously complies with several requirements for injection compression molding. A fundamental requirement was the modularity for using different mold inserts. Furthermore, the compliance with the desired surface contour and quality is important to reach the optical function of the lens. The positioning of the mold inserts in the mold base is achieved with a cone-alignment. This mold base design can be used for both injection molding and injection-compression molding. For injection-compression molding, it is necessary to seal the cavity before the polymer is injected. Figure 5.16 shows the centering principle of the mold base. In the moving mold half, a conical centering ring 1 is mounted with a spring support and touches the opposite side. As the mold closes, the two mold halves are automatically aligned against each other. Subsequently, the spring supported sealing ring 2 contacts the fixed half and seals the cavity. In this position, the mold is not closed completely, and it is possible to mold polymer melt using both processing techniques, injection molding and injection-compression molding.

5.3.3 Processing Parameters

In this section, optical lenses are taken as an example of precision plastic parts, and the effects of the major processing parameters on the quality of the molded lenses are discussed. There are many different technical parameters to evaluate the quality of optical lenses, including surface contour error, light transmission, surface waviness, surface finish, refractive index variation, etc. Different conclusions can be drawn if different assessment criteria concerning the physical, mechanical, and geometrical properties are employed. In the present evaluation, we focus on optical properties and geometrical accuracies.

Since there are many process parameters involved in an injection molding process, and more importantly, an optical lens needs precisely controlled surface contours, determination of the processing conditions for lens molding is very complicated. Earlier, Lu and Khim [19] investigated experimentally some effects of the molding conditions on the surface contours of injection molded lenses. A spherical lens was molded using polycarbonate. Statistical methods were employed in the experimental studies in order to systematically analyze the effects of various process parameters on the lens contour errors. The process parameters studied include

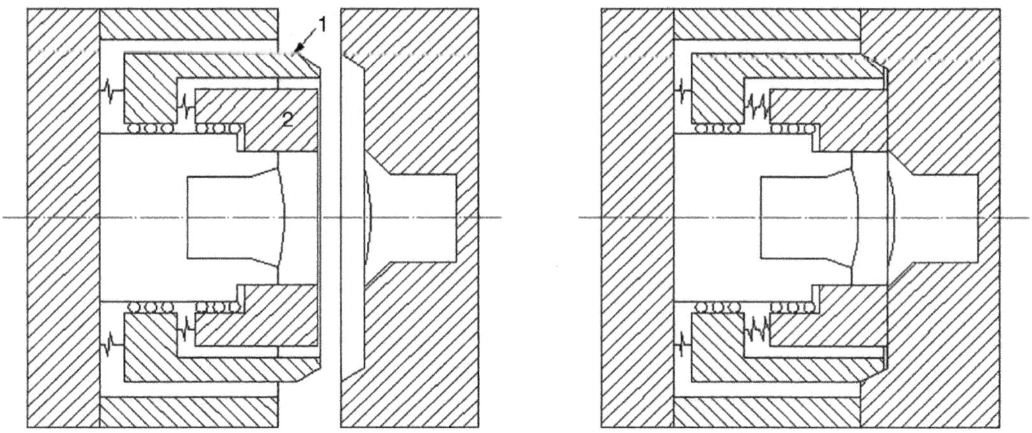

Figure 5.16 Cross-section of mold base for injection compression molding.

injection speed, holding pressure, and mold temperature. The result indicates that the mold temperature was the most important process parameter affecting the contour errors. The holding pressure, somehow, showed a less influence on the contour errors. Behind these process parameters, the ultimate fundamental parameters affecting dimensional accuracy and stability are mold shrinkage and residual stress. The effect of the injection speed on the residual stress was mainly to change the stress distribution. The use of two-stage holding pressure was found to provide better surface replication for the lenses.

Precision fabrication of optical components by injection molding and injection compression molding poses a serious challenge for tooling and machinery technology as well as for process control. Recently, Michaeli et al. [18] analyzed the accuracy of the molded geometry as well as the optical performance. They summarized that injection molding offers a good capability to produce high precision parts with tolerances in the micron range. If dimensional accuracy needs to be further improved, the injection compression molding technique with an appropriate mold can be called in. Overall, injection compression molded lenses showed a better optical performance than injection molded lenses. A further improvement of the quality of polymer optics can be achieved by local shrinkage compensation in the mold.

More recently, Tsai et al. [20] designed a two-stage experimental process to investigate the effects of process parameters on quality characteristics of injection molded optical lenses. In the first stage, significant factors were identified through the Taguchi screening procedure. The significant factors were then used to implement the full factorial experiments in the second stage. The quality characteristics chosen were light transmission, surface waviness, and surface finish. Through empirical and theoretical analysis, the most significant process parameters affecting surface waviness was determined to be the melt temperature, followed by mold temperature, injection pressure, and packing pressure. On the other hand, these process parameters were found to have little effect on light transmission and surface finish of lenses. Regression approach was then implemented based on surface waviness data from full factorial experiments to formulate regression models. The results showed the correlations were highly nonlinear and a nonlinear regression model was needed for high prediction accuracy.

The Six Sigma approach has also been used to improve the quality of injection-molded lenses with the implementation of a "define, measure, analyze, improve, and control" (abbreviated as DMAIC) procedure [21]. At first, critical-to-quality factors (CTQ) were determined according to customer requirements for quality. The causal factors for the lens properties were then identified within the processing window being tested. Next, the Taguchi method for design of experiments (DOE) was employed for screening pertinent process parameters in the injection molding process. After completing the DOE procedure, verification experiments were conducted with selected combinations of factors and levels. With this methodology, the quality characteristics of the lenses were significantly improved. Furthermore, the Taguchi method adopted in the analysis step successfully identified the optimal combination of process parameters as well as the most significant factors affecting the surface qualities. The final results suggested that among all process parameters, the most significant factors affecting surface accuracy are packing pressure, melt temperature, injection pressure, and packing time, accordingly.

Refractive index variation is another important quality attribute for optical lenses, which is also significantly affected by the process conditions. Yang et al. [22] recently investigated the refractive index variation in injection-molded PMMA optical lenses under different levels of packing pressure using a Shack–Hartmann wavefront sensor-based metrology system. The experimental results showed that refractive index variation exhibited a large difference under different packing pressures used in injection molding. Higher packing pressure was discovered to have led to much more uniform refractive index distribution as compared with lower packing pressure. Generally, it is believed that refractive index variation results from density variation, which itself is caused by shrinkage variation under different packing conditions.

5.4 Injection Molding of Thin Wall Parts

Thin wall plastic parts are a particular kind of miniature plastic parts. It is conventionally defined as the part that has a nominal wall thickness of 1 mm or less and a surface area of at least 50 cm^2 with a flow length/thickness ratio above 100. Thin-wall injection molding has become increasingly important due to the explosive growth of wireless telecommunication and portable electronic devices that require thinner and lighter plastic housings. This trend is partially caused by the increasing demand for product portability and material savings. Nowadays, the part thickness of cellular phone components, laptop and notebook computer components, and similar telecommunication items has already reached 1 mm or less. Especially for some electrical connectors, electric component packages, and power supply items, manufacturers require even thinner wall thickness.

However, thin-wall injection molding presents several technical challenges. First of all, process physics and material behavior under the extreme processing conditions are not fully understood. Second, existing commercial computer simulation tools, whose solutions are based on questionable input material properties, often fail to predict the experimental observations for thin-wall parts. Third, thin-wall parts have a more restrictive flow path compared to conventional injection molded parts. This leads to a much narrower processing window and poorer production yield. Finally, thermal degradation and material degradation from shear also become significant during thin-wall molding. How can one successfully and economically injection mold long and thin parts? Therefore, in this section, concepts of scalable filling and low-speed filling and a RTR mold will be introduced. Then, the effects of processing parameters on the molding process for thin-wall parts are discussed.

5.4.1 Scalable Filling

The filling length of a thermoplastic material can best be understood as a ratio that compares the length and the thickness of the wall section that the material travels through [23]. For a part with flow length of 300 mm and wall thickness of 3 mm, the length to thickness (L/T) ratio is 100. This L/T can be easily reached with standard tools and process. However, the once easy-to-hit L/T becomes more difficult to reach as wall sections drop to 1.0 mm because the ratio of frozen layer to molten core increases. It is considered difficult with the current thin-wall molding technology to reach L/T of 50 when wall thickness drops to 0.5 mm. Therefore, the filling length is not scalable as the wall thickness decreases and the curve of L/T vs. injection pressure is thickness dependent.

The scalability of the filling length can be retained in isothermal filling, which can be shown by scaling analysis as follows. The flow in strip type or spiral cavities that are frequently used to determine the flow length for thermoplastics can be considered as one-dimensional. By neglecting the inertia effects, isothermal filling can be described by the following momentum equation:

$$\frac{\partial}{\partial z}\left(\eta(T_0,\dot{\gamma},p)\frac{\partial u}{\partial z}\right)-\frac{\partial p}{\partial x}=0 \tag{5.1}$$

where x denotes the streamwise direction, z the gapwise (transverse) direction, u the velocity in the x direction, T_0 the isothermal temperature, η the shear rate and pressure dependent viscosity, and $\dot{\gamma} = |\partial u / \partial z|$ the shear rate. Normalizing the dimensions, the velocity, and the shear rate with the half gap thickness, we can obtain a new set of variables:

$$x^* = \frac{x}{b}, z^* = \frac{z}{b}, u^* = \frac{u}{b}, \text{ and } \dot{\gamma}^* = \frac{\partial u^*}{\partial z^*} = \frac{\partial u}{\partial z} = \dot{\gamma} \tag{5.2}$$

The normalization does not change the shear rate and consequently the viscosity does not change. Substituting these normalized variables in Eq. (5.1) results in

$$\frac{\partial}{\partial z^*}\left(\eta \frac{\partial u^*}{\partial z^*}\right) - \frac{\partial p}{\partial x^*} = 0 \tag{5.3}$$

The above thickness-independent equation indicates that L/T will be the same for strips with different thicknesses if an identical pressure history is applied at the polymer entrances. Hence, the flow length in isothermal filling is scalable with respect to the part thickness.

5.4.2 Rapid Thermal Response Molding

The difficulty in molding thin-wall parts lies in the fact that the ratio of frozen layer thickness to part thickness during filling will drastically increase with the decrease in part thickness. The problem becomes worse when the frozen layer thickness and the part thickness are comparable. The frozen skin of each side is approximately 0.25 mm thick regardless of the nominal wall thickness of the part in question. A part with a nominal wall thickness of 1 mm will, as a result, have a true flow path of 0.5 mm.

In order to overcome the filling difficulty in thin-wall molding, specialized molding conditions were recommended: higher injection pressure and speed, lower viscosity materials, and strengthened mold plates. These conditions, however, result in increasing demands for machine, material, and tooling resources. Another solution is to increase the mold temperature so as to reduce the amount of frozen layer during filling. For this purpose, Yao and Kim [24] developed the RTR molding process with active heating and cooling.

In the RTR molding process, the mold surface can be rapidly heated to the injection temperature prior to the injection stage so that a hot mold cavity can be obtained. The hot mold has many advantages in injection molding such as improvement of flow characteristic, reduction of flow-induced molecular orientation and residual stress, increase of weld line strength, etc. The resulting process is especially useful for thin-wall injection molding where short-shot and excessive molecular orientation are among main concerns. The schematic setup for the RTR molding is shown in Figure 5.17. The heating source can be turned on and off during the molding cycle. A data acquisition unit synchronizes the heating cycle with the molding cycle in order to control the mold temperature [25].

Two RTR mold inserts were constructed first. The RTR is accomplished by a thin metal heating layer, below which an insulation layer provides thermal and electrical insulation from the steel mold base. Figure 5.18 shows a top view of the RTR mold containing side insulation elements. Closure of the two mold plates forms a rectangular mold cavity, 80 mm long and 30 mm wide, with the thickness adjustable. The RTR mold can raise the surface temperature from 25 to above 200 °C within 5 seconds and cool to 50 °C in about 20 seconds, as shown in Figure 5.19.

Bayer polycarbonate CD2000 was used as the material in the molding experiment. The injection temperature was set to 265 °C. Two levels of machine injection speed were used: 30% and

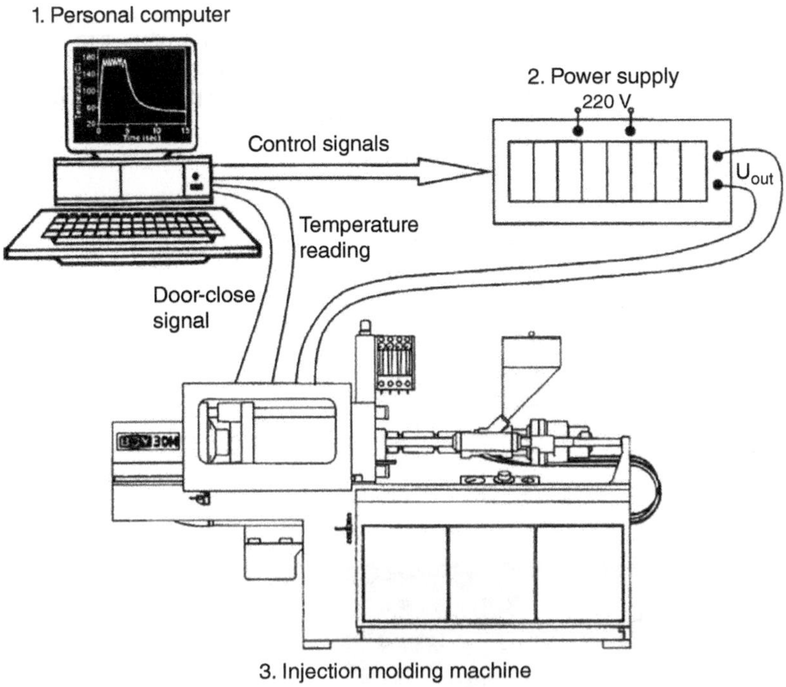

Figure 5.17 The schematic setup for the RTR molding process. *Source:* Yao and Kim 2002 [24]. Reproduced with permission of Taylor & Francis.

Figure 5.18 RTR mold containing side insulation elements. *Source:* Park et al. 2006 [25]. Reproduced with permission of Taylor & Francis.

100%. The actual flow rates determined from short shot experiments are about 9 and 30 cm^3/s, respectively. The conventional molding condition was carried out at mold temperature of 25 °C, using the same mold without turning on the heating source. Typical experimental flow patterns of 0.5 mm thick polycarbonate strips are shown in Figures 5.20 and 5.21. It can be seen that RTR molding provides much longer flow length for both filling speeds.

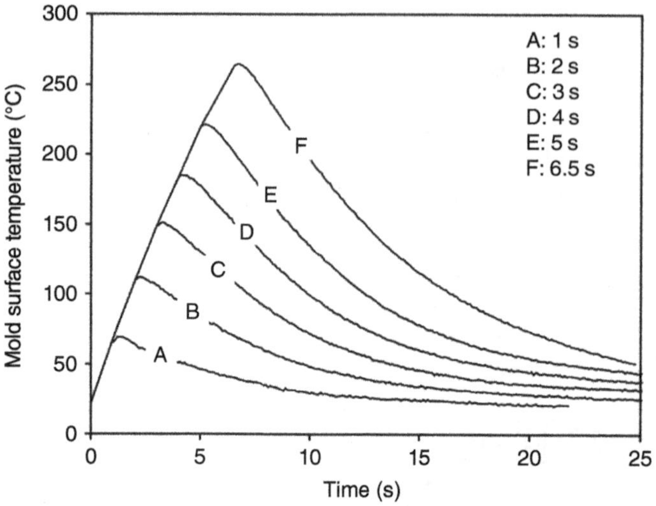

Figure 5.19 Experimental mold surface temperature of the RTR mold. *Source:* Park et al. 2006 [25]. Reproduced with permission of Taylor & Francis.

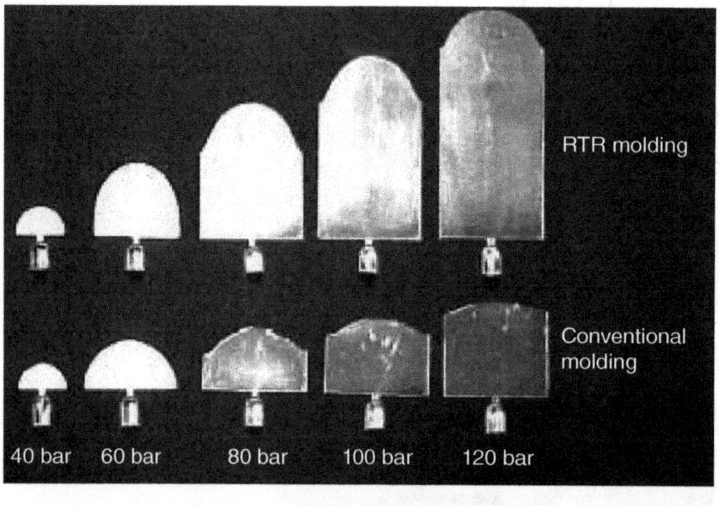

Figure 5.20 Experimental flow patterns with flow thickness of 0.5 mm at 30% machine injection speed. *Source:* Yao and Kim 2002 [24]. Reproduced with permission of Taylor & Francis.

For quantitative comparison, the results were also plotted as injection pressure vs. flow length, as shown in Figures 5.22 and 5.23. From Figures 5.22 and 5.23, it is clearly seen that the pressure used in hot molding is much lower than that in cold molding. At 30% speed, the pressure required in cold molding for filling the same L/T ratio is more than twice as high as that in hot molding. For example, at an injection pressure of 150 MPa, L/T about 160 can be achieved in hot molding, which is about three times as long as that in cold molding. The slope of the L/T vs. pressure curve changes about 50% in cold molding when the speed decreases from 100% to 30%, while it only changes about 30% in hot molding.

5.4.3 Effect of Processing Parameters

The filling difficulty of thin-wall parts lies in the fact that the ratio of frozen layer thickness to part thickness will drastically increase with the decrease in part thickness. Therefore, it is hard to fill cavities for thin-wall parts under the conventional processing condition. It is difficult to

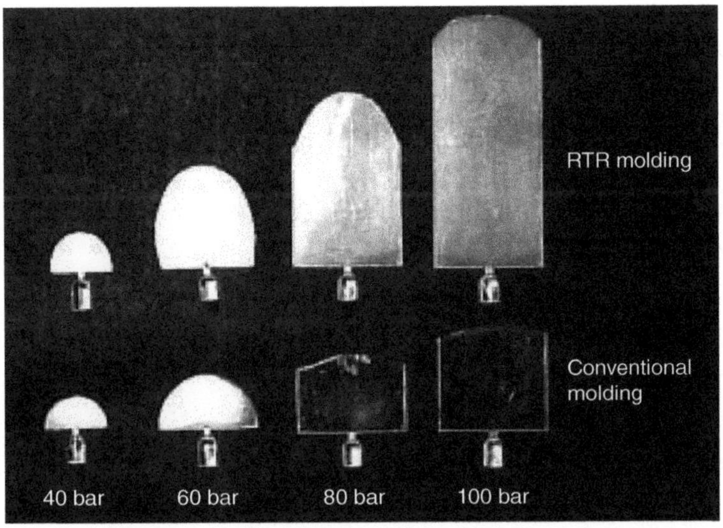

Figure 5.21 Experimental flow patterns with flow thickness of 0.5 mm at 100% machine injection speed. *Source:* Yao and Kim 2002 [24]. Reproduced with permission of Taylor & Francis.

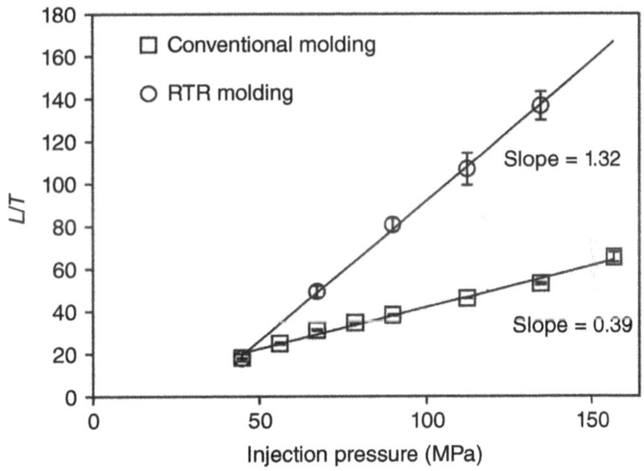

Figure 5.22 Experimental flow lengths with flow thickness of 0.5 mm at 30% injection speed. *Source:* Yao and Kim 2002 [24]. Reproduced with permission of Taylor & Francis.

select process parameters correctly because there are more influential factors involved in thin-wall injection molding than in conventional injection molding. In this section, the roles of different process parameters in the thin-wall molding process are discussed.

As an example, the work on process study of thin-wall molding by Song et al. [26] is discussed herein forward. The cavities and runners of that thin-wall part are shown in Figure 5.24. One cavity is round and the other is rectangular. Both thin-wall parts have the same nominal thickness of 0.2 mm (or 0.1 mm) and the same volume nearly. The material used in this experiment was polypropylene (PP). The injection machine used in this experiment was BOY 12A with the minimal injection volume of 0.1 cm^3, maximal injection pressure of 180 MPa, and maximal injection rate of 240 mm/s.

The orthogonal experimental method (Taguchi method) was employed to design experiments. With this method, experimental results can be analyzed scientifically, and relations between primary and secondary factors and relations between the target value and input factors, as well as further research directions, can all be worked out. There are a good number of

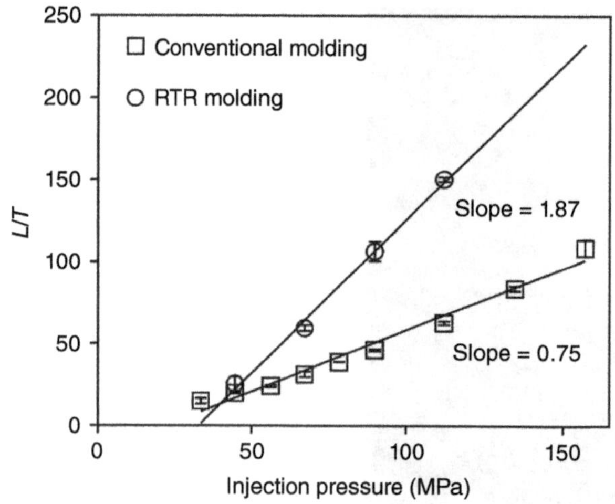

Figure 5.23 Experimental flow lengths with flow thickness of 0.5 mm at 100% injection speed. *Source:* Yao and Kim 2002 [24]. Reproduced with permission of Taylor & Francis.

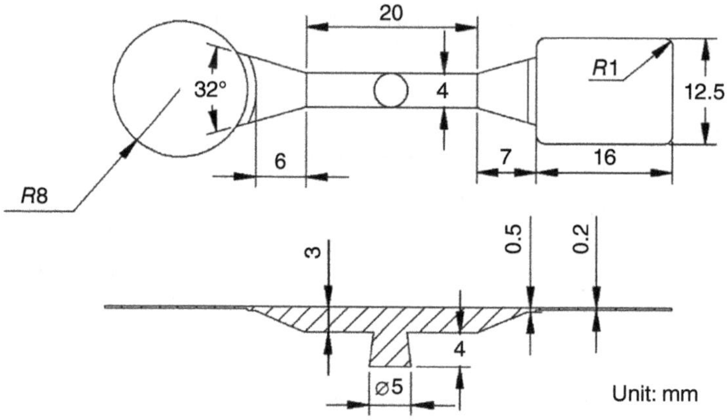

Figure 5.24 Cavities and runners of a thin wall mold. *Source:* Song et al. 2007 [26]. Reproduced with permission of Elsevier.

factors that can influence the process of thin-wall injection molding. Injection pressure, injection rate, melt temperature, metering size, and part thickness were regarded as recommendable factors, disregard of man-made and circumstantial factors. Filling volume was selected as the target variable. Since the two cavities have the same uniform thickness, filling area can also be selected as the target variable instead of filling volume. The metering size of the BOY injection molding machine was calculated on the basis of the distance between the nozzle and the top of the screw. The levels of factors employed in the experiments are shown in Table 5.4.

Orthogonal DOEs were conducted to investigate the effects of different process parameters in injection molding of thin-wall parts. The results are given in Table 5.5. Based on the calculated values of dispersion R, some useful conclusions were made

1) The part thickness is the decisive parameter in molding of thin-wall plastic parts; the polymer moldability decreases rapidly with the reduction of part thickness.
2) The metering size and the injection rate are two other principal factors in thin-wall injection molding. An appropriate metering size must be determined and employed in thin-wall molding. Further, higher injection rates are desired in filling longer flow length.

Table 5.4 The levels of factors.

Factors	Levels		
	Level 1	Level 2	Level 3
A: Injection rate (mm/s)	60	84	108
B: Injection pressure (MPa)	85	95	105
C: Melt temperature (°C)	220	230	240
D: Metering size (mm)	7.0	7.5	8.0
E: Part thickness (mm)	0.2	0.1	—

Source: Song et al. 2007 [26]. Reproduced with permission of Elsevier.

3) The melt temperature and the injection pressure are secondary factors, but higher melt temperature and higher injection pressure are also necessary in molding longer and thinner flow length.

Table 5.5 Experimental design and results.

No.	A	B	C	D	E	Area (mm^2)
1	1	1	2	2	1	149.5
2	2	1	1	1	1	143.3
3	3	1	3	3	1	172.5
4	1	2	1	2	1	160.3
5	2	2	3	1	1	111.0
6	3	2	2	3	1	176.2
7	1	3	3	1	1	86.0
8	2	3	2	3	1	191.7
9	3	3	1	2	1	172.3
10	1	1	1	3	2	52.2
11	2	1	3	2	2	64.3
12	3	1	2	1	2	46.9
13	1	2	3	3	2	54.5
14	2	2	2	2	2	61.7
15	3	2	1	1	2	55.8
16	1	3	2	1	2	35.1
17	2	3	1	3	2	59.1
18	3	3	3	2	2	56.6
Sum of level 1	537.6	600.8	643	478.1	1362.8	
Sum of level 2	631.1	619.5	661.1	664.7	486.2	
Sum of level 3	680.3	628.7	544.9	706.2	—	
Dispersion R	142.7	27.9	116.2	228.1	876.6	

Source: Song et al. 2007 [26]. Reproduced with permission of Elsevier.

5.5 Injection Molding of Microstructured Parts

In recent years, more attention has been paid to fabrication of microstructured polymer components and devices, especially for optical and biomedical applications. The weight of a microstructured part may be more than 1 g, but the accuracy of microstructure (holes, slots, ribs, etc.) is in microns or sub-microns. Polymer materials are favored in such applications because of their low cost, good biocompatibility, high optical clarity, and high impact strength, compared with silicon or glass. For microstructured parts, injection molding has the potential for economical mass-production. Microstructured parts suitable for injection molding include diffractive optical elements, grating optical elements, multi-fiber connector, microlens arrays, high-density optical disc, light-guided plates, microfluidic devices, among others. However, some technical challenges caused by the microscale feature sizes need to be overcome for successful molding of microstructured parts. This section aims at presenting the main significant developments that have been achieved in different aspects of injection molding of microstructured parts. Aspects covered include material developments, mold manufacturing technologies, and process enhancements.

5.5.1 Materials

Injection molding of microstructured parts involves more severe operation conditions than conventional injection molding, in terms of temperature and pressure required for successful molding. In addition, being a viscoelastic material, the polymeric melt experiences shear-thinning, i.e. a decrease in viscosity with increase in shear rate. Thus, radical changes in the material properties are expected due to high shear-rates resulting from flow into microcavities. Therefore, the polymer selected should be appropriate for injection molding of microstructured parts. Generally speaking, the polymer should possess low viscosity and hence good flow properties.

High mechanical strength is also recommended in order for the molded part to resist mechanical stresses associated with demolding frictions or ejector forces. This is especially important for high-aspect ratio structures, where a larger surface of contact between the mold and the polymer imposes higher frictional resistance during demolding.

The polymer should be compatible with the mold material, as during processing, polymers may have different effects on the mold material. Also, the material should not leave deposits on the mold surface. For example, nickel mold inserts are not affected by molding PC even after 10 000 molding cycles. On the other hand, polymers containing aggressive chemicals can lead to corrosion in the inserts. This causes rough mold surfaces leading to aggressive demolding and damage to structural details in the submicron range.

Table 5.6 summarizes polymers that have been reported as being moldable by the microinjection molding process [7]. Most of the known engineering plastics appear on this list. Among the listed materials, some polymers including POM, PPS, polybutylene terephthalate (PBT), and liquid crystal polymer (LCP) have been specifically recommended for medical applications, because they comply with the approval criteria of the European regulatory agencies. In addition to the listed materials, thermoplastic elastomers, such as polyurethanes, were recently reported to be injection moldable with high aspect ratio features.

5.5.2 Mold Manufacturing Technologies

Several manufacturing techniques are used to produce inserts with microcavities, including micromechanical machining, electroforming, electron beam lithography (EBL), micro EDM,

Table 5.6 Polymers for injection molding of microstructured parts.

Class	Polymer	Full name
Amorphous	PMMA (acrylic)	Polymethylmethacrylate
	PC	Polycarbonate
	PSU	Polysulfon
	PS	Polystyrene
	COC	Cyclic olefin copolymer
	COP	Cyclic olefin polymer
	PPE (PPO)	Polyphenylene oxide
	PEI	Polyetherimide
	PAI	Polyamide imide
	MABS	Methylmethacrylate acrylonitrile-butadienestyrene
	SAN	Styrene acrylonitrile
	SBS	Styrene-butadiene-styrene
	ABS	Acrylonitrile-butadiene styrene
Semicrystalline	LCP	Liquid crystal polymer
	PP	Polypropylene
	PE	Polyethylene
	POM (acetal)	Polyoxymethylene
	POM-C	Polyoxymethylene (carbon filled)
	PBT (polyester)	Polybutylene terephthalate
	PBT-HI	Polybutylene terephthalate (filled with 15% glass fiber)
	PA 6 (nylon)	Polyamide 6
	PA 12	Polyamide 12
	PA 12 C	Polyamide 12 (carbon filled)
	PVDF	Polyvinylidene fluoride
	PFA (Teflon)	Perfluoroalkoxy
	PEEK	Polyetheretherketone
	PLA (polyester)	Polylactic acid (polylactide)

Source: Attia et al. 2009 [7]. Reproduced with permission of Springer.

LIGA/UV-LIGA, deep reactive ion etching (DRIE), anisotropic silicon etching, etc. Some complex molds require the combination of different processes. The choice of a particular technique depends on the dimension/geometry of the structure, the accuracy requirements, and the fabrication costs. In this section, micromechanical machining, electroforming, EBL, and DRIE are discussed regarding their capabilities in fabrication of microstructured molds.

5.5.2.1 Mechanical Micromachining

One representative work on mold fabrication with mechanical micromachining is due to Jung et al. [27] who used a three-axis micromachining stage to form microchannels on lab-on-a-chip (LOC) molds. A diameter of the micro end-mill they used was 200 μm, and its length of cutter is 400 μm. Mold steel NAK80 with a uniform hardness of 40 HRC was chosen as a testing mold material. In their research, a series of machining experiments using micro end-mills were performed to determine optimum machining conditions for improving surface finish and shape accuracy of the microchannels.

Another work is due to Sha et al. [28, 29] who micromachined two microstructured mold inserts using a KERN HSPC micromachining center. The two mold inserts were made of tool steel (AISI 01). The microfeatures assumed a pattern of circular "pins" and "fingers." There were two types of pins having diameters of 100 and 150 μm, heights of 350 and 400 μm, and in-between distances of 330 and 660 μm, respectively. The fingers also had two in-between distances of 50 and 150 μm.

Murakami et al. [30] also fabricated a mold insert with micro V-grooves by a precision micromachining center. The dimension of the micro V-groove was 20 μm in depth and 50 μm in width. In their process, tool steel Stavax HRC52 was first electroformed with a nickel layer of 200 μm thickness. Then micro V-grooves were milled on to the mold insert using a precision micromachining center.

Weng et al. [31] also fabricated an aluminum alloy mold insert with a microlens array by an ultraprecision freeform machine Nanotech® 350FG equipped with a Fast Tool Servo. The diameter of the microlens was 2 mm. The root-mean-square roughness is around 50 nm for the machined mold insert surface.

More recently, Yang et al. [32] manufactured a mold insert with microchannels using combined diamond tools and carbide tools on the Nanotech 350FG machine. After a flat surface was created on the mold insert by diamond turning, microchannels were machined on to the mold surface by high-speed micro milling using an ultra-precision high-speed air bearing spindle designed and built by Professional Instrument. A 50 μm diameter carbide flat end mill was used, and the spindle speed was set at 20 000 rpm. Finally, the machined insert with the main cavity and microchannels was mounted into a two-platen mold.

5.5.2.2 Electroforming

In recent years, electroforming has taken on new importance in the fabrication of micro and nanoscale metallic devices and in producing precision injection molds with micro and nanoscale features for production of nonmetallic micromolded objects. As shown in Figure 5.25, electroforming is a deposition process for depositing a metallic layer on a conductive object. It is an electrolyte process where the electrolyte is a solution containing ions of the metal to be deposited. The ions from the anode are attracted to the cathode when applying an electric field between the anode and the cathode. The anode is the metal to be deposited and the cathode is the surface to be coated. As the metallic ions move from the anode to the cathode, the plating

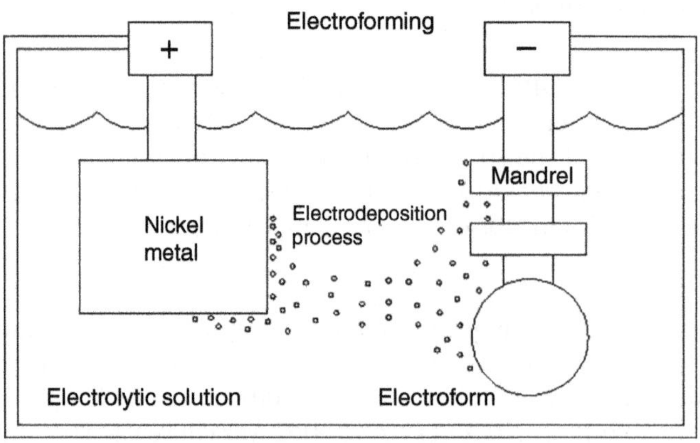

Figure 5.25 Schematic of electroforming process.

material builds up to a thickness where it can be separated from the original surface and form the desired part. The advantage of electroforming is that it produces replicas of the original pattern without shrinkage and distortion as encountered in other metal forming processes. Another benefit of using electroformed material is its high purity and the suitability for experimental control. Typical materials used in the electroforming process include gold, silver, nickel, copper, and iron. The reason of choosing nickel for most mold insert applications is its relatively robust mechanical properties compared to other materials. Kim et al. [33] created a nickel master stamp with a nanograting pattern by the electroforming method. The scanning electron microscope (SEM) image of the master stamp is shown in Figure 5.26. The pitch and depth of the nanogratings were 540 and 320 nm, respectively.

5.5.2.3 Electron Beam Lithography (EBL)

EBL is a technological tool for achieving nanoscale accuracy in the definition of nanopatterns. Lin et al. [34] have fabricated a mold insert with nanostructures using EBL. Figure 5.27 shows a schematic diagram of the fabrication steps used to produce the mold insert. First, a nanometer thick layer of photoresist was spun onto a silicon wafer substrate. Second, the photoresist was selectively exposed to radiation by an E-beam Writer. After developing, some surface of the wafer will expose to the air. Third, a chromium layer was evaporated on top of the photoresist and the wafer. After lifting off, the remaining layer of chromium served as the mask for the following dry etching. The substrate was then etched by Inductively Coupled Plasma Etcher. Finally, the mold insert with nanostructures was electroformed with nickel on the patterned substrate.

5.5.2.4 Deep Reactive-ion Etching (DRIE)

Deep reactive-ion etching (DRIE) is a highly anisotropic etching process used to create deep penetration, steep-sided holes and trenches in wafers/substrates, typically with high aspect ratios. In this process, etch depths of hundreds of microns can be achieved with almost vertical sidewalls. The primary technology is based on two different gas compositions alternated in the reactor. The first gas composition creates a polymer on the surface of the substrate, and the second gas composition etches the substrate. The polymer is immediately sputtered away by the physical part of the etching, but only on the horizontal surfaces and not the sidewalls. Since the polymer only dissolves very slowly in the chemical part of the etching, it builds up on the

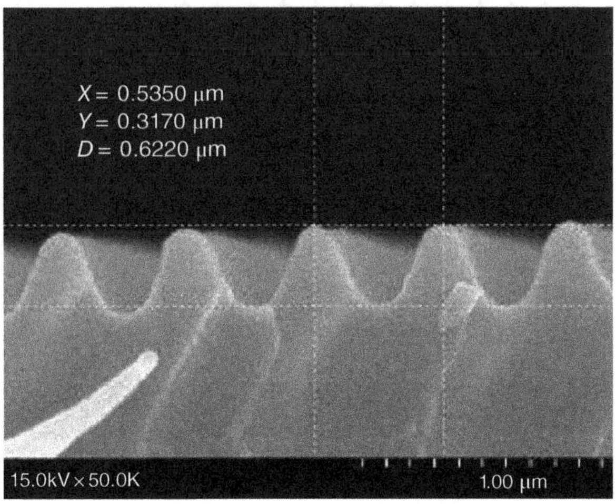

Figure 5.26 SEM image of nanogratings on the nickel stamp. *Source:* Kim et al. 2007 [33]. Reproduced with permission of Taylor & Francis.

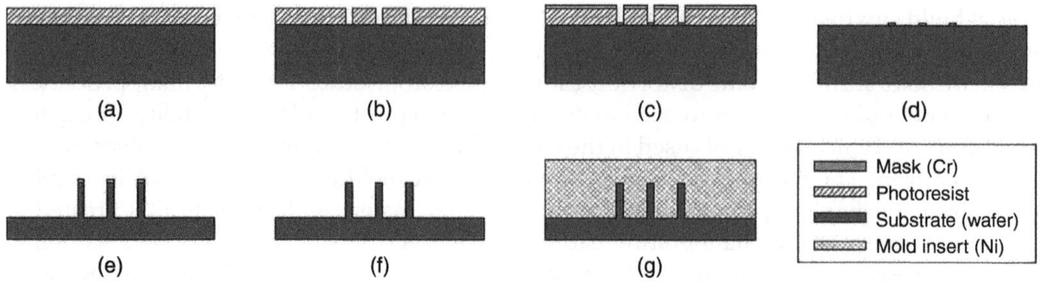

Figure 5.27 Schematic of the fabrication steps employed to produce the nanostructured mold insert: (a) spin coating, (b) lithography, (c) e-beam evaporation, (d) lift off, (e) etching, (f) mask remove, and (g) electroforming.

sidewalls and protects them from etching. As a result, etching aspect ratios of 50 to 1 can be achieved. The process can easily be used to etch completely through a silicon substrate, and etch rates are three to four times higher than wet etching. In particular, Fu et al. [35] manufactured a mold insert with microfluidic channels by DRIE.

5.5.2.5 Combination of Mechanical Micromachining and LIGA

Some complex molds require the combination of different processes. For example, microstructured mold inserts may be fabricated by a combination of mechanical micromachining and LIGA. In fact, such a method has been successfully applied to fabricate a mold insert for multifiber connectors [36]. The first layer of the latter ferrule was micromilled into a copper substrate. Afterward, the substrate was coated with a thick layer of PMMA in which the rippled alignment structures were patterned by X-ray lithography followed by electroforming of nickel. The nickel block was released from the substrate and the mold inserts were cut to the final shape by wire-EDM, as shown in Figure 5.28.

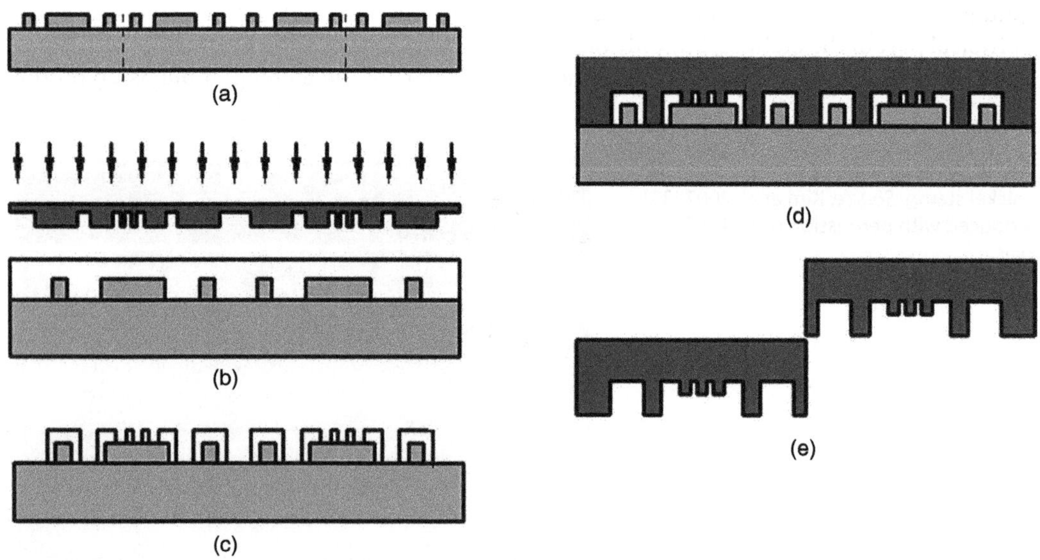

Figure 5.28 Schematic of the combinatorial technology: (a) micromilling, (b) X-ray lithography, (c) development, (d) electroforming, and (e) wire-EDM.

5.5.3 Molding Machine and Mold Design

The performance of the injection molding machine and the mold design affect significantly the quality of microstructured parts. Higher pressure injection molding machines may be needed to fill particular parts. For extreme cases, injection compression molding may be required. In this process, the melt is injected with a small amount of daylight in the mold. The mold is then rapidly closed to finish filling and packing.

Molds made for injection molding of microstructured parts are in principle similar to molds made for conventional injection molding. They usually consist of a fixed part and one or more moving parts, depending on the design. Finished parts are demolded with ejector pins that are usually controlled hydraulically or electrically. However, molds made for microstructured parts have special features. For example, gates may have to be relatively larger than for conventional parts to delay freeze-off. Mold temperature can be controlled by the Variotherm process [7]. This process permits heating rates, typically by heated oil or induction heating, of the order of several tens of Kelvin degrees per second, and hence imposes no significant change in the cycle time compared to conventional injection molding. The Variotherm process raises the mold temperature instantly above the polymer T_g during injection to prevent early freezing. This process is similar to the RTR molding process previously mentioned, only differing in the method used for mold heating. During cooling, the mold temperature is decreased below the T_g to assist in part solidification.

In addition to the Variotherm process, evacuation of the mold is done by air evacuation with evacuation rates up to 25 m^3/h in some systems (SMS Group 2006). Air evacuation is preferred over air vents as used in conventional injection molding. This is because the vents used in conventional injection molding are larger in size than some of the cavity features in micro injection molding, i.e. they will simply be clogged with polymer melt. In addition, microcavities may form air pockets that cannot be released while the polymer is flowing in the mold, so the mold has to be evacuated just before injection [15].

Due to the size of the mold cavity and features, modified mold sensors with miniature sizes are used, because conventional sensors are not suitable to microfeatured mold cavities. For example, pressure sensors with front diameters down to 1 mm and pressure range up to 2000 bar are currently available for measuring cavity pressures in micro-injection molding.

Part ejection becomes more difficult, since higher pressures are required and thinner features are more easily damaged. Special ejector design is important so that microfeatures are not deformed due to the induced friction between the mold and the part. For larger parts with microfeatures, more ejector pins may be required.

5.5.4 Process Parameters

5.5.4.1 Effect of Process Parameters

The process parameters involved in microstructure injection molding are nearly the same as those in conventional molding. The major process parameters are melt temperature, mold temperature, injection speed, injection pressure, holding pressure, holding time, and cooling time. The part quality can be measured in terms of cavity and microstructure fill, dimensional accuracy and stability, mechanical properties, or any other properties related to the desired performance of the molded article.

So far, most studies made to investigate the effects of the major process parameters have mainly focused on changing one parameter at a time and observing its effect on the part quality. This method was useful for injection molders in making some basic conclusions about the role of each parameter by its own. For example, Despa et al. [37] found that low mold temperature offers a short cycle time but can cause premature freezing. Thus, the injection rate has to be high enough to allow for complete filling before freezing. In addition, because of the viscoelasticity of polymeric melts, high melt temperatures allow for complete penetration without the need for high speeds. Thus, it was deduced that the injection speed and the melt temperature at various stages within the process are the most important parameters for the injection process.

In another experiment [38], the parts were not completely filled using the conventional setting for the mold temperature, unless the mold was heated up to a temperature above the T_g of the polymer, implying that the mold temperature significantly affected the filling depth of the polymer. This relation between the mold temperature and the filling depth was also found in another experiment involving molding of nanoscale features [33]. The nanoscale features were successfully filled only when the mold surface temperature was set very close to the melt temperature.

The effect of holding pressure was also found to be important, especially for optical parts. Birefringence was often used to evaluate the effect of process parameters on the residual stresses in the molded parts. A higher holding pressure was shown to provide better replication by eliminating material shrinkage, but at the expense of increased residual stresses and birefringence. Recent experimental results showed that the mold temperature is the predominating parameter affecting the optical properties of microstructured PC parts [39]. In general, a higher mold temperature results in better optical properties of molded parts.

5.5.4.2 Design of Experiments Approach

In fact, the interaction of more than one parameter may affect both the filling behavior and the part quality. However, testing of all possible combinations of different parameters is not practical in terms of time and resources. Hence, the DOE approach has become more widely accepted in experimental assessments of micro-injection molding.

Table 5.7 compares a number of DOE studies conducted by different research groups. The experiments presented indicate that many parameters can affect the part quality either independently or interactively. The table also shows that the main results from each study can be quite different. This is understandable considering many different factors such as the material used, the geometry of the mold, the selected response, the surface finish of the mold, and the performance of molding machine simultaneously affect the yield of the molding process. Further study and investigation is, therefore, required whenever an element is changed, such as the mold geometry or the polymeric material.

5.5.4.3 In-process Monitoring of Process Parameters

Multiple-sensors and data-acquisition systems have been used to monitor the change in processing parameters during the different stages of injection molding. Several additional sensors can be used to monitor the change in, for example, injection pressure, cavity pressure, displacements, and velocity of the injection pin, temperature in both halves of the mold and the injection force. In addition, in-process rheometry has been used to monitor the change in the strain rates during processing. Ultrasonic probes are currently used to evaluate the part quality in terms of filling incompletion, polymer degradation, and melt flow speed [47].

5.5.4.4 Multiple Quality Optimization of Process Parameters

In design of experiments, the Taguchi method is mainly used for optimizing a single quality attribute. However, this optimized process setting is only applicable to that particular quality attribute and do not represent the optimal solution for the overall quality. For this reason, Fung [48] adopted both the gray relational analysis and the Taguchi method to optimize the

Table 5.7 A summary of different DOE studies on microinjection molding.

Response	Materials	Main results	References
Filling quality of microfeatured channels	PC, SBS, MABS, COC, and PMMA	Melt temperature and mold temperature are most significant parameters	Mönkkönen et al. [40]
Part weight and dimensions	PC and POM	Metering size and holding pressure are most significant. The interaction between both is also important	Zhao et al. [5]
Surface profile of microlens	PS, PMMA, and PC	Packing pressure and flow rate significantly affect the final surface profile of the injection molded product	Lee et al. [41]
Complete filling of doughnut-shaped parts	PS and PC	Injection speed and holding pressure are the most influential, while melt temperature and mold temperature have less influence	Aufiero [42]
Complete filling of high aspect ratiorods	PP, POM, and ABS	Melt temperature and injection speed are key factors for PP and ABS. Mold temperature is also significant in case of POM	Sha et al. [29]
Flow length along a microchannel into a flat cavity	PP, ABS, and PC	The high levels of all processing parameters result in better filling. Surface finish is related to level of turbulence in melt flow.	Griffiths et al. [43]
Weld-line formation	PS	Injection speed and mold temperature have the main effect on weld-line placement and orientation	Tosello et al. [44]
Filled volume fraction of microfilters	COC	Flow rate was found to be the most important processing parameter	Lee et al. [45]
Maximum residual stress of microlens arrays	PC	Melt temperature, mold temperature, and packing pressure are the major factors affecting maximum residual stresses	Weng et al. [31]
Flatness of microfluidic component	PMMA	Cooling time and ejection system design are the key factors influencing the flatness of microfluidic components	Marson et al. [46]

process parameters in injection molding. Using different sliding directions as the target qualities, optimal process conditions were obtained through experiments and the wear properties of molded parts were improved. Chang et al. [49] used the gray relational analysis and the Taguchi method to optimize parameters for injection molding of PC/acrylonitrile-butadiene styrene (ABS) blends. Using tensile properties as the target qualities, optimal process conditions were obtained through experiments. Effective improvements in part quality were also achieved in their experiments. In addition, Liao and Hsieh [50] used the back-propagation neural network (BPNN) to predict shrinkage and warpage in injection molding and investigated the capability of the BPNN method in process modeling and optimization. The optimal process conditions determined by BPNN also resulted in minimized shrinkage and warpage in the molded parts.

Kuo and Su [51, 52] also conducted a process optimization study for the injection molding process of a microstructured light-guide plate by a combined implementation of the Taguchi method, the gray relational analysis, and the BPNN method. There were two quality characteristics of the light-guide plate concerned in this study: the depth and the angle of V-grooves. The authors chose the Taguchi method and the gray relational analysis for determining the optimal processing conditions and applied BPNN to establishing a quality prediction system for molded light-guide plates. The Taguchi method was primarily used for experimental planning, while the gray relational analysis was mainly used for integrating multiple qualities and identifying the optimal process conditions. The Taguchi method was also used to design the learning parameters for the BPNN, and this speeded up network convergence in determining preferable combinations of learning parameters. As the experimental results revealed, the prediction system so established was effective in accurate prediction of the qualities of molded light-guide plates.

5.6 Injection Molding of Microparts

Injection molding has also been employed to fabricate a variety of polymer microparts. In contrast to microstructured parts previously discussed, the weight of a micropart is much smaller, in milligrams. The dimensional accuracy is similar, in microns or submicrons. Most applications of microparts are in the field of microoptics, microelectronics, micromechanical devices, and microfluidics. These microparts all have an extremely small volume and desire supper high accuracy in fabrication. Therefore, some new problems arise regarding mold design, mold fabrication, and precision microcavities replication.

For mold fabrication, generally speaking, the methods previously mentioned for making mold inserts for microstructure injection molding are equally useful since in both cases precision microcavities need to be fabricated. LIGA, UV-LIGA, mechanical micromachining, micro EDM, high energy beam ablation, lithography, etc., can all be used for fabricating microcavity mold inserts. Cell container mold inserts is shown in Figure 5.29.

However, some changes on the mold design are often needed. In particular, a tiny gate needs to be precision fabricated on the mold insert to connect the microcavity to a relatively large runner system. In contrast, in microstructure injection molding, the microstructures are attached to a thick substrate and therefore no additional gates are needed. Also, clever ejection schemes need to be implemented since the part is so small. Besides mounting precision tiny

Figure 5.29 SEM images of cell container mold inserts.

ejection pins directly in the microcavity, another possible method for ejection is to install ejection pins to nearby runners and pull the micropart out of the cavity by the force transferred from the runners.

The more important problem to be tackled with in micropart injection molding is on precision metering and delivering an extremely small volume of plastic melt to the microcavity. Some precision injection molding machines for microstructure injection molding or thin-wall injection molding, as previously mentioned, particularly with a small shot size, can be used for micropart injection molding. In general, any precision injection molding machines with a separation metering screw, e.g. a screw-plunger hybrid machine, is suitable for micropart injection molding. Plunger-type machines like the sesame machine from Medical Murray, Inc., also become a feasible solution in micropart machine due to the small shot size required. On the other hand, because of an extremely small shot size is needed, some unconventional plasticization methods may now be possible. Examples are the CO_2-assisted plasticization, the X-melt process, and ultrasonic plasticization. As examples, CO_2-assisted plasticization and the X-melt process are discussed in detail as flows. We also discuss some recent experimental results on micropart injection molding.

5.6.1 CO_2-assisted Plasticization

Katoh et al. [53] developed a special injection molding process called AMOTEC (Asahi molding technology with CO_2) where CO_2 is used to plasticize polymer. Figure 5.30 shows a schematic of the injection molding system equipped with a heating cylinder, a CO_2 supplying system to the barrel and the cavity, and a gaseous-sealing mold. The CO_2 dissolved into the spaces between the molecules of molten resin inside the barrel with a heating cylinder acts as a plasticizer that improves the flowability of the resin. Using this property of CO_2 gasses, good transcription can be achieved when molding products require ultra-fine replication. One main benefit of using CO_2 is that no solidification layer is formed at the wall or at the melt front during filling. After molding, the CO_2 whose contents is less than several percent of the total amount of the resin, evaporates and does not significantly change the properties of the resin. Last but not least, since the glass transition temperature (T_g) of the resin decreases as the CO_2 pressure increases, molding can be performed with both the mold and the resin at lower temperatures. Less time is required for cooling so that the cycle time can be shortened.

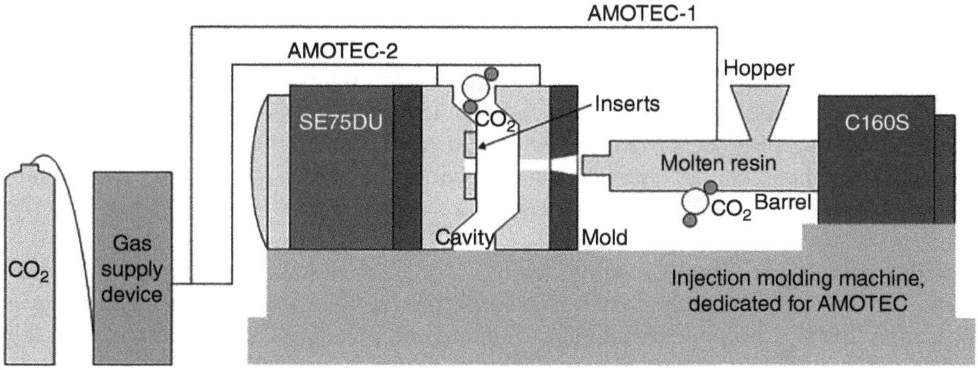

Figure 5.30 Experimental setup for micro injection molding using AMOTEC with CO_2.

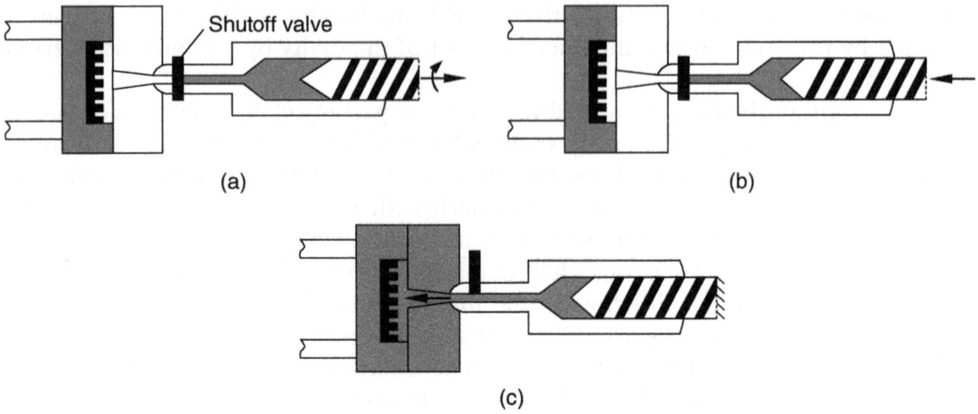

Figure 5.31 Stages involved in expansion injection molding: (a) dosing, (b) compression, and (c) expansion injection.

5.6.2 X-melt Process

A commercially available process called X-melt, developed by Engel Machinery [54], uses the energy stored in the polymer melt to force the polymer melt into the mold cavity. The process sequence is shown in Figure 5.31. Simply to put, the polymer melt is compressed to build up the pressure. When the shutoff valve is opened, the statically pressurized polymer melt expands into the mold cavity, driven by the energy stored in the melt. Due to this characteristic, the process may be named expansion injection molding, but one should not confuse this process with expansion processes for foaming, involving internal gas pressure in the polymer melt. The main advantage of this method over the high-speed method is that no injection ram is used and thus the inertia effect is minimal. Further, the highest injection pressure is attained at the beginning of the injection stage. This allows the polymer to flow at the highest speed at the beginning of the injection stage, thus most effectively suppressing the premature freezing problem. Note that in the speed control scheme, the injection pressure is near zero at the startup. Because of the limited compressibility of the polymer melt, a possible limitation of the expansion molding process is the large cushion material used and thus an increased potential for thermal degradation.

5.6.3 Process Study on Micropart Injection Molding

Zhao et al. [55] investigated the effects of process parameters on quality of precision molded microparts. A series of micro gears were molded using a POM resin in a set of statistically designed experiments. Micro component inspection, characterization, and data analysis were carried out to study the quality characteristics of the molded gears. It was found that the metering size and the holding time were the two most significant factors affecting part quality. The process was also significantly affected by the interaction of these two parameters. It is understandable that there must be a range of metering size within which the holding pressure is effective.

Attia and Alcock [56] used the DOE approach to correlate the quality of the molded microparts to the processing parameters. Five processing parameters were investigated using a screening half-factorial experimentation plan to determine their possible effect on the filling quality of the molded parts. The part mass is used as an output parameter to reflect the filling of the parts. The experiments showed that the holding pressure was the most significant processing parameter for all different shapes. In addition, the experiments showed that the geometry of the parts played a role in determining the significant processing parameters. For a more complex part, injection speed and mold temperature became statistically significant. A desirability function approach was successfully used to improve the filling quality of each part.

Micropart injection molding, with advantages of easy mass production and low cost, is a key technology for producing micro components. Nevertheless, a low yield rate of high-quality molded parts is common due to problems associated with geometric precision, molecular orientation, and optical properties. Solutions to such problems must consider the machine, mold design, and process parameter settings. However, optimal performance becomes relatively less attainable when process parameters deviate due to inevitable process tolerances and change in an operation environment. Huang et al. [57] therefore developed a robust parameter search method for meeting the requirements of multiple quality characteristics in molded microparts. Two gear mold inserts with outside diameters of approximate 6 and 4 mm are used to investigate two key geometrical dimensions of a molded gear, outside diameter and tooth thickness. The UV-LIGA process was employed in fabricating the Ni gear mold inserts to take advantage of its easy access and low cost. The micropart injection molding experiments suggest that mold temperature, holding pressure, and injection speed had significant effects on dimensional quality characteristics of molded gears. The robust parameters derived from the proposed method increased yield rates of the 6 mm gear from 10% to 91%, and those of the 4 mm gear from 38% to 93% in comparison with the initial point, thereby demonstrating application effectiveness.

The above experimental results indicate that many parameters can play a role in the part quality either independently or interactively. The main results of each set of experiments can be largely different. This infers that the processing parameters studied were not the only parameters affecting the part quality. Other factors play a significant role include the material used, the geometry/dimension of the mold, the selected quality goal, and the surface finish of the mold. Further study and investigation is required in this area before conclusions can be drawn.

5.7 Simulation of Injection Molding

The quality and performance of injection molded parts depend not only on the material, shape, and function of the part design but also on how the material is processed during molding. By employing scientific-based simulation tools, engineers can quickly qualify final design and material selection without physically committing real material and machine time [58]. The ability to numerically simulate injection molding would allow for the following major goals to be achieved:

(a) To visualize the flow and predict the last-filled sections of the mold. This is useful to identify defects that are usually associated with the last filled parts like incomplete filling, weld lines, and voids.
(b) To economically optimize the design of the mold. It would be very useful to simulate different geometrical designs, sprue and gating systems, and flow-paths to determine the optimum mold design before manufacturing.
(c) To simulate the thermal conditions of the flow during filling and cooling which would be useful in estimating the cycle time and determining the processing bottlenecks.
(d) To assist in designing more effective experiments and determining the most influential processing parameters on the part quality.
(e) To identify postprocessing properties, such as residual stresses, shrinkages, and warpage.

In this section, the standard injection molding simulation is introduced first. Then, special simulation needs for miniature molding processes are discussed. Some representative simulation results are presented, and potential areas for improvement are identified.

5.7.1 Standard Injection Molding Simulation

5.7.1.1 Flow Model

Three different types of flow models have been used in 3D filling simulation: midplane models, surface models, and solid models [59].

Most injection molded parts have the characteristic of being thin but generally of complex shape. Therefore, theoretical studies of filling flow in a thin cavity based on the Hele-Shaw flow formulation have been conducted. In midplane models, the flow is assumed to be quasi-steady state and the inertia terms are neglected due to the low Reynolds numbers encountered in the flow of molten polymers. The filling of a mold cavity becomes a 2D flow problem for the gapwise averaged velocity which is related to the pressure gradient through a quantity called "fluidity" representing the sum of the effect of changing temperature and rheology across the gap. The arbitrary planar midplane (middle-plane/center plane) with a defined thickness is used to represent the three-dimensional geometry of the part, so that it is called a "midplane" model. Although the industry has referred to this as 3D, the midplane models are actually 2.5D models. A midplane model works well when the part is thin-walled; however, the complexity of creating the midplane mesh has been a leading factor in limiting the use of such tools.

The surface model for molding simulation represents a three-dimensional part with a boundary or skin mesh on the outside surfaces of a solid model, instead of the midplane. This allows users to analyze solid geometries of thin-walled parts directly, resulting in a significant decrease in model preparation time. The surface model creates a shell mesh on the outside surfaces of the 3D solid geometry. The elements on the opposite surfaces of a wall are matched and aligned. The flow and thermal calculations are then conducted on the two surfaces. It should be noted that the surface model still adopts the Hele-Shaw approximation and therefore its formulation is very similar to that of the mid-plane model. The surface model was introduced by Moldflow® and commercialized as a fusion method and dual domain technology.

Several situations occurring during mold filling, however, cannot be accurately predicted using the Hele-Shaw approximation. Among the most important factors, we can cite the fluid behavior at the free surface (flow front), the fluid behavior near and at the solid walls, the phenomenon occurring at the merging of two or more fluid streams, and the kinematics in areas where shear and extensional deformations contribute significantly to the stress field (gates, ribs, sudden thickness changes, etc.).

With the rapid development of computer hardware, full 3D simulation has become a reasonable and promising solution to these problems. In general, a true, three-dimensional simulation should be performed on geometries that do not conform to the criteria defining traditional, thin-wall designs. This solid model (full 3D model) works well especially with thick and solid parts such as electrical connectors, thick structural components, and those that have extreme thickness changes. If the polymer melts are assumed to be incompressible during filling, the conservation of mass and momentum equations reduce to the Navier–Stokes equations:

$$\nabla \cdot \mathbf{u} = 0 \tag{5.4}$$

$$\rho \frac{D\mathbf{u}}{Dt} = -\nabla P + \nabla \cdot \bar{\bar{\tau}} + \rho \mathbf{g} \tag{5.5}$$

where **u**, **g**, and $\overline{\overline{\tau}}$ denote the velocity vector, body force vector, and stress tensor, respectively. The energy equation is expressed as

$$\rho C_p\left(\frac{\partial T}{\partial t}+u\cdot\nabla T\right)=\nabla\cdot(k\nabla T)+2\eta\overline{\overline{\dot{\gamma}}}:\overline{\overline{\dot{\gamma}}} \tag{5.6}$$

where the strain rate tensor $\overline{\overline{\dot{\gamma}}}$ is defined as

$$\overline{\overline{\dot{\gamma}}}=\frac{1}{2}\left(\nabla\mathbf{u}+(\nabla\mathbf{u})^{\mathrm{T}}\right) \tag{5.7}$$

To date, several numerical methods for solving these equations in injection molding have been developed, including the finite element method (FEM), the finite volume method (FVM), the control-volume-based finite-element-method (CVFEM), and the boundary element method (BEM). In the three-dimensional filling simulation, accurate tracking of the melt fronts (polymer/air interfaces), as well as the representation and evolution of their complex topology, is very important. Depending on the selection of the reference frame, the moving boundary problems can be tackled by Lagrangian or Eulerian algorithms.

5.7.1.2 Simulation Example

To demonstrate the procedure, the function, and the application of injection molding simulation, we here discuss one case study due to Wu et al. [60].

This example dealt with the analysis of the precision molding process of a 3D biodegradable polymeric scaffold [60]. The FEM was adopted to compute the kinematic and dynamic quantities in a 3D inertia-free incompressible flow. A control volume scheme with a fixed finite element mesh was employed to predict flow front advancement. Poly(lactic acid) (PLA) was chosen as a plastic material for the scaffold.

In summary, this example illustrates that, through a capable injection molding simulation, the flow dynamics, processing window, and optimal processing conditions can all be predicted without actually fabricating a testing mold.

5.7.2 Challenges for Simulating Miniature Injection Molding Processes

Due to some simplifications assumed in the formulation, the capability of the standard injection molding simulation is limited. This is partly because the difficulty in realistically modeling the non-Newtonian effects in a 3D filling simulation. This weakness will be magnified when flow problems with large Weissenberg number are encountered. In miniature molding processes, additional issues caused by scaling can arise, as such some assumptions used in the standard simulation now become questionable [61]. Several particular challenges affecting the simulation accuracy of miniature injection molding processes are elaborated below.

1) In standard injection molding, shell modeling is common, which neglects the effects of sides and edges. In comparison, the geometry involved in miniature molding processes, particularly micropart molding, are typically quite three dimensional. Thus, the Hele-Shaw approximation, although used several times in the literature to simulate micro injection molding, may not indeed be suitable.
2) Some effects that are neglected in conventional injection molding simulation can become significant in the microscale due to the increased surface-to-volume ratio, such as surface roughness, surface tension, heating of the melt by viscous friction, and cooling of the melt front due to increased heat loss. In addition, models should account for the differences in

dynamics of heat and mass transfer in the microscale. The heat transfer coefficient at the polymer-mold contact, for example, was shown to be significant in the microscale [62].

3) The viscoelastic nature of the polymeric melt typically becomes more significant at the microscale because of the high shear rates involved in, for example narrow gates. Particularly, it has been mentioned in the literature that the melt viscosity is different from that classically known.
4) Meshing elements should be chosen with particular care. For example, for meshing of microfeatures sitting on a relatively large substrate, special considerations must be given. A smooth transition of meshing from macrofeatures to microfeatures is important in reaching a converged numerical solution.
5) Special processing conditions, such as variable mold temperature and air vacuuming need to be considered in modeling and simulation.

5.7.3 Simulation of Miniature Injection Molding

Although it is difficult to completely solve the aforementioned challenges in modeling and simulation of miniature injection molding processes, a number of studies have been performed and some progress has been made. At the moment, this area is still highly research oriented despite the high industrial significance. Most studies attempted to address the microscale effects including size-dependent rheological properties and boundary conditions on the polymer filling process [63]. The distinctive features of some representative studies in this area are compared in Table 5.8.

Most studies employed commercial computer aided engineering (CAE) tools for simulating miniature filling processes. Several earlier studies, in fact, focused on applying the Hele-Shaw

Table 5.8 Comparison of the simulation studies for miniature injection molding processes.

Works	Macroscale G.E.	Microscale G.E.	Microscale feature
Yao and Kim [64]	Hele-Shaw (2.5D) In-house code	Modified Hele-Shaw (2.5D) In-house code	Slip, Surface tension Size-dependent viscosity
Yu et al. [62]	Hele-Shaw (1D) In-house code	Navier–Stokes 2D In-house code	None
Kim and Turng [58]	Navier–Stokes 3D In-house code	Navier–Stokes 3D In-house code	None
Tofteberg and Andreassen [65]	Hele-Shaw (2.5D) MoldFlow	Navier–Stokes 2D Ansys CFX	None
Shen et al. [66]	Navier–Stokes (3D) MoldFlow	Navier–Stokes (3D) MoldFlow	None
Park and coworkers [67]	Hele-Shaw (2.5D) MoldFlow	Navier–Stokes (2D) In-house code	Surface tension
Lin and Young [68]	Navier–Stokes 3D MoldFlow	Analytical (1D)	Air-trapping
Cao et al. 2011 [69]	Navier–Stokes 2D In-house code	Navier–Stokes 2D In-house code	Slip, Surface tension
Choi and Kim [63]	Navier–Stokes 3D MoldFlow	Navier–Stokes (2D axisymmetric) Comsol, Fluent	Slip, Surface tension

flow model and obtained some reasonable predictions, at least at a qualitative level. The driver behind this election was the relatively low computational load compared with a full 3D implementation. However, with the rapid advancement in computer hardware and computing power over the past two decades, the tendency is being shifted toward more capable 3D simulations. It is known that both midplane and dual domain approximations have significant limitations in predicting 3D flow phenomena such as fountain flow and flow imbalance in runner splitting. Furthermore, inertial effects that are important in high-speed injection and 3D geometries are neglected in the Hele-Shaw flow model or alike. In microcavity filling, the fountain flow phenomenon plays an important role since the out-of-plane velocity is expected to affect the filling flow near the inlet of the microcavity. Although the macroflow can be assumed independent of the microflow, the viscosity of the inlet flow into the microcavity is determined by the shear rate in the macro-regime.

Since most microparts and microstructured parts cannot be simply considered to be thin walled or shell like, 3D flow formulations are expected to be more suitable. Edge effects and other boundary effects that cannot be handled by the shell formulation can now be better treated. Microscale effects have also been discussed in the literature. In particular, Kim and Turng [58] argued that the surface tension is negligible since the capillary number is quite big. However, they commented that wall slip could more significantly affect the filling flow. Kim et al. [67] have employed the Brackbill's model to account for the capillary effect by converting the surface force into a volume force in terms of the phase boundary curvature. Lin and Young [68] proposed an analytical model that can consider air trapping in the microcavity. Cao et al. [69] used the Phan-Thien–Tanner (PTT) model to represent the rheological behavior of viscoelastic fluids. They also included the slip boundary condition and surface tension in the mathematical model for microcavity filling in order to effectively describe the microscale effects. Choi and Kim [63] presented a multiscale model for simulating the microcavity filling process. Their results show that wall slip and surface tension both play important roles in the microregime.

5.8 Summary and Outlook

Recent developments in injection molding are moving into a multiscale arena where thin and long parts, microstructured parts with large substrates, and microparts with weight less than a few milligrams have all showed a strong economical relevancy to the modern society. Although these areas have been developed rapidly in the past two decades, there are still some technical challenges, particularly regarding materials, mold fabrication, process optimization, and modeling and simulation, remaining to be solved. In this chapter, we have therefore focused on special considerations on materials, equipment, and processing suitable for successful molding at these different size scales. The standard injection molding technology was introduced in good detail first, and the other newer processes were described more in terms of their uniqueness or difference from the conventional counterpart. We particularly highlighted some proven solutions or strategies useful for molding thin and long parts, for filling a microcavity, and for replicating microstructures from a mold insert. Moreover, the simulation of miniature injection molding processes has been discussed. Based on the understanding of the state-of-the-art, we proposed the following work for possible future studies in the related area:

1) At the moment, filling into ultrathin wall sections and microcavities still remains to be a technical challenge. New molding materials with enhanced flowability or moldability and yet low cost are highly demanded. The mechanical properties of the molding material also

need to be improved since most known composite formulations for macroscale applications are not suitable for miniature sizes. Development of cost-effective formulations for strong and easy-flow materials is therefore an important direction for future work.

2) Although several manufacturing techniques are currently available for fabricating mold inserts with microcavities, some key problems have not yet been solved. For example, it is difficult to make a real 3D microcavity on stainless steel. The combination of different processes may provide a new capability, for example the combination of UV-LIGA and micromachining or micro-EDM as discussed in the previous sections. Furthermore, creations of super surface finishes and appropriate draft angles on hard and strong metallic tooling materials remain to be a challenge. Some kinds of new methods for polishing microstructured mold surface in a noncontact manner, e.g. by electrochemical erosion or ultrasonic erosion, are particularly of interest.
3) The literature shows that the results from different process studies are often different or even contradictory to each other. This may be attributed to the different part geometry involved and different material used in different studies. A standard geometry-based processing database is therefore highly desired. This database may be integrated into simulation packages, for example Moldflow.
4) Modeling and simulation of injection molding at miniature scales is still in its infancy. Current studies have demonstrated the importance of incorporating size effects into the filling simulation in microcavities. However, fundamental understanding on such size effects is still primitive, and the predications from existing simulations and analyses are quite qualitative. Important phenomena that still lack in understanding include dynamic wetting or de-wetting at the tiny melt front, size-dependent rheological properties, microscale wall slip, among others. Formulation of proper constitutive models on the related physical processes is critical to the development of more capable simulation capabilities for microscale molding processes.

References

1 Chen, Z.B. and Turng, L.S. (2005). A review of current developments in process and quality control for injection molding. *Advances in Polymer Technology* 24 (3): 165–182.

2 Bryce, D.M. (1996). *Plastic Injection Molding: Manufacturing Process Fundamentals*. Society of Manufacturing Engineers.

3 Yao, D.G. and Kim, B. (2004). Scaling issues in miniaturization of injection molded parts. *Journal of Manufacturing Science and Engineering-Transactions of the ASME* 126 (4): 733–739.

4 Osswald, T.A. and Hernández-Ortiz, J.P. (2006). *Polymer Processing*. Munich: Hanser.

5 Zhao, J., Mayes, R.H., and Chen, G. et al. (2003). Polymer micromould design and micromoulding process. *Plastics Rubber and Composites* 32 (6): 240–247.

6 Piotter, V., Mueller, K., and Plewa, K. et al. (2002). Performance and simulation of thermoplastic micro injection molding. *Microsystem Technologies* 8 (6): 387–390.

7 Attia, U.M., Marson, S., and Alcock, J.R. (2009). Micro-injection moulding of polymer microfluidic devices. *Microfluidics and Nanofluidics* 7 (1): 1–28.

8 Bibber, D.M. (2004). Micro molding challenges. In: *ANTEC Conference Proceedings*, vol. 10(3), 3703–3711. Society of Plastics Engineers.

9 Yao, D.G., Chen, S.C., and Kim, B.H. (2008). Rapid thermal cycling of injection molds: an overview on technical approaches and applications. *Advances in Polymer Technology* 27 (4): 233–255.

10 Demirer, A., Soydan, Y., and Kapti, A.O. (2007). An experimental investigation of the effects of hot runner system on injection moulding process in comparison with conventional runner system. *Materials and Design* 28 (5): 1467–1476.

11 Michaeli, W. and Opfermann, D. (2006). Micro assembly injection moulding. *Microsystem Technologies* 12 (7): 616–619.

12 Zhu, T., Dong, H.-J., and Liu, Y. (2012). In-mold assembly injection mold for micro plastic hinge. *Molding Engineering* 38 (1): 50–54.

13 Kaneton, Y. and Yokoi, H. (2011). Visualization analysis of side-edge flow phenomena in different thickness/width rectangular cavities using a rotary runner exchange system. *Polymer Engineering and Science* 51 (4): 721–729.

14 Fathi, S. and Behravesh, A.H. (2008). Real-time measurement of flow front kinematics using quantitative visualization in injection molding process. *Polymer Engineering and Science* 48 (3): 598–605.

15 Heckele, M. and Schomburg, W.K. (2004). Review on micro molding of thermoplastic polymers. *Journal of Micromechanics and Microengineering* 14 (3): R1–R14.

16 Mayer, R. (2007). Precision injection molding: how to make polymer optics for high volume and high precision applications. *Optik and Photonik* 2: 46–51.

17 Sano, T., Lyoda, Y., Shimazu, T. et al. (2010). Injection molded optical lens using a heat resistant thermoplastic resin with electron beam cross-linking. *Japanese Journal of Applied Physics* 49 (5): 075–083.

18 Michaeli, W., Heβner, Klaiber, F., and Forster, J. (2007). Geometrical accuracy and optical performance of injection moulded and injection-compression moulded plastic parts. *CIRP Annals* 56 (1): 545–548.

19 Lu, X.H. and Khim, L.S. (2001). A statistical experimental study of the injection molding of optical lenses. *Journal of Materials Processing Technology* 113 (1–3): 189–195.

20 Tsai, K.M., Hsieh, C.Y., and Lo, W.C. (2009). A study of the effects of process parameters for injection molding on surface quality of optical lenses. *Journal of Materials Processing Technology* 209 (7): 3469–3477.

21 Lo, W.C., Tsai, K.M., and Hsieh, C.Y. (2009). Six Sigma approach to improve surface precision of optical lenses in the injection-molding process. *International Journal of Advanced Manufacturing Technology* 41 (9–10): 885–896.

22 Yang, C., Su, L., Huang, C. et al. (2011). Effect of packing pressure on refractive index variation in injection molding of precision plastic optical lens. *Advances in Polymer Technology* 30 (1): 51–61.

23 Cosma, L. (1996). Molding engineering resins into thin wall applications: issues and answers. In: *SPE ANTEC Technical Papers*, 466–469.

24 Yao, D.G. and Kim, B. (2002). Increasing flow length in thin wall injection molding using a rapidly heated mold. *Polymer-Plastics Technology and Engineering* 41 (5): 819–832.

25 Park, K., Kim, B., and Yao, D.G. (2006). Numerical simulation for injection molding with a rapidly heated mold, Part I: flow simulation for thin wall parts. *Polymer-Plastics Technology and Engineering* 45 (8): 897–902.

26 Song, M.C., Liu, Z., Wang, M.J. et al. (2007). Research on effects of injection process parameters on the molding process for ultra-thin wall plastic parts. *Journal of Materials Processing Technology* 187: 668–671.

27 Jung, W.C., Heo, Y.M., Yoon, G.S. et al. (2007). Micro machining of injection Mold inserts for fluidic channel of polymeric biochips. *Sensors* 7 (8): 1643–1654.

28 Sha, B., Dimov, S., Griffiths, C. et al. (2007). Investigation of micro-injection moulding: factors affecting the replication quality. *Journal of Materials Processing Technology* 183 (2–3): 284–296.

29 Sha, B., Dimov, S., Griffiths, C. et al. (2007). Micro-injection moulding: Factors affecting the achievable aspect ratios. *International Journal of Advanced Manufacturing Technology* 33 (1–2): 147–156.

30 Murakami, O., Kotaki, M., and Hamada, H. (2008). Effect of molecular weight and molding conditions on the replication of injection moldings with micro-scale V-groove features. *Polymer Engineering and Science* 48 (4): 697–704.

31 Weng, C., Lee, W.B., and To, S. (2010). A study of the relevant effects on the maximum residual stress in the precision injection moulding of microlens arrays. *Journal of Micromechanics and Microengineering* 20 (3): 033–035.

32 Yang, C., Huang, H.X., Castro, J.M., and Yi, A.Y. (2011). Replication characterization in injection molding of microfeatures with high aspect ratio: influence of layout and shape factor. *Polymer Engineering and Science* 51 (5): 959–968.

33 Kim, S.H., Shiau, C.S., Kim, B.H., and Yao, D.G. (2007). Injection molding nanoscale features with the aid of induction heating. *Polymer-Plastics Technology and Engineering* 46 (10–12): 1031–1037.

34 Lin, H.Y., Chang, C.H., and Young, W.B. (2010). Experimental and analytical study on filling of nano structures in micro injection molding. *International Communications in Heat and Mass Transfer* 37 (10): 1477–1486.

35 Fu, G., Tor, S.B., Hardt, D.E., and Loh, N.H. (2011). Effects of processing parameters on the micro-channels replication in microfluidic devices fabricated by micro injection molding. *Microsystem Technologies* 17 (12): 1791–1798.

36 Wallrabe, U., Dittrich, H., Friedsam, G. et al. (2002). Micromolded easy-assembly multi fiber connector: RibCon®. *Microsystem Technologies* 8 (2–3): 83–87.

37 Despa, M.S., Kelly, K.W., and Collier, J.R. (1998). Injection molding using high aspect ratio microstructures mold inserts produced by LIGA technique. *Materials and Device Characterization in Micromachining* 3512: 286–294.

38 Wimberger-Friedl, R. (1999). Injection molding of sub-mu m grating optical elements. In: *ANTEC '99: Plastics Bridging the Millennia, Conference Proceedings*, vol. I–III, 476–480.

39 Murakami, O., Yamada, K., and Kotaki, M. (2007). Replication and optical properties of injection moldings with microstructures. In: *SPE ANTEC Technical Papers*, 2040–2044.

40 Monkkonen, K., Pakkanen, T.T., Hietala, J. et al. (2002). Replication of sub-micron features using amorphous thermoplastics. *Polymer Engineering and Science* 42 (7): 1600–1608.

41 Lee, B.K., Kim, D.S., and Kwon, T.H. (2004). Replication of microlens arrays by injection molding. *Microsystem Technologies* 10 (6–7): 531–535.

42 Aufiero, R. (2005). The effect of process conditions on part quality in microinjection molding. In: *SPE ANTEC Technical Papers*, Boston, 36–40.

43 Griffiths, C.A., Dimov, S.S., Brousseau, E.B., and Hoyle, R. T. (2007). The effects of tool surface quality in micro-injection moulding. *Journal of Materials Processing Technology* 189 (1–3): 418–427.

44 Tosello, G., Gava, A., Hansen, H. et al. (2007). Influence of process parameters on the weld lines of a micro injection molded component. In: *ANTEC-Conference Proceedings*.

45 Lee, B.K., Hwang, C.J., Kim, D.S. et al. (2008). Replication quality of flow-through microfilters in microfluidic lab-on-a-chip for blood typing by microinjection molding. *Journal of Manufacturing Science and Engineering* 130 (2): 010–021.

46 Marson, S., Attia, U.M., Lucchetta, G. et al. (2011). Flatness optimization of micro-injection moulded parts: the case of a PMMA microfluidic component. *Journal of Micromechanics and Microengineering* 21 (11): 1–9.

47 Ono, Y., Cheng, C.C., Jen, C.K. et al. (2005). Ultrasonic technique and probes for monitoring surface imperfections of microfluidic plastic devices during injection molding. In: *ANTEC-Conference Proceedings*.
48 Fung, C.P. (2003). Manufacturing process optimization for wear property of fiber-reinforced polybutylene terephthalate composites with grey relational analysis. *Wear* 254 (3–4): 298–306.
49 Chang, S.H., Hwang, J.R., and Dong, J.L. (2000). Optimization of the injection molding process of short glass fiber reinforced polycarbonate composites using grey relational analysis. *Journal of Materials Processing Technology* 97 (1–3): 186–193.
50 Liao, S.J. and Hsieh, W.H. (2004). Shrinkage and warpage prediction of injection-molded thin-wall parts using artificial neural networks. *Polymer Engineering and Science* 44 (11): 2029–2040.
51 Kuo, C.F.J. and Su, T.L. (2007). Multiple quality characteristics optimization of precision injection molding for LCD light guide plates. *Polymer-Plastics Technology and Engineering* 46 (5): 495–505.
52 Kuo, C.F.J. and Su, T.L. (2007). Optimisation of precision injection moulding processing parameters and implementation of quality prediction system for LCD light guide plate. *Polymers and Polymer Composites* 15 (1): 17–28.
53 Katoh, T., Tokuno, R., Zhang, Y.P. et al. (2008). Micro injection molding for mass production using LIGA mold inserts. *Microsystem Technologies* 14 (9–11): 1507–1514.
54 Heckele, M. and Durand, A. (2001). Microstructured through-holes in plastic films by hot embossing. In: *EUSPEN: European Society for Precision Engineering and Nanotechnology. International Conference*, 196–198.
55 Zhao, J., Mayes, R.H., Chen, G. et al. (2003). Effects of process parameters on the micro molding process. *Polymer Engineering and Science* 43 (9): 1542–1554.
56 Attia, U.M. and Alcock, J.R. (2009). An evaluation of process-parameter and part-geometry effects on the quality of filling in micro-injection moulding. *Microsystem Technologies* 15 (12): 1861–1872.
57 Huang, M.S., Li, C.J., Yu, J.C. et al. (2009). Robust parameter design of micro-injection molded gears using a LIGA-like fabricated mold insert. *Journal of Materials Processing Technology* 209 (15–16): 5690–5701.
58 Kim, S.W. and Turng, L.S. (2006). Three-dimensional numerical simulation of injection molding filling of optical lens and multiscale geometry using finite element method. *Polymer Engineering and Science* 46 (9): 1263–1274.
59 Cardozo, D. (2008). Three models of the 3D filling simulation for injection molding: a brief review. *Journal of Reinforced Plastics and Composites* 27 (18): 1963–1974.
60 Wu, T.L., Ou, K.L, Cheng, H.C. et al. (2008). Analysis for biodegradable polymeric scaffold of tissue engineering on precision injection molding. *International Communications in Heat and Mass Transfer* 35 (9): 1101–1105.
61 Tolinski, M. (2005). Macro challenges in micromolding. *Plastics Engineering* 61 (9): 14–16.
62 Yu, L., Lee, L.J., and Koelling, K.W. (2004). Flow and heat transfer simulation of injection molding with microstructures. *Polymer Engineering and Science* 44 (10): 1866–1876.
63 Choi, S.J. and Kim, S.K. (2011). Multi-scale filling simulation of micro-injection molding process. *Journal of Mechanical Science and Technology* 25 (1): 117–124.
64 Yao, D.G. and Kim, B. (2002). Simulation of the filling process in micro channels for polymeric materials. *Journal of Micromechanics and Microengineering* 12 (5): 604–610.
65 Tofteberg, T. and Andreassen, E. (2008). Simulating injection moulding of microfeatured components. In: *Proceedings of the Polymer Processing Society 24th Annual Meeting*.

66 Shen, Y.K., Chang, C.Y., Shen, Y.S. et al. (2008). Analysis for microstructure of microlens arrays on micro-injection molding by numerical simulation. *International Communications in Heat and Mass Transfer* 35 (6): 723–727.

67 Kim, S.M., Park, S.H., and Lee, W.I. (2008). Multi-scale simulation of molding process to achieve micro-features on the large surfaces. In: *Proceedings of KSPE Fall Meetings.*

68 Lin, H.Y. and Young, W.B. (2009). Analysis of the filling capability to the microstructures in micro-injection molding. *Applied Mathematical Modelling* 33 (9): 3746–3755.

69 Cao, W., Kong, L., Li, Q. et al. (2011). Model and simulation for melt flow in micro-injection molding based on the PTT model. *Modelling and Simulation in Materials Science and Engineering* 19 (8): 003–008.

6

Manufacturing Techniques of Bulk Metallic Glasses

Mustafa Bakkal[1], *Umut Karagüzel*[2], *and Ali T. Kuzu*[3]

[1] *Faculty of Mechanical Engineering, Istanbul Technical University, Istanbul, Turkey*
[2] *Mechanical Engineering Department, Isik University, Istanbul, Turkey*
[3] *Yeditepe University, Istanbul, Turkey*

6.1 Introduction

Glass is any material that can be cooled from a liquid to a solid without crystallizing. Most metals do crystallize as they cool, arranging their atoms into a highly regular spatial pattern called a lattice. But if crystallization does not occur, and the atoms settle into a nearly random arrangement, the final form is a metallic glass. In recent years, production of new metallic glasses in bulk form is getting popular because of their superior strength, elasticity, and magnetic properties [1]. Metallic glasses were discovered over 40 years ago when rapid quenching methods were first applied to Au–Si system. This discovery is carried out the first worked on the technique known as "splat cooling" that achieves cooling rates in the range 10^6–10^8 °C/s as liquid alloy is solidified. In this technique, liquid droplets of the alloy are shot from a carbon dioxide gun onto a cooled Cu or Ag block to produce solid "splats." This amorphous structure was metastable and could be converted into a metastable crystalline phase by annealing [2, 3]. Suppressing the crystallization, one of the crucial step to get low critical cooling rate for bulk metallic glasses (BMGs) has enabled with the discovery of new compositions. The glass forming ability (GFA), combination of large differences in atomic size of the constituent elements, and negative heat of mixing hinder local atomic rearrangements in the undercooled liquid state are other important thermodynamic and kinetic criteria to get BMGs. During the past several years, advances have been made in this field as a result of the discovery and development of several families of alloys with substantially improved glass-forming ability. Whereas previous metallic glasses were generally formed by cooling the melt at rates of 10^5–10^6 K/s, the more recently developed alloys require cooling rates of only 1–100 K/s or less. As such, the new materials can be cast from the molten state into glassy objects with dimensions up to several centimeters as compared with maximum thickness or diameter of 10–100 μm for rapidly quenched ribbons and powders. These new alloys are referred to as "bulk glass formers" or "bulk metallic glasses."

The work of the David Turnbull and his group in the early 1960s was another critical contribution to the subject [4]. This work illustrated the similarities between metallic glasses, ceramic glasses, and silicates. Specifically, Turnbull, Chen, and other later collaborators clearly

Modern Manufacturing Processes, First Edition. Edited by Muammer Koç and Tuğrul Özel.

demonstrated the existence of a glass transition in rapidly quenched Au–Si glasses as well as the other Pd–Si and Pd–Cu–Si glass forming alloys synthesized initially by the Duwez group [5].

The field of metallic glasses gained momentum in the early 1970s when Allied Chemical Corporation developed continuous casting processes for commercial manufacturing of metallic glass ribbon and sheets [6]. During the same period, Chen and collaborators used simple suction casting method to form millimeter diameter rods of ternary Pd–Cu–Si alloys at significantly low cooling rates in the range of 10^3 K/s [7]. The synthesis of metallic glasses in a bulk form (with thickness >1 mm) was successfully achieved in Ni40Pd40P20 alloys in 1976 [8]. During the late 1980s, Inoue et al. in Sendai, Japan, found exceptional GFA in the rare-earth rich alloys [9], e.g. La–Al–Ni and La–Al–Cu. By casting the alloys into copper molds, they fabricated fully glassy rods and bars with cast thickness of several mm. From there, they studied similar quaternary and quinary materials (e.g. La–Al–Cu–Ni) and developed alloys that formed glass at cooling rates under 100 K/s with critical casting thickness ranging upward toward 1 cm [10]. A similar family of alloys, with the rare-earth metal partially replaced by the alkali earth metal Mg were also developed along with the parallel family of multicomponent Zr-based alloy (e.g. Zr–Cu–Ni–Al) [11]. The work opened the door for the development of the other broad classes of BMGs. Building on the Inoue work, Peker and Johnson [12] developed a family of ternary, quinary, and higher-order glass formers based on higher-order alloys of Zr, Ti, Cu, Ni, Be (also combined with other transition metals [TM]). One extensively studied example, referred to as Vitreloy 1, has a composition of $(Zr_3Ti)_{0.55}$ $(Cu_5Ni_4)_{0.225}{\cdot}Be_{0.225}$. Critical cooling rate for glass formation are 1 K/s. The alloys were cast in the form of fully glassy rod with diameters ranging up to 5–10 cm. The alloys require no fluxing or special processing treatments and form bulk glass by conventional metallurgical casting methods [5].

Figure 6.1 shows the Zr-based BMG button and ingot prepared by arc melting and drop casting in inert gas atmosphere in Oak Ridge National Laboratory (ORNL).

Since 1988, bulk glass formation has been reported in various alloys based on Cu, Ti, Fe, Nd, and Pr. But these alloys generally have small supercooled regions and their critical sizes for glass formation are generally smaller than that (30 mm) of Zr-based alloys. However, with attention back to Pd-based alloys recently, large glass formation by water quenching using B_2O_3 flux has been reported in $Pd_{40}Cu_{30}Ni_{10}P_{20}$ alloy with diameter up to 72 mm [14], which is the

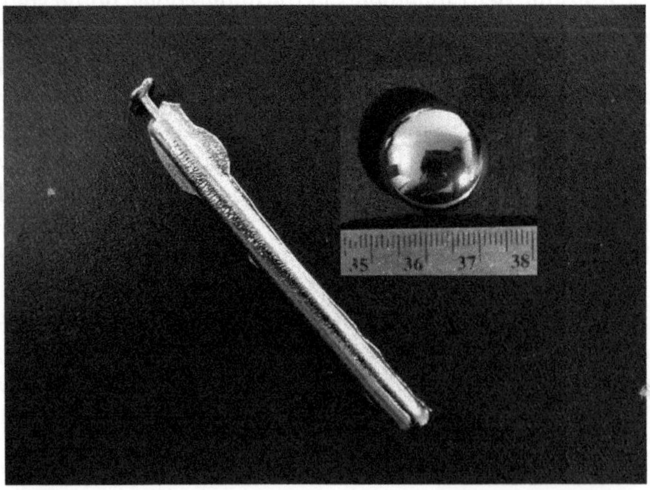

Figure 6.1 Zr-based BMG button and ingot prepared by arc melting and drop casting in inert gas atmosphere. *Source:* From Bakkal 2004 [13].

Table 6.1 Critical cooling rates and section thicknesses for glass formation of bulk glass forming alloys in chronological order.

Year	Alloy content	Thickness (mm)	Critical cooling rate (K/s)	Production method
1969	Pd–Cu–Si	1	$>10^6$	Splat quenching
1974	$(Pd_{1-x}M_x)_{0.835}Si_{0.165}$	1–3	$<10^3$	Splat quenching
1974	$(Pd_{1-x}T_x)_{1-xP}P_{xP}$	1–3	$<10^3$	Splat quenching
1974	$(Pt_{1-x}Ni_x)_{1-xP}P_{xP}$	1–3	$<10^3$	1974
1982	$Pd_{40}Ni_{40}P_{20}$	5–6	~1	Casting with B_2O_3 flux
1984	$Pd_{40}Ni_{40}P_{20}$	10		Casting with B_2O_3 flux
1993	$Zr_{65}Al_{7.5}Ni_{10}Cu_{17.5}$	<16	1.5	Water quenching
1993	$Zr_{41.2}Ti_{13.8}Cu_{12.5}Ni_{10}Be_{22.5}$	14–50	<10	Suction casting
1996	$Pd_{40}Cu_{30}Ni_{10}P_{20}$	40	1.57	Casting with B_2O_3 flux
1997	$Pd_{40}Cu_{30}Ni_{10}P_{20}$	50–72	0.1	Casting with B_2O_3 flux
1999	$Pd_{40}Ni_{40}P_{20}$	7	100	Water quenching
1999	$Pd_{40}Ni_{32.5}Fe_{7.5}P_{20}$	7	100	Water quenching
1999	$Pd_{40}Ni_{20}Fe_{20}P_{20}$	7	100	Water quenching
2001	$Fe_{67}Co_{9.5}Nd_3Dy_{0.5}B_{20}$	0.5		Powder consolidation
2003	$Y_{36}Sc_{20}Al_{24}Co_{20}$	25		Water quenching
2004	$Pt_{57.5}Cu_{14.7}Ni_{5.3}P_{22.5}$	16		Water quenching
2004	$Cu_{46}Zr_{44}Al_7Y_5$	<4	10	Copper mold casting
2006	$Cu_{40}Zr_{44}Ag_8Al_8$		6	Copper mold casting
2007	$La_{65}Al_{14}(Cu_{5/6}Ag_{1/6})_{11}(Ni_{1/2}Co_{1/2})_{10}$	30		Water Quenching
2008	$Fe_{68.7-x}C_7Si_{3.3}B_{5.5}P_{8.7}Cr_{2.3}Mo_{2.5}Al_2Co_x$	6		Casting

largest size of BMGs reported so far [15]. Table 6.1 summarizes the critical cooling rates section thickness and production methods for glass formation in chronological order of different alloy systems, respectively.

Some of the most common metallic glass sheets, ribbons, and BMGs rod, bar, and plate manufacturing methods are discussed in next sections.

6.2 Mechanical Properties and Usage of Bulk Metallic Glasses

Having no crystalline phase makes BMGs a potential engineering material in a wide variety of applications due to their distinctive mechanical, physical, and chemical properties. For example, as shown in Table 6.2, BMG possesses a high-ultimate tensile stress (1900 MPa), high elastic strain limit (2%), low thermal conductivity (4 W/m K), and high hardness (534 kg/mm^2) compared to other conventional crystalline metals and alloys. These properties indicated that metallic glasses could be better than conventional metals in many applications.

Some of the potential and already commercialized application of BMG is given in Table 6.3.

Table 6.2 Comparison of properties of Zr-based bulk metallic glass (BMG) and other metals.

Material	Elastic modulus (GPa)	Poisson ratio	Ultimate tensile stress (MPa)	Percent of elongation to fracture	Thermal conductivity (W/m K)	Fracture toughness (MPa m$^{1/2}$)
BMG ($Zr_{41.2}Ti_{13.8}Cu_{12.5}Ni_{10}Be_{22.5}$)	*96*	*0.36*	*1900*	*2 (Elastic)*	*4*	*40–55*
Ti–6Al–4V	114	0.34	880	14	6.7	75
Inconel 718	200	0.28	1370	25	11.4	20
Aluminum 6061-T6	69	0.33	310	12	167	29
AISI 4340 steel Q+T	205	0.32	1250	14	45	50
316 stainless steel	200	0.30	550	60	16	80

Source: Bakkal et al. [16]. Reproduced with permission of Elsevier.

Table 6.3 Already commercialized and possible applications of BMG.

Applications	Properties
Golf clubs[a)]	High strength
Soft magnetic materials	Good soft magnetism
Armacor coating[a)] (Liquidmetal)	High wear, temperature, and corrosion resistance
Jewelry (Pd–Pt based alloys)	High strength, high workability
Machinery structural material	High hardness, wear and corrosion resistance, precious metal content, high workability
Cutting materials	High hardness
Knives for ophthalmology[a)]	High hardness, wear and corrosion resistance
Sporting goods materials	High elastic energy
Materials for digital master discs	Wear resistance, no grain structure
Writing appliance materials	High wear resistance
Optical precision materials	High reflection ratio
High magnetorestrictive materials	High frequency permeability
Composite materials	High viscous flowability
Acoustic absorption materials	High acoustic attenuation
Penetrator	Self-sharping property
Medical devices materials	High wear resistance and manufacturability
Precision mechanical elements	High strength, high hardness, wear and corrosion resistance, good surface finishing

a) Already commercialized products.
Source: From Liquidmetal Technologies, 2007 [17].

6.3 Rapid Quenching Methods

The first amorphous metallic alloy, $Au_{75}Si_{25}$, was produced by Duwez et al. [3] by applying a rapid quenching method with a cooling rate of ~10^8 °C/s. Since the first synthesis of Au–Si amorphous alloy system by a rapid solidification technique in 1960 [18], a number of studies

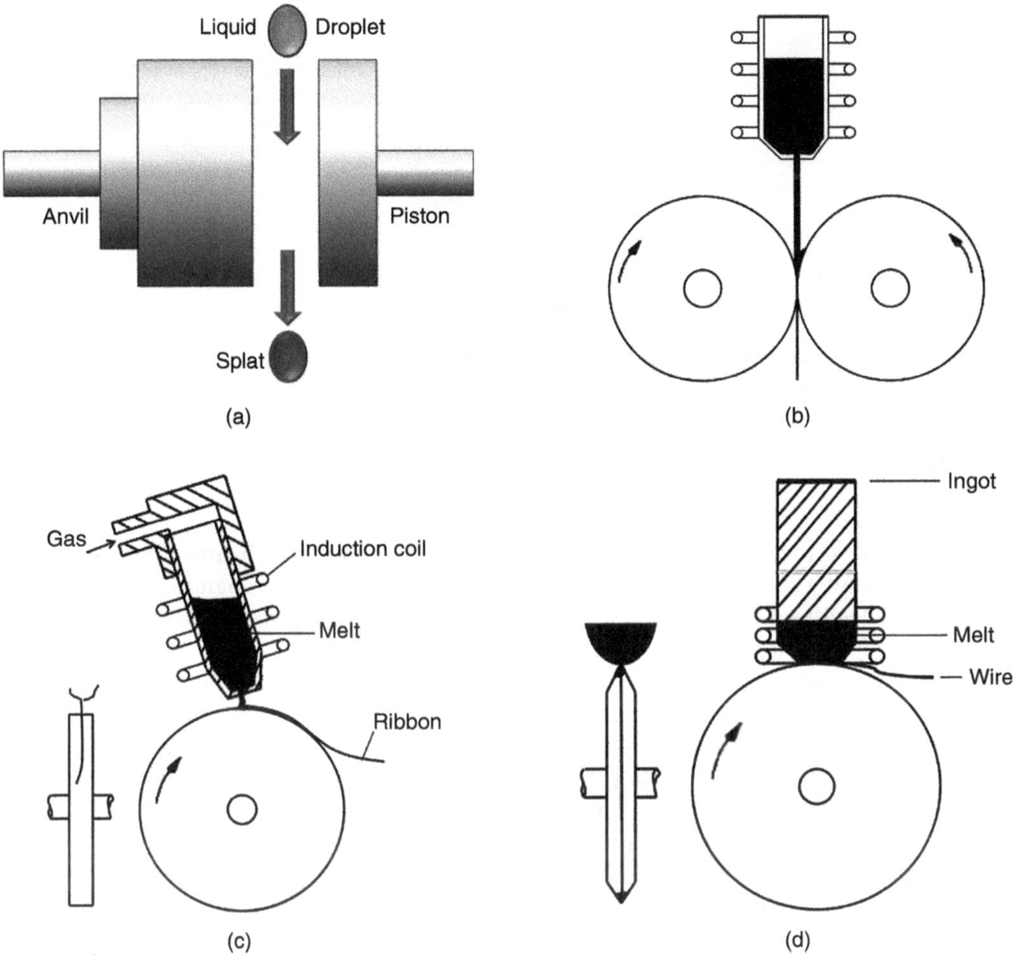

Figure 6.2 Some common rapid quenching methods. (a) Piston and anvil technique. (b) Twin roller quenching technique. (c) Melt spinning technique. (d) Pendant-drop melt extraction technique. *Source:* From Anantharaman 1984 [19].

have been carried out on the formation of the metallic glasses. This method was discovered by them and known as splat quenching or piston and anvil technique shown in Figure 6.2a.

To understand the mechanical properties of metallic glass, the first amorphous metallic alloys were not proper on account of shape and size. The following years, new rapid quenching methods such as twin roller, melt spinning, and pendant-drop melt extraction techniques were developed and achieved restrictions on dimensions partly. Figure 6.2 shows schematic plots of these four techniques. Especially, melt spinning method enabled mass production of ribbon metallic glass. All these rapid quenching techniques used in production of metallic glass were based on rapid quenching with a cooling rate range of 10^4–10^8 °C/s.

6.4 Water-Quenching Method

Water quenching method is one of the easiest methods to obtain amorphous microstructure for some BMG alloys. Method includes a sequence of heating and quenching processes.

Figure 6.3 This Pd-based ($Pd_{40}Cu_{30}Ni_{10}P_{20}$) sample one of the largest BMG sample ever recorded with its 72 mm in diameter. *Source:* From Suryanarayana and Inoue 2011 [20].

The specimen is held for predetermined time at certain temperature at heating step, and then soaked into quenching medium very fast for hindering the stable phase change. Average cooling rates achieved with is about 10–100 K/s. This range of cooling rate is enough for some BMG alloys to obtain glassy state.

Most of the early BMG alloys, Pd and Pt based, were fabricated with water quenching method. The precisely calculated and measured multiple alloy content is mixed as an alloy by vacuum arc melter and induction furnaces at the first stage. Then prepared alloy is heated until molten state inside of a quarts tube. The last step is direct quenching of full molten alloy under suitable quenching medium. If prepared alloy has large glass-forming ability, it is high possibility to obtain fully amorphous microstructure with shiny and almost no casting shrinkage on sample surface. A sample of water quenched BMG is shown in Figure 6.3. One of the key advantages of water quenching is low residual stress due to the low cooling rates [20].

6.5 Arc Melting Drop/Suction Casting Method

Another BMG manufacturing method is arc melting method with a special combination of different casting methods. Arc melting and drop casting methods are well established and known processes for BMG alloys. Drop casting is typically used to process materials with diameters larger than 6 mm, while suction casting is used for casting of materials with diameters under 6 mm. The authors of [21] is to combine and optimize these technologies to create an integrated drop and suction-casting, which may be used to quickly and efficiently fabricate BMG alloys in various sizes and geometries. A schematic picture of arc melting method is given in Figure 6.4.

Arc melting method is mostly used for a low critical cooling rate required alloys. Arc between W tip and copper heart melts alloy under extremely pure environment and inert gas atmosphere. After alloy preparation, molten metal is transferred into water-cooled drop/suction casting molds, which sits under a vacuum bell. Arc melting drop/suction casting system consists of water cooled heart system and three types of mold, as shown in Figure 6.5. Mold types include an alloying crucible (a), in which elemental raw materials are combined to form an alloy of a predetermined composition, a two-piece split die (b) for drop casting, and a four-piece split die (c) for suction casting. The choice of mold material is generally copper.

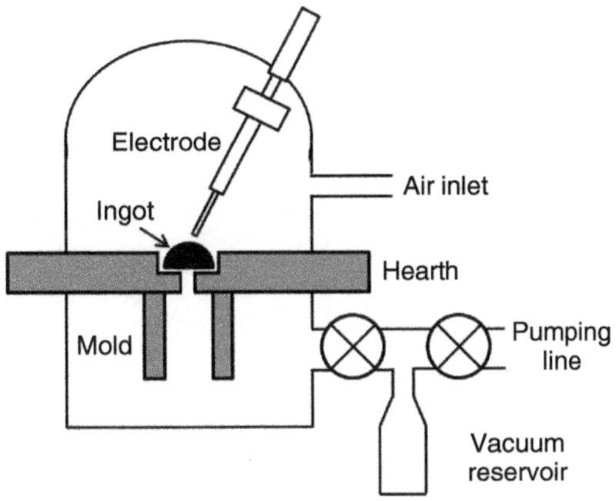

Figure 6.4 Schematic picture of the arc melting drop/suction casting system. *Source:* Gu et al. 2002 [22]. Reproduced with permission of Elsevier.

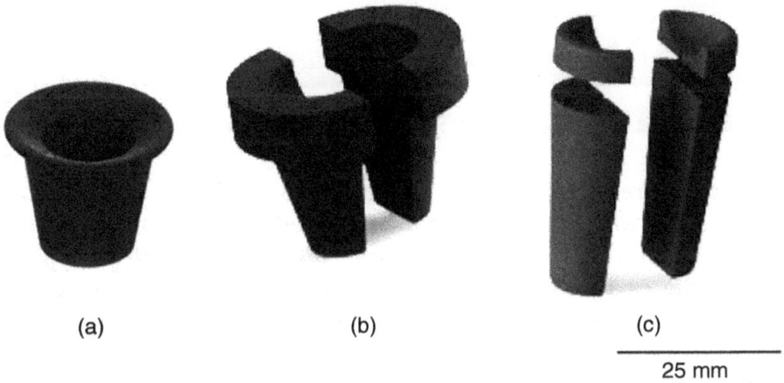

Figure 6.5 A set of whole arc melting and suction casting process tooling (a) the alloying crucible, (b) the two-piece drop-casting split die (c) the four-piece suction casting split die. *Source:* Wall et al. 2006 [21]. Reproduced with permission of AIP publishing.

The copper as the mold material serves dual purposes. The copper has very good electrical conductivity, and it is an ideal anode for the arc melting process. Besides, the copper hearth acts like a heat sink and extracts the heat from the melt facilitating the rapid vitrification of the alloy melts into amorphous alloys [20].

With arc melting and drop/suction casting method, different BMG alloys with complex geometries can be successfully fabricated. However, a serious drawback of this method is that it is very difficult to completely avoid formation of a small amount of the crystalline phase, at least on the surface. This is because of the ease of heterogeneous nucleation due to incomplete melting of the alloy at the bottom side that is in contact with the copper hearth.

6.6 High-Pressure Die Casting Method

Due to its higher cooling and solidification rates and easy to obtain complicated geometries of metal molds, die casting method is one of the common method for production of BMG alloys and BMG parts. The applied pressure in die casting helps higher cooling rate between mold and molten metal. Higher the applied pressure, better the contact and higher the cooling rate, about 1000 K/s. Higher pressure also compensates solidification shrinkage. Selecting

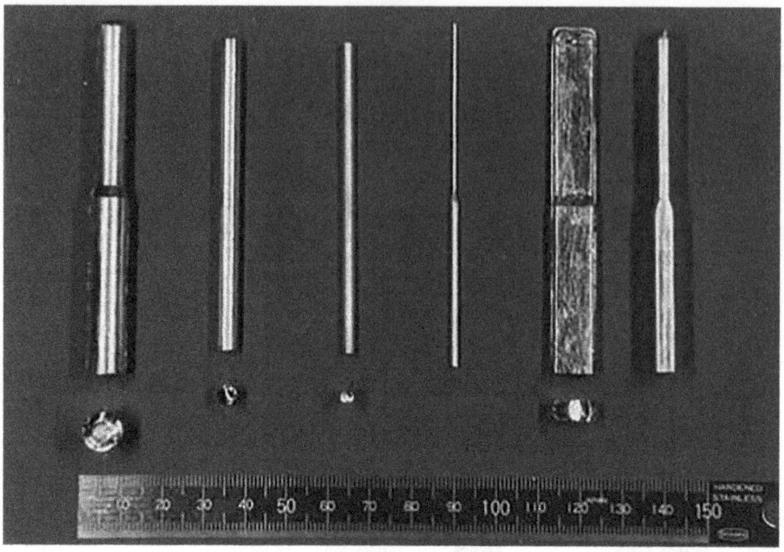

Figure 6.6 Some die-casting product of the $Mg_{65}Cu_{25}Y_{10}$ parts. *Source:* From Inoue et al. 1992 [10].

the copper mold material also ensures very high cooling rates. Process in die casting starts with one of the fast melting method which is induction melting method under protective gas atmosphere to protect molten metal from contaminant gases. Next step is transferring pressurize molten alloy to copper die cavity via plunger system. Solidification process takes a couple milliseconds to complete. One important issue for die casting is trapped gas evacuation system for preventing porosity problem. With this method, large size of Mg-(Figure 6.6), La-, Cu-, and Zr-based BMG parts is produced by Inoue et al. [23].

6.7 Copper Mold Casting Method

Copper mold casting is another common casting technique to produce different BMG alloy parts due to its simple structure. System includes induction melting system, pressurized quartz tube for alloying and melting elements, copper mold for fast heat extraction, inert atmospheric condition for hinder crystallization, and vacuum chamber for clean atmosphere. Formerly, prepared and precisely measured alloy content is placed into a quartz tube and melted several times for being sure about the homogenized mixture by high-frequency induction coils. The prepared molten alloy is transferred to argon-protected copper crucible via pressurized injection gas. Crucible may have different shapes, but most common geometry is foil, cylindrical bar, cone, or wedge shape with predetermined length [20]. A schematic diagram of cooper mold casting technique is shown in Figure 6.7.

6.8 Cap Casting Method

One of the most significant problem about BMGs is very limited part dimensions. So as to increase the maximum diameter of the amorphous alloys, more homogeneous and faster heat extraction is needed. To resolve this problem, Yokoyama et al. [25] have developed the cap-cast method. Using this method, achieved one Zr-based BMG sample that has about

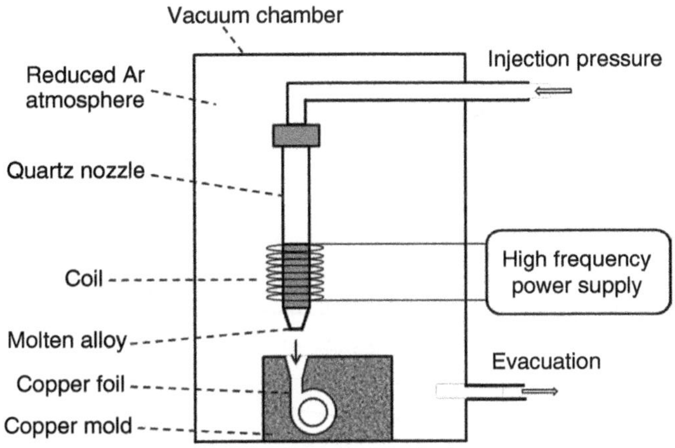

Figure 6.7 Schematic illustration of a copper mold casting equipment. *Source:* Ikarashi et al. 2001 [24]. Reproduced with permission of Elsevier.

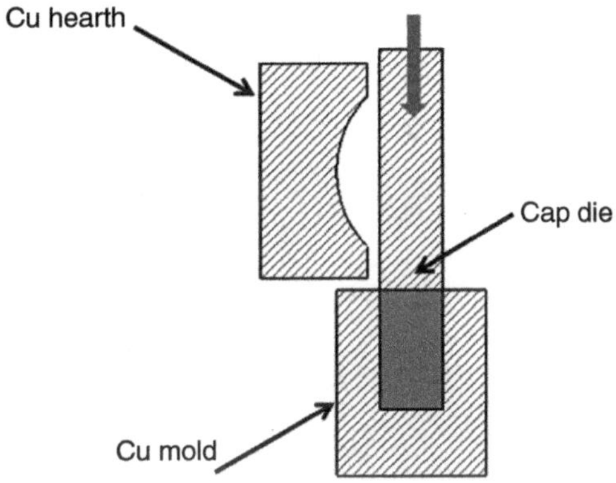

Figure 6.8 A schematic diagram of cap casting method. *Source:* Yokoyama et al. 2007 [25].

30 mm diameter. Method is schematically shown in Figure 6.8. Before starting the cap-casting method, predetermined content is alloyed in induction furnace or arc melting furnace. Prepared alloy is gotten to molten state and poured into a copper mold. Finally, as shown in Figure 6.9, copper top cap is brought to contact with molten metal and forced in a compressive direction. All these close contact of copper surfaces and applied force causes rapid and equal solidification.

6.9 Centrifugal Casting Method

The studied centrifugal casting system consists of two main parts: casting apparatus and injection system of molten alloy. The casting apparatus includes a cylindrical copper mold, which is rotated by a motor with a belt transmission. The transmission allows to changing the speed of rotating mold. Moreover, the injection system consists of quartz container where the alloy ingot is melted by inducing melting. Process of ingot melting is realized by using high-frequency power supply [18].

Figure 6.9 shows a schematic system diagram of the centrifugal casting method for the production of ring shape BMG parts. The process parameters about the solidification BMG alloys depend on many parameters such as duration, metal density, thermal conductivity, cooling rate, and melting temperature. Figure 6.10 shows the two example rings with 25 mm in diameter and 1 mm in thicknesses.

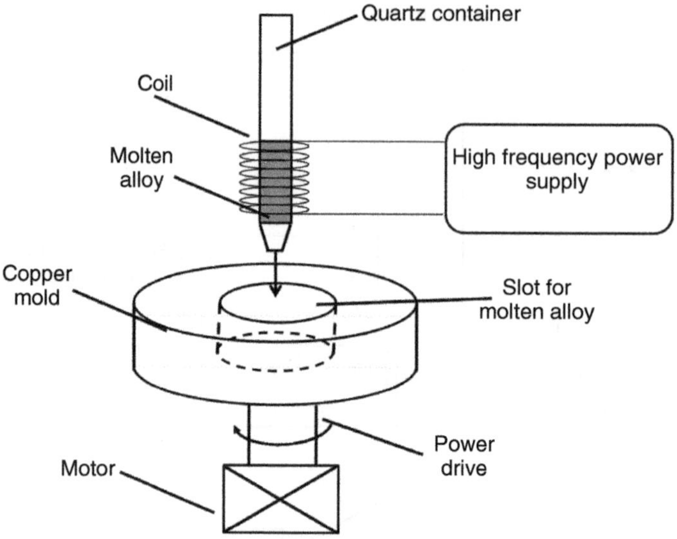

Figure 6.9 Schematic diagram of the centrifugal casting setup with injection system of molten alloy. *Source:* Nowosieski and Babilas 2007 [18].

Figure 6.10 Ring shape BMG samples, $Zr_{55}Al_{10}Ni_5Cu_{30}$ alloy produced by centrifugal casting method. *Source:* Zhang et al. 2002 [26]. Reproduced with permission of Elsevier.

6.10 Metal Foaming Method

Because of high ductility in compression of amorphous metal foams, many researches were focused on to fabricate porous BMG parts [27–29]. Qiu and coworkers [29] developed a process, for Super Plastic Foaming Method to fabricate BMG alloys foams. According to this method, a mixture of NaCl and $CoCl_2$ salts and prealloyed amorphous powder were mixed at specified ratios. About 1% ethanol is added as a binder into mixture for better binding of amorphous alloy and salts. Then mixture is placed into a copper crucible which was sealed and outgassed. After that copper crucible is heated isothermally, and hot pressed at high compressive forces to got a preform. Finally, the mixture is quenched into room temperature and then removed the copper coat. The salt is removed with flowing water, leaving the BMG foam (Figure 6.11).

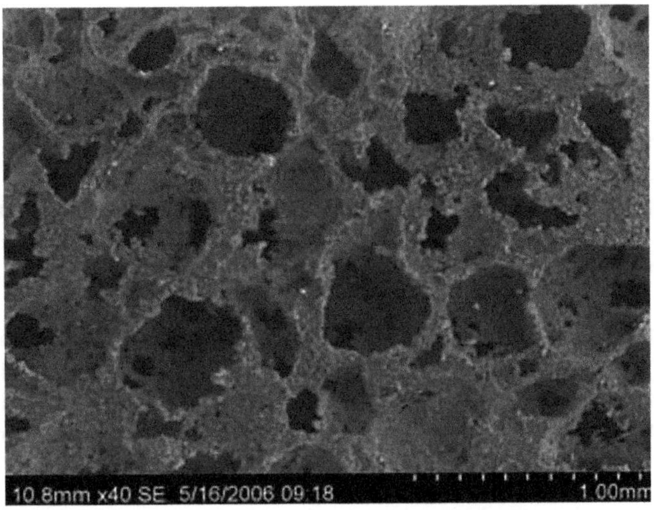

Figure 6.11 The cellular structure of the BMG foam. *Source:* Suo et al. 2008 [29].

6.11 Concluding Remarks

Some of the most common BMG manufacturing technique have been discussed and described above. All these methods allow us to produce sheets, films, ribbons, bar, rod, and plate shapes that are most essential structural geometries of BMG parts. Every day, a new and mostly unconventional technique has been developing by number of researchers to obtain more complex BMG geometries. But still the main problem of maximum size of BMGs has not been solved yet.

References

1 Sneiderman, P. (1998). Metallic glass: material of the future?. Headlines of Hopkins, Johns Hopkins University News Release.
2 Klement, W., Willens, R.H., and Duwez, P. (1960). Non-crystalline structure in solidified gold-silicon alloys. *Nature* 187: 869–870.
3 Duwez, P., Willens, R.H., and Klement, W. (1960). Continuous series of metastable solid solutions in silver-copper alloys. *Journal of Applied Physics* 31: 1136–1137.
4 Chen, H.S. and Turnbull, D. (1968). Evidence of a glass-liquid transition in a Au-Ge-Si Alloy. *Journal of Chemical Physics* 48: 2560–2571.
5 Johnson, W.L. (1999). Bulk glass-forming metallic alloys: science and technology. *MRS Bulletin* 24 (10): 42–56.
6 Kasvesh, S. (1978). *Metallic Glasses*. Metals Park, OH: ASM International.
7 Chen, H.S. (1974). Thermodynamic consideration of formation and stability of metallic glass. *Acta Metallurgica* 22: 1505–1511.
8 Chen, H.S. (1976). Glass temperature formation and stability of Fe, Co, No, Pb and Pt based glasses. *Materials Science and Engineering* 23: 151–154.
9 Inoue, A., Yamaguchi, H., and Zhang, T. (1990). Al–La–Cu amorphous alloys with a wide supercooled region. *Materials Transactions, JIM* 31: 104–109.
10 Inoue, A., Nakamura, T., Nishiyama, N., and Masumoto, T. (1992). Mg–Cu–Y bulk amorphous-alloys with a wide supercooled region. *Materials Transactions, JIM* 33: 937–945.

11 Inoue, A., Zhang, T., and Nishiyama, N. (1993). Preparation of 16mm diameter rod of amorphous $Zr_{65}Al_{7.5}Ni_{10}Cu_{17.5}$ alloy. *Materials Transactions, JIM* 34: 1234–1237.
12 Peker, A. and Johnson, W.L. (1993). A highly processable metallic glass$Zr_{41.5}Ti_{13.8}Cu_{12.5}Ni_{10}Be_{22.5}$. *Applied Physics Letters* 63: 2342–2344.
13 Bakkal, M. (2004). Machinability of bulk metallic glasses. Ph.D thesis. North Carolina State University, USA.
14 Inoue, A., Nishiyama, N., and Matsuda, Y. (1996). Preparation of bulk glassy $Pd_{40}Ni_{10}Cu_{30}P_{20}$ alloy of 40 mm in diameter by water quenching. *Materials Transactions, JIM* 37 (2): 181–184.
15 Lu, Z.P. (2000). Glass forming ability and lass transition study of rare-earth based bulk metallic glasses. Ph.D thesis. National University of Singapore.
16 Bakkal, M., Liu, C.T., Watkins, T.R. et al. (2004). Oxidation and crystallization of Zr-based bulk metallic glass due to machining. *Intermetallic* 12: 195–204.
17 Liquidmetal Technologies. (2007). http://www.liquidmetal.com.
18 Nowosieski, R. and Babilas, R. (2007). Fabrication of bulk metallic glasses by centrifugal casting method. *Journal of Achievements in Materials and Manufacturing Engineering* 20 (1–2): 487–489.
19 Anantharaman, T.R. (1984). *Metallic Glasses: Production, Properties and Applications*. Zurich: Trans Tech SA.
20 Suryanarayana, C. and Inoue, A. (2011). *Bulk Metallic Glasses*, 145–186. Boca Raton, FL: CRC Press https://doi.org/10.1201/9781420085976.
21 Wall, J.J., Fan, C., Liaw, P.K. et al. (2006). A combined drop/suction casting machine for the manufacture of bulk-metallic glass materials. *Review of Scientific Instruments* 77: 033902–033902-4.
22 Gu, X., Xing, L.Q., and Hufnagel, T.C. (2002). Glass-forming ability and crystallization of bulk metallic glass $(Hf_xZr_{1-x})_{52.5}Cu_{17.9}Ni_{14.6}Al_{10}Ti_5$. *Journal of Non-Crystalline Solids* 311: 77–82.
23 Inoue, A., Nakamura, T., Nishiyama, N., and Masumoto, T. (1992). Mg–Cu–Y bulk amorphous alloys with high tensile strength produced by a high-pressure die casting method. *Materials Transactions, JIM* 33: 937–945.
24 Ikarashi, K., Mizushima, T., Makino, A., and Inoue, A. (2001). Preparation of the bulk Fe-Al-Ga-P-C-Si-B glassy alloys in a ringed form by copper mold casting. *Journal of Materials Science and Engineering A* 304–306: 763–766.
25 Yokoyama, Y., Mund, E., Inoue, A., and Schultz, L. (2007). Production of $Zr_{55}Cu_{30}Ni_5Al_{10}$ glassy alloy rod of 30 mm in diameter by a cap-cast technique. *Materials Transactions, JIM* 48: 3190–3192.
26 Zhang, Q.S., Guo, D.Y., Wang, A.M. et al. (2002). Preparation of bulk $Zr_{55}Al_{10}Ni_5Cu_{30}$ metallic glass ring by centrifugal casting method. *Intermetallics* 10: 1197–1201.
27 Qiu, K., Yu, B., and Ren, Y. (2007). Porous bulk metallic glass fabricated by powder hot pressing. *Journal of University of Science and Technology Beijing* 14-I: 59.
28 Brothers, A.H. and Dunand, D.C. (2006). Amorphous metal foams. *Scripta Materialia* 54 (4): 5–13.
29 Suo, Z.Y., Liu, S.W., Zhang, L. et al. (2008). Porous bulk metallic glass fabricated by powder consolidation. *Journal of Minerals & Materials Characterization & Engineering* 7 (2): 97–104.

7

Micromanufacturing

Ömer N. Cora[1] and Muammer Koç[2]

[1] Department of Mechanical Engineering, Karadeniz Technical University, Trabzon, Turkey
[2] College of Science and Engineering, Hamad bin Khalifa University, Doha, Qatar

7.1 Introduction

Micromanufacturing and microforming, sometimes used interchangeably, refer to production of parts, or part features with at least two dimensions of those are in submillimeter range. Part dimension may go up to a few millimeters, whereas features on the part are mostly in micrometer range. Microengineering, on the other hand, is regarded as systematic approach to produce functional parts, surfaces, and machines through multiple micromanufacturing techniques in microscale. It usually requires multidisciplinary team work from concept and design phase to realization of final product. The fundamental characteristics of micromanufacturing are ultraprecise dimensions, ability to produce very complicated three-dimensional (3D) surfaces through utilization of variety of materials. Micromanufacturing is seen as critical to aeronautics, telecommunication, defense, energy, and health-care industries.

The technology that developed the micromanufacturing concept was first regarded as Microstructures or micro systems technology (MST) in Europe. It is also named as micro engineering technologies (MET). A more specific term "micro-electro mechanical systems (MEMS)" was used initially by US companies in semiconductor and microelectronics industry and adopted by rest of the world later on. Recently in Europe, the concept of "micro nano manufacturing technologies (MNMT)" has been introduced as there is no clear boundary between in micro and nanotechnology. Similarly, many MEMS researchers have directed their focus on nano-electro mechanical systems (NEMS) research instead of MEMS in the United States. Meanwhile, concept of microfactory, small desktop-size production systems suitable for manufacturing of small-size parts and products was introduced and first realized in 1980s, by Japanese scientists.

Micromanufacturing can be categorized as silicon-based manufacturing (or MEMS-based) and nonsilicone-material-based manufacturing depending on the material to be dealt with. The technology for MEMS products has been well established and verified while nonsilicone-material-based micromanufacturing is an emerging field [1].

7.1.1 Micromanufacturing Market Projections

Although it is difficult to determine exact values, there are several estimations regarding the market size for MET; however, it is unanimously accepted that market is continuously

Modern Manufacturing Processes, First Edition. Edited by Muammer Koç and Tuğrul Özel.

growing thanks to developments in materials, electronics, and micromanufacturing industries. According to the NEXUS market analysis report published in 2005 the MEMS and MST market size was estimated as $25 billion by 2009. The report was based on the 28 most common first-level packaged products including read-write heads (51% among all), microdisplays, inkjet, pressure sensors, RF MEMS, microfluidics chips, gyroscopes, accelerometers, microphones, flow sensors, MEMS fingerprint sensors, drug delivery systems, IR sensors, optical MEMS for telecom, inclinometer, micro energy sources, microreaction, chip cooler, inclinometers, MEMS memories, MEMS fingerprints, liquid lenses, microspectrometers, wafer probes, micro-mirrors for optical processing, micro-pumps, micromotor, and chemical analysis systems [2].

Another report, published in 2012 by Yole Dévelоppement, forecasted $10 billion market size for MEMS by 2012, and $21 billion by 2017 [3]. MEMS market for consumer and medical products are expected to grow significantly. As a specific application, market size for MEMS for cell phones and tablets is expected to reach $6.4 billion market by 2018 [4]. Meanwhile, pacemakers, which fully consist of microsystem technology components, are $2 billion worldwide business among $50 billion medical devices market [5, 6]. In next 10 years, micro and nanotechnologies are expected to change almost every product on the current market resulting in $2.6 trillion global market [7].

7.1.2 Trends in Micromanufacturing

Microsystem technologies have been the trend since 1950s with the discovery of piezoresistive and semiconductor effects at Bell Laboratories. Since then, microsystems and parts continue to shrink in terms of dimensions along with the increasing demand for functionality, precision, reduced energy consumption, and improved performance. Starting from 1990s, the MEMS-oriented research has boomed, and endless efforts continue to move the current technology forward. Compared to MEMS technologies, which are regarded as mature technologies, microforming/manufacturing is in its crawling stage, and there are still many issues that need to be addressed. Figure 7.1 shows a comparison of scientific research papers published on "MEMS", and "Microforming." As it can be noticed, more than 4000 papers have been published annually on MEMS since 2006, whereas about 35 research papers per year have been published on microforming in the same period. Some of the key points in micromanufacturing that need to be addressed are noted as improvement on quality and efficiency, low-cost/volume production, tool design and fabrication, manufacturing precision, material handling, microassembly, measurement of products and product properties, etc. [1, 8–11].

7.2 Classification of Micromanufacturing Processes

Several categorizations have been suggested for micromanufacturing processes by researchers in literature. The noteworthy ones are summarized below.

7.2.1 Classification Based on Manufacturing Approach (Top-down vs. Bottom-up)

Even though it is mostly used in nanotechnology process categorization and in nanomanufacturing, top-down manufacturing methods refer to the processes that start with a larger piece of material and etching, milling or machining or any other material removal process to get smaller structure. This can be performed via precision engineering and lithography in semiconductor industry. Bottom-up manufacturing processes, on the other hand, involve the

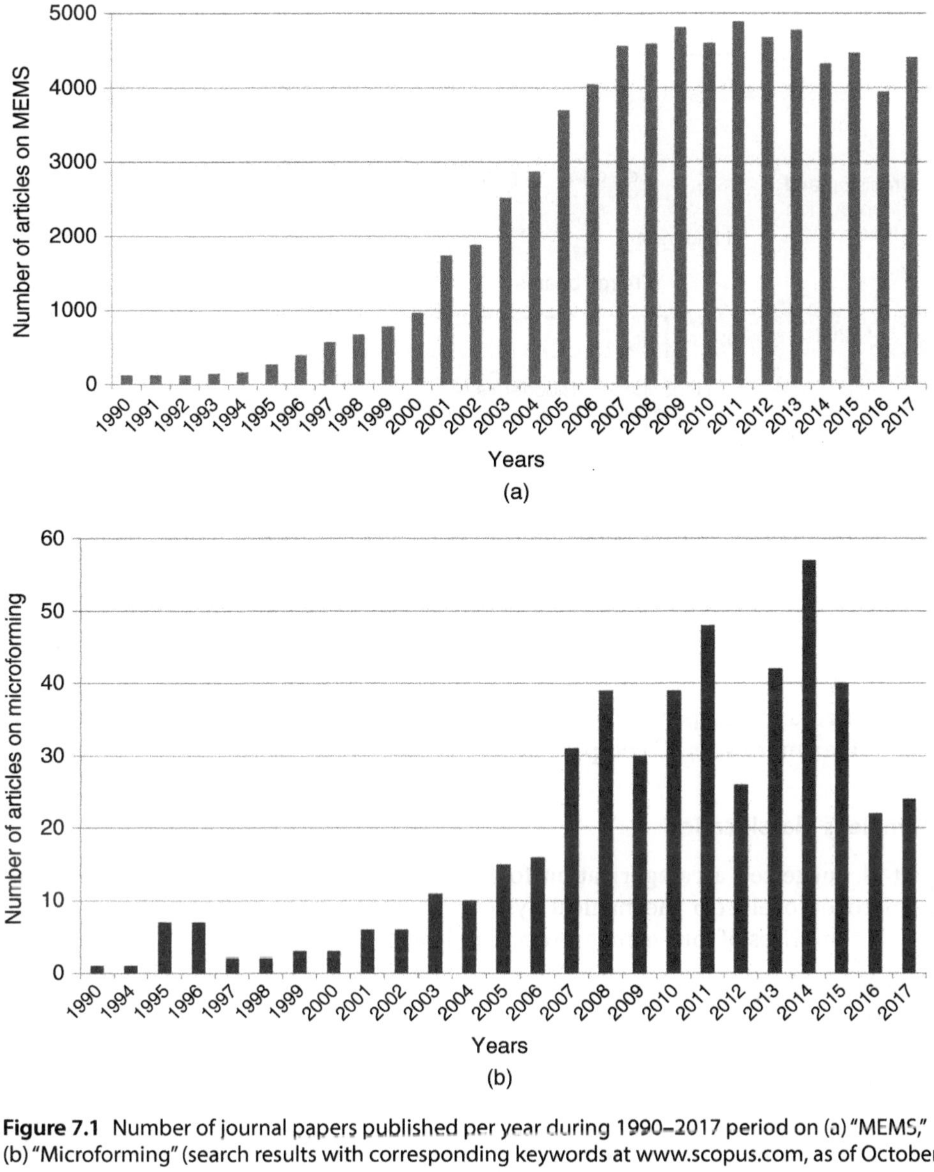

Figure 7.1 Number of journal papers published per year during 1990–2017 period on (a) "MEMS," (b) "Microforming" (search results with corresponding keywords at www.scopus.com, as of October 2018).

building of structures atom-by-atom, or molecule-by-molecule [12]. The recent trend is to synthesize these two different approaches as can be seen in Figure 7.2. This approach is also dubbed as hybrid manufacturing.

7.2.2 Classification Based on Lithography Technique Involvement

Madou suggested a classification, mainly for MEMS processes, miniaturization methods are divided into categories as follows:

(1) Nonlitogpraphy-based (traditional),
(2) Lithography-based (nontraditional) processes [14].

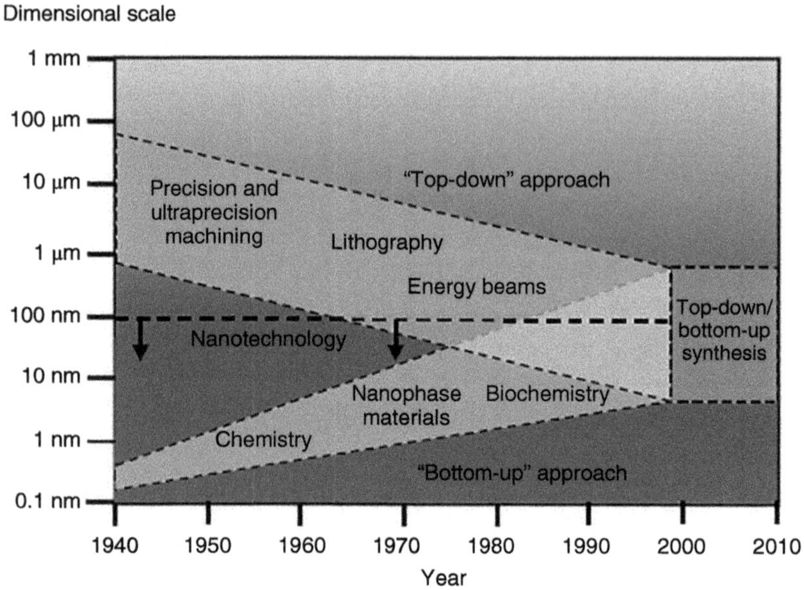

Figure 7.2 Top-down and bottom-up manufacturing trends in nanotechnology. *Source:* From Whatmore 2001 [13].

He also exemplified and compared these methods in terms of used material/application, typical minimum/maximum size, IC compatibility, tolerance, etc. [15].

7.2.3 Brinksmeier's Classification

Brinksmeier et al. suggested a categorization for machining of mechanical parts as well as MEMS and micro opto electro mechanical systems (MOEMS) products as illustrated in Figure 7.3 [16]. It comprises of four main groups as follows:

(1) MEMS processes,
(2) Energy-assisted processes,
(3) Mechanical processes,
(4) Microreplication techniques.

As it can be understood from Figure 7.3 that there is no certain process type that is dedicated to certain technology. Brinksmeier, in addition to abovementioned techniques, included one more step that covers handling, mounting, metrology, quality control, etc. in realization of precision parts and microstructures. Gandarias noted that Brinksmeier's listing is one of the commonly accepted one in micromanufacturing categorization [17].

7.2.4 Classification Based on the Effect of Process

Alternatively, Masuzawa grouped the micromachining processes based on the effect of process and compared their capabilities [18]. Type of effect-based categorization and corresponding manufacturing techniques are given in Table 7.1. This approach was also adopted by Alting et al. later on [19].

Qin used and modified this approach and divided micromanufacturing processes into five categories and categorized more than 40 processes as listed in Table 7.2 [20]. In another

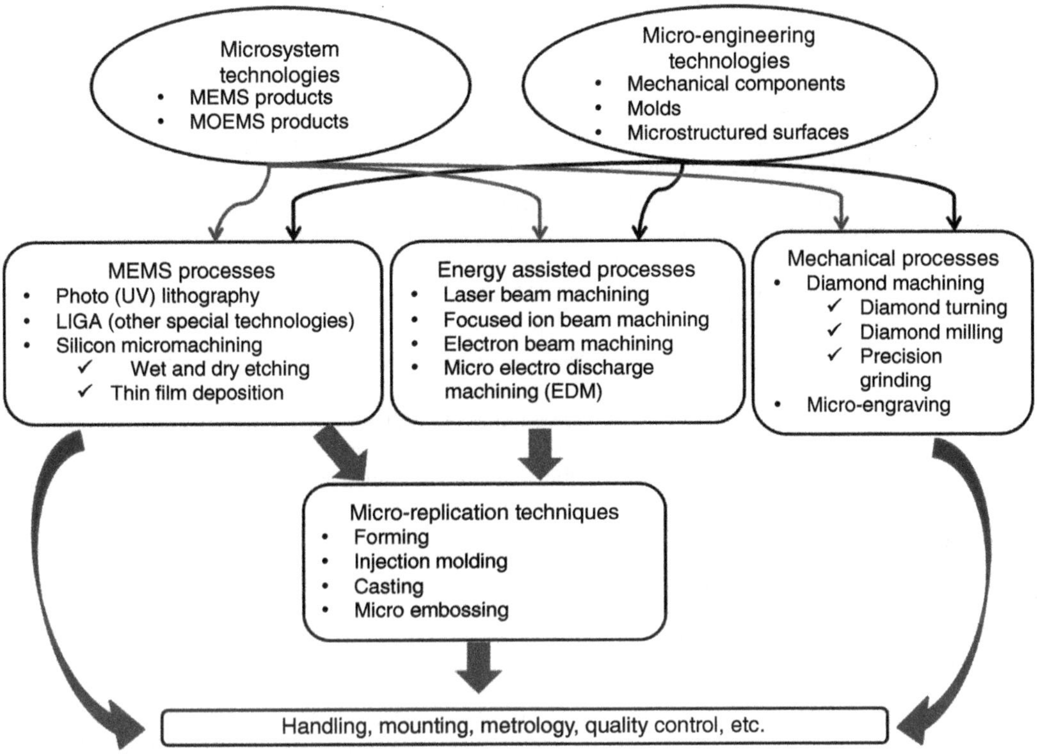

Figure 7.3 Process categorization for microsystem and micro-engineering technologies. *Source:* Brinksmeier et al. 2002 [16]. Reproduced with permission of Springer.

Table 7.1 Micromachining categorization based on the effect of process.

Process effect		Processes
Material removal	Mechanical force	Micro-cutting (drilling, milling, turning, and fly cutting), grinding, sandblasting, punching, ultrasonic machining (USM)
	Melting and vaporization	Micro-electro discharge machining, laser beam machining (LBM), electron beam machining (EBM)
	Ablation	LBM (Excimer, femtosecond lasers)
	Dissolution	Chemical dissolution, electrochemical dissolution (machining)
Plastic deformation		Micro-punching, press
Solidification		Micro-injection (metal, ceramic) molding, die casting
Lamination		Stereolithography
Recomposition		Photoetching, electroplating, LIGA

Source: Masuzawa 2000 [18]. Reproduced with permission of Elsevier.

extensive categorization by Dimov et al. 42 micromanufacturing processes are listed under three main categories [8]. The categorization is based on the combination of Brinksmeier approach and effect of process type classification as given in Table 7.3.

Table 7.2 Micromanufacturing process categorization.

Process type	Processes
Subtractive	Micro-mechanical cutting (milling, turning, grinding, polishing etc.), micro-EDM, micro-electrochemical machining (ECM), LBM, EBM, photo-chemical machining, etc.
Additive	Surface coating (CVD, PVD), direct writing (ink-jet, laser guided), micro-casting, micro-injection molding, sintering, photo-electro-forming, chemical deposition, polymer deposition, stereolithography, etc.
Deforming	Micro-forming (stamping, extrusion, forging, bending, deep drawing, incremental forming, superplastic forming, hydro-forming, etc.), hot-embossing, micro-/nano-imprinting, etc.
Joining	Micro-mechanical assembly, laser-welding, resistance, laser, vacuum soldering, bonding, gluing, etc.
Hybrid	Micro-laser-ECM, LIGA and LIGA combined with laser-machining, micro-EDM and laser assembly, shape deposition and laser machining, Efab, laser-assisted-micro-forming, micro assembly injection molding, combined micro-machining and casting, etc.

Source: From Qin 2010 [20].

Table 7.3 Micromacturing classification.

Process type		Processes
Energy assisted and mechanical	Beam-based	E beam, focused ion beam (FIB), laser ablation, laser hardening, plasma machining, projection mask-less patterning (PMLP)
	Electrical/chemical	Etching, EDM, ECM, electrochemical polishing, electroforming
	Mechanical machining	Abrasive water jet, drilling, milling, grinding, lapping, polishin g, turning/diamond turning
	Prototype/layer based manufacturing	3D printing, 3D lithography, selective laser sintering
Surface modification and structuring	Lithography	E beam lithography, ion beam lithography, laser lithography, nanoimprint lithography (NIL), photo/UV lithography, X-ray lithography
	Coating	Physical vapor deposition (PVD), chemical vapor deposition (CVD), electroplating, spin coating
Replication	Replication	Casting, direct LIGA, hot/UV embossing, injection molding, metal forming, multi-component injection molding, nanoimprinting, screen printing, powder injection molding, reel to reel embossing, self-assembly

Source: From Dimov et al. 2006 [8].

As it can be noted from the classification suggested by Dimov et al., not only effect of the process but also process technology is considered in their categorization. Therefore, this approach is found to be more definitive and will be adopted through rest of the chapter.

7.3 Micromanufacturing Processes

This chapter is intended to provide brief information on some of the micromanufacturing techniques. As the MEMS-based processes are given in a detail elsewhere [14, 15], other forms of micromanufacturing methods are the focus.

7.3.1 Mechanical Processes

7.3.1.1 Micromachining Processes

Methods will be described in this category are mostly down-scaled versions of conventional machining processes in which the forming tool is in contact with the workpiece and remove material by different actions such as turning, drilling, grinding, milling. Micromachining methods are the pioneering processes in micromanufacturing field. They are superior to their corresponding conventional techniques in terms of energy savings, scrap rate, tool speed, tolerances, and surface finish. Moreover, microfactory concept was first realized with micromachining processes in 1990s. Developments in micromachining processes were mainly driven by watchmaker, and jewelry industries as well as optical device manufacturing; however, medical, aeronautics, automotive, and printed circuit industries implemented these technologies later on.

7.3.1.2 Micromilling

Micromilling is a chip removal process in microscale, and it is appropriate technique to produce parts with high-aspect ratios and complex geometries. Therefore, it is preferred for mold-making applications in microinjection [21], blood diagnostic devices [22], microchannels in fuel cells, heat exchangers, and MEMS [23]. The main difference between micromilling and conventional milling is noted to be scale of operation, though they are kinematically same [24].

Micro end mills are mainly produced with grinding process and the quality of micro-cutting tools is reported to be strongly depending on the properties of grinding wheel properties such as wheel size, cutting edge radius, and wear. Substrate material that mainly preferred in micro end mills is ultrafine grain tungsten carbide with cobalt as binder. The grain size is usually smaller than 0.5 μm [25]. The commonly used coatings on the substrates are TiN, TiCN, TiAlN, TiAlCrN, etc. Typical end mill size changes from 100 to 500 μm in diameter with edge radii of 1–10 μm [24]. As an ultimate example, end mills with Ø 5 μm are commercially available [26]. Figure 7.4 shows a micro end mill and a micromachining of a watch component.

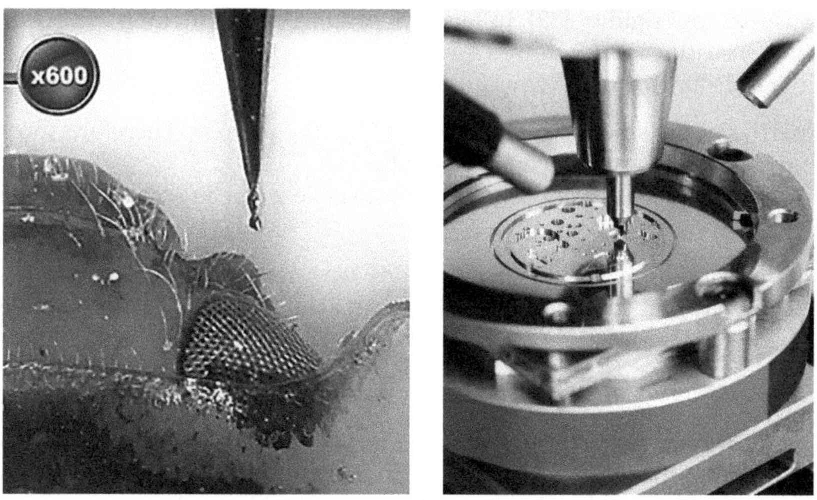

(a) (b)

Figure 7.4 (a) A 10 μm diameter end mill shown next to the eye of a ladybug. *Source:* Courtesy of Performance Micro Tool, Inc., USA. (b) Micromachining of a watch component. *Source:* Courtesy of Willemin-Macodel SA, Switzerland.

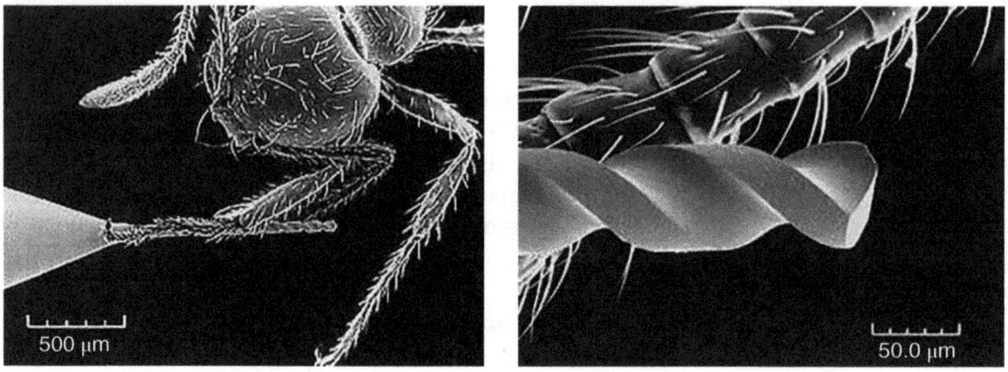

Figure 7.5 (a) Drill with 50 μm diameter shown next to a ant leg and (b) close-up view. *Source:* Courtesy of Performance Micro Tool, Inc., Janesville, WI, USA.

7.3.1.3 Microdrilling

Microdrilling refers to making holes smaller than 3 mm in diameter in practice, and it is used in manufacturing of many parts including aerospace, and electronic parts, mold and dies, diagnostic tools, implants, joint replacements that require deep holes in addition to intricate details [27]. Flute length is critical and limits the depth of the hole to be drilled. Nowadays, depending on their many microdrills are available with flute lengths for hole applications ranging from 10 to 15 times diameters [28]. Figure 7.5 shows an example to microdrills.

Tool material selection is critical as the flute should resist the bending effects during the drilling process. The main materials for drills are solid carbides and high-speed steels (HSS). The material to be drilled is another concern. For example, medical parts are mostly produced from difficult-to machine medical-grade materials including stainless steel and heat-resistant superalloys. To overcome the problems encountered in machining of the metals, microdrills have advanced tip geometries, flute designs, and coatings. With these advanced designs, cutting forces are reduced and tool life is extended. Moreover, chip formation and evacuation is enhanced along with ensured tool rigidity [28]. In terms of coating, ultrathin, long-life TiAlN diamond-like carbon (DLC) coatings are preferred for wear resistant tool surfaces.

Main advantages of microdrilling process are as follows:

(1) The electrical properties of workpiece material do not affect the process. Therefore, most metals, plastics, and composites can be machined easily.
(2) Drilling time can be controlled easily since the process is stable when an appropriate feed rate per rotation is used.

Disadvantages of microdrilling, on the other hand, are reported as straightness problem of the machined product, and insufficient quality in machining very hard or brittle materials [18].

7.3.1.4 Microturning

Also referred as "diamond turning," microturning is suited for generating small parts with 3D convex shapes [29]. In addition, grooves, threads in micron level can be produced with microturning. Micro diamond turning is considered as one of the most suitable machining processes for micro-optical device manufacturing. Figure 7.6 shows a microturning tool and a high aspect ratio shaft produced through microturning.

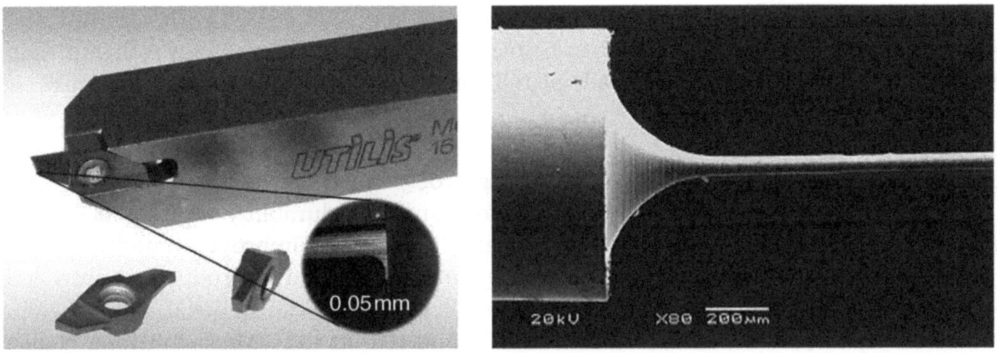

(a) (b)

Figure 7.6 (a) A grooving and microturning tool with insert width of 50 μm. *Source:* Courtesy of Genevieve Swiss Industries Inc., USA. (b) Shaft with aspect ratio of 188 (80 μm in diameter and 15 mm in length) produced by microturning. *Source:* Courtesy of Prof. Mustafizur Rahman, Mikrotools PTE Ltd., Singapore.

7.3.1.5 Microgrinding

Microgrinding, also called as precision/ultraprecision grinding, ductile-regime grinding, ductile-regime finish machining, etc., is an alternative approach to microstructuring that eliminates the problems of micromilling such as surface defects, poor edge finish and comparatively large burr formation, and inability to machine brittle materials. Therefore, instead of cutting action, hard and brittle materials such as ceramics, glasses, hardened steels, and semiconductor materials have to be machined by abrasive action as in microgrinding. The process generates high-surface quality, high-form accuracy, and surface integrity parts that are vitally important in electronics, optic, and astronomical applications (Figure 7.7). Microgrinding is increasingly adopted in fabricating of clock resonators, ferrite parts for electronic data recording devices, ceramic parts for engines and bearings, germanium aspherics, and infrared windows for optics [30–32].

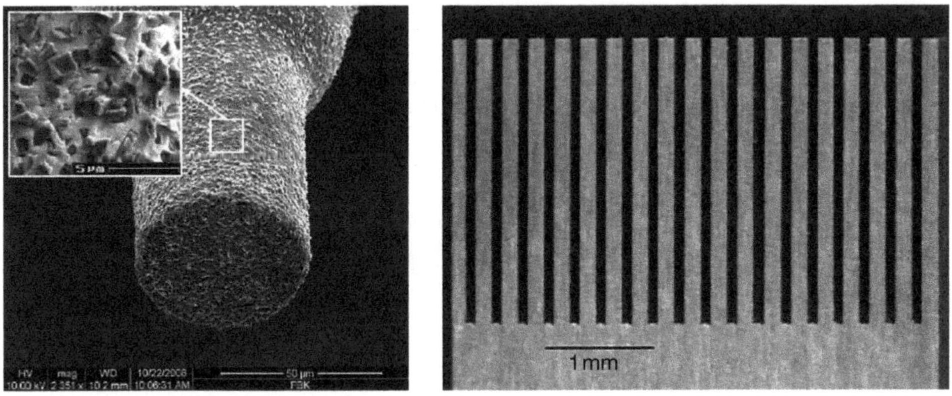

(a) (b)

Figure 7.7 (a) Scanning electron microscope (SEM) image of a grinding tool with 45 μm in diameter and 1–3 μm grit size surface. *Source:* Aurich et al. 2009 [30]. Reproduced with permission of Elsevier. (b) Ground slots with an aspect ratio of 15 on WC material. *Source:* Ramesh et al. 2004 [31]. Reproduced with permission of Elsevier.

From tool manufacturing point of view, microgrinding tools are coated with very fine and hard particles for abrasive action. For example, micropin shown in Figure 7.7a was first ground out of a carbide shank, and then electrically plated with nickel-embedded diamond grains.

7.3.1.6 Micropolishing

Micropolising is another abrasive surface finishing process in which material removal from surface is precisely performed. It enhances the performance and durability of products such as bearings, camshafts, crankshafts, transmission shafts by means of improved roundness, flatness, and reduced lobing. Besides, micropolishing can be benefited in producing controlled cross-hatches, straight-lines, tangential fillet radius blend, ferrite cap removal, and gauging. Typical material removal thickness is between 0.25 and 0.5 μm [33]. Micropolishing is also utilized as final surface finishing operation of high-precision structured molds. It improves the surface roughness, removes feed marks resulted from pre-machining. These issues are highly critical in manufacturing of optical elements made of glass and polymers. A surface roughness of less than 1 nm and form accuracy in the nanometer range can be achieved through polishing of plane spherical surfaces while the 5 nm surface roughness was achieved for structured surfaces shown in Figure 7.8 [34]. Suzuki et al. developed an ultrasonic vibration assisted polishing machine for molds of micro glass lenses as illustrated in Figure 7.9. Surface roughness of 7 nm (R_y) was reported to be achieved by the researchers [35].

(a) (b)

Figure 7.8 Abrasive polishing of V-grooved surfaces made of optical mold steel using (a) pin type (b) wheel type polishing tools made of polyamide PA6. *Source:* Brinksmeier et al. 2006 [34]. Reproduced with permission of Elsevier.

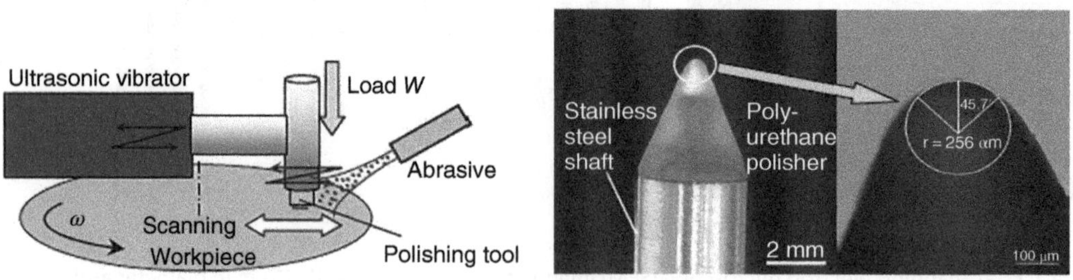

(a) (b)

Figure 7.9 (a) Ultrasonic vibration-assisted polishing method and (b) polyurethane tool with 250 μm. *Source:* Suzuki et al. 2006 [35]. Reproduced with permission of Elsevier.

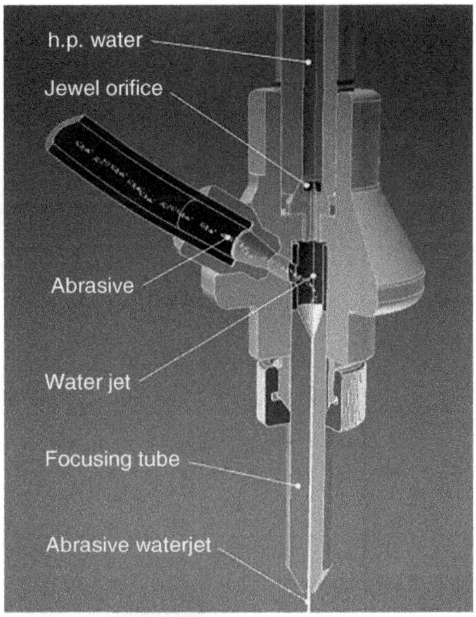

Figure 7.10 Schematic of water jet system. *Source:* Courtesy of Finepart Sweden AB, Sweden.

7.3.1.7 Micro Abrasive Waterjet Machining (μ-AWM)

Micro abrasive water jet machining (μ-AWM) is one of the non-contact, subtractive, and top-down micromachining processes. The cutting mechanism is erosion rather than the process name implies. The working principle of the process is based on abrasive water jet process in which a mixture of water and fine-grade abrasives called as "cutting stream" are impacted on the material to be cut at velocities up to twice the speed of sound. Abrasive particles constitute less than 1% of cutting stream volume, while water droplet 10% or so with the remainder air (Figure 7.10). Cutting accuracy of 10 μm, and accuracy of 3 μm can be achieved. Hard stainless steel with a 1 mm thickness can be cut with this tolerance level remembering that material thickness to be cut depends on material type, and cutting accuracy desired.

μ-AWM process have particular advantages in manufacturing of miniature parts (Figure 7.11) and features that are made of difficult to cut materials and thickness levels that are not economical to cut using other cutting methods. Other benefits of μ-AWM can be listed as follows:

Ability to produce fine details: Although it is not commercially practiced, the cutting stream with abrasive particles at nanometer range could theoretically machine smaller features than any laser can does. Laser cutting diameter is dictated by the wavelength and current laser practices cannot approach to the level that nanoparticles hypothetically can erode. However, particles smaller than 10 μm are not reliably classified that leads to minimum cutting jet diameter of 40 μm in industrial practice [37]. In no-particle water jet machining (pure water cutting stream), on the other hand, cutting diameter of 25 μm was achieved; however, the cutting power is as low as to cut soft materials such as foam, only.

No clamping fixtures are required: Since the mechanical load on the workpiece at a few Newton levels and local cutting forces are in micronewton range, expensive clamping fixtures are no longer needed [38].

No structural change: As the process is water based, the cutting stream provides an extreme cooling effect and lowers the temperature increase at zones particles impacted. Moreover, minimal kerf width results in mechanical stress-free cutting. Hence, the process does not cause neither thermal nor mechanical stress effect that can change the microstructure of

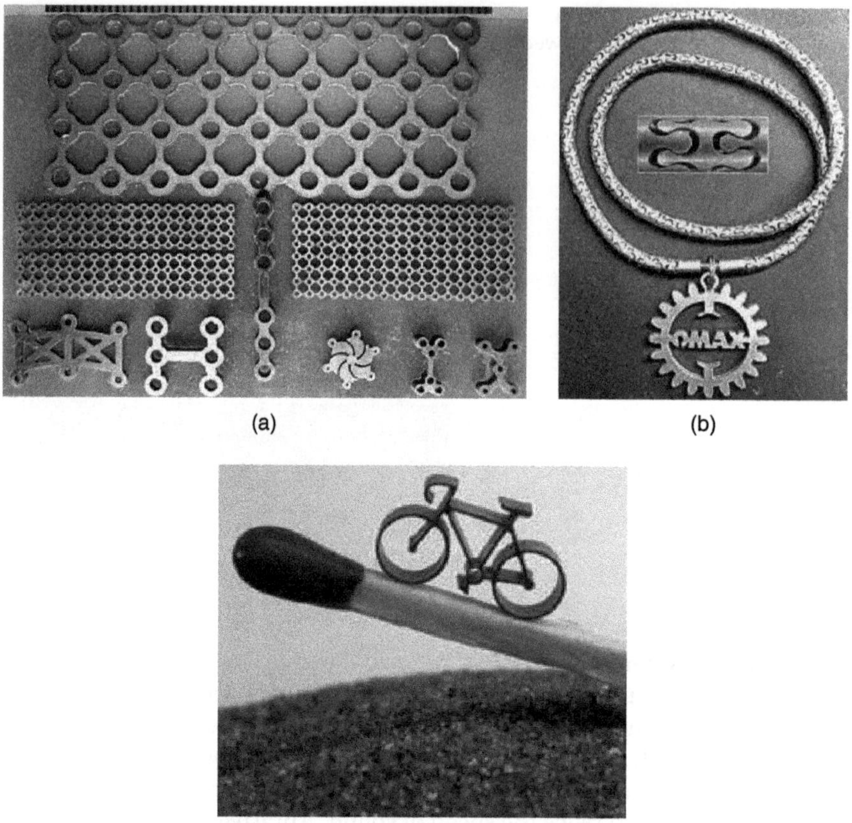

(a) (b) (c)

Figure 7.11 Examples of micro abrasive water jet machined parts: (a) Titanium and stainless steel orthopedic parts and (b) Interlocking link made from a solid titanium tube (outer diameter is 0.64 mm). *Source:* Liu and Schubert 2012 [36]. Licenced under CC BY 3.0. (c) Micro abrasive water jet machined bicycle on a match. *Source:* Courtesy of Finepart Sweden AB.

material. Cut surfaces are free of heat-related effects. These features make μ-AWM process a prominent choice in manufacturing of heat-sensitive and adhesive-backed materials in medical, aircraft, and electronics industries. For example, titanium and nitinol implants remain microstructurally unchanged after the cutting process that is challenging with other machining methods such as laser micromachining. Apart from thermal stress absence, process is also mechanical stress-free with minimal kerf width and does not yield work-hardening effect. Therefore, delicate materials and composite materials that contain carbon and high-tech fibers can successfully be cut with this technique.

Suitable for wide spectrum of materials: In contrast to wire electro discharge machining (WEDM) process that is also capable of producing fine details, the material to be cut does not need to be electrically conductive in μ-AWM. Metals, plastics, and glass, composites can be cut by means of abrasive water jet. Materials that are much harder than the abrasive particles are the only exemptions.

No need for post-processing: In addition to cutting accuracy and free-of-defects at cut surfaces, micro water jet can achieve a surface roughness up to $R_a = 0.8\,\mu m$ so that most parts are produced with minimal burr that eliminated post-processing.

Economical and fast way of micromachining: Especially for small production runs, prototyping purposes, or when production is urgently needed, the μ-AWM is fairly useful as it does not require tooling as conventional micromachining does. Processing speed, on the other hand, is 5–10 times faster than WEDM for similar cut surface roughness values.

7.3.2 Microforming Processes

This category of micromanufacturing processes refers to downscaled version of conventional forming techniques including forging, extrusion, heading, punching, blanking, deep-drawing, stamping, hydroforming, etc.

7.3.2.1 Microforging

Forging is one of the oldest bulk metal forming processes, and its downscaled version enables fabricating microparts with good mechanical properties at low cost and high productivity. Production of heat sink cooling fins and microblades for microturbines with a high aspect ratio [39], lens array molds [40] can be made possible through microforging. As in many microforming processes, size effect is reported for heading of pins made of copper [41]. Figure 7.12 shows variety of microheaded parts in different dimensions while Figure 7.13 shows cooling fins with high aspect ratio that can be realized through MicroForging® process developed by Alpha Comp. Ltd. (Japan).

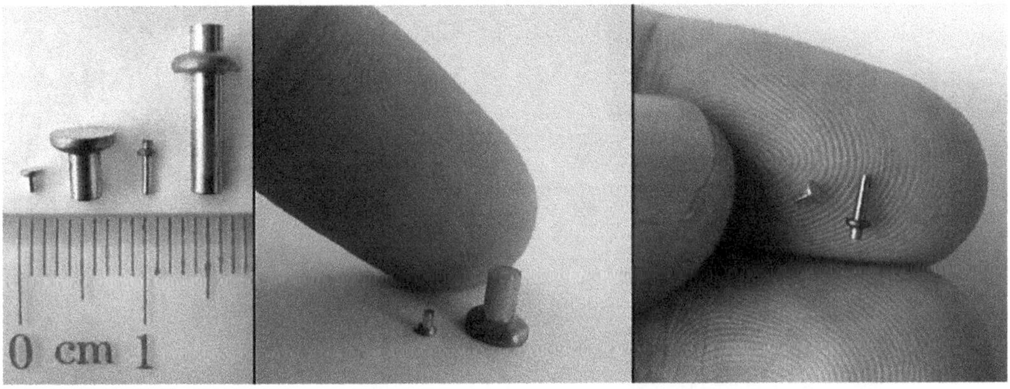

Figure 7.12 Microheaded parts in different scales. *Source:* Chan and Fu 2012 [41]. Reproduced with permission of Elsevier.

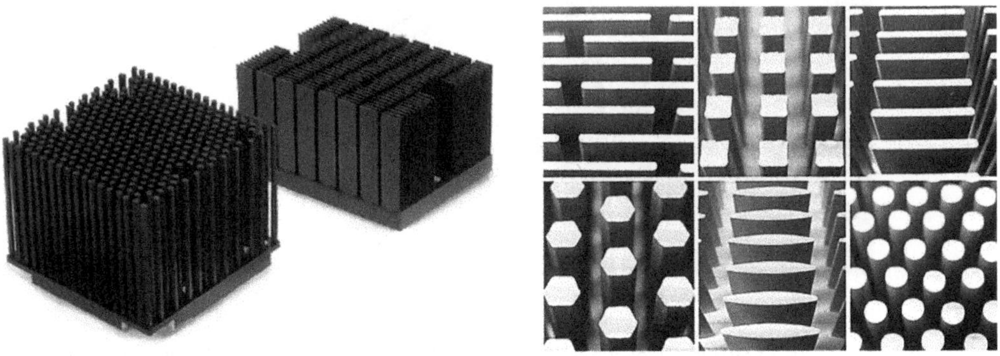

(a) (b)

Figure 7.13 (a) Cooling fins with high aspect ratio for heat sinks by *MicroForging*® process and (b) fins with different cross-sections in detail. *Source:* Courtesy of Alpha Company Ltd., Japan.

7.3.2.2 Micro Extrusion

Micro extrusion is used to produce parts such as micro gears for MEMS, micro extruded tubes that are used in micro-fluidics, minimally invasive surgery as well as in micro-analytics; micro catheters for pediatrics, kink resistant, and flexible tubes for angiography as seen in Figure 7.14. Tubes with 100 μm in diameter and 50 μm in wall thickness can be realized with this technology. High temperature thermoplastics such as polyether ether ketone (PEEK), polyether imide (PEI), polytetrafluoroethylene (PTFE), and fluorinated ethylene propylene (FEP) in addition to commonly used materials polyurethane and polyamide are exploited in microtubing applications [43].

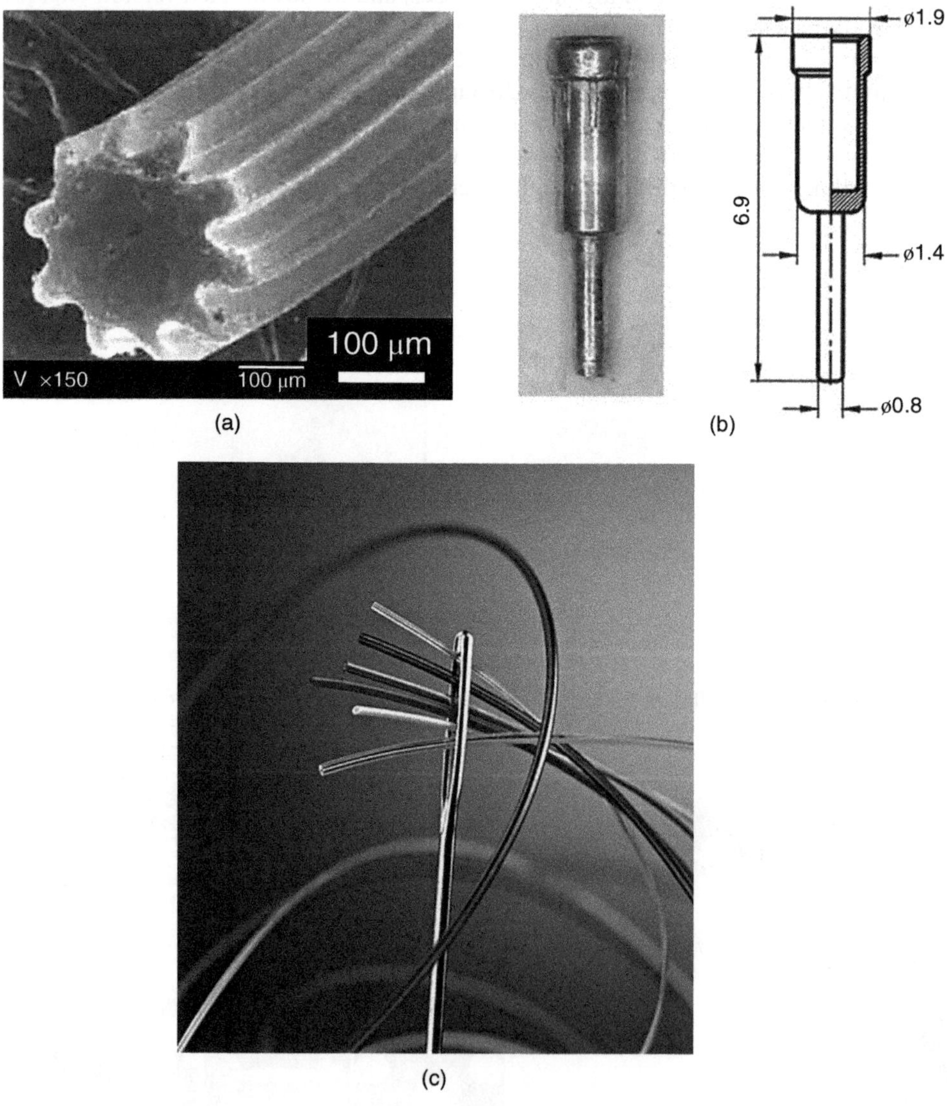

Figure 7.14 (a) SEM picture if extruded gear shaft and (b) copper pin with a shaft diameter of 0.8 mm and a wall thickness 125 μm produced with forward rod and backward extrusion. *Source:* Geiger et al. 2001 [42]. Reproduced with permission of Elsevier. (c) Various micro tubing examples. *Source:* Courtesy of Raumedic AG, Germany.

Figure 7.15 Surface detail of extruded polymer jacket for wire. *Source:* Courtesy of Hoowaki LLC, USA.

Micro extrusion is also used to modify surface structure of materials. Properties including coefficient of friction, liquid wettability, appearance, tactile feel, and heat transfer can be lowered or increased through micro extrusion. Many extruded products, for example polymer-coated wires or medical tubing, require low-friction surfaces for installation and use as friction often damages the surface of products, making them unusable or unsafe. In the case of medical tubing, lowered coefficient of friction results in a more precisely delivered stent, reduced damage to a passage, or faster rotating speed of a plaque removal device. In current practice, such medical devices are coated with expensive coatings, exotic polymers, and lubricants to lower the coefficient of friction. On the other hand, microstructured extruded surfaces made of polypropylene, polyethylene, nylon, and fluoropolymers reduce or eliminate lubricant and coating use (Figure 7.15). It is noted that 20% in dry sliding while 80% reduction in lubricated sliding for friction can be achieved through microstructured surface [44].

7.3.2.3 Progressive Microforming

Progressive microforming refers to series of operations with different actions to produce final part of distinct form compared to initial shape. Fu and Chan demonstrated the fabrication of meso-scaled bulk cylindrical and flanged parts (Figure 7.16a) by using sheet metals through blanking (for cylindrical part), and progressive punching, extrusion, and blanking (for flange region) [45]. In another example, Ghassemali et al. manufactured pins with different diameters (0.3, 0.5, 0.8 mm) using copper sheets (Figure 7.16b) by means of (i) indentation, (ii) upsetting, (iii) extrusion effects [46].

7.3.2.4 Micro Deep-drawing

Manufacturing of small cups, housings, shells for batteries, resistor end caps, shells for valves in antilock braking system (ABS), eyelets, etc., can be realized by micro deep-drawing process. Medical, communications, aerospace, automotive, and electronic industries take advantage of micro deep-drawing process where unique size, tolerance, and surface finishing are required [47, 48]. In academia, therefore, micro deep-drawing is studied extensively by researchers around the world [49–54]. Various kind materials such as copper, as commonly used material in electronic engineering, stainless steel, magnesium or molybdenum alloys, brass, bronze, gold, Inconel, platinum and its alloys, silver, titanium can be drawn (Figure 7.17).

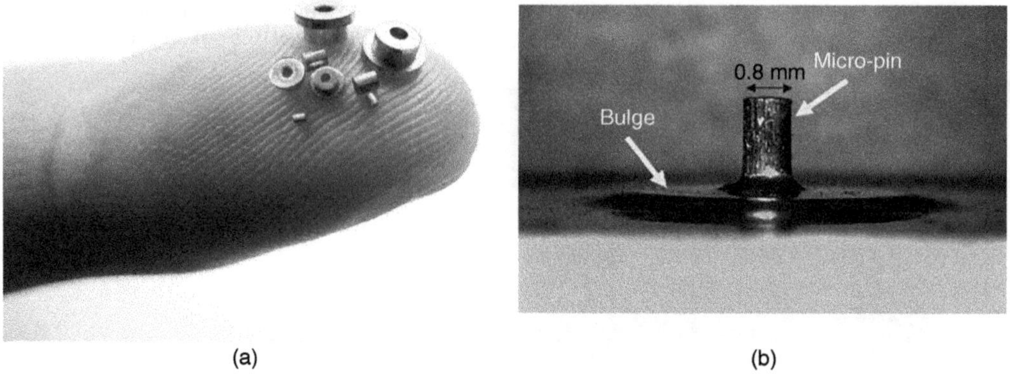

Figure 7.16 Examples of progressive microforming. (a) Various parts made from pure copper sheets. *Source:* Fu and Chan 2013 [45]. Reproduced with permission of Elsevier. (b) Micro pin with 0.8 mm in diameter. *Source:* Ghassemali et al. 2013 [46]. Reproduced with permission of Springer.

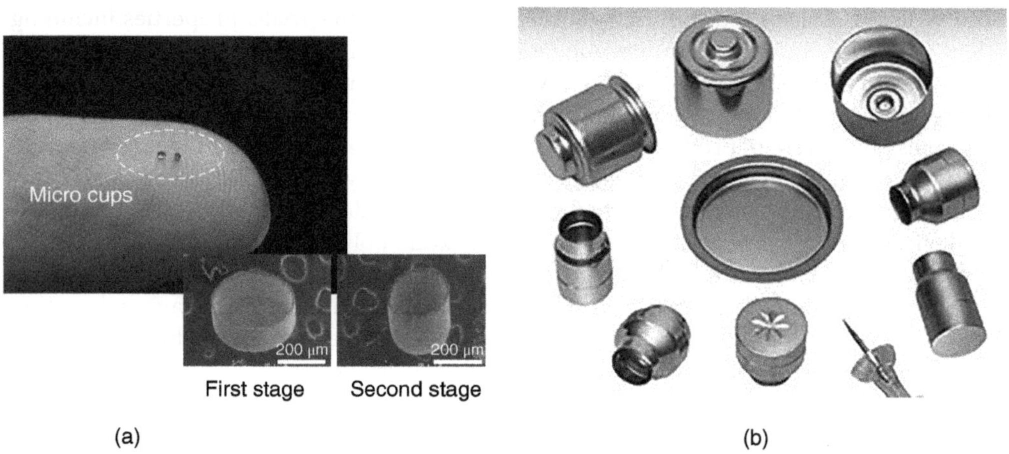

Figure 7.17 (a) Micro deep drawn cup with a initial blank of 20 μm in thickness and 1.1 mm in diameter (first stage: 700 μm, second stage: 500 μm in diameter). *Source:* Shimizu et al. 2012 [49]. Licenced under CC BY 3.0. (b) Various industrial parts produced with micro deep drawing process. *Source:* Courtesy of Braxton Manufacturing Co. Inc., USA.

7.3.2.5 Micro Stamping

The need for microfeatures on such as channels, bumps, and fins on large surfaces have been increasing particularly in the field of electronics cooling, energy generation/storage, medical devices, and MEMS industry. Fabrication of such microfeatures is usually desired in confined and ever-decreasing volumes due to miniaturization trend in these industries. Stamping, among the other sheet metal forming processes, is well-known for its high productivity rates, near-net shapes, excellent mechanical properties, and close tolerances etc. Therefore, it is well-suited to part production with high volume at low cost. In micro stamping, a sheet blank is formed against the female die cavity using a male die under the effect of applied load. Figure 7.18 shows the schematic of micro stamping approach used to produce metallic bipolar plates with channels from 51 μm thick SS 316L stainless steel blanks. Blanks were formed into 250 μm deep

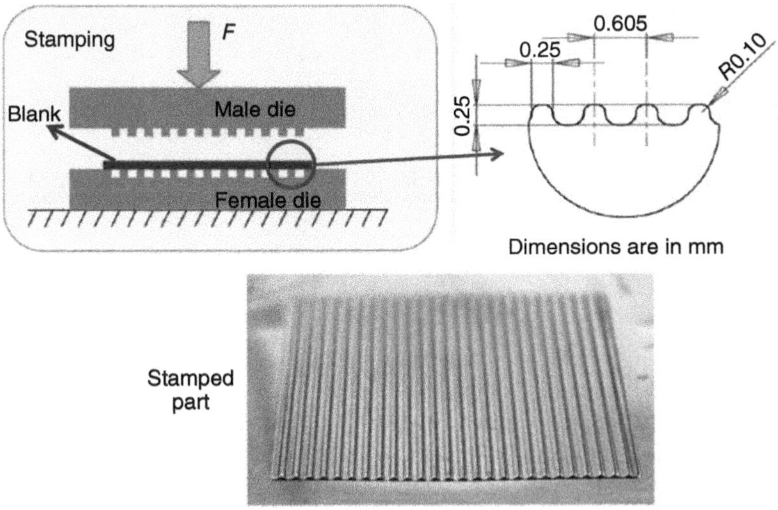

Figure 7.18 Schematic description of micro stamping process, die dimensions, and stamped part with channels (channeled surface area is 40 mm × 40 mm) [55, 56].

channels; however, the stamped part height might be lower depending on the load level applied [55, 56]. A similar process, forming pure copper sheets into channeled surfaces, is referred as micro embossing by Chan and Fu [57].

7.3.2.6 Micro Hydroforming

Micro hydroforming, similar to micro stamping, is another replication process in which sheet blank or tube is formed inside a die. Contrary to stamping, male die is not used in hydroforming and forming effect is provided with pressurized fluid and both sheet blanks and tubes can be formed. Some of the advantages of hydroforming processes are reduced tooling cost (due to fewer parts) and setup time, elimination of direct contact (punch-workpiece) related problems and wear, and hence excellent surface quality, precise control of material thinning and uniform sheet thickness distribution, less scrap, and ability to produce complex-shaped parts in one operation. Figure 7.19 illustrates the schematics of sheet and tube hydroforming approaches, while Figure 7.20 shows some micro hydroformed tube parts.

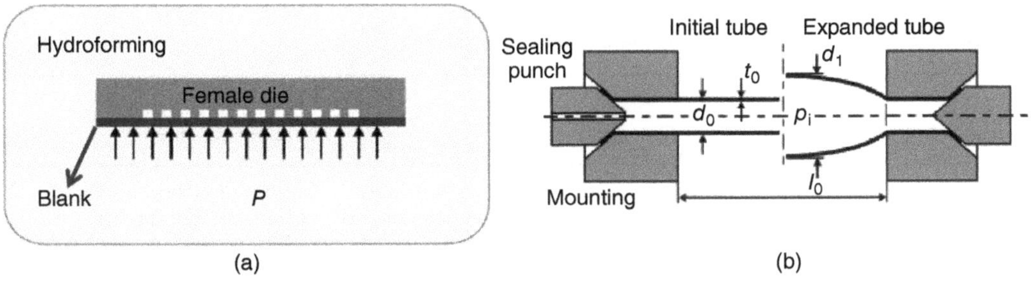

Figure 7.19 Schematics of (a) sheet hydroforming. *Source:* Mahabunphachai et al. 2010 [55]. Reproduced with permission of Elsevier. (b) Tube hydroforming. *Source:* Hartl et al. 2011 [58]. Reproduced with permission of AIP Publishing.

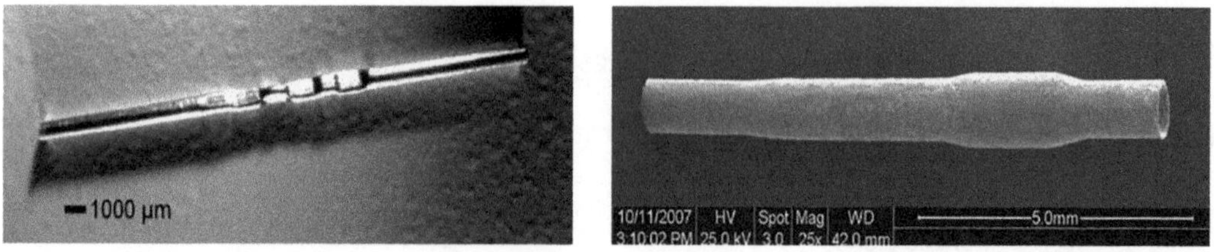

Figure 7.20 Examples of tube microhydroforming [58, 59].

7.3.2.7 Micro Injection Molding

Micro injection molding (or moulding), or powder injection molding (PIM), refers to medium or high volume production of metal, ceramic, or plastic complex parts or surface features/ structures measured in microns and with volumes usually less than 10 mm^3. The smallest part achieved by company Accu-Mold has dimensions of 800 μm × 300 μm × 380 μm [60]. The weight of the micromolded part, on the other hand, can be as low as 0.001 g [61]. Besides, micromolded parts have complex geometries and require very small tolerances. Therefore, micromolding machines with high injection speed and pressure, precise shot control, uniform melt temperature and fine resolution servo-electric drives, and sophisticated control techniques are required. MIM parts can be found in micro-electronics, cell phones, antennas, connectors, surgical tools, hearing aids, drug delivery, glucose monitoring, transdermal wearable drug patches, etc. Some specific applications are listed in Table 7.4.

Depending on the main powder material, process is named metal injection molding (μ-MIM), ceramic injection molding (μ-CIM), or plastic/polymer injection molding. Schematic of micro metal injection molding (μ-MIM) process is illustrated in Figure 7.21 along with an example part. Before injection molding, metal powders are blended with polymer, waxes, and further additives to get a feedstock. Different from conventional injection molding, powders with smaller particle size have to be used in μ-MIM process. In addition, the applied binder systems need to be more flowable for injection and very stiff for green part ejection to guarantee a precise molding of the micro geometry. The injection molding process resembles to the polymer process. Nevertheless, the ejected part called green compact still contain binder that needs to be removed. The part after debinding process is dubbed as brown compact. The last step is known as sintering in which brown compact is densified, and metallurgical bonds are obtained between powders. Depending on the binding content of the feedstock %10–15 shrinkage can be experienced when comparing the green compact and finished part [65].

Table 7.4 Micro molding applications in different industries [60–64].

Medical industry	Electronics, IT	Military, aerospace, automotive, other
Staples, needles, sharps, anchors, forceps, blade holders, cannulas/ catheters, bone screws, heart valve ring, dental tooth implants, abutments, orthodontic brackets, medical endoscopic devices, nozzles, catheter tips, hearing aid sensor housing	Fiber optic connectors/housings, micro optic lenses, semiconductor wire bonding capillaries, guides, gearwheels, parts with screw threads, receptical guides, grippers, dot matrix printer heads, sensor housings	Rotor pinion, lead frame, cogwheels, micro gears, nozzles, stepper motor axle for watches, gaskets, seals, rollers, micro switches, connectors, small screws

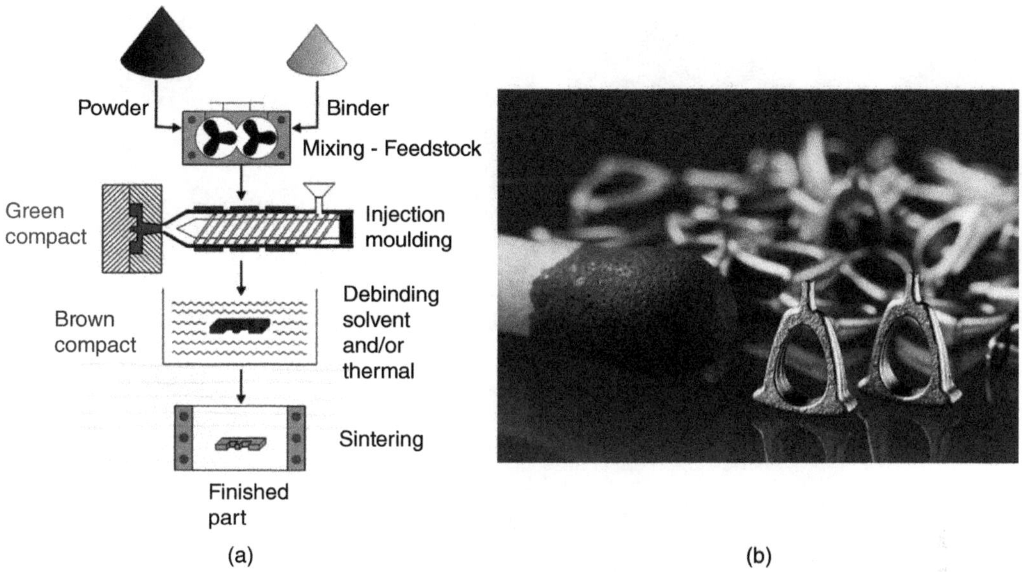

Figure 7.21 (a) Schematic of μ-MIM process. *Source:* From Haack and Imgrund 2012 [65]. (b, c) Replica of smallest bone (stapes) in human body made of surgical steel through micro metal injection molding. *Source:* Courtesy of Fraunhaufer IFAM, Germany.

Injection molding is increasingly used to manufacture plastic medical devices owing to its ease of use, recyclability, disposability, and cost-effectiveness. Figure 7.22 shows examples to micro plastic injection molding.

To satisfy requirements such as bio-compatibility, corrosion resistance, electrical insulation properties, wear resistance, surface quality, thermal stability, high modulus of elasticity with low weight ceramic powders are molded after blending of appropriate binder and additives. This process is known as micro ceramic injection molding (μ-CIM). The most common ceramic materials used in industry are aluminum oxide (Al_2O_3) and zirconium oxide or zirconia (ZnO). Aluminum oxide is preferred when high strength and hardness, high wear resistance, corrosion resistance, high thermal conductivity, superior electrical insulation properties, and high-temperature resistance features are required. Zirconia, on the other hand, is selected when exceptional bending strength, high modulus of elasticity (comparable to steels), high temperature resistance, and good tribological characteristics are demanded. Dental tooth implants, abutments, orthodontic brackets, medical endoscopic devices, and biological fluid analysis nozzles in medical field; semiconductor wire bonding capillaries, receptable guides, and grippers in electronics sector; and guides, gearwheels, parts with screw threads in manufacturing industry are some of the examples to μ-CIM products [66]. Figure 7.23 shows some μ-CIM parts manufactured by Small Precision Tools Roth Ltd. (SPT Roth, Switzerland).

7.3.2.8 Powder Compaction

Powders can be compacted between two dies under the effect of pressure and temperature to fabricate micro-featured surfaces in relatively simple manner compared to PIM process. Sintering is applied subsequently to obtain metallurgical bonds between powders. Such surfaces can be used for advanced heat transfer purposes, and as interconnect plates in solid-oxide fuel cells. Cora et al. investigated the powder compaction process using spherical copper powders at different process conditions (e.g. pressure, temperature, powder size, and sintering

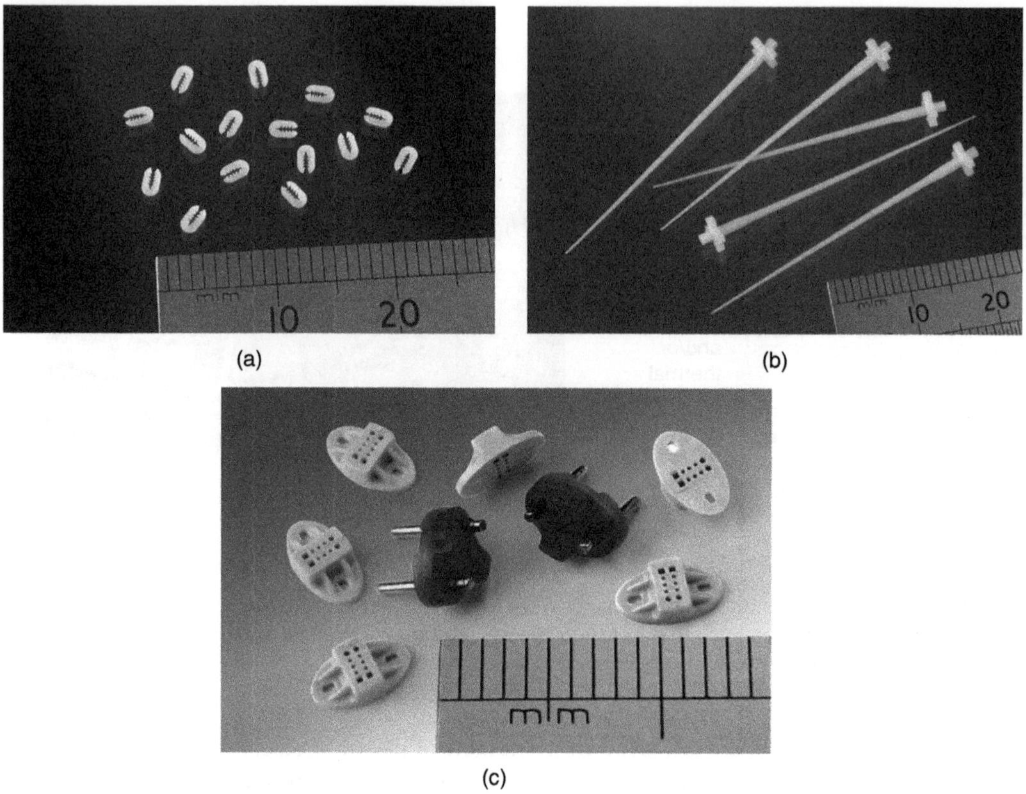

Figure 7.22 Micro plastic injection molding applications: (a) medical staples made of bioresorbable materials (with 0.3 mm pitch of the teeth), (b) dental surgery device for root canal treatment with a tip diameter of 0.18 mm, and (c) micro-molded fiber optic connector with 0.2 mm through holes. *Source:* Courtesy of Micro Systems UK Ltd, UK.

temperature), surface geometry (flat, large, small channels; two dimensional [2D], 3D modulations), and aspect ratio and with different process routes (compaction + sintering vs. sintering + compaction). The main idea behind these efforts was to obtain sound porous surfaces enabling two-phase flow and heat transfer enhancements through mass-production compatible compaction process. It was revealed that such surfaces made of coppers can increase the heat transfer rate by 300% [67, 68]. Upon successful results on 2D modulations, 3-D surface geometries were also realized by the same process [69, 70]. Figure 7.24 shows some of those surfaces realized by means of copper powder compaction process with different dies.

7.3.3 Electrical/Chemical Micromanufacturing Processes

Processes in this group use top-down approach and form the workpiece with different material removal effects including electric discharge melting, laser melting, laser ablation, chemical dissolution, etc.

7.3.3.1 Micro Wire Electro Discharge Machining (Micro WEDM, μ-WEDM)

WEDM is considered as near-contact cutting method with a tensioned wire moving between a top and a bottom guide through a workpiece. The cutting area is either submerged or flooded in a dielectric fluid, which can be deionized water or oil. Oil is preferred as it has

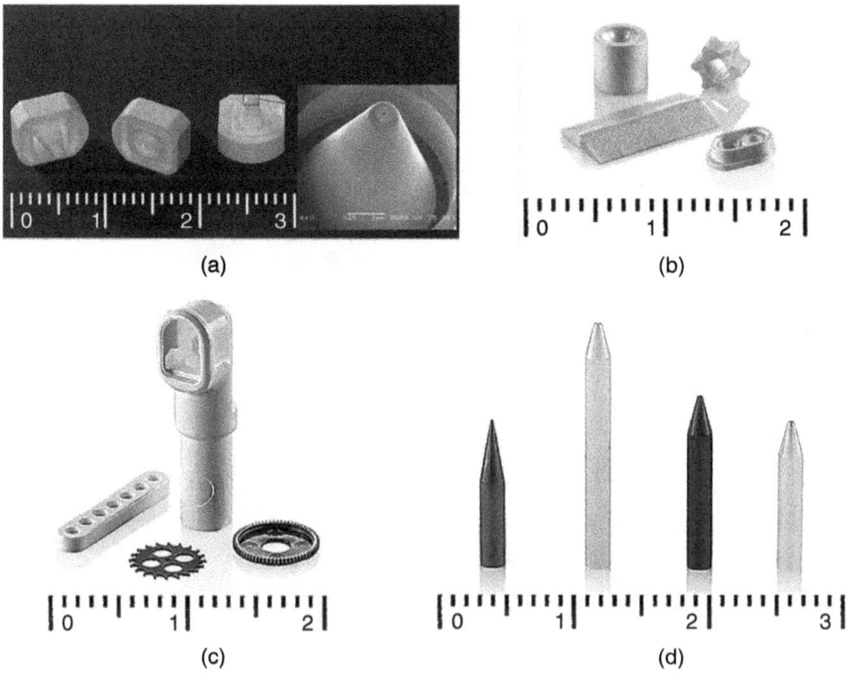

Figure 7.23 Examples to microceramic injection molded parts: (a) pharmaceutical jet nozzle made of zirconia toughened alumina with 50 μm orifice size, (b) industrial parts made of alumina, zirconia, or zirconia-toughened alumina, (c) medical parts made of zirconia-toughened alumina and gears made of black zirconia or polycrystalline ruby, and (d) ceramic capillaries with inner diameters down to 0.015 ± 0.001 mm. *Source:* Courtesy of SPT Roth, Switzerland.

higher insulating capability and improves the accuracy and surface finish. Oil use also greatly reduces wire breakage problem. When a timed electrical pulse is applied current begins to flow at the spot along a wire where the electric field is strongest, a plasma channel forms, and a spark occurs between wire and the workpiece. Temperatures as high as 10 000 °C can be experienced, and this results in immediate melting and ejection of small volume of workpiece material from its surface and the wire. At the end of electrical pulse, the gas-filled plasma channel implodes and melted material solidifies to form spheres a few microns in diameter that are carried away by dielectric fluid [71].

The requirement and development of WEDM were driven by the mold and tool industry needs in cutting 3D parts made of hard and super hard materials, precisely. WEDM has been used for most of electrically conducting materials for +25 years. It is known as slow machining process when compared with conventional machining operations; however, use of automated wire threading enables 24-hour operation.

7.3.3.2 Electro-chemical Micromachining (ECM)

Electro-chemical micromachining (ECM) process is a hybrid process that combines the effects of electropolishing and ECM. It is regarded one of the precision cold forming technologies, and mainly used for very precise shaping and micropatterning of metallic workpieces and surfaces. Microfluidics (Figure 7.25), high-quality aesthetic engraving (watch industry, security marking), biomedical applications, and all kind of replication tools, in general.

The process provides superior smoothness compared to laser machining, chemical etching, electric discharge machining, and micro milling. Precise dimensioning of engraved patterns is

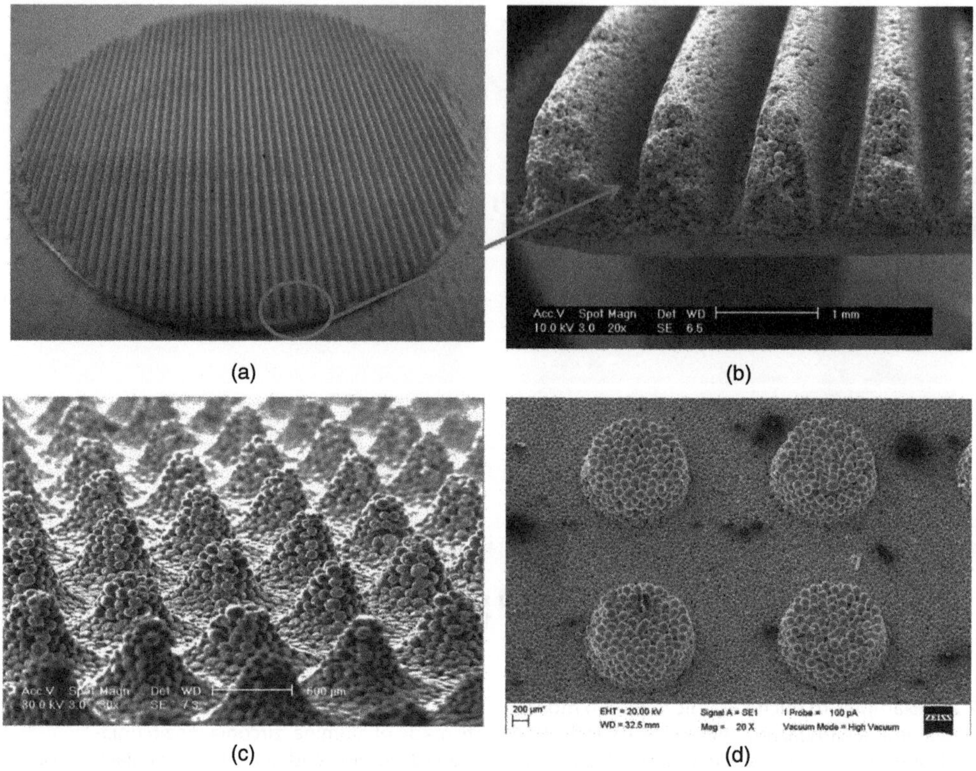

Figure 7.24 The 2D and 3D functional surfaces through powder compaction + sintering, (a) Actual photo of 55 mm in diameter 2D small channel design sample with aspect ratio of 1.68 (channel height 1.68 mm, channel with 1 mm), (b) SEM image of channels (20× magnification), (c) SEM image of 3D pyramid-like surface geometry (scale bar: 500 μm), and (d) SEM image of 3D bumps-like design (feature height 500 μm, base diameter: 1 mm).

obtained by monitoring the electric charge flown through the system. This control is very precise and, unlike for chemical etching, is independent of the aging and temperature of the bath. Micropatterning of 3D solid objects such as biomedical implants shown in Figure 7.26 is one of the important applications. Round cavities or cylindrical channels can be precisely etched with diameters in the range of 30–300 μm. Positioning and dimensional accuracy can be as low as 1 μm over the surface area as large as 150 mm × 150 mm. From metallurgical point of view, the process does not cause oxidation, darkening, and corrosion resistance loss on the surface. Nonetheless, the metallic surface is left in a passive state without residual stress [72].

7.3.4 Electron Beam-Based Micromanufacturing Processes

The continuously increasing level of integration in micromechanical products requires precise machining and joining technologies. Electron beam machining (EBM) is thermal machining process in which use of a focused beam of high-velocity electrons result in instant heating, melting, or vaporization of material. It has a variety of applications including drilling, cutting, annealing, surface treatment, and welding. Marking, labeling, and sample characterization can also be performed through EBM. A variety of materials can be machined, and material properties such hardness, ductility, electrical conductivity, thermal conductivity, and melting points are usually not limiting factor [73]. The first electron beam processing machine was built by physicist Dr. Karl-Heinz Steigerwald in 1952.

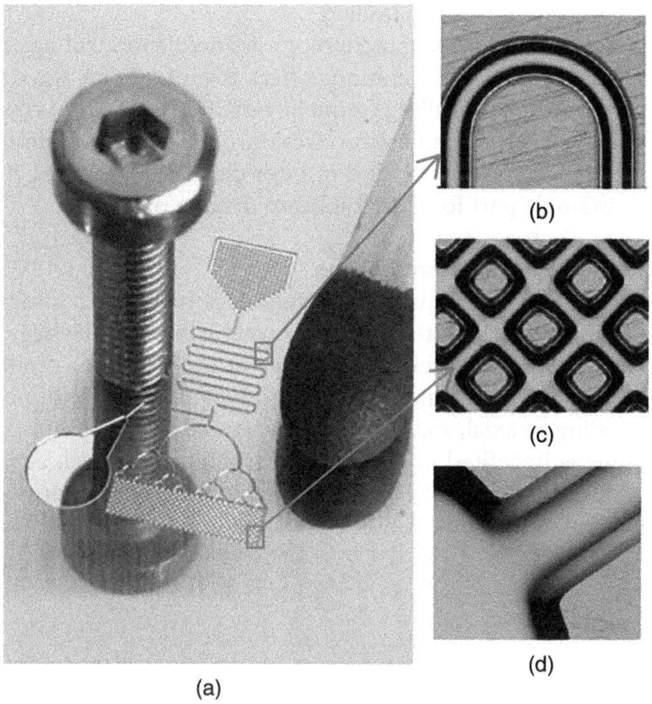

Figure 7.25 Microfluidic channel on stainless steel by means of electrochemical micromachining process, (a) General view showing micro-macro assembly, (b) curved channel with 60 μm width and 25 μm depth, (c) 60 μm wide high-density channel array, and (d) smooth transitions between different sections. *Source:* Courtesy of Micropat SA, Switzerland.

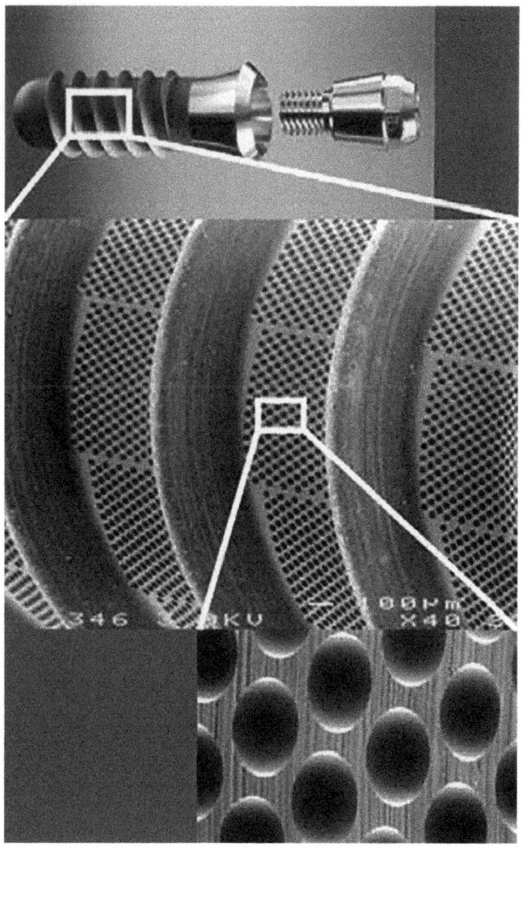

Figure 7.26 Titanium dental implant microstructured with 27 000 hemispherical cavities 50 μm in diameter. *Source:* Courtesy of Micropat SA, Switzerland.

7.3.4.1 Electron Beam Drilling

Electron beam drilling technology is suitable for drilling as the power density of beam is as high as 10^8 W/cm^2. The technique offers advantages for the workpieces and requires large number of micro-holes. Drilling frequency of 3000 holes per second is possible with the current technology. Typical application areas are drilling of spinning heads for fiber glass production, inclined holes, and all kind of filter plates [74]. Figures 7.27 and 7.28 show the schematic and an example part for electron beam drilling process.

7.3.4.2 Electron Beam Welding

Electron beam welding is well suited for high-precision joining of metallic materials due to its localized heat input and narrow weld geometry. Hence, it is regarded as an effective joining technique in the vicinity of temperature sensitive components [75]. In addition, it provides joining for critical metallic material combinations (bimetallic joints), complex-shaped parts. Welding in axial, radial directions as well as plane and curved surfaces is possible. The technique is benefited in automotive, machine, medical, aerospace, power engineering, and vacuum industries. As the power density increases deep welding effect can also be realized as illustrated in Figure 7.29.

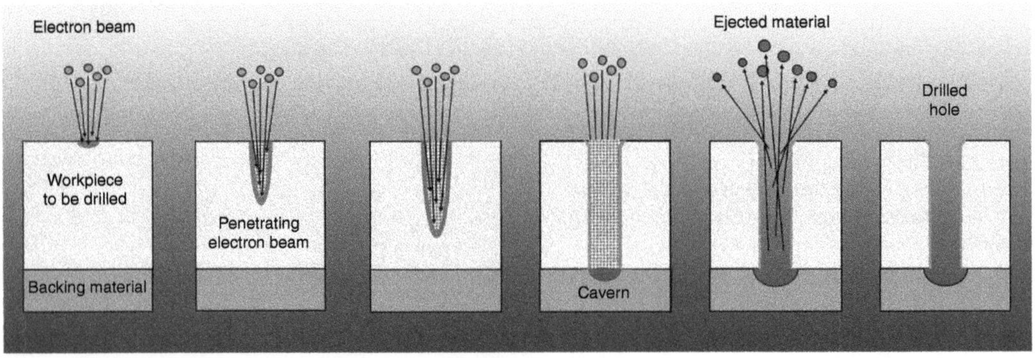

Figure 7.27 Schematic of electron beam drilling process. *Source:* Courtesy of Steigerwald Strahltechnick GmbH, Germany.

Figure 7.28 Micro holes on a spinning head for glass fiber production. *Source:* Courtesy of Steigerwald Strahltechnick GmbH, Germany.

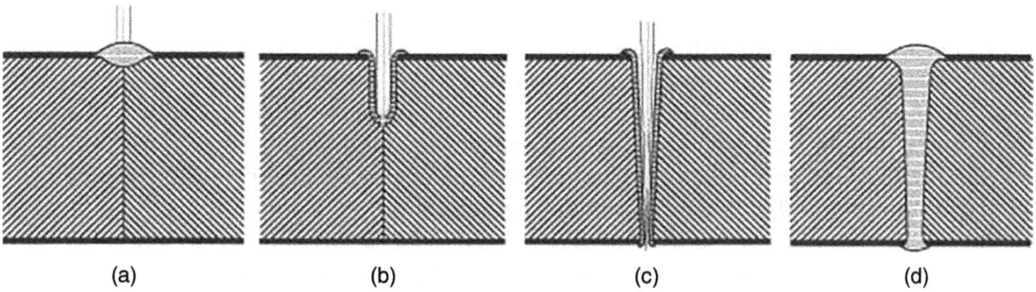

Figure 7.29 Process sequence in deep welding through electron beam (a) High-energy concentrate at the beam spot and melts the material, (b) material at the center vaporizes, (c) beam penetrates into deeper points in workpiece through the vapor channel formed, and (d) melted material flows from bottom to up and solidifies. *Source:* Courtesy of Steigerwald Strahltechnick GmbH, Germany.

7.3.5 Laser-assisted Microforming

Although it is detailed in Chapter 19, laser-assisted forming or in other words laser beam machining (LBM) processes will be discussed briefly throughout this section. The term "laser" stands for Light Amplification Stimulated Emission of Radiation, and its foundations was established by Einstein who provided the concept of stimulated emission. First laser (ruby laser) was realized in 1960s by Townes and Shawlow [76]. Since then, it has been increasingly adopted by several industries due to its high accuracy, repeatability, and speed. In addition, laser-assisted forming does not results in tool wear as in traditional milling operations or in EDM. It is environmentally clean process as no solvent chemical is used as in chemical etching and less amount of scrap material is produced.

In terms of process capability, wide range of operations such as micro cutting, milling, grinding, drilling, welding, 3D machining, as well as annealing, marking/scribing, engraving, surface texturing, surface treatment, selective layer removal, and perforating can successfully be performed with lasers. Micro holes with a few microns in diameter and submicron tolerances. Different from conventional machining, laser-assisted micromachining is a noncontact type microforming operation that does not involve mechanical tools and capable of producing parts and/or features with intricate details, high aspect ratios, and strict tolerances. Table 7.5 lists some of common applications in different areas. Wide range of materials from metals, alloys, polymers, composites, ceramics, glasses to organic, and inorganic compounds can be machined with lasers.

Table 7.5 Common laser micromachining application in various sectors, after [77–79].

Medical	Aerospace	Automotive	Electronics	Life sciences	Energy
Surgical tools, stents, artificial heart valves, hypotubes, aerosols, needles, catheters, medical implants, angioplasty devices, endoscopes, artificial pacemakers, micro fluidic sensor, metered-dose inhalers, etc.	Turbine parts, turbine blades, engine components, cooling holes, etc.	Fuel injection nozzles for gasoline and diesel engines, filters and graders, gas sensors, fuel cells, etc.	Circuit boards, electronic components, solar cells, inkjet heads, optical circuits, IC repair, various types of sensors, etc.	Jewelry, clockwork mechanisms	Photovoltaic (PV) cells

Traditionally, Neodymium-doped yttrium-aluminum-garnet (Nd:YAG) and CO_2 are the most widely used lasers in industry. CO_2 lasers, which belong to gas lasers group, with 10.2 μm wavelength are high-efficient, long-penetration depth (5–10 μm or more), and inexpensive lasers. Nd:YAG lasers, the most conventional solid-state lasers, on the other hand, are selected for machining thinner and high reflective materials. In general, CO_2 laser is used in continuous mode, and Nd:YAG laser is operated in pulsed mode. Nd laser outputs can be frequency doubled, tripled, or quadrupled through harmonic generation.

Excimer lasers are regarded as the "third" commercial laser along with CO_2 and solid-state lasers operating in ultraviolet wavelength band (157–351 nm). Lasing medium is a mixture of rare gas (krypton, argon, and xenon) and a halogen gas (fluorine or hydrogen chloride). This mixture determines the laser beam output wavelength. Excimer lasers are characterized with their short wavelength providing higher resolution (~1 μm) on the target surface. Their high peak power enables ablation of the target material with no or small heat affected zone (HAZ). In addition to its applications in medical and electronics industry, excimer lasers are well suited to eye surgery operations (LASIK), precision micromachining of organic materials including some polymers and plastics. Their shallow absorption depth allows tight control of feature depth by manipulating the number of pulses produced. Typical pulse rate is around 100 Hz, and pulse length is 10–20 ns resulting in peak powers of tens of megawatts [78, 79]. Although it is not common practice to machine metals with excimer lasers, drilling holes in aircraft wings is performed with this mature technology. High-power excimer lasers can drill multiple holes simultaneously as larger area can be illuminated. In these hole array applications, excimer laser beam illuminates a noncontact mask that contains array of holes. A downstream lens images the hole array onto the part, allowing multiple holes to be drilled at once [80]. Apart from their advantages, excimer lasers have high initial cost and frequent maintenance issues. Laser gas is toxic and corrosive, and its consumption is high and expensive.

Fiber lasers operate at near infrared wavelength, typically at 1070 nm, and offer superior features over traditional laser sources. Their power, usually in the range of 50–400 W, can be focused to a much smaller spot. They can also support kilowatt levels of power output owing to the fiber's high surface area to volume ratio, which enables efficient cooling. They are compact compared to solid-state and gas lasers as fiber can be bent and coiled. Moreover, with their high vibrational stability, fiber lasers are regarded as robust laser processes. Fiber lasers are used in precision cutting of stents, welding and drilling of medical products such as hypotubes, needles, endoscopes, and canulae [81]. Different from excimer lasers, their relatively lower initial cost, maintenance-free operation, and extended life time make them more preferable in microforming field.

Known also as "ultrafast lasers," ultrashort pulse lasers are the most recent laser technology that are commercially available. This group of lasers cut, drill, and structure virtually any material with micron-scale precision and without thermal damage. Those also do not require postprocessing as the surfaces are burr-free. Among those, picosecond lasers are increasingly used in semiconductor and display production. Picosecond laser with up to 100 W power, and repetition rates of 1 MHz are becoming standard micromachining tools in various sectors. Femtosecond lasers, as second group under ultrashort pulse lasers, are suitable for almost any material and have great potential in medical technology, sensors, and aeronautics applications [81]. Machining polymers with low melting temperatures such as bioabsorbables and fluropolymers is a very recent application of picosecond and femtosecond lasers [80].

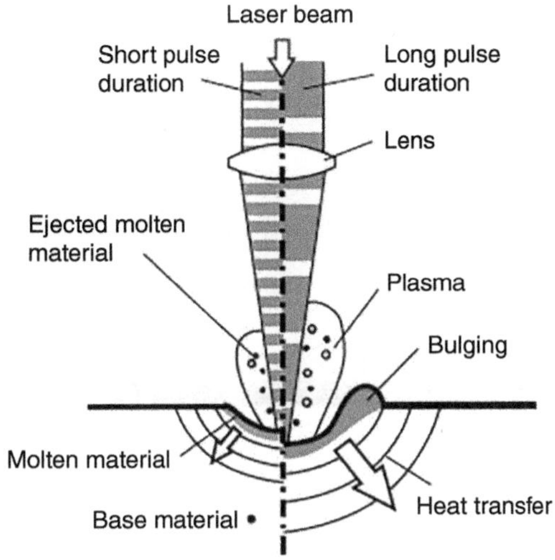

Figure 7.30 Schematic of laser ablation process. *Source:* Schubert et al. 2011 [83]. Reproduced with permission of AIP Publishing.

Despite these advantages, ultrafast lasers are much more expensive than any other laser system, currently. This kind of lasers also operates at low power levels and pulse energies for superior surface quality that results in slower processing speeds. Therefore, it should be noted that selecting the right laser application is not an easy process as many factors (such as material, initial cost, process speed, and machining quality) are involved, and an unique laser system that can be used for any application may never be developed [82].

7.3.5.1 Laser Ablation

Laser ablation is the process in which laser radiation reaches the substrate material, the absorbed laser energy melts the target surface, first and then heats it to the vaporization temperature (Figure 7.30).

Laser ablation is not only used for material removal purposes but also as to produce micro-/nanostructure from a bulk material immersed in a liquid. In this top down method, known also as pulsed laser ablation in liquid (PLAL), laser is used to fragment a solid structure of macroscopic size down to micro-nano sized objects [84]. It has been reported that the method can be applied to various materials predominantly metals, mainly gold and silver, as well as polymers, semiconductors, oxides, etc., with more or less the same protocol [85].

7.3.5.2 Laser Cutting

Micro cutting of intricate complex parts is a conventional laser process. Laser cutting of coronary stents, drug-eluting, and bare metal stents are the most common applications. Special stainless steel, cobalt-chromium, or shape memory alloy Nitinol can be material of choice in stent production. Most metal stents are cut from tubes with wall thicknesses that are often just a few tenths of a millimeter. The stent design needs to be optimized in a way that best blood flow is guaranteed and maximum flexibility is retained. The stent should also withstand the expansions. These requirements can be realized by means of laser cutting very fine structures with extreme precision and preventing HAZs and microcracks. Cutting speed of 3000 mm/min can be achieved with the geometry quality of ±1 μm can be achieved with current technology.

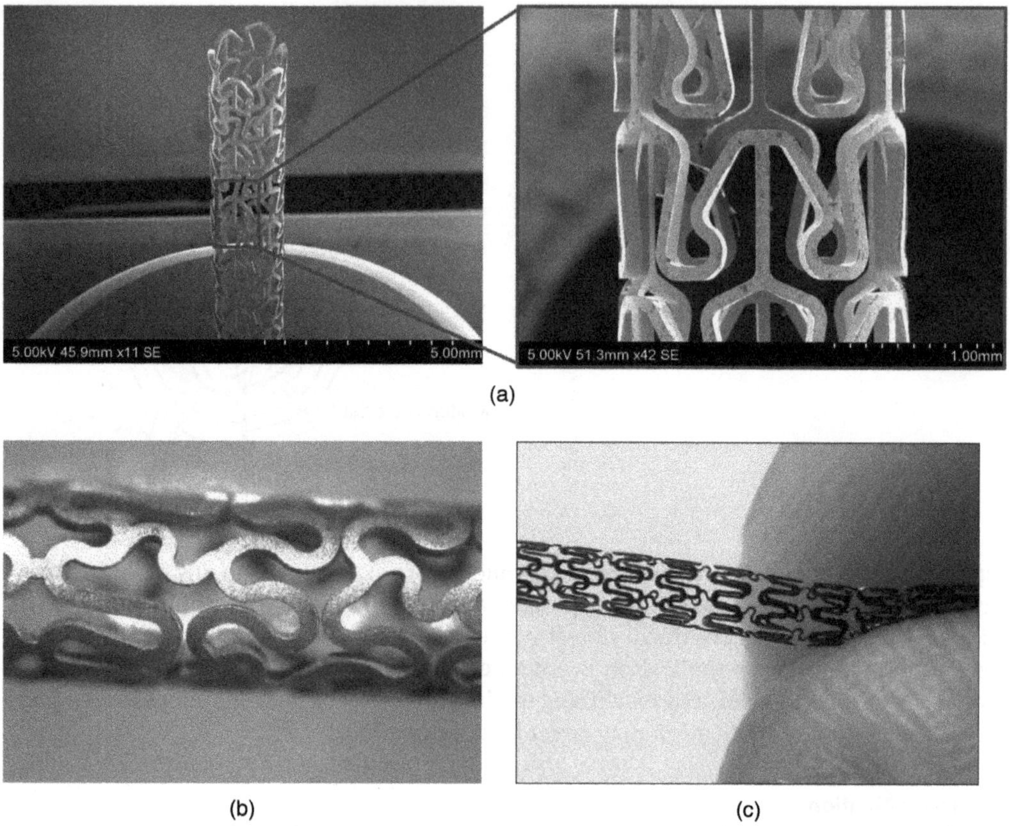

Figure 7.31 Laser cut stents (a) Stainless steel stent cut with single mode 100 W fiber laser. *Source:* Courtesy of Dr. Mohammed Naeem. (b) Magnesium stent. *Source:* Courtesy of Swisstec AG, Switzerland. (c) Polyether-ether-ketone (PEEK) stent. *Source:* Courtesy of Blueacre Technology Ltd, Ireland.

Instead of stainless steel and Nitinol stents, a recent trend is the use of biodegradable magnesium and polymer as stent material. Figure 7.31 shows stents made of different materials. PEEK and polyimides offer advantages over the metallic stents as those can be extruded with smaller tubes prior the laser processing and require less postprocess operation. These new materials, on the other hand, require new laser technologies such as ultrashort pulse lasers with picosecond (10^{-12}) or even femtosecond (10^{-15}) range. Extensive amount of energy is packed into these short pulses, converting the material instantly into gaseous state instead of melting it, first. Hence, material is removed pulse by pulse without generating HAZ. Polymer stent cutting does not usually require postdeburring process as the laser process vaporizes the polymer material, leaving no flash. In some cases, the only postprocessing required for laser-machined plastics might be the ultrasonic bath cleaning [81, 86].

7.3.5.3 Laser Micro Welding

Laser welding is used to join high value parts in medical and electronics industry such as catheters, implants, orthodontic brackets, watch parts, and hard disk drive flexures. During the manufacture of hard disk drives (Figure 7.32a), three or four 20–150 μm thick SS 300 stainless steel foils must be spot welded to make the flexure arm that will hold the ferrite reader over the

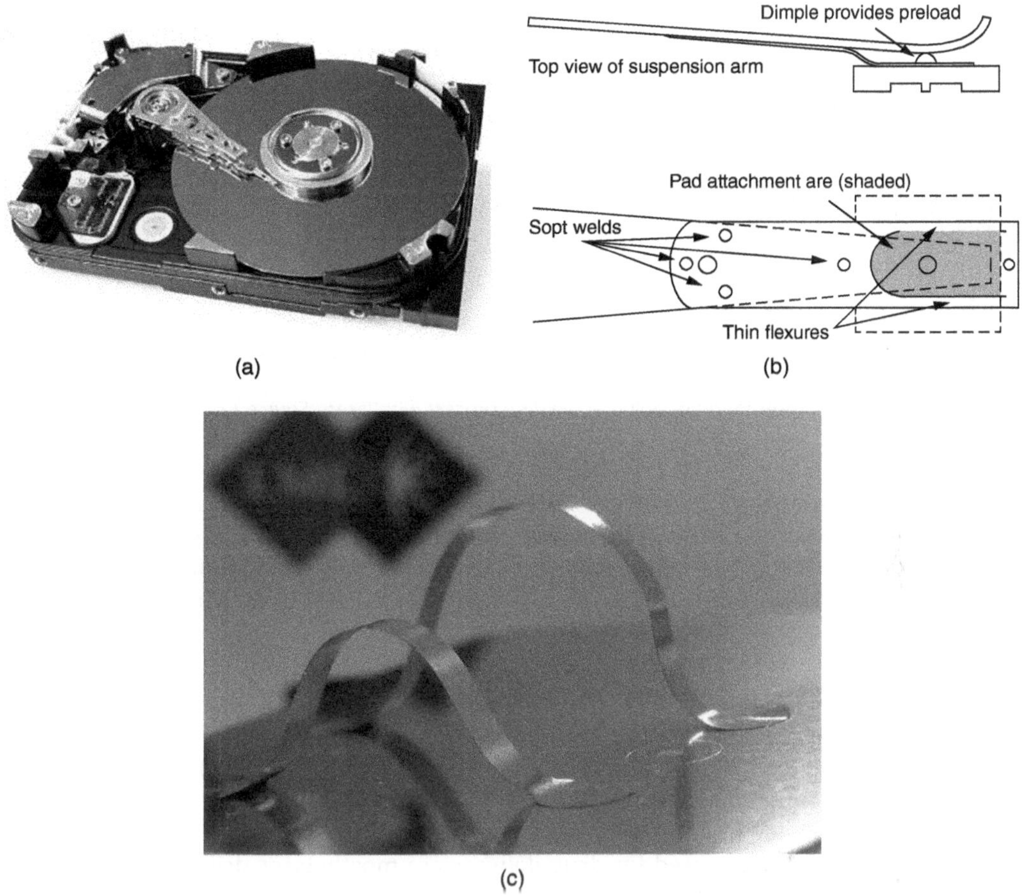

Figure 7.32 (a) Hard disk drive and (b) flexure arm schematic showing laser spot weld positions. *Source:* Courtesy of Dr. Mohammed Naeem. (c) Micro welding of 100 μm thick stainless steel foil onto stainless steel plate. *Source:* Courtesy of Dr. Nadeem Rizvi, Laser Micromachining Ltd., United Kingdom.

spinning magnetic media. Spot weld dimensions shown in Figure 7.32b vary between 76 and 175 μm [87]. Figure 7.32c shows micro welding of 100 μm thick SS foil onto SS plate.

7.3.5.4 Laser Microwelding of Polymers

In addition to metals, various kinds of fusible polymers can be welded by lasers. It is important to note that a minimal chemical compatibility has to exist between the polymers to be joined and their melting temperatures should be in proximity. The laser wavelength is also critical as some of the polymers do not absorb laser radiation from the near ultraviolet and near infrared range of the spectrum. They are transparent for these wavelengths. The frequently used laser types in polymer processing are ultraviolet and ultrashort-pulse lasers.

The schematic of polymer welding process using lasers is given in Figure 7.33. The joining process is performed by the use of heat, with or without additional material, and application of externally acting compressive force. The result is industrially preferred lap joint arrangement. If the joint is made of one polymer type, each part has to exhibit different optical behaviors against the wavelength used. As it can be seen from the Figure 7.33, one component has to be

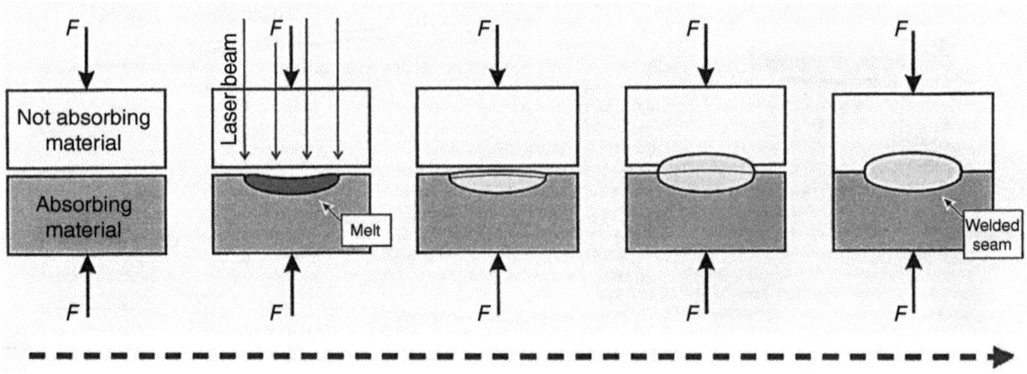

Figure 7.33 Schematic representation of polymer welding process with lasers. *Source:* Courtesy of 3D-Micromac AG, Germany.

transparent for the laser beam, while the other one is absorbing. The beam penetrates the transparent component and connects to the absorbing one at the contact interface. A conversion of radiation into heat is possible almost without any loss. In case of joining of two similar polymers, the absorption in one part is improved by relevant additives. These additives are usually carbon black, pigment, or absorber layers [88].

7.3.5.5 Laser Drilling

As a commonly accepted rule, lasers can drill holes or slots from millimeter range to as small as to 1 μm diameter. The maximum drilling depth is around 1 mm. One of the biggest advantages of the lasers is to drill through holes in polymer materials one hole at a time or drill multiple holes at once [80].

Liquid and gas-flow devices used for drug delivery are mostly laser-drilled. They include catheters that are extruded, and flexible tubes inserted into the body to allow for drainage, localized drug delivery and to provide surgical instruments access to internal organs. Catheters require holes that are precisely sized and positioned to ensure that the drug reaches the specific area [89]. Figure 7.34 shows two different applications of laser drilling on plastic materials.

7.3.5.6 Laser Etching

The process is also referred to as micro stripping or micro etching. Different from the mechanical processes, lasers can remove material layer-by-layer that results in selective stripping or etching of a material and producing features such as blind holes and channels, surface modification, tube notching, or stripped wires and cables [80].

7.3.5.7 Other Laser-Assisted Microforming Processes

Laser involved micromachining operations are becoming increasingly popular in industry due to several advantages including accuracy, speed, and cost-effectiveness. Apart from abovementioned laser micromachining (also referred as laser beam machining) processes, there are some other micromanufacturing processes in which laser is involved or combined with other techniques (hybrid micromanufacturing). Driven by specific needs, there are some laser-assisted processes such as indirect laser shock micro embossing [90], laser cladding [78], laser peen forming [91], laser shock micro deep-drawing [92].

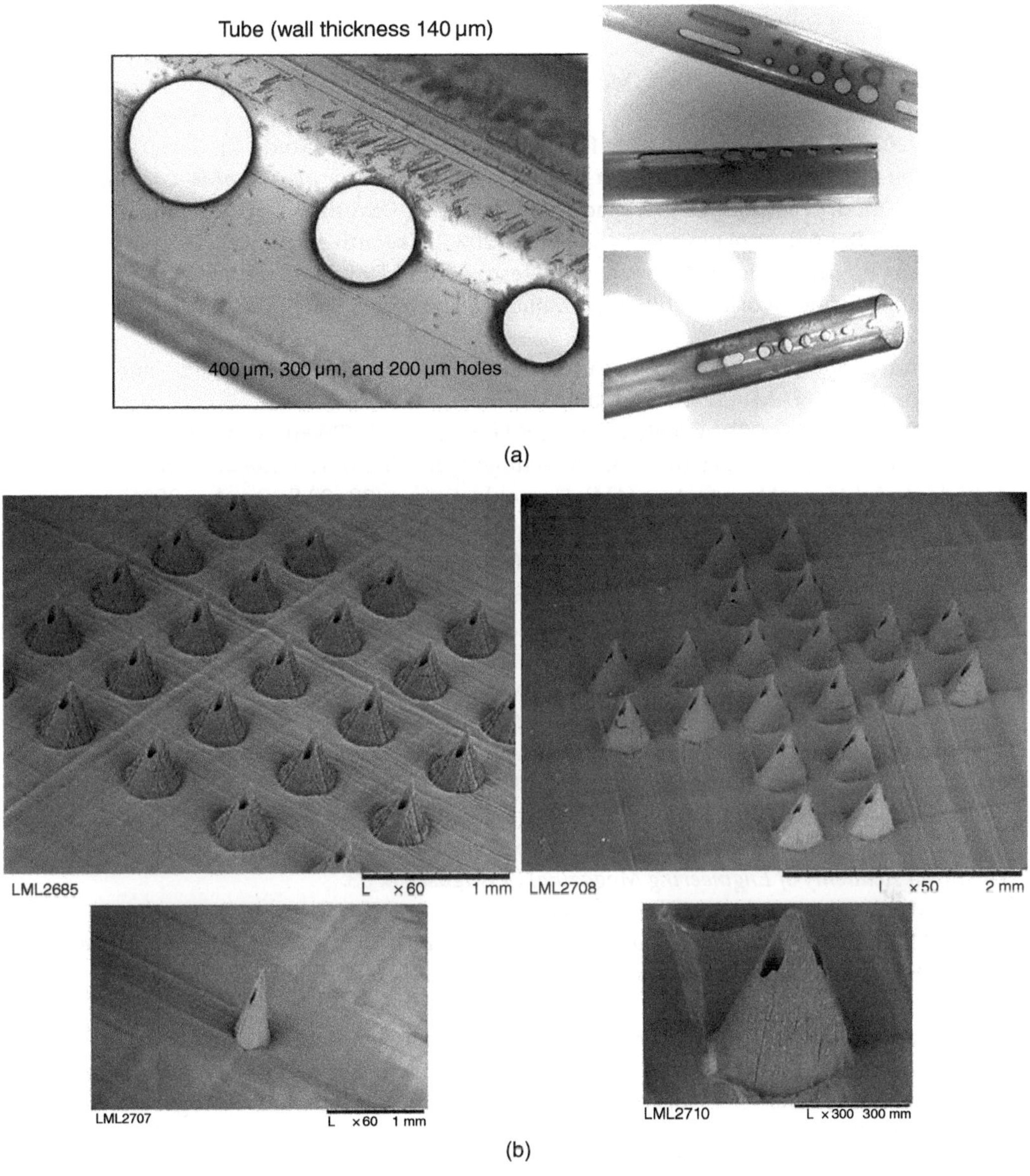

Figure 7.34 (a) Hole drilling on Kapton tube with a wall thickness of 140 µm and (b) laser micromachined micro-needle arrays with central through holes made in polyimide. *Source:* Courtesy of Dr. Nadeem Rizvi, Laser Micromachining Ltd., United Kingdom.

References

1 Qin, Y., Brockett, A., Ma, Y. et al. (2010). Micro-manufacturing: research, technology outcomes and development issues. *International Journal of Advanced Manufacturing Technology* 47: 821–837.

2 NEXUS Market Analysis for MEMS and Microsystems III: 2005–2009 (September 2005). Grenoble Cedex, France.

3 Status of the MEMS Industry. Market Analysis Report. Yole Développement (July 2012). Lyon, France.
4 MEMS for Cell Phones and Tablets. Market Analysis Report. Yole Développement (19 June 2013). Lyon, France.
5 Audet, S. (2010). Micro-manufacturing for medical devices: a case study. In: *Keynote presentation in ICOMM/4M 2010 – 5th International Conference on MicroManufacturing and Multi-Material Micro Manufacture*, Madison, Wisconsin, USA (5–8 April 2010).
6 Ehmann, K.F., Bourell, D., Culpepper, M.L, et al. (2005). Executive Summary Presentation of Panel Report on International Assessment of Research and Development in Micromanufacturing. World Technology Evaluation Center (WTEC), Baltimore, MD, USA (October 2005).
7 MNT Future Vision Newsletter. European Technology Platform for Micro- and Nano Manufacturing (MINAM) (February 2009).
8 Dimov, S.S., Matthews, C.W., Glanfield, A., and Dorrington, P. (2006). A roadmapping study in Multi-Material Micro Manufacture. In: *Proceedings of the 2nd International Conference on Multi-Material Micro Manufacture (4M)*, Grenoble, France (20–22 September 2006).
9 Ehmann, K.F. (2007). A synopsis of U.S. micro-manufacturing research and development activities and trends. In: *Proceedings of the 3rd International Conference on Multi-Material Micro Manufacture (4M)*, Borovets, Bulgaria (3–5 October 2007).
10 Razali, A.R. and Qin, Y. (2013). A review on micro-manufacturing, micro-forming and their key issues. *Procedia Engineering* 53: 665–672.
11 Waurzyniak, P. (2013). Micro manufacturing keeps shrinking the envelope. *Manufacturing Engineering, Society of Manufacturing Engineers Magazine* (January 2013), pp. 65–73.
12 The Royal Society and Royal Academy of Engineering (2004). *Nanoscience and Nanotechnologies: Opportunities and Uncertainties"*, Final Report. London: The Royal Society and Royal Academy of Engineering.
13 Whatmore, R.W. (2001). Nanotechnology: big prospects for small engineering. *Ingenia, (The Royal Academy of Engineering Magazine)* (9 August), p. 33.
14 Madou, M.J. (2002). *Fundamentals of Microfabrication*, 2e. CRC Press.
15 Madou, M.J. (2012). *Fundamentals of Microfabrication and Nanotechnology*, From MEMS to Bio-MEMS and Bio-NEMS: Manufacturing Techniques and Applications, 3e, vol. 3. CRC Press.
16 Brinksmeier, E., Riemer, O., and Stern, R. (2002). Machining of precision parts and microstructures. In: *Initiatives of Precision Engineering at the Beginning of a Millennium – 10th International Conference on Precision Engineering (ICPE)* (ed. I. Inasaki). Springer.
17 Gandarias, E. (2007). MICROM: a revolutionary monitoring system to detect tool breakages & collisions, enhance machine cycles and introduce a new probing concept in micro-milling. Ph.D. Dissertation. Mondragon University, Mondragon, Spain.
18 Masuzawa, T. (2000). State of the art of micromachining. *CIRP Annals – Manufacturing Technology* 49 (2): 473–488.
19 Alting, L., Kimuara, F., Hansen, H.N., and Bissacco, G. (2003). Micro engineering. *CIRP Annals – Manufacturing Technology* 52 (2): 635–657.
20 Qin, Y. (2010). Overview of micro-manufacturing. In: *Micro-manufacturing Engineering and Technology* (ed. Y. Qin). Glasgow. William Andrew Books – Elsevier.
21 Bissacco, G., Hansen, H.N., and Chiffre, L.D. (2005). Micromilling of hardened tool steel for mould making applications. *Journal of Materials Processing Technology* 167 (2–3): 201–207.
22 Sodemann, A.A., Li, M., Mayor, R., and Forest, C.R. (2009). Micromilling of molds for microfluidic blood diagnostic devices. In: *Proceedings of 24th Annual Meeting of the American Society for Precision Engineering*, Monterey, CA (4–9 October 2009).

23 Monroy-Váazques, K.P., Attanasio, A., Ceretti, E. et al. (2013). Evaliation of superficial and dimensional quality features in metallic micro-channels manufactured by micro-end-milling. *Materials* 6: 1434–1451.

24 Cardoso, P. and Davim, J.P. (2012). A brief review on micromachining of materials. *Reviews on Advanced Materials Science* 30: 98–102.

25 Li, P. (2009). Micromilling of hardened tool steels. Ph.D dissertation. Delft University of Technology, The Netherlands.

26 Performance Micro Tool Inc., Janesville, WI, USA. http://www.pmtnow.com/ (accessed 20 October 2018).

27 Müller, P. and Gschrey, H. (2014). The elements of successful microdrilling. *Micro Manufacturing*. John Wm Roberts & Associates 7 (1): 16–19.

28 Lauman, S. (2013). Drilling into the medical market. *Micro Manufacturing*. John Wm Roberts & Associates 6 (2): 25–27.

29 Dornfeld, D., Min, S., and Takeuchi, Y. (2006). Recent advances in mechanical micromachining. *CIRP Annals – Manufacturing Technology* 55 (2): 745–768.

30 Aurich, J.C., Engmann, J., Schueler, G.M., and Haberland, R. (2009). Micro grinding tool for manufacture of complex structures in brittle materials. *CIRP Annals – Manufacturing Technology* 58: 311–314.

31 Ramesh, K., Huang, H., Yin, L., and Zhao, J. (2004). Microgrinding of deep micro grooves with high table reversal speed. *International Journal of Machine Tools and Manufacture* 44: 39–49.

32 Brinksmeier, E., Mutlugüneş, Y., Klocke, F. et al. (2010). Ultra-precision grinding. *CIRP Annals – Manufacturing Technology* 59: 652–671.

33 Barton, K. (2007). *Beyond the Shine – A Micropolishing Primer*. QPAC Inc http://www.micropolishing.com/home/index.html (accessed 20 October 2018).

34 Brinksmeier, E., Riemer, O., and Gessengarter, A. (2006). Finishing of structured surfaces by abrasive polishing. *Precision Engineering* 30: 325–336.

35 Suzuki, H., Moriwaki, T., Okino, T., and Ando, Y. (2006). Development of ultrasonic vibration assisted polishing machine for micro aspheric die and mold. *CIRP Annals – Manufacturing Technology* 56 (1): 385–388.

36 Liu, H.-T. and Schubert, E. (2012). Micro abrasive-waterjet technology. In: *Micromachining Techniques for Fabrication of Micro and Nano Structures* (ed. M. Kahrizi), 225–227. InTech. ISBN: 978-953-307-906-6. Available at: https://www.intechopen.com/books/micromachining-techniques-for-fabrication-of-micro-and-nano-structures/-micro-abrasive-waterjet-technology.

37 Miller, D. (2013). Micro machining – abrasive waterjets. *Commercial Micro Manufacturing International* 6 (2): 40–42.

38 Parette, S. (2012). Micro manufacturing utilising waterjet technology – a brief history. *CMM International* 5 (2): 22–24.

39 Wang, C., Wang, C., Shan, D. et al. (2013). Manufacturing high aspect ratio microturbine by isothermal microforging process. *Materials and Manufacturing Processes* 29 (1): 42–45.

40 Forest, C.R., Saez, M.A., and Hunter, I.W. (2007). Microforging technique for rapid, low-cost fabrication of lens array molds. *Applied Optics* 46 (36): 8668–8673.

41 Chan, W.L. and Fu, M.W. (2012). Experimental studies of plastic deformation behaviors in microheading process. *Journal of Materials Processing Technology* 212: 1501–1512.

42 Geiger, M., Kleiner, M., Eckstrein, R. et al. (2001). Microforming. *Annals of CIRP* 50 (2): 445–462.

43 Raumedic AG, Helmbrechts, Germany. https://www.raumedic.com/technologies/extrusion/micro-tubing/ (accessed 20 October 2018).

44 Hulseman, S. (2011). Manufacturing micro scale features. *Commercial Micro Manufacturing* 4 (6): 37–39.
45 Fu, M.W. and Chan, W.L. (2013). Micro-scaled progressive forming of bulk micropart via directly using sheet metals. *Materials and Design* 49: 774–783.
46 Ghassemali, E., Tan, M.-J., Jarfors, A.E.W., and Lim, S.C.V. (2013). Progressive microforming process: towards the mass production of micro-parts using sheet metal. *International Journal of Manufacturing Technology* 66: 611–621.
47 Witulski, N., Justinger, H., and Hirt, G. (2004). Validation of FEM simulation for micro deep drawing process modeling. *AIP Conference Proceedings* 712: 952–957.
48 Braxton Manufacturing Co. Inc. Watertown, CT, USA. http://www.braxtonmfg.com/parts.htm (accessed 20 October 2018).
49 Shimizu, T., Yang, M., and Manabe, K. (2012). Impact of surface topography of tools and materials in micro-sheet metal forming. In: *Metal Forming* (ed. M. Kazeminezhad), 117. IntechOpen Available from: https://www.intechopen.com/books/metal-forming-process-tools-design/impact-of-surface-topography-of-tools-and-materials-in-micro-sheet-metal-forming.
50 Irthiea, I., Green, G., Hashim, S., and Kriama, A. (2014). Experimental and numerical investigation on micro deep drawing process of stainless steel 304 foil using flexible tools. *International Journal of Machine Tools and Manufacture* 76: 21–33.
51 Justinger, H. and Hirt, G. (2009). Estimation of grain size and grain orientation influence in microforming processes by Taylor factor considerations. *Journal of Materials Processing Technology* 209: 2111–2121.
52 Saotome, Y., Yasuda, K., and Kaga, H. (2001). Microdeep drawability of very thin sheet steels. *Journal of Materials Processing Technology* 113: 641–647.
53 Vollertsen, F., Hu, Z., Niehoff, H.S., and Theiler, C. (2004). State of the art in micro forming and investigations into micro deep drawing. *Journal of Materials Processing Technology* 204: 70–79.
54 Molotnikov, A., Lapovok, R., Gu, C.F. et al. (2012). Size effects in micro cup drawing. *Materials Science and Engineering A* 550: 312–319.
55 Mahabunphachai, S., Cora, Ö.N., and Koç, M. (2010). Effect of manufacturing processes on formability and surface topography of proton exchange membrane fuel cell metallic bipolar plates. *Journal of Power Sources* 195: 5269–5277.
56 Turan, C., Cora, Ö.N., and Koç, M. (2011). Effect of manufacturing processes on contact resistance characteristics of metallic bipolar plates in PEM fuel cells. *International Journal of Hydrogen Energy* 36: 12370–12380.
57 Chan, W.L. and Fu, M.W. (2012). Experimental and simulation based study on micro-scaled sheet metal deformation behavior in microembossing process. *Materials Science and Engineering A* 556: 60–67.
58 Hartl, C., Anyasodor, G., and Lungershausen, J. (2011). Formability of micro tubes in hydroforming. *AIP Conference Proceedings* 1353: 529–534.
59 Hartl, C., Anyasodor, G., Ptaschlik, T., and Lungershausen, J. (2010). Investigation into reduction of die-cavity deflection in micro-hydroforming process using FEA. *International Journal of Advanced Manufacturing Technology* 47: 853–858.
60 Accu-Mold LLC. Ankeny, IA, USA. https://www.accu-mold.com/micro-molding-design-guidelines/ (accessed 20 October 2018).
61 Shah V., "Micro Injection Molding". http://www.consultekusa.com/pdf/Consulting/Process/MICRO%20INJECTION%20MOLDING.pdf (accessed 27 October 2018).
62 Bibber, D. (2012). The science of compliance: micro moulding needles and sharps. *Commercial Micro Manufacturing International* 5 (5): 35–37.
63 Micro Systems (UK) Limited. http://www.microsystems.uk.com/ (accessed 20 October 2018).

64 Shah, V. *Micro Injection Molding*. Colsultek Consulting Group http://www.consultekusa.com/processing.html (accessed 20 October 2018).

65 Haack, J. and Imgrund, P. (2012). Quality control of micro metal injection moulding via neural networks. *Commercial Micro Manufacturing Europe* 5 (1): 37.

66 Sutter, M. and Whitlock, B. (2013). Micro injection moulding of fine ceramics. *Commercial Micro Manufacturing International* 6 (1): 8–11.

67 Cora, Ö.N., Usta, Y., and Koç, M. (2009). Micro-manufacturing of micro-scale porous surface structures for enhanced heat transfer applications: an experimental process optimization study. *Journal of Micromechanics and Microengineering* 19, 045011, 12pp.

68 Cora, Ö.N., Min, D., Koç, M., and Kaviany, M. (2010). Microscale-modulated porous coatings: fabrication and pool-boiling heat transfer performance. *Journal of Micromechanics and Microengineering* 20, 035020, 9pp.

69 Öztürk, B., Cora, Ö.N., and Koç, M. (2011). Effects of sintering temperature on the porosity and microhardness of the micro-scale 3-D porous gradient surfaces. In: *The Proceedings of The 8th International Multi-Material Micro Manufacture (4M-2011) Conference*, Stuttgart, Germany (8–10 November 2011) (ed. H. Kück, H. Reinecke and S. Dimov), 176–179.

70 Öztürk, B., Cora, Ö.N., and Koç, M. (2012). Micro-manufacturing of 3-D porous surfaces-investigation on porosity and microhardness" , Oral presentation. In: *The Proceedings of 7th International Conference on MicroManufacturing*, Evanston, IL, USA (11–14 March 2012), 520–525.

71 Miller, D. (2011). Non-contact micro cutting methods: lasers, wire EDM and abrasive waterjets. *Commercial Micro Manufacturing Europe* 4 (8): 13–16.

72 Diserens, M. and Chauvy, P.-F. (2012). An art of fine surface tuning: electro-chemical micro machining. *Commercial Micro Manufacturing International* 5 (6): 10–12.

73 Schneider, R.W. (1989). Electron beam machining. In: *ASM Handbook*, Machining, vol. 16 (ed. ASM Handbook Committee), 568–571. Materials Park, OH: ASM International.

74 Steigerwald Strahltechnick GmbH. http://www.steigerwald-eb.de/en/application-areas/drilling.html (accessed 20 October 2018).

75 Tanasie, G., Boehm, S., Baertle, J. et al. (2008). Electron beam machining of microsystem products. In: *proceedings of 2nd European Conference & Exhibition on Integration Issues of Miniaturized Systems –MOMS, MOEMS, ICS and Electronic Components (SSI)*, Barcelona, Spain (9–10 April).

76 Dubey, A.K. and Yadava, V. (2008). Laser beam machining – a review. *International Journal of Machine Tools and Manufacture* 48: 609–628.

77 Swisstec AG. http://www.swisstecag.com/ (accessed 20 October 2018).

78 Meijer, J. (2004). Laser beam machining (LBM), state of the art and new opportunities. *Journal of Materials Processing Technology* 149: 2–17.

79 Ramanathan, D. *Micromachining*. Resonetics Inc https://resonetics.com/wp-content/uploads/2016/04/article_micromachining_dec2012.pdf (accessed 20 October 2018).

80 Ogura, G. (2013). Fundamentals of laser micromachining polymers in the life science industry. *Commercial Micro Manufacturing International* 6 (1): 24.

81 Fassbind, E. (2012). Laser technology produces the perfect stent. *Commercial Micro Manufacturing International* 5 (5): 16–19.

82 Rizvi, N. (2013). Choosing the laser and processing approach. *Commercial Micro Manufacturing International* 6 (2): 12–18.

83 Schubert, A., Edelmann, J., Gross, S. et al. (2011). Micromanufacturing of hard to machine materials by physical and chemical ablation processes. In: *Proceedings of International Conference on Advanced in Materials and Processing Technologies*, Paris, France (24–27 October 2010).

84 Yan, Z. and Chrisey, D.B. (2012). Pulsed laser ablation in liquid micro-nanostructure generation. *Journal of Photochemistry and Photobiology C: Photochemistry Reviews* 13 (3): 204–223.
85 ALPhANOV (2012). Micromachining news. *Commercial Micro Manufacturing International* 5 (4): 28.
86 Moore, D. and Gillen, D. (2013). Laser micromachining of PEEK and polyimides: advanced tube geometries. *Medical Plastics News* (12): 13.
87 Naeem, M. (2012). Micro welding hard disk drive flexure arms using fibre lasers. *Commercial Micro Manufacturing International* 5 (4): 30–33.
88 John, B., Gebhardt, M., and Wulst, G. (2012). Laser welding of polymer materials. *Commercial Micro Manufacturing International* 5 (5): 12–15.
89 Schaeffer, R.D. (2012). Drilling microholes in plastic medical devices. *Micro Manufacturing Magazine* 5 (1): 24–25.
90 Wang, X., Shen, Z., Gu, C. et al. (2014). Laser indirect shock micro-embossing of commercially pure copper and titanium sheet. *Optics and Lasers in Engineering* 56: 74–82.
91 Wang, Y., Fan, Y., Vukelic, S., and Yao, Y.L. (2007). Energy-level effects on the deformation mechanism in microscale laser peen forming. *Journal of Manufacturing Processes* 9 (1): 1–12.
92 Vollertsen, F. and Sakkiettibutra, J. (2010). Different types to use laser as a forming tool. *Physics Procedia* 5: 193–203.

Part II

Thermal and Energy-assisted Manufacturing Processes

8

Warm Stamping

Fahrettin Ozturk[1,2], *Serkan Toros*[3], *and Ilyas Kacar*[4]

[1] *Department of Mechanical Engineering, Ankara Yıldırım Beyazıt University, Ankara, Turkey*
[2] *Strategy and Technology Management, Turkish Aerospace Industries Inc., Ankara, Turkey*
[3] *Department of Mechanical Engineering, Nigde Omer Halisdemir University, Nigde, Turkey*
[4] *Department of Mechatronics Engineering, Nigde Omer Halisdemir University, Nigde, Turkey*

8.1 What Is Stamping?

Stamping is the general name of sheet-metal forming manufacturing processes, such as blanking, punching, embossing, bending, drawing, stretching, flanging, and coining. All of the sheet-metal forming processes are mostly carried out by machines called presses. The term *stamping press* is used to distinguish these presses from forging and extrusion presses. The tooling that performs sheet metalworking is called *punch*-and-*die* or stamping die. The sheet-metal products are also called as "*stampings*." Stamping process involves clamping a sheet blank at its edges and forcing it into a die cavity with a punch. The working components are the punch and die, which perform the operation. Stretching is more effective rather than squeezing in sheet metal forming process. Material quantity to be drawn into the die is determined by pressure applied on punch. The pressure is used to create not only for cutting force but also for other forces acting at the same time, such as the blank holding force exerted by a die cushion. A *blank* is a starting sheet work piece in plane geometry. It is held between blank holder and drawing die. The *die holder* is assembled to the base of the press, and the punch holder is assembled to the cylinder-piston. Actuation of the piston provides the press working operation. Sheet metal to be processed is in the continuous form of coils or strips in general. The device to stop the sheet metal as it moves through the die when punch is in operation cycle is called as *stop*. Stops block strip's forward motion. In addition to these components, a simple plate named *striper* is used to prevent the sheet metal from sticking to the punch when it moves upward after the operation. As the newly created hole in the coil is the same size as the punch, it tends to put on to the punch when the punch returns back. Stripper reverses the movement of material that tends to put on to the punch after the operation.

8.2 Benefits and Usage Areas of Warm Stamping

Warm stamping method provides as much as possible elimination for disadvantages of the cold or hot forming. In the optimal deformation temperature range, the warm forming allows for maximum deformation ratios. Warm forming offers a viable alternative to cold and hot

Modern Manufacturing Processes, First Edition. Edited by Muammer Koç and Tuğrul Özel.

forming processes. When warm forming is compared to cold forming, it needs lower press work and process results in lower strain hardening, higher ductility. It provides the following advantages over cold forming:

- Power requirement for warm forming operation is less than cold forming. Also forces acting on the tool equipment and strain hardening are less.
- Production of more complex work geometries is possible.
- In some application, annealing need may be reduced or eliminated. Mild steels may be formed without a spheroidization treatment.
- Increased material ductility.

As we compare warm forming to hot forming, warm forming includes the following benefits:

- Oxidation layer and decarburizing is less than hot forming.
- More accurate dimensioning and surface quality can be obtained.
- Finish machining requirements on product are less and very small scrap.
- Because of the finer structures and the presence of some hardening, there is no final heat treatment.
- Tool life is longer since there is less thermal shock and thermal fatigue.
- Heating up to warm forming temperature; generally, requires less energy than hot working temperature. Less press power is required.

Warm stamping is not a subject only in the materials science but also its further development intensively depends on the development in the other fields. The warm stamping of sheet metal forming is an integrated technology of a variety of fundamental techniques. The forming of some high-strength metals in the sheet form is difficult because the surface-to-volume ratio is sufficiently large to rapid heat transfer and the heat in the sheet is transferred rapidly to cold and much massive dies. Successfully, warm formation of the temperature-sensitive materials is obtained under isothermal conditions by a thermostatic chamber or the passing of a large electrical current between tooling besides the blank.

A closed tooling is used for the warm stamping process, producing near-net components almost without springback. Warm stamped parts' weight and dimensional ranges depend on press capacity, e.g. the maximum length of the parts is limited by the available slide stroke. Most of warm formed components are for the automotive industry, e.g. automatic transmission, flanges, alternator pole pieces, bevel gears. Also, some components of ball bearings and parts for the hand tool industry can be manufactured by warm forming.

The warm stamping process is ideally used with aluminum and magnesium alloy material for considerably larger elongation. So many companies can develop new ways of forming aluminum, magnesium, and steel panels into innovative shapes that will help augmented designs. Using warm stamping, much larger parts can be formed on a smaller press due to the lower forming pressures requirement. It makes warm forming a cost-effective production method.

When aluminum, magnesium, and steel are warm stamped, the stamped parts can be painted directly after operation. It makes total production time decreased.

8.3 Warm Stamping and Recent Developments

Sheet metal forming processes are generally performed at room temperature (RT) due to the adopting difficulties of the heating system to the production lines and some other economic concerns. Materials with reasonable ductility are formed at room temperature. However, the materials with low ductility such as aluminum alloys, magnesium alloys, and titanium alloys have premature failure at room temperatures. The fact is that production of a complex shaped part from these alloys is extremely difficult even impossible. In recent years, a warm forming process (temperature ranges from $0.3T_m$ to $0.5T_m$) is proposed in order to overcome the difficulties at room temperature. Warm stamping or forming is a method for improving the formability of low ductility materials at room temperatures. Warm forming operation was first studied in the late 1970s and 1980s by Shehata et al. [1] and Wilson [2] with increasing attention dedicated to this subject, and for the last two decades, it has continued with an increasing trend. It was demonstrated that the formability of the materials was improved at elevated temperatures. In warm stamping, the die and the blank holder usually are heated to a temperature in the required range using electrical heated rods that are located in the dies. At some applications, all dies and punches are kept at room temperature. Also, the blank is heated by several heating methods such as furnace heating, induction heating, and electric resistance heating, etc. If the whole system is kept at same temperature, that process is called *isothermal heating* if not it is called *nonisothermal heating.* Isothermal heating is very rare in production line due to the necessity of the bigger heating system which must be able to cover all die and punch system. Besides, its grandeur, it is difficult to satisfy homogeny temperature distribution.

A typical nonisothermal warm forming setup is shown in Figure 8.1. In the warm forming setup, dies and blank holders are heated up to the required temperature that depends on material. It must be taken into consideration that there will be the risk of necking during heating and cooling processes. As it is described before, the warm stamping is one of the important methods that are used in the metal forming operations in order to increase the formability of the material without creating any change in microstructure contrary to hot stamping operations. In the mechanical properties aspect, it is desired that the general microstructural features of the materials should be kept like before or obtain better features after operation. The selected temperature for the warm forming operations is very important due to the microstructural transform and its effects on the general mechanical properties after cooling of the finished products. In a general sense at elevated temperatures, while formability is facilitating and ductility is increasing; strain hardening rate and strength of the finished product may decrease drastically when a specific temperature level that varies from material to material is exceed. However, at elevated temperatures, some undesired properties in the finished product have started to arise, and after forming process, cooling conditions have to be taken into consideration too due to the undesired effects on the microstructure of the materials. Besides the advantages in which the microstructural features are the same as the materials after forming operations at warm stamping process, lower heating costs and inexpensive tooling are another important opportunity provided to sheet metal manufacturers.

Formability in warm working is usually better than in cold working because of the lower stresses. This method is convenient for the sheet that cannot be formed at room temperature without additional forming or joining working as can be seen in Figure 8.2. Warm stamping

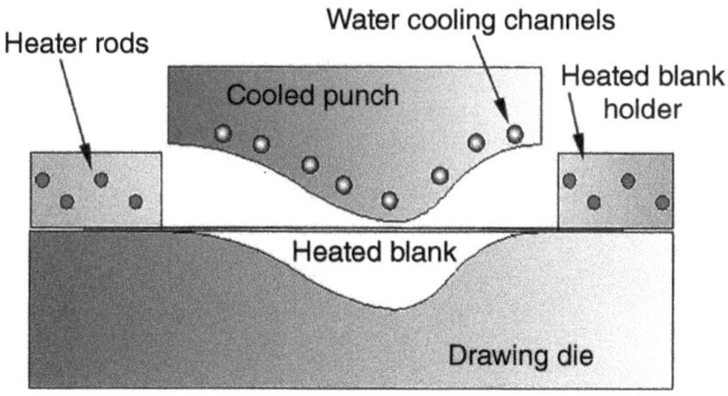

Figure 8.1 A typical nonisothermal warm forming setup.

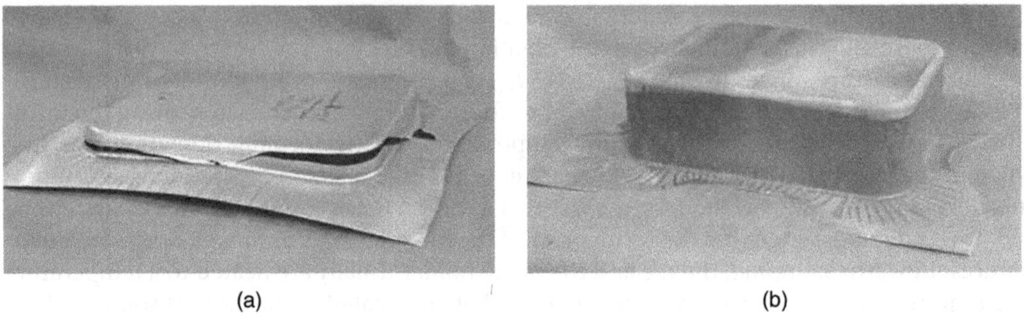

Figure 8.2 Formability at room and warm temperatures. (a) Aluminum – formed at room temperature and (b) aluminum – warm formed. *Source:* Courtesy of US Department of Energy/NETL – National Energy Technology Laboratory.

process may be treated roughly as a method with some advantages of cold working and without some limitations of hot working.

Since all processing is done in the solid-state condition under recrystallization temperature of metal in warm forming operations, there is no need to handle molten material or deal with the complexities of solidification. The material's grains are just moved (or rearranged) to produce the new shape. Starting temperatures of warm stamping process are above room temperature but below the recrystallization temperature. The boundary line between cold forming and warm forming is often expressed in terms of the melting point for a metal. In practical aspect, temperature limits are $0.3T_m < T < 0.5T_m$, where T_m is the melting point (as absolute temperature) for a particular metal. Even today, material behavior is not well-characterized for the warm forming temperatures. For steel, cold forming temperature is 25 °C, warm forming is in 750—900 °C and hot forming is in 1150—1250 °C [3].

The warm stamping operations are generally conducted via heated dies and cold material or heated materials and cold dies. These processes cause nonisothermal condition, which means that there is the heat transfer between materials and die surfaces. In order to obtain isothermal conditions in which there is no heat transfer between working materials and tools, very big thermally isolated volume like environmental chamber with a temperature control unit is required. Application and control of these kinds of systems are difficult for a wide range of stamping operations.

In Figure 8.3, the general heating methods used for warm stamping operation is illustrated schematically. As can be seen in the figure, the work piece (blank) is heated in external chamber and then carried and put into the dies by the transfer system or operator. In the operation, the temperature variations are obtained due to the heat transfer between blank, dies, and

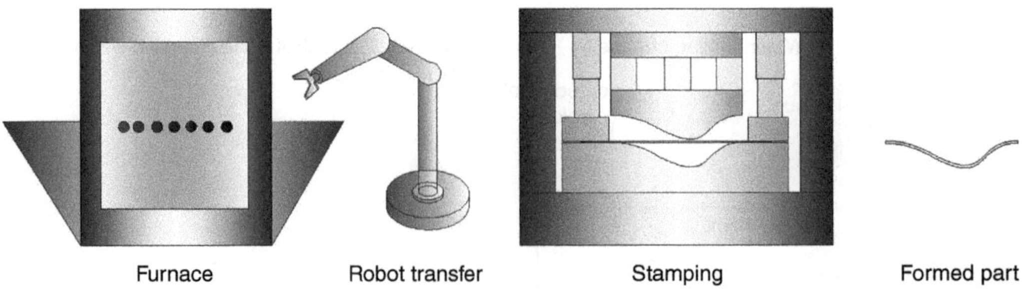

Figure 8.3 A schema of typical warm stamping process.

atmosphere. Therefore, reaching the set temperature value at all points on the materials' surface is impossible. Due to this variation, materials' mechanical properties under various temperature points have to be known very well.

In recent years, there has been significant interest in aluminum and magnesium alloy sheets because of their high strength to weight ratio, excellent thermal conductivity, excellent corrosion resistance, and their weight and energy saving potentials. Although these alloys offer the advantages of a high strength to weight ratio and corrosion resistance, their low formability characteristics limit their usage in many applications with complex shape especially in automotive industry. If aluminum and magnesium alloy sheets are stamped using dies designed for steel working without considering temperature requirements, there have been usually challenging formability problems in part such as *fracture, wrinkling,* or *splitting* to be overcame. Therefore, warm stamping operation is generally applied to these kinds of light-weight materials in order to alleviate this problem by using the elevated but below recrystallization temperature. In literature, there have been numerous studies about determining of the formability characteristics of aluminum and magnesium alloys under various temperature and strain rate. The warm forming method has improved the formability of the aluminum alloys such as 5082 and 5005 due to the increased strain rate hardening. Also partial or localized heating techniques are an other important factor to make comparison on nonisothermal formability. The formability with a partial heating in the holder or matrices area is much better when compared with the homogeneously heated tools. A significant increase in the limiting drawing ratio (LDR) for the aluminum alloy AlMg4.5 Mn0.4 can be achieved by a heated and lower strain rated hydro-mechanical stamp [4, 5].

Another determiner of material formability is the *limiting drawing ratio,* defined as the ratio of the successfully drawn largest blank radius to the punch radius. In a general aspect, for aluminum and magnesium alloys, the grandeur of LDR is increased with temperature and decreased with increasing strain rate [6, 7]. Besides that, LDR increases mostly with increasing die temperature, the deformation resistance in flange shrinkage decreases with an increase in temperature. Additionally, LDR becomes smaller with increasing forming speed at all temperatures since the flow stress of the heated blank (at the flange) increases with increasing strain rate. Moreover, the cooled blank at the punch corner becomes less ductile.

The *forming limit diagram* (FLD) is another approach to evaluate formability of sheet metals. When the effects of temperature and forming speed on the FLD for fine grain Al–Mg alloy (5083-O) sheet were investigated by performing stretch-forming test at various forming speeds (0.2–200 mm/min) at several temperatures from 20 to 300 °C, it was shown that the forming limit strain increased drastically with decreasing speed for any strain paths at a high temperature ranging from 150 to 300 °C [8]. It is known that the FLD is not sensitive to speed at room temperature for numerous metals. The improvement in formability at 300 °C at low forming speed is specifically due to the high strain rate hardening characteristic of the material, but below 200 °C, the formability is also affected strongly by strain hardening [6, 8]. Figure 8.4 shows the effects of temperature on FLDs of 5754, 5182, and 6111-T4 aluminum alloys. As seen in the figures, the formability of aluminum alloys increases with increasing forming temperature. Temperature sensitivity of Al 5754 is greater than other aluminum alloys [9].

The number of available 5XXX series Al-based alloys for passenger vehicles is very limited. At the present time, 5052 and 5456 are the most commonly used alloys. Although 5052 offers a good combination of mechanical properties, corrosion resistance, and formability, it is unsuitable for use at temperatures above 120 °C due to its poor creep resistance and low strength at elevated temperatures. In order to get a better overall understanding of alloys and to identify the most promising compositions, most researchers have examined and evaluated the micromechanical features.

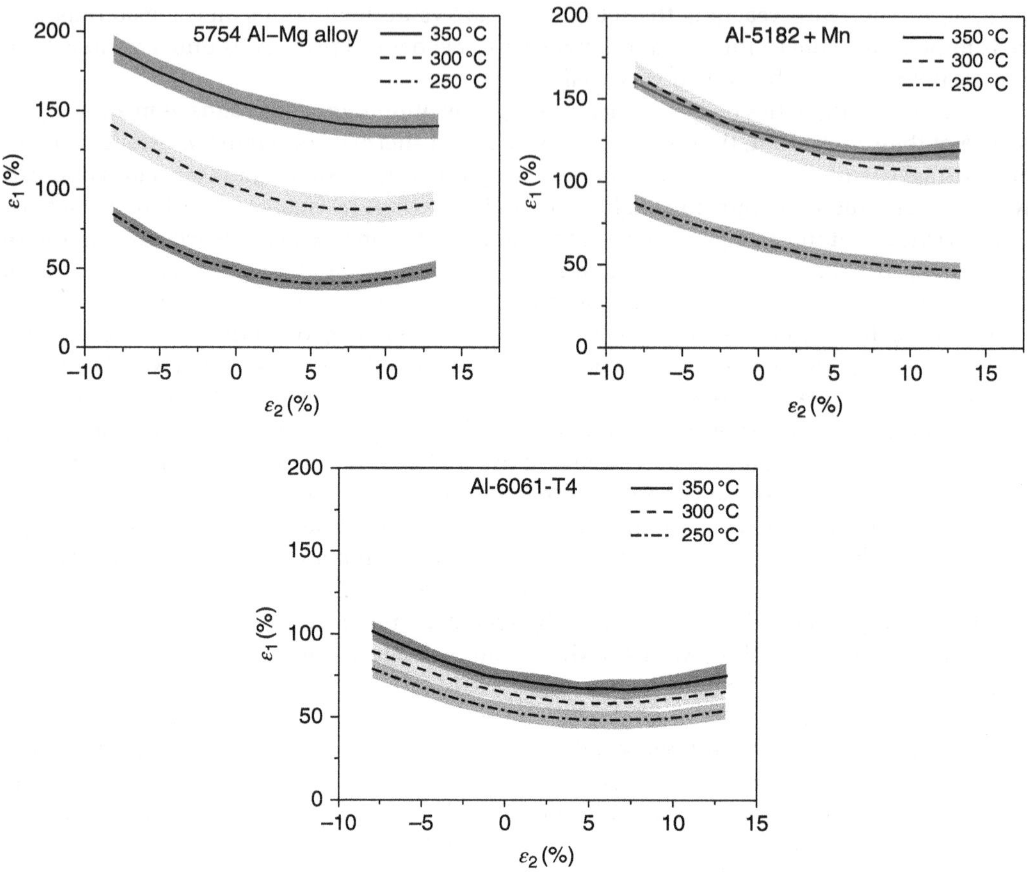

Figure 8.4 The effect of warm temperatures on FLDs. *Source:* From Li and Ghosh 2004 [9].

When the formability was compared for 1050, 5754, and 6016 type aluminum alloys from 100 to 250 °C, it was shown that the minimum die temperature of 6016 alloy on the die process limits was lower than that of the 5754 alloy [10]. Also, when the strain rate sensitivity of 5754 and 5182 type aluminum alloy sheets is investigated at several temperature ranges from room temperature to elevated temperatures and at high strain rates like 600, 1100, and 1500 s^{-1}, the quasistatic and dynamic stress–strain responses are low for both alloys. AA5754 exhibits a mild increase in flow stress with strain rate, while AA5182 is insensitive. The ductility of the alloys shows little differences in the temperature range between 23 and 150 °C at a strain rate of 1500 s^{-1}. However, the final elongation is decreased for both aluminum alloys at 300 °C and a strain rate of 1500 s^{-1} when compared to that at lower temperatures [11]. For the tensile properties of 5052 aluminum–magnesium alloy in the range between room temperature to 300 °C and in the strain rate range of 0.0083–0.16 s^{-1}, the uniaxial tensile elongation of the material increases with increasing temperatures and decreases with increasing strain rates. The formability of this material at warm temperatures is better than that of at room temperature and the most suitable forming conditions are obtained at 300 °C and 0.0083 s^{-1} [12]. The effects of strain rate and temperature on the deformation behavior of hardened 5083-H111 aluminum–magnesium alloy sheet had been studied by performing uniaxial tensile tests at various strain rates from 0.0083 to 0.16 s^{-1} and temperatures from −100 to 300 °C. The obtained improvement in formability at cold temperatures is principally due to the strain hardening of the material. However, the improvement at warm temperature and low strain rate is specifically due to the high strain rate sensitivity characteristic of the material. This alloy should be formed at temperatures higher than 200 °C and at low strain rates [12, 13]. Another most important material property for the plasticity used for determining the optimum forming condition is the yield surfaces. A material's yield surface is generally described as the initiation of the plastic deformation under several stress combinations that occur during the forming operation. This yield locus is directly related with the yield stress of the materials and therefore, this feature also depends on temperature and strain rate. For the fine grained Al–Mg alloy (5083-O), the size of the yield locus drastically decreases with increasing temperature and increased strain rate [14]. This feature also shows the possible springback characteristic of the material. If the yield locus is large, the springback will increase. The formability of the materials is also affected from isothermal or nonisothermal forming conditions. According to a study, for biaxial formability characteristics of Al 5754, Al 5182 containing 1%Mn and Al 6111-T4 at warm temperature levels, it can be seen that the formability for all these three alloys improves at elevated temperatures. However, the strain hardened alloys Al 5754 and Al 5182+Mn show considerably greater improvement than the precipitation hardened alloy Al 6111-T4. It is also shown that if the die temperature is set slightly higher than punch temperature, formability increases remarkably [9]. For warm deep drawing process of AA5754-O aluminum alloy, the temperature in the blank center has a strong influence on the process feasibility and thus on the material formability [15]. Also, in deep drawing process with localized heating and cooling technique with variable blank holder pressure control, deep drawing performance of the materials is enhanced using the appropriate temperature distribution for the local heating and cooling technique [16–18]. For the fine grained Al–Mg alloy (5083-O) sheet in cylindrical deep drawing tests at various forming speeds (0.2–500 mm/min) at die temperatures of 20–180 °C (the die was heated, while the punch was water cooled during the tests), the LDR mostly increases with increasing die temperature, because the deformation resistance in flange shrinkage decreases with temperature rise and the LDR also becomes lower with increasing forming speed at all temperatures because of the flow stress of the heated blank at the flange increases with increasing strain rate. Moreover, the cooled blank at the punch corner becomes less ductile [6].

Different test methods performed are very remarkable to make comparison on deformation behaviors. In most materials, the compression and tensile properties shows difference and in general aspects, the strength of the alloy is slightly greater in compression than tension. Particularly at elevated temperature levels, the difference gradually disappears as strain rate decreases [19].

Another most common forming process that is used for manufacturing of dome-shaped pressure vessels for passenger and aeronautical vehicles is *spin forming*. In this process, high plasticity is expected from the materials, and thus the application of this process to aluminum and magnesium alloys is so rare. A large amount of deformation can be obtained by the application of the temperature to these kinds of lightweight materials [18].

Besides the low formability characteristics at cold temperature levels, another disadvantage of aluminum alloys is that these alloys suffer from stretcher lines, which gives an uneven surface after deformation. Because of this reason, 5XXX series aluminum is used for inner panels of vehicles. These undesired surface defects can be eliminated by the forming processes at the elevated temperatures [20]. The aluminum, which contains 6% magnesium could give a 300% total elongation at about 250 °C, is used in more applications in industry [21].

In the investigation on the mechanical behavior of the commercial aluminum alloy AA5182-O, the dynamic strain aging effect was observed at all temperature between −80 and 110 °C and at strain rates lower than $10^{-1}\,s^{-1}$ [22]. In uniaxial tensile deformation behavior investigation of 5754 aluminum–magnesium alloy sheet at cold (−60 to 0 °C) and warm (RT to 250 °C) forming temperatures and a strain rate range of 0.0016–0.042 s^{-1}, the formability of this material at cold and warm temperatures is better than at room temperature. Stretching lines are eliminated at both cold and warm temperatures especially above 175 °C and 0.0016 s^{-1} strain rate. The most suitable forming conditions are obtained at 250 °C and 0.0083 s^{-1} or 175 °C and 0.0016 s^{-1} [23].

8.4 Effects of Temperature on Strain Hardening for Warm Stamping

That metals are deformed plastically at temperatures lower than recrystallization temperature of that metal causes an increment on to the strength and hardness. It is named as work/strain hardening. Strain hardening is simply based on the new orientation of the grains done by presswork.

The amount of formability is directly related with hardening ability of a material and it increases with increasing strain hardening ability "n" which is determined by using power law (Hollomon) equation ($\sigma = K\varepsilon^n$) in the simplest way. This value can be calculated by curve fitting methods to experimental tensile test data for the uniform plastic deformation region. Although heated materials' hardening ability decreases with increasing temperature, the uniform elongation increases remarkably. What might be the main reason of this behavior? Initially, it is necessary for us to describe the hardening phenomena. The number of the dislocations increases with the plastic deformation, and these increments make their movements locked and fall in to the grain boundaries. These intersections force the dislocations to move on different slip planes during deformation. The dislocations on different slip planes move and then start to lock the others and a rapid hardening mechanism occurs.

During the forming operation at elevated temperatures, the recrystallization is not seen in the microstructure, however, with the recovery mechanism, the dislocations that occur during the forming operation, can be rearranged. Increasing and locking of them with each other are delayed by the applied forming temperature. Therefore, although the strength level of the materials decreases with the temperature, the uniform elongation, thus the formability increases. The numbers of the slip planes have an important role on the work hardening mechanisms of a material. Besides the recovery mechanism, the number of slip planes and directions increase with the temperature and applied stress level. So the dislocation can move more easily with the deformation. As the movement of dislocations become easier, important material parameters on stamping operation will change. Materials' yield, ultimate tensile strength, and hardening exponent values will decrease with increasing temperature.

In Figure 8.5, the variation of the mechanical properties of 5754-O aluminum–magnesium alloy with applied temperature is depicted. As can be seen in the Figure 8.5a, strain hardening ability decreases with increasing temperature. At cold temperature (−60 °C), strain hardening exponent has the highest value, about 0.43 and at 250 °C this value decrease to 0.17. However, the graphics in Figure 8.5b,c show some fluctuations on the yield and ultimate tensile curves with increasing temperature. Both of yield and maximum stresses increase with increasing temperature up to 100 °C and then softening starts after that point. In 5XXX series aluminum–magnesium alloys, that behavior is seen for both 5754 (121 °C) and 5052 (93 °C) [24], and 5083 (100 °C) alloys at a specific strain rate [6]. The amount of the total elongation decreases up to 100 °C but then increases remarkably.

The need for fuel savings by structural weight reduction in vehicles is driving the replacement of steel by lightweight metals. The usage of aluminum and magnesium alloys

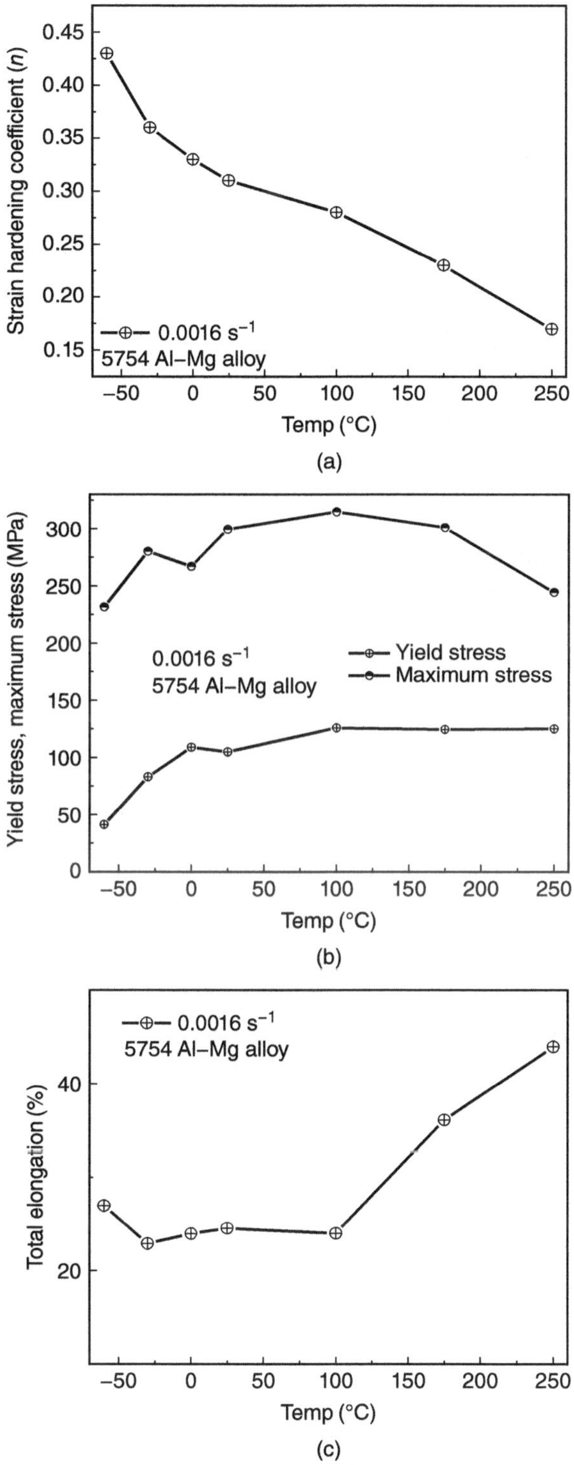

Figure 8.5 Variation of some mechanical properties with applied temperature. (a) Strain hardening coefficient, (b) strength, and (c) elongation.

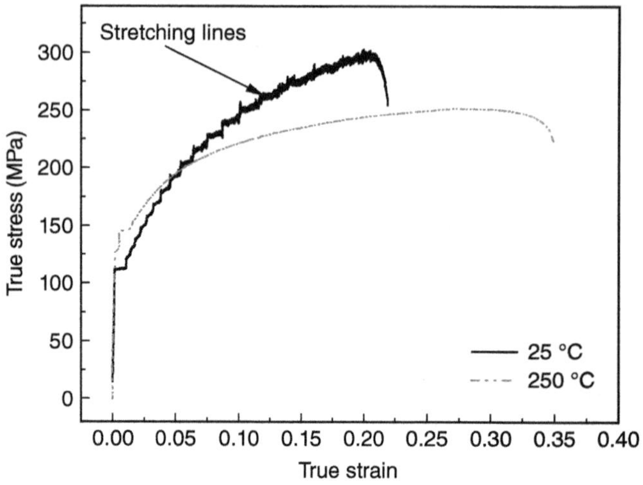

Figure 8.6 Stretching strain marks (Lüder's band).

in automotive applications requires overcoming formability and surface finish problems, for example the stretcher lines defects on the product surfaces. The stretching strain marks are similar to the Lüder's bands, which are observed in steel. These problems usually appear at room temperature and low strain rates. These bands can be visible even after painting the parts. During the stamping of parts, Lüder's band is undesirable since they affect the surface quality of the finished parts. If the Lüder's band does not traverse to the whole surface, its boundaries will create a surface roughness, and it will create some trouble during the painting operation. As can be seen from the Figure 8.6, the 5754 Al–Mg alloy exhibits serrated flow curves (as other types of Al–Mg alloys like 5052 and 5083) at lower temperatures from RT to 100 °C. This is due to dynamic strain aging, which leads to stretch marks in a product after the forming process. Dynamic strain aging can be explained as the interaction between dislocations and solute atoms. Particularly in this aluminum group (5XXX series), the magnesium atoms block dislocation movement, and this leads to a higher initial yield stress. If dislocations move slowly (at low strain rates), the solute atoms can migrate to the dislocations while arresting at other obstacles or solutes. That migration also constrains the dislocation movement. As temperature increases, flow stress of the material decreases with corresponding increase in the elongation up to failure. This is due to the increment in the mobility of the solutes that eliminates the serrated flow behavior. Correspondingly, no Lüder's band develops at elevated temperatures.

This phenomenon is similar to discontinuous yielding in steels. The yield point effect in steels can be also explained by the hindering of dislocations with interstitially dissolved C and N atoms. To eliminate Lüder's band for the steel sheets, stamping speed can be increased as shown in Figure 8.7.

8.5 Interrelation of Temperature and Strain Rate

Deformation speed has great effects on the formability of the materials. In the ideal conditions, it is hoped that hardening exponential $n = 0$, and yielding remains uniform. In practice, some other unwanted cases arise during deformation process. Especially elevated temperatures, a condition that is frequently encountered in metal forming process, make this process more sensitive to the speed of deformation. For these speed-sensitive materials may require more

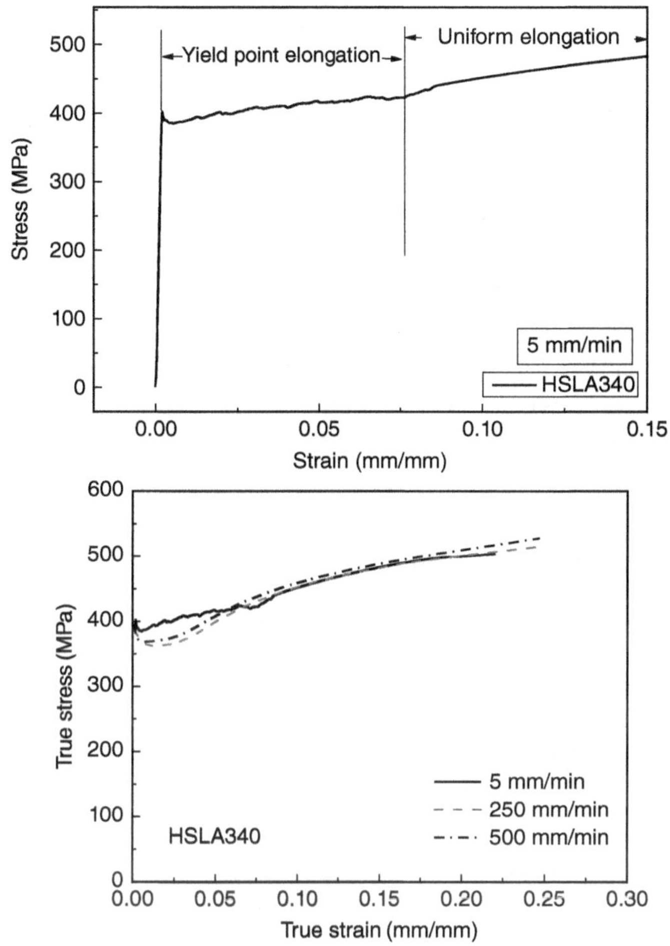

Figure 8.7 Discontinuous yielding and Lüder's band.

energy and stronger tools. The selection of press for the warm forming of a small-sized product may depend on the speed sensitivity of the material being formed.

In the case of the significantly greater rate of deformation, tensile test data obtained from slow strain rates may be totally useless. If impacted type or higher level loading is implemented on the strain rate sensitive materials, it may shatter or crack but will be formed plastically when subjected to slow-speed loadings. Strain rate is directly related with deformation speed, in metal forming process. Yield stress depends on strain rate and temperature, usually increases with strain rate and decreases with temperature.

For most of the materials, when the strain rate sensitivity increases, the flow stress rises and this case depends on the material and temperature. When such materials have the positive strain rate sensitivity, the necking delays and formability increases. When a blank starts to be formed gradually, a specific region on the blank may thin faster than the remaining portion. Therefore, in this region, the strain rate will be much higher than the remaining portion. If the strain rate sensitivity of the material is high enough, additional strengthening can be obtained on the deformed region. Then the remaining portion that are close to thinner portion will start to be deformed, respectively. Thanks to positive and high strain rate sensitivity, fractures are delayed during the forming operations.

The effects of temperature and strain rate are interrelated, and generally, at room or negative temperature points, the mechanical properties change as if the strain rate increases. As aforementioned before, with the increasing temperature, the dynamic recovery increases and lower value of steady-state stress level is obtained. Similarly, when the deformation speed increases, the rate of increments in dislocation density increases and larger stress saturation can be obtained. At low temperatures, the strain rate effect essentially disappears because the dynamic recovery is much smaller than the higher temperatures, and it is negligible when compared to the effects of hardening rate.

Although deformation stress increases with strain rate, hardening capability decreases. The strain rate sensitivity may significantly increase with increasing temperature for the material which has negative strain rate sensitivity at RT shows. Superplastic deformation depends on strain rate sensitivity. The material that has higher strain rate sensitivity than 0.5 shows the superplastic deformation behavior at high temperatures. The elongation can be reached up to 1000% (one thousand percent) in the superplastic deformation case. This feature also provides high production rate that affects the production cost.

In Figure 8.8, stress–strain curves obtained at various temperatures and strain rates are depicted for AZ31 and 5083 alloys.

As can be seen from the Figure 8.8, the flow curves of two different alloys show similar behavior with increasing temperature and strain rate. For AZ31 magnesium alloy, the strain rate sensitivity does not change with increasing strain rate at room temperature. Therefore, no stress increment is obtained. These results are also valid for 5083 Al–Mg alloy. However, the stress levels of both materials show more sensitivity to the increased deformation speed at elevated temperatures and while the stress value increases, total elongation decreases. Additionally, the stretcher strain marks (Lüder's bands) disappear with increased temperature and strain rate for the 5083 Al–Mg alloy.

8.6 Effect of Temperature and Deformation on Elasticity Modulus

Nowadays, the capability of the simulations and analysis in manufacturing industry has been improved significantly by using developed models that describes the materials' behavior under various conditions such as temperature, strain rate, and amounts of deformation. Understanding of the materials' behaviors is one of the key components in providing the reliable results of capability predictions, and the compatibility of the simulation results with the experiments or real part forming operations increasing remarkably day by day.

Another most important material parameter used for describing the stiffness capability is elasticity modulus. This property strongly depends on the temperature selected for forming operations. If a forming operation is performed under isothermal conditions (no heat transfer between blank and dies), the constant elasticity (Young's) modulus measured at prescribed temperature is required in order to obtain good agreement with simulation and experiments. However, since the temperature gradient is obtained in nonisothermal forming operations, the general variation of elasticity modulus with temperature should be known in order to obtain well-predicted simulation results, particularly for springback simulations. The variation of elasticity modulus with temperature for some metals is depicted in Figure 8.9. As it is a well-known phenomenon, the elasticity modulus decreases with increasing temperature. Besides the elasticity modulus, the absorbed strain energy decreases with increasing temperature due to remarkably decreased yield strength. Recently, in addition to temperature, some researches show that the elasticity modulus decreases with not only temperature but also deformation. Since the elongation increases with increasing temperature, it is expected that the elasticity modulus decreases more than those in lower temperature levels.

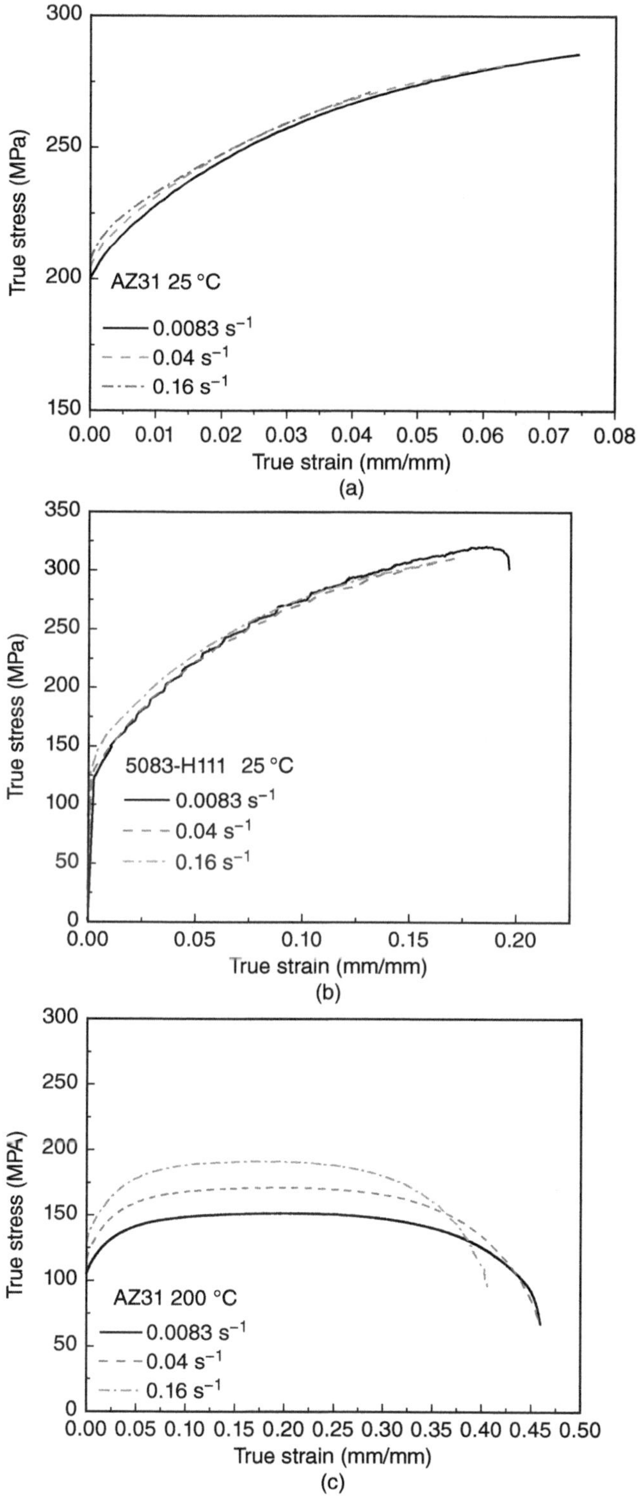

Figure 8.8 Effects of temperature and strain rate on flow curves of AZ31 magnesium and 5083 Al–Mg alloys. (a) Stress–strain curve of AZ31 alloy at 25 °C, (b) stress–strain curve of 5083 alloy at 25 °C, (c) stress–strain curve of AZ31 alloy at 200 °C, (d) stress–strain curve of 5083 alloy at 200 °C, (e) stress–strain curve of AZ31 alloy at 300 °C, and (f) stress–strain curve of 5083 alloy at 300 °C.

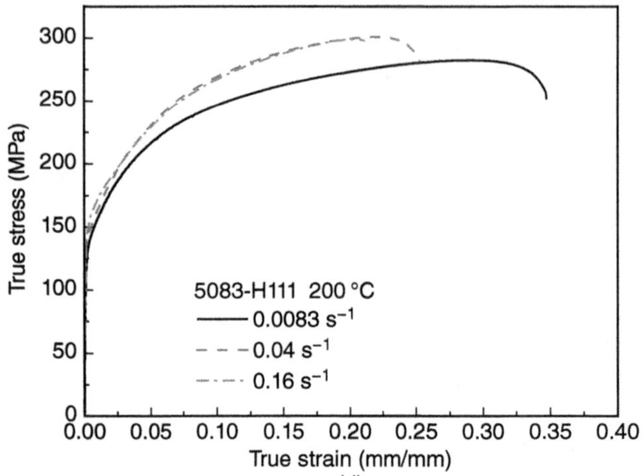

(d)

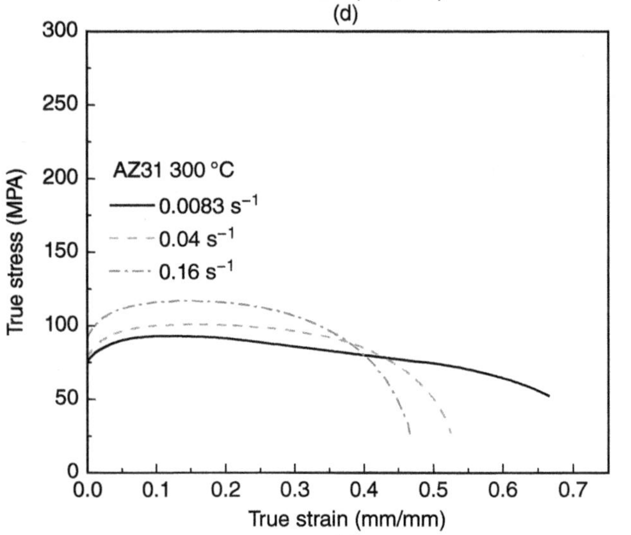

(e)

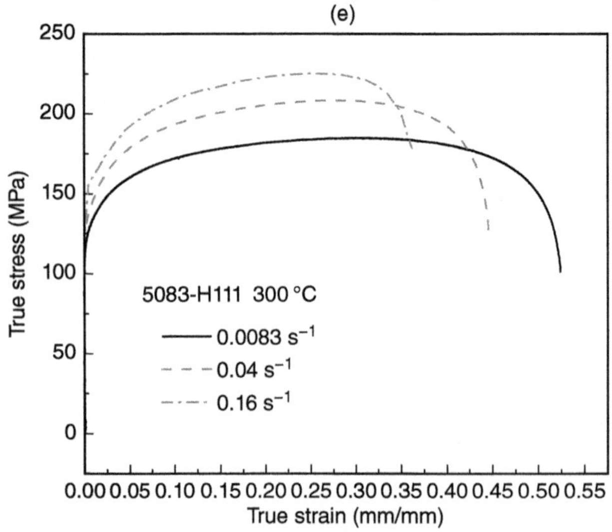

(f)

Figure 8.8 (*Continued*)

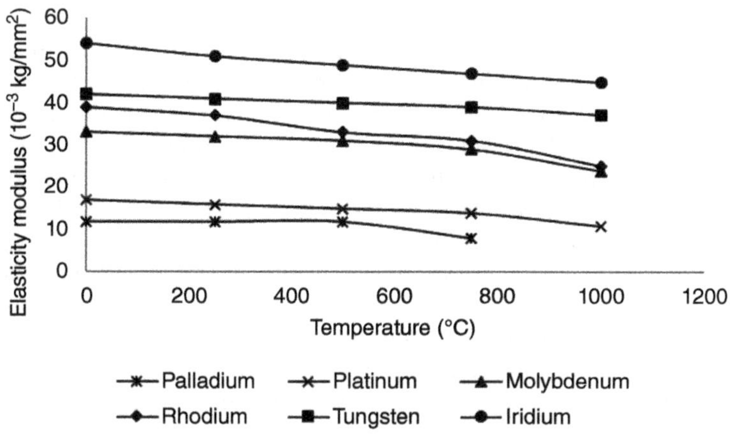

Figure 8.9 The variation of elasticity modulus of several materials with temperature. *Source:* From Darling 1966 [25].

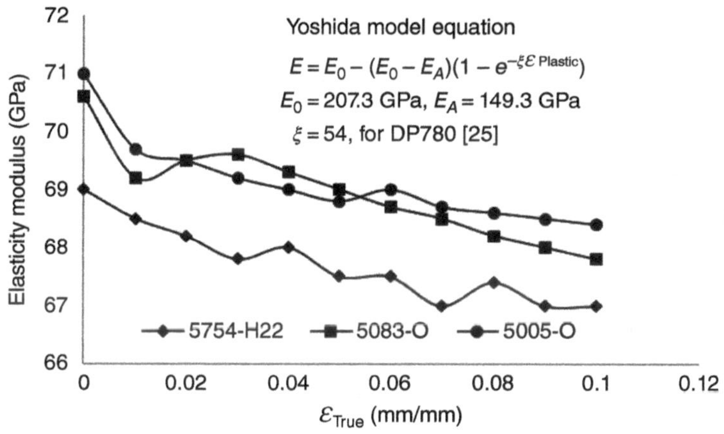

Figure 8.10 The variation of elasticity modulus with plastic deformation.

In Figure 8.10, the variation of the elasticity modulus with respect to the deformation is depicted. As can be seen from the figure, the decrease of the elasticity modulus can be defined as in exponential form. In literature, this variation is generally defined the best by Yoshida model [26]. These kinds of property changes must be considered during modeling and simulations of the processes.

8.7 Effect of Temperature on Springback

In most metal forming operations, the target geometrical shape cannot be obtained exactly due to the material and process parameters. The final shape of the material depends on the tool design, process parameters, and material type of the blank. These parameters and their influences must be considered and taken into account prior to manufacturing of the dies in order to obtain desired shape and features. Otherwise, some kinds of defects such as fracture, wrinkling, and orange peel might be encountered.

Another most common defect in metal forming operations is *springback*. The springback can be defined as geometrical distortion caused by "elastic recovery." Right after forming operation, the shape of the deformed product closely conforms to the geometry of tools. However, as soon as the tools are retracted, an elastically-driven recovery is seen and deformed blank tries to return its original shapes a little. Besides the low elasticity modulus, springback is related to many parameters such as forming conditions, tool geometry, sheet thickness, yield stress, work hardening coefficient, and strain rate sensitivity. As a forming condition, most of the studies show that the applied temperature is a key factor to eliminate the springback. As aforementioned before, principally the applied temperature affects the material properties such as its yield stress, elongation, elasticity modulus, and absorbed strain energy. Generally, springback is associated with elastic energy, and this value decreases with increasing temperature. In Figure 8.11, a typical forming tool and obtained geometry are shown. As can be seen in the figure, the target and real part shapes show some differences at various regions.

In the literature, there have been numerous studies about the eliminating of springback. Two –dimensional (2D) drawing and *V-shaped* die bending test methods are the most common ways for determining of springback under different conditions as temperature,

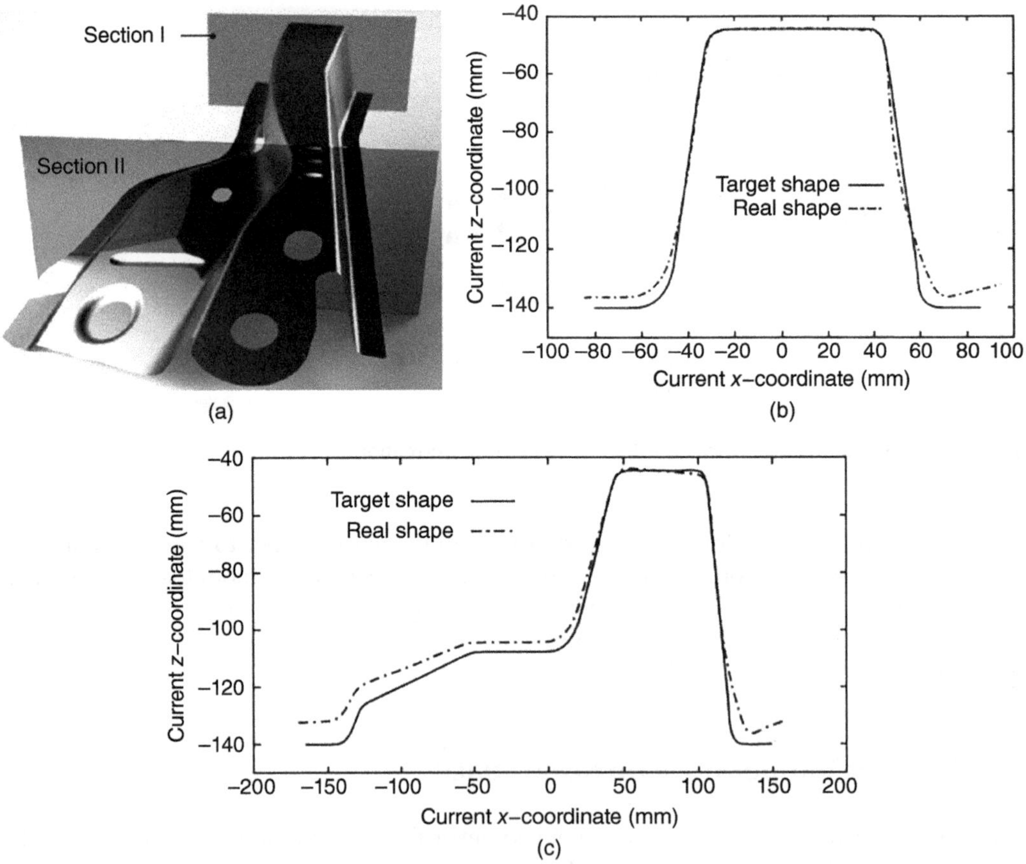

Figure 8.11 Obtained geometries have some differences from desired geometry. (a) Measuring planes, (b) section I, and (c) section II. *Source:* From Burchitz 2008 [27].

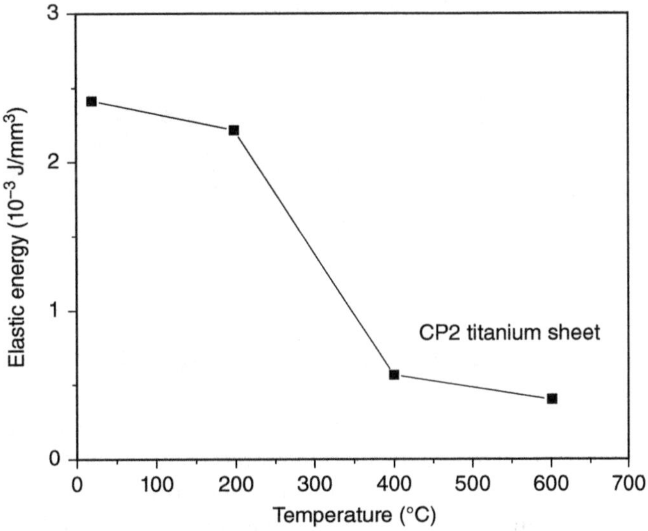

Figure 8.12 Variation of the elastic absorbed energy with temperature.

deformation speed, and blank holder force (BHF). The experimental studies about springback characterization are generally consist of both experimental methods. In Figure 8.12 variation of the absorbed elastic energy with temperature is depicted. The elastic energy can be calculated by using the area up to yield point under stress–strain curve for each temperature.

As shown in the Figure 8.12, the elastic energy of commercially pure (CP grade) Ti alloy reduces remarkably with temperature. The springback angle also reduces with temperature as can be seen in Figure 8.13. Springback on the CP2 titanium alloy sheet can be eliminated at approximately 578 °C as can be seen in Figure 8.14.

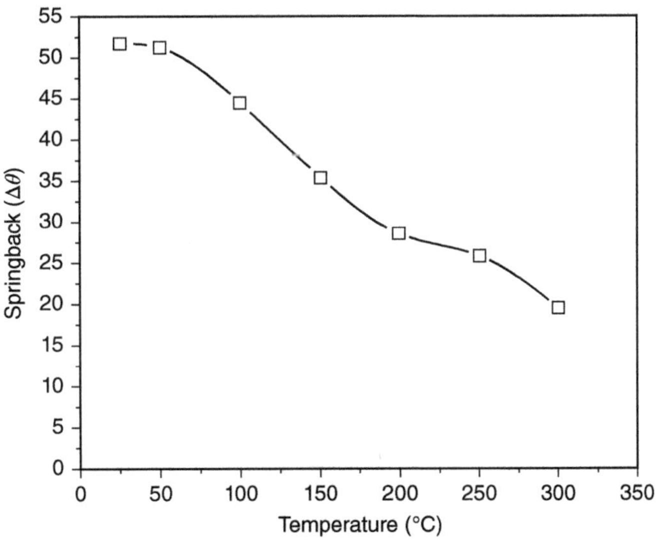

Figure 8.13 Variation of the springback angle with temperature.

Figure 8.14 Comparison of springback behaviors of CP2-Ti alloy sheets at RT and 578 °C.

8.8 Effect of Temperature on Forming Limit Diagrams (FLD)

In the real forming application, there are some deformation modes that occur during the forming process. These deformation modes completely depend on the dies' complexity and materials' characteristics prior to forming operation, and formability limits at all points on the blank are not similar according to the deformation modes. In order to determine the forming limits under different deformation modes, FLD was proposed firstly by Keeler and Backhofen [28] and then Goodwin [29]. Forming limit diagram is a representation of critical combination of two principal surface strains named as *major* and *minor*. It has a border curve that localized necking is observed in upper side. For varying strain ratios, from pure shear to equi-biaxial tension, the forming limit curve is plotted. When the strain ratio is positive, in other words while minor strain is positive, the process is *stretching*. In the case of negative strain ratio, when minor strain is negative, the process is *deep drawing*. It should also be noticed that the strains plotted are true strains. The different straining modes that can occur during the forming operations can be seen in Figure 8.15.

Because the elongation capability increases with increasing temperature for most materials, forming limit curves change with temperatures. These curves give more freedom to designers for more complex parts. In Figure 8.16a,b, the FLDs obtained under various temperature are shown for 5754 and AZ31 alloys. As seen in the figure, the formability limits at all deformation

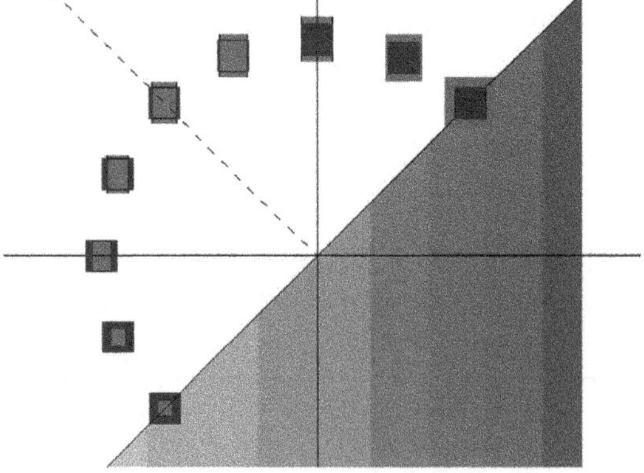

Figure 8.15 Deformation modes on a sheet blank during forming operations.

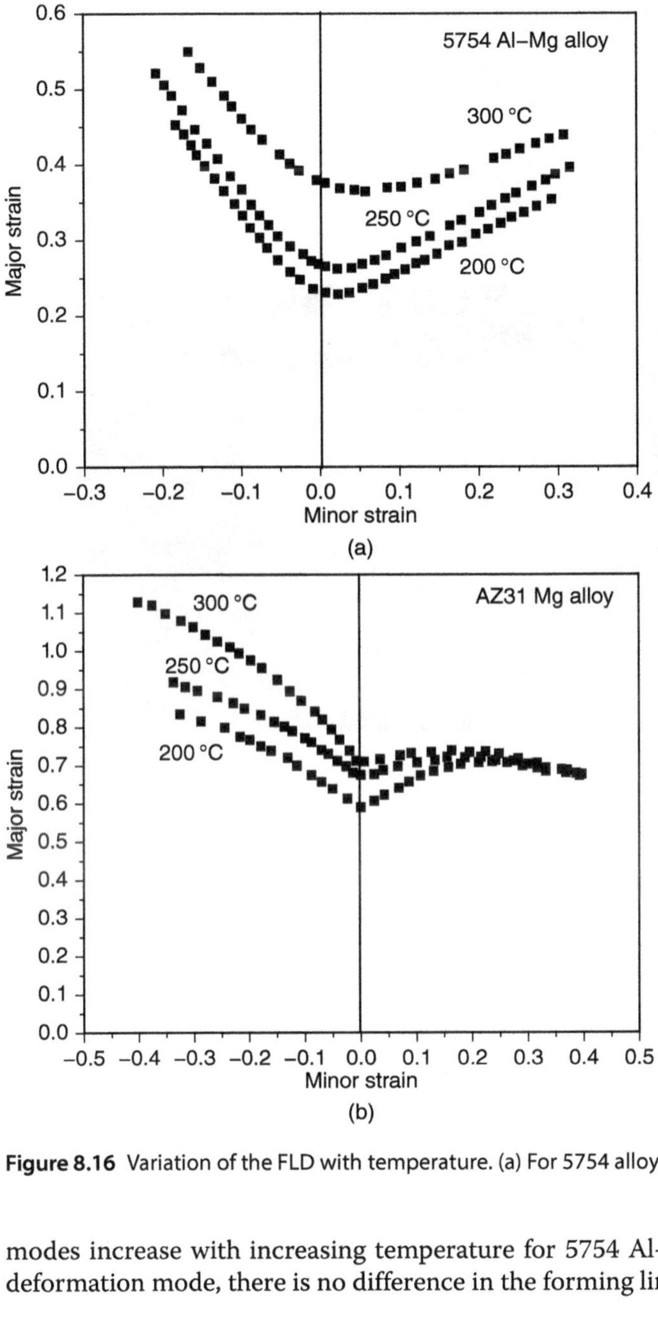

Figure 8.16 Variation of the FLD with temperature. (a) For 5754 alloy and (b) For AZ31 alloy.

modes increase with increasing temperature for 5754 Al–Mg alloy. However, for the biaxial deformation mode, there is no difference in the forming limit for AZ31 Mg alloy.

8.9 Analyze Techniques on Formability at Warm Stamping

Evaluation of grids on the surface of a deformed sheet is an analysis method that is used for determining the strains. In this technique, grids are painted or etched on sheet surface before deformation. After the deformation, quantities and directions of the deformation are measured

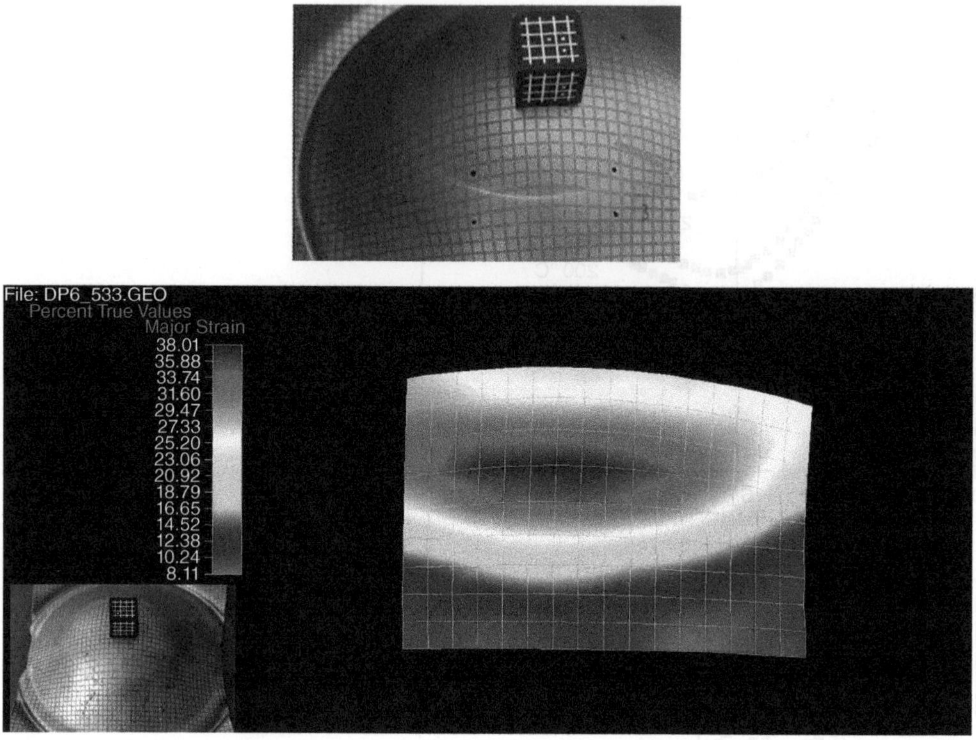

Figure 8.17 Measurements of deformation by using the grid technology.

grid by grid to compare undeformed one as in Figure 8.17. The magnitude and direction of principal strains are calculated by evaluating these measurements taken over necking zone on stamping. Measurements are plotted on the FLD. Each sample corresponds to a point on the diagram.

As another analysis technique, the finite element modeling method has become popular since 1980s for metal forming simulations and the continued development of computing power and speed; process simulations are now quick, inexpensive, and quite accurate. Advanced simulation techniques can provide a clear and thorough understanding of the process, eliminating costly trial-and-error development cycles. Finite element analyze techniques have capabilities on modeling, part design, manufacturing process design, heat treatment, and surface treatment optimization, and others of the complex 2D or 3D continuum problems in metal forming. However, the prediction capability of the simulation is limited by used mathematical models that represent the material behavior. The prediction results will be more convenient when right material model and its parameters are used. The mathematical models should predict how a material will respond to warm forming operation and how the final properties at all locations within the product will be.

Designing optimal temperature distribution on the blank is directly affecting the overall success of the warm forming process. Numerical simulations of the stamping were performed for the temperature distribution of 5083-O Al–Mg alloy at various die and blank initial temperature conditions. The main goal of these numerical simulations is to perform thermal transient analyses and obtain the optimum warm forming condition. Using symmetric boundary conditions on the model in Figure 8.18, temperature distributions on the blank

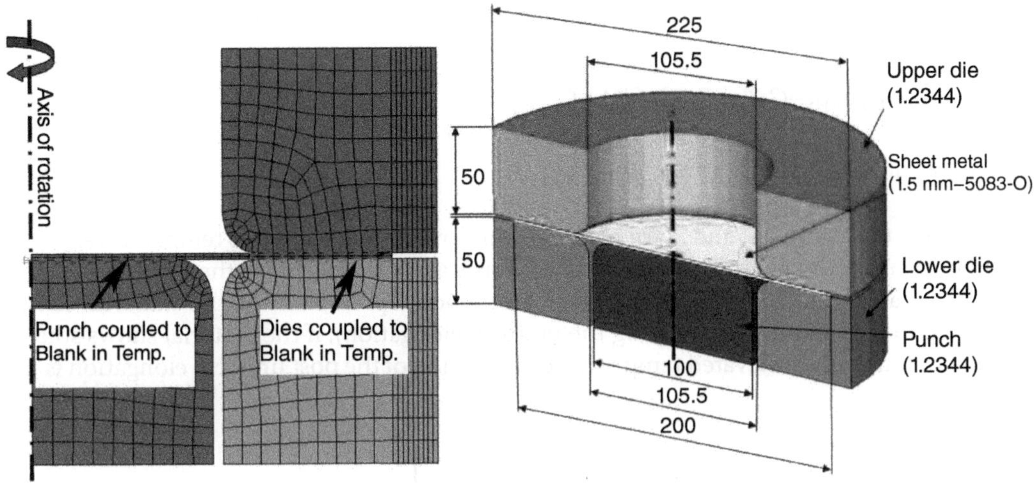

Figure 8.18 A 2D axial-symmetric finite element model.

surface were predicted for different temperatures, RT, 200, 250, and 300 °C, respectively, as in Figure 8.19.

A hemispherical dome test simulation model is also illustrated as is in Figure 8.20. Results for a 5083 Al alloy blank show the effect of temperature on formability as in Figure 8.21. The finite element results show that formability increases with temperature.

Particularly, parameters for finite element analysis (FEA) model are obtained by using curve fitting technics on data from tensile tests performed at various temperatures and strain rates. Description of the flow curve for a certain condition can be done by three methods. The first one is to use an *experimental flow curve* in software. For any case of deformation, the stress and strain values are calculated by various linear or other kinds of interpolation methods. The second one is *static formulation* in which the stress depends on the strain. The most common formulation of the flow curve is Power law ($\sigma = K\varepsilon^n$) where K is the strength coefficient and n is the strain hardening parameter. Both of these parameters can be determined by curve fitting technique to the experimental data obtained for a certain condition. Another most common method is to use a *dynamic model*. This method generally includes both temperature and strain rate effects or only one of them. The flow curve is given as a function of temperature and strain rate according to given equations.

In literature, many dynamic models can be found, but the most commons are Johnson–Cook (JC) [32] and Zerilli–Armstrong [33] models due to their simplicity in the view of computational effort. The JC model is given in Eq. (8.1).

$$\sigma_Y = \left[A + B\left(\varepsilon^p\right)^n\right]\left(1 + C\ln\dot{\varepsilon}\right)\left[1 - \left(T_h\right)^m\right] \tag{8.1}$$

where A is the initial yield strength of the material at RT. The equivalent plastic strain rate $\dot{\varepsilon}$ is normalized with a reference strain rate $\dot{\varepsilon}_0$. T_h is the homolog temperature, and it can be defined as $((T - T_0)/(T_m - T_0))$. T_0 is RT, T_m as the melting temperature of the material, and T is the test temperature. While the parameter n takes into account the strain hardening effect, the parameter m models the thermal softening effect and C represents strain rate sensitivity.

Zerilli–Armstrong developed a constitutive model based on dislocation-mechanics theory and considering crystal structure of materials. The model makes a distinction between body cubic centered (BCC) and face cubic centered (FCC) lattice structure as given in Eqs. (8.2) and (8.3), respectively.

$$\sigma = C_0 + C_1 \exp\left(-C_3 T + C_4 T \ln(\dot{\varepsilon})\right) + C_5 (\varepsilon)^n \tag{8.2}$$

$$\sigma = C_0 + C_2 (\varepsilon)^n \exp\left(-C_3 T + C_4 T \ln(\dot{\varepsilon})\right) \tag{8.3}$$

Where C_0, C_1, C_2, C_3, C_4, C_5, and n are material constants and T is the test temperature.

Another dynamic model that was derived by Toros and Ozturk [34] is the softening model. Although JC and Zerilli–Armstrong models are good at prediction of the material tensile behavior up to maximum stress (during the uniform elongation), if the material shows softening behavior tendency at elevated temperature, prediction of the post uniform elongation is not

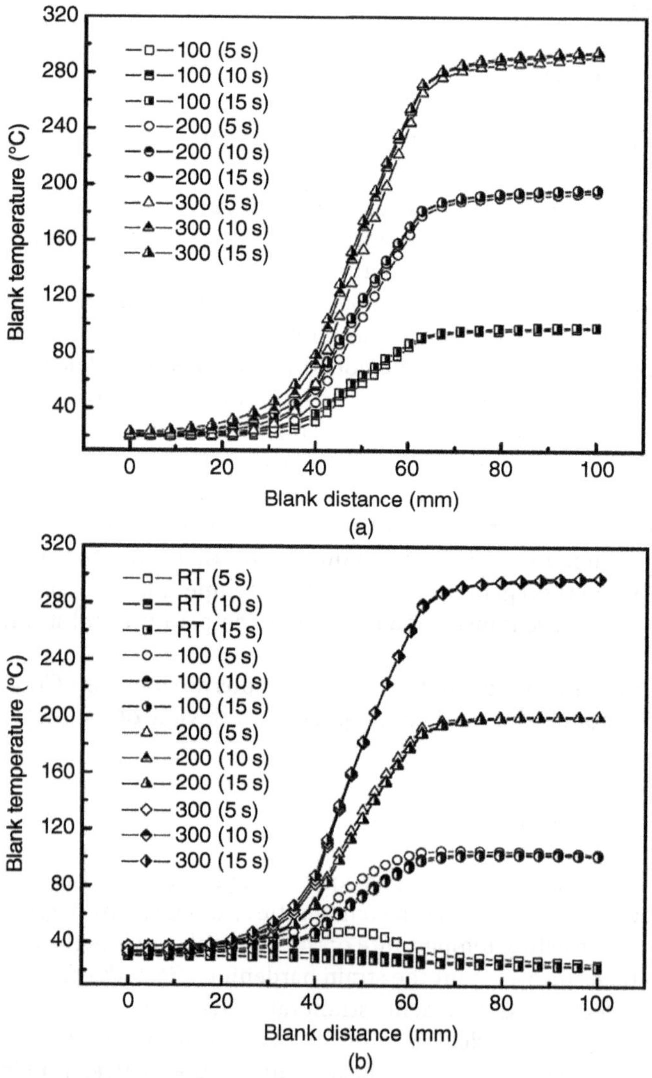

Figure 8.19 Temperature distributions on the blank surface (a) RT, (b) 200 °C, (c), 250 °C, and (d) 300 °C. *Source:* From Ozturk and Arslan 2011 [30].

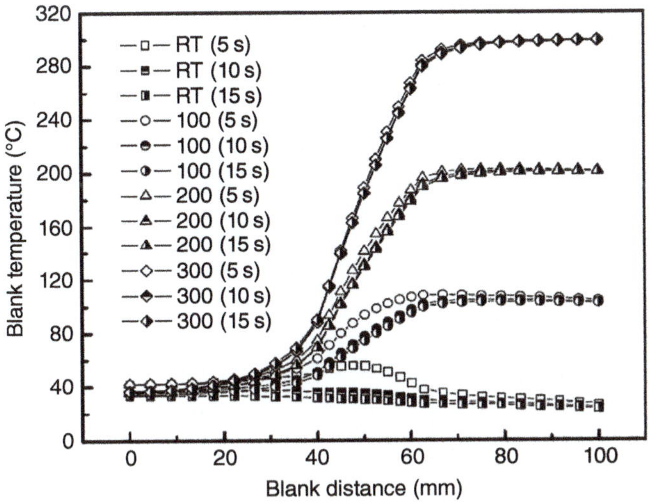

(c)

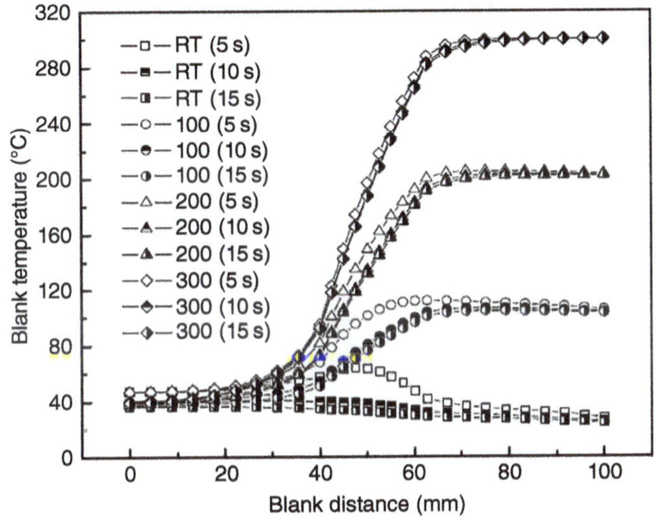

(d)

Figure 8.19 (*Continued*)

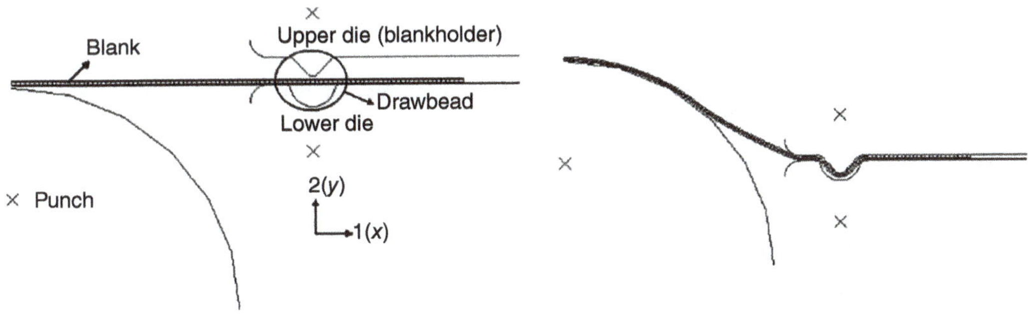

(a) (b)

Figure 8.20 (a) Finite element modeling of the hemispherical dome test and (b) finite element simulation of the deformed blank.

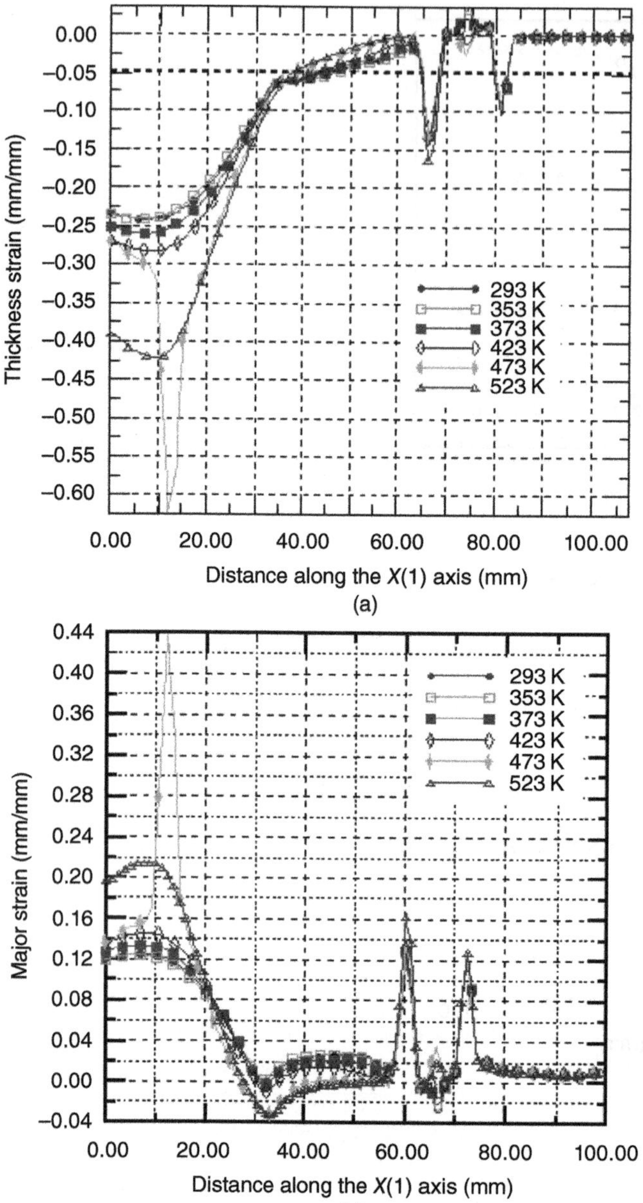

Figure 8.21 (a) Thickness strains vs. blank length at various temperatures and (b) major strains vs. blank length at various temperatures. *Source:* Ozturk et al. 2004 [31].

satisfactory. Therefore, the softening model is superior than these models. The softening model (the flow curve equation) is given in Eq. (8.4).

The flow curve equation is

$$\sigma_{\mathrm{f}} = C(T, \dot{\varepsilon})\varepsilon^{n(T, \dot{\varepsilon})} \exp\left(-\lambda^{*} \log(T) - T_{\mathrm{h}}\varepsilon\right) \tag{8.4}$$

where C and n are strength coefficient and strain hardening parameter, respectively, and λ is fitting parameter of the equation.

$$C(T,\dot{\varepsilon}) = C_0 + a_1\left[1-\exp\left(a_2\frac{T-273}{T_\mathrm{m}}\right)\right] + a_3\log\left(1+\frac{\dot{\varepsilon}}{\dot{\varepsilon}_\mathrm{ref}}\right) \tag{8.5}$$

$$n(T,\dot{\varepsilon}) = n_0 + b_1\left[1-\exp\left(b_2\frac{T-273}{T_\mathrm{m}}\right)\right] + b_3\log\left(1+\frac{\dot{\varepsilon}}{\dot{\varepsilon}_\mathrm{ref}}\right) \tag{8.6}$$

In the model, $a_1, \ldots, a_n$ and $b_1, \ldots, b_n$ are material parameters, $\dot{\varepsilon}$ is strain rate of experiment and $\dot{\varepsilon}_\mathrm{ref}$ is initial strain rate. C_0 and n_0 are the results from room temperature tests.

The performances of the models are depicted in Figure 8.22 for room temperature and $0.0017\,\mathrm{s}^{-1}$ strain rate for an advanced high strength steel (AHSS), TRIP800. Most materials show temperature increments which is called as adiabatic heating during the plastic deformation. It is possible to implement the adiabatic heating formulation in Eq. (8.7) to the dynamic flow curve models.

$$\Delta T = \frac{1}{\rho C_p}\int \sigma\, \mathrm{d}\varepsilon_\mathrm{p} \tag{8.7}$$

where ρ is the mass density and C_p is the specific heat capacity of the materials. In the models, the increasing temperature with the plastic deformation is taken into account by adding the temperature on the previous one.

$$T_{i+1} = T_i + \Delta T \tag{8.8}$$

Both cases' results are shown in Figure 8.22 as experimental vs. predicted stress values. Softening model showed the best performance when the adiabatic heating was taken into account. However, for no adiabatic heating implementation, JC model has the best performance.

For the variable temperature and strain rate, the softening model's performance is depicted in Figure 8.23 for AZ31 Mg alloy. As can be seen from the figure the predicted results have good agreement with the experimental results in all cases.

Abedrabbo et al. [24] developed a temperature-dependent anisotropic material model for FEA and formability simulation for two automotive aluminum alloys, AA5182-O and AA5754-O. Multiple temperatures to simulate the formability of more complex automotive parts, where the temperatures of the different sections will be determined automatically, can be found by this model. The forming speed controlling, the strain rate, the die and stamp corner radii, and other geometric parameters of the die set-up determine the forming characteristics of aluminum alloy sheets in addition to the temperature.

When thermo-mechanically coupled FEA is performed for forming of rectangular cups at elevated temperatures in the view of applicability, accuracy, and repeatability by considering the several failure criteria (maximum load, minimum load, and thickness ratio) to identify the onset of failure, the thickness ratio criterion is preferable since it results in accurate prediction of necking-type failure when compared with experimental measurements obtained under a variety of warm forming conditions. When predicted part depth values from FEA at various die-punch temperature combinations and blank holder pressures conditions are compared with experiments, results indicate that they are in good agreement. In addition, strain distributions on the formed part obtained under different die-punch temperature combinations are better in terms of the formability when a high temperature gradient between die and punch

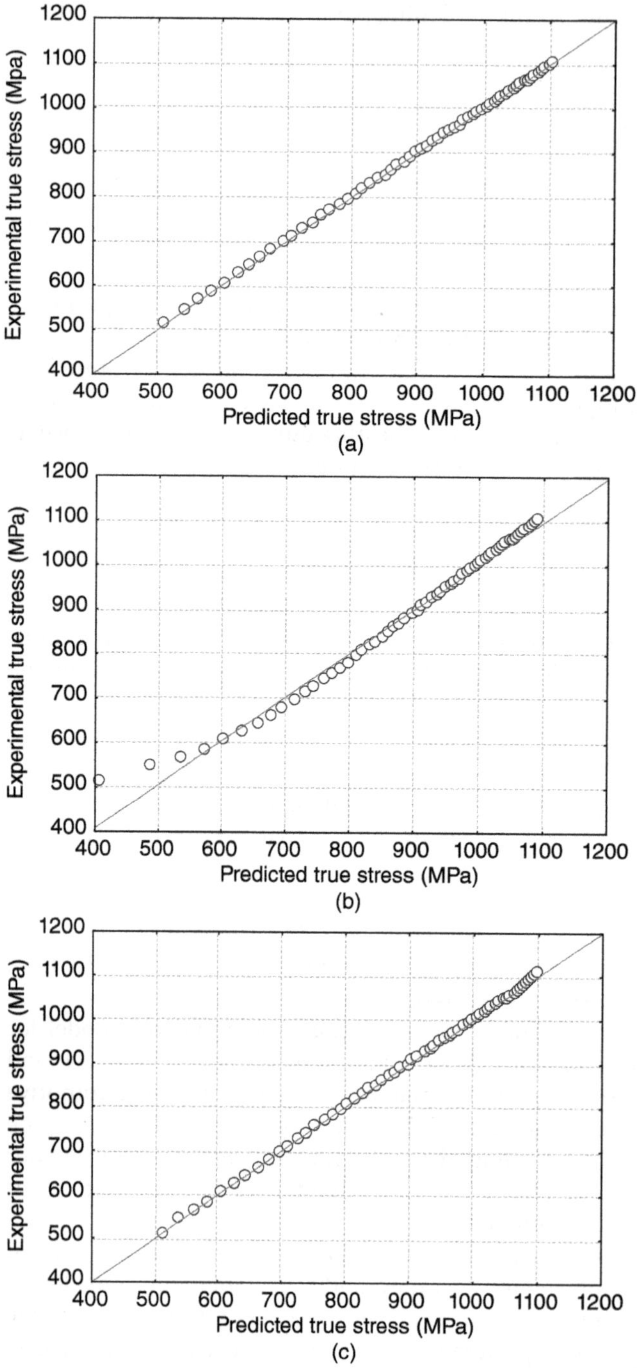

Figure 8.22 Experimental vs. predicted values for softening model. (a) Softening model with adiabatic heating, (b) Softening model with no adiabatic heating, (c) JC model with adiabatic heating, (d) JC model with no adiabatic heat, (e) Zerilli–Armstrong model with adiabatic heat, and (f) Zerilli–Armstrong model with no adiabatic heat.

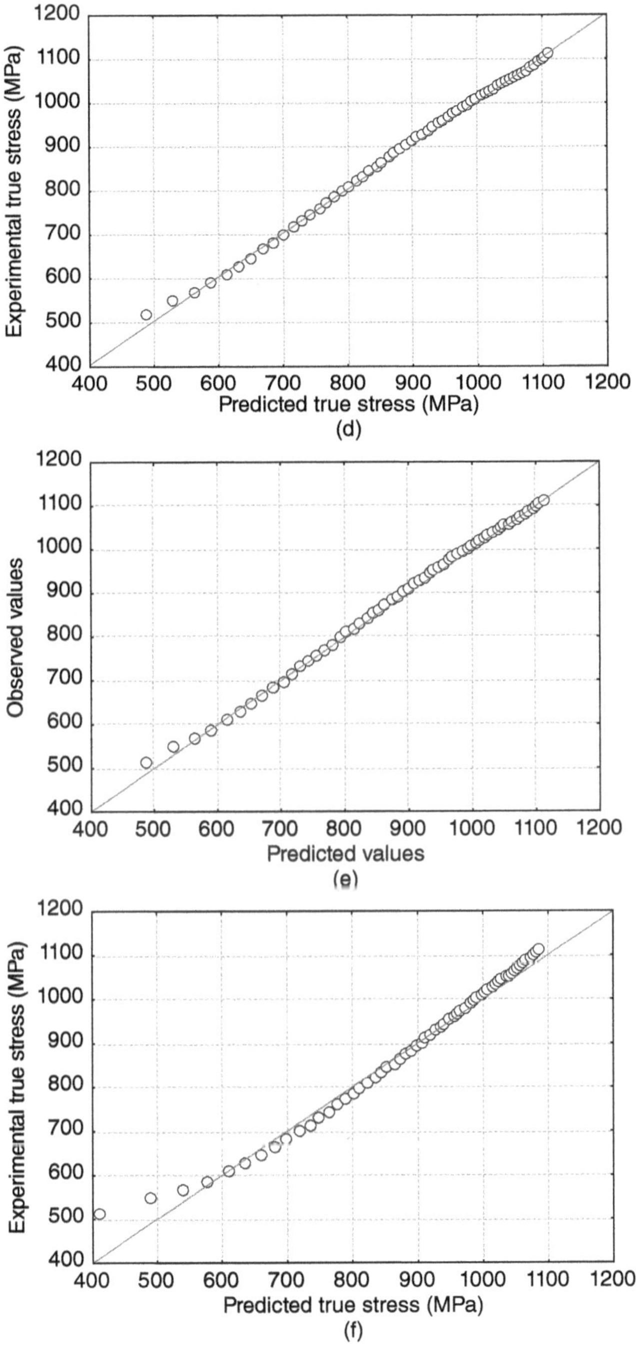

Figure 8.22 *(Continued)*

$T_{die} > T_{punch}$ is provided [35]. Also, an analytical model for hydro-mechanical deep drawing tests to investigate the effects of process conditions such as temperature, hydraulic pressure, BHF, and forming speed was developed and the experimental results show good agreement with the finite element (FE) models [36].

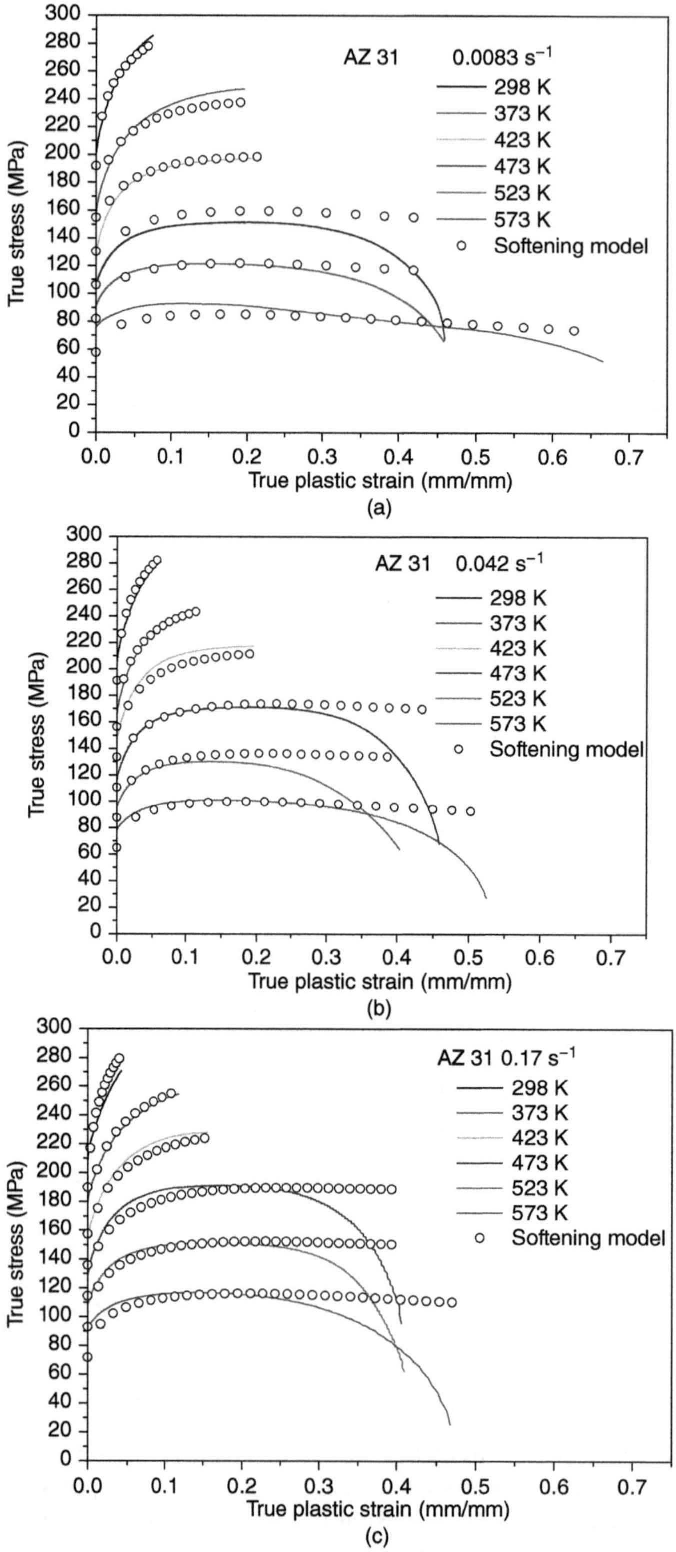

Figure 8.23 Comparison of experiments and the softening model.

For accurate simulation of the warm forming of aluminum sheet, a material model is required that incorporates the temperature and strain-rate dependency [37, 38]. Because of this, the effect of temperature distribution on warm forming performance is very important. The formability of the Al–Mg alloy sheets can be improved by increasing the temperature in some parts of the sheet and cooling the other parts when simulated by the cylindrical cup deep drawing at the different temperature gradients of the tools and the blanks. In another FEA, an isothermal/nonisothermal combined model was analyzed with a design of experiments tools to predict appropriate warm forming temperature conditions for 5083-O (Al–Mg) sheet metal blanks, deep drawing and two-dimensional stamping cases. To achieve satisfactory forming results at the end of the analysis, different temperature levels should be assigned to the corner and body of the die and punch. A 25–250% elongation ranges were obtained. The formability of Al-5083 alloy is greatly dependent on the temperature distribution of the die and punch. Also, the optimal temperature distributions for warm deep drawing and warm 2D stamping are not identical [39].

8.10 The Effects of Lubrication

One of the important parameters for the forming of sheet metals is lubrication. Surface quality is largely dependent on the effectiveness of the lubrication. Material flow in the die cavity is influenced by frictional conditions at the die/work piece contact surfaces. Therefore, a good understanding of the parameters that affect friction is essential for selecting lubricants and producing good-quality sheet metal parts. A suitable lubricant is used during the forming process to get better surface quality and to decrease the friction on contact surfaces. This contributes to an increase in the die lifetime by reducing wear. When dry film lubricants are compared with other types of lubricants on aluminum sheet metal forming, it is seen that dry film lubricants show advantages over conventional oil lubricants because of their high deep drawing performance, especially on complex shaped auto body panels. The formability is increased as a consequence of reduced friction, and it is possible to get more homogeneous sheet thickness distributions [40]. When a super plastic 5083 Al alloy was formed under biaxial deformation with or without lubrication, it is seen that reducing the interfacial friction by use of a lubricant altered the metal flow after the deformed sheet had made contact with the die surface. In addition, changes of the metal flow during forming operation not only developed a better thickness distribution of the formed part but also improved cavity distribution [41, 42].

8.11 Future Directions

Warm stamped parts of Al and Mg alloys sheets have wide acceptance by the automotive component industry, but further development is still needed to improve forming technology of parts for industrial applications. Unfortunately, this method is not convenient for every material types, is optimum for only specific application areas when we consider energy consumption and product quality requirements. Recent studies show that forming at elevated temperatures provides enhancements on the constitutive structure of aluminum–magnesium alloys by means of increment in the material's strain rate sensitivity.

In spite of heating costs and limited application areas, warm stamping process will be cost-effective in some applications where without any additional requirements or limitations as in cold/hot forming. Interest in warm forming of lightweight materials that have potential in automotive manufacturing will increase much more with respect to further improvement on forming technology and design innovations.

References

1 Shehata, F., Painter, M.J., and Pearce, R. (1978). Warm forming of aluminium/magnesium alloy sheet. *Journal of Mechanical Working Technology* 2: 279–291.
2 Wilson, D.V. (1988). Aluminium versus steel in the family car – the formability factor. *Journal of Mechanical Working Technology* 16: 257–277.
3 Pieja, T., Stachowicz, F., and Trzepieciński, T. (2010). Warm forming of stainless steel sheet. *Archives of Civil and Mechanical Engineering* 10 (4): 85–94.
4 Liebler, B.C., Schmoeckel, D., and Speck, F.D. (1995). Temperaturgeführter stofffluß beim tiefziehen von Al-blech-realversuche. In: *Bänder, Bleche, Rohre*, vol. 36 H. 7/8, 24–28.
5 Schmoeckel, D., Liebler, B.C., and Speck, F.D. (1995). Temperaturgeführter stofffluss beim tiefziehen von Al-blech-grundlagen und modellversuche. In: *Bänder, Bleche, Rohre*, vol. 36 H. 6, 14–21.
6 Naka, T. and Yoshida, F. (1999). Deep drawability of type 5083 aluminum-magnesium alloy sheet under various conditions of temperature and forming speed. *Journal of Materials Processing Technology* 89/90: 19–23.
7 Lee, H.M., Kim, Y.H., Kim, K.H. et al. (2008). Non-isothermal simulation of warm circular cup deep drawing processing of an AZ31 magnesium alloy sheet. *Materials Transactions* 49 (5): 1120–1123.
8 Naka, T., Torikai, G., Hino, R., and Yoshida, F. (2001). The effect of temperature and forming speed on the forming limit diagram for type 5083 aluminum-magnesium alloy sheet. *Journal of Material Processing Technology* 113: 648–653.
9 Li, D. and Ghosh, A. (2004). Biaxial warm forming behaviour of aluminium sheet alloys. *Journal of Materials Processing Technology* 145: 281–293.
10 Bolt, P.J., Lamboo, N.A.P.M., and Rozier, P.J.C.M. (2001). Feasibility of warm drawing of aluminium products. *Journal of Materials Processing Technology* 115: 118–121.
11 Smerd, R., Winkler, S., Salisbury, C. et al. (2005). High strain rate tensile testing of automotive aluminum alloy sheet. *International Journal of Impact Engineering* 32: 541–560.
12 Ozturk, F., Toros, S., and Kilic, S. (2008). Evaluation of tensile properties of 5052 type aluminum-magnesium alloy at warm temperature. *Archives of Materials Science and Engineering* 34: 95–98.
13 Ozturk, F., Toros, S., and Kilic, S. (2010). Tensile deformation behavior of AA5083-H111 at cold and warm temperatures. *International Journal of Materials Research* 101 (9): 1172–1179.
14 Naka, T., Nakayama, Y., Uemori, T. et al. (2003). Effects of temperature on yield locus for 5083 aluminium alloy sheet. *Journal of Materials Processing Technology* 140: 494–499.
15 Palumbo, G. and Tricarico, L. (2007). Numerical and experimental investigations on the warm deep drawing process of circular aluminum alloy specimens. *Journal of Materials Processing Technology* 184: 115–123.
16 Yoshihara, S., Nishimura, H., Yamamoto, H., and Manabe, K. (2003). Formability enhancement in magnesium alloy stamping using a local heating and cooling technique: circular cup deep drawing process. *Journal of Materials Processing Technology* 142: 609–613.
17 Yoshihara, S., Yamamoto, H., Manabe, K., and Nishimura, H. (2003). Formability enhancement in magnesium alloy deep drawing by local heating and cooling technique. *Journal of Materials Processing Technology* 143/144: 612–615.

18 Yoshihara, S., Mac Donald, B., Hasegawa, T. et al. (2004). Design improvement of spin forming of magnesium alloy tubes using finite element. *Journal of Materials Processing Technology* 153/154: 816–820.
19 Spigarelli, S., Ciccarelli, D., and Evangelista, E. (2004). Compressive deformation of an Mg–Al–Si-RE alloy between 120 and 180 °C. *Materials Letters* 58: 460–464.
20 van den Boogaard, A.H., Bolt, P.J., and Werkhoven, R.J. (2001). Modeling of Al–Mg sheet forming at elevated temperatures. *International Journal of Functional Programming* 4: 361–375.
21 Altan, T. (2002). Warm forming of aluminium alloys academic exercise or practical opportunity. *Stamping Journal* 14: 58.
22 Picu, R.C., Vincze, G., Ozturk, F. et al. (2005). Strain rate sensitivity of the commercial aluminum alloy AA5182-O. *Materials Science and Engineering: A* 390: 334–343.
23 Ozturk, F., Toros, S., and Pekel, H. (2009). Evaluation of tensile behavior of 5754 aluminum–magnesium alloy at cold and warm temperature. *Materials Science and Technology* 25 (7): 919–924.
24 Abedrabbo, N., Pourboghrat, F., and Carsley, J. (2007). Forming of AA5182-O and AA5754-O at elevated temperatures using coupled thermo-mechanical finite element models. *International Journal of Plasticity* 23: 841–875.
25 Darling, A.S. (1966). The elastic and plastic properties of the platinum metals. *Platinum Metals Review* 10 (1): 14.
26 Yoshida, F. and Uemori, T. (2002). A model of large-strain cyclic plasticity describing the Bauschinger effect and workhardening stagnation. *International Journal of Plasticity* 18: 661–686.
27 Burchitz, I.A. (2008). *Improvement of Springback Prediction in Sheet Metal Forming*. Enschede: University of Twente.
28 Keeler, S.P. and Backhofen, W.A. (1964). Plastic instability and fracture in sheet stretched over rigid punches. *ASM Transactions Quarterly* 56: 25–48.
29 Goodwin, G.M. (1968). Application of strain analysis to sheet metal forming in the press shop. Society of Automotive Engineers-SAE International. Technical Paper 680093.
30 Ozturk, F. and Arslan, M.A. (2011). Analysis of temperature distribution on warm forming of aluminum–magnesium alloy. *Advanced Materials Research* 264–265: 329–336.
31 Ozturk, F., Donmez, O., and Sisman, A. (2004). Analysis of temperature effects on formability of Al alloys. In: *CMES1-04 Proceedings of the First Cappadocia International Mechanical Engineering Symposium*, Cappadocia, Turkey.
32 Johnson, G.R. and Cook, W.H. (1983). A constitutive model and data for metals subjected to large strains, high strain rates and high temperatures. In: *Proceedings of 7th International Symposium on Ballistics*, 541–547.
33 Zerilli, F.J. and Armstrong, R.W. (1987). Dislocation mechanics based constitutive relations for material dynamics calculations. *Journal of Applied Physics* 61 (5): 1816–1825.
34 Toros, S. and Ozturk, F. (2010). Modeling uniaxial, temperature and strain rate dependent behavior of Al–Mg alloys. *Computational Materials Science* 49: 333–339.
35 Kim, H.S., Koç, M., and Ni, J. (2006). Finite element modelling and analysis of warm forming of aluminium alloys – validation through comparisons with experiments and determination of a failure criterion. *Journal of Manufacturing Science and Engineering* 128: 613–621.
36 Choi, H., Koç, M., and Ni, J. (2007). A study on the analytical modeling for warm hydro-mechanical deep drawing of lightweight materials. *International Journal of Machine Tools and Manufacture* 47: 1752–1766.
37 van den Boogaard, A.H. and Huétink, J. (2004). Modelling of aluminium sheet forming at elevated temperatures, modeling, simulation and applications, NUMIFORM 2004. *Langmuir* 712: 893–898.

38 van den Boogaard, A.H. and Huétink, J. (2006). Simulation of aluminium sheet forming at elevated temperatures. *Computer Methods in Applied Mechanics and Engineering* 195: 6691–6709.
39 Chen, P., Lin, Z., Chen, G., and Koç, M. (2006). Parametric analysis of warm forming of aluminum blanks with FEA and DOE. *Transactions of Nonferrous Metals Society of China* 16: 267–273.
40 Meilera, M., Pfestorf, M., Geiger, M., and Merklein, M. (2003). The use of dry film lubricants in aluminium sheet metal forming. *Wear* 255: 1455–1462.
41 Kelly, J.F. and Cotterell, M.G. (2002). Minimal lubrication. machining of aluminium alloys. *Journal of Materials Processing Technology* 120: 327–334.
42 Wu, H., Chio, H.C., Wang, Y.J., and Lee, S. (2006). Effect of lubrication on deformation characteristics of a super plastic 5083 Al alloy during bi-axial deformation. *Materials Science and Engineering: A* 427: 268–273.

9

Warm Hydroforming

Muammer Koç[1], Ömer N. Cora[2], Hüseyin S. Halkacı[3], and Mevlüt Türköz[3]

[1] *College of Science and Engineering, Hamad bin Khalifa University, Doha, Qatar*
[2] *Department of Mechanical Engineering, Karadeniz Technical University, Trabzon, Turkey*
[3] *Department of Mechanical Engineering, Konya Technical University, Konya, Turkey*

9.1 Introduction

In order to reduce the fuel consumption and hazardous tail pipe emissions in cars, various researches have been conducted toward the use of lightweight alloys, such as aluminum and magnesium, instead of low carbon steel (LCS). Schultz presented that 10% reduction of body weight could improve the fuel efficiency by 6–8% [1]. The mass reduction by replacing steel with aluminum can be as high as 40–60% and, with magnesium, it can go up to 60–75%. However, the cost of a part is expected to increase because of the raw material and production costs. In the case of aluminum alloys, the cost increases by 30–100%, whereas for magnesium, it is a 50–150% increase [2]. However, even with such significant price increases in initial purchase, lightweight alloys are still found to be cost saving when the life cycle of a passenger car is considered due to the savings in the fuel consumption. Benefits gained in terms of remarkable reductions in tail pipe emissions are priceless as it is difficult to calculate the savings when multiplying effects on health, environment, and global warming are taken into account.

Lightweight alloys have to be manufactured into functional lightweight parts and components in order to realize the abovementioned benefits. But there are some formability problems for lightweight materials especially at room temperature conditions. These problems have been under investigation by means of developing novel manufacturing methods.

One of such methods is hydroforming, which has been increasingly used in automotive and aerospace industries since 1990s and that used to deform sheet or tube materials via a pressurized fluid medium. In sheet hydroforming, hydraulic pressure is used to deform metal blanks into desired shapes. Hence, it offers various advantages over conventional stamping plus welding approach such as an increase in forming limits due to biaxial state of deforming stresses, savings in the die/mold due to elimination of one of the dies (female or male depending on the part and process design), consolidation of parts and process steps (hence, savings in further die/mold and equipment usage), improved tolerances, surface finish, and robustness. In tube hydroforming, a pressurized fluid is applied to the inside of a tube that is held by dies with the desired cross sections and forms. When the dies are closed, the tube ends are sealed by axial punches, and the tube is filled with hydraulic fluid. The internal pressure can go up to a few thousands of bars, and it causes the tube to calibrate against the dies. Some complex parts manufactured by tube hydroforming cannot be manufactured by another forming process.

Modern Manufacturing Processes, First Edition. Edited by Muammer Koç and Tuğrul Özel.

Another manufacturing process that has been heavily investigated since 1960s is warm forming, in which metal blanks are heated up to below their recrystallization temperature and deformed up to higher strains with considerably less amount of forces without any significant oxidation or changes in the microstructure. So it offers increasing formability advantage with selective heating at elevated temperature levels. One of the main concerns in warm forming is determination of optimal temperature distributions in tooling.

Warm hydroforming is an innovative forming technology and defined as forming of the heated tube or sheet materials by fluid pressure into the complex shapes in one step. It is a hybrid approach that integrates the advantages of both hydroforming and warm forming and offers even further benefits and opportunities. In warm hydroforming, the attained temperatures are below the recrystallization temperatures of the materials. So the problems such as oxidation of the parts, the poorness of the dimensional accuracy can be overcome, and the need of the heat treatment after forming process disappears. It has been investigated since the beginning of 2000s with premises of further increases in formability, part consolidation, reductions in equipment size and capacity, and process step reductions.

9.2 Warm Sheet Hydroforming

There are two main sheet hydroforming types. One of them is sheet hydroforming with die and the other one is sheet hydroforming with punch or hydromechanical deep drawing. This classification is also valid for warm hydroforming. When female die is used in the process, it is known as warm sheet hydroforming with die (Warm SHF-D) which is mainly used in production of shallow but larger parts such as hoods, door panels, fuel tanks. With warm hydromechanical deep drawing (Warm HMD), male die is used with punch acting against the hydraulic pressure to deform the blank into the desired deeper shapes.

9.2.1 Implementation of Warm Sheet Hydroforming

Warm SHF-D is conducted as following:

- First, the sheet is located on the die, and it is compressed at a particular force by blank holder.
- The blank holder and die are heated up to a particular temperature before or after the blank holder force (BHF) applied.
- The heated and pressurized fluid is applied on the sheet so the sheet is deformed into a female die (Figure 9.1).

The BHF is generally fixed at a high value in order to prevent leakage of the pressurized fluid in the most of the studies conducted on this topic. The sealing is generally provided by metal sealing materials like copper or aluminum because of rubber materials that cannot be used in warm temperatures. The sealing is provided by compressing of the metal sealing material at a high BHF. So it is not allowed the sheet to be drawn between the tools, and the desired part shape is obtained by stretching of the sheet. Therefore, this process is suitable for manufacturing of shallow parts. Even if the BHF can be optimized and the sheet slides between the tools, the friction force occurring on the wall of the die obstructs sliding of the sheet. When the die is higher, the center of the sheet again stretches until it contacts the bottom of the die. So the sheet can be failure before it does not take the desired shape.

In order to manufacture more quality parts by warm SHF-D, the BHF should be optimized, and it is allowed to slide of the sheet between the tools. The BHF coming on the sheet should

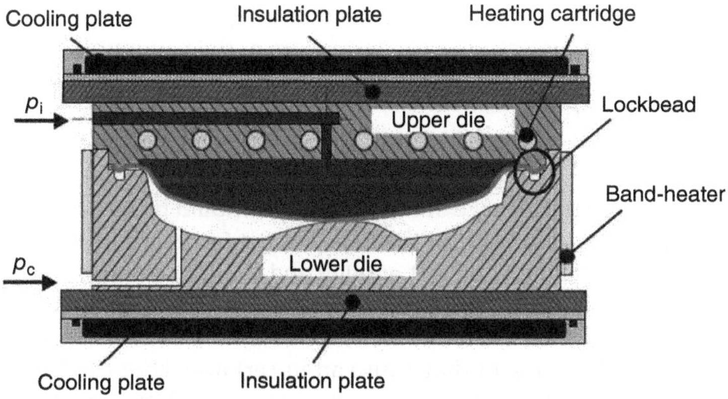

Figure 9.1 Warm sheet hydroforming with die. *Source:* Siegert and Jager 2004 [3].

be known to do this. Therefore, the sealing material must be very soft, and the applied force should directly act on the sheet. Otherwise, the sheet is thinned too much.

The sheet is heated by the tools, and it is desired that all areas of the sheet have high temperature. So the fluid must also be heated in warm SHF-D process. Hence, the fluid should be suitable for using high temperatures. Dynalene 600, synthetic oil, and marlotherm SH are among the some of the suitable fluids.

If it is desired to manufacture deeper parts, the warm HMD process should be preferred. There are three basic differences between the warm hydromechanical deep drawing and the warm sheet hydroforming with die.

- First, the male die namely the punch is used in warm HMD and the punch acts against the hydraulic pressure to deform the blank into the desired shapes (Figure 9.2).
- Second, the heating strategies are different between the two methods. Although the whole area of the sheet is heated in warm SHF-D, only the flanges of the sheet should be heated and the radius of the pressure cap and the areas contact to the punch (side wall and punch corner of the cup) should have relatively less temperature levels. Because most of the deformation occurs in the flange of the sheet thus heating of the flange region increases the ductility of the material and permitted more material to be drawn into the pressure cap. The lower temperature level of the radius of the pressure cap and side wall of the cup increases the strength of the blank and delay the localized thinning, resulting in the improved part depth [4].

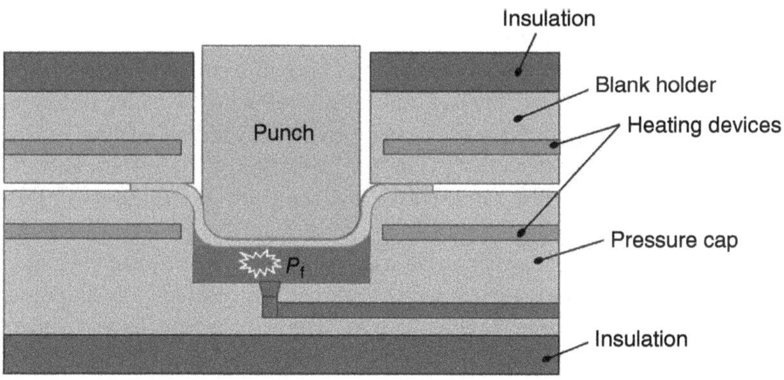

Figure 9.2 Warm hydromechanical deep drawing process (Warm HMD).

- Third, the BHF in warm HMD should not be applied as fixed at a high force. If it is not allowed to be drawn, then the blank into the pressure cap is a failure at the beginning of the forming process. In order to obtain the desired deeper parts, the BHF acting on the blank should be optimized and arranged properly.

The sealing is one of the most important problems in process design of the warm HMD. The BHF should be high enough for preventing the leakage of the fluid and it should be less enough for not causing tearing of the blank. In order to optimize the BHF, the net force coming on the blank must be known. So the sealing material must be soft enough and should not carry much force on itself. In addition, it must have high temperature resistance. The sealing rings manufactured from the soft metals like cooper and aluminum can be a candidate for solving the sealing problem, but they carry much load, and they cause an uncertainty about BHF. So rubber based seals having high temperature resistance can be used for sealing with their softness property. Fluorocarbon (Viton®, FKM) elastomers and silicon compounds (VMQ) can resist the temperatures up to 200 °C. If the forming process will be performed at higher temperatures, perfluoroelastomer (FFKM) compounds can be used up to 300 °C temperatures. But their prices are nearly a hundred times higher than the FKM or VQM seals. Another solution of the sealing is the metal seals having different shape designs such as U, E, C, O. These sealing materials can resist to very high temperatures like 870 °C and very high pressures (530 MPa). The C-Ring seals are the most convenient one in the metal seals for warm HMD process with their flexible design.

The flange of the sheet should be heated between the blank holder and pressure cap in warm HMD. There is no need to heat the fluid medium in warm HMD process. Because it is desired to be cooled, the blank areas contact to the punch. The tools are generally heated by cartridge heaters ([4–6]), band heaters [7], and tubular heating elements [8] in both process types of the warm sheet hydroforming. The tubular and band heaters do not carry much current so heating time is long and heating temperatures not so much (around 250 °C) with these devices. Cartridge heaters can carry more current, and it is possible to attain higher temperatures in a fewer times. Another solution about heating of the tools can be induction heating method. In this method, the tools constitute an electromagnet, through which a high-frequency alternating current (AC) is passed and this causes heating of the tools. The method is a noncontact heating process, and the heat occurs in the tool so it is possible to attain very high temperatures in a few times.

The tools should be insulated from the press block with the insulation plates in order to attain desired temperatures. The plates should have high compression strength and temperature resistance. Calcium silicate matrix composites and fiber cement boards are suitable materials for the insulation with their desired properties.

The punch should be cooled in warm HMD process in order to keep the side wall and punch corner of the part at low or medium temperature levels. Generally, cool water is recirculated in the punch for cooling process. The proper punch temperature can be decided according to the flow behavior of the material at different temperatures. The temperature of the punch should be held under the temperature ranges that do not cause the flow strength to decrease so much.

Appropriate lubricants should be used in the flange region for enhancing the process efficiency. It was specified in a study that dry lubricants can be used satisfactorily, but the viscosity of the mineral oils and synthetic drawing lubricants is being reduced dramatically at elevated temperatures, thus considerably higher friction forces are produced according to the room temperature [9]. In a study [10], three groups of lubricants were found as suitable. Ester oils (e.g. Oemeta – HTV 36) had been determined as suitable with their low viscosity due to good lubrication at room temperature and due to the easy application and removal of the lubricant before and after the forming process. High viscosity lubricants (e.g. Oemeta – VP 420) and solid graphite based lubricants (Fuchs Lubritech GmbH – Gleitmo HFL 9040) had been found suitable due to good lubrication even at high temperatures.

9.2.2 Parameters in Warm Sheet Hydroforming Process

The warm hydroforming parameters that affect the part quality and formability can be grouped under four main categories. These are material, die and tool, tribological and process and loading variables (Figure 9.3).

In warm hydroforming process after the part and its tools are designed and its material and suitable lubricants are determined, the achievement of the process mostly depends on the loading and process variables. For every part shape and material, different optimized levels of these variables should be selected. Hence, these variables should be controlled and changed easily for every process. Therefore, these variables are also called as controllable process variables. Choi et al. [11] grouped the controllable process variables into two major categories according to their response times as shown in Figure 9.4. The first group includes the temperature of the

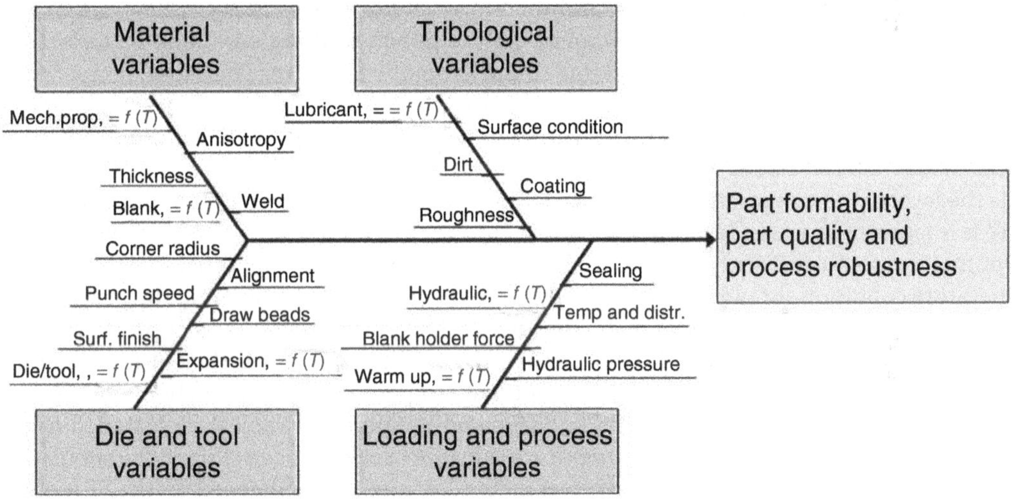

Figure 9.3 Warm hydroforming parameters [7].

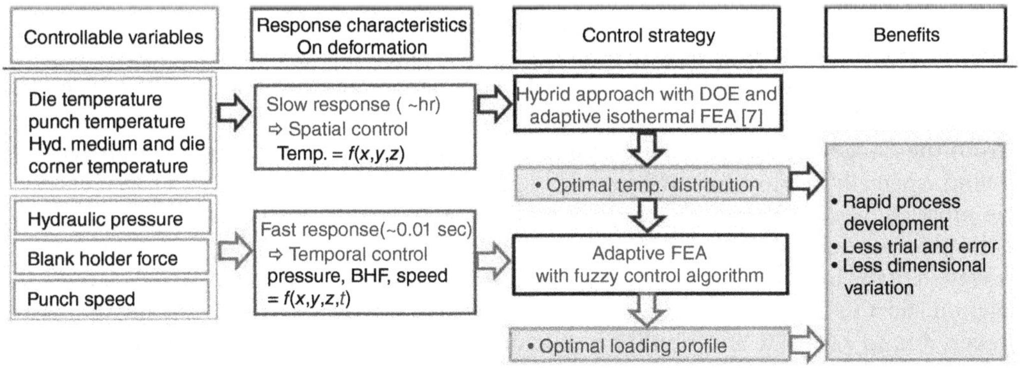

Figure 9.4 Controllable process variables and control strategies [11].

tooling and hydraulic medium, which in turn affects temperature distribution of the blank material, and can be controlled slowly on a time basis. Generally, the tooling and the hydraulic medium have a large thermal capacity that results in very slow responses, in ranges of hours. Therefore, the temperature levels of the tooling and the hydraulic medium cannot be controlled on a time basis during a forming cycle, which is usually in the range of 10–30 seconds depending on the part size. To ensure increased part formability, an appropriate way would be to control the temperature levels spatially with selective heating or cooling devices. The second group consists of the variables that can be controlled rapidly on a time basis in response to the part forming process. For example, BHF, pressure of the fluid medium, and speed of the punch can be controlled rapidly in response to the forming process. These variables can be controlled during each forming cycle very quickly (i.e. on the order of 10^{-2} seconds). Therefore, these variables can be controlled temporally on the foundation of the predetermined optimal temperature distribution [12].

The temperature level of the blank is one of the most important variables in warm SHF process. While the whole areas of the blank are desired to have the same high enough temperature in warm SHF-D process, the flanges of the blank should have high temperature and the areas that contact the punch should have less temperature in warm HMD process. Because if the material at the flange region is heated, and therefore, its flow stress is decreased remarkably leading to high ductility and easy movement. At the punch wall region, blank material is in continuous contact with the punch wall due to the hydraulic pressure. Therefore, temperature level of the blank at the punch wall and die corner regions is low because of contact heat transfer with the punch wall (conduction) and with the hydraulic medium (convection) resulting in an increased material strength providing additional resistance to drawing force [12]. Hence, while the level of the temperature is important in warm SHF-D, the temperature gradient of the blank is important in warm HMD process.

Optimal temperature distribution results in better formability in the materials. So the optimal temperature condition should be investigated before the forming process. First, the proper temperature ranges should approximately be decided by knowing the mechanical properties and flow behavior of the material at different temperatures. But this is not sufficient for obtaining optimal temperature condition. In addition, optimal temperature condition should be investigated by finite element analysis (FEA). The estimated temperature values are assigned to the blank or tools in FEA. Then the best temperature values are selected according to the simulation results of the forming process. Several types of methodologies were enhanced for the FEA of the warm SHF process with respect to calculation method of the temperature. These are isothermal, thermomechanical, hybrid isothermal/thermomechanical, and adaptive isothermal FEA.

In the isothermal analysis, the temperatures are directly assigned to the blank, the heat transfer is not calculated, and the areas of the blank keep the assigned temperatures through the analysis. Hence, if a high temperature level is assigned to the flange region of the blank, when the forming process proceeds and the material in the flange region moved into the die corner or punch region its temperature should be decreased, but in the isothermal analysis, the material from the flange region still maintain its initial temperature level. This is not an accurate modeling especially at a low forming speed.

The most convenient approach for the warm SHF is fully coupled thermomechanical FEA with the properly defined contact and material models. In this approach, the temperature levels are assigned to the tools and heat transfer between the blank and tools is calculated through the simulation. So the most real like temperature distribution of the blank can be estimated. However, due to complex and nonlinear iterations and calculations, the thermomechanical FEA takes a very long time and a great amount of resources, depending on the size of the parts being simulated.

Hybrid isothermal/thermomechanical approach was suggested by Kim et al. [4] to reduce long calculation time and resources for investigation of warm SHF process. In this approach, it is assumed that a constant but spatially nonuniform temperature distribution on blank and no heat transfer between tooling and material (isothermal FEA) exists. After determining the optimal blank temperature distribution for improved formability (i.e. high part depth, minimal thinning) by the design of experiment (DOE) method, the concluded temperature distribution on the blank was mapped onto the tooling followed by few (one to three) thermomechanical FEA (nonisothermal FEA) for verification and further tuning [12].

To overcome the limitations mentioned in other methods, adaptive isothermal analysis was proposed by Choi et al. [12]. In this method, the temperatures are assigned to the tools, and the blank temperature is determined by checking material position and assigning neighboring tooling temperature to the blank. Advantage of the method is determining rapidly and accurately the proper temperature conditions for a given tooling, part, and process configuration as opposed to lengthy and costly fully coupled thermomechanical simulations. The accuracy of the method was checked by fully coupled thermomechanical analysis, and it was concluded that the method can predict the proper temperature condition rapidly and accurately [12]. The isothermal, adaptive, isothermal, and thermomechanical approaches are shown in Figure 9.5.

After determination of the most appropriate temperature condition with FEA, the arranging of the blank temperature in the experiments is performed by controlling the temperatures of the neighboring tooling and the fluid medium. Because the whole region of the blank is to be heated in warm SHF-D process, the fluid must be heated. But there is no need to heat the fluid in warm HMD process even if it is needed to be cooled by additional equipment. It was specified in the study of Choi et al. [12] that the effect of hydraulic fluid temperature was found to be insignificant because it was specified that convective heat transfer coefficient (30–180 W/m^{2}K) between the hydraulic fluid (oil or water) and the blank was much less than the contact heat transfer coefficient (1400 W/m^{2}K) between the tooling and the blank. Hence, the

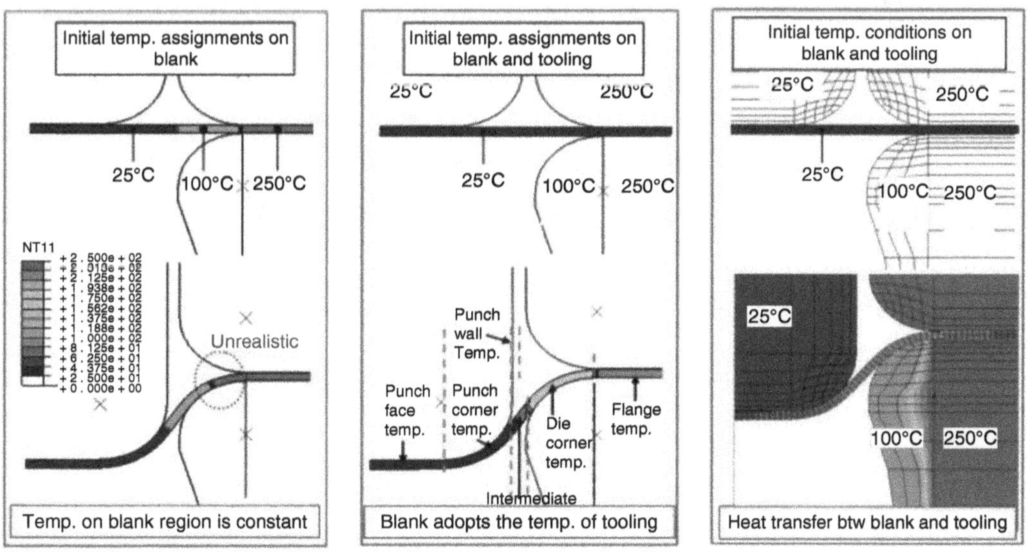

Figure 9.5 Three different thermal models at low forming speed: (a) isothermal approach, (b) adaptive-isothermal approach (checking material position and assigning neighboring tooling temperature to blank temperature), and (c) fully coupled thermomechanical approach [12].

temperature distribution of the blank substantially depends on the temperatures of the blank holder, die, and punch. In addition, the temperature distribution is also dependent on the speed of the punch. At high forming speeds, the blank material that draws from the flange region is in contact for only a short time increment. Therefore, the conduction to neighboring tooling and fluid medium is negligible. As a result, temperature level of the material mostly depends on the upstream flange temperature. However, at low punch speeds, which are typical in warm HMD process, the blank material stays longer on the die corner. Therefore, the temperature level of the blank material highly depends on the neighboring tooling and hydraulic medium temperatures due to the conduction effects [12].

The pressure of the fluid medium is an important variable for both of the process types which are warm SHF-D and warm HMD. The blank is formed at a particular pressure by the fluid and the level of the pressure is generally increased as linear in warm SHF-D process. When the strain rate sensitivity of the material is important, the flow rate of the fluid may be controlled for maintaining a constant strain rate. So the fluid pressure path can be obtained in a different form. But in warm HMD process, not only pressure level of the fluid medium but also the path of the fluid pressure according to the position of the punch is so important. The value of the fluid pressure should be different at the beginning, progressive, and final stages of the forming process.

In both process types, the value of the fluid pressure is based on the material and thickness of the blank, the minimum radius of the part, and draw ratio. As the yield strength, thickness of the blank, and the draw ratio increase and the minimum radius of the part decrease, the necessary fluid pressure increases. But the effect of the radius of the part is more pronounced in warm SHF-D process or rather in the parts having concave radiuses. Because the necessary force for bending of the blank and overcoming from the friction are provided by the fluid pressure. But in warm HMD process, the punch force forms the radius and fluid pressure only assists the blank against to the punch. So the necessary fluid pressure can be less in this process. The necessary fluid pressure according to the material and the thickness of the blank and the radius of the part in stretch formed parts is given in a study as shown in Figure 9.6.

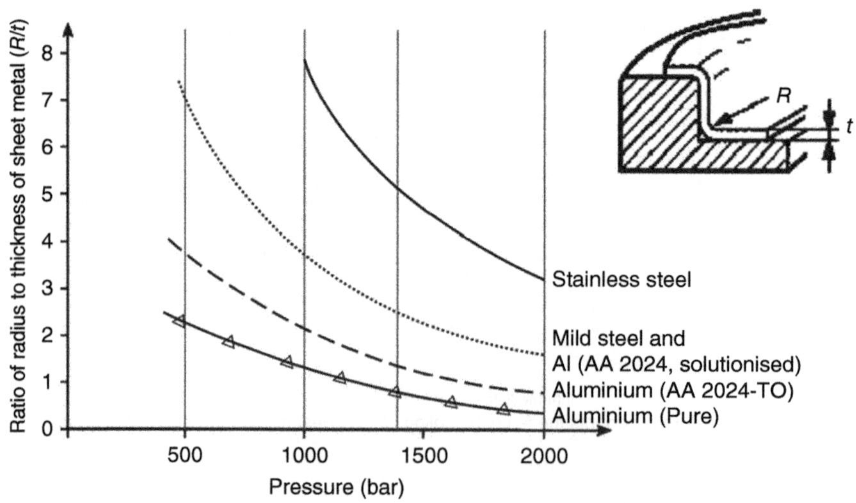

Figure 9.6 The necessary fluid pressure according to the material, the thickness of the blank, and the radius of the part in stretch formed parts [13].

As the path of the fluid pressure is important in warm HMD, to obtain the optimal fluid pressure path, different methods are applied. Obtaining of the optimal fluid pressure path by trial and error is a hard duty. Therefore, FEA should be used to avoid expensive and time consuming experimental trials. For a convenient path, the trials can be performed in the FEA of the warm HMD process. But this method is also not adequate because of the deficiency of a systematic approach. Therefore, a systematic approach that determines the optimal path whose name is adaptive FEA coupled with a fuzzy logic control algorithm was proposed by Choi et al. [11]. In this method, a fuzzy control algorithm written as a user subroutine works with the finite element software, and it determines the optimum levels of the fluid pressure through the simulation of the forming process according to selected criteria such as thinning, wrinkling, and contact to the punch. The optimal BHF can also be determined with the same technique. After the simulation optimal paths are obtained. An example of optimized fluid pressure and BHF paths obtained with this method are shown in Figure 9.7.

It is also possible to use some analytical models for obtaining an optimal path. Choi et al. [14] developed some analytical models for warm HMD process that can predict success/failure of the forming process, proper levels of process variables, and the process windows rapidly. It was specified that while the calculation speed of the thermo-mechanical FE simulation was about 45 min/case, it was only 30 s/case for the analytical model. In this analytical model, the fluid pressure path had determined for the floating condition of the blank on the die corner with the developed equations. Similarly, the failure was determined by comparing the maximum drawing force with the critical drawing force calculated by some equations. The determination of the process windows of the warm HMD process had been based on the analytical model that can predict failure, wrinkling, and floating/nonfloating conditions. An example of the process window for hydraulic pressure and BHF can be seen in Figure 9.8.

The BHF can be applied on the blank as constant or variable in sheet hydroforming process. It has generally being applied as a constant in warm SHF-D process as more shallow parts are manufactured, and the flange portion of the blank is smaller according to warm HMD process. The BHF should be high enough to prevent leakage of the pressurized fluid and wrinkle of the blank. But it should not be so much that it causes tearing of the blank.

The BHF can be applied excessively for preventing the blank to draw-in in this case the blank is formed via stretching by the fluid pressure, so this process is called as stretch forming. If the BHF is optimized as permitting the blank to draw-in between the blank holder and die, this process is called as warm SHF-D and more complex and deeper parts can be formed according to the stretch forming.

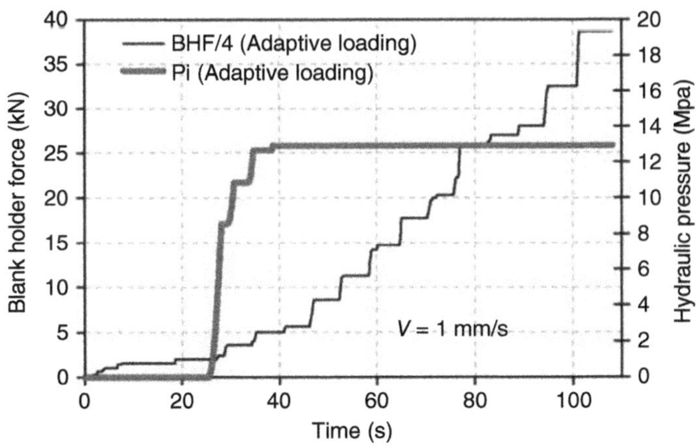

Figure 9.7 Optimized fluid pressure and blank holder force paths obtained by the adaptive FEA coupled with a fuzzy control algorithm [11].

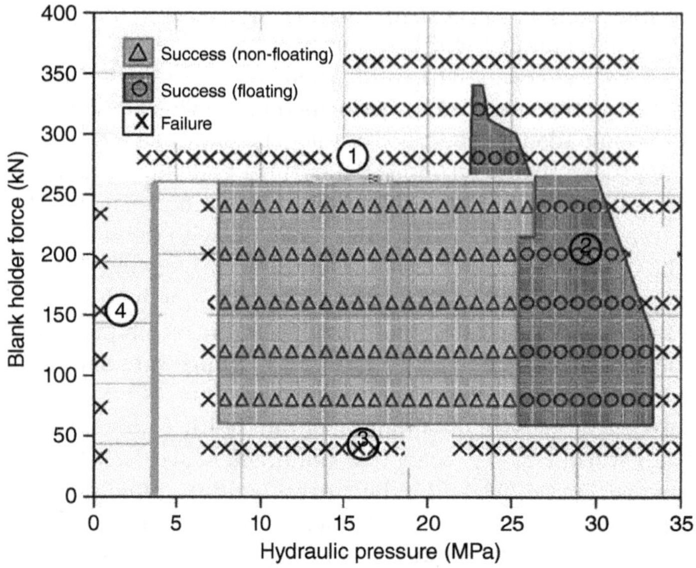

Figure 9.8 Process windows for hydraulic pressure and blank holder force at 3 mm/s punch speed [14].

While the BHF is generally applied as constant in warm SHF-D, it should have a profile that changes according to the position of the punch in warm HMD process. Because deeper parts have more flange area and for a successful forming that is free from wrinkles and failures, the BHF profile should be optimized. The same optimization methods with that of the fluid pressure are used for the optimizing of the BHF.

Most of the materials have less strain rate sensitivity values in the room temperature. But as the temperature increases, the strain rate sensitivity is also increased, hence strain rate becomes more important. So the forming speed is also an important variable and should be considered in warm hydroforming process. As the temperature increases, the flow stress of the materials generally decreases, and the total elongation increases. Contrary to the temperature, as the strain rate increases, the flow stress generally increases and the total elongation decrease. So at low forming speeds, the formability of the material becomes better in warm hydroforming process.

9.2.3 Studies on Warm Sheet Hydroforming

The studies on warm SHF were classified as following, and an evaluation was carried out according to their purposes, methods, results, and imperfections.

- Material characterization
- The optimization of the process parameters
- The types of the warm sheet hydroforming
  - Warm sheet hydroforming with die (Warm SHF-D)
  - Warm hydromechanical deep drawing (Warm HMD)

The first studies on warm SHF started in Germany [9, 15]. The literature is not abundant in terms of warm SHF, and process is in its development stage.

The studies were carried out as experimental or experimental + numerical except of a numerical and an analytical one. In the numerical studies, the effects of different parameters on warm SHF process were investigated. The experimental studies were performed on two different types of warm SHF process that are warm SHF-D and warm HMD processes. The studies on warm SHF-D are much, and it was generally concerned about temperature and fluid pressure parameters. Conducting of warm HMD process is more difficult as the BHF must be arranged sensitively besides the fluid pressure and the blank must have a temperature distribution at different levels. In addition, there are studies conducted for determining the mechanical behaviors of the materials at different temperatures. The effects of the parameters were determined by DOE methods in some of the experimental and numerical studies.

9.2.4 Studies on Material Characterization

In order to develop appropriate warm hydroforming strategies, first, the behavior of the materials to be formed at high temperatures should be determined. Therefore, the behaviors of the materials at high temperatures are determined by tensile and bulge tests. As well as there are studies on only material characterization at high temperature [16–18], this issue was investigated in the studies performed on warm hydroforming [5, 9, 15]. It was found from the tensile tests of the aluminum and magnesium alloys that the ductility had increased considerably and the yield stress of the material decreased in the manner that to eliminate the hardening at the elevated temperatures. In addition, the yield curves of the materials could be determined up to elevated strains from the bulge test according to the tensile test.

9.2.5 Studies Conducted on the Optimization of the Parameters

The studies on determination of the optimal parameter levels and investigation of the effects of the parameters how and which rate on the process were conducted generally on the warm SHF-D process by DOE methods. One of the main studies on warm HMD process about the topic of the optimization of the parameters is the study of Choi et al. [11]. In the study, they determined the optimal fluid pressure and BHF profiles at different punch speeds by a developed methodology in warm HMD process. First, 2D and 3D FE models had been generated and verified by experiments conducted in other studies. Then the optimal fluid pressure and BHF profiles had been determined by developed adaptive FEA with fuzzy control algorithm based on previously determined temperature conditions. The thinning and wrinkling of the material, contact to the punch and cavity filling ratio had been selected as the criteria that should be satisfied in the algorithm. After determining the optimal profiles, the effects to the thinning of the material, stress and strain values and temperature distributions as well as success/failure of the process had been presented. It had been concluded that by using the adaptive FEA, time and many resources to determine proper loading profile can be reduced significantly. It had been proved that the adaptive FEA with fuzzy control algorithm enables the forming process and can be performed in an efficient way.

The effects of the parameters to the process were determined in some studies by DOE methods [5–7]. In these studies, the effects of the different values of the temperatures and fluid pressures on thinning of the materials and cavity filling ratios were determined by response surface method.

9.2.6 Warm Sheet Hydroforming with Die

Generally, 5xxx group of the aluminum alloys that were used in automotive was used in the investigations. In the studies, the process was investigated between the room temperature and maximum 300 °C. Generally, the effects of the temperature and the fluid pressure on the thinning of the blank and cavity filling ratios were investigated [5–7, 19]. Only in the study of Gedikli et al. [19], it was tried to increase the accuracy of the warm SHF simulations rather than the effects of the parameters. In the study, the appropriate combinations of the numerical modeling conditions like as solver, element type, and material model had been determined by comparing the numerical results with the warm SHF-D experimental findings.

While the materials are generally strain rate insensitive at the room temperature, strain rate may be more important at high temperatures. Thus, also the effects of the strain rate to the mentioned response variables had been investigated in some studies [5, 7]. In addition, the effects of the material and the part geometry to the forming process had been investigated in a study [6]. In the mentioned studies, as the blank was being formed at a particular temperature with a particular pressure into a die, the knowledge of the used BHF was not given. In this case, it was thought that the blank was formed by stretching as it was not permitted to slide below the blank holder. Hence, in these studies, the effect of the BHF to the process was not investigated.

9.3 Warm Hydromechanical Deep Drawing

The studies on warm sheet hydroforming with punch or in another words warm hydromechanical deep drawing is limited. The aluminum and magnesium alloys were further used in the investigations. It was found that in some experimental studies that the formability of the materials was increased considerably by warm HMD process (Table 9.1) [8, 9, 15].

While the effects of the temperature variable to the limiting drawing ratio (LDR) were investigated in these studies, the optimal levels of the fluid pressure, BHF, friction, and the other geometrical parameters did not get investigated, and the effects to different performance criteria did not get determined. Only Groche et al. [9] had estimated the appropriate fluid pressure value according to the position of the punch as a variable by the numerical analysis of the process for every analysis step. The fluid pressure was used as fixed in the other experimental investigations [8, 15].

While the BHF is an important parameter for warm HMD process, there is not any experimental study for the optimization of the BHF. Although there is a study that the effects and optimal values of the BHF and fluid pressure parameters besides the temperature were investigated [11], this study was performed as numerical and the effects of the parameters to the LDR of the materials were not introduced.

9.3.1 Current/Future Trends in Warm Sheet Hydroforming

The conducted studies on warm sheet hydroforming are limited and still on the onset. The studies conducted on warm SHF-D are more extensive and the effects of the main variables like temperature and fluid pressure were further investigated experimentally. The studies conducted on warm HMD process are shallow. Although there are some numerical studies focused on the optimal temperature conditions and determining the effects of the fluid pressure and BHF, there is no study to investigate these variables experimentally. Only in a few studies, the effects

Table 9.1 Increasing of the limiting drawing ratio.

Study	HMD	Warm forming	Warm HMD
Kurz [8]	—	2.8	3.6
Groche et al. [9]	2.5	2.7	3

of the temperature variable on the process were investigated experimentally. Especially the optimization and effect of the BHF were not investigated in both of the process types. The reason of this may be to provide the sealing at high temperatures by controlling the BHF. As a future trend, the BHF and fluid pressure profiles should be optimized and should be controlled in warm SHF process.

Generally, aluminum alloys especially 5xxx and 6xxx series are investigated in the studies conducted until now. There are a few studies on other materials. Therefore, forming characteristics of the different materials by warm hydroforming should be investigated more. In addition, simple components were manufactured in the conducted studies until now. Real industrial components can be manufactured by developing the warm SHF process.

Warm sheet hydroforming is still a lab scale manufacturing method, and it is not used for manufacturing of industrial parts currently. By conducting investigations on process design and process optimization, the process will be a candidate for manufacturing of sheet parts in industrial plants. To do this, the problems like sealing, insulation, heating, and lubrication should be solved with an efficient way and investigations should be carried out to determine optimal process conditions (e.g. variable loading profiles: pressure, BHF, punch speed, and temperature).

9.4 Warm Tube Hydroforming

To extend the capabilities of tube hydroforming, the process can also be carried out at elevated temperatures. Hence, materials with higher strength such as ultrahigh strength steels, and materials with poor formability at room temperature (e.g. Al and Mg alloys) can successfully be formed via this technique.

Warm tube hydroforming also provides producing complex hollow shapes, minimize the need for welding, highly accurate part dimensions, lower tooling costs, and/or simplify the assembly of components leading to improved structural integrity. Improved part strength, stiffness, and less springback are other advantages [20–23].

9.4.1 Implementation of Warm tube Hydroforming Process

Figure 9.9 shows the schematic of warm tube hydroforming process steps. Different from the tube hydroforming operations conducted at room temperature, water-based pressure media is no longer appropriate to be used in warm hydroforming operations as their boiling temperature is in the range of 100 °C. Heat stable liquids (e.g. heat transfer oils) are common choices up to the temperature levels of 300 °C. Beyond that point, gases can be used as pressure medium. Heating can be performed in two different ways: (i) heating of dies only, (ii) heating of dies and tubes. Figure 9.9 shows the latter of those. In both techniques, the pressurized gas does not

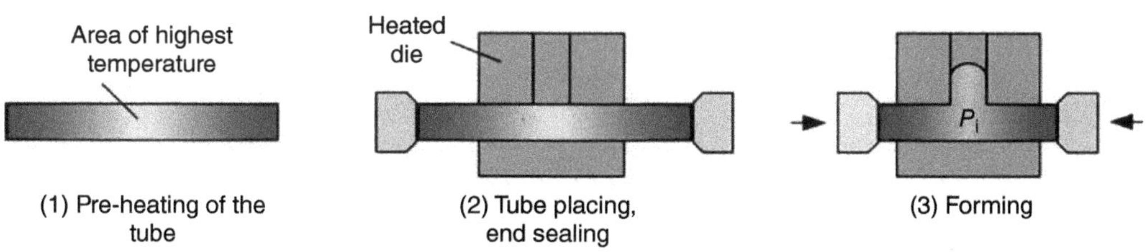

Figure 9.9 Schematic representation of warm tube hydroforming steps [24].

need to be heated as gases provide little specific heat and therefore cooling effect of gases on the tube material is neglected [24].

9.4.2 Parameters in Warm Tube Hydroforming Process

The most critical parameters in warm tube hydroforming are regarded as temperature, pressure, and feed loading paths. In addition, material properties, die and tube geometry, friction, and lubrication are noted other important parameters. Each of these parameters that play has a key role in the successful forming process and has to be addressed during the process development stage [25].

In particular, friction affects the part surface quality such as wall thickness distribution and process formability. Figure 9.10 illustrates the different friction zones in a hydroforming of T-shaped tube [26].

9.4.3 Studies on Warm Tube Hydroforming

Particular interest for warm tube hydroforming comes from the automotive and aerospace industries to reach the vehicle weight reduction and fuel-saving goals. Nevertheless, it is necessary a well-established procedure, accurate material behavior knowledge, and precise testing and/or measurement equipment, etc., to carry out warm tube hydroforming.

In warm hydroforming, the thickness uniformity is mostly poor, and high thinning ratio is experienced in forming zone, while a high thickening ratio is obtained in feeding zone. In order to eliminate this problem, Gang et al. implemented nonuniform temperature field in tube hydroforming of magnesium AZ31 alloy. They realized temperature gradient on the tube blank by controlling the die temperature in feeding, and forming zones as lower, and higher, respectively (Figure 9.11).

Effects of loading paths, forming zone temperature on formability, and effect of temperature difference on thickness distribution, and effect of preform shape on limit expansion were investigated. In the effect of temperature difference on thickness distribution study, temperature of the forming zone was kept constant at 240 °C, and temperature differences of 130, 150, and 170 °C were provided between forming and feeding zones. It was concluded that the

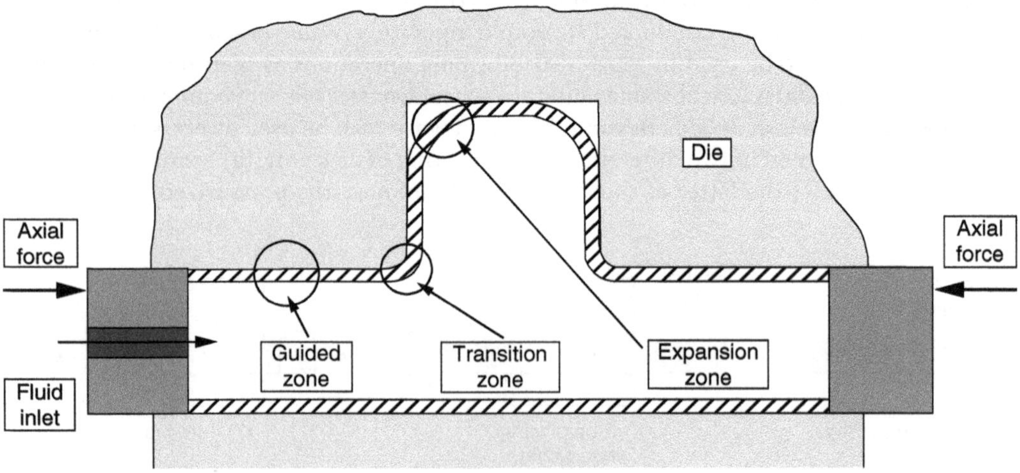

Figure 9.10 Friction zones in tube hydroforming for T-shape [26].

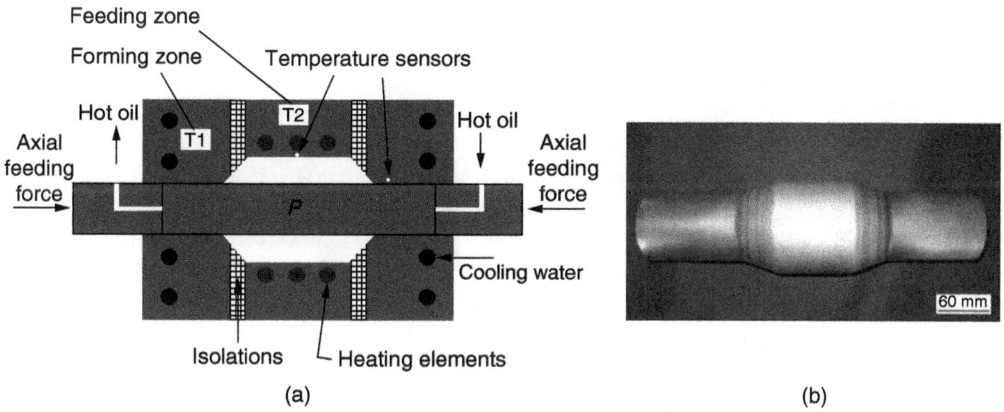

Figure 9.11 (a) Warm hydroforming setup with nonuniform temperature fields, (b) formed AZ 31 sample with 35% expansion ratio (forming zone temperature 240 °C, feeding zone temperature 90 °C) [27].

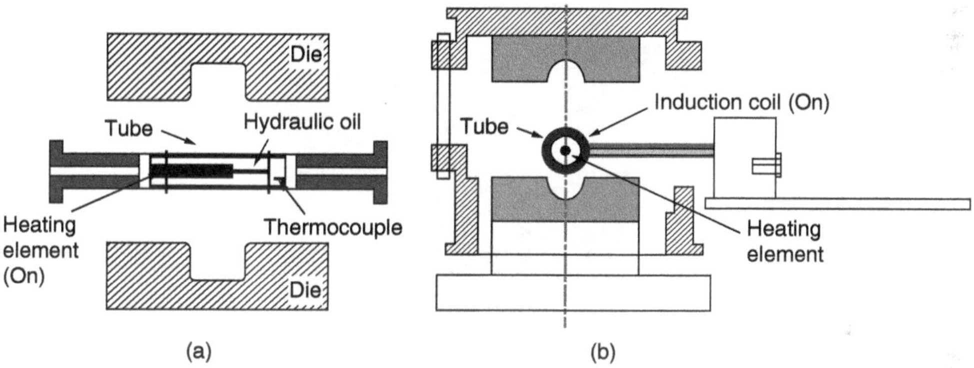

Figure 9.12 Integration of heating system to a warm tube hydroforming setup (a) heating element inside the tube, (b) induction coil system around the tube [20].

optimum temperature difference is 150 °C as more uniform thickness distribution was obtained. Larger temperature distributions resulted in nonuniform deformation in forming zone, whereas lower temperature difference led to thickening in feeding zone near tube ends [27].

Yi et al. on the other hand, aimed for developing a heating system for tube hydroforming process that provides uniform temperature distribution. The heating system devised comprises of an induction coil on the outer of the tube as a main heating source and a heating element embedded in the tube as a secondary source (Figure 9.12). Experiments were performed on AA 6061 aluminum tubes. It was noted that the heating system developed ensures faster heating of the tube and enhances the uniformity of temperature distribution in comparison with only using induction heating or a heating element. Moreover, free bulge hydroforming tests were conducted with the heating system developed, and it was shown that circumferential bulge heights are more uniform due to the increased temperature uniformity.

As the material data is critical in both process designs of hydroforming and numerical analyses, especially in elevated temperature, Elsenheimer and Groche developed a new testing technique referred to as "tube bulge test" in which high process temperature and constant strain rates can be realized (Figure 9.13). They reported flow curves for an aluminum alloy AlMgSi0.5 (EN-AW 6060) tubular material at a strain rate of 0.09 s^{-1}, and at temperatures 480–575 °C [24].

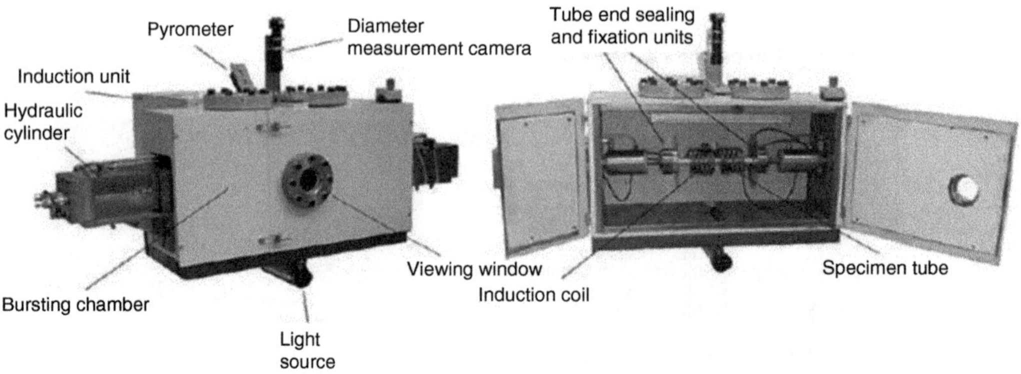

Figure 9.13 Hot bulge testing equipment developed by Elsenheimer and Groche [24].

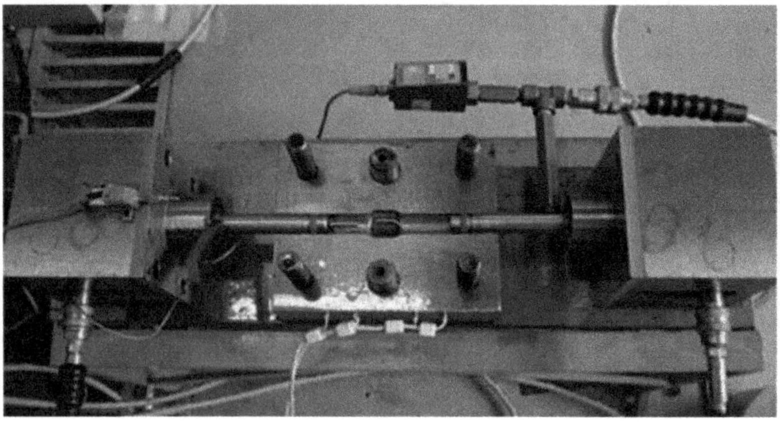

Figure 9.14 Warm tube hydroforming setup used by Hashemi et al. [28].

Numerical analyses are essential tools significantly reducing the number of trial–error experiments to optimize the warm hydroforming process conditions. Besides, it is also of critical importance that the feasibility of process should be checked before manufacturing of expensive tooling. Nevertheless, the numerical analysis of hydroforming is not easy to conduct as the process is highly nonlinear due to large deformations, material nonlinearity, and contact interactions. Three-dimensional fully coupled thermo-mechanical analyses are required in most cases to properly reflect actual conditions and reveal the defects in warm tube hydroforming.

Hashemi et al. investigated the effect of warm hydroforming temperature on thickness distribution of AA 1050 tubes with had initial diameter of 24, and thickness of 1 mm using FEA + experimental approach. Figure 9.14 shows the experimental setup utilized in their study. The tubes were bulged at different temperatures up to 230 °C, and it was reported that if both the ends of the tube are free, increasing temperature causes softening the material and therefore leads to more uniform thickness distribution and less thinning. On the contrary, if the both ends are constrained, increasing temperature results in increased thinning. Nonuniform temperature distribution led to nonhomogeneous bulging of the tubes, and it was reported to be the major cause of bursting [28].

Bielak et al. performed FEA analyzes for tube hydroforming of an aluminum alloy having two different diameters. They noted that THF process should be solved in three steps to avoid the common convergence problems. Those steps are as follows: (i) fixing the initial and boundary conditions, (ii) heating the die to the quasi-static thermal equilibrium level, (iii) applying pressure to the inner surface of the tube. Coulomb coefficient of friction (μ) for the contact surfaces was imposed in the range of 0.022–0.035 [29].

As the magnesium alloys offer great reduction in weight, and those are of particular interest in warm tube hydroforming researches, Chen and Kot investigated the effect of loading paths on formability of quadrilateral tubular components made of magnesium based alloy AZ31B. The formability of AZ31B can significantly be improved at temperature from 200 to 300 °C, and forming limit was found to be six times higher than that of room temperature according to literature. Moreover, it is reported that tube hydroformed AZ31B components exhibit good damping properties, electromechanical shielding, and acceptable component quality. The researchers selected internal pressure and axial feeding as key parameters to optimize the process and tried nine different loading paths in numerical analyses. Forming temperature was $T = 270\,°C$, and Coulomb coefficient of friction of $\mu = 0.05$ was defined for contacting surfaces. Upon, numerical analyses, the optimum paths were further verified with experiments (Figure 9.15). It was noted that formability of hydroformed tubular components can be predicted accurately, good agreement for thickness distribution was obtained between numerical simulation, and tube hydroforming experiments [23].

Another study on AZ31B alloy was performed by He et al. They studied the effect of forming temperature and flattening displacement on the maximum expansion ration and bursting pressure. Ring hoop tension test, and hydro-bulge test were employed to determine formability behavior. Based on the mechanical properties and formability measurements, a tubular part with curved axis and different cross-sections was manufactured using warm tube hydroforming system developed in-house. The system uses heat transfer oil as forming medium and can provide 100 MPa of forming pressure at up to 300 °C. The researchers reported that total elongation in hoop direction is significantly different from the one along axial direction that shows highly anisotropic behavior of the material of interest [30].

Along with the flow curves at specified temperature and strain rates, fracture criterion in a material is one of the most important design properties in numerical analyses. Lee et al. developed failure criterion for warm hydroforming using AA 6061 aluminum tubes with different treatments (e.g. as extruded, fully annealed, T6 tempered). They proposed a forming-limit-based ductile fracture failure criterion based on experimental results and FEA [31] (Figure 9.16).

Accurate prediction and control of warm tube hydroforming parameters are vital factors in process success. As the nature of warm tube hydroforming process is quite complicated, it is relatively difficult to determine the optimum process parameters. Most of the literature concerned about the effect of single criterion on the output parameters, yet efforts should be concentrated on multiobjective optimization studies so that optimized values of multiple parameters can obtained simultaneously [32, 33].

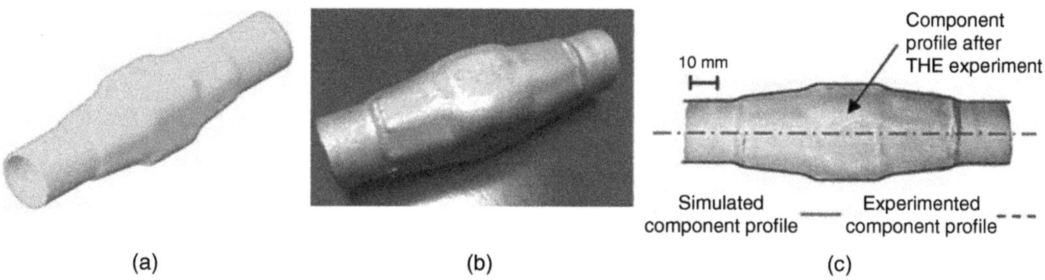

Figure 9.15 Results with optimum profile (a) FEA, (b) experimental, and (c) formed tube profile comparison [23].

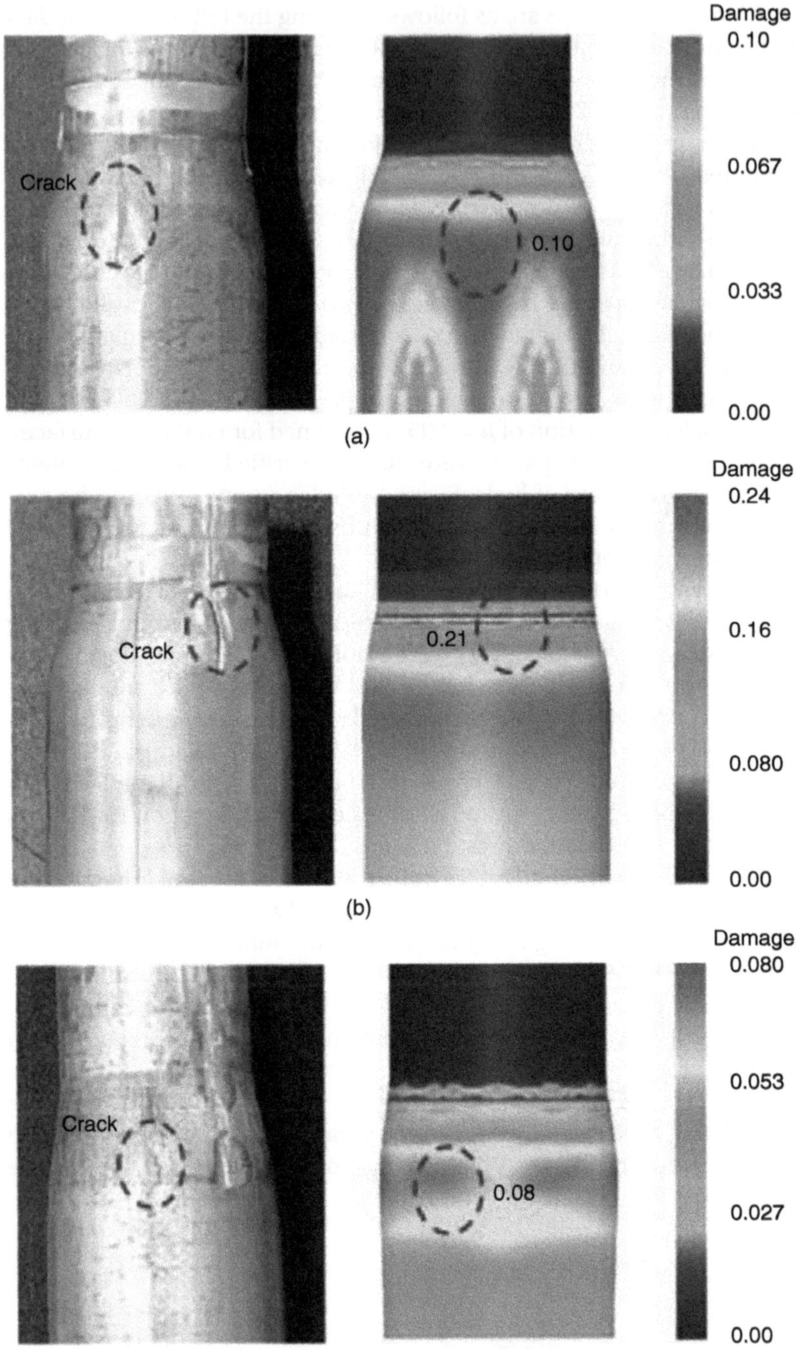

Figure 9.16 Verification of critical damage value obtained with the developed failure criterion. (a) As-extruded, (b) full-annealed, and (c) T6-temper [31].

An example to this approach was performed by Seyedkashi et al. They implemented an adaptive simulated annealing technique in optimization of warm tube hydroforming to determine the optimal pressure loading paths for AA6061 alloy tubes with different wall thicknesses and corner fillets. Material properties were derived from tensile tests conducted at 300 °C, and then those were used in numerical analyses and optimization of 40 different cases. An optimized pressure loading path was found for each case and compared with theoretical predictions. Optimal pressure profile is accepted as fully forming the final geometry with minimum thinning and without any defects. To validate the numerical results, experiments were also performed. It was noted that corner fillet, and tube thickness have great effect on internal pressure [33].

References

1 Schultz, R.A. (1999). Aluminum for light vehicles an objective look at the next ten to twenty years. Ducker Research, Presented at 14th International Aluminum Conference, Montreal, Canada.

2 Koç, M. (2008). *Warm Hydroforming of Lightweight Materials, Hydroforming for Advanced Manufacturing* (ed. M. Koç), 352–380. Cambridge, England: Woodhead Publishing Ltd.

3 Siegert, K. and Jager, S. (2004). Pneumatic bulging of magnesium AZ31 sheet metal at elevated temperatures. *Magnesium Technology* 2004: 87–90.

4 Kim, H.S., Koc, M., and Ni, J. (2004). Determination of proper temperature distribution for warm forming of aluminum sheet materials. *Journal of Manufacturing Science and Engineering* 128 (3): 622–633.

5 Mahabunphachai, S. and Koç, M. (2010). Investigations on forming of aluminum 5052 and 6061 sheet alloys at warm temperatures. *Materials and Design* 31: 2422–2434.

6 Koç, M., Mahabunphachai, S., and Billur, E. (2011). Forming characteristics of austenitic stainless steel sheet alloys under warm hydroforming conditions. *The International Journal of Advanced Manufacturing Technology* 56: 97–113.

7 Koç, M., Agcayazi, A., and Carsley, J. (2011). An experimental study on robustness and process capability of the warm hydroforming process. *Journal of Manufacturing Science and Engineering* 133: 021008.

8 Kurz, G. (2004). Heated hydro-mechanical deep drawing of magnesium sheet metal. *Magnesium Technology* 2004: 67–71.

9 Groche, P., Huber, R., Dorr, J., and Schmoeckel, D. (2002). Hydromechanical deep-drawing of aluminium alloys at elevated temperatures. *CIRP Annals-Manufacturing Technology* 51: 215–218.

10 Geiger, M., Merklein, M., Celeghini, M. et al. (2003). Sheet and tube hydroforming at elevated temperatures. *Hydroforming of Tubes, Extrusions and Sheet Metals* (ed. K. Siegert) 3: 273–292.

11 Choi, H., Koc, M., and Ni, J. (2007). Determination of optimal loading profiles in warm hydroforming of lightweight materials. *Journal of Materials Processing Technology* 190: 230–242.

12 Choi, H., Koc, M., and Ni, J. (2008). A study on warm hydroforming of Al and Mg sheet materials: mechanism and proper temperature conditions. *Journal of Manufacturing Science and Engineering* 130: 1–14.

13 Johanissson, T. (2001). *Low Volume Production of Sheet Metal Parts,' in: Hydroforming Tubes, Extrusions and Sheet Metals* (ed. K. Siegert), 159–179. Frankfurt: Werkstoff Informationsgesellschaft mbH.

14 Choi, H., Koc, M., and Ni, J. (2007). A study on the analytical modeling for warm hydro-mechanical deep drawing of lightweight materials. *International Journal of Machine Tools and Manufacture* 47: 1752–1766.

15 Siegert, K. and Jager, S. (2004). Warm forming of magnesium sheet metal. SAE 2004 World Congress and Exhibition, Detroit, MI, USA, (8–11 March 2004).

16 Novotny, S. and Geiger, M. (2003). Process design for hydroforming of lightweight metal sheets at elevated temperatures. *Journal of Materials Processing Technology* 138: 594–599.
17 Kaya, S., Altan, T., Groche, P., and Klopsch, C. (2008). Determination of the flow stress of magnesium AZ31-O sheet at elevated temperatures using the hydraulic bulge test. *International Journal of Machine Tools and Manufacture* 48: 550–557.
18 Koç, M., Billur, E., and Cora, Ö.N. (2011). An experimental study on the comparative assessment of hydraulic bulge test analysis methods. *Materials and Design,* 32: 272–281.
19 Gedikli, H., Cora, Ö.N., and Koç, M. (2011). Comparative investigations on numerical modeling for warm hydroforming of AA5754-O aluminum sheet alloy. *Materials and Design* 32: 2650–2662.
20 Yi, H.K., Pavlina, E.J., Van Tyne, C.J., and Moon, Y.H. (2008). Application of a combined heating system for warm hydroforming of lightweight alloy tubes. *Journal of Materials Processing Technology* 203: 532–536.
21 Hang, L.H., Wang, Z.R., Kang, D.C. et al. (2004). Hydroforming highlights: sheet hydroforming and tube hydroforming. *Journal of Materials Processing Technology* 151: 165–177.
22 Esnaola, J.A., Torca, I., Galdos, L., and Garcia, C. (2009). Determination of the optimum forming conditions for warm tube hydroforming of ZM21 magnesium alloy. *Journal of Achievements in Materials and Manufacturing Engineering* 32 (2): 188–195.
23 Chan, L.C. and Kot, W.K.R. (2014). Determination of loading paths in warm hydroforming reinforced quadrilateral tubular components. *Materials and Manufacturing Processes* 29: 32–36.
24 Elsenheimer, D. and Groche, P. (2009). Determination of material properties for hot hydroforming. *Production Engineering – Research and Development* 3: 165–174.
25 Alaswad, A., Benyounis, K.Y., and Olabi, A.G. (2012). Tube hydroforming process: a reference guide. *Materials and Design* 33: 328–339.
26 Plancak, M., Vollertsen, F., and Woitschig, J. (2005). Analysis, finite element simulation, and experimental investigation of friction in tube hydroforming. *Journal of Materials Processing Technology* 170: 220–228.
27 Gang, L., Wen-da, Z., Zhu-bin, H. et al. (2012). Warm hydroforming of magnesium alloy tube with large expansion ratio within non-uniform temperature field. *Transactions of Nonferrous Metals Society of China* 22: 408–415.
28 Hashemi, S.J., Naeini, H.M., Liaghat, G. et al. (2013). Numerical and experimental investigation of temperature effect on thickness distribution in warm hydroforming of aluminum tubes. *Journal of Materials Engineering and Performance* 22: 57–63.
29 Bielak, R., Liedl, G., and Behulova, M. (2010). Contribution to the lightweight alloy warm tube hydroforming. *The proceedings of International Scientific Conference on Mechanical Engineering COMEC 2010*, Santa Clara, Cuba (2–4 November 2010).
30 He, Z., Yuan, S., Liu, G. et al. (2010). Formability testing of AZ31B magnesium alloy tube at elevated temperature. *Journal of Materials Processing Technology* 210: 877–884.
31 Lee, Y.S., Yi, H.K., Van Tyne, C.J., and Moon, Y.H. (2011). Failure criterion for aluminium tubes during warm hydroforming and in crash application. *Proceedings of the Institution of Mechanical Engineers Part B Journal of Engineering Manufacture* 225: 357–366.
32 Abdessalem, A.B. and El-Hami, A. (2014). Global sensitivity analysis and multi-objective optimization of loading path in tube hydroforming process based on metamodelling techniques. *The International Journal of Advanced Manufacturing Technology* 71: 753–773.
33 Seyedkashi, S.M.H., Naeini, H.M., Liaghat, G. et al. (2012). Experimental and numerical investigation of an adaptive simulated annealing technique in optimization of warm tube hydroforming. *Proceedings of the Institution of Mechanical Engineers Part B Journal of Engineering Manufacture* 226: 1869–1879.

10

Hot Stamping

Fahrettin Ozturk[1,2], *Ilyas Kacar*[3], *and Muammer Koç*[4]

[1] *Department of Mechanical Engineering, Ankara Yıldırım Beyazıt University, Ankara, Turkey*
[2] *Strategy and Technology Management, Turkish Aerospace Industries Inc., Ankara, Turkey*
[3] *Department of Mechatronics Engineering, Nigde Omer Halisdemir University, Nigde, Turkey*
[4] *College of Science and Engineering, Divison of Sustainable Development, Hamad Bin Khalifa University, Doha, Qatar*

10.1 Introduction

Recent environmental laws and regulations force automotive industry to try and/or develop new materials and processes in order to meet the strict reduced fuel consumption and emission requirements. Prospective candidate lightweight materials are aluminum alloys, magnesium alloys, advanced and ultrahigh strength steels (A/UHSSs), composites, and plastics. A/UHSSs have become an attractive choice of material, particularly for body in white structural parts such as pillars, due to their high strength to weight ratio as well as their low cost, availability, weight reduction capability, and collision safety when compared to other options in the short and mid-terms. Several structural automotive parts can be manufactured using thinner advanced strength steel (AHSS) sheet metal blanks when compared to the current sheet metals without sacrificing any requirements including strength, safety, and environmental. However, in stamping operations, some of the A/UHSS do not have good formability at room temperature (RT). Besides, they tend to fail when producing complex shaped parts at the RT. As complexity of parts has been increasing due to part consolidation and several other design needs, forming at the RT becomes more difficult and complex failure mechanisms occur (such as edge cracking). Thus hot forming becomes an optimum process for an acceptable formability. It is well known that press hardening steels are used for hot deformation to obtain desired shapes and materials properties. Deformation above the recrystallization temperature enables the material to crystallize again during formation. The more strain hardening indicates the higher yield strength and hardness. Recrystallization process makes strain hardening decreased and ductility increased, which helps achieving high formability. With a consequent quenching, high strength in the part is ensured. For quenchable steels, sheet material is heated up to austenitization temperature in the furnace and transferred to the stamping presses to be deformed and quenched. Stamping of the new generation steel alloys with high strength and low formability is a challenge for auto body manufacturing technology, and hot stamping is still not a fully developed technology for the production of complex geometries using all types of A/UHSSs.

RT is preferred for any stamping operations, if possible. Because elevated temperatures need any additional process that affects part tolerances and increases costs. Although hot forming is a difficult and expensive process, it has very distinct advantages such as low press force, easy material flow for complicated surfaces and regions and high strength on the final

Modern Manufacturing Processes, First Edition. Edited by Muammer Koç and Tuğrul Özel.

product, which leads to thinner gauges and reduced weight. In some situations, the heating of the entire sheet is neither efficient nor cost effective, and local heating is considered for this fact. Sometimes pressured gases at 60–70 MPa are used for forming force. This type of forming provides easiness to form for both tubular-shaped profiles and sheet metals.

10.2 Process Description and Motivation

The hot press forming process was originally developed by Carl-Erik Ridderstrale of Plannja Hard Tech [1]. It has been seen that yield and tensile strength of material are increased up to 2.5–3 times after hot stamping process. This technique is preferable when material can be formed neither at RT nor by warm forming without any additional costly forming or joining procedures. Especially, UHSSs have some limited formability, but significant springback at RT. Hot stamping process can overcome the typical difficulties seen in cold and warm process [2]. It is almost possible to get springback-free parts after process. Figure 10.1 depicts temperature ranges with respect to melting point (T_m). The hot forming temperature starts about at $0.5T_m$ temperature. If a comparison is made between hot-stamped and cold-formed parts, it can be seen that hot-stamped parts have better deformation capability and no springback. In the hot deformation, very high plastic deformation is achieved. Therefore, elastic energy in the part is almost negligible. At the hot stamping, the elastic energy reduction is the main effect on springback compensation.

Hot stamping requires temperatures above the recrystallization point. This leads to that strength coefficient of flow curve of the hot formed metal is less than that at RT and the strain hardening exponent is zero (at least theoretically), so metal ductility increases remarkably. But, in addition to temperature, deformation itself produces further heating locally exerted by strain work. Sometimes this may cause melting in these regions unwittingly. Also the higher temperature causes scale layer formation on the stamping surface.

Recrystallization of the metal grains involves atomic diffusion that depends on time. Metal forming processes are generally performed at high strain rates where time is not enough to complete the recrystallization. However, during cooling period, recrystallization continues due to higher temperatures, even after hot stamping.

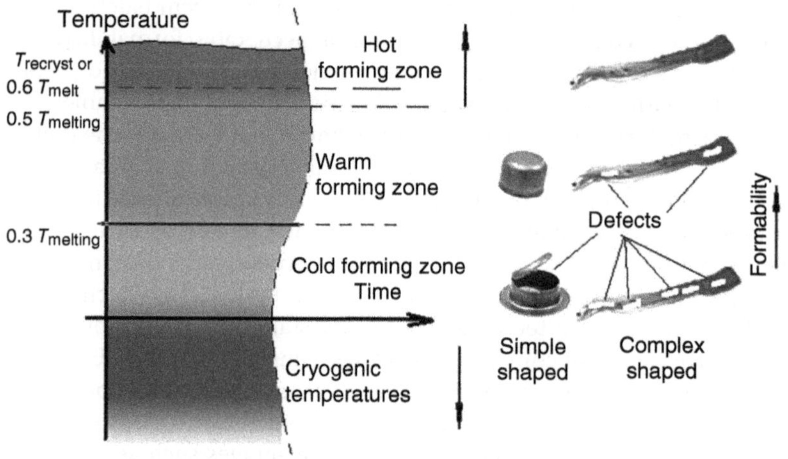

Figure 10.1 Temperature ranges for forming processes.

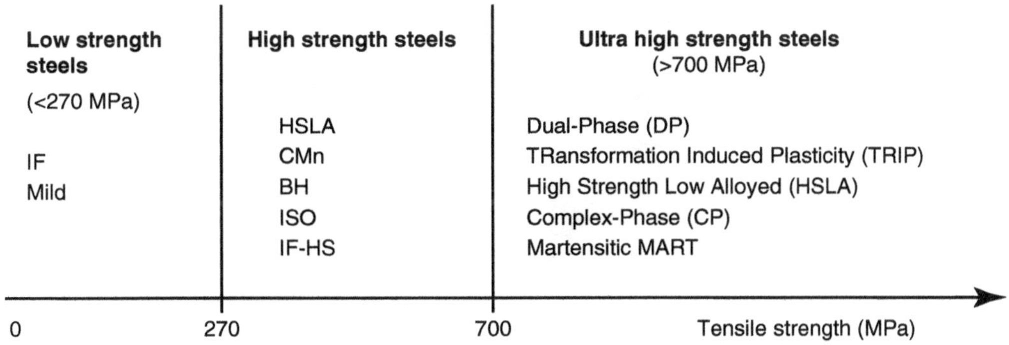

Figure 10.2 Steel grades.

10.3 Why Hot Stamping?

It is inevitable to use more strengthened materials for new generation car bodies and some complex-shaped parts such as roof reinforcement, pillars, side member, tunnel reinforcement, crash box, roof rail, rocker panel, and rear rails. The process makes yield strength of the material rise from 500–600 to 1000–1200 MPa and put superior mechanical properties to the part stamped. A classification with respect to strength for different steel grades was shown in Figure 10.2.

While mild steels have excellent formability (e.g. for deep drawing), HSS has a good balance among strength, formability, energy absorption, and durability, and AHSS has a high strength property. Dual phase (DP) steels that are a considerable portion of the AHSS applications have higher energy absorption ability in any crash case. Martensitic steels (MSs) or UHSS are typically used to provide collision protection by minimizing the deformation from side impacts. DP steels can provide both enough ductility and strength, although it has high work hardening properties. Hot stamping of boron-alloyed steel has better properties than that of the DP and other AHSS steels.

10.4 Automotive Parts by Hot Stamping and Potentials

Some simple shaped parts that ultimate tensile strength (UTS) is up to 1200 MPa such as side impact beams can be formed by cold stamping. There are no heating and cooling costs in cold forming, but it limits formability and causes significant springback even at warm temperatures especially for parts in planar geometries like sheet metals. These difficulties can be overcome by stamping performed at hot temperatures. Some parts in a typical auto body manufactured by hot stamping are seen in Figures 10.3 and 10.4.

The AHSS tubes are mainly used as side impact door beams, seat structures, and instrument panel beams in the automobiles. The shapes of the tubes are common circular, square, and oval. Welding of tapered tubes to other tubes may be necessary in application. Excellent strength and hardness with minimal defects have been obtained with very favorable welding characteristics up to 1.5 mm thickness by the hot forming of TRIP780 and DP980 sheet metals welded together.

In the forming of the HSS blanks, some defects known as *splitting* and *wrinkling* can be seen. These types of defects can be predicted by means of forming limit diagrams (FLDs). For AHSS, defect prediction is not easy task. Although recent developments and challenges in stamping of

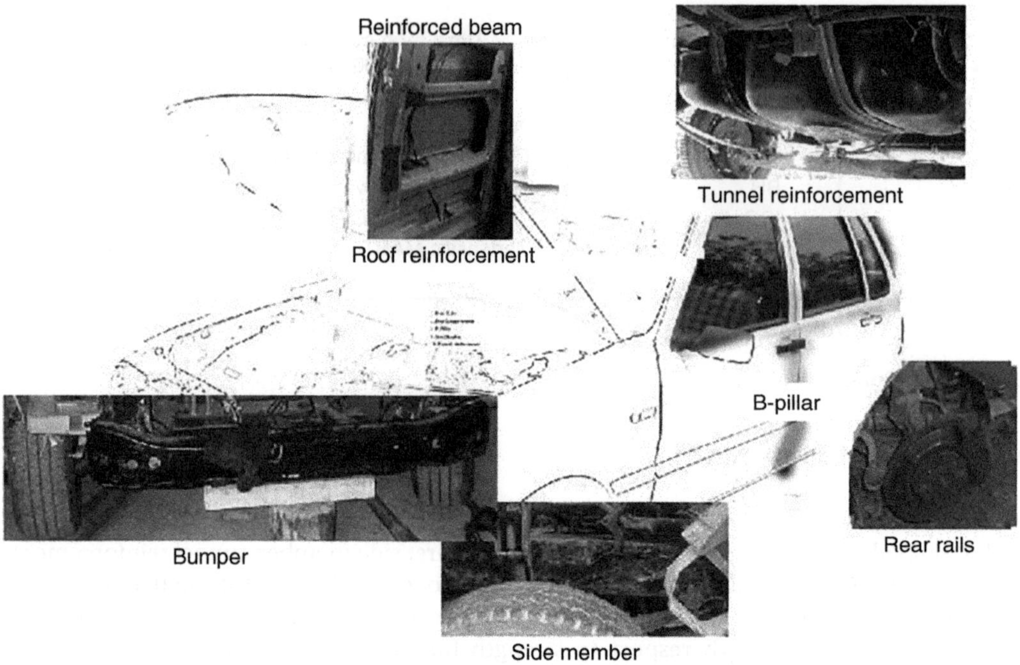

Figure 10.3 A typical middle-class car body with typical hot stamping parts.

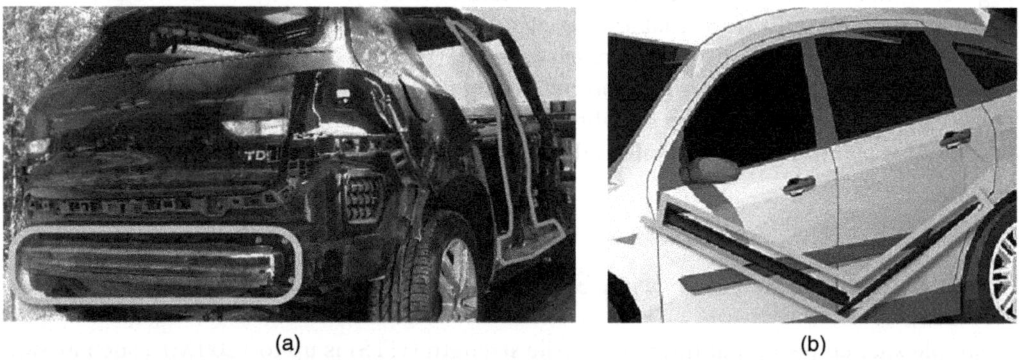

Figure 10.4 Hot stamping applications in automotive industry. (a) Bumper, B-pillar and (b) side impact beams.

the AHSS are satisfactory on new levels of strength and formability, the fracture mode known as *shear fractures* has still been a more general failure mechanism. These types of failures occur while bending and tension are together as in the stretching case and sometimes without localized necking. There are a number of factors leading to fractures in the AHSS. The ratio of the punch radius to the material thickness, called as "r/t ratio," is one of those factors. While a value of that ratio is sufficient for defect-free deformation in pure bending case, it may remain too small for combination of bending and tension. Because the failure originated from r/t ratio is either on the sidewall of the sample or at the punch nose radius in the stamping for AHSS blanks in general sense, the FLD is the most convenient way to predict the failure for bending with tension applications. One of the structural parts known as front beam in auto bodies is formed by stamping process, and due to security requirements it has to be more resistive to impact. An application is illustrated in Figure 10.5.

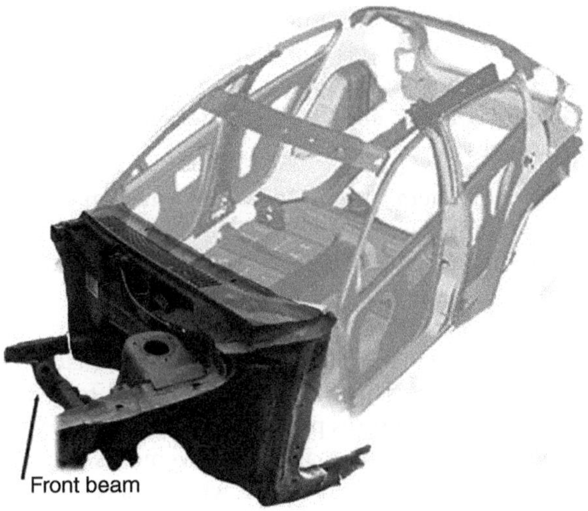

Figure 10.5 A hot stamping part made from DP steel in an auto body.

Figure 10.6 Hot stamped front windshield beam member.

Another factor is microstructure. Especially, failure behavior in multiphase materials is not predictable because of complex phase behavior. In DP steel stampings, homogeneity in microstructure provides more resistance to failure occurrences.

While a fuel tank guard consists of five pieces, weights 4.41 kg, and has 545 mm weld line by conventional manufacturing process of mild steel with UTS = 270 MPa, it may consist of three pieces, weights 4.28 kg, and has 120 mm weld line by direct hot stamping of AHSS with UTS = 1300 MPa. When the stamping process is completed, the flash is trimmed off by means of a laser cutting process after hot stamping. By means of AHSS, thinner sheet metals can be used without sacrificing strength requirements. Due to bigger surface area, windshield beams are another structural part having great effect on weight reduction strategies. Reducing the thickness of the beam in its construction can significantly provide weight savings opportunities. A windshield beam member is seen in Figure 10.6.

10.5 Advantages and Disadvantages

Hot stamping is a forming process that has a worldwide acceptance and accomplishes as much as possible elimination for the disadvantages of the cold or warm forming. In the optimal deformation temperature range, the hot forming allows maximum deformation ratios without springback and makes improvement on microstructure due to recrystallization of deformed grains. Hot forming is sometimes inevitable for especially forming of complex shaped parts. When the hot stamping is compared to cold stamping, during operation sheet material's strength is lower and strain hardening is less, ductility is higher. It provides the following advantages [3]:

- Force or energy required to deformation is small due to decrease in yield strength and increase in ductility.
- The higher temperatures lead to the more diffusion resulted in chemical homogeneities in microstructure.
- New crystallization leads to reduction of porosities in size or removing totally due to reorienting of inclusions (impurities). While inclusions and pores are randomly oriented before heating, those porosities are reoriented to a new string-like shape during cooling. These string-like structures prevent crack propagation.
- New crystallization makes improvements on the work piece due to fine spherical grain distributions in microstructure which increase the strength, ductility, and toughness of the material.

Ferritic steel with body-centered-cubic (BCC) lattice at RT is stronger than austenitic steel with face-centered-cubic (FCC) lattice at higher temperatures. So the austenite phase is more ductile and deformation of austenitic steel is easier than ferritic steel. Main advantage of hot stamping of metals is the ability to produce more plastic deformation than that of cold/warm stamping. A comparison between cold and hot-deformed tubes is seen in Figure 10.7. Hot stamping provides following advantages over cold stamping:

1) Productivity of many different and complex-shaped blanks is possible.
2) Lower force and power are required.
3) Metals often resulted in breakage during cold stamping may hot-stamped.
4) Isotropic strength properties that are the lack in cold stamping due to oriented grain structure can be obtained.
5) Strain hardening does not occur. Strain hardening is undesirable due to decreased ductility.

Besides aforementioned advantages, there are some limitations in hot stamping. Heating causes scaling or rapid oxidation on the blank surface. Dimensional accuracy is lower because

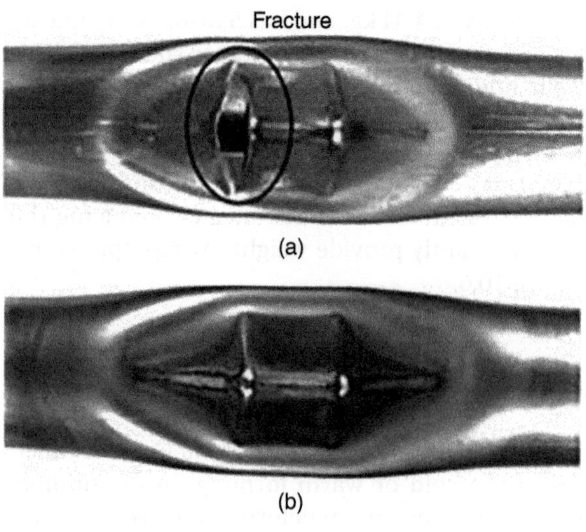

Figure 10.7 V-shaped cold and hot-formed tubes using resistance heating: (a) cold forming and (b) hot forming. *Source:* Mori 2012 [4]. Reproduced with permission of Elsevier.

contraction or distortion from insufficient cooling conditions can cause less precise tolerances on blank. Thermal or cross-sectional variations may cause undesired grain structures. Heating unit consumes energy and eventually causes higher cost. Severity in the hot working conditions leads to shorter tool life.

10.6 Process Description and Methods

Main idea in hot stamping is that disadvantages of conventional manufacturing processes on workpiece can be compensated by controlling the temperature of the workpiece during deformation. Hot stamping consists of both deforming and cooling of the blank just in die cavity. So hot stamping process does not include only a hot press deforming but also a hardening process simultaneously in-die cooling. In the hot stamping process, the die and punch are usually cooled up to RT at the required cooling range to make *press hardening (die quenching)*.

10.6.1 Hot Stamping Process Types

Hot stamping can be classified mainly into two types: *direct* and *indirect* hot stamping as demonstrated in Figure 10.8. Direct method was developed first.

The hot stamping process is explained briefly as follow:

1) Blanks are heated up to recrystallization temperature to convert the microstructure into more ductile one. The ferritic-pearlitic microstructure of steel is converted to austenite by heating up to ~950 °C and held during approximately five minutes.
2) A conveying system transfers the heated blank into an internally cooled die set. During transfer, the temperature of the part drops as $\Delta T = 100–80\,°C$. Between heating and press working, the hot blanks are needed to be transferred and handled properly (fast with minimizing heat transfer). In the *direct process*, blanks are transported to furnace first. In the *indirect process*, blanks are transported to press machine to be pre-formed in cold conditions. Then the pre-shaped blanks are transferred to furnace.
3) After the heating, blanks are transferred to press machine equipped with cooling ducts for both direct and indirect hot stamping. The blanks between die set are simultaneously

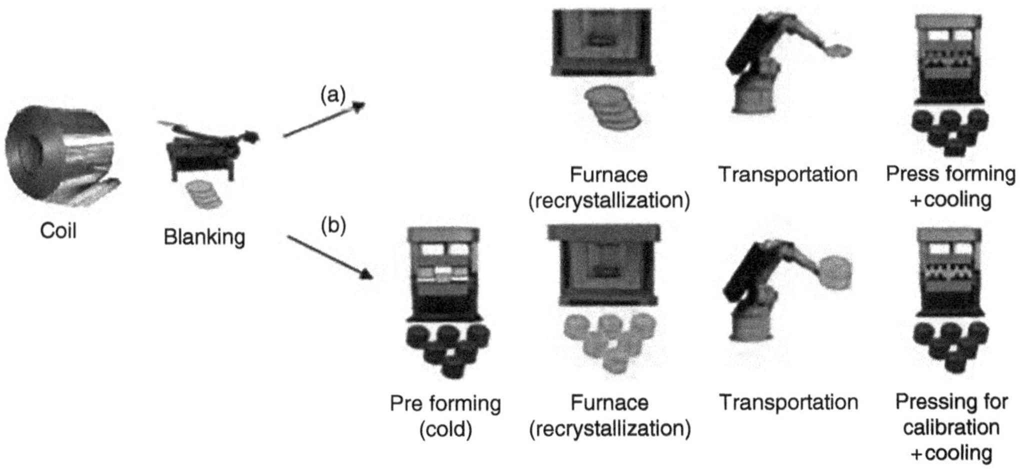

Figure 10.8 Basic hot stamping process chains: (a) direct and (b) indirect hot stamping.

stamped and quenched in the die cavity simultaneously. While press progress gives both *deforming* and *quenching* to blanks in direct hot stamping, press progress provides fine tuning (*calibration*) for previously deformed part and *quenching* in the indirect hot stamping. Cooling rate for quenching is important to eliminate the occurrences of other soft phases besides hardened phase (e.g. bainitic phase instead of martensite in steel). Cooling rate of 30–100 °C/s is preferable for steels to avoid bainitic transformation.

4) Quenching causes microstructural transformation to hardened phase, martensite (UHSS).

Those steps are illustrated in Figure 10.9 according to temperature, formability, microstructure, and strength aspects.

Transfer between furnace and press is usually performed by robots, and hence, temperature drop is about 60 °C unavoidable. In order to reduce and control the temperature drop during transferring of the blank, the heated part must be transferred as fast as possible. Forming process in the press must be completed before the martensitic transformation starts. Otherwise, the desired material properties cannot be achieved. The speed of the tool closure stroke is another important parameter for the process of hot stamping. Cooling process must begin as soon as possible after tool closes. Cooling is performed by means of water channels inside the tool. A certain clearance must be left between blank and blank holder to prevent the hardening of the blank holder. The strain rate in addition to the strain itself has a great effect on the deformation characteristics.

Forming and hardening are combined in a single process in the direct hot stamping method, whereas forming is conducted separately in cold conditions in the indirect hot stamping. When they are compared, it is clear that the direct hot stamping cycle is much shorter than the indirect stamping. In both methods, quenching operation transforms steel phases to satisfy the strength and hardness requirements. Through quenching, it has been possible to increase the strength from 400 MPa up to 1500 MPa for 22MnB5 steel [5]. The 22MnB5 steels are deformed in the hot stamping process. They have good formability and reasonable hardenability. Additional processes such as hole blanking, trimming, piercing, or cutting are completed in the pre-forming step because deformation is not easy at the last stage anymore. Therefore, the indirect method has been developed as an alternative to direct hot stamping.

It is mentioned previously that the starting temperature of hot stamping process are above the recrystallization temperature. Cooling rate is a very important parameter to obtain the desired microstructure and the consequent mechanical properties in the final product. Slower

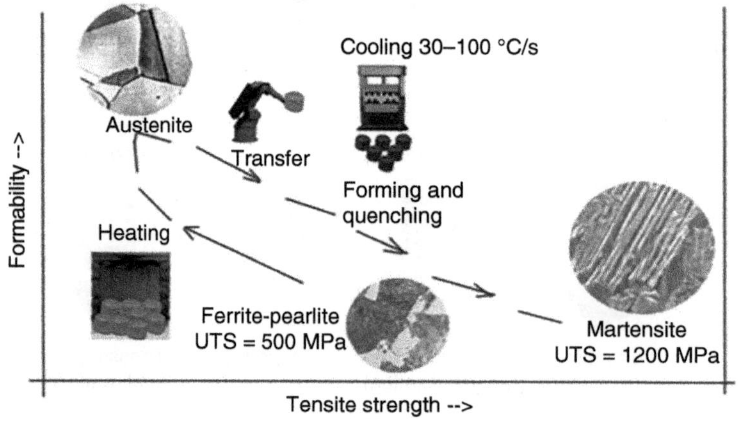

Figure 10.9 Hot stamping on steel material.

cooling may cause occurrences of undesired phases instead of martensitic phase in the microstructure. Another important fact is that material's chemical composition is a critical factor in hot stamping process to control microstructure and obtain desired phases at the final stage. The upper limit of hot stamping temperature depends on various factors such as oxidation, grain size, or unwanted phase occurrences. Table 10.1 lists hot forming parameters of some most commonly used sheet materials. Heating rate is almost the same as 27 °C/s for all materials. After quenching, die is opened when the part has an average temperature of 80 °C.

After pressing and quenching, last processes are applied on the stamped part such as trimming, cleaning, and welding. In automotive industry, lasers are generally used to trim and weld. Sand blast and shot peening are suitable to clean stamped parts.

10.6.2 Materials for Hot Stamping

Steel industry has started to develop the dual-phase steels (ferritic–martensitic) and other types advanced high strength steels due to fuel economy, collision safety, and environmental requirements of automotive industry since the oil crisis in 1970s. The alloys of the lightweight metals are widely used in motor vehicles, aircraft, and electronics.

One of the easy and efficient ways to reach the goal of light weighting cars is the usage of traditional lower density materials such as Al, Mg, and Ti alloys or plastics and polymer composites. However, these lower density materials often require bigger thicknesses to compensate for their lower strength. Therefore, required weight reduction is not possible every time due to bigger thickness. On the one hand, any suitable material to be used to meet the demands for vehicle weight reduction for more fuel economy and less harmful gas emissions, on the other hand, it has to meet the customer demands for high strength requirements. A balance among material costs, performance, and manufacturing techniques must be satisfied.

High performance steels and new materials can be used as an optional way. The new materials require new manufacturing techniques to overcome the deformation difficulties and defects during their forming. Forming and hardening are combined within a single manufacturing step in the hot stamping process. Hot stamping is a fundamental and essential technology for the manufacturing of sheet metals. By combination of both into one progress, thinner walled sheets can be used to manufacture the stampings due to the fact that the full martensitic transformation in the microstructure causes an increase of the tensile strength up to three times as much as 1500 MPa for 22MnB5 [12]. So, the thinner sheets mean the lightener structures.

Table 10.1 Optimal hot stamping temperature values of various materials.

	Optimal hot forming temperature (°C)	Cooling rate (°C/s)
Steel in general [6]	1150–1250	25
AISI 4140 [7]	1000–1100	water quenched
22MnB5 [8]	>900	30
Tungsten	2200	20–30
Tin (Sn)	Room temperature	—
Aluminized CMnB [9]	930	25
Ti–6Al–4V [10]	875 and 925	25
22SiMn2TiB [11]	950	25–75

Table 10.2 The mechanical characteristics of some AHSS materials.

Grade	Yield stress, R_p (MPa)	Tensile strength, R_m (MPa)	Elongation (%) min
TM1400 [13]	1287	1431	4.7 (A_{50})
CP800 GI [14]	600–750	780–950	≥10 (A_{80})
DOCOL 1400 [15]	1150	1400–1600	≥3 (A_{80})
TWIP [16]	450	900	60.3 (A_t)

Thus the new material concepts (e.g. ultra-lightweight HSSs (One grade is the manganese boron alloyed steels), metal foam, AHSSs, ceramic products, polymer composites, etc.) are very common in automobile body manufacturing. Usage of new manufacturing techniques (such as hot stamping) in the automotive industry allows the new designs with new materials. Generally, high strength materials are used for hot stamping processes to accomplish the advantage of the hot stamping. AHSS that offers up to five times the strength of the mild steels is a viable alternative to satisfy the balance between weight reduction and high strength demands. AHSS is popular steel used in light-weight automobiles with high strength for safety and production of critical parts, especially for safety of the passenger compartments. Some of them are seen in Table 10.2.

By hot stamping, flow stress decreases with increasing forming temperature that gives superior property to uniqueness of the hot stamping as in Figure 10.10.

10.6.3 Coating and Lubrication for Hot Stamping

Friction force increases in metal forming processes since the surface of the tool and blank is in close contact and subjected to high pressures and high temperature during hot stamping operation. Tool wear results in the loss of dimensional accuracy. Tools in metal forming are generally expensive. Therefore, tool wear is an important issue that needs to be addressed. Friction and tool wear become more troublesome due to high temperature and pressure of the hot stamping environment. Lubricant reduces the harmful effects of friction on the tool-blank interface for the most of metal forming process. Metal flow is sometimes delayed due to residual stress caused by friction and defects in the product. Friction in hot stamping is different from those of in mechanical systems, e.g. gearbox, shaft and bearings, and machine parts

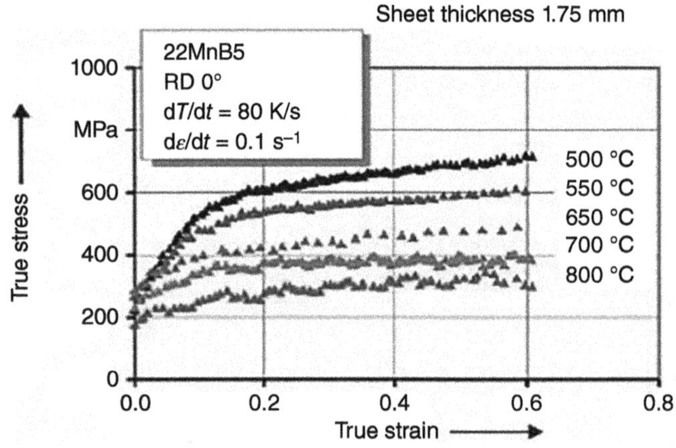

Figure 10.10 Influence of forming temperature on the flow stress. *Source:* Merklein et al. 2006 [17]. Reproduced with permission of Elsevier.

Table 10.3 Typical values of temperature, strain-rate sensitivity, and coefficient of friction in cold, warm, and hot working.

Category	Temperature range	Strain rate sensitivity exponent	Coefficient of friction
Cold working	$<0.3T_m$	$0.000 < m < 0.05$	0.1
Warm working	$0.3T_m$–$0.5T_m$	$0.05 < m < 0.1$	0.2
Hot working	$0.5T_m$–$0.75T_m$	$0.05 < m < 0.4$	0.4–0.5

Source: Groover 2010 [18]. Reproduced with permission of John Wiley & Sons.

moving relative to each other. In those cases, the friction is usually occurring due to metal-to-metal contact with low contact pressure and low temperature. It can be reduced by ample lubrication easily. But, in hot stamping case, higher pressure between soft blank and hard tools and more plastic deformation of soft materials at high-temperatures are present. Thus relatively high friction coefficient is created in hot stamping processes, even if lubricants are used. Friction coefficient values, for the three different forming types are given in Table 10.3.

If the coefficient of friction was too big, *sticking* can occur. Sticking is their tendency to adhere together instead of sliding of two surfaces relatively moved with respect to each other. When stresses from the friction exceed shear stress, one of the defects called *splitting* (tearing) will be present on blank due to sticking rather than sliding. Decreasing in the sticking, friction, power requirement, and tool wear leads to better surface quality in the final product. It can be obtained by proper lubricant usages. Tool wearing is a function of temperature and lubricant types in hot stamping. The following considerations are important when the selection of the appropriate metal processing lubricant:

- Blank's material.
- Chemical reactivity between tools and the blank materials (for most effective friction reducing, lubricants are coated to the whole surfaces generally).
- Ease of usage.
- Whether toxic.
- Whether flammable.
- Cost.

Sometimes hot stamping is carried out as dry for certain processes and materials (e.g. some types of steel and aluminum). Lubricants used for hot stamping operations are generally mineral oils, graphite, and glass. Another effective way for effective lubrication is usage of protective coatings for shielding and lubricating aimed. Blanks are pre-coated by Al–Si based elements as in Figure 10.11 to protect the blank from oxidation during heating and transfer from the furnace to the press. Coating provides protection from oxidation during the heating process. But coatings limit the heating range because it must not evaporate accidentally by rapid heating [19]. Proper heating time should be selected.

The hot-stamped parts are ready to be painted directly after the stamping operation without occurrences of any scaling as long as protective coating is applied before heating, e.g. Al–Si coating on 22MnB5. Coating also can be performed on the tool surfaces to reduce the abrasive wearing. Another metallic coating is galvanizing (zinc coating) and provides cathodic protection. If a material includes aluminum inherently such as aluminized CMnB, and aluminum in the microstructure provides protection naturally. Shielding gases like argon can be used as a protective medium for uncoated sheets. Shielding gases prevent the scale layer (oxidation) that increases stiffness and acts on fragile layer leading to undesired mechanical properties.

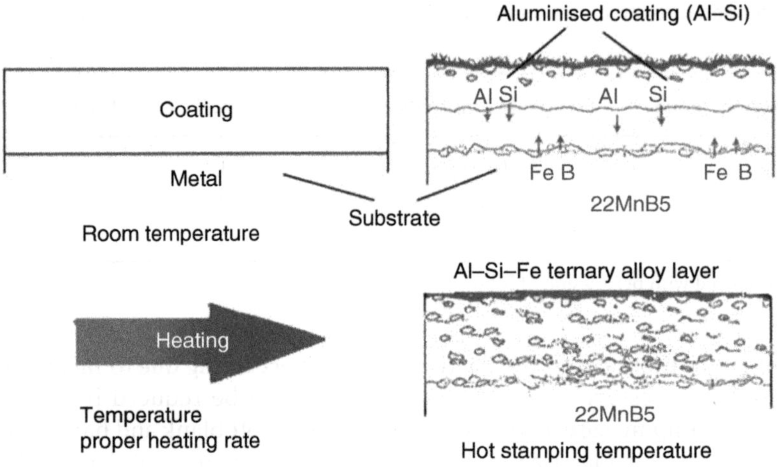

Figure 10.11 Usage of Al–Si alloying layer as a lubricant and oxidation protector.

Shielding gases prevent the contact between material and air. Lubricant usage for preventing the oxidation is not common in the hot stamping process due to their lower melting points. Also sol–gel process provides protective coatings as well as lubrication for cold forming or indirect hot stamping.

Although melting point of the Al–Si coating layer is 600 °C, which is under the austenitizing temperature of blank material, diffusion of Fe into Al–Si rises to a new layer consisted of Al and Fe on the surface and Al–Fe layer prevents evaporation of Al–Si coating in a certain heating speed. To prevent the melting and evaporating of the Al–Si coat, and to ensure sufficient diffusion, a special heating curve is used as depicted in Figure 10.12. Dwell time of blanks inside the furnace is limited and determined by the Al–Si–Fe ternary alloy layer thickness. Dwell time depends on the blank thickness.

Since coating may be damaged during the cold forming stage in the pre-forming progress of the indirect hot stamping, coating is not protective enough. Thus, direct hot stamping process is usually preferred for auto body industry in order to take advantage of the Fe diffusion into

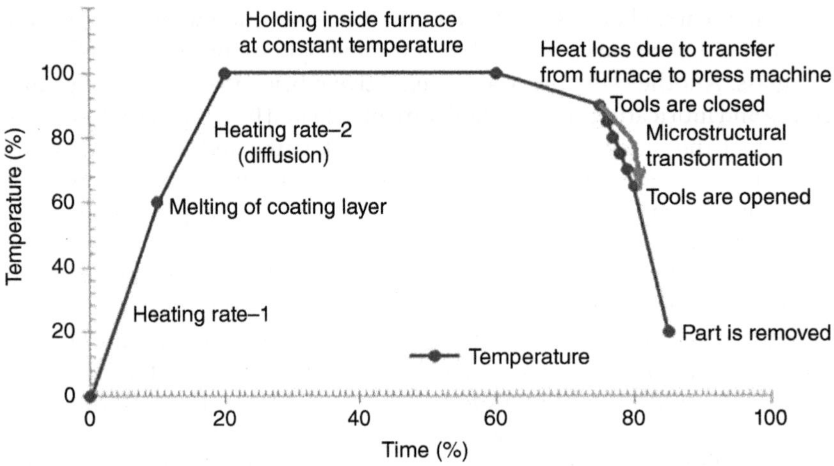

Figure 10.12 Schematic representation of the process sequence in relation to the heat treatment.

the Al–Si layer and higher deformation speed. Zn coating reduces friction between blank and die. It also has higher corrosion resistance, but Zn causes porosities during welding because of vaporization due to the lower melting point of Zn. Zinc in galvanized dual-phase steels splashes during welding and reduces the weld quality [20]. Improvements on weldability are obtained by removing of the Zn coat before welding [21].

10.6.4 Tools for Hot Stamping

In hot stamping, dies have two main functions, forming of blanks as expected and transferring the heat rapidly. Tools wear away due to high friction between blank and tool surfaces at higher temperatures especially. Coefficient of friction decreases as the temperature increases [22].

Deformation of blanks is easier at higher temperatures because strength and strain hardening are both reduced and deformation forces are going to be lower during hot stamping. In addition, ductility increases at elevated temperatures, which allows greater plastic deformation. These severe environments under high temperature and contact pressures lead to excessive wear. When designing a new die, selection of a tool material with 1500 MPa strength to prevent plastic deformations will be enough. Most hot stamping dies are commonly made from hot work steel (HRC 45). Tool wear leads to loss of dimensional accuracy, resulting in defective parts and requiring replacement of the expensive tooling. Die materials with better wear resistance is important to maintain tolerances and heat transfer efficiency. Die designing, manufacturing, and replacement costs are high, longer die life is an important issue that needs to be studied carefully.

The heat removing efficiency and cooling rate (cycle time) depend on contact pressure. Full contact between dies and blank surfaces also depends on tool surfaces roughness. When pressure is low, contact decreases and heat is concentrated on the contactless zones on blank called as *hot spot* if heat is not transferred properly enough. Heat transfer is reduced regionally in hot spot regions. One of the elimination methods is to use tool materials with high heat transfer coefficients.

If die material's thermal conductivity is increased, heat transfer will increase. Thermal barriers (coatings, corrosion layers) at blank/die interface need to be cleaned before stamping process in order to improve heat removing. In this context, designing of segmented cooling channels compared to conventional tool cooling techniques are more effective for heat transfer but so expensive in machining and prevention from coolant leakage in sealing.

10.6.5 Heating for Hot Stamping

Heating of the blanks up to the recrystallization temperature is the important step in the hot stamping process. It is expected that the basic functions of heating systems are to provide a homogeneous heating and short heating time. Dwell time of the blank depends on the thickness of sheet metal. Various research activities on the cooling performance of the press tooling focus on reducing cycle time. Higher energy costs and environmental friendly products lead to less energy-consuming concepts. So new heating-cooling technologies in hot stamping are still continuing research areas.

Heating type may be critical in terms of avoiding corrosion layer. Heating system should be able to provide desired heating sequences in a selected heating rate. It is important especially in the coating stage. Heating technique should be able to present phase transformation completely at the recrystallization temperature on the microstructure. Blank in the heating system should be deformed at homogeneous temperature distribution. Deviation from set value of temperature should not exceed required interval. Energy consumption should be less. Available furnaces in the hot forming line have length up to 30–40 m. Height is based on the process requirements. A lot of volumetric spaces are needed in the factory environment. These requirements make heating and furnace foundation cost to increase. New and cheap furnace designs have been a major issue. Some of the heating systems are briefly explained below.

10.6.6 Conventional Chamber Furnace (Radiation)

In the conventional continuous furnaces, heat is transferred by radiation and convection methods. Electricity or gas (natural gas, LPG) is commonly used to heat these furnaces. Blanks are transported to the furnace chamber by means of rollers or a closed-loop continuously moving conveyor. The heating rate is adjusted by the rate of this movement. Some furnaces have multiple chambers so that different heating sequences may be applied sequentially. One of these furnaces is shown in Figure 10.13.

The thicknesses of the blank, the coating thickness of the surface, and emission coefficient of the blank surface are important heating parameters influencing efficiency. If there are coating layers on blanks that behave as a barrier on the blank surface, heating rate is the most important parameter to avoid these layers by evaporation.

Later, heated blanks must be transported to the stamping presses. During transportation, temperature starts to drop and blank surfaces are oxidized, causing extra labor, cost, and loss of time. Also, initial warming-up is slow by this method. The method is not usually suitable for automation.

10.6.7 Conduction Heating

As an alternative heating technique, conduction heating (or electric resistance heating) can also be used in the hot stamping operation. A blank to be heated is connected to both ends of an electrical power supply. Heating occurs proportional to the current flow on the blank due to electrical resistance of the blank material. The temperature depends on the amount of current passed through the blank length. Heating of the material is based on the Joule's law. It is an appropriate method for fast and cheap heating. The method is flexible and suitable for various shaped and limited blank sizes. Efficiency depends on the blank metal's resistance and resistance of the cable used for connections.

Figure 10.13 A conventional laboratory scaled heating furnace.

Installation and operation costs are significantly lower than conventional heating systems. Varieties of blank materials are suitable for heating. Warming-up is fast. The heating unit can be combined with the deforming press. Therefore, there is no need to transport the blank. The cycle time can also be reduced.

Since the electrical resistance increases with increasing length of the sheet, this method is efficient for only certain ratios (length/section). It cannot provide uniform heating throughout blank thickness and industrial applications is not yet widespread. Complex shaped blanks cannot be heated uniformly. Special temperature measurement device is also needed to measure the blank temperature correctly. This method is used in the laboratory scale heating yet. One of these systems is shown in Figure 10.14.

10.6.8 Induction Heating

In this method, electrically conductive or semi-conductive materials can be heated. Induction heating has already been used for the purpose of pre-heating, melting, and heat treatment. Hot stamping has not been widely used as an industrial process yet. The use of the method for hot stamping operation is not common for industrial sized blanks. It has only been used in the laboratory scale hot stamping process so far.

An induction heating system consists of a high frequency generator and an inductor (induction coil). While a blank is passing through the windings, magnetic field from windings generates another current called as *eddy current* inside of the blank. Eddy current is exposed to short-circuit throughout the blank body due to electrical conductivity of the blank material. This short-circuit generates heat in blank. In the magnetic materials, the magnetic field at the working frequency causes an effect called as "hysteresis loss."

The operating frequency of the high-frequency generator affects the penetration depth of magnetic field. The lower frequency causes the higher penetration and vice versa. When the surface is hardened by this system, the layer thickness is determined by this property of the method.

Inductor type, distance to blanks, and blank's geometry are key parameters affecting the efficiency of heating. Also winding shape, part shape, direction, and amount of magnetic field

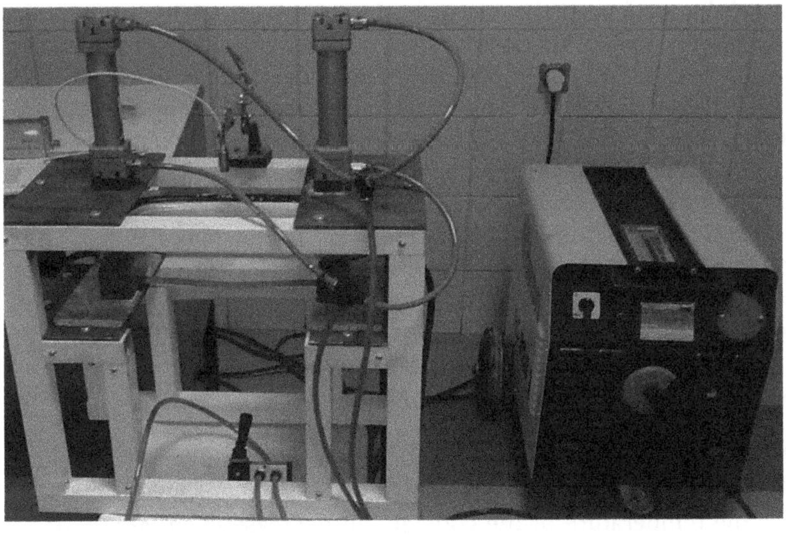

Figure 10.14 A laboratory-scaled electric conduction heating system.

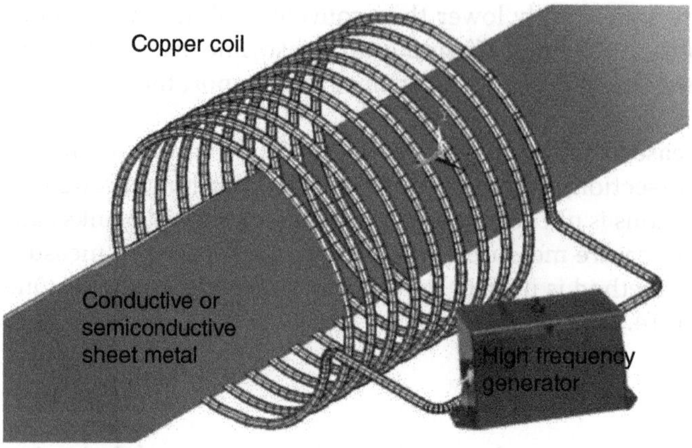

Figure 10.15 An illustration of the induction heating system.

influence the heating process. Three types of inductor can be used in the sheet metal processing; longitudinal field, cross field, and face inductors. Conveyor or material transporting speed does not influence the temperature but the rate of heating. If winding number in inductor is increased, the temperature will increase. For uniform heat transfer, the distance between material and winding should not be changed during the movement.

Induction heating is a fast method for heating of metals. It is reported that it takes 8–10 seconds to heat up to 900 °C for 22MnB5 (Al–Si coated). The heater is not in contact with the material in this method. The use of this method is new for hot stamping process. An induction heating system is illustrated in Figure 10.15.

10.7 Cooling for Hot Stamping

For optimum heat transfer in the cooling progress, contact between tool and blank surfaces must be set-up properly. There should be no spaces or oxide layers. Convective heat transfer depends on the tool-blank contact interface. In the design of the cooling channels; numbers, location, and sizes of channels have effects on heat transfer. Some restrictions arising from the manufacturing methods, for example drilling limit the design of cooling channel. By casting, places to be the channel holes can be formed by inserting pipes before casting.

The blank microstructure depends on material type and the cooling rate. Using the continuous time–temperature transformation curves (TTT) as in Figure 10.16 is more utilized to be able to determine the cooling rate to avoid bainitic transformation. Required curve is determined from TTT diagrams by means of successive experiments on several samples [23].

Accurate prediction of temperature distribution on blank and tools is a phenomenon that includes significant plastic deformation, phase transformations, and heat transfer by radiation and convection. Heat transfer coefficients are responsible in thermal behavior (heating and cooling). The greater the contact pressure between tool and blank causes to the bigger heat transfer [24].

Coolant fluids may flow through an open or a closed circuit. In closed circuit, contamination from particle concentration can be under control to prevent clogging channel holes up. In closed circuits, cooling tower, chiller unit, etc., can be used to cool the coolant itself. Increment in the yield strength up to 50–65 MPa was reported in the case of liquid nitrogen as coolant for 22MnB5.

With the cooling tower, the temperature of the coolant can decrease from 900 to 20 °C. Water cooled by a chiller unit may be used to cool the dies. If cooling process is supported by

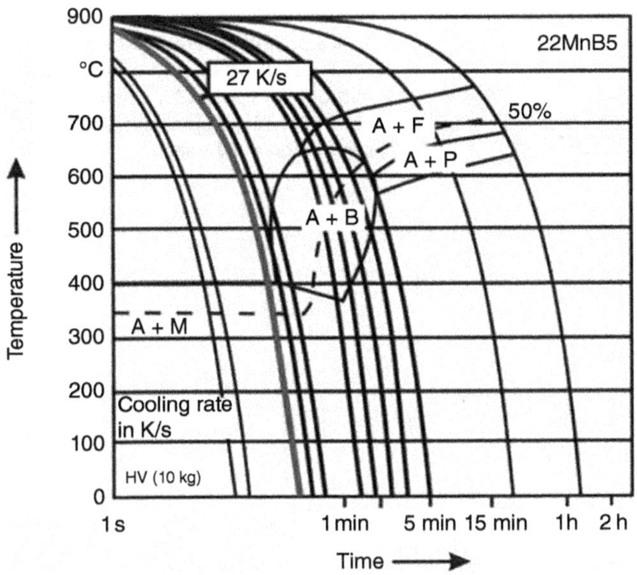

Figure 10.16 TTT-diagram of 22MnB5. *Source:* Merklein et al. 2006 [17]. Reproduced with permission of Elsevier.

another water cooling machine like a chiller unit, both cooling time and electricity consumption decrease.

10.8 Process Control

Forming and cutting operations are performed on metal sheets, strips, and coils. Press working (or stamping) is the term often applied for sheet metal because the machines used to perform these operations are presses (various types of presses are also used in other manufacturing processes). Today's press machines are equipped with advanced control systems. Conventional presses are driven by hydraulic cylinders or electrical motors with reducer. Mechanical servo presses (AC servo motor and hydraulic servo press) have been used in hot stamping operations. With the direct drive technology, programming of the press machine has been easier. A proportional-integral-derivative (PID) type control is applied on press machines. Punch position and velocity can be set simultaneously during stamping operation. When supported with a feedback unit, instantaneous action during process can be done immediately by a closed loop control. The load oscillations can be applied to reduce the load requirement by means of reduction of friction in compression as in Figure 10.17. Pulsation is obtained by control of load steps.

Digital linkages with other machines can be done by servo systems. Formability, accuracy, and productivity can be controlled by servo machines. A smart servo system can provide localized heat control as seen in Figure 10.18. An improvement in productivity and cost of the equipment are obtained.

10.9 Modeling and Analysis

There have been numerous numerical simulation and optimization studies on the failure predictions but none of them includes real stamping conditions totally. If there is no localized necking detected, it is almost impossible to predict failure. This phenomenon is a process

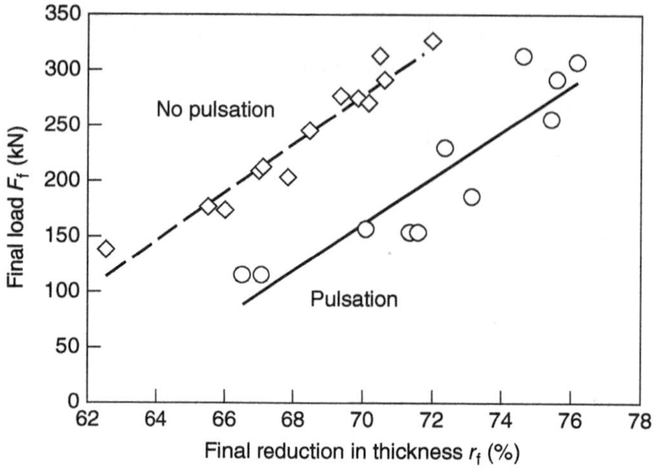

Figure 10.17 Relationships between final load and final reduction in thickness using servo press with and without load pulsation. *Source:* Maeno et al. 2011 [25]. Reproduced with permission of Elsevier.

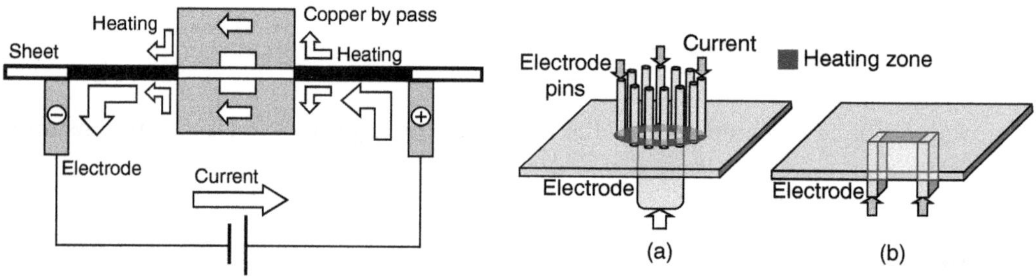

Figure 10.18 A resistance heating developed and local resistance heating of hole region for punching: (a) Large hole and trimming and (b) Small hole. *Source:* Mori 2012 [4]. Reproduced with permission of Elsevier.

that includes thermo-mechanical-phase transformation features at the same time as seen in Figure 10.19.

10.10 Design and Optimization in Hot Stamping

The design considerations for hot stamping process such as die design (cooling channels, die life improvements, and tool wearing), dwell time and temperature, heating (rate and method), cooling (distance and location of channels to the contact surface, the cavity designing, and cooling rate), strain and strain rate during hot stamping are important issues because it has still been difficult issues to be controlled or estimated satisfyingly especially for complex shaped parts.

Heat is transferred from blank surfaces to tools cooled. There are numerous numerical simulations and optimization studies on them. Quenching brings a microstructural transformation from austenite to martensite due to cooling during deforming. One of the methods to obtain homogenous heat transfer is the heating of both die and punch. But it decreases die life due to wearing of tool surfaces. Surface treatments are often used to improve tool life. Main concern on hot stamping is how to shorten the process cycle. The cycle time mainly depends on heating time required to recrystallization of the sheet blanks and also waiting time to diffusion of coating layer if any available coating and waiting time inside die cavity when tools are closed. FE-based methods are used for the proper designing and optimization of hot stamping operations.

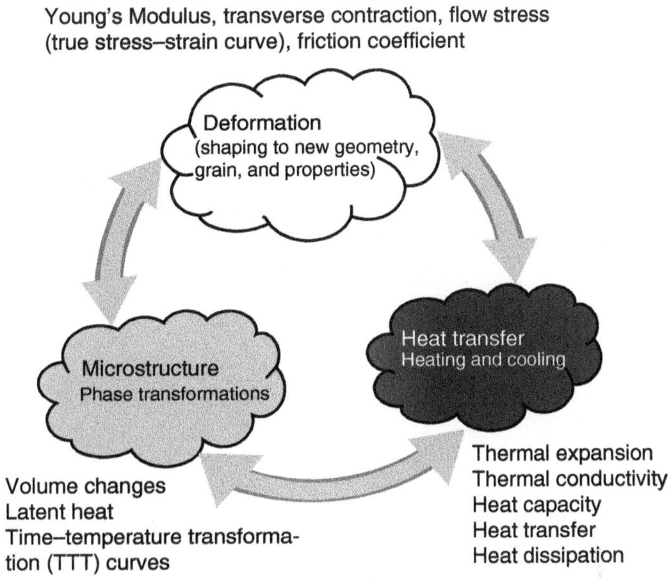

Figure 10.19 Main thermomechanical-phase transformation trilogy in hot stamping.

10.11 FEA in Hot Stamping

Finite element analyses (FEA) are most widely used methods for hot stamping analyses. Not only macro-scale modeling is important but also micromechanical modeling by finite element simulations is present to determine the formability issues in micro-scale. The FEA of hot stamping is a more complex and time-consuming process when compared to that of cold process. While tools can be modeled as a rigid body in cold or warm conditions, in hot stamping case, mechanic, thermal (temperature and hardness dependency with time), and microstructural properties have to be put into analysis. For this reason, tool and blank interaction has to be modeled adequately. As a crucial phenomenon, extensive research and engineering work are done on cooling systems. Due to nonuniform thickness of the crosssection, temperature distribution varies on the surface during deformation. The thinner sections are the cooler. Cooler sections have bigger tendency to cracking. A fully equipped model with thermal, mechanical, and microstructural conditions is needed to analyze. It is an important point that FE simulations do not include all forming conditions because of assumptions such as neglecting viscoplastic strains, the thermal contact resistance between steel sheet and die [26]. For this reason, experiences and trial and error investigations are still necessary to improve analyses. Before a mechanical analysis, a flow rule, a yield criterion, and a hardening model must be selected properly.

Analysis software need to know the *direction* and *amount* of the *plastic strain increment* to be able to determine the plastically deformed shape in every increment of calculation steps. Flow rule provides both the direction and amount of the plastic strain increment from each load step.

A yield criterion defines the material state at the transition from elastic to elastic-plastic behavior expressing a relationship between stress components at the moment when plastic yielding occurs. Since sheet materials show anisotropic behavior due to their crystal structure and rolling process, anisotropic yield functions should be employed in the plasticity model to determine the elastic strain limit from every load step. In a hot stamping operation, plastic anisotropy is a function of temperature, strain rate, material formability limits, friction, and heat transfer coefficients.

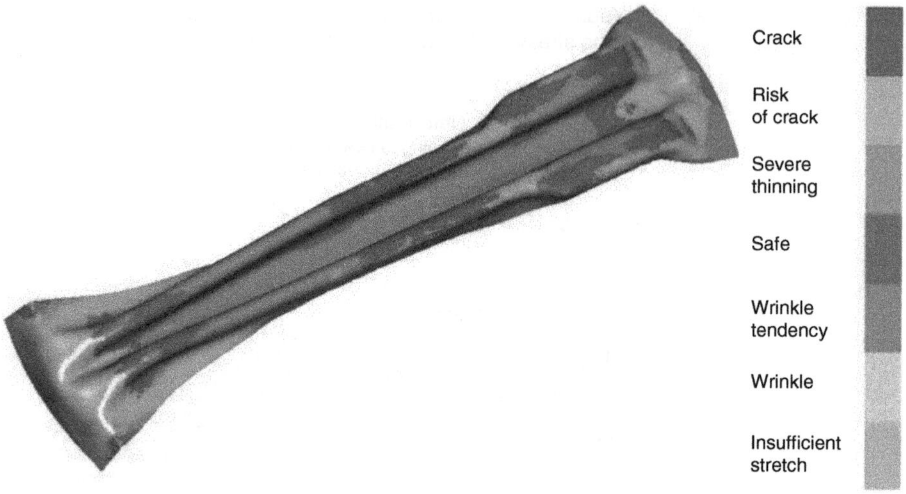

Figure 10.20 An isothermal FE analysis result. *Source:* Wang et al. 2014 [27]. Reproduced with permission of Elsevier.

Realistic modeling of sheet materials requires using of nonlinear hardening laws. Actually, due to plastic strain, material's yield strength changes at the end of every load step due to hardening (and softening). This new yield strength value must be used in the next load step as the yield strength. Yield strength depends on the strain hardening. During a stamping process, the blank undergoes lots of bending, unbending, and rebending cycles when passing on tool edges that lead to cyclic load and hardening in the blank. Hardening is effective on yield strength value. The hardening rule specifies the change in the yield strength value during the course of plastic flow. The better prediction of those hardening results in the more reliable finite element simulations.

In order to set-up an accurate model for FE-based numerical simulation of hot stamping, prominent issues on the analyses can be summarized as follows:

- Heat transfer between blank and die (and die equipment) by considering temporal and spatial changes.
- Phase transformations by evaluating temperature, time, and deformation.
- Material models by considering temperature, strain rate, and microstructure.
- Interaction between thermo-mechanical-microstructural trilogies called *thermal coupling* for setting a finite element model.
- Computational time consumption.

A simple FE simulation result of a thermo-mechanical isothermal hot forming process is seen in Figure 10.20.

10.12 Research and Development Trends and Needs

Various R&D activities have been continued on coatings, heating techniques, design, cycle time reduction efforts, and developing new steel grades. Many R&D efforts are ongoing issues on development of processes and tools for more accurate and rapid models, including mechanical equipment development.

By new design, as in Figure 10.21, the target product could be successfully hot stamped, satisfying the specification of high hardness and good dimensional accuracy [28]. A cast conform-

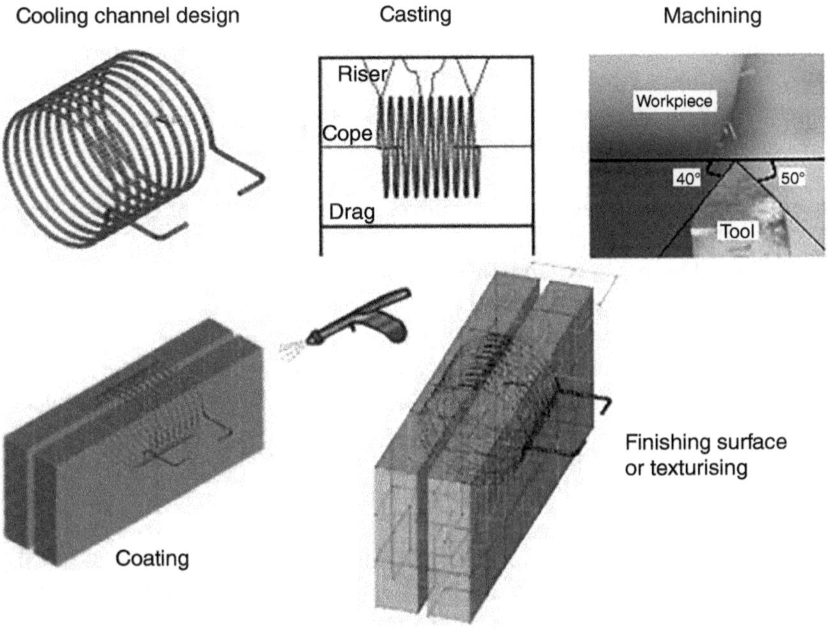

Figure 10.21 Cast conformable cooling channels.

able cooling channels are seen in Figure 10.22. One of the new trends on the tool design in hot stamping is partial heating and partial quenching of blank to obtain variation in properties.

Segmented tools with drilled cooling channels are new with respect to milled-machined cooling channels. Machined cooling channels are more effective for heat transfer but have sealing problems. Also machining/manufacturing costs are higher.

Research and development facilities on alternative heating and cooling strategies are still ongoing process such as heating one side of the blank to higher temperature while other side remains at a relatively low temperature. Similarly, cooling only on certain zones during deforming and other zones after deforming to avoid the martensitic transformation in certain zones. Also, nonferrous alloys with much more heat conductivity are new challenges.

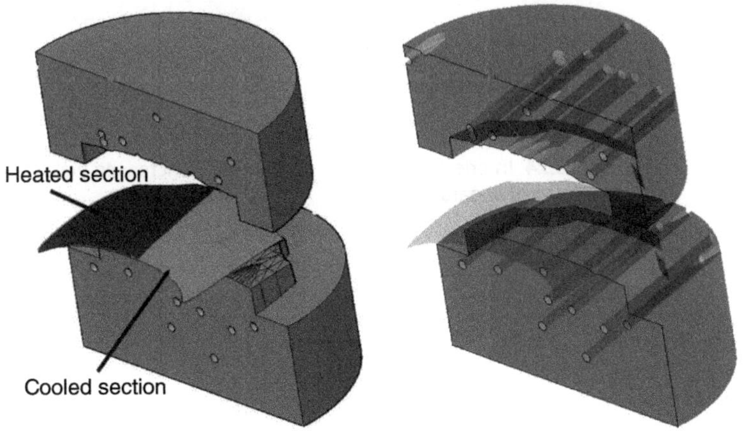

Figure 10.22 A new trend on the tool design in hot stamping.

Nowadays, press machines with servo drive has been started to be used on finishing processes, e.g. cutting, trimming, and coating to obtain process flexibility and controllability.

Alternative coatings are another center of interest. Hot-dip galvanizing is inexpensive and easy to apply or remove. During austenization, Zn coating transforms to a Zn–Fe layer which protects surfaces against scaling in addition to cathodic protection maintained even if the Zn-coating is damaged.

Nanotechnology-based coatings have been recently developed for many application including the hot stamping. Main idea in nano coating is that all materials are porous and contain microscopic peaks and valleys. The nano-coated particles are in laminar shapes, which mean they overlap each other like the scales of a fish. When these platelets are deposited on the capillary wall and closely follow the contour of its surface, it reduces the permeability of the treated surface [29]. A paint based on nanoparticles has been developed to apply for direct hot stamping. This coating protects the blank against corrosion and improves the lubrication between sheets and dies [30].

In Figure 10.23, it is seen that the nanocolumnar coating has sufficiently high wear resistance even in severe bending and ironing step of progressive stamping sequence at the narrowed clearance situations.

Another interest is in combination of hot stamping and Tailor welded blanks (TWBs). TWBs are used to improve the crash performance and a strategy on weight reduction as in Figures 10.24 and 10.25. Both of sheet metals made of different steel grades with/without different thick-

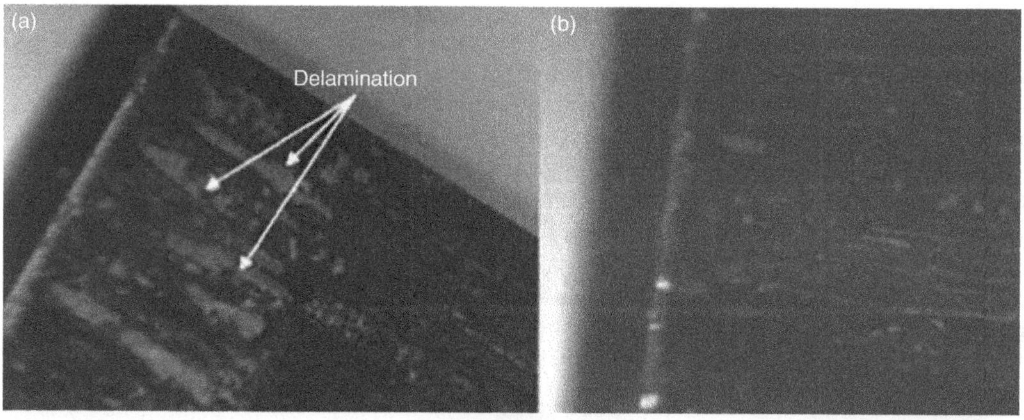

Figure 10.23 Comparison of contact surface after loading and unloading (a) uncoated (b) coated with nanocolumnar carbon coating. *Source:* Aizawa et al. 2007 [31]. Reproduced with permission of Elsevier.

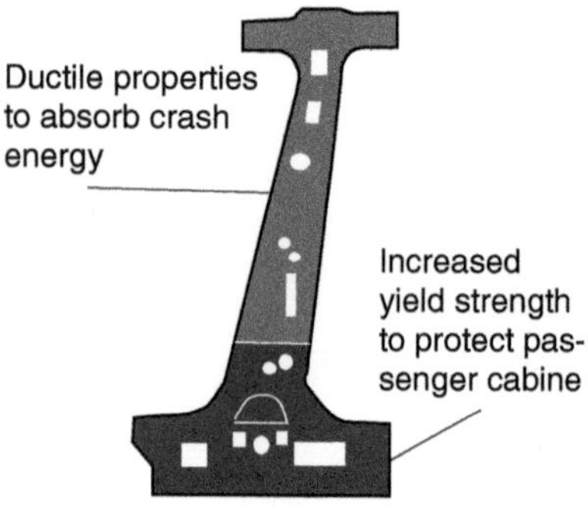

Figure 10.24 In one product, both ductility and increased yields can be desired separately.

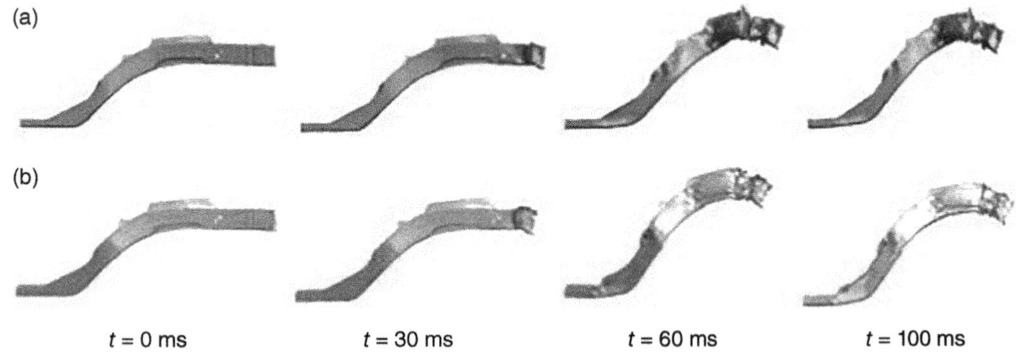

Figure 10.25 The deformation history of the front side rail using (a) simple structure and (b) TWB structure. *Source:* Shi et al. 2007 [32]. Reproduced with permission of Elsevier.

nesses are welded together before stamping process. This allows adjusting the strength and elongation properties along the part, leading to different deformation modes under crash situations.

Higher strength (greater than 1500 MPa) is not desired on an auto body in terms of energy absorption during the crash. B-pillar is a vertical part between door sill and roof frame. Nowadays, the B-pillars have consisted of lower and upper sections with/without different thicknesses, materials. B pillar's lower section has ductile properties to absorb crash energy by starting to yield initially, while upper section has got increased yield strength to protect and maintenance the stabilization of the passenger compartment in any side impact case [33].

TWBs are so deterministic welding style in automotive industry especially in terms of strength. The usage of the TWB in the automotive industry allows to new designs with new materials. Selection of a suitable partner material effects mechanical and thermal response of blank. Tailor welded high strength low alloyed (HSLA)/boron steels shows good hot stamping behavior as in Figure 10.26.

HSLA steel's yield strength is about 400–500 MPa. The strengthening mechanisms mainly depend on fine dispersion of small amounts of alloying elements such as titanium, vanadium, and niobium. Hot forming of TWB results in high-strength martensitic steel combined with a ductile ferritic, pearlitic, and bainitic microstructure.

Laser welding equipment is preferred for tailor welding so that welding has very narrow line and heat affected zone (HAZ) as shown in Figure 10.27. It is seen that after stamping, necking occurs far enough from the weld line and HAZ.

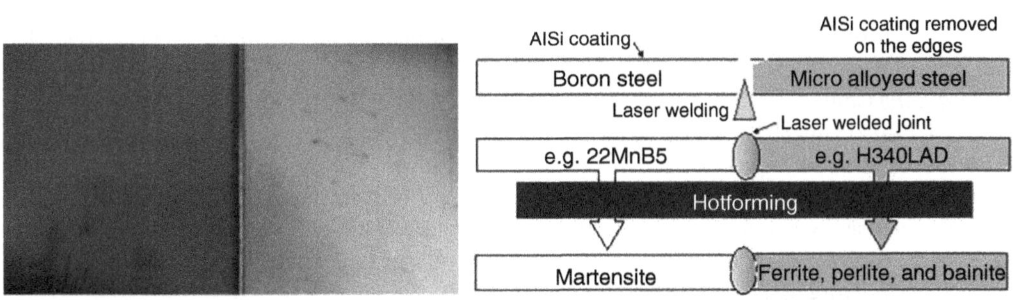

Figure 10.26 TWBs of low alloy high strength steel and boron steel. *Source:* Tang et al. 2013 [34]. Reproduced with permission of Elsevier.

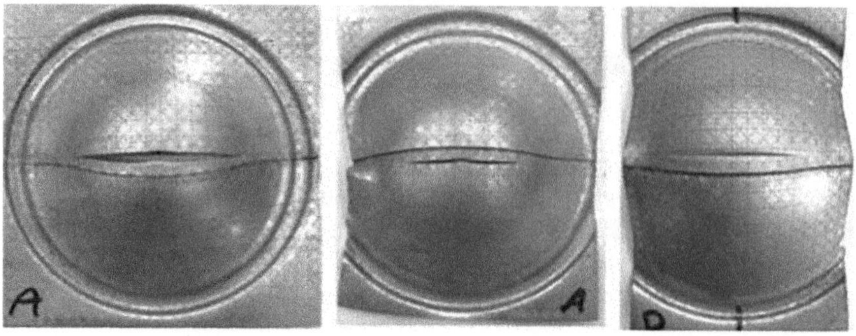

Figure 10.27 Deformations on TWBs. *Source:* Mohebbi and Akbarzadeh 2012 [35]. Reproduced with permission of Elsevier.

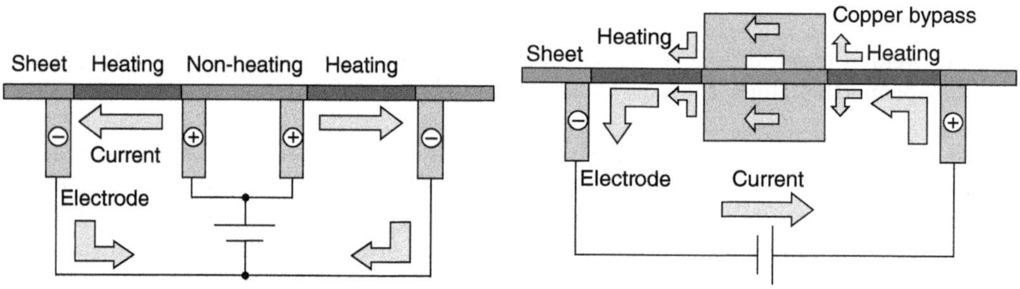

Figure 10.28 New and alternative heating concepts. *Source:* Mori et al. 2013 [36]. Reproduced with permission of Elsevier.

The other trend in hot stamping is designs of alternative faster, economic, and selective heating concepts as in Figure 10.28.

Eventually, hot stamping is a process that includes thermal, mechanical, microstructural, and technological fields. Hot stamping process has great potential for further investigations and innovations for car bodies.

References

1 Norrbottens, J.A.B. (1977). *GB1490535 (A) – Manufacturing a Hardened Steel Article*. Sweden: Norrbottens Jaernverk.

2 Mori, K., Maki, S., and Tanaka, Y. (2005). Warm and hot stamping of ultra high tensile strength steel sheets using resistance heating. *CIRP Annals – Manufacturing Technology* 54 (1): 209–212.

3 Koç, M. and Altan, T. (2008). Hot stamping of ultra high strength steels – a review. In: *AP&T Hot Stamping Conference*, Dearborn, MI (2 October 2008).

4 Mori, K.I. (2012). Smart hot stamping of ultra-high strength steel parts. *Transactions of Nonferrous Metals Society of China* 22 (2): s496–s503.

5 Altan, T. (2006). Hot-stamping boron-alloyed steels for automotive parts. Part I: process methods and uses. *Stamping Journal – An FMA Publication* 40–41.

6 Stachowicz, F., TrzepieciŇSki, T., and Pieja, T. (2010). Warm forming of stainless steel sheet. *Archives of Civil and Mechanical Engineering* 10 (4): 85–94.

7 Samadian, P., Parsa, M.H., and Shakeri, A. (2014). Determination of proper austenitization temperatures for hot stamping of AISI 4140 steel. *Journal of Materials Engineering and Performance* 23: 1–8.

8 Önal, A.S. and Güler, H. (2012). Hot forming process. *Union of Chambers of Turkish Engineers and Architects, Chamber of Mechanical Engineers Journal of Engineer and Machine* 53 (626): 66–70.

9 Fan, D., Kim, H.S., and Cooman, B.C.D. (2008). *Thermo-mechanical Analysis of Hot Press Forming of AHSS.* Association for Iron & Steel Technology.

10 Filice, L., Gagliardi, F., Lazzaro, S., and Rocco, C. (2010). FE simulation and experimental considerations on TI alloy superplastic forming for aerospace applications. *International Journal of Material Forming* 3 (1): 41–46.

11 Shi, Z., Liu, K., Wang, M. et al. (2012). Effect of non-isothermal deformation of austenite on phase transformation and microstructure of 22SiMn2TiB steel. *Materials Science and Engineering: A* 535: 290–296.

12 Turetta, A., Bruschi, S., and Ghiotti, A. (2006). Investigation of 22MnB5 formability in hot stamping operations. *Journal of Materials Processing Technology* 177 (1–3): 396–400.

13 Rehrl, J., Mraczek, K., Pichler, A., and Werner, E. (2014). Mechanical properties and fracture behavior of hydrogen charged AHSS/UHSS grades at high- and low strain rate tests. *Materials Science and Engineering: A* 590: 360–367.

14 Steel, T. (2013). CP800 GI, Highly bendable roll forming steel. In: *NLAMLTDS:750/ENG:11/2013* (ed. T.S.E. Limited). The Netherlands: Tata Steel Automotive Commercial Headquarters.

15 SSAB (2007). Docol 1400. Data Sheet: 12-11-07 GB8207 DOCOL. 4.

16 Chen, L., Voswinckel, H., Kim, H.S. et al. (2008). *Stretch-flangeability of TWIP Steel.* Association for Iron & Steel Technology.

17 Merklein, M., Lechler, J., and Geiger, M. (2006). Characterisation of the flow properties of the quenchenable ultra high strength steel 22MnB5. *CIRP Annals – Manufacturing Technology* 55 (1): 229–232.

18 Groover, M.P. (2010). *Fundamentals of Modern Manufacturing: Materials, Processes, and Systems*, 4e. Wiley.

19 Frutos, E., Adeva, P., González-Carrasco, J.L., and Pérez, P. (2013). Oxidation behavior of AISI 316 steel coated by hot dipping in an Al–Si alloy. *Surface and Coatings Technology* 236: 188–199.

20 Chen, H.C., Pinkerton, A.J., Li, L. et al. (2011). Gap-free fibre laser welding of Zn-coated steel on Al alloy for light-weight automotive applications. *Materials and Design* 32 (2): 495–504.

21 Pennington, E.J. (1987). Laser welding of galvanized steel. Patent US4642446 A.

22 Stöhr, T., Merklein, M., and Lechler, J. (2008). Determination of frictional and thermal characteristics for hot stamping with respect to a numerical process design. In: *First International Conference on Hot Sheet Metal Forming of High-Performance Steel*, Kassel, Germany, 2093–2300.

23 Karbasian, H. and Tekkaya, A.E. (2010). A review on hot stamping. *Journal of Materials Processing Technology* 210 (15): 2103–2118.

24 Lechler, J. and Merklein, M. (2008). Hot stamping of ultra high strength steels as a key technology for lightweight construction. In: *Materials Science and Technology (MS&T).* Steel Product Metallurgy and Applications, Pittsburgh, PA, 1698–1709.

25 Maeno, T., Osakada, K., and Mori, K. (2011). Reduction of friction in compression of plates by load pulsation. *International Journal of Machine Tools and Manufacture* 51 (7–8): 612–617.

26 Ying, X. and Zhong-de, S. (2013). Design parameter investigation of cooling systems for UHSS hot stamping dies. *The International Journal of Advanced Manufacturing Technology* 70 (1–6): 257–262.

27 Wang, G.-F., Wu, X.-S., Sun, C. et al. (2014). Auxiliary current hot forming of high-strength steel for automobile parts. *Procedia Engineering* 81: 1701–1706.

28 Lee, M.-G., Kim, S.-J., Han, H.N., and Jeong, W.C. (2009). Application of hot press forming process to manufacture an automotive part and its finite element analysis considering phase transformation plasticity. *International Journal of Mechanical Sciences* 51 (11–12): 888–898.

29 Deardorff, J. (2011). Permanon: nanotechnology-based surface protection reduces factors that cause coating failures. *Metal Finishing* 109 (1–2): 29–31.

30 Altan, T. and Tekkaya, A.E. (2012). *Sheet Metal Forming: Processes and Applications.* ASM International.

31 Aizawa, T., Iwamura, E., and Itoh, K. (2007). Development of nano-columnar carbon coating for dry micro-stamping. *Surface and Coatings Technology* 202 (4): 1177–1181.

32 Shi, Y., Zhu, P., Shen, L., and Lin, Z. (2007). Lightweight design of automotive front side rails with TWB concept. *Thin-Walled Structures* 45 (1): 8–14.

33 Kolleck, R., Steinhöfer, D., Feindt, J.-A. et al. (2004). Hot forming and cold forming – two complementary processes for lightweight auto bodies. In: *Proceedings of The International Conference of New Developments in Sheet Metal Forming Technology*, Stuttgart/Germany, 235–244.

34 Tang, B., Yuan, Z., Cheng, G. et al. (2013). Experimental verification of tailor welded joining partners for hot stamping and analytical modeling of TWBs rheological constitutive in austenitic state. *Materials Science and Engineering: A* 585: 304–318.

35 Mohebbi, M.S. and Akbarzadeh, A. (2012). Prediction of formability of tailor welded blanks by modification of MK model. *International Journal of Mechanical Sciences* 61 (1): 44–51.

36 Mori, K., Maeno, T., and Mongkolkaji, K. (2013). Tailored die quenching of steel parts having strength distribution using bypass resistance heating in hot stamping. *Journal of Materials Processing Technology* 213 (3): 508–514.

11

High-Speed Forming (Electromagnetic, Electrohydraulic, and Explosive Forming)

Brad Kinsey[1] *and Yannis Korkolis*[2]

[1] *Department of Mechanical Engineering, University of New Hampshire, Durham, NH, 03824, USA*
[2] *Department of Integrated Systems Engineering, The Ohio State University, Columbus, OH, 43210, USA*

11.1 Introduction

High-speed forming (HSF) processes are historically associated with the space race of the 1950s and 1960s [1]. A variety of novel alloys (e.g. nickel and titanium alloys) and very complex aerospace components (e.g. integrally-ribbed, thin-walled missile shells for the Apollo missions) could not be fabricated by traditional forming techniques (e.g. stamping in a press). As a result of this aerospace push, a plethora of forming techniques were created and pursued to the extent that numerous HSF machine tools became commercially available. However, despite the initial interests, these techniques failed to gain widespread acceptance, and those that survived into the 1970s and beyond remained confined to niche applications. The main reason for that was the lack of fundamental understanding of (i) the science of each process and (ii) the material behavior at high strain rates. This led to inefficient tooling/machine design and limited process parameter understanding, tied to expensive and time-consuming trial-and-error implementation. Thus, manufacturing engineers were unable to exploit these techniques fully, oftentimes resulting in only marginal improvement over well-established conventional forming techniques for mass-produced components (e.g. automotive, appliance, etc.). However, with current day advances, HSF processes are poised to become more widely used in production applications.

The term HSF is, to some extent, a misnomer. Indeed, it clearly describes a basic feature of these processes, i.e. that the actual forming of the specimen is completed on the order of milliseconds. But given the fact that the workpiece needs to be loaded in and then removed from the forming machine, the overall time savings of HSF over conventional processing might not be enough to justify the additional investment. (Note that this loading–unloading delay may be necessary to e.g. provide sufficient time to recharge capacitor banks for electromagnetic forming [EMF] processes.) On the other hand, the term HSF fails to disclose that an almost universal feature of these processes is the significantly enhanced formability they offer over the conventional techniques [2]. For example, see Figure 11.1 for the forming of a cone-shaped geometry. Thus, not only is a component formed faster but it can also be formed in a single operation. Eliminating the multiple forming steps; the warm-/hot-working or possible annealing steps in-between; the loading, unloading, and transferring of semifinished workpieces to and from processing stations; the work-in-progress inventories; and the capital equipment and

Modern Manufacturing Processes, First Edition. Edited by Muammer Koç and Tuğrul Özel.

Figure 11.1 Example of enhanced formability in electromagnetic forming (EMF) of a cone shaped geometry. *Source:* Courtesy of G. Daehn.

tooling needs for multiple stations have brought the HSF processes back to the research forefront. This trend is exacerbated by the emergence and adoption of modern automotive alloys (such as advanced high strength steels and aluminum and magnesium alloys), which generally suffer from reduced ductility in comparison to the mild steels they are meant to replace. Hence processes that hold the promise to enhance the ductility are becoming increasingly relevant. In the case of larger components and smaller production runs, HSF can be attractive as it often requires modest investment in capital equipment and tooling. Lastly, in recent years, the concept of hybrid processes has emerged. For example, a large panel with local sharp features can be stamped conventionally and then, in the same forming die, a HSF process might be used locally to reproduce these features [3].

11.1.1 Advantages

There are numerous advantages of HSF processes compared to conventional methods, e.g.:

- HSF processes are highly repeatable with reduced springback and wrinkling. These processes can be free-forming (without a female die), but calibration of the workpiece against a die can improve the dimensional consistency in the process. The reduced springback is due to the more uniform distribution of strains.
- Improved formability is achieved during HSF processes compared to conventional metal-forming processes (see Figure 11.1). These can be contributed to inertial stabilization of the deformation, a more uniform strain distribution in the material, and large compressive stresses when forming into a female die. Kinetic energy, which can be ignored during conventional forming, is significant during HSF processes and is responsible for this improved formability.
- Dissimilar materials can be joined together. While different metal alloys are of interest, even composites, polymers, glasses, and ceramics can be joined to metals using these techniques. Various joints are achievable, from interference fits to solid state welds. For metal welding, this prevents the formation of intermetallic phases, heat-affected zones, and thermal distortion which are all detrimental to weld quality.
- Free-forming without die contact assures no damage to, e.g. a coating on the workpiece. Also, no lubricant is required providing environmental and cost benefits, e.g. since a processing step to remove the lubricant is not required.
- If female tooling is used, only light tooling systems are required as the inertia of the tooling is often sufficient to constrain the deforming workpiece, and the forces generated will dissipate quickly due to the short timeframe of the process.
- Components that are not achievable by traditional processes are possible, e.g. coining over a significant surface area due to the large pressures produced.

11.1.2 Disadvantages

Of course, there are disadvantages to HSF processes as well, e.g.:

- The modeling of the process is difficult due to the multiple phenomena occurring, the impulse required to initiate the process (whether electrical or chemical), the modified material behavior at high strain rates, etc. Thus coupled, physics-based simulation techniques are required. The equations for these various phenomena are beyond the scope of this book chapter. But references are provided throughout. For example, information on the wave behavior in solids which dominate during high-rate forming processes can be found in classic books such as, Johnson [4].
- The classic method to predict failure in sheet and tube metal-forming operations, i.e. the forming limit diagram, cannot be readily used for HSF processes as such criteria were created without inertial effects which are dominant at high strain rates.
- The energy efficiency of HSF processes is low as some energy is also converted to heat, sound, motion of other entities, etc. not just deforming the workpiece. Efficiencies typically reported are in the range of 2–25% depending on the application [5].
- The cost of HSF processes can be difficult to assess due to the lack of physical understanding with respect to how long coils and electrical components will last and process efficiency.
- In the case of electromagnetic processes, the materials used need to possess reasonably high electrical conductivity, e.g. above 15 $\mu\Omega$ cm [5]. However, another workpiece can be used as a sacrificial driver material in order to achieve deformation.
- Safety concerns must be addressed when implementing HSF processes due to the high currents, sometimes in the presence of water, explosives, and deformation rates used. However, there is no reason that the processes cannot be conducted in a safe and reliable manner in an industrial setting. Note though that additional regulations may be required.

11.1.3 Process Variations Mentioned Briefly

There are three main variations of HSF processes that will be covered in this chapter:

i) *Electromagnetic forming and magnetic pulsed welding (MPW)*: Electrical energy is stored in a capacitor bank and quickly dissipated into a specially designed coil to create a magnetic field that repulses the workpiece away at a high deformation rate.
ii) *Electrohydraulic forming (EHF)*: Electrical energy is driven across a wire or a spark gap in a fluid medium that generates a shock wave which deforms the workpiece.
iii) *Explosive forming*: A shock wave is produced in a fluid medium through the use of a chemical reaction.

As is evident by these brief definitions, there are similarities but key differences between these HSF processes that will be explored in more detail throughout the chapter. An additional method to produce high deformation rates which is used in experimental setups, e.g. Kolsky bars, is a pneumatically driven hammer. But as this is used more for test purposes than a mass production process, this will not be covered in this chapter.

11.2 Electromagnetic Forming and Magnetic Pulsed Welding

11.2.1 Methodology/Basics

In EMF, electrical energy is stored in a capacitor bank (on the order of $10–10^2$ kJ and $10^2–10^3$ V) and released into a low inductance circuit and specially designed coil. A magnetic field is created in close proximity to the coil. This induces eddy currents, based on Maxwell's principles, and a magnetic field in the workpiece, which is opposite in sign to the applied field. These opposing magnetic fields produce Lorentz forces that launch the workpiece away from the coil at a high deformation rate (on the order of 10^2 m/s). The workpiece geometry can either be tubular (for compression or expansion of the diameter) or a sheet. See Figure 11.2 for schematic representation of process variations. The distance between the workpiece and the coil, i.e. the stand-off distance, is an important parameter in the process as this effects the eddy currents induced in the workpiece. As the workpiece travels away from the coil, the magnetic pressure is reduced.

Due to the need to induce eddy currents, the workpiece material is preferably highly electrically conductive (e.g. copper and aluminum alloys). However, a sacrificial flier (i.e. a driver material) or conductive coatings can be used as well to achieve deformation in less or nonconductive workpieces. Note that the process is noncontact in nature (i.e. there is not a punch as in traditional forming operations). The deformations achieved are typically well beyond the quasi-static forming limits of the material. The proposed rationale for this behavior is that this high-rate process prevents the localization of deformation from occurring, which would lead to a failure.

Potential sizes of workpieces in the process can vary from 10^{-4} to 0.02 m^2 for sheets and from 3 mm to 2 m in diameter for tubes with thicknesses from 0.5 to 5 mm. In order for the current to be effectively induced in the material, a skin depth (see e.g. [5] for skin depth equation) of approximately half the thickness is desirable. Due to various losses such as heating of the supply lines/coil/workpiece, the magnetic field that is not used to create deformation, etc. the energy efficiency of the process is on the order of 5–25%. Of course, for larger sheets and tube diameters, higher charging energies are required to achieve deformation, thus creating more energy loses.

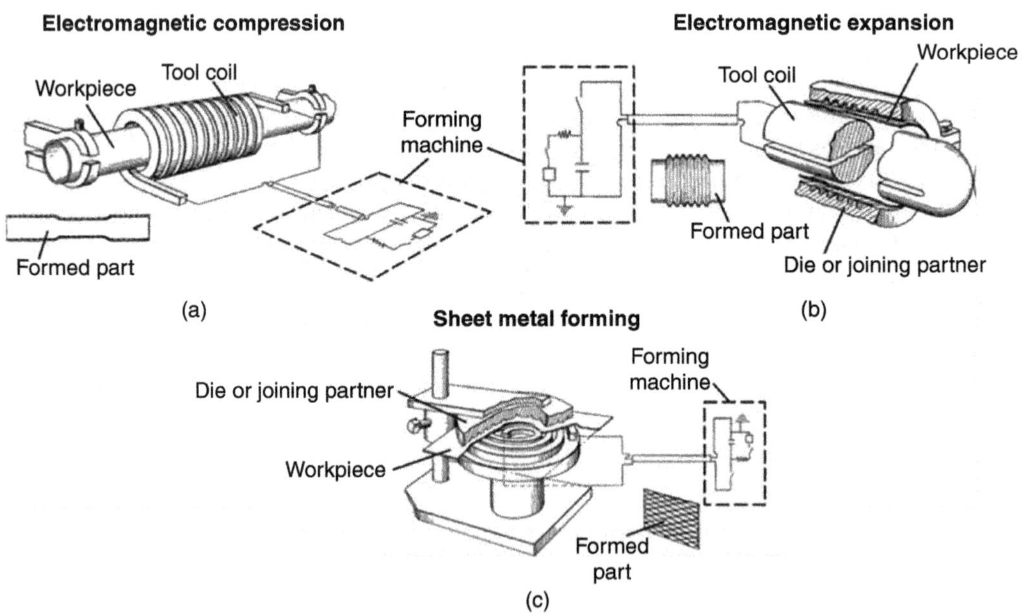

Figure 11.2 Variations on electromagnetic forming (a) compression, (b) expansion, and (c) sheet forming [6]. *Source:* From Harvey and Brower, 1958.

During MPW, which is similar to EMF, the deforming workpiece (e.g. compressed tube) impacts a stationary workpiece (e.g. a shaft) thus creating a solid state weld. This allows dissimilar materials to be joined together. See Figure 11.3 for a schematic representation of the MPW process [7]. A field shaper (i.e. a field or flux concentrator) can also be used in the process in order to focus the magnetic field and thus the eddy currents that are induced in the workpiece. The magnetic field in the initial coil induces eddy currents in the field shaper which in turn produces eddy currents in the workpiece. However, there are significant losses when inducing these eddy currents (i.e. the eddy currents induced are significantly less than the applied current into the coil). A benefit of using a field shaper is that this considerably reduces the mechanical loading on the coil thus increasing the tooling life. Also various field shaper geometries can be used with the same coil thus increasing process flexibility for different products. Note that field shapers could be nonsymmetric and used with EMF processes other than tube compression, as well.

While MPW is considered a solid state joining process, melting and solidification at the interface between the two workpieces have been shown to occur (e.g. [8]). In addition, a wavy pattern is induced at the interface between the workpieces as shown in Figure 11.4. This phenomenon can be predicted using a shear instability analysis based on the relative motion between the two components [10].

The materials also have to impact at a proper angle (e.g. 5°–15°) in order to create a welded interface. For sheet-to-sheet welding, this angle needs to be incorporated into the tooling

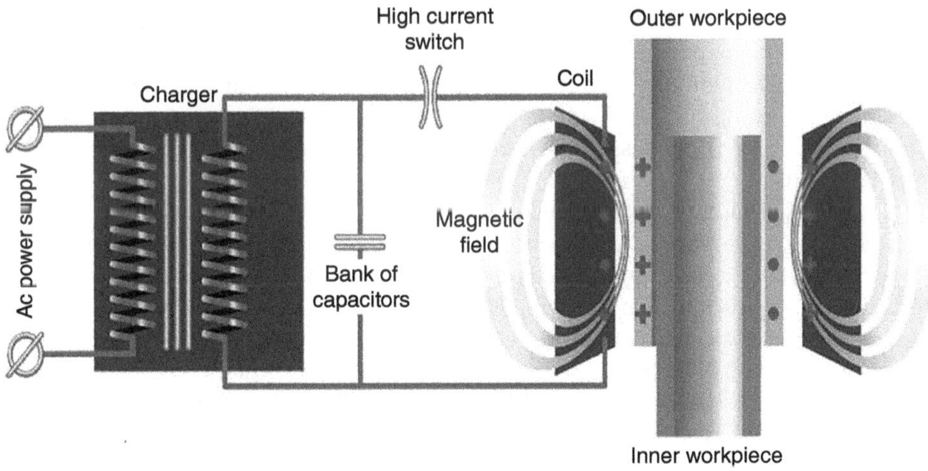

Figure 11.3 Schematic diagram of MPW process [7]. *Source:* From Blakely, M., 2008.

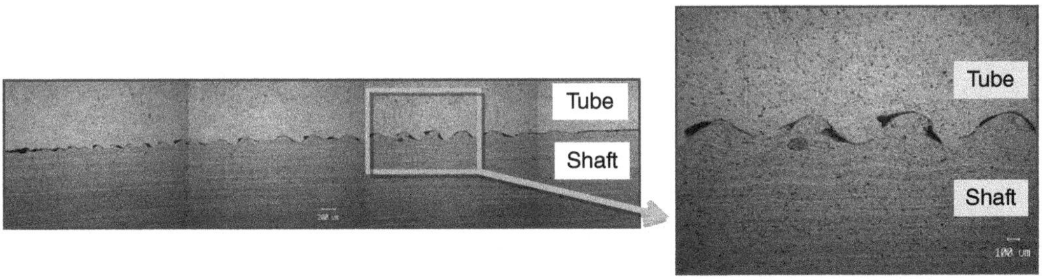

Figure 11.4 Wavy pattern observed at the interface during MPW [9]. *Source:* From Thibaudeau, E., 2013.

design. For tube-to-shaft welding, the proper angle will be created from the deformation process with no welding occurring at the center of the coil, but welding occurring away from the center of the process as the deforming workpiece achieves the desired angle. But this is dependent on the gap between the tube and the shaft and the distance that the peak velocity is obtained in the process.

Another benefit of this high-rate method of creating a weld is that the significant plastic deformation at the interface between the two materials will "clean" any oxide layer from the workpieces and thus allow for a better weld with the base materials. Note that in addition to welded connections, mechanical crimps, etc. can be created with a magnetic pulse in order to join materials together mechanically. This again is particularly useful for the joining of dissimilar materials.

The basic components of the EHF and MPW processes are as follows:

(a) the workpiece,
(b) the means to generate the high current pulse,
(c) the coil (and field shaper where applicable), and
(d) optional female forming die or other tooling.

The elements that are unique to EHF, i.e. (b), (c), and (d), are examined in detail in the following, starting with the generation of the current discharge.

11.2.2 Generation of the Current Pulse

As previously mentioned, a capacitor bank is used to store the electrical energy (on the order of $10–10^2$ kJ). Typical peak currents are on the order of $10–10^3$ kA. Various methods are available to dissipate the electrical energy from the capacitor bank into the coil, e.g. mercury-filled ignitrons, spark gap switches, solid-state silicon–based switches, etc. A key process parameter to achieve a more efficient EMF process is a faster rise time for the first current peak, typically on the order of $10–10^2$ μs.

Figure 11.5 shows a typical damped sinusoidal current response from the short circuit of a machine which uses ignitron switches. A high-capacity diode can be used in parallel in the circuit to only allow the first half of the sine wave to be outputted in order to prevent multiple peaks and thus multiple deformation pulses from occurring. From the current trace, key process parameters such as the inductance and resistance of the circuit can be determined assuming a ringing inductance-resistance-capacitance (LRC) circuit. These are essential for electrical modeling of the process. A low machine inductance is desirable to produce a more efficient EMF process.

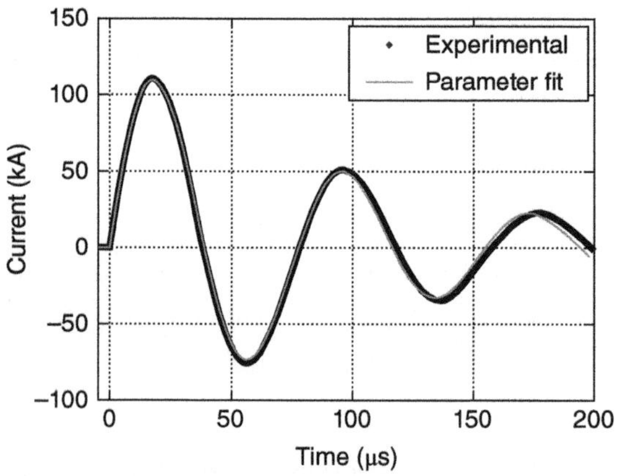

Figure 11.5 Current trace from a capacitor bank discharge measured using a Rogowski coil [9]. *Source:* From Thibaudeau, E., 2014.

Note that, increasing the electrical energy typically improves the forming process initially. However, this does not always occur as excessive energy levels may lead to nonuniform deformation and rebounding of the workpiece off of the die and thus less accurate final part geometries. Therefore, care must be taken when determining process parameters. While the physics regarding EMF is not presented here due to space limitations, classic texts provide such information, e.g. ASTME (1968) and Daehn [2] just to reference two.

11.2.3 Coil (and Field Shaper Where Applicable)

The coil, field shaper, and workpiece combination have a significant influence on the efficiency and effectiveness of the EMF process. There are several geometries for the coil. For sheet forming, flat spiral coils have been used (see Figure 11.2c). Because of the magnetic field generated, the largest magnetic pressure is observed in-between the center of the coil and the outer edge Al-Hassani [11]. While often used, these coils are not optimal for sheet applications as the spiral nature causes a nonsymmetric feature, and edge effects can exist which do not lead to efficient deformation [12]. See Figure 11.6 where the forming angle is less on one side of the coil and also decreases with smaller specimen widths due to edge effects [13].

A helical coil design (see Figure 11.2a) is typical for either tube compression or expansion. This in essence works like a solenoid when creating the magnetic field. For a MPW process, tube compression is typical in order to collapse the outer tube onto e.g. a shaft (see Figure 11.3). Alternatively, tube expansion can also be used to create joints, such as mechanical connections, or features on a bellowed tube. The coil in Figure 11.2b is a single turn coil which is structurally sound, but is electrically and magnetically less efficient.

Daehn's research group at The Ohio State University created a uniform pressure actuator [14] for sheet forming. This coil design is unique as the workpiece is a part of an outer loop so that the induced eddy currents are able to flow uniformly through the sheet. Figure 11.7 shows pictures of this coil design and the process setup [15]. The coil material used for this application is a copper–beryllium alloy, which is a good compromise between strength and conductivity.

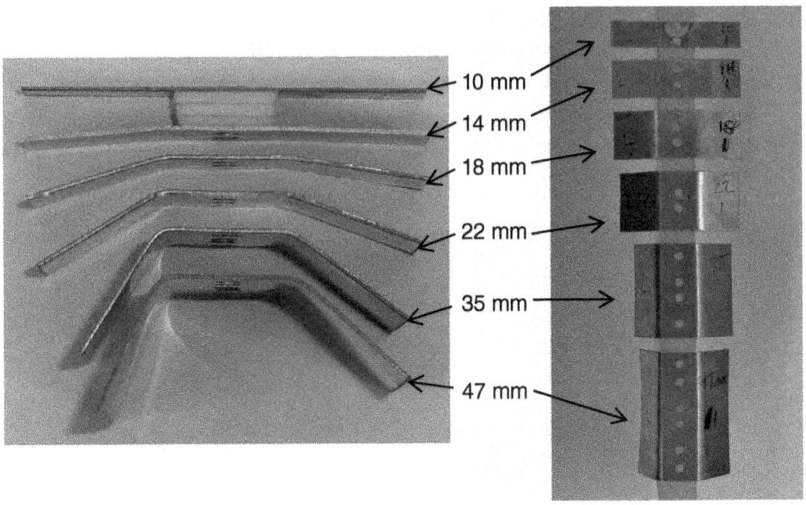

Figure 11.6 Formed components using a flat spiral coil showing nonsymmetric and edge effects with less deformation on one side and for the narrower specimens. *Source:* VanBenthysen et al. 2013 [13]. Reproduced with permission of Elsevier.

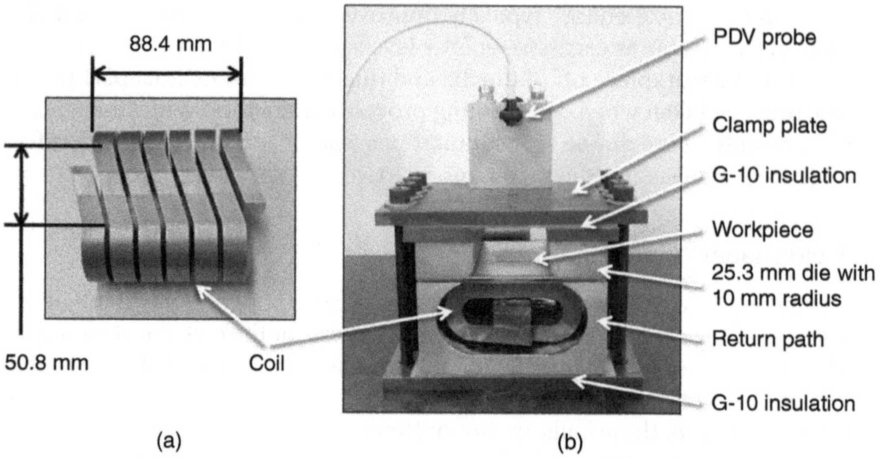

Figure 11.7 Pictures of (a) a uniform pressure actuator coil and (b) entire process setup. *Source:* Thibaudeau and Kinsey 2015 [15]. Reproduced with permission of Elsevier.

Using this type of coil, microscale features can be embossed [14]. Therefore, features such as those on bipolar plates are feasible.

For all of these coil designs, there are several key parameters. A wound coil is in essence a solenoid, and increasing the number of turns increases the magnetic pressure generated. However, increasing the number of turns of the coil will also increase the inductance which decreases the rise time of the current trace and thus decreases the forming efficiency. Therefore, there is a trade-off for these parameters. Also, the cross-section of the wire must be sufficient to ensure the physical strength of the coil. The large magnetic pressures not only deform the workpiece but also are applied to the coil which could cause permanent plastic deformation. In addition, sufficient insulation is required between the turns of the coil to prevent the electrical current from jumping between leads. Finally, there are edge effects so the coil geometry should extend beyond the workpiece. These physical coil parameters affect the resistance and inductance of the coil in the electrical current and thus the EMF process. Iterations are often required in order to effectively design the coil for these competing factors in order to achieve a long coil life. The severe mechanical and electrical stresses make it difficult to determine the life of a coil, but not impossible. For example, see Figure 11.8 from Golovashchenko et al. [16] which exceeded over 50000 discharges. Note that an alternative to creating a durable coil that will withstand several cycles is to have an

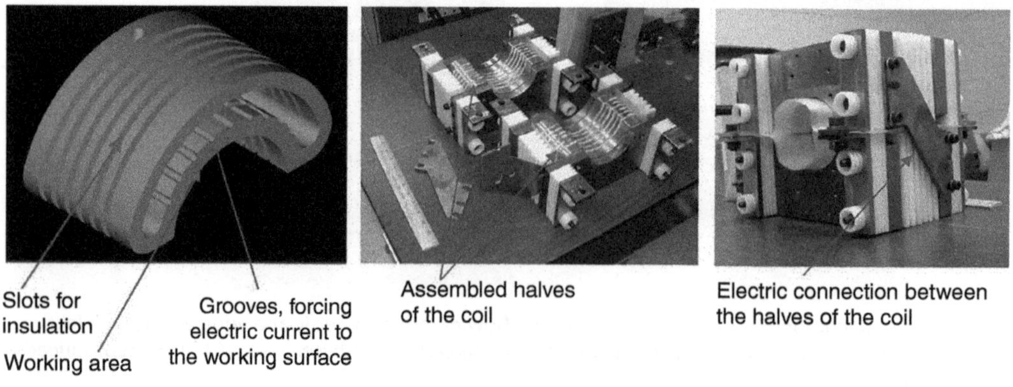

Figure 11.8 Coil design with two halves [16]. *Source:* From Golovashchenko et al., 2006.

expendable coil for the process that will only last one cycle. The economics and other requirements of the given application will determine what the most desirable option is.

In addition to the main coil geometry, as previously mentioned, a field shaper can be used with flat spiral and helical coils in order to concentrate the magnetic pressure. The electrical efficiency of the process significantly decreases; however, the increases in the concentrated magnetic pressure at a specific location may offset this. In addition, a field shaper can be interchangeable allowing a single coil to be used for various applications. Finally, the strength and rigidity of the tooling system can be increased using a field shaper.

As for materials for the coil and field shaper, the goal is for a strong and conductive material. Some typically used materials include copper-beryllium alloys, brasses, high-strength aluminums, etc. Kapton tape (a polyamide film) is useful for electrical insulation purposes as well as urethane or reinforced epoxy when potting the coil geometry.

11.2.4 Optional Female Forming Die

In order to achieve reverse features in the direction opposite to the magnetic pressure, a female die is required. The significant magnetic pressures and impact forces require that tooling be sufficiently rigid to prevent permanent plastic deformation. Due to the high velocities in the process, the material may rebound off of the female die leading to less accurate final part geometries. Also, the air would need to be evacuated out of the die cavity, either through a pump or more simply holes, to assure trapped air does not prevent the workpiece from conforming to the die geometry.

As previously mentioned, in some applications, a particularly sharp feature is desired in only a certain location in the formed part. In order to achieve this, a deep drawing process can be used to produce the majority of the deformation, while a magnetic coil can be imbedded in the punch or die for the localized deformation. See Figure 11.9 for an example of such a hybrid process. Such a configuration takes advantage of some of the key benefits of EMF while allowing a majority of the deformation to be achieved with conventional processing. In addition, instead of a single high-energy electrical pulse, multiple lower energy pulses can be used to incrementally deform the material into the desired geometry [17].

11.2.5 Typical Applications and Equipment

A typical application for MPW is the dissimilar material welding of tubes and shafts (see Figure 11.3 again for a schematic diagram of the process, and see Figure 11.10 for several components which were produced by MPW). Figure 11.11a shows a 12-kJ capacitor bank, a forming

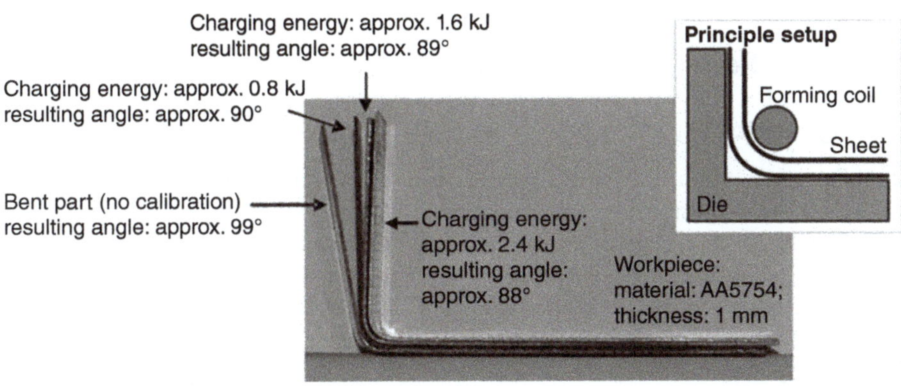

Figure 11.9 Schematic representation of EMF used to create a local feature with large deformations [3]. *Source:* From Iriondo et al. 2006.

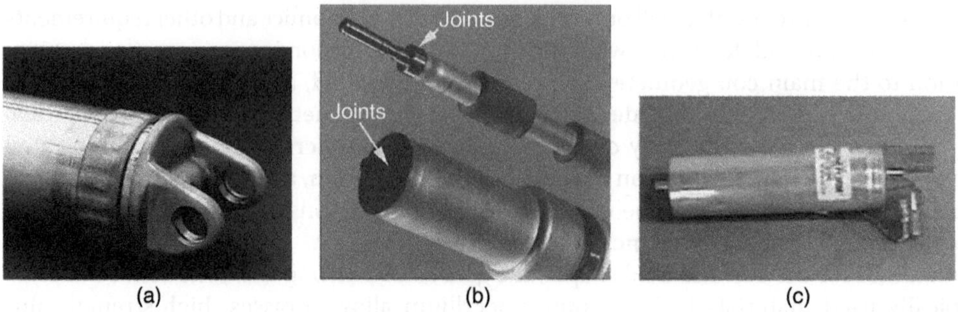

Figure 11.10 Example MPW components from (a) Ford, (b) Xerox and (c) Johnson & Johnson (medical inhaler, Al to plastic weld). *Source:* Courtesy of Magneform.

chamber, and Figure 11.11b shows a cross-section of a commercial coil, including field shaper, used for axisymmetric welding. With the correct impact angle and impact velocity, the weld strengths, confirmed through, e.g. peel tests, can be as strong as the base materials. In addition to axisymmetric geometries, sheets can be impacted together to create weld seams or spot welds.

11.2.6 Other Applications

In addition to forming and joining components, EMF can be used to shear edges as well. Figure 11.12 shows an example of this process. Note that since there is not a shear and break surface on the final edge, the profile of the component is significantly different, e.g. a reduced burr and less rollover. The parameters of interest in this process are a combination of those from the conventional shearing process (e.g. die edge radius, shear gap, etc.) and those from EMF (e.g. charge energy, stand-off distance, etc.). This process may have particular applications for shearing high-strength materials, e.g. advanced high-strength steels. Finally, note that a combined process of forming and cutting in a single discharge operation is possible.

Finally, another potential opportunity for EMF is powder compaction and consolidation. With the increased interest in rapid prototyping, in particular for metallic components, EMF could be used in addition to static compaction in order to achieve denser structures compared to traditional compacting alone [19]. This method could even replace classic sintering as local heating to melt the particles together can be achieved with the magnetic pressure and induced eddy currents.

11.3 Electrohydraulic Forming

11.3.1 Methodology/Basics

EHF is another HSF process, applicable to both sheets and tubes. The basic concept is that the workpiece is deformed by pressure waves that are generated and propagate inside a quiescent fluid medium. Similar to EMF, electrical energy is stored in a capacitor bank. But here the electrical energy is transferred through an exploding wire or over a spark gap inside the fluid, generating the pressure wave that deforms the material. The fluid is contained in an open or closed pressure vessel (or chamber), with the deformable workpiece forming part of the chamber wall. As the wave impacts the workpiece, it deforms it plastically. This deformation can be

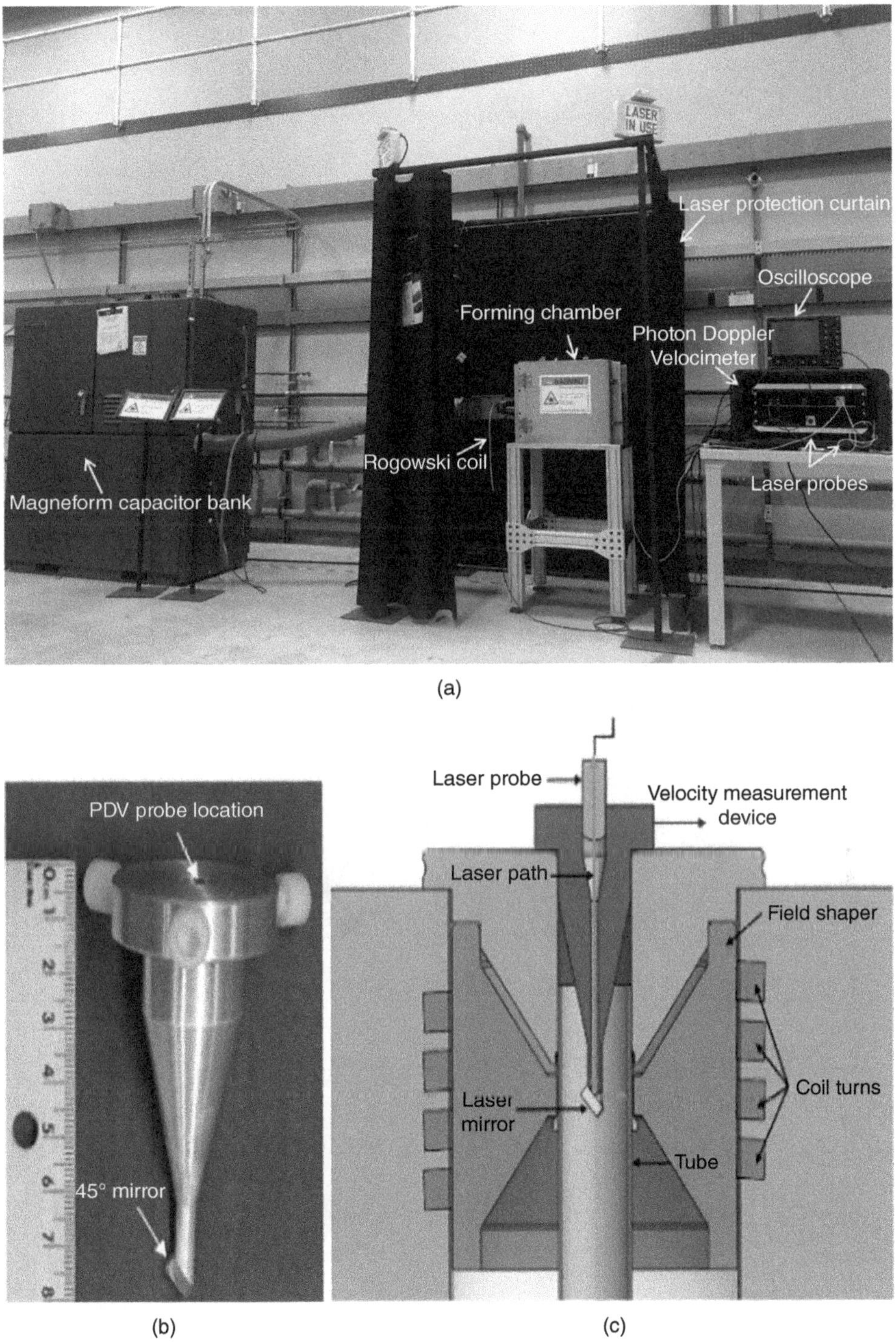

Figure 11.11 Equipment used for MPW: (a) Capacitor Bank and forming chamber, (b) measuring probe, and (c) cross-section of commercial coil with field shaper and measuring probe.

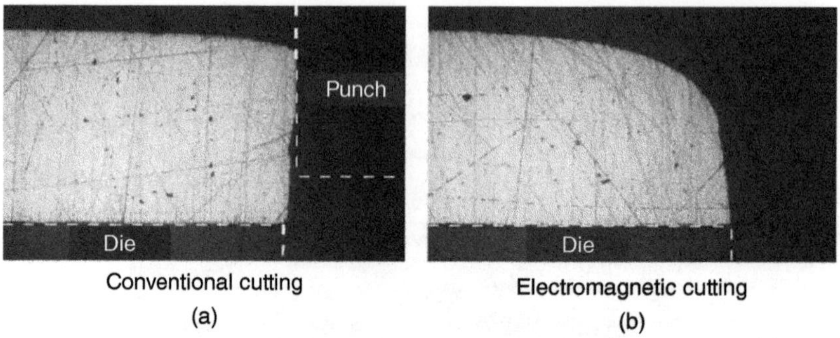

Figure 11.12 Comparison between (a) conventionally and (b) electromagnetically sheared edges [18]. *Source:* From Uhlmann and Scholz, 2003.

either free or against a female die with the desired final part geometry. At the same time, the pressure wave impacts and is reflected off of the remainder of the pressure chamber and can often restrike the workpiece a number of times with intensity comparable to that of the initial impact [1]. See Figure 11.13 for schematic representation of the EHF process.

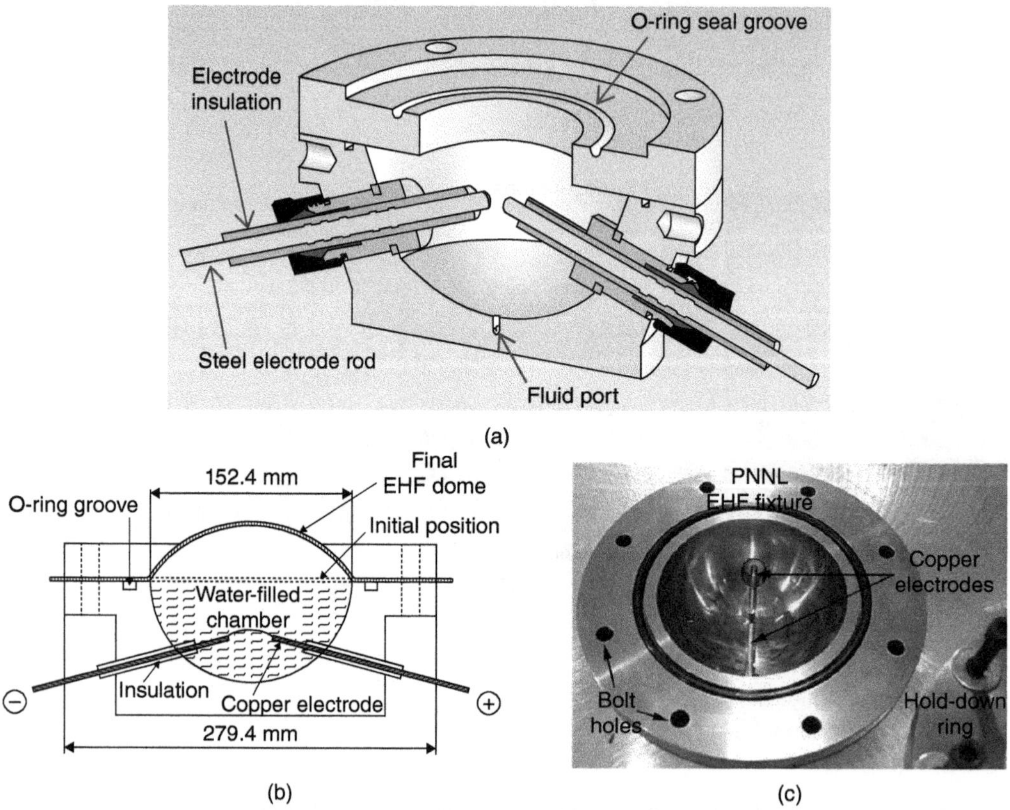

Figure 11.13 Schematics and photographs of EHF chambers for free bulging at (a) Ford Motor Company. *Source:* Gillard et al. 2013 [20]. Reproduced with permission of Elsevier. (b, c) Pacific northwest National Lab. *Source:* Rohatgi et al. 2011 [21]. Reproduced with permission of Elsevier. Notice that (a) does not include the specimen or the hold-down ring.

In comparison to the EMF process described earlier in this chapter, EHF is slower by an order of magnitude or more. Also, EHF can be used on materials irrespectively of their electrical conductivity. The presence of the fluid usually provides sufficient damping to the system so that the inertial effects which often lead to wrinkling and self-folding in EMF are less pronounced here. On the other hand, the multiple reflections of the pressure pulse on the pressure chamber and the resulting multiple impacts on the specimen may render the shape- or dimensional-control impossible. Hence, depending on the desired component geometry, EHF or EMF might be more suitable than the other for the given component. Lastly, EHF involves high voltages in close proximity to a wet environment, so the safety issues are more critical here than in EMF. In fact, in large-scale applications, the handling of workpieces is done robotically, and the EHF manufacturing cell is operated remotely.

The basic components of the EHF process are as follows:

(a) the workpiece,
(b) the remainder of the pressure chamber, which is stiff and possibly shaped,
(c) the means to generate the pressure pulse,
(d) optionally, a wave shaper, and
(e) optionally, a female forming die.

The elements that are unique to EHF, i.e. (b), (c), and (d), are examined in detail in the following, starting with the generation of the pressure pulse.

11.3.2 Generating the Pressure Pulse and the Exploding Wire Technique

For EHF applications, a typical pressure profile is shown in Figure 11.14. The pressure rises to 2500 psi (172 bar) in a span of 50 μs. Ideally, this is the only peak of significant intensity in the pressure wave that impacts the workpiece. In reality, reflections of that initial wave off of the chamber walls may result in multiple restrikes. This is seen in Figure 11.14 as subsequent spikes in the decaying part of the pressure profile. Also shown in Figure 11.14 is the effect of the

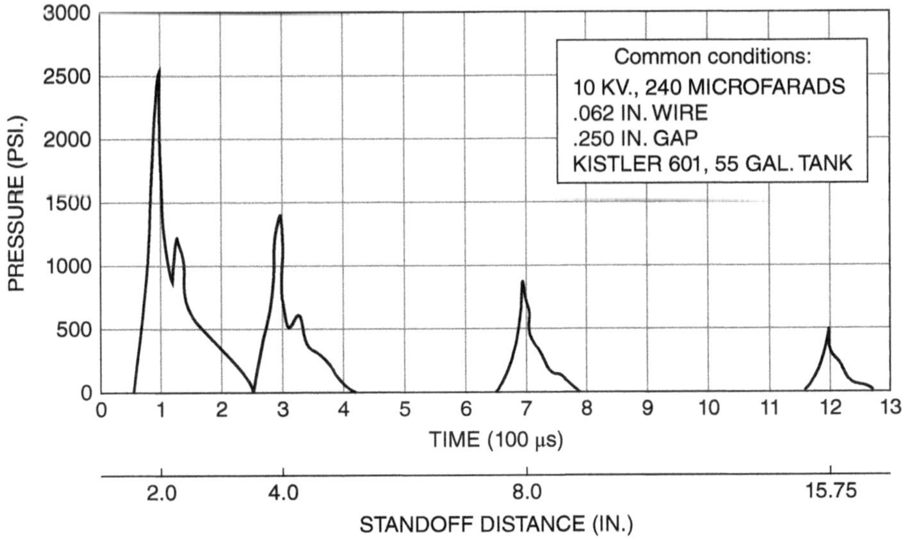

Figure 11.14 Pressure pulses generated by 12-KJ energy release at various stand-off distances [22]. *Source:* From Bruno, E.J. 1968.

stand-off distance (i.e. the distance between the exploding wire or the spark gap and the specimen).

The most common way of generating a pressure pulse in a quiescent fluid medium is the exploding wire technique, which was used in the measurements of Figure 11.14. This involves immersing a conducting wire in the fluid. A high electrical charge, stored in a capacitor bank as in the EMF process, is then discharged through the wire. The resulting increase in temperature due to ohmic heating (e.g. 10^{11} °C/s, [22]) vaporizes the wire. This phase change of the metal results in a sudden volume expansion, which sets off a pressure wave in the quiescent fluid. In industrial EHF applications, the wire is often omitted and only two electrodes in close proximity (i.e. the spark gap) are used, see Figure 11.12a [20]. This allows for very short cycle times, as there is no need to replace the wire after each forming. Also, while the electrodes progressively wear out, the cost per part is less than that of a wire. A disadvantage that the process designer has to take into account is that the pressure pulse from the spark gap is less clean and sharp than the one from the exploding wire [1]. In industrial installations, this is often mitigated by providing a mechanism to advance the electrodes as they wear out, aiming to keep the gap constant.

Other ways of generating a pressure pulse include the use of a plunger in the pressure chamber, that is impacted from the outside, e.g. with a Hopkinson bar or some other type of projectile. This method has been used in academic installations [23], but it is not favored in industrial settings as the discharge energies available are difficult to scale-up for forming larger components. An exception is the forming of very small tubes outside diameter (OD < 1/4″) in which fitting the wire and/or electrode inside the tube is not feasible [1, 24].

11.3.3 The Pressure Chamber

The pressure chamber serves to contain the fluid and the pressure waves, as well as to support/clamp the deforming workpiece. A unique feature of the chamber is that it has to be able to withstand repeated impact loads from the pressure pulses, while its steady-state operating pressure is zero. As a result of that, the die is often made from high-strength steels, such as AISI 4340 [23].

In many applications, the back side of the workpiece is held at a vacuum to improve the dimensional accuracy by avoiding trapped air between the workpiece and the female die. According to Lange [25], the optimal back-pressure is between 100 and 2500 Pa. Larger back pressure (less vacuum) does not avert the danger of dimensional deviations, while back pressure less than 100 Pa tends to adhere the workpiece to the die, with the effect that there is pronounced springback when it is removed from the die.

11.3.4 The Wave Shaper

The wave shaper is used to focus the incident and reflected pressure pulses as per the requirements of the target geometry. A vivid illustration of the role of the wave shaper is given in Figure 11.15 [1], which shows a combustor tube successfully formed in a single EHF process with and without using a shaper.

Generally, the shaper design is done empirically, following some rules-of-thumb [1]. This is not surprising, as a rational design of the shaper would require solving a coupled fluid-structure interaction, thermomechanical problem, with a finitely deforming, rate-dependent elastoplastic solid; pressure waves generated, transmitted, and reflected in the fluid chamber; and possible phase changes of the fluid medium. Such analysis is at the limit of current state-of-the-art of industrial forming simulations [26].

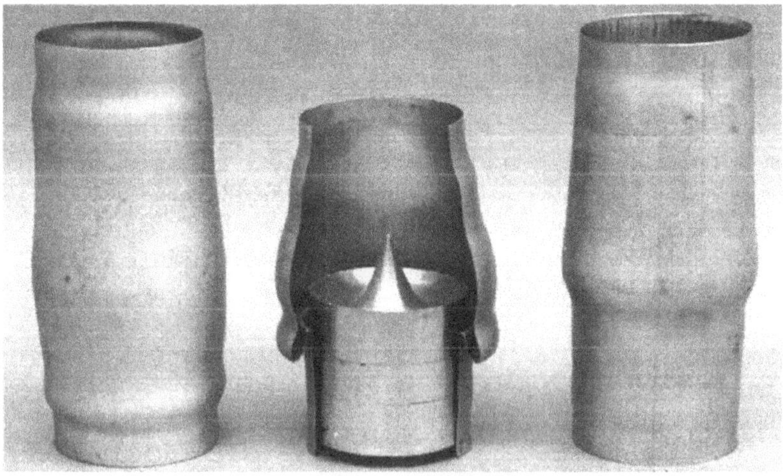

(a) (b) (c)

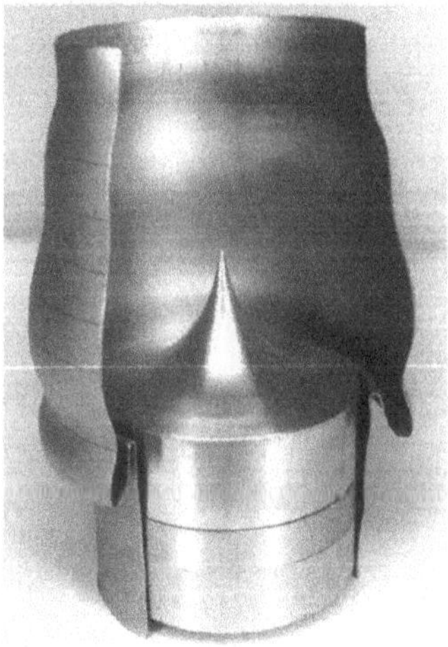

(d)

Figure 11.15 Aluminum tube formed with different techniques: (a) EHF without reflector, (b) EHF with reflector, (c) explosively, and (d) close-up of the component [1]. *Source:* From Noland et al., 1967.

11.3.5 Typical Applications and Equipment

EHF is applicable to both sheets and tubes, irrespectively of their electrical conductivity. Some typical aerospace components are shown in Figure 11.16.

The development of EHF processes in the 1950s and 1960s led to a variety of commercially available forming machines, such as the Cincinnati Electroshape and the Rohr SoniForm, Figure 11.17. None of these machines is currently marketed.

11.4 Explosive Forming

11.4.1 Methodology/Basics

In contrast to EHF where a shock wave is produced by vaporizing a wire with electrical energy, in explosive forming, a shock wave is produced by a chemical detonation mechanism. But the

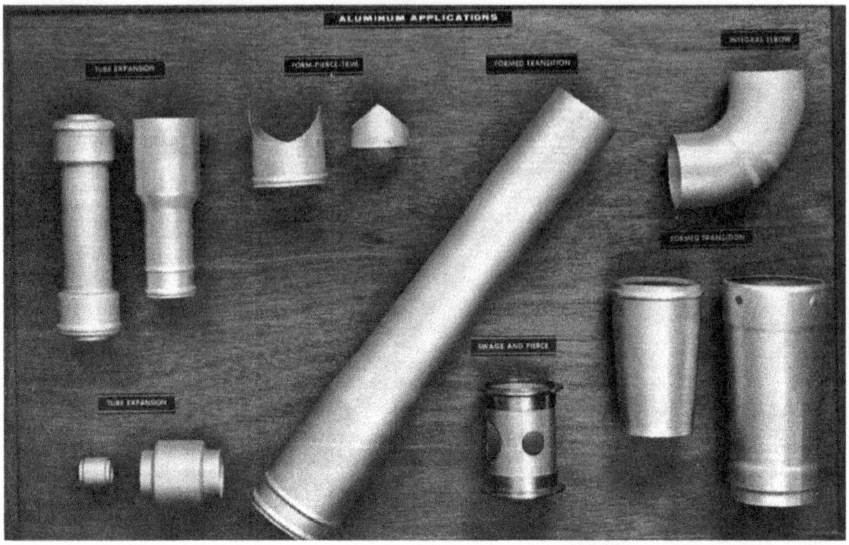

(a)

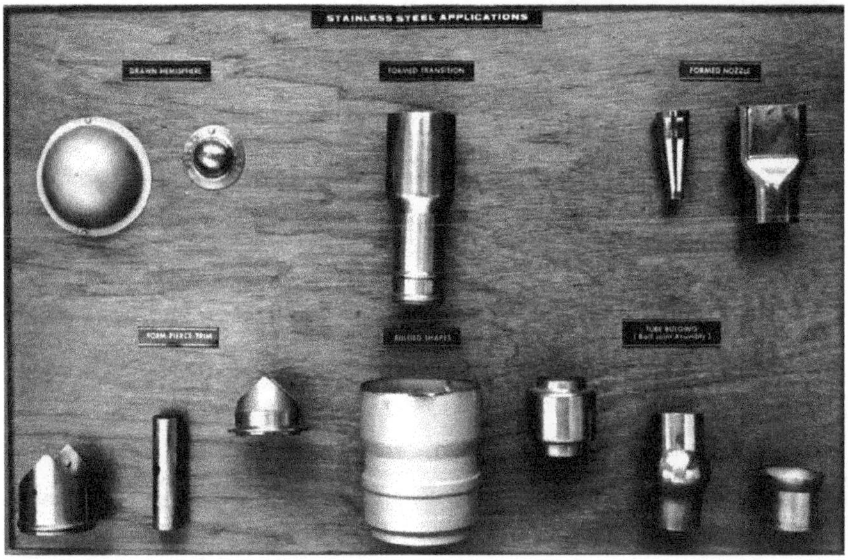

(b)

Figure 11.16 Components formed electrohydraulically from (a) aluminum and (b) stainless steel [1]. *Source:* From Noland et al., 1968.

end result is the same, with the shock wave from the gas generated during the explosion propagating through an energy-transmitting, quiescent medium (e.g. water or sand) to deform the workpiece. Much higher energies can be achieved in explosive forming than in EHF, thus producing larger deformation pressures. Figure 11.18 shows a schematic diagram of the process. Potential sizes of workpieces in the process can be on the order of $10\,\text{m}^2$ for sheet area and on the order of 1 m in diameter for tubes with thicknesses in excess of 25 mm.

In a typical explosive forming setup, the sheet is clamped over a die and the air is evacuated out so the material can be formed to match the female cavity. The forming speeds are typically

(a) (b)

Figure 11.17 Electrohydraulic forming machines. (a) Cincinnati Electroshape and (b) Rohr SoniForm [1]. *Source:* From Noland et al., 1969.

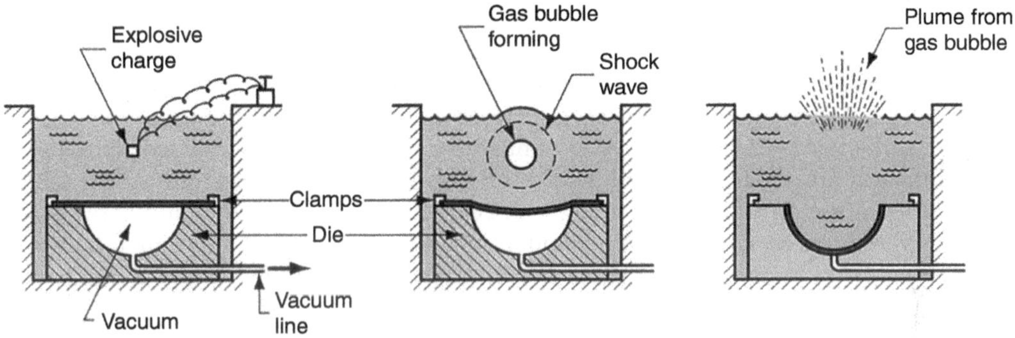

(1) (2) (3)

Figure 11.18 Schematic representation of an explosive-forming process. *Source:* Groover 2007 [27]. Reproduced with permission of John Wiley & Sons.

similar to those observed in EMF (i.e. on the order of $10–10^2$ m/s), but for high explosives, velocities on the order of 10^3 m/s can be achieved. Since the energy-transmitting medium transfers the shock wave in the process, two key parameters in explosive forming are the compressibility and the acoustic impedance (i.e. the mass density multiplied by the sound velocity) of the fluid. Water is a commonly used fluid medium in explosive forming. Other key process parameters include the stand-off distance between the explosive and the workpiece and the detonation charge (which can be on the order of 10 kg of an explosive substance such as TNT). The explosive can be placed directly on the workpiece, but utilizing a stand-off distance is more common so the wave has a distance over which to propagate before deforming the workpiece. Typical tolerances achievable in explosive forming are ±0.025 mm.

11.4.2 Setup and Equipment

Depending on the number of components to be produced, the tooling (i.e. the female die) can be fabricated from simple materials such as reinforced concrete, hard plaster, or wood.

However, often the tooling is fabricated from metal (e.g. steel) in order to have the required strength to withstand the high velocity impact without plastically deforming. This allows the tooling to be used to produce multiple components. Similar tooling alternatives can be used for the other high-rate deformation processes discussed (i.e. EMF and EHF). However, these other processes typically use metal dies as the production quantities are large enough to justify the cost of producing the tooling. Note that the stiffness and damping coefficient of the tooling significantly affects the energy dissipation and thus the forming process.

Other basic components in explosive forming are similar to those of EMF and EHF (see previous sections for further details). For example, as in EHF, the wave shaper and the chamber geometry can have a considerable effect on the deformation due to the shock waves reverberating off of the chamber walls. One technique that is used in explosive forming to mitigate these reflections is to generate an air curtain on the walls of the forming chamber, e.g. by passing compressed air on a perforated tube located at the bottom circumference of the chamber. Since explosive forming is often used for very large components and low production runs, the forming chamber is oftentimes open to the atmosphere, and the forming is taking place outdoors. In the past, explosive forming has been also attempted under the sea, e.g. to fabricate the end-domes of the SEALAB vessel [1].

11.4.3 Typical Applications

Explosive forming is a desirable process for fabricating large pressure vessels and sections of ships/vehicles out of plate material as other processes cannot achieve the forming pressures required (e.g. the SEALAB vessel [1]). Also, similar to the MPW process, cladded plates can be produced that are created from this solid-state welding process over a large surface area. Refer to classical literature on explosive forming, e.g. Pearson and Rinehart [28] and Ezra [29], or society-based references, e.g. ISEE [30], for additional details. Note that there are other variations on this process, e.g. exploding gases and burning propellants. But these are in essence equivalent to explosive forming.

11.5 Emerging Technologies

In addition to the major variations of HSF processes presented in the previous three sections, there are two emerging technologies that achieve high deformation rates using plasma as the driving force. First, vaporizing foil actuator welding (VFAW) [31] is similar to EMF in that the electrical energy in a capacitor bank is quickly discharged. However, instead of flowing into a coil to create a magnetic field, the energy is used to vaporize a thin foil of metal, e.g. an Aluminum alloy. This creates a plasma that drives the workpiece. See Figure 11.19a for a schematic diagram of a VFAW process [32]. Velocities on the order of 800 m/s are achieved which allows material combinations, such as Titanium–Copper, to be welded with complex vortices at the interface. See Figure 11.19b for experimental results, as well as modeling results to be discussed later in this chapter, for impact welding of commercially pure Titanium to a Copper alloy (Cu 110) for various thicknesses of the deforming workpiece [33]. Benefits of VFAW over MPW are the higher velocities achieved, and the fact that the electrical conductivity of the deforming workpiece is not required as the plasma is the driving mechanism.

Similar to VFAW, in laser impact welding, a laser is shined through a transparent medium, e.g. polycarbonate, to vaporize, e.g. ink on the workpiece in a localized area. This creates a plasma whose pressure plastically deforms the workpiece, which is typically a thin foil of metal.

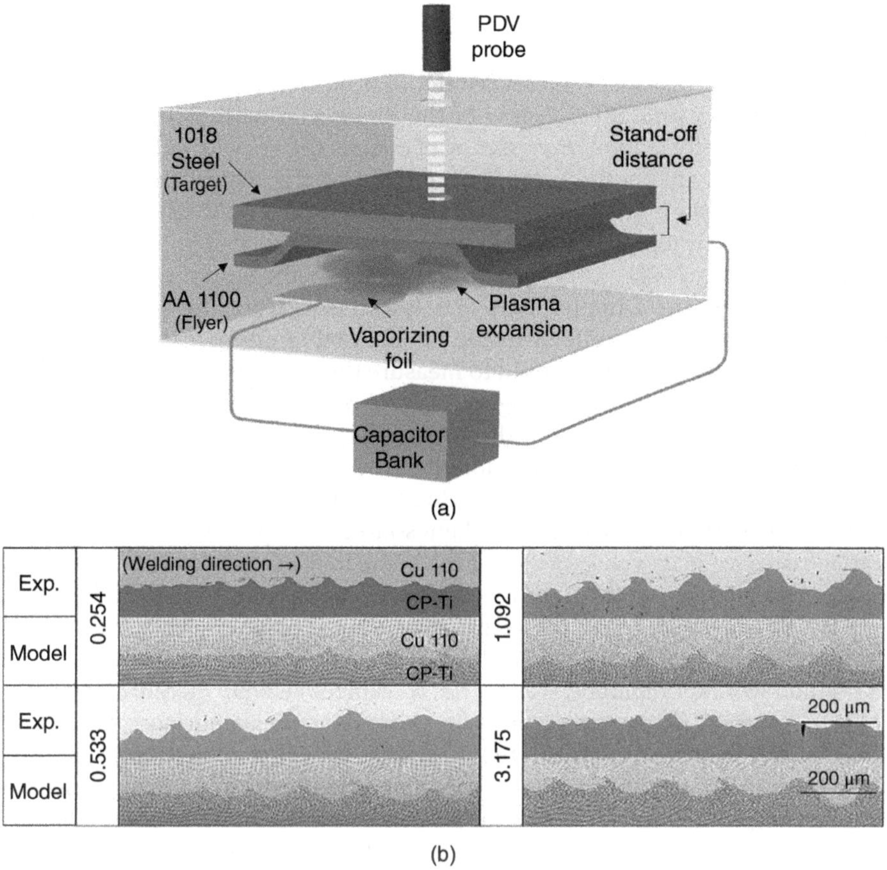

Figure 11.19 (a) Schematic representation of vaporizing foil actuator welding apparatus. (b) Impact of welding results, both experimental and modeled by smooth particle hydrodynamics, for various deforming workpiece thicknesses.

See Figure 11.20 for a schematic diagram of this process [34]. Laser ablation is a localized effect so used for smaller scale forming and joining processes. These emerging technologies will continue to expand possible applications for HSF processes.

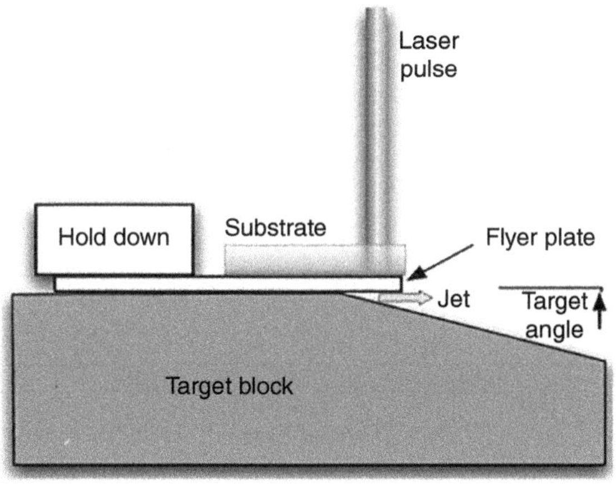

Figure 11.20 Schematic diagram of laser impact welding process.

11.6 Metrology and Measurements

As with various manufacturing processes, measurement of process parameters is essential in order to evaluate the efficiency and effectiveness of the process. A parameter of particular importance during high-rate deformation processes is, of course, the velocity of the workpiece. During MPW, this parameter affects the solid state joining of the materials. Velocities of approximately 300 m/s are required in order to create the desired final weld strength. In order to measure velocity, various technologies are used.

In photon Doppler velocimetry (PDV), the velocity of a point on the workpiece is measured using laser interferometry (i.e. the Doppler shift in a laser beam from the source and the reflected beam from the surface) [35]. Reflective tape can be used to improve the reflection of the laser beam from the surface. PDV can be used to measure the velocity during the deformation or welding at varying discharge energies. Figure 11.21 shows the maximum velocity data for four different materials and three different discharge energies. As expected, the velocity increases with increasing discharge energy. Also note that the electrical conductivity, density, and other material parameters affect the maximum velocity achievable in the process.

A similar but alternative method uses a fiber optic light source, as opposed to the laser source in PDV, in order to measure velocity values. Some of the fibers in the sensor emit infrared light while others measure the intensity of the light received after being reflected off of the target surface. As the workpiece moves away from the sensor, less light is received. Thus, the relative position of the workpiece can be determined and by differentiating with time, a velocity can be resolved. This method works comparably well for flat workpiece velocity measurements (see e.g. Figure 11.22 [36]). However, for curved surfaces (in particular concave surfaces if reflective tape is used to increase the light intensity reflected from the surface), the additional light may provide erroneous values. Also, surface roughness can affect the results if not using reflective tape, and the effective range of the sensor is only on the order of a few centimeters. Thus, care should be taken when using this fiber optic sensor method. However, there are distinct advantages for this technique over PDV, i.e. the cost of the system (approximately 1/10), safety as the laser beam has both eye and skin hazards, and ease of implementation.

For both PDV and fiber optic sensor applications, a key component in the data acquisition system is the oscilloscope, as the sampling rate determines if the desired velocity can be detected accurately. For both of these systems, reasonable specifications are a 1-GHz bandwidth with a 4–5 Gigasamples/s sampling rate and sufficient data storage capacity for the application (e.g. 2 MB). This allows velocity measurements on the order of 800 m/s. Another key is that these sensors are nonconductive so their use in high-voltage and large magnetic field environments, e.g. during EMF, is possible.

There are other means to measure velocity as well. The workpiece can be located between a laser sheet and a photodiode detector. As the deformation progresses, the displacement of the workpiece, and thus velocity, can be detected and measured. See Figure 11.23 for a schematic representation of this experimental setup [37]. Alternatively, for MPW or similar processes, a voltage difference between, for example, the shaft and the tube can be applied. Once the

Figure 11.21 Velocities measured using PDV system during electromagnetic tube compression.

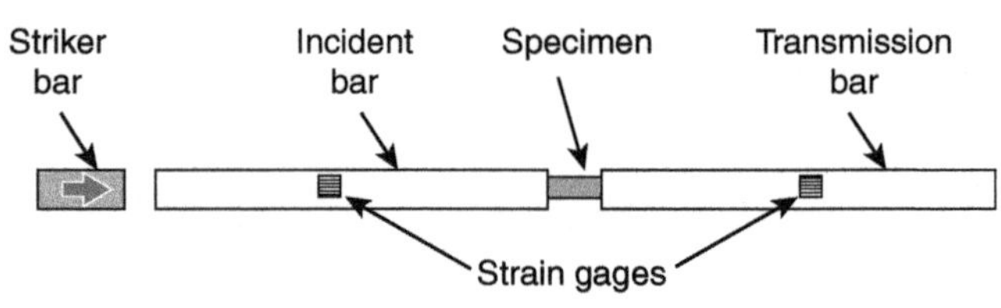

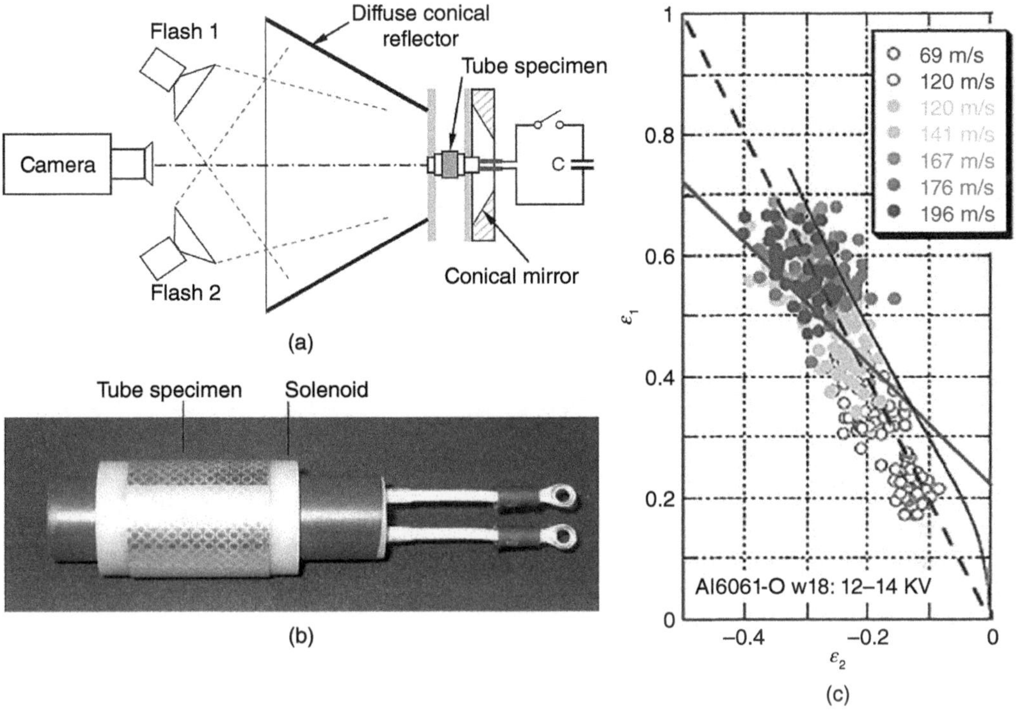

Figure 11.22 Comparison of velocities measured using PDV and a fibre optic system at three levels of discharge energy: (a) 3.6 kJ, (b) 6.0 kJ, and (c) 8.4 kJ. *Source:* Thibaudeau et al. 2015 [36]. Reproduced with permission of ASME.

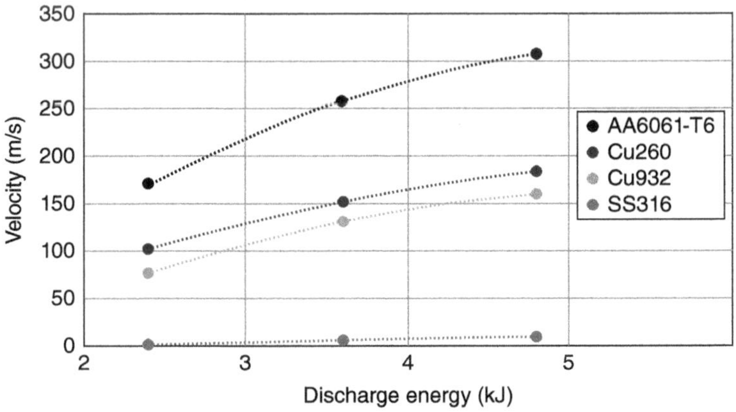

Figure 11.23 Schematic representation of laser sheet and photodiode detector means to measure velocity. *Source:* Kleiner et al. 2005 [37]. Reproduced with permission of Elsevier.

workpiece is deformed and the two are in contact due to the high-rate deformation, the velocity can be approximated based on the initial gap and the time measured. Again, the sampling rate of the data acquisition system is a key factor for the velocity measurements which can be obtained for any of these alternative systems.

Finally, high-speed imaging has also been implemented to photograph the workpiece deformation and velocity (e.g. [38]). See Figure 11.24 of a sheet deforming during EMF. These

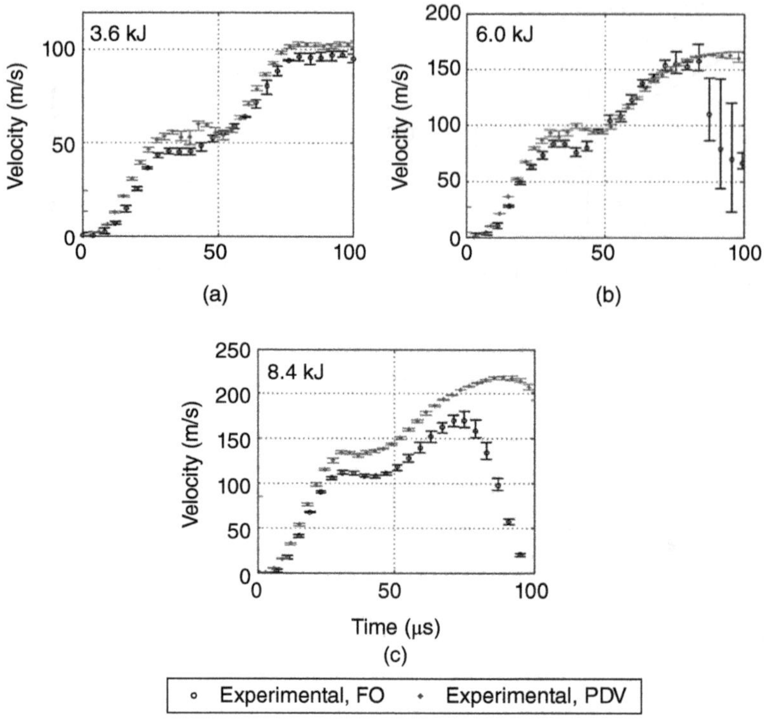

Figure 11.24 High-speed imaging (Every 30 Ms) of an electromagnetic forming process. *Source:* Courtesy of G. Daehn.

systems are limited by the shutter speed of the camera. Current technology allows 1 M frames/second; however, the cost of these systems can be prohibiting, and the pixel resolution at these high sampling rates is significantly reduced. In addition to measuring velocity, the high-speed imaging can be coupled with digital image correlation in order to measure the strain and strain rate in the process [39]. In theory, the final strain in a specimen can be obtained by taking an image of the undeformed workpiece and an image of the deformed workpiece. However, if the deformation is excessive, as is typically the case in high-rate deformation processes, then the correlation may not be successful.

11.7 Material Characterization

Wood [40] conducted standard tension test experiments with varying deformation velocities. He found that a slight increase in the failure strain occurred until a critical velocity was achieved. This von Karman velocity is related to the plastic wave speed of the material and is between 30 and 150 m/s for most useful engineering alloys [2]. In contrast to quasi-static deformation processes, failure during high-rate processes typically occurs at multiple locations due to the inertial stabilization of the deformation preventing a single localization from occurring. In other words, as soon as the deformation begins to localize in a quasi-static tension test, the rest of the specimen unloads elastically and thus the localization of deformation is promoted further. However, in a high-rate process, the unloading wave is not sufficiently fast and hence multiple localized necks may form. This leads to enhanced formability over the single neck, quasi-static case.

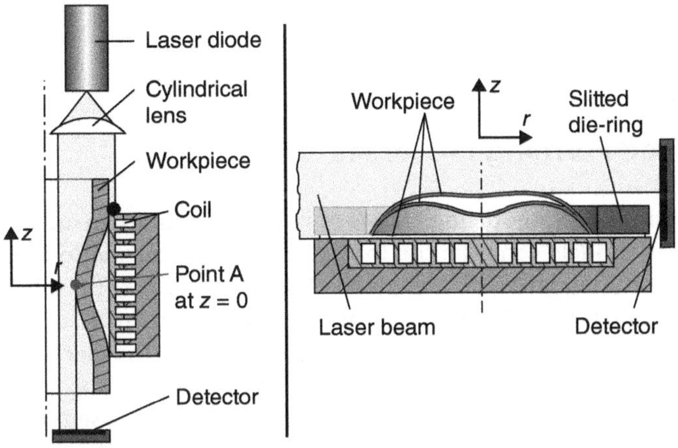

Figure 11.25 Schematic diagram of a split Hopkinson tension bar (Aka Kolsky bar).

Typical experiments for assessing the material properties at the high strain rates that correspond to HSF are the Hopkinson/Kolsky bar [41, 42] and the expanding ring [43–45] techniques.

In the Hopkinson bar (see Figure 11.25), a pneumatically operated projectile impacts the incident bar and creates a longitudinal wave. The wave then passes through the specimen to the transmission bar and is reflected a number of times before decaying. A momentum arrestor is used to reduce the impact of these reflected waves. Each bar is instrumented with strain gages that measure the strain pulses from which the stress applied to the specimen can be inferred. Typically, Hopkinson bar data can be quite noisy, and special postprocessing of the data is required [41, 42, 46–48].

An alternative method for high-rate testing is the expanding ring technique that is more suitable for testing tubular materials (or sheets that have been rolled into tubes), see Figure 11.26. This experiment is in essence identical to the EMF/expansion process described in Section 11.2. A ring or tube specimen is placed around a coil/inductor. A transient current is applied to the inductor circuit, which leads to a transient magnetic field. This induces secondary currents in the specimen, which lead to their own magnetic field. The two fields oppose each other, and the resulting Lorentz force accelerates the specimen. A high-speed camera system can be used to acquire the shape evolution of the specimen as it deforms. Furthermore, a circle or random pattern can be etched on the specimen prior to the deformation and can be used to assess the strains at failure (Figure 11.26c). Analytical modeling of the failure limits during high-rate

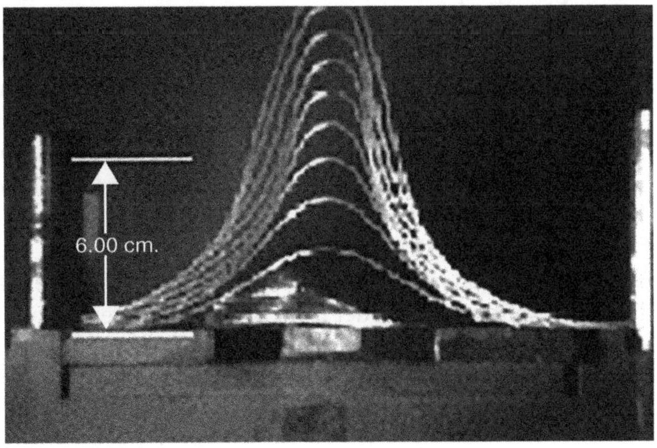

Figure 11.26 (a) Schematic representation of the experimental setup, including the camera viewpoint. (b) Photograph of a tube specimen on the solenoid. Note the circular grid pattern applied on the tube. (c) Forming limit diagram, where the open circles denote safe, the closed circles denote fail, and the blue line is the assumed uniaxial tension path. *Source:* Zhang and Ravi-Chandar 2010 [45]. Reproduced with permission of Springer.

deformation reveals that the dynamic forming limit diagram is higher than the quasi-static one and correlates well with measurements from the expanding ring experiment [43].

11.8 Modeling of High-Speed Forming Processes

As with other metal-forming applications [49], modeling efforts of HSF operations can aid the design process considerably [50]. The models can take the form of analytical calculations as well as numerical simulations using commercially available finite element analysis (FEA) software. When performing FEA simulations, the physical mechanism used to create the high deformation rates can be incorporated into the analyses. For example, for EMF, the electrical discharge of the capacitor bank into the coil can be modeled in order to determine the magnetic field generated and accompanying Lorenz forces [51, 52]. In addition, the increased temperature due to induced eddy currents in the workpiece and the high deformation rate can be simulated. Thus, an electrical-magnetic-mechanical-thermal model is achieved. This is essential in order to accurately capture the physical phenomena occurring. For example, as the deformation progresses, the magnetic field that is created varies based on the current discharge trace vs. time, as well as the physical position of the workpiece, which is in turn determined by the constitutive behavior of the material deforming at elevated temperatures. Various commercially available software packages have the capability to incorporate all these physical effects into the numerical analysis.

Of course, there are considerable assumptions within such models, e.g. the constitutive behavior of the material, effect of surrounding materials on the magnetic field generated, current trace, yield criterion, etc. Models that are simply loosely coupled, i.e. where all these physical parameters are not determined simultaneously in the simulations at each time increment, have been shown to provide reasonable results when comparing simulations to physical experiments. See Figure 11.27 [53].

In addition to classical Lagrangian numerical simulations of the HSF processes, alternative modeling techniques have been implemented, e.g. smoothed particle hydrodynamics (SPH) [54, 55], Eulerian [56], and Arbitrary Lagrangian–Eulerian (ALE) formulations [57]. SPH is a meshless method where the continuum body is represented by a collection of particles with specific spatial locations. The physical properties of each particle are determined by averaging the relative characteristics of the neighboring particles. A kernel function determines the smoothing length and the number of neighboring particles included with nearby particles affecting the properties and deformation more than distant ones. The kernel function corresponds to the shape function in the traditional Lagrangian finite element method. If the severe distortion exceeds a threshold based on the material behavior, then the elements create a separate body in the simulations, e.g. jetting of surface material. The SPH technique works well for high deformation phenomena such as ballistic impacts or in this case HSF (e.g. the wavy pattern observed at the interface of an impact weld). See Figure 11.19b for results from SPH for impact welding commercially pure Titanium to a Copper alloy (Cu 110) for various thicknesses of the deforming workpiece.

In Eulerian numerical simulations, the mesh is fixed in space as opposed to be defined with respect to the material being deformed. This technique is typically used in computational fluid dynamics software to model fluid flows. However, it can be used in HSF as well, to model the wavy pattern interface observed during impact welding. Finally, as the name implies, ALE is a hybrid between Lagrangian and Eulerian formulations where only a small area of interest is defined with the more computationally expensive Eulerian approach.

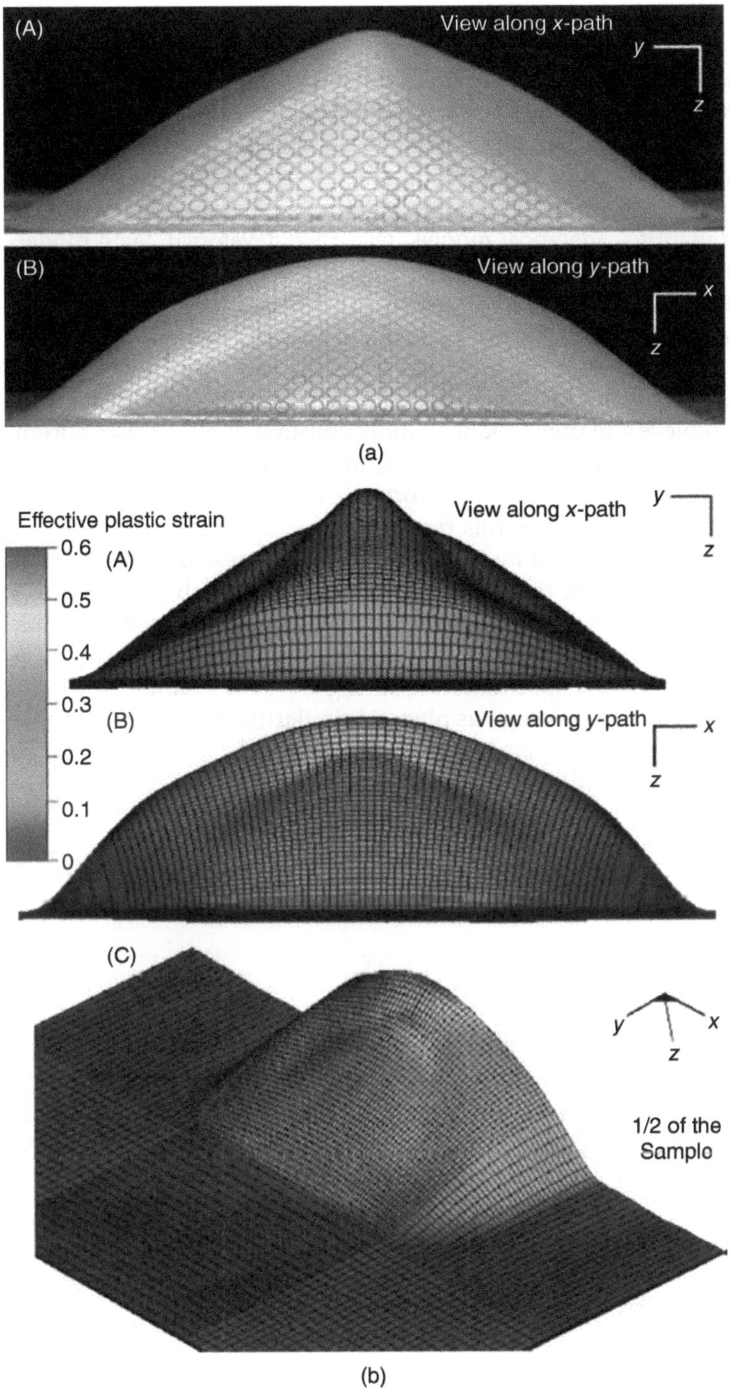

Figure 11.27 Comparison between (a) experiments and (b) loosely coupled FEA simulations. *Source:* Oliveira et al. 2005 [53]. Reproduced with permission of Elsevier.

A critical component for such modeling efforts that has been studied extensively is the constitutive behavior of the material. The model used must account for the high strain rate and temperature increases that occur during HSF. A classical constitutive behavior that is often used is the Johnson–Cook [58] model, which has been extended to account for active mechanisms and void formation during deformation [59] or for a specific material, such as glass [60]. FEA itself has been used as a means to assist in the accurate determination of constitutive behavior of the material [61].

In addition to numerical simulations, analytical models have been developed. Such models can aid in the design process of the coil. Al-Hassani [11] created a model to predict the nonuniform magnetic field and resulting pressure distribution for two coil geometries, flat spiral and solenoidal coils, by summing the contributions of small coil areas. Furthermore, Kamal and Daehn [14] developed a uniform pressure actuator and an accompanying analytical model to aid the design process. This coil geometry is innovative in that the eddy currents induced flow uniformly through the workpiece and thus produce a more constant pressure distribution. These models by Al-Hassani [11] and Kamal and Daehn [14] were combined and extended in the work by Thibaudeau and Kinsey [15] to predict the forming velocity of various materials during HSF deformation. See Figure 11.28 from this research. Thus, accurate predictions of velocity are possible from modeling efforts. An interesting result of this research was that for these materials, magnetic coupling between the coil and workpiece effectively induced eddy currents down to a skin depth to thickness ratio of 0.66. Previous research estimated this value as 1.5–2.0 [5], thus demonstrating that thinner workpieces could be effectively used during HSF.

Finally, modeling efforts can investigate a specific effect during HSF. For example, the wavy interface pattern observed during impact welding has physical similarities to instabilities that

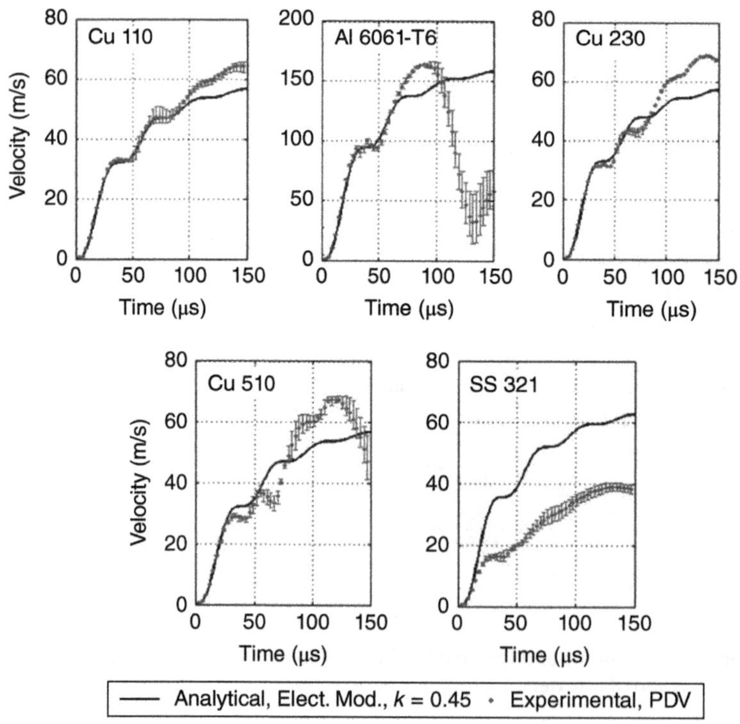

Figure 11.28 Workpiece velocity with varying workpiece conductivity and analytical model prediction [15]. Discharge energy was 6 KJ. *Source:* Thibaudeau et al. 2015 [15]. Reproduced with permission of ASME.

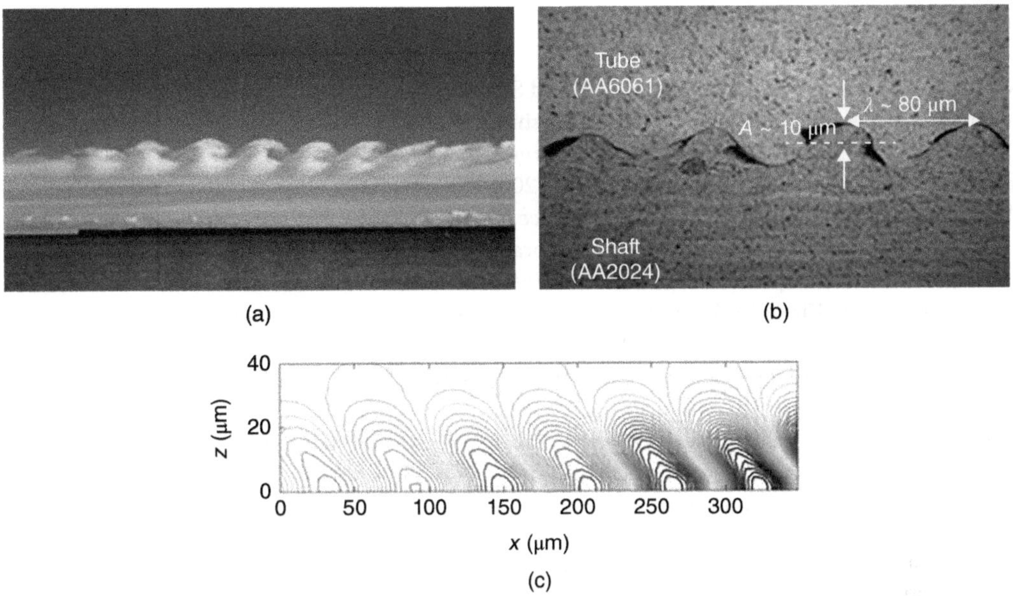

Figure 11.29 Wavy patterns in (a) cloud formations [6], (b) MPW [2], and (c) stream function from instability analysis.

occur in fluid dynamic systems. See Figure 11.29 for images from (a) cloud formations and (b) MPW showing such similarities. This lead to the hypothesis that the wavy pattern could be caused by a Kelvin–Helmholtz instability [62]. Through modeling efforts, Nassiri et al. [10] showed that the wavy pattern could indeed be the result of a shear instability through the solution of an eigenvalue problem with a symmetric velocity profile at the interface. See Figure 11.29c) for images of the wavy pattern predicted, as well as the interface of a MPW joint. Note that some researchers believe that melting and solidification occurs at the interface which causes the wavy pattern [8]. Alternatively, others mention stress waves reflecting off backing surfaces and causing the interface wavy pattern [63, 64].

11.9 Summary and Future Work

HSF processes have the potential to significantly advance the current state-of-the-art for component fabrication, e.g. joining of dissimilar materials to create a lightweight structure, shearing of high strength/advanced materials, and exploiting the enhanced formability to create a complex geometry in a single operation. In this chapter, details regarding electromagnetic, electrohydraulic, and explosive forming processes were discussed. However, the adaption of these technologies into industrial practice has been slow. A possible reason for this is that HSF processes require various knowledge-bases, i.e. mechanical, electrical, physics, chemistry for explosives, etc. to fully understand the fundamental mechanisms involved. With increases in computation power, multiscale simulations are feasible that will allow accurate prediction of processing parameters and thus address these concerns to some extent. This in turn will require characterization of material parameters at high strain rates and the ability to measure key process parameters, e.g. the material velocity in the process. Finally, further research and development efforts at higher technology readiness levels will allow the broader implementation of HSF processes in industrial applications in order to fully take advantage of their unique benefits.

References

1 Noland, M.C., Gadberry, H.M., Loser, J.B., and Sneegas, E.C. (1967). High-Velocity Metalworking. A Survey, NASA SP-5062, Washington, DC.
2 Daehn, G.S. (2005). *High Velocity Metal Forming*, vol. 14B, 405–418. ASM Handbook.
3 Iriondo, E., Gonzalez, B., Gutierrez, M. et al. (2006). Electromagnetic springback reshaping. In: *Proceedings of the 2nd International Conference on High Speed Forming – ICHSF 2006*, 153–160. Dortmund: Available at: https://eldorado.tu-dortmund.de/handle/2003/27064.
4 Johnson, W. (1970). *Impact Strength of Materials*. London: Edward Arnold.
5 Psyk, V., Risch, D., Kinsey, B.L. et al. (2011). Electromagnetic forming – a review. *Journal of Materials Processing* 211 (5): 787–829.
6 Harvey, G.W. and Brower, D.F. (1958). Metal Forming Device and Method. US-Patent Nr. 2976907.
7 Blakely, M. (2008). *Filler Metal is for Wimps*. Fabricating and Metalworking, April.
8 Ben-Artzy, A., Stern, A., Frage, N., and Shribman, V. (2008). Interface phenomena in aluminum-magnesium magnetic pulse welding. *Science and Technology of Welding and Joining* 13 (4): 402–408.
9 Thibaudeau, E. (2013). Development of Analytical and Experimental Tools for Magnetic Pulsed Welding. Master's Degree Thesis. University of New Hampshire.
10 Nassiri, A., Chini, G.P., and Kinsey, B.L. (2014). Investigation of spatial shear instability theory to predict wavy interfacial pattern during magnetic pulsed welding. *Annals of CIRP* 63: 245–248.
11 Al-Hassani, S.T.S. (1975). Magnetic pressure distributions in sheet metal forming. In: *Conference on Electrical Methods of Machining, Forming, and Coating*, 1–10. London, UK.
12 Imbert, J. and Worswick, M. (2011). Electromagnetic reduction of a pre-formed radius on AA 5754 sheet. *Journal of Materials Processing Technology* 211: 896–908.
13 VanBenthysen, R., Thibaudeau, E., and Kinsey, B.L. (2013). Effect of specimen planar area on electromagnetic flanging. *Journal of Manufacturing Processes* 15: 194–200.
14 Kamal, M. and Daehn, G.S. (2007). A uniform pressure electromagnetic actuator for forming flat sheets. *Journal of Manufacturing Science and Engineering* 129 (2): 369–379.
15 Thibaudeau, E. and Kinsey, B.L. (2015). Analytical design and experimental validation of uniform pressure actuator for electromagnetic forming and welding. *Journal of Materials Processing Technology* 215: 251–263.
16 Golovashchenko, S., Dmitriev, V., Canfield, P. et al. (2006). Apparatus for electromagnetic forming with durability and efficiency enhancement. US Patent 2006/0086165.
17 Vohnout, V.J., Shang, J., and Daehn, G.S. (2004). Improved formability by control of strain distribution in sheet stamping using electromagnetic impulses. In: *Proceedings of the 1st International Conference on High speed Forming – ICHSF 2004*, 211–220. Dortmund, Online available at: https://eldorado.tudortmund.de/handle/2003/27036.
18 Uhlmann, E. and Scholz, M. (2003). Zerteilen von Aluminiumblechen durch Impulsmagnetfelder. In: *Proceedings of the 2. Kolloquium Elektromagnetische Umformung*, 87–94. Dortmund, Germany: Technische Universität Dortmund.
19 Mamalis, A.G., Manolakos, D.E., Kladas, A.G., and Koumoutsos, A.K. (2004). Electromagnetic forming and powder processing: trends and developments. *ASME Applied Mechanics Review* 57 (4): 299–324.
20 Gillard, A.J., Golovashchenko, S.F., and Mamutov, A.V. (2013). Effect of quasi-static prestrain on the formability of dual phase steels in electrohydraulic forming. *Journal of Manufacturing Processes* 15 (2): 201–218.
21 Rohatgi, A., Stephens, E.V., Soulami, A. et al. (2011). Experimental characterization of sheet metal deformation during electro-hydraulic forming. *Journal of Materials Processing Technology* 211 (11): 1824–1833.

22 Bruno, E.J. (ed.) (1968). American society of tool and manufacturing engineers. In: *High Velocity Forming of Metals*. NJ: Prentice Hall Inc.

23 Espinosa, H.D., Lee, S., and Moldovan, N. (2006). A novel fluid structure interaction experiment to investigate deformation of structural elements subjected to impulsive loading. *Experimental Mechanics* 46 (6): 805–824.

24 Chu, G. and Liu, W. (2013). Experimental observations of 5A02 aluminum alloy in electromagnetically assisted tube hydroforming. *JOM* 65 (5): 599–603.

25 Lange, K. (ed.) (1985). *Handbook of Metal Forming*. Dearborn, MI: Society of Manufacturing Engineers.

26 Mamutov, A.V., Golovashchenko, S.F., Mamutov, V.S., and Bonnen, J.F. (2015). Modeling of electrohydraulic forming of sheet metal parts. *Journal of Materials Processing Technology* 219: 84–100.

27 Groover, M. (2007). *Fundamentals of Modern Manufacturing*. Hoboken, NJ: Wiley.

28 Pearson, J. and Rinehart, J.S. (1963). *Explosive Working of Metals*. New York: Macmillan.

29 Ezra, A.A. (1973). *Principles and Practices of Explosive Metallurgy*. London: Metal Working, Industrial Newspapers, Ltd.

30 International Society of Explosives Engineers (2011). *ISEE Blaster's Handbook*. Cleveland, OH: International Society of Explosives Engineers.

31 Vivek, A., Hansen, S., Liu, B., and Daehn, G.S. (2013). Vaporizing foil actuator: a tool for collision welding. *Journal of Materials Processing Technology* 213 (12): 2304–2311.

32 Lee, T., Zhang, S., Vivek, A. et al. (2018). Flyer thickness effect in the impact welding of aluminum to steel. *Journal of Manufacturing Science and Engineering* 140: 121002-1-7.

33 Lee, T., Zhang, S., Vivek, A. et al. (2019). Wave formation in impact welding: study of the Cu–Ti system. *CIRP Annals – Manufacturing Technology* 68 (1).

34 Zhang, Y., Babu, S.S., Prothe, C. et al. (2011). Application of high velocity impact welding at varied different length scales. *Journal of Materials Processing Technology* 211 (5): 944–952.

35 Zhang, Y., Daehn, G.S., L'Eplattenier, P., and Babu, S. (2009). Experimental study and numerical simulation on magnetic pulse welding for pre-flanged AA6061-T6 and Cu101 sheets. In: *ASM Proceedings of the International Conference: Trends in Welding Research – Proceedings of the 8th International Conference*, 715–720. Materials Park, OH: ASM International.

36 Thibaudeau, E., Turner, B., Gross, T., and Kinsey, B.L. (2015). Analysis of fiber optic sensor to measure velocity during electromagnetic forming and welding. *ASME Journal of Manufacturing Science and Engineering* 137 (031013): 1–7.

37 Kleiner, M., Beerwald, C., and Homberg, W. (2005). Analysis of process parameters and forming mechanisms within the electromagnetic forming process. *CIRP Annals – Manufacturing Technology*: 225–228.

38 Watanabe, M. and Kumai, S. (2009). High-speed deformation and collision behavior of pure aluminum plates in magnetic pulse welding. *Materials Transactions* 50 (8): 2035–2042.

39 Tiwari, V., Sutton, M.A., McNeill, S.R. et al. (2009). Application of 3D image correlation for full-field transient plate deformation measurements during blast loading. *International Journal of Impact Engineering* 36 (6): 862–874.

40 Wood, W.W. (1967). Experimental mechanics at velocity extremes – very high strain rates. *Experimental Mechanics* 7 (10): 441–446.

41 Field, J.E., Walley, S.M., Proud, W.G. et al. (2004). Review of experimental techniques for high rate deformation and shock studies. *International Journal of Impact Engineering* 30 (7): 725–775.

42 Gama, B.A., Lopatnikov, S.L., and Gillespie, J.W. (2004). Hopkinson bar experimental technique: a critical review. *Applied Mechanics Reviews* 57 (4): 223–250.

43 Thomas, J.D., Seth, M., Daehn, G.S. et al. (2007). Forming limits for electromagnetically expanded aluminum alloy tubes: theory and experiment. *Acta Materialia* 55 (8): 2863–2873.

44 Zhang, H. and Ravi-Chandar, K. (2006). On the dynamics of necking and fragmentation–I. Real-time and post-mortem observations in Al 6061-O. *International Journal of Fracture* 142 (3–4): 183–217.
45 Zhang, H. and Ravi-Chandar, K. (2010). On the dynamics of localization and fragmentation-IV. Expansion of Al 6061-O tubes. *International Journal of Fracture* 163 (1–2): 41–65.
46 Frew, D.J., Forrestal, M.J., and Chen, W. (2002). Pulse shaping techniques for testing brittle materials with a split Hopkinson pressure bar. *Experimental Mechanics* 42 (1): 93–106.
47 Mates, S.P., Rhorer, R., Whitenton, E. et al. (2008). A pulse-heated Kolsky bar technique for measuring the flow stress of metals at high loading and heating rates. *Experimental Mechanics* 48 (6): 799–807.
48 Vecchio, K.S. and Jiang, F. (2007). Improved pulse shaping to achieve constant strain rate and stress equilibrium in split-Hopkinson pressure bar testing. *Metallurgical and Materials Transactions A* 38 (11): 2655–2665.
49 Tekkaya, A.E. and Martins, P.A.F. (2009). Accuracy, reliability and validity of finite element analysis in metal forming: a user's perspective. *Engineering Computations* 26 (8): 1026–1055.
50 Unger, J., Stiemer, M., Schwarze, M. et al. (2008). Strategies for 3D simulation of electromagnetic forming processes. *Journal of Materials Processing Technology* 199: 341–362.
51 Mamalis, A.G., Manolakos, D.E., Kladas, A.G., and Koumoutsos, A.K. (2005). Physical principles of electromagnetic forming process: a constitutive finite element model. *Journal of Materials Processing Technology* 161: 294–299.
52 Stiemer, M., Unger, J., Svendsen, B., and Blum, H. (2006). Algorithmic formulation and numerical implementation of coupled electromagnetic-inelastic continuum models for electromagnetic metal forming. *International Journal of Numerical Methods in Engineering* 68: 1301–1328.
53 Oliveira, D.A., Worswick, M.J., Finn, M., and Newmanc, D. (2005). Electromagnetic forming of aluminum alloy sheet: free-form and cavity fill experiments and model. *Journal of Materials Processing Technology* 170: 350–362.
54 Johnson, G.R., Stryk, R.A., and Beissel, S.R. (1996). SPH for high velocity impact computations. *Computer Methods in Applied Mechanics and Engineering* 139: 347–373.
55 Wang, X., Zheng, Y., Liu, H. et al. (2012). Numerical study of the mechanism of explosive/impact welding using smoothed particle hydrodynamics method. *Materials & Design* 35: 210–219.
56 Akbari Mousavi, A.A. and Al-Hassani, S.T.S. (2005). Numerical and experimental studies of the mechanism of the wavy interface formations in explosive/impact welding. *Journal of Mechanics and Physics of Solids* 53: 2501–2528.
57 Fenton, G. and Daehn, G.S. (1998). Modeling of electromagnetically formed sheet metal. *Journal of Materials Processing Technology* 75: 6–16.
58 Johnson, G.R. and Cook, W.H. (1985). Fracture characteristics of three metals subjected to various strains, strain rates, temperatures and pressures. *Engineering Fracture Mechanics* 21 (1): 31–48.
59 Lesuer, D.R., Kay, G.J., and LeBlanc, M.M., TMS (2001). Modeling large strain, high rate deformation in metals. In: *Modeling the Performance of Engineering Structural Materials II* (eds. D.R. Lesuer and T.S. Srivatsan). The Minerals, Metals & Materials Society.
60 Johnson, G.R. and Holmquist, T.J. (2011). A computational constitutive model for glass subjected to large strains, high strain rates and high pressures. *Journal of Applied Mechanics* 78 (051003): 1–9.
61 Kleiner, M. and Brosius, A. (2006). Determination of flow curves at high strain rates using the electromagnetic forming process and an iterative finite element simulation scheme. *Annals of CIRP* 55 (1): 267–270.
62 Hunt, J.N. (1968). Wave formation in explosive welding. *Philosophical Magazine* 17 (148): 669–680.
63 Blazynski, T.Z. (1983). *Explosive Welding, Forming and Compaction*. Springer.
64 Godunov, S.K., Deribas, A.A., Zabrodin, A.V., and Kozin, N.S. (1970). Hydrodynamic effects in colliding solids. *Journal of Computational Physics* 5 (3): 517–539.

Part III

Advanced Material Removal Processes

12

High-Speed Machining

Elisa Vázquez[1] *and Guillem Quintana*[2]

[1] *Department of Mechanical Engineering and Civil Construction, Universitat de Girona, Girona, Spain*
[2] *ASCAMM Technology Centre, Barcelona, Spain*

12.1 High-Speed Machining Overview

Productivity and quality improvement are undoubtedly two of the main demands of the productive sector; key factors for competitiveness and survival. Within this sector, material removal processes play, still nowadays, a principal role despite the emergence of additive manufacturing techniques. Industries such as aerospace, automotive, molds, and dies, or energy largely depend on machine tools' performance.

Technologies involved in machining operations have advanced greatly the last decades, and machines have experienced significant changes such as the incorporation of numerical control. Every year, it is possible to observe in fairs, conferences, and of course, in the market, how production capabilities increase thanks to the development of new concepts, devices, materials, tools, coatings, structures, etc. Accuracy, flexibility, and productivity are enhanced constantly with innovative solutions that permit to achieve the market demands and raise them into higher levels. At the end, all these improvements are possible thanks to the generation of knowledge. The understanding of the metal cutting fundamentals has become a key factor in the last decades, but this is not an easy task due to complexities of the chip formation mechanism.

In general, high speed machining (HSM) consists in the production of tools and single parts or small series in their final material, metal, or plastic, through metal removal technologies. This technology derives from conventional machining, but with the technological philosophy of applying different cutting conditions. The process involves high cutting speeds, combined with extremely rapid feed rates and small passes.

Therefore, the evolution of the state of science, technology and technique, and the market pressure in terms of productivity, cost, quality, and time have stimulated the emergence of new machining concepts that are overcoming the conventional machining. Consolidation within the business reality of these new concepts, such as high-speed machining or high-performance cutting among others, has been made possible thanks to the advances in the process understanding and the constant innovations in materials, coatings, mechanics, and computers.

The first relevant research in the field of metalworking was published by Taylor in 1907 [1]. However, the concept of high-speed machining was initially conceived by Dr. Eng Carl. J. Salomon during a series of experiments, conducted between 1924 and 1931 in the German city of Hannover (German Patent No. 523,594, [2]). Despite the long journey of the high-speed machining since 1931 until today, some authors are reluctant to define concisely and bounded

Modern Manufacturing Processes, First Edition. Edited by Muammer Koç and Tuğrul Özel.

the concept. Others, however, are more unambiguous to determining the frontier between the HSM and conventional machining:

- *The term "High Speed Machining" has been used for many years to describe end milling with small diameter tools at high rotational speeds, typically 10 000–100 000 rpm* [3]
- *High speed machining is vaguely defined as the use of higher spindle speeds and higher feed rates to achieve higher material removal rates without diminishing the accuracy and quality of the part. An exact definition covering speeds and feeds is difficult to achieve because it depends quite the type of material being machined* [4].
- *There is no definition HSM except that associated with overcoming a certain cutting speed, angular speed, feed rate ..., dependent on the process and the combination material-part/material-cutting tool.*
- *High Speed Machining, HSM, follows an idea contrary to the traditional concept. Maintaining the chip section, i.e. the feed per tooth and the conditions of immersion (depths of cut) even greatly reduced, but the cutting speed rises above the usual* [5].
- *High Speed Machining (HSM) generally refers to flat milling or ball end milling at high rotation speeds (10 000–100 000 rev./min)* [6].
- *The term High Speed Machining (HSM) is generally used to describe milling with small diameter tools (10 mm) at high rotational speeds (≥10 000 rpm)* [7].
- *High Speed Machining is defined as machining using medium higher spindle speeds to 1047 rad/s (10 000 rpm) that can generate cutting speeds of 6 m/s. This high speed process produces plastic flow heterogeneous severe local tensions that provoke high ratios of plastic deformation (up to 106 s^{-1}), thermal gradients high (100 °C/mm), and ratios of generation of high heat (105°C/s)* [8].

It has to be mentioned that, in some cases, HSM acronym is also applied for high-speed milling operations.

12.2 High-Speed Machining Processes and Capabilities

Machining is applied to a wide range of materials to create a great variety of geometries and shapes, practically without restrictions on complexity. Typical workpiece materials are aluminum alloys, cast iron, titanium, austenitic stainless or hardened steels, copper, carbon graphite, and also plastics, woods, and plastic composites. Regarding the complex geometries and shapes, two types of surfaces are usually defined: ruled surfaces, (e.g. for blades) and sculptured surfaces or free-form surfaces (e.g. for molds and dies) (Figure 12.1). This enormous number of combinations is able to meet the specific manufacturing requirements of a wide range of manufacturing industries, such as automotive, aerospace, or dies and molds [9].

The recent years' trends, thanks to the advances in computers and sensors, seem to be focused on the online monitoring, measuring, and controlling the machining process. Several kinds of sensors and signal processing techniques have been developed for direct or indirect diagnosis of aspects such as the workpiece condition (surface roughness, integrity, and dimensional accuracy), detection of tool wear and breakage, chatter identification, analysis of machine components condition, etc.

12.3 Machine Tools for High-Speed Machining

Globalization has brought new challenges and requirements to the manufacturing sector. Industries are continuously pushed to improve productivity and quality and to reduce costs by increasing material removal rates and machine tool flexibility, capabilities, and performance.

Figure 12.1 Example of sculptured surfaces by HSM.

Naturally, it becomes necessary to consider machine tools with new and improved technologies when attempting to cater to increased levels of quality and production system parameters [10]. Selecting the most appropriate machining center from among the wide range of types and models available is a difficult but very important decision for a manufacturing company (Figure 12.2) [11]. An improper selection can negatively affect productivity, precision, flexibility, and company's responsive manufacturing capabilities [12]. Machine tools are most important for the manufacture of discrete products.

Figure 12.2 Vertical machine tool.

This issue affects both manufacturers of machined products and manufacturers of machine tools. Manufacturers of machined products have to acquire new machine tools and new technologies to keep on competing in the market. Manufacturers of machine tools have to design and produce competitive machines that fulfill the needs of their customers. Therefore, manufacturers of machined products have to decide which of the machines of the market have to be acquired to increase the performance of the company, and manufacturers of machine tools have to decide which machines have to be designed and developed to launch to the market the machine tool that better suits the needs of their customers. The decision-maker should be an expert, or at least be very familiar with the machine properties, to select the most suitable machine among the alternatives. Selecting the most suitable machine from the increasing number of available machines can be highly demanding [12]. For the base machine selection, the first factor to be considered is the cost of the machine, with the main goal being to satisfy production system design requirements and the budgetary constraints [10]. As the machine-tool industry provides a large range of machine-tool options and configurations to satisfy the particular requirements of each sector, selecting the most suitable spindle is an iterative process that must take into account the final characteristics of the end product and the requirements of the production system. Aspects such as, the maximum spindle speed, effective power, torque, drive method, type of bearings, stiffness and dynamic behavior, tool holder technology, and thermal properties should be taken into consideration and evaluated to get the best final product specifications in terms of tolerances, surface finish, etc., and meet the production system requirements in terms of productivity, flexibility, etc. In general, the most important characteristics that should be considered in a process of machinery selecting can be classified into three groups:

Geometric characteristics: work volume, work area, table-loading capacity, weight, linear X axis travel, linear Y axis travel, and linear Z axis travel.

Axis characteristics: positioning accuracy, positioning repeatability, number of axis, rapid traverse in X axis, rapid traverse in Y axis, rapid traverse in Z axis, and feed force speed also in the three axis.

Spindle characteristics: maximum spindle speed, power, and torque.

The lack of standard format in machine catalogues, the large number of factors to be considered, and introduction of new machine tools together with the advancements in the technology complicate the comparison between machines of different companies [12]. In some cases, not all the characteristics of the machine are specified in its catalogue; sometimes, suppliers use different standards or modes to indicate the same characteristic.

12.4 Tools for High-Speed Machining

12.4.1 Cutting Tool Materials

One of the most important advancements in the field of the cutting tools is related with the improvement of the material science technology. Tool manufacturers have faced the challenge to machine new materials characterized by being lighter and stronger, adapting their capacities, and looking to maintain high productivity and reduced costs. The main influence factors that the toolmakers work in order to improve the cutting tools are the materials, coatings, and geometries. The first requirement to be met by the material of the tool is that it must be harder than the material to be machined. Particularly, in the cutting tools for high-speed machining, the tools should be able to support high cutting speeds (V_c) and feed rates (f) [13]. Furthermore, due to the plastic deformation between the workpiece and the tool, the temperature increases until reaching the 1000 °C; this phenomenon allows that the surfaces of the work material and the tool are activated chemically and produce wear. The tool also is exposed to cyclic periods of high mechanical and thermal stresses. Then, the cutting tool material should be able to resist high temperatures, wear and thermal shock, and impact and should not interfere chemically with the workpiece and cutting fluids. The life tool depends on the machining conditions and the combination of tool and work material used [14]. The diversity of materials includes high-speed steel (HSS), carbides, ceramics, and diamonds; however, some authors group the last two in one category called the ceramic and the superhard materials; these entire materials offer a wide range of properties and capabilities.

The HSSs appeared around 1900 and were replaced to high carbon steel because as the name says this material allows higher speeds. There are a broad range of grades of HSSs that are differentiated for the alloy element added to a high content of carbon. The alloy elements more used are tungsten, molybdenum, vanadium, chromium, and cobalt. This material is characterized by mean hardness of 75 HRC (hardness Rockwell C), and the working temperature of HSS is about 500 °C. One of the most important improvements in the history of the high steels is the technology of the power metallurgy. This technology has allowed to increase the toughness, the hardness, the hot hardness and to improve the wear behavior [15]. However, due to the working temperature limit, having without a coating is not the best option as cutting tool material for high-speed machining.

For this reason, the carbide tools have taken the space formerly occupied by HSSs. Furthermore, the carbide tools reach faster speeds than the HSSs. The process of manufacturing of the carbide tools are basically by power metallurgy and use as primarily material carbides; characterized as a hard material; and tough metal of the binders group as a cobalt. The most commonly used materials as carbide are tungsten, titanium, and tantalum, among others. Usually, the tungsten carbide (WC) powder is mixed with powered cobalt at high temperature and pressure. The ratio between the two elements and the particle size of the tungsten carbide determines the final properties of the cutting tool. For example, if the content of tungsten is increased, then the wear resistance is also increased; however, the tool strength is reduced. Associated with the particle size and increasing the hardness, the advances have allowed obtaining submicron and ultra-fine grained alloys with WC grain sizes of 0.5–0.8 μm and 0.2–0.5 μm, containing between 6% and 16% of cobalt by mass [16]. There are an extensive variety of grades, and the selection should be according to the application and the manufacture recommendations.

The classification of the carbides can include the cermets because these contain a similar microstructure, and they include nickel–chrome binder. The main element is carbonitrides of titanium (Ti) in combination with tantalum (Ta), tungsten (W), and molybdenum (Mo).

Ceramics are characterized by high temperatures (>1500 °C), chemically stable, and harder than carbides. Ceramic materials are based, basically, of two kinds of matrix materials: the alumina (Al_2O_3) and the silicon nitride (Si_3N_4). Commonly, the alumina matrix is combined with hard materials in order to improve the mechanical properties [15]. The silicon nitride is typically used for rougher and, particularly, cast iron and super alloys, as it has good heat resistance, creep resistance, and oxidation resistance. The reinforcement of alumina ceramics with silicon carbide whisker allows important improvements in the mechanical properties as increasing hardness, fracture toughness, thermal conductivity, thermal shock resistance, and high-temperature creep resistance [16].

The super hard materials include the polycrystalline diamond (PCD) and the cubic boron nitride (CBN). The CBN ranks the second place as hardest material after to diamond, and this material is characterized by hardness up to 4500 Hk [17]. The synthetic material contains CBN grain that is sintered with a specific ceramic binder, has high abrasive wear resistance, and presents chemical stability when is subjected to high temperature. CBN is used for machining hardened tool steels by high-speed machining [18]. PCD is manufactured by sintering at high pressure and temperature in combination of a solvent/catalyst metal, commonly cobalt or cobalt/nickel alloy [19]. This produces a tough material that is preferred as a tool material for cutting compared to monocrystalline diamond (MCD). MCD is a natural diamond and the hardest of all known materials; this material can produce defined cutting edges with precisely notch-free flutes [15]. Diamond properties include low coefficient of friction and thermal expansion, resistance to corrosion, and high strength. PCD tools are used mainly in aluminum–silicon by high-speed machining, composites, and nonferrous materials.

No single cutting tool material meets all requirements. Selecting the suitable tool material supposes trade-offs appear among the range of materials. In the case of the high-speed machining with higher rates of metal removal exists the need for heat resistant because this is one of the largest sources of error in the machining process.

12.4.2 Coating Application and Materials

One of the most common surface treatments of the tip tool is known as the coating. Coatings are used to improve the mechanical and thermal properties of tool, thereby increasing the tool life. In HSM, hard coatings are critical and extensively used to increase the tool productivity [20]. Normally, the tool coatings are between 2 and 15 μm of thickness [15]. The physical vapor deposition (PVD) and chemical vapor deposition (CVD) are mostly used as technologies to apply the coatings. The main difference lies in the temperature at which each process is performed; in the case of PVD, the coating material is applied at less than 500 °C, thereby achieving stress-reduction on the deposited layer of material and also avoids the reduction of the endurance. On the other hand, the CVD process is carried out between 800 and 1200 °C resulting in residual stresses due to the thermoelastic component between the film of material and the tool as well as the inherent thermoelastic component of the film. Typical coating materials include the titanium nitride (TiN), the titanium carbide (TiCN), the TiN adding aluminum (TiAlN), and the chromium nitride (CrN). When dry high-speed machining is performed, the recommended coating is the titanium aluminum nitride because it provides a better heat insulation for high temperatures. Additionally, the multilayer system is preferred because it allows compensating the mechanical stress, thermal conductivity, and could provide the reduction of diffusion processes for high tool temperatures. It should be emphasized that it is possible to recondition worn coated HSS and cemented carbide tools considering as these must be recoated after regrinding and taking into account that the mechanical properties could be affected by these procedures [21].

12.4.3 Tool Microgeometry

One of the limitations in order to achieve microfeatures machined by micromilling is the size of cutting edge radius and diameter tool. Currently, the capabilities of the tool manufacturing process are limited as concerns to the minimum cutting edge radius obtainable, which is in the order of 1–4 μm [22]. For instance, in the case of micromilling of hardened steels, the tools used are micrograin tungsten sintered carbide, which limit the minimum allowable cutting edge radius (2–3 μm) (Figure 12.3). It is usual for the chip thickness value in micromilling of hardened steels to be smaller than 1 μm.

Therefore, the cutting area corresponds to the radius, producing a large negative rake angle [23]. Likewise, due to the limitations of the current micro-tool–manufacturing methods, the

Figure 12.3 Cutting edge radius of micro end mill tool.

cutting edge radius cannot be considered perfectly sharp having considerable nonuniformities of the value radius along the flutes of the cutter [24]. If tool diameter is smaller than the nominal value, then the possibilities to breakage are bigger. Commonly, the diameters of commercial micro-tools are in the range of 100–500 µm. Smaller tool diameters can be obtained with customized order, however, the cost increases.

Another problem is the commercial availability of micro-tools. Actually, the production of micro-tools is limited for geometry types. Only few models are available in the market, and most of the current cutting micro-tools are designed for steel machining [25].

In tool customization, electro-discharge machining (EDM) and ion beam machining (IBM) are commonly used to fabricate micro-tools with smaller diameter than commercial diameters or with different cutting geometry. Using IBM, the authors Adams et al. [26] realized micro-milling of trenches in aluminum with customized tools. Another different method to produce custom microtool is EDM. Chern et al. [27] fabricated micro-tools of less than 100 µm in diameter using wire electro discharge machining (WEDM) technology.

One important consideration for tooling is the coating. For conventional and micromilling, coatings are usually made by appropriate vapor (PVD) deposition techniques. Commercially, the market offers micromills with different coatings: TiAl, TiAlN, CBN, etc. However, high tool wear appears when hard materials are machined, changing drastically the tool condition and consequently the process performance [25].

After some experimental trials using microcoated mills in steel, Aramcharoen et al. [28] realized some important conclusions about coatings:

- TiN, TiCN, TiAlN, CrN, and CrTiAlN coatings help reduce cutting-edge chipping and edge radius wear as compared to uncoated ultra-fine grain carbide end mills.
- TiN and CrTiAlN coatings reduced flank wear compared to uncoated tools.
- Wear patterns of CrN- and TiCN-coated micromills were identified as coating delamination, while localized flank wear was found on TiAlN-coated micromills.
- TiN appears to be the best coating in terms of least flank wear and edge chipping. This coating had the lowest hardness coupled with the highest adhesion and fairly good surface finish.

- At the beginning of cutting (burn-in-period), none of the coatings showed any improvements in surface finish compared to the uncoated tool. This implies the need to produce wear-resistant coatings that have an ultra-fine surface finish compared to the uncoated tool. Additionally, the use of self-lubricating, low friction topcoats should be pursued.
- Most coatings, with the exception of TiAlN, lead to reduced burr size compared to uncoated tools.
- The tests showed that TiN consistently performs the best in micromilling tool steel based on flank wear, chipping, edge radius wear, surface finish, and burr size.

12.4.4 Tool Wear and Failure

As mentioned in Section 12.4.1, one of the challenges in the technology of HSM is to find the best combination of tool material, coating, workpiece, and the process parameters in order to achieve efficient operations. In fact, the only way to ensure the productivity is to extend the tool life and avoid the tool wear and the posterior failure. Wear could be defined as volume loss of tool material due to the interaction phenomena generated, as temperature and stresses, between the tool material and workpiece material.

Cutting of titanium alloys has a relevant interest for industrial applications; however, it is considered as difficult-to-machine material. Coated carbide tools at lower cutting speed machining Ti–6Al–4V present critical mechanical damage caused by mechanical loading conditions. Whereas at the higher cutting speed, the thermal damage dominates the progressive tool wear [29]. The cutting speed of 628 m/min is suitable for milling Ti6Al6V2Sn alloy [30]. The lubrication conditions are a critical factor that affects significantly the tool wear. Under dry conditions, the flank wear increases rapidly while when the machining of Ti6Al4V is performed under lubrication by applying compressed in cold nitrogen gas conditions. This achieved smallest flank wear improving the tool life (Figure 12.4). The coated cemented carbide tools primarily through diffusion wear mechanism undergo all conditions of cooling/lubrication [31]. The Inconel 718 super alloy is widely used in industries as aviation, nuclear reactors, and many fields where high temperatures are used [32, 33]. Nanocomposite ceramic tools are used in the milling of Inconel 718, these tools showed higher tool life than the homogeneous ones, due to their improved mechanical properties, superior abrasive wear, and fracture resistance [34]. Concerning to micromilling of Inconel 718, experiment was performed

Figure 12.4 High-speed milling under cooling conditions.

comparing cutting tools, the results shown that the micro tools filmed with AlTiN, TiAlN+AlCrN presented improved performance compared to those coated with TiAlN+WC/C and diamond-like carbon (DLC). However, the last two evince goodness performance avoiding built-up edge formation [35]. The tool steels are broadly used to dies and molds due to the hardening characteristics. Koshy et al. [36] investigated the end milling of AISI D2 comparing indexable insert ball nose end mills using carbide and their solid carbide equivalents. The results indicated that adhesion, chipping, and attrition wear are the main forms for tool wear.

12.5 High-Speed Machining Applications and Future Trends

The aerospace industry is one of the most traditional applications in the high-speed milling. Aluminum is being replaced by titanium alloy and is a typically material used by this industry (Figure 12.5). However, this material is also used in others fields as the automotive field.

On the other hand, the global trends of replication manufacturing techniques causes that die and mold industries (Figure 12.6) as well as machining of castings take the place of the major consumers of high-speed machining [37, 38].

Figure 12.5 Motor block by milling.

Figure 12.6 Example of mold cavity.

Figure 12.7 High-speed milling using a micro end mill tool of 200 µm.

The complex geometries characterizing the forging dies make the manufacturing process of high-speed machining to be the best option for manufacturing.

High surface quality is one of the characteristics achieved with the use of the HSM technology; for this reason, one of the application fields is the precision machining. A study by Schulz, H. [39] is cited as an example for the optical industry and particularly fine mechanical parts.

The global tendency toward the miniaturization of products demands to make a bridge between the macro and micro world. In this regard, important works have been realized in order to increase base-knowledge of micromilling (Figure 12.7) ([22, 23, 40, 41];).

Biomedical applications are one of the emerging opportunities for manufacturing processes. Magnesium–calcium is seen as a potential material for use in orthopedic prostheses due to their mechanical properties similar to bone tissue. For this reason, the high-speed dry machining of Magnesium–Calcium (MgCa) alloy has become a topic of interest for research [42, 43].

The evolution of manufacturing processes demand for time and cost reduction. High-speed machining technology guarantees superior metal removal rate, improved machined surface quality, and dimensional accuracy. Benefits of high-speed machining can be achieved in so far as it can be ascertained the best combination of tool, workpiece, machine tool, and process parameters, among others.

References

1 Taylor, F.W. (1907). *On the Art of Cutting Metals*. New York: The American Society of Mechanical Engineers.

2 Salomon, C. (1931). Process for the machining of metals or similarly acting materials when being worked by cutting tools. German Patent, 523594, 1931–1904.

3 Dewes, R. and Aspinwall, D. (1997). A review of ultra high speed milling of hardened steels. *Journal of Materials Processing Technology* 69 (1–3): 1–17.

4 Arnone, M. (1998). *High Performance Machining*. Hanser Gardner Publications.

5 López de Lacalle, L.N., Sánchez Galíndez, J.A., and Lamikiz, A. (2004). *Mecanizado de alto rendimiento: Procesos de arranque*. Ediciones Técnicas Izaro.

6 Urbanski, J., Koshy, P., Dewes, R., and Aspinwall, D. (2000). High speed machining of moulds and dies for net shape manufacture. *Materials and Design* 21 (4): 395–402.

7 Toh, C. (2005). Design, evaluation and optimisation of cutter path strategies when high speed machining hardened mould and die materials. *Materials & Design* 26 (6): 517–533.

8 Campbell, C., Bendersky, L., Boettinger, W., and Ivester, R. (2006). Microstructural characterization of al-7075-T651 chips and work pieces produced by high-speed machining. *Materials Science and Engineering: A* 430 (1): 15–26.

9 López de Lacalle, L.N. and Lamikiz, A. (2008). *Machine Tools for High Performance Machining.* Springer.

10 Gopalakrishnan, B., Yoshii, T., and Dappili, S. (2004). Decision support system for machining center selection. *Journal of Manufacturing Technology Management* 15 (2): 144–154.

11 İç, Y.T. and Yurdakul, M. (2009). Development of a decision support system for machining center selection. *Expert Systems with Applications* 36 (2): 3505–3513.

12 Arslan, M., Catay, B., and Budak, E. (2004). A decision support system for machine tool selection. *Journal of Manufacturing Technology Management* 15 (1): 101–109.

13 Kopač, J. (1998). Influence of cutting material and coating on tool quality and tool life. *Journal of Materials Processing Technology* 78 (1): 95–103.

14 Komanduri, R. (2000). *Tool Materials. Encyclopedia of Materials: Science and Technology,* 1–13. Wiley Online Library.

15 López de Lacalle, L.N., Lamikiz, A., Fernández de Larrinoa, I., and Azkona, I. (2011). Machining of hard materials. In: *Advanced Cutting Tools* (ed. J. Paulo Davim), 35–36. London: Springer.

16 Byrne, G., Dornfeld, D., and Denkena, B. (2003). Advancing cutting technology. *CIRP Annals-Manufacturing Technology* 52 (2): 483–507.

17 Davis, J.R. (1995). *Tool Materials.* Novelty, OH: ASM International.

18 Özel, T. (2003). Modeling of hard part machining: effect of insert edge preparation in CBN cutting tools. *Journal of Materials Processing Technology* 141 (2): 284–293.

19 Philbin, P. and Gordon, S. (2005). Characterisation of the wear behaviour of polycrystalline diamond (PCD) tools when machining wood-based composites. *Journal of Materials Processing Technology* 162: 665–672.

20 Kopač, J. (2003). Advanced tool materials for high-speed machining. In: *Paper Presented at the Proceedings of the 12th Scientific International Conference Achievements in Mechanical and Materials Engineering,* 1119–1128. AMME.

21 Bouzakis, K., Michailidis, N., Skordaris, G. et al. (2012). Cutting with coated tools: coating technologies, characterization methods and performance optimization. *CIRP Annals-Manufacturing Technology* 61 (2): 703–723.

22 Bissacco, G., Hansen, H.N., and De Chiffre, L. (2005). Micromilling of hardened tool steel for mould making applications. *Journal of Materials Processing Technology* 167 (2): 201–207.

23 Uriarte, L., Azcárate, S., Herrero, A. et al. (2008). Mechanistic modelling of the micro end milling operation. *Proceedings of the Institution of Mechanical Engineers, Part B: Journal of Engineering Manufacture* 222 (1): 23–33.

24 Özel, T. and Liu, X. (2009). Investigations on mechanics-based process planning of micro-end milling in machining mold cavities. *Materials and Manufacturing Processes* 24 (12): 1274–1281.

25 Uriarte, L., Herrero, A., Ivanov, A. et al. (2006). Comparison between microfabrication technologies for metal tooling. *Proceedings of the Institution of Mechanical Engineers, Part C: Journal of Mechanical Engineering Science* 220 (11): 1665–1676.

26 Adams, D.P., Vasile, M.J., Benavides, G., and Campbell, A.N. (2001). Micromilling of metal alloys with focused ion beam–fabricated tools. *Precision Engineering* 25 (2): 107–113.

27 Chern, G., Wu, Y.E., Cheng, J., and Yao, J. (2007). Study on burr formation in micro-machining using micro-tools fabricated by micro-EDM. *Precision Engineering* 31 (2): 122–129.

28 Aramcharoen, A., Mativenga, P., Yang, S. et al. (2008). Evaluation and selection of hard coatings for micro milling of hardened tool steel. *International Journal of Machine Tools and Manufacture* 48 (14): 1578–1584.

29 Li, A., Zhao, J., Luo, H. et al. (2012). Progressive tool failure in high-speed dry milling of ti-6Al-4V alloy with coated carbide tools. *The International Journal of Advanced Manufacturing Technology* 58 (5): 465–478.

30 Maekawa, K., Kitagawa, T., and Kubo, A. (1997). Temperature and wear of cutting tools in high-speed machining of inconel 718 and ti-6Al-6V-2Sn. *Wear* 202 (2): 142–148.

31 Su, Y., He, N., Li, L., and Li, X. (2006). An experimental investigation of effects of cooling/lubrication conditions on tool wear in high-speed end milling of ti-6Al-4V. *Wear* 261 (7): 760–766.

32 Dudzinski, D., Devillez, A., Moufki, A. et al. (2004). A review of developments towards dry and high speed machining of inconel 718 alloy. *International Journal of Machine Tools and Manufacture* 44 (4): 439–456.

33 Ezugwu, E., Wang, Z., and Machado, A. (1998). The machinability of nickel-based alloys: a review. *Journal of Materials Processing Technology* 86 (1): 1–16.

34 Zheng, G., Zhao, J., Zhou, Y. et al. (2013). Performance of graded nano-composite ceramic tools in ultra-high-speed milling of inconel 718. *The International Journal of Advanced Manufacturing Technology* 67 (9–12): 2799–2810.

35 Ucun, İ., Aslantas, K., and Bedir, F. (2013). An experimental investigation of the effect of coating material on tool wear in micro milling of inconel 718 super alloy. *Wear* 300 (1–2): 8–19.

36 Koshy, P., Dewes, R., and Aspinwall, D. (2002). High speed end milling of hardened AISI D2 tool steel (~58 HRC). *Journal of Materials Processing Technology* 127 (2): 266–273.

37 Elbestawi, M., Chen, L., Becze, C., and El-Wardany, T. (1997). High-speed milling of dies and molds in their hardened state. *CIRP Annals-Manufacturing Technology* 46 (1): 57–62.

38 Fallböhmer, P., Rodríguez, C., Özel, T., and Altan, T. (2000). High-speed machining of cast iron and alloy steels for die and mold manufacturing. *Journal of Materials Processing Technology* 98 (1): 104–115.

39 Schulz, H. (1997). State of the art of high-speed machining. In: *1st French and German Conference on High Speed Machining*, 1–7. France: Metz University.

40 Aramcharoen, A. and Mativenga, P. (2009). Size effect and tool geometry in micromilling of tool steel. *Precision Engineering* 33 (4): 402–407.

41 Vázquez, E., Rodríguez, C.A., Elías-Zúñiga, A., and Ciurana, J. (2010). An experimental analysis of process parameters to manufacture metallic micro-channels by micro-milling. *The International Journal of Advanced Manufacturing Technology* 51 (9): 945–955.

42 Salahshoor, M. and Guo, Y. (2011). Cutting mechanics of high speed dry machining of magnesium-calcium biomedical alloy using internal state variable plasticity model. In: *Paper presented at the ASME 2011 International Manufacturing Science and Engineering Conference*, vol. 1, 85–92. Oregon.

43 Salahshoor, M. and Guo, Y. (2011b). Cutting mechanics in high speed dry machining of biomedical magnesium–calcium alloy using internal state variable plasticity model. *International Journal of Machine Tools and Manufacture* 51 (7): 579–590.

13

Hard Machining

Durul Ulutan and Tuğrul Özel

Manufacturing and Automation Research Laboratory, Department of Industrial and Systems Engineering, School of Engineering, Rutgers University, Piscataway, NJ 08854, USA

13.1 Introduction

The basic definition of hard machining can be given as material removal operations such as turning, milling, and drilling that are applied to heat-treatable metal alloys, mostly steels, in their hardened state in which the workpiece hardness is significantly higher than 45 HRC (hardness Rockwell C) [1]. Hard machining operations are typically final manufacturing operations executed on the parts and components, and desired dimensional and surface quality is subjected to hard machining parameters and conditions. In order to machine these hard workpiece materials, machine tools have to be rigid enough to withstand high cutting forces and vibrations, and cutting tool material and cutting tool geometry must be carefully selected with attention to tool wear/life concerns. Tool edge preparation and addition of coating material also affect the outcome in both aspects of tool life and machining accuracy/quality.

The harder materials frequently utilized in the industry are steels (hardened, heat treated, bearing steels) and cast iron. These materials are often utilized in advanced industries such as automotive, rail transportation, machine tool, and aerospace industries. Superior mechanical properties of these materials at their hardened state improve their service life when they are manufactured with hard machining processes. For this reason, hard machining–based process chains offer advantageous outcomes in terms of productivity and quality when compared to conventional machining processes and subsequent heat treatment and finish grinding–based traditional process chains in producing value-added hardened parts and components.

13.1.1 Hard Machining Operations

Machining operations that process materials harder than 45 HRC are considered hard machining operations and can include practically any operation. The major manufacturing processes that utilize hard machining are turning, milling, broaching, and boring operations [2]. It is important to understand how hard machining applies to these processes and its advantages and disadvantages compared to traditional machining operations.

In hard turning, the major goal is to eliminate the grinding process afterward, by improving the accuracy of turning [2]. With the correct selection of tool material and geometry suited for the specific workpiece material and machining parameters, it is possible to reach surface roughness (R_a) levels less than 1 μm [2]. This level of surface roughness would lower the

Modern Manufacturing Processes, First Edition. Edited by Muammer Koç and Tuğrul Özel.

possibility of a stress concentration location, thus a crack initiation and end-part failure. Coolant usage and selection can also be an important decision, but for cost and effectiveness reasons, coolant usage is not advised for most cases of hard turning. Both tailstock support and workpiece geometry are important factors in terms of the rigidity of the system; therefore, tailstock should be used to support the workpiece whenever possible to reduce the possibility of chatter. In any case, the length to diameter ratio of the workpiece should be selected as low as possible (smaller than 4) to improve the rigidity of the system; hence, the quality of the final surface. Although machines used for traditional machining can be utilized for hard turning if rigid enough, machines specifically designed for hard machining provide more rigidity and reliability necessary to improve process outputs. Most commonly, hard turned materials are steels (bearing, tool, etc.), and with their hardness level of 60 HRC and above, precision hard turning of these materials require high rigidity machine tools with multitasking to consolidate subsequent operations.

In hard milling, there are more parameters to consider than hard turning. In addition to the cutting tool, the toolholder, as well as the machine selection is more important, and the computer-aided design and manufacturing of the part plays a bigger role. Hard milling requires specifically designed milling machines with properties similar to high speed milling machines. The machine and its components such as the spindle and tool holding system should have high rigidity to decrease the effects of vibrations; and this improved rigidity can be provided by fixturing considerations and by material selection. Lower cutting speeds might be needed as well as higher speeds, so the spindle should be able to provide high torque at all cutting speeds [3]. Carbide inserts and end mills are usually the choice of cutting tools, while ceramic inserts are also utilized in some cases. During selection of the cutting tool system, the difficult nature of the hard milling process should be considered, and less expensive inserts that would offer decreased tool life, hence efficiency, should be avoided.

Hard broaching process is usually used when the heat brought by finish broaching process creates excessive distortion of the part geometry, and postprocessing is required. End-part quality, in terms of surface finish, changes mainly with different tool selection and coating [2]. Compared to conventional broaching, uncut chip thickness should be kept small to keep the tool wear low, while the cutting speed is higher to reduce machining time and improve machining economy. Due to this higher speed requirement, specific hard broaching machines are needed to have the support for high axial (thrust) forces.

On the other hand, hard boring is used to eliminate internal grinding process to achieve required part accuracy and tolerance levels, in terms of size and concentricity of the hole. This process also ensures the straightness of the hole, compared to grinding finish. Mostly, polycrystalline cubic boron nitride (PCBN) and ceramic tools are used for these processes, and single-blade tools are used to increase tool rigidity to machine bigger diameter holes [4].

13.1.2 Characteristics of Hard Machining

When machining harder materials, process requirements and outputs are different than those of conventional machining. In particular, the differences in power requirements and tool requirements cause changes in force outputs, tool wear, chatter vibrations, and surface integrity, which result in lower process reliability.

Since the material hardness is higher, cutting tool edges are treated to increase edge strength. Therefore, mechanics of cutting in hard machining processes differ than conventional machining. Although the power required for plastic deformation (shearing) of the material is the highest in conventional machining processes, this requirement decreases in hard machining as the parts do not deform as much. On the other hand, the power required for ploughing due to larger edge geometry increases greatly because of the hardness of the material; therefore, the axial (thrust) component of the machining force increases compared to conventional machining. Increasing axial force generates excessive heat, creates chatter vibrations, and causes rapid tool wear, which might lead to catastrophic tool failure. Hence, the choice of tool material and geometry is very critical for successful hard machining. Although sintered carbide tools are still utilized, cutting tool choice for hard machining changes toward ceramics, PCBN, and polycrystalline diamonds (PCDs) [2]. Tool coating is also known to improve machinability in hard machining; therefore, coatings such as TiN and TiAlN are utilized for better results [5]. Most hard machining tools have negative rake angle to increase the strength of the cutting edge and make the tool more reliable against higher cutting forces. To create a similar effect, tools are also produced with bigger edge radii or with chamfered and wiper geometries as well (Figure 13.1). The use of nontraditional insert geometries such as round and rhombic shapes is also common to increase strength and durability [2].

Formation of new machined surface also requires considerable power, so the power usage during the process shifts from conventional to hard machining. With increasing cutting speed, cutting force decreases, but for this reason, machining systems need to be very rigid. In most cases, specific machining centers are needed because the regular machining centers are not rigid enough to support for hard machining.

In hard machining, when higher cutting speed and lower feed values are utilized for aforementioned reasons, higher machining temperatures (up to 1200 °C) than conventional machining are induced [1]. In addition to the increase in material removal rate, this is also preferred because of its thermal softening effect that reduces the power requirements and force outputs. However, with increased temperature, thermal effects during machining increase, which

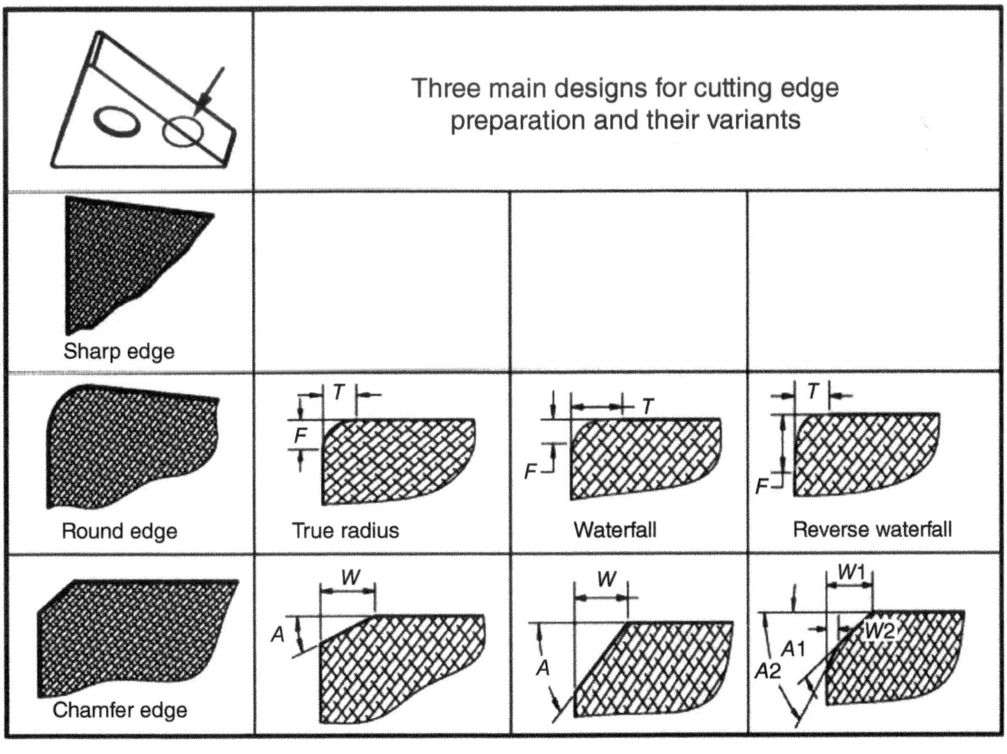

Figure 13.1 Common tool geometries. *Source:* Yen et al. 2004 [6]. Reproduced with permission of Elsevier.

adversely affect the surface finish of the final part in terms of white layer formation, microhardness, and residual stresses [7].

13.2 Mechanics of Hard Machining

Mechanisms of hard machining in terms of chip formation, heat generation, machining forces, energy and power, tool wear, and lubricant usage are different from those of conventional machining. Therefore, it is important to understand these mechanisms of hard machining before designing a manufacturing process utilizing a hard workpiece material.

While machining hardened steels, chip usually forms in a saw-tooth shape. This is considered to be due to the crack formation in the primary shear zone, and the propagation of these cracks creates sharp edges of the chip (Figure 13.2). Due to the cyclic behavior of these cracks, the chip resumes a saw-tooth shape, indicating concentrated shear zones. Because of this cyclic crack formation in the primary shear zone, and the excessive rubbing between the flank and the workpiece, machining temperature is higher than conventional machining, which creates higher thermal stresses [1]. High temperatures can also cause the material to approach or even achieve transition temperatures, which changes the nature of the machining process and can deteriorate the surface quality of the final product through microstructural alterations and white layer formation.

It is understandable that all components of cutting forces increase when machining materials with hardness greater than 45 HRC because these materials provide higher resistance against machining. Moreover, particularly due to the tool geometry required to machine these materials, i.e. edge radius, chamfer, and negative rake angle, the axial (thrust) component of the cutting force increases significantly. It is also known that while machining hard materials,

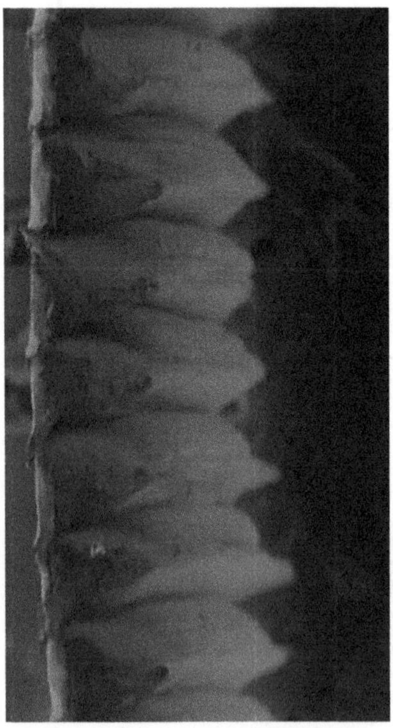

Figure 13.2 Saw-tooth shape chip formation in machining hardened steel. *Source:* Karpat and Özel 2008 [8]. Reproduced with permission of Elsevier.

increasing cutting speed compared to conventional machining speeds decreases the cutting force, which is also an advantage in terms of machining rate. This is due to the significant decrease in radial forces, as the axial (thrust) component of forces actually increases with harder materials, mostly due to the necessary tool geometry required for machining. Different tool geometries (chamfered, wiper, round edge, etc.) can also affect the cutting forces, but these effects vary depending on the workpiece material, cutting speed, and feed. Increasing the depth of cut increases the load on the cutting tool, which causes an increase in the cutting forces, but excessive tool wear can change the characteristic of the process and eliminate this effect.

In terms of energy and power usage, hard machining is substantially more consumptive [2]. This is expected, as the cutting forces are greater than in conventional machining. In addition, with increasing cutting speed and feed, specific cutting energy decreases. Also, negative rake angles are needed for hard machining, but extreme negative angles or high edge radii that cause largely negative angles at the lower uncut chip thicknesses cause the specific cutting energy to increase greatly. For the same reason, largely chamfered or wiper tools might require more energy, and thus power input.

Furthermore, tool wear is more pronounced in hard machining because hard materials cause the cutting tools to wear very quickly and face catastrophic tool failure. Higher cutting forces and higher thermal loads due to increased friction and heat generation develop tool wear on the cutting tool faster in hard machining compared to conventional machining. When the tool wears faster, it causes surface integrity problems such as surface defects, roughness, white layer formation, and residual stresses. Therefore, it is important to either find methods to decrease tool wear or frequently change the cutting tool. In order to reduce the tool wear, lubrication can be utilized. However, the lubricant must be selected with care, as some lubricants adversely affect the process results because of their incompatibility with the workpiece material [9]. Researchers developed ways to utilize minimum quantity lubrication (MQL) to eliminate these problems, while also reducing tool wear and hence improving surface quality of the final product. With MQL, cutting forces can be reduced comparably to both dry machining and wet machining, thus improving tool life.

13.3 Cutting Tools

In hard machining, mechanical loads and high temperature occur at the same time, and cutting tools must be able to withstand this combination. Since hard machining requires different cutting tools that can support and enhance the mechanics of the process, it is important to understand the options in terms of tool materials, coating materials and application methods, tool microgeometry, and tool wear and failure characteristics. Different materials may require different tool geometries for better performance, or some materials may not respond well to some coatings, and an operator must be aware of these differences before machining a hard workpiece material.

13.3.1 Cutting Tool Materials

Machining materials that are harder than 45 HRC requires special cutting tool materials that can withstand the challenge of elevated temperatures. The chemical composition of the tool material must be stable and nonreacting with the workpiece material (and the lubricant, if applicable). The material should also possess physical and thermal properties that do not create problems at high temperatures. Especially, the cutting tool material hardness must not decrease too much at elevated temperatures and should be able to machine the hard workpiece material and should possess high thermal conductivity to dissipate the heat generated in the cutting

Table 13.1 Comparison of cutting tool materials and key properties.

	Hardness (HV)	Transverse rupture strength (MPa)
WC	1500–3000	2000–5000
PCBN	2800–4200	85–115
PCD	8000–12000	110–220

Source: Lopez de Lacalle et al. 2011 [9]. Reproduced with permission of Springer.

zone. In addition, especially in the absence of tool coating, tool material should have good wear resistance to enhance performance in the long run. To avoid premature catastrophic tool failure, tool materials with higher toughness are also preferred. Materials that can satisfy these requirements are mainly sintered carbides (tungsten carbide with cobalt binder), ceramics, PCBN, and PCDs (Table 13.1).

Sintered carbides are still the majority of the cutting tools in hard machining due to their low cost and recyclability. They are usually a combination of tungsten or vanadium carbides with cobalt. Cobalt is usually around 10% in the mix, and it helps to increase the melting point of the cutting tool to withstand higher machining temperatures, but decreases the hardness of the cutting tool, so higher percentages are not advised. They can be found in both as integral (single cutter) tools and as inserts. Although integral tools provide better rigidity, they are significantly more expensive compared to inserts. If inserts are used, they are replaced frequently to have better surface finish. When integral tools are used, they are sharpened several times to reduce new tool costs as much as possible [9].

Ceramics are also used to machine hard materials, mainly due to their combination of hardness and chemical resistance at elevated temperatures. Their main disadvantage is their fragility that prevents them from being utilized for intermittent machining [9]. Silicon nitride ceramics with silicon carbide reinforcements are the most frequently seen type in the industry because of their high thermal resistance, and improved fracture toughness. Because of their good properties at elevated temperatures, hot machining is advised, so lubricant usage should be avoided with these tools.

PCBN tools benefit from the good mechanical properties of the crystalline structure of boron nitride materials. It is extremely hard (>80 HRC) [10], although still not as hard as diamond tools. As with sintered carbide tools, PCBN tools are also bound with cobalt, or similar binder elements. With higher cubic boron nitride (CBN) percentage, toughness of PCBN is increased for better intermittent cutting performance while reducing thermal conductivity, but lower CBN percentage decreases the cost of the cutting tool. PCBN tools cannot be used with ferritic irons because of the reaction between the ferritic matrix and boron that leads to rapid degradation of the CBN tool, but it can be used for machining pearlitic irons. Thanks to the superior properties of these materials, and very high cutting speeds can be achieved in machining. Since this is also a very expensive material, PCBN tools mostly emerge as PCBN layers on metal tool bodies to decrease costs, while not compromising from rigidity.

PCD tools are sintered diamond grains bound with a similar binding element such as cobalt. By altering the grain size of the diamond, it is possible to compromise between surface quality and tool life. When extremely well surface quality is required, natural diamond (monocrystalline diamond) can also be utilized, but these are very expensive options and are limited to very strict requirements of the manufacturing process. However, PCD tools cannot be used with ferrous materials because diamond reacts with ferrous materials under high machining temperatures.

13.3.2 Coating Application and Materials

After the tool is prepared, it is sometimes coated with a layer of material so that tool properties (hardness, thermal conductivity, wear resistance, etc.) are enhanced, and the performance of the cutting tool is improved. However, each layer of added coating creates additional parameters to consider in a cutting tool. The coating thickness is an important parameter, and depending on the process requirements, the tool coating thickness should be determined. When more tool wear is anticipated, coating thickness might be higher so that tool endures the process longer, but increased coating thickness also increases the resultant edge radius, which may induce higher cutting forces [9]. If this is not a major issue, the thickness might be kept less to avoid adhesion and ploughing. There are two major methods of coating application: Physical vapor deposition (PVD) and chemical vapor deposition (CVD). Especially, PVD coatings are helpful at high temperatures; therefore, hard machining is an important application for PVD coatings. Most of the coating materials available for the industry are nitrides, such as titanium nitride (TiN), titanium carbo-nitride (TiCN), and titanium aluminum-nitride (TiAlN) [5]. These advanced coating materials increase the hardness of the cutting tool and provide mechanical and thermal wear resistance. For even better thermal resistance, aluminum content of the TiAlN coatings can be increased when higher machining temperatures are anticipated. Addition of silicon (Si) to the coating to make it a TiAlSiN coating improves hardness properties at elevated temperatures. Other elements such as chromium, cobalt, tantalum, vanadium, ytterbium, and zirconium are also added with small percentages to improve the balance or chemical adhesion of the coating material. Application of multilayers of coating also helps in combining the advantages of multiple layers. Moreover, applying the coating in multiple layers rather than at once helps fighting against crack propagation. The thickness of the coating could be varying through the coating region, thinner on the back edge and thicker on the cutting edge. This way, reduction of friction can be achieved on the cutting edge, while the disadvantages of coating can be avoided on the regions where coating is not needed. The effectiveness of the coating and the duration it stays on the cutting tool can also be improved by preparing the edge with an edge radius rather than sharp [9,11,12].

13.3.3 Tool Microgeometry

Cutting tools have many geometrical properties, and different machining processes (end milling, bar turning, etc.) require different geometries for better efficiency. The main geometrical properties of a cutting tool are its rake angles, clearance angle, nose radius, edge radius, and chamfer angle if applicable. Also, when coating is applied on a cutting tool, the geometry of the tool is altered slightly, changing the characteristics of the machining process.

Orthogonal turning is the simplest form of machining; however, it is not commonly utilized in the industry because it lacks practicality. In orthogonal turning, the edge of the cutting tool is perpendicular to the cutting velocity direction. Then, rake angle is the angle between feed direction and tool cutting edge, and clearance angle is the angle between cutting velocity direction and tool flank. Edge radius and chamfer define how initial contact between tool and workpiece will be, and if coating is applied, it increases the edge radius by its thickness. In oblique cutting processes, the orthogonality between the cutting velocity vector and the cutting edge no longer exists. In hard turning, in order to provide enough strength to support the process, rake angle is preferred to be negative. Also, including a chamfer in the tool edge preparation acts as a negative rake angle while machining at low feed rate. Edge radius acts as a variable rake angle in such cases, also helping with the machining process.

In hard machining, negative back and side rake angles can be preferred. This is a comparably aggressive approach that helps in machining very difficult-to-machine materials such as nickel-based alloys. However, if such extreme measures are not required, a negative side rake angle and a small positive back rake angle provide enough support for most hard machining processes. When coating is applied, it decreases the friction coefficient between the cutting tool and the workpiece (as well as the flowing chip) but increases the tool edge radius by its applied thickness [9].

13.3.4 Tool Wear and Failure

During machining processes, the cutting tool is worn by the flowing chip, as well as the rubbing mechanism of the machined workpiece surface. The chip has its maximum impact at the tool rake face, but not at the tip, causing a crater shape. The machined workpiece also contacts the clearance face of the cutting tool, abrading the flank region. These two wear types are both affected by the heat generation at the primary shear zone. Because of the increased temperature at the chip–tool interface and the tool tip contacting the workpiece, thermal effects are significant on tool wear.

Crater wear is a process where small pieces from the tool rake face material are adhered to the flowing chip and thus removed from the surface [9]. Especially, when the thermal conductivity of the workpiece material is low, the tool–chip interface will be very hot, which will ease the adhesion process as well as enabling diffusion between the tool rake surface and the flowing chip. Flank wear is a process where abrasion between the cutting tool tip and the machined workpiece surface removes small pieces of cutting tool material from the flank face. With the harder workpiece materials, this mechanism plays more of a major role in determining the tool failure. Especially, if the chemical reactivity between the tool and the workpiece ceases to exist, adhesion and diffusion are not observed, and the tool life considerations only include flank wear. Since these two mechanisms require high temperatures and hard workpiece materials, they are the two most common types of tool wear observed in hard machining.

Another mechanism effective in hard milling, or intermittent machining of hard materials, is chipping. In this mechanism, the small pieces of the cutting tool break away from the cutting edge during interrupted cutting due to transient thermal and mechanical impacts. It usually causes catastrophic tool failure, which means that the tool is deteriorated heavily and rapidly and/or it broke down, and further usage of the tool will not be possible or beneficial. If the size of the chipped tool is small, it can be considered as an intermediate step between fresh and worn tool. Further wear might correct the chipped portion of the tool, and it might be used for more machining, but its geometry will be different, so the cutting parameters such as width of cut and feed might be different. Alternatively, chipped portion can rapidly deteriorate, and further machining may be harmful for the surface finish of the machined workpiece. Catastrophic tool failure can also occur with extreme flank wear, usually not more than 0.5 mm. In most cases, tool wear types affect each other to increase the rapidity and severity of each other, so once the cutting tool becomes somewhat worn, it takes a short time to reach catastrophic failure. Therefore, especially in hard machining, frequent changing of cutting tools is advised.

13.4 Surface Quality and Integrity

Surface integrity of a machined product is composed of different types of modifications to the surface and immediate subsurface, such as surface defects, microstructural alterations, microhardness changes, surface roughness, and residual stresses [7, 13]. Surface integrity of a machined part directly influences the end use and performance of that part, so these effects should be categorized, analyzed, understood, and altered in favor of the specific practical use. The effects can be on postprocessing such as on frictional and wear behavior and lubrication, on mechanics of the post process, or on end use such as lubrication (such as in bearings), thermal and electrical conductivity, and fatigue life and corrosion characteristics. All these influences might come from different types of surface integrity issues, so categorizing the surface problem is important.

13.4.1 Surface Defects

Surface defects are observed practically on every machined part (Figure 13.3). Feed marks, which occur on every machined workpiece, are one type, but obvious feed marks can be avoided using lower feed rates or eliminated using postprocessing. Also, because of the contact between the tool and the workpiece, surface drag, material cracking, tool material particles adhering on workpiece surface, surface plucking and hence cavities, and formation of chip layer are some other examples of common surface defects [7]. Some of these defects may be easier to eliminate, whereas some of them may not. When it is harder to eliminate surface defects, it is important to quantify these defects to understand whether or not they accumulate to a substantial amount that requires postprocessing or remanufacturing.

13.4.2 Microstructural Alterations

Machining processes, especially machining of hard materials induce high thermal and mechanical loads that may lead to phase changes in the material [7]. For machining of hardened steels,

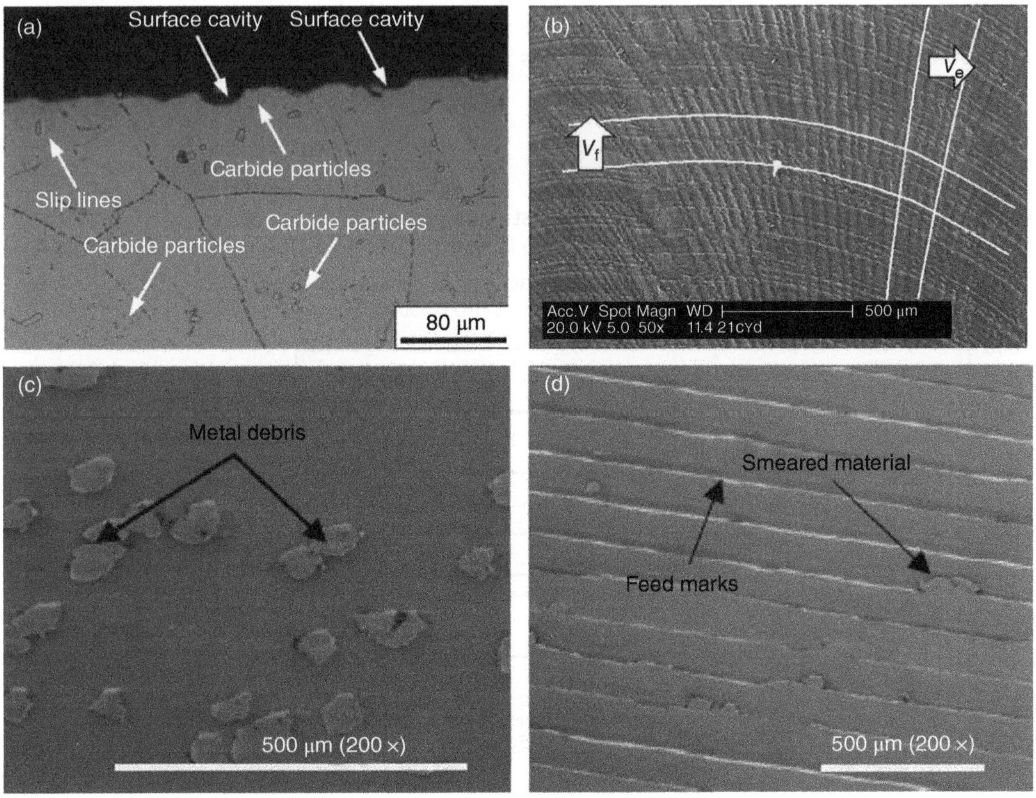

Figure 13.3 Examples of surface defects in machining. *Source:* Ulutan and Özel [7]. Reproduced with permission of Elsevier.

high friction and plastic strain during machining causes austenite to be formed, and rapid cooling by the ambient air, bulk material, or coolant induces quenching that forms "white layer" on the surface [13]. This layer is usually a very brittle martensitic structure that possesses high residual stress, which causes early fatigue failure due to crack initiation and propagation under dynamic loading. Similar layers of plastic deformation with different characteristics were also observed in Titanium-based alloys (although these are usually not considered hard materials), and similar mechanisms may be effective in other materials as well [7, 14, 15, 16, 17, 18]. Most of the work on microstructural alterations and white layer formation after hard machining has been done on bearing and tool steels, and formation of nanograins (100–300 nm) following dynamic recrystallization is observed [13]. In another study, honed edge geometries were found to create deeper subsurface plastic deformation compared to sharp and chamfered tools while machining hardened AISI 52100 bearing steel [19, 20].

13.4.3 Microhardness

While machining, because of the hardening effect of the machining process especially for materials that exhibit work-hardening behavior, the surface of the workpiece becomes harder than the bulk material [7]. This increase in hardness is measured by the microhardness profile from surface to the bulk material. The diminishing effects of thermal and mechanical loads into the material cause less change in microhardness into the depth of the material, but the surface experiences heavy thermal and mechanical loading, causing the difference. It was shown on hardened AISI 52100 bearing steel that these changes in microhardness are related to the phase transformations as well as residual stress distributions [21]. In hard turning of nickel-based alloy IN-718, it was found that the surface hardness measurements are approximately 50% higher than bulk material hardness values at 0.2-mm deep into the material and above [7, 13, 22], as depicted in Figure 13.4.

13.4.4 Surface Roughness

Surface roughness is one of the easiest and most common ways to quantify and compare surface integrity of a machined workpiece after machining. The dimensional tolerances and accuracy can be achieved by minimizing the roughness of the surface. The traditional parameters

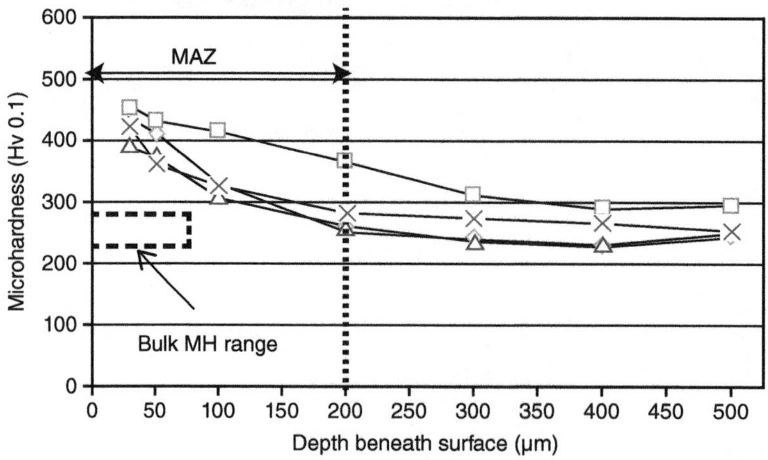

Figure 13.4 Example of microhardness profile and machining-affected zone.

that are known to increase the roughness of the machined surface are feed rate and tool nose radius, but rapid tool wear has also been identified as a major effect on surface roughness during hard machining [13]. Solid tools are usually considered to provide better surface quality compared to indexable inserts because of the higher rigidity they possess.

Since the thermal loading during machining is considered to have a considerable role in tool wear, it also affects surface roughness adversely. Hard machining creates higher cutting temperatures due to the nature of the process, which is a significant factor in the high surface roughness values observed if the cutting parameters are not adjusted for hard machining. The accumulation of built-up edge on the tool-chip-workpiece contact point also drifts the tool away from the cutting route, which contributes to the surface roughness [23].

13.4.5 Residual Stresses

During machining processes, the workpiece experiences heavy thermal and mechanical loading [13]. Especially in hard machining, these loads cause plastic deformation on the workpiece, which leads to irreversible deformation. The remaining stress and strain on the workpiece after the thermal and mechanical loads are removed and are called residual stress and strains [7]. These residual stresses are usually tensile (positive) on the surface due to phase transformations within the material, particularly when the machining temperatures are higher than certain phase-transformation temperatures [13]. The residual stress profile becomes negative after a short distance through the material, where the thermal effects are lower than that of surface. After making a compressive peak at approximately 50–100 μm deep into the material, the profile settles to a near-zero value at approximately 200 μm depth. Here, the effects of thermal and mechanical loads are assumed to have vanished or minimized because of the distance (see Figure 13.5).

Residual stresses are considered to be the primary reason for crack initiation and propagation and resultant fatigue failure. As a result, postprocessing is utilized to remove or alter the portion with tensile residual stresses. It is known that with increasing tool wear, residual stresses at the surface tend to become more tensile, which elevates the need for more frequent tool change [7]. It was also reported that with increasing workpiece hardness, residual stresses may tend to shift from tensile to compressive, while the magnitude and depth of the compressive peak residual stress may increase [25].

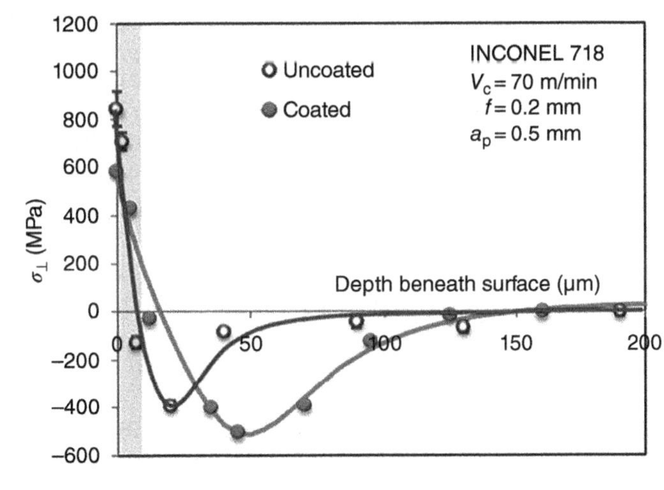

Figure 13.5 Typical residual stress profile. *Source:* Outeiro et al. 2008 [24]. Reproduced with permission of Elsevier.

13.5 Summary and Conclusions

Hard machining operations such as are turning, milling, broaching, and boring operations are typically final manufacturing operations executed on the parts and components, and desired dimensional and surface quality is usually high demanding robust hard machining parameters and conditions. The choice of tool material and geometry is very critical for successful hard machining. As tool materials, sintered carbide tools, ceramics, PCBN, and PCDs are often utilized. Coating of TiN, TiCN, TiAlN, and TiAlSiN is often applied to sintered carbide tools in order to improve the performance of cutting tool and machinability. Edge preparation and custom microgeometry fabrication has been applied to cutting tools to improve the edge strength, tool life, and finished surface quality. Finished surface quality including surface finish, microstructural alterations, subsurface residual stress formation, and surface microhardness is highly affected by tool material and coating, tool microgeometry, and hard machining conditions.

References

1 Chen, L., Elwardany, T.I., Nasr, M., and Elbestawi, M.A. (2006). Effects of edge preparation and feed when hard turning a hot work die steel with polycrystalline cubic boron nitride tools. *Annals of the CIRP* 55 (1): 88–92.

2 Klocke, F., Brinksmeier, E., and Weinert, K. (2005). Capability profile of hard cutting and grinding processes. *Annals of the CIRP* 54 (2): 557–580.

3 Dewes, R.C. and Aspinwall, D.K. (1997). A review of ultra high speed milling of hardened steels. *Journal of Materials Processing Technology* 69: 1–17.

4 Astakov, V.P. (2011). Chapter 1: Definitions and industrial applications. In: *Machining of Hard Materials* (ed. J.P. Davim), 1–31. Springer.

5 Coelho, R.T., Ng, E., and Elbestawi, M.A. (2007). Tool wear when turning hardened AISI 4340 with coated PcBN tools using finishing cutting conditions. *International Journal of Machine Tools and Manufacture* 47: 263–272.

6 Yen, Y.C., Jain, A., and Altan, T. (2004). A finite element analysis of orthogonal machining using different tool edge geometry. *Journal of Materials Processing Technology* 146: 72–81.

7 Ulutan, D. and Özel, T. (2011). Machining induced surface integrity in titanium and nickel alloys: a review. *International Journal of Machine Tools and Manufacture* 51 (3): 250–280.

8 Karpat, Y. and Özel, T. (2008). Mechanics of high speed machining with curvilinear tools. *International Journal of Machine Tools and Manufacture* 49: 195–208.

9 Lopez de Lacalle, L.N., Lamikiz, A., Fernandez de Larrinoa, J., and Azkona, I. (2011). Chapter 2: Advanced cutting tools. In: *Machining of Hard Materials* (ed. J.P. Davim), 33–86. Springer.

10 Harris, T.K., Brookes, E.J., and Taylor, C.J. (2004). The effect of temperature on the hardness of polycrystalline cubic boron nitride cutting tool materials. *International Journal of Refractory Metals & Hard Materials* 22: 105–110.

11 Klocke, F. and Kratz, H. (2005). Advanced tool edge geometry for high precision hard turning. *Annals of the CIRP* 54 (1): 47–50.

12 Özel, T., Hsu, T.-K., and Zeren, E. (2005). Effects of cutting edge geometry, workpiece hardness, feed rate and cutting speed on surface roughness and forces in finish turning of hardened AISI H13 steel. *International Journal of Advanced Manufacturing Technology* 25: 262–269.

13 Abrao, A.M., Riberio, J.L.S., and Davim, J.P. (2011). Chapter 4: Surface integrity. In: *Machining of Hard Materials* (ed. J.P. Davim), 115–141. Springer.

14 Che-Haron, C.H. (2001). Tool life and surface integrity in turning titanium alloy. *Journal of Materials Processing Technology* 118: 231–237.

15 Che-Haron, C.H. and Jawaid, A. (2005). The effect of machining on surface integrity of titanium alloy Ti-6%Al-4%V. *Journal of Materials Processing Technology* 166: 188–192.

16 Sharman, A.R.C., Hughes, J.I., and Ridgway, K. (2008). Surface integrity and tool life when turning Inconel 718 using ultra-high pressure and flood coolant systems. *Proceedings of Institute of Mechanical Engineering Part B: Journal of Engineering Manufacturing* 222: 653–664.

17 Sharman, A.R.C., Amarasinghe, A., and Ridgway, K. (2008). Tool life and surface integrity aspects when drilling and hole making in Inconel 718. *Journal of Materials Processing Technology* 200: 424–432.

18 Matsumoto, Y., Hashimoto, F., and Lahoti, G. (1999). Surface integrity generated by precision hard turning. *Annals of the CIRP* 48 (1): 59–62.

19 Thiele, J.D. and Melkote, S.N. (2000). Effect of tool edge geometry on workpiece subsurface deformation and through-thickness residual stresses for hard turning of AISI 52100 steel. *Journal of Manufacturing Processes* 2 (4): 270–276.

20 Smith, S., Melkote, S.N., Lara-Curzio, E. et al. (2007). Effect of surface integrity of hard turned AISI 52100 steel on fatigue performance. *Material Science and Engineering A* 459: 337–346.

21 Warren, A.W., Guo, Y.B., and Weaver, M.L. (2006). The influence of machining induced residual stress and phase transformation on the measurement of subsurface mechanical behavior using nanoindentation. *Surface Coating Technologies* 200: 3459–3467.

22 Pawade, R.S., Joshi, S.S., and Brahmankar, P.K. (2008). Effect of machining parameters and cutting edge geometry on surface integrity of high-speed turned IN-718. *International Journal of Machine Tools and Manufacture* 48: 15–28.

23 Mantle, A.L. and Aspinwall, D.K. (2001). Surface integrity of a high speed milled gamma titanium aluminide. *Journal of Materials Processing Technology* 118: 143–150.

24 Outeiro, J.C., Pina, J.C., M'Saoubi, R. et al. (2008). Analysis of residual stresses induced by dry turning of difficult-to-machine materials. *CIRP Annals – Manufacturing Technology* 57: 77–80.

25 Matsumoto, Y., Barash, M.M., and Liu, C.R. (1986). Effect of hardness on the surface integrity of AISI 4340 steel. *Journal of Engineering for Industry* 108: 169–175.

14

Advances in Material Modeling for Manufacturing Analysis and Simulation (Deformation and Cutting Processes)

Elisabetta Ceretti[1], Claudio Giardini[2], and Antonio Fiorentino[1]

[1] Department of Mechanical and Industrial Engineering, University of Brescia, Via Branze 38, 25123, Brescia, Italy
[2] Department of Engineering, University of Bergamo, Via Pasubio 7b, 24044, Dalmine (Bg), Italy

14.1 Introduction on Material Characterization and Modeling

A model is not the exact representation of the real world, but it is a tool to better understand the real systems according to some simplifications introduced for conducting several studies in a virtual environment so to reduce the costs with respect to carry out experiments. The field of modeling operations is extremely wide, and the application of these models covers a lot of areas.

In order to understand if a model has been developed in a correct way, it is necessary to compare the effective results with the expected ones: if the two results are similar, the model is considered as a good model; on the contrary, it is necessary to refine it or to build another one.

There are many different reasons to use a model in the cutting and forming operations fields, such as the following:

- processes design
- processes optimization
- processes control
- processes simulation
- equipment design

In this chapter, how the features to build a good simulative and analytical model will be described.

In order to realize a reliable simulative and/or analytical model for cutting and forming environments, it is important to know some parameters since they strongly influence the final results. The most significant parameters are as follows:

- Material characterization (known also as flow stress law)
- Friction model and value
- Failure criteria of the material

Modern Manufacturing Processes, First Edition. Edited by Muammer Koç and Tuğrul Özel.

14.2 Material Models and Applications

The most important variables affecting the material flow during cutting and deformation processes are the following:

- *Strain*, defined as the variation in shape or size of an object due to an applied force or a change in temperature with respect to the initial ones;
- *Stress*, defined as the internal forces acting within a deformable body;
- *Strain rate*, defined as the rate of change in strain with respect to time; and
- *Temperature*, defined as the local temperature of the material during the process.

All these variables (or some of them) are linked together according to the type of model used. The simplest one correlates only the stress and the strain (typical of elastic problem). The most complete one correlates the stress and the strain as the strain rate and the temperature change (typical of plastic-deformation process carried out at high temperature).

As a consequence, the most general flow stress law can be written as a function of strain ε, strain rate $\dot{\varepsilon}$, and temperature T, as reported in (14.1):

$$\sigma = f(\varepsilon, \dot{\varepsilon}, T) \tag{14.1}$$

As it can be easily seen in Figure 14.1, at room temperature, the flow stress depends mainly on strain and subsequently on strain rate. If the temperature grows up to higher value (e.g. above the recrystallization temperature), the stress decreases while the effect of temperature and strain rate increases.

After an analysis of literature, attention has been focused on the following material models:

- Johnson–Cook
- Oxley
- Maekawa
- El-Magd

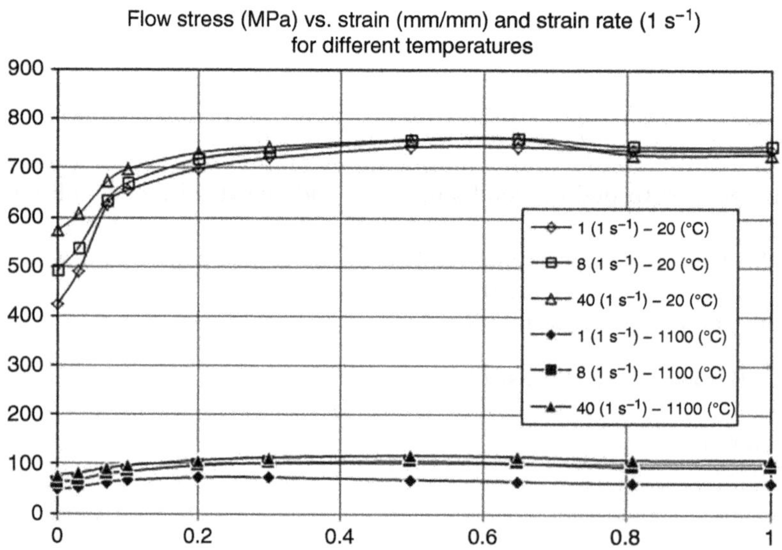

Figure 14.1 Flow stress distributions for different strain, strain rate, and temperature for a 15CrNi6 steel.

which are briefly explained in the following sections.

14.2.1 Johnson–Cook Model

The Johnson–Cook's model [1] is normally known in its general expression as follows (14.2):

$$\sigma=\left(A+B\cdot\varepsilon^{n}\right)\cdot\left(1+C\cdot\ln\left(\dot{\varepsilon}/\dot{\varepsilon}_{0}\right)\right)\cdot\left(1-\left(\frac{T-T_{\mathrm{r}}}{T_{\mathrm{m}}-T_{\mathrm{r}}}\right)^{m}\right) \tag{14.2}$$

where

σ is the true stress
ε is the true strain
$\dot{\varepsilon}$ is the strain rate
$\dot{\varepsilon}_0$ is the reference strain rate in quasistatic test used to determine A, B, and n
T is the temperature of the material
T_{r} is the reference temperature
T_{m} is the melting temperature of the material
A, B, and n are respectively yield stress and hardening parameters in quasistatic condition
C and m are numerical coefficients depending on the material that can be identified by means of experimental tests.

14.2.2 Oxley Model

Oxley defined a model [2] based on the Eq. (14.3):

$$\sigma=\sigma_1\cdot\varepsilon^{n} \tag{14.3}$$

where

σ is the true stress
σ_1 is the flow stress at a strain ε = 1.0.

it is evident how this model does not take into account the effect of temperature and strain rate. Therefore, it is a simple model able to represent process at constant temperature where the influence of the strain rate can be neglected. This model can be used for cold-forming processes where temperature and strain rate are not affecting the material behavior.

14.2.3 Maekawa Model

Another important model is represented by the Maekawa's one [3] (14.4):

$$\sigma=A\cdot\left(\frac{\dot{\varepsilon}}{1000}\right)^{M}\cdot\varepsilon^{N} \tag{14.4}$$

where:

A, M, and N are material constants
$\dot{\varepsilon}$ is the strain rate

14.2.4 El-Magd Model

Finally, there is the El-Magd's model [4] (14.5), where the base equation is

$$\sigma = \left[K \cdot (B + \dot{\varepsilon})^n + \eta \cdot \dot{\varepsilon}\right] \cdot e^{\left[-\beta \cdot \left(\frac{T - T_0}{T_{melt}}\right)\right]} \tag{14.5}$$

where:

σ is the true stress,
ε is the strain,
$\dot{\varepsilon}$ is the strain rate,
T is the temperature of the material,
T_0 is the reference temperature,
T_{melt} is the melting temperature of the material, and
K, B, n, η, and β are coefficients depending on the material.

The choice between the different material models can be done as a function of the working technology, the type of material, and the process parameter (temperature, strain, and strain rate). As an example, when considering orthogonal cutting of AISI 1045 steel, it is possible to define that at 500 °C, Maekawa et al. model is the best for prediction flow stress using an error of only around 7%, followed by Johnson–Cook and Oxley models where the deviations are 13% and 15% values, respectively (Table 14.1).

In case of higher temperatures, (800 °C as a reference), for higher strains of 0.5, the minor variation of 0.5% is presented by the Oxley model. Using this model, it can be possible to underline the trend of increasing accuracy for higher strains.

If the temperature is higher than 800 °C, similar to 1000 °C, the best model to use is Johnson–Cook's one, which presents an error of 5%, followed by Oxley model (15% error) and Maekawa et al. and El-Magd et al. models (21% error in both cases). It is important to underline that Johnson–Cook model gives an increasing accuracy trend followed by a deterioration at the higher strain.

So, thanks to this analysis, it appears that (Table 14.2) Oxley and Johnson–Cook models have the potential to be the representative of the high temperature deformation and the high cutting

Table 14.1 Comparison of the accuracy of different material models.

Temperature	Johnson–Cook	Oxley	Maekawa	El-Magd
500 °C (AISI 1045 steel, orthogonal cut)	13%	13%	**7%**	n.a.
800 °C ($\varepsilon > 0.5$)	n.a.	**0.5%**	n.a.	n.a.
1000 °C	**5%**	15%	21%	21%

Table 14.2 Preferred material models for different processes.

Process	Johnson–Cook	Oxley	Maekawa	El-Magd
High cutting speed	x	x		
Low cutting speed			x	
Cold forming		x		
Hot forming	x			

speed (corresponding to high temperature), while Maekawa model is representative for a lower cutting speed (lower temperature).

Regarding the forging operation, the process appears to be less critical than the cutting one because the achieved temperatures are more uniform within the material, and the strain rates are lower than the respective values typical of cutting.

In particular, in forming, we can divide between hot and cold processes. Johnson–Cook is preferable for hot forging processes such as extrusion, hot stamping, or hot drawing where the main effect on stresses are caused by the increase of temperature and strain rate. In cold-forming processes such as deep drawing, stamping, etc. the Oxley model represents a good solution for material modeling.

14.3 Failure Models

14.3.1 Prediction of Ductile Fracture

One of the basic problems dealing with ductile deformation is to predict the occurrence of the ductile fracture for a given material, at different stress and strain states and history, and for different strain rates, temperature and friction according to the different working conditions.

What is needed is a general criterion related with deformation in multiaxial stress system, which is able to rationalize all the data available about material properties and process parameters.

Studies of ductile fracture can be divided into three categories:

- *Experimental works*, investigating the forming limits, fracture sites, and types of cracks (internal cracks caused by severe and nonhomogeneous deformations and external cracks mainly due to high friction);
- *Theoretical modeling* of ductile fracture criteria;
- *Numerical analysis*, such as upper bound solution, slip line method, or finite element analysis.

Many experimental studies have been conducted to establish relevant testing methods to determine formability and fracture limit diagrams, and several ductile fracture criteria have been suggested based on experimental observations and results [5–10].

Some of the fracture criteria quoted previously seem to be good for predicting surface cracks, while others seem to be good for predicting internal cracks. Knowing the validity of the ductile fracture criteria is crucial when predicting fracture, especially when the choice of the criterion affects the results of the simulation model.

In order to find the best solution for the application of fracture criteria, two conditions must be satisfied:

- the validity of the criterion for different types of cracks and
- the possibility of its implementation in commercial FE codes.

After a detailed analysis of the literature available on this subject, attention has been focused on the following criteria:

- Effective strain [11, 12],
- Cockroft and Latham [13], and
- McClintock [8].

In particular, Cockroft and Latham and McClintock criteria seemed to be reliable in the study of damage because they deal with both stresses and strains for the calculation of the accumulated damage.

14.3.2 Ductile Fracture Criteria

During the years, different models have been developed for representing this phenomenon. The most important damage criteria will be briefly explained in the following.

14.3.2.1 Effective Strain

The effective plastic strain is not considered a real damage criterion; however, many authors have used this value in predicting when the fracture occurs [11, 12]. In fact, because of the simplicity of calculation, the effective plastic strain can be easily applied to the studies of plastic-deformation processes with finite element codes.

The formula for calculating the effective strain is (14.6):

$$\dot{\bar{\varepsilon}} = \sqrt{\frac{2}{3}\left(\dot{\varepsilon}_1^2 + \dot{\varepsilon}_2^2 + \dot{\varepsilon}_3^2\right)} \tag{14.6}$$

where

$\dot{\varepsilon}_1$, $\dot{\varepsilon}_2$, and $\dot{\varepsilon}_3$ are the strains along the principal directions of deformations.

The value of the accumulated effective strain can be evaluated as the integral of the effective strain along all the process duration as follows:

$$\varepsilon_{\text{fracture}} = \int_{t_0}^{t_1} \dot{\bar{\varepsilon}} \cdot \mathrm{d}t \tag{14.7}$$

where

$\varepsilon_{\text{fracture}}$ is the strain prior to fracture.

14.3.2.2 Cockroft and Latham

Cockroft and Latham [13] suggested that fracture occurs when the cumulative energy due to maximum tensile stress exceeds a certain value, defined as in Eq. (14.8). Oh et al. [9] modified the criterion in a nondimensional form obtaining the expression reported in Eq. (14.9):

$$C_{3a} = \int_0^{\varepsilon_f} \sigma^* \, \mathrm{d}\bar{\varepsilon} \tag{14.8}$$

$$C_{3b} = \int_0^{\varepsilon_f} \left(\frac{\sigma^*}{\bar{\sigma}}\right) \cdot \mathrm{d}\bar{\varepsilon} \tag{14.9}$$

where

ε_f is the effective fracture strain,
$\bar{\varepsilon}$ is the effective strain,
$\bar{\sigma}$ is the effective stress,
$\overset{\bullet}{\sigma}$ is the maximum tensile stress, and
C_{3a} and C_{3b} are constants depending on material and must be determined by experiment.

Oh et al. applied the Cockroft and Latham's criterion to free surface cracks in simple upsetting as well as internal cracks in the tensile test, extrusion, and drawing, in conjunction with rigid-plastic FE simulations [9]. They claimed that the Crockroft and Latham's criterion reasonably agrees with the experimental results of Bridgeman [14].

Yoshida and Wanheim conducted block upsetting tests using aluminum 6061-T6 alloy and concluded that the Crockroft and Latham's criterion reasonably predicts both surface fracture site and the fracture strain [15]. Cockroft and Latham's criterion has been applied in many other experimental studies for various cold-forging processes such as internal defects in opposed extrusion [16], cracks in plane strain bar forging [17], free surface cracks in the collar test [18], and internal cracks in tensile tests with notched specimens. In these studies, the validity of the Cockroft and Latham's criterion is acceptable.

14.3.2.3 McClintock

McClintock et al. proposed a mathematical formulation assuming that fracture occurs because of the presence of voids in the material that can be considered as long cylinders [8]. The cylindrical voids are assumed to grow in the transverse direction, and fracture occurs when the voids coalesce (Figure 14.2).

Oh et al. modified McClintock's criterion by assuming that smaller voids than the primary voids are distributed throughout the slip band, between the primary voids, and that the smaller voids are touching together and connecting the primary voids at fracture [9] as in Eq. (14.10):

$$C=\int_0^{\varepsilon_f}\left[\frac{2}{\sqrt{3(1-n)}}\sinh\left(\frac{\sqrt{3(1-n)}}{2}\cdot\frac{\sigma_a+\sigma_b}{\bar{\sigma}}\right)+\frac{\sigma_b-\sigma_a}{\bar{\sigma}}\right]\cdot d\bar{\varepsilon} \tag{14.10}$$

where

C is a material constant to be determined by experiment,
n is the work-hardening constant (assuming that flow stress curved is represented by $\sigma = K\varepsilon^n$,
ε_f is the effective fracture strain,
$\bar{\varepsilon}$ is the effective strain,
$\bar{\sigma}$ is the effective stress, and
σ_a and σ_b are the maximum and minimum principal stresses.

Oh et al. [9] showed that the modified McClintock's criterion almost coincides with Cockroft and Latham's criterion under various hydrostatic stress when work hardening is neglected

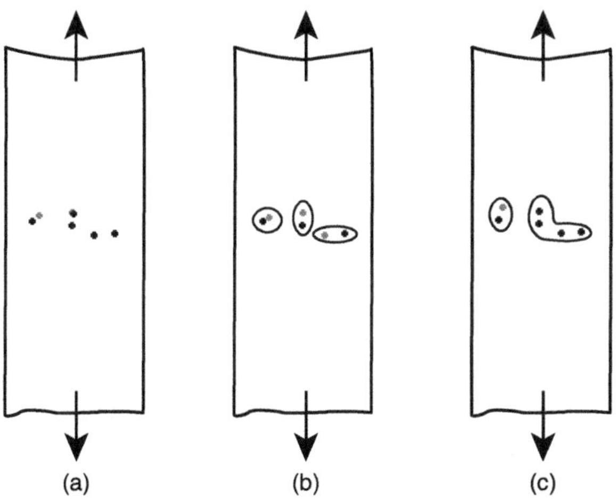

Figure 14.2 Stage of the ductile fracture: (a) void initiation, (b) void growth, and (c) holes coalescence.

($n = 0$). They claimed that McClintock's criteria, similar Cockroft and Latham's criterion, reasonably agrees with Bridgeman's data in tensile tests [14], but the validity of this criterion is not conclusive for predicting chevron cracks in extrusion and drawing.

Dung and Mahrenholtz [18] attempted to compare various criteria proposed by Cockroft and Latham [13], Oyane [10], and McClintock based on the results of experimental work and rigid-plastic FE simulations. They reported that McClintock's criterion most successfully predicted the location of the free surface fracture as well as fracture strain in the collar test, the punch test, and the upsetting test, while the other models showed good agreement only in the upsetting test.

In cutting operation, the most used criterion is the Cockroft and Latham one based on the cumulative energy due to maximum tensile stress. Another simple criterion is based on the reaching of the shear strength value for the cut material in the shearing zone of the chip.

14.3.2.4 Other Criteria

Gurson [19] suggested a yield criterion and flow rules for porous ductile media, based on the continuum theory of ductile rupture by void nucleation and growth [19]. Gelin and coworkers [6] modified Gurson's model and applied the model to various forging processes such as simple upsetting, forward extrusion, wire drawing, color test, tapered disk compression, and the plain strain tensile test in conjunction with elastic-plastic FE simulation. The results of FE simulations showed higher values of void volume fraction at the locations where cracks are prone to occur. The studies qualitatively concluded that the modified Gurson's model can predict ductile fracture in various forging processes. However, no quantitative verification of these models was found in literature, possibly because of difficulties in practical application.

Dawson et al. [20] proposed another void growth model, which calculated void volume fraction. They implemented the model using elastic-plastic FE formulation. They compared the results of FE simulations with experimental results of the drawing operation, from literature, and claimed good agreement in the location of the fracture site.

Although the concept of void volume fraction may be capable of predicting occurrence of ductile fracture, as well as its site, practical use of the models is restricted because of the difficulties in experimental verification and the requirement of separate FE formulations other than commercial FE packages.

Vujovic and Shabaik suggested a stress formability index, $\beta = 3\sigma_m/\sigma$ (where σ_m is the mean stress and σ is the effective stress) and proposed that equivalent fracture strain (ε_f) is a function of β, which can be represented as a forming limit curve for bulk and sheet metal working processes [21]. The forming limit curves can be determined by several standard tests such as the tensile test, the simple upsetting test, the torsion test, thin wall tube under internal pressure, and the profile rolling test for given material. Vujovic and Shabaik reported forming limit curves for 0.15%, 0.25%, and 0.35% steels [21]. Bolt [22] obtained similar forming limit curves (σ_m/σ vs. ε_f) for low and medium carbon steel, copper, brass, and aluminum from the notch tensile test and the torsion test. Bolt also simulated the tensile test using an elastoplastic FE code and claimed the validity of the curves. Drawbacks of using forming limit curves in cold forging would be the need of much experimental work to obtain the curve for a given material and the difficulties in implementing commercial FE packages.

Lamaitre derived a model of isotropic ductile plastic damage based on a continuum damage variable, by assuming that both tensile and compressive stresses induce the same mechanical effects in void and micro cracks [7]. Lemaitre claimed that the criterion agreed with McClintock's and Rice and Tracey criteria. Bolt applied Lemaite's criterion to the notch tensile test, but the validity of the criterion was not conclusive [22]. However, the assumptions made in this model limit the application within small coessive stress conditions.

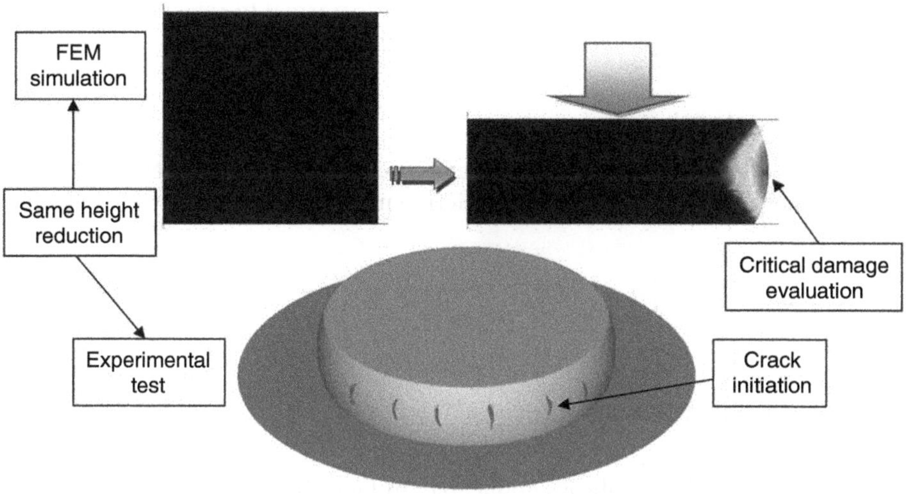

Figure 14.3 Critical damage value evaluation using FE analysis.

14.3.3 Critical Damage Value

The common hypothesis of the ductile fracture criteria, previously described, is that ductile fracture occurs when the maximum damage value in a workpiece exceeds a critical value or "critical damage value" [7, 17].

The critical damage value can be mainly considered as a material constant similar to the yield stress or the tensile strength even if it is affected by more process parameters (velocity of deformation, temperature, etc.).

Since different ductile criteria lead to different damage values, critical damage values corresponding to different damage criteria are different for a given material. Oh, Chen, and Kobayashi in 1979 have found a relation between the critical damage value determined by Cockroft and McClintock in uniaxial tensile test [9]. The critical damage value found using McClintock is twice that found using Cockroft ($K=2C$).

Critical damage value is difficult to be determined. Its value can be evaluated by experiments in the case of simple forming processes, using tensile or compression tests, but, for all the cutting processes, the identification of the critical damage value is a critical issue [9, 23]. In fact, the results and the prediction of finite element simulations are deeply affected by the value chosen for this parameter.

A typical coupled simulative-experimental test is carried out to evaluate the critical damage value corresponding to the compression test of a cylindrical body until cracks occur on the external surface of the cylinder. By simulating this process and calculating the damage value according to the different criteria, it is possible to know for which value of the damage parameter the breakages take place (Figure 14.3).

14.4 Modeling of Contact, Friction, and Wear

Friction can be defined as the force resisting the relative motion of material surfaces sliding against each other. In manufacturing processes, the knowledge of the friction conditions between a workpiece and a tool or die is very important for determining the material flow and the part feasibility.

In order to quantitatively express the interface friction, it is normal to refer to two models: the Coulomb model and the Schley model (or Shear model or constant friction model).

14.4.1 Coulomb Friction Model

The Coulomb friction model is the most well-known and easiest friction model to be implemented; the equation that represents the Coulomb friction model is expressed by the following Eq. (14.11):

$$\tau = \sigma_{\mathrm{n}} \cdot \mu \tag{14.11}$$

where

τ is the frictional shear stress,
σ_n is the normal stress at the interface, and
μ is the coefficient of friction.

The Coulomb friction model is not always applicable. In fact, the linear relationship is not valid for high normal stress levels because the value of shear stress τ cannot exceed the shear strength value of the material, defined as $k = \sigma_0 / \sqrt{3}$ (see Figure 14.4). In this situation, the Schley friction law can be used instead Eq. (14.13).

From this, a simple model can be derived taking into account in a very straightforward way the existence of the two regions observed on the rake face in cutting, namely the sliding and the sticking one. Thus, a constant coefficient of friction, according to the Coulomb model, is utilized in the sliding region, while a constant frictional stress, equal to the shear flow stress of the chip material, is applied in the sticking region. Typically, Coulomb law is suitable for friction in cold forming (i.e. sliding sheet drawing), while Schley models the condition of soft materials that undergo severe normal stresses (i.e. sticking in hot forging or aluminum stamping). In cutting, where the normal stress on tool rake face decreases with the distance from the tool tip, there can be a transition from Schley to Coulomb friction behaviors (Figure 14.5). Such a model can be expressed by means of the following mathematical formulation (14.12):

$$\begin{aligned} \tau(\sigma_{\mathrm{n}}) &= \mu\sigma_{\mathrm{n}} \quad \text{when} \quad \tau < \tau_{\mathrm{f}} \\ \tau &= \tau_{\mathrm{f}} \quad \text{when} \quad \tau \geq \tau_{\mathrm{f}} \end{aligned} \tag{14.12}$$

14.4.2 Schley Friction Model or Shear Model

The Schley friction model is described by the following Eq. (14.13):

$$\tau = \frac{m}{\sqrt{3}}\sigma_0 = m \cdot k \tag{14.13}$$

where

τ is the frictional shear stress,
m is the shear or friction factor,
($m = 0$ represents a frictionless interface, while $m = 1$ represents sticking conditions)
σ_0 is the yield strength of the deforming material, and
k is the shear strength of the material.

Another expression for modeling the particular friction conditions existing between chip and tool (which is a very important situation for determining the chip formation) is reported in

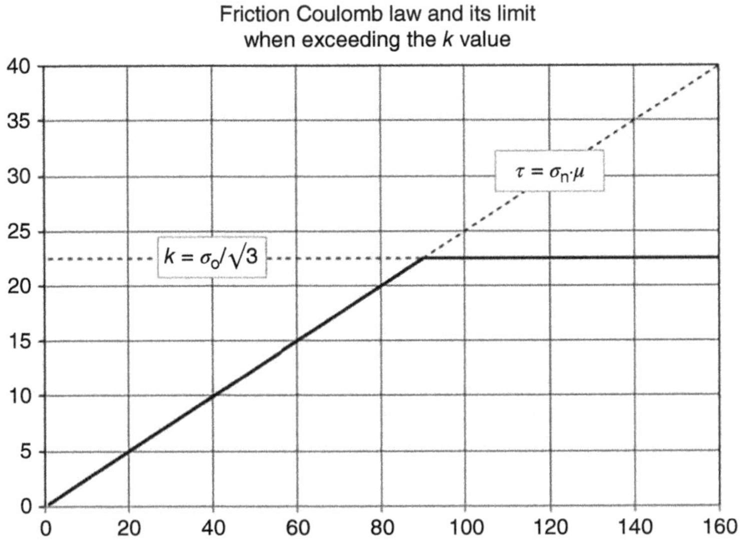

Figure 14.4 Coulomb law limited by the shear strength value.

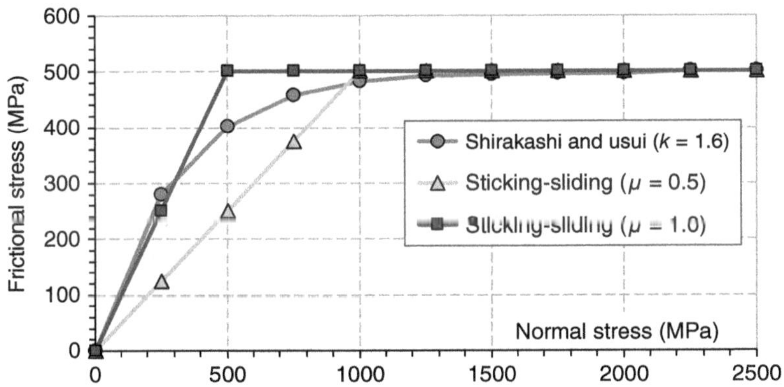

Figure 14.5 Frictional stress vs. normal stress distributions obtained utilizing the Shirakashi and Usui model and the sticking-sliding model and assuming $\tau_f = 500$ MPa [24].

the following formula where nonlinear friction characteristic is taken into account. This model was proposed by Shirakashi and Usui [25], which relates the frictional stress τ to the normal stress σ at the chip–tool interface:

$$\tau = \tau_f \cdot \left[1 - e^{-\frac{k \cdot \sigma}{\tau_f}} \right] \tag{14.14}$$

where

τ and σ are the frictional and the normal stress on the tool,
τ_f is the shear flow stress of the chip material, and
k is a material constant.

14.4.3 Parameters Influencing Friction

The friction conditions in forging or cutting processes are influenced by a wide number of factors, namely

1) Process parameters,
2) Workpiece and tool materials,
3) Workpiece and tool surface finishing, and
4) Lubricant.

14.4.3.1 Process Parameters

The following parameters are classified as process parameters:

1) Contact normal stresses (i.e. normal pressure acting between the tool and the workpiece),
2) Sliding velocity of the workpiece material on the tool surface,
3) Contact area, and
4) Temperature in the contact area.

In general, when these parameters increase, the friction between the tool and the workpiece interface increases. The temperature influences the viscosity of the lubricant.

14.4.3.2 Workpiece and Tool Materials

Regarding the workpiece and the tool materials, the most influencing parameters on friction are the following:

1) Mechanical properties of the two materials,
2) Geometrical features, and
3) Surface topology and finishing.

Depending on the hardness of the tool/die and workpiece materials, the workpiece material behavior at the contact interface may vary. This affects the way friction is imposed. A good surface finish may cause welding of asperities and lead to higher friction and galling, which can be prevented by the presence of a lubricant layer. Welding of asperities may also take place if the tool and workpiece material are both made of the same material. Therefore, the die is normally realized using a harder material than the one of the workpiece. This can prevent the tool wear too.

14.4.3.3 Lubricants

The lubricant properties are influenced by the following parameters:

1) *Temperature.*
2) *Pressure*: The pressure at the tool–workpiece interface can reach very high values especially in cutting operations where the lubricant has to be able to form a limit lubrication layer in order to avoid direct contact between surfaces.
3) *Shear* strength of the involved material surfaces.
4) *Viscosity* that has to be chosen according to the sliding velocity.

14.4.4 Lubricant

Lubricant is one or more substances (liquid or solid) interposed between tool and workpiece in order to decrease their direct contact and so reducing both friction and wear. Lubricants have to simultaneously fulfill various requirements. The most important one are listed as follows [26]:

1) *Constant low friction*: The friction stresses should be low and uniform in order to guarantee a homogenous deformation.
2) *Separation of surfaces*: The formation of a thin layer of lubricant avoids the direct contact between tool and part surfaces so reducing friction and wear mechanisms. Moreover, the lubricant film on the tool must be durable in order to withstand continued or repeated contacts and separations with the workpiece surface.
3) *Thermal insulation*: The lubricant should insulate tool and part in hot and heat-generating processes so to reduce workpiece cooling and to protect the tools from excessive heating.
4) *Chemical inertness*: The lubricant must not chemically interact with die, tool, and workpiece surfaces and with the surrounding atmosphere during part process or storage. In fact, residues left on workpiece and tool should neither cause long-time changes in chemical or metallurgical properties (i.e. staining) nor interact during following heat treatment. Moreover, the lubricant has to protect against oxidation or gas absorption at high (furnaces or deformation induced) temperatures.
5) *Rapid response*: The lubricant must be effective in short times since the changes involved in many manufacturing processes occur in few millisecond time lapses. In other words, it has to rapidly respond to the process changes.
6) *Controlled stability*: The lubricant properties should remain stable. In particular, it has to withstand against repeated encounters, oxidation, bacteriological attack, and contamination, which are often present in industrial practice.

Moreover, the lubricant has to respect the following constrains:

1) *Storage, application, and removal*: Characteristics such as a long shelf life or ease of controlled application and removal are all favorable.
2) *Safe and disposable*: The lubricant has to withstand the standards in force regarding its environmental impact. At the end of its life, it should be completely reclaimed and treated for reuse. Moreover, the effluents resulting from the treatment must be environmentally acceptable.
3) *Safety*: Lubricants must not be toxic, carcinogenic, or cause irritation to skin. Moreover, they should be odorless.
4) *Cost*: This factor becomes an important factor when the lubrication costs (material and reconditioning for reuse and disposal) represent a significant portion of the product total cost.

14.4.5 Tool Wear

Wear can be defined as the progressive loss of material when two or more surfaces are in contact and in relative motion. The main cause of wear is friction, and the following effects are some of the wear consequences:

1) Alteration of the size of the pieces and
2) Degradation of surface characteristics.

As said for friction, the wear of cutting tools or dies depends on tool material and geometry, workpiece material, cutting or forming parameters (cutting speed, feed rate and depth of cut for cutting, pressure, sliding velocity for forming), lubricants, and machine-tool characteristics.

There are several wear's mechanisms; the most important are the following:

1) Abrasive wear,
2) Adhesive wear,

3) Diffusion wear, and
4) Chemical wear.

The following sections describe some models for each wear mechanism; only the chemical wear does not present a specific model.

14.4.5.1 Abrasive and Adhesive Tool Wear Model

Two separate models can be found in literature for describing the tool wear due to abrasion and to adhesion, but some authors proposed a model taking into account the adhesive and the abrasive wear together, so obtaining a single model to determine the total wear.

The following wear models have been suggested in the literature [27], for tool wear prediction. These models were found to give reasonable results.

$$Z_{\text{ab}} = \int K_1 \frac{p^{a_1} v_c^{b_1}\,dt}{H_{\text{d}}^{c_1}} \qquad Z_{\text{AD}} = \int K_2 \frac{p^{a_2} v_c^{b_2}\,\text{d}t}{H_{\text{w}}^{c_2}} \tag{14.15}$$

where

Z_{ab} and Z_{AD} are abrasive wear depth and adhesive wear depth,
K_1 and K_2 are the coefficients determined by experiments,
p is the local contact pressure orthogonal to workpiece and tool surfaces,
v is the local sliding velocity,
$\text{d}t$ is the incremental time interval,
H_{d} is the die (or tool) hardness,
H_{w} is the workpiece hardness, and
a_1, b_1, c_1, a_2, b_2, and c_2 are the experimental constants.

Estimation of wear using FE simulations requires the appropriate wear models be available in a discretized form. For wear prediction, the exponent values a_1, b_1, a_2, and b_2 are commonly taken to be equal to 1. Abrasive wear prediction commonly uses $c_1 = 2$ (for steel); adhesive wear prediction commonly uses $c_2 = 1$ [28]. When implementing the model (Eq. (14.15)), the exponents can be modified for fine tuning purposes. It is desirable to keep the constants a_1, b_1, a_2, b_2 quite close to unity. The constants c_1 and c_2 allow more flexibility but should remain in the range of 0.5–2.5. The coefficients K_1 and K_2 simply scale the magnitudes of the predicted wear profiles and are experimentally found by comparing the predicted wear profiles with the actual ones.

It is important to highlight that the abrasive wear model always predicts material removal from the tools, whereas the adhesive wear model always predicts material built-up (edge) on the tool.

In case of wear modeling in cutting process, a further abrasive wear model is available. This model was developed by Usui et al. [29] based on the following equation:

$$\frac{\partial W_{\text{a}}}{\partial t} = A \cdot p \cdot v_{\text{s}} \cdot \text{e}^{-\frac{B}{T}} \tag{14.16}$$

where

$\partial W_{\text{a}}/\partial t$ is the tool wear rate calculated according to the abrasive model,
p is the contact pressure at the interface between tool and chip (MPa),
v_{s} is the sliding velocity at the interface between tool and chip (mm/s),
T is the tool temperature at the interface between tool and chip (K), and
A and B are two constants to be found from the comparison with experiments.

By means of this expression, it is possible to forecast both flank and crater wear for the cutting tool.

Some observations by the application of the aforementioned model can be derived:

1) The abrasive tool wear rate increases with increasing cutting speed,
2) The tool wear and the length of the crater increase with increasing value of the friction, and
3) The maximum value of the tool wear and the length of the worn area increase with increasing tool rake angle.

14.4.5.2 Diffusion Tool Wear Model

Another important cause of tool wear is diffusion. According to this mechanism, atoms diffuse from the tool toward the chip and from the chip toward the tool and are removed by the flow of work material along the contact surfaces. Diffusion wear is characterized by a smooth worn surface with no plastic deformation. The material transferred toward the chip leads eventually to the formation of a crater on the tool rake face.

The equation describing the model of the wear diffusion is represented by the following expression [30] (14.17):

$$\frac{\partial W_d}{\partial t} = D(T) \cdot e^{-\frac{E}{RT}} \tag{14.17}$$

where

$\partial W_d / \partial t$ is the tool wear rate calculated according to the diffusive model;
E is the activation energy (7535 kJ/mol);
R is the gas constant (8314 kJ/mol K);
T is the tool temperature at the interface between tool and chip (K); and
$D(T)$ is defined according to ([31]) as a third order polynomial function of the temperature (T).

The two models representing abrasive (Usui) and diffusive (Takeyama) wear can be used together so representing different tool material behavior depending on the tool temperature. In fact, abrasive wear takes place under the so called "activation temperature," while both the wear mechanism are normally present above the activation temperature [32].

14.4.6 Applications of Material Modeling in 2D and 3D Simulations

In the following paragraph, some examples of the simulation of forming and cutting processes by means of the implicit Finite Element Code Deform 2D and 3D will be given.

The first set example refers to the study of material fracture in orthogonal cutting, blank shearing, and forward extrusion. In particular, a commercial FEM code DEFORM 2D has been modified, and fracture criteria have been implemented to study material separation [33]. In the FE model, ductile fracture is simulated deleting the mesh elements, and cracking starts when the element damage value reaches the "critical value," determined by experiments.

The second set refers to wear simulation in 2D stamping process and in 3D cutting tool simulations. In both cases, abrasive and diffusive wear models are used.

14.4.6.1 Orthogonal Cutting

The first simulated process is orthogonal cutting with discontinuous chip flow. In this case, the chip is segmented, and, as the tool advances, it breaks. The damage criterion utilized for fracture modeling was Cockroft and Latham [13]. The model of the simulation is plane strain and nonisothermal. In the simulation, the workpiece is assumed to be plastic and has a rectangular shape. The density of the mesh is not constant in order to have more elements in the area affected by the cutting process and to save computational time and disk memory. The tool is assumed to be rigid.

The simulation parameters are as follows (Figure 14.6):

Workpiece material: AISI 1045 steel (flow stress law as a function of strain rate and temperature [33]),
Workpiece dimensions: Height 5 mm, length 20 mm,
Initial tool and workpiece temperature: 20 °C,
Tool geometry: Rake angle −6° and 30°, clearance angle 6° and 10°, and tip radius 0.1 mm.
Tool speed: 15 200, and 300 m/min,
Depth of cut: 0.5 mm,
Lubrication: No lubricant, $\tau = m \cdot \tau_{\text{lim}}$ $(m = 0.5)$,
Damage criterion: Cockroft and Latham ($C_{cr} = 0.4$) [13],
Grooved tool: Land = 0.5 mm
Groove radius = 1 mm,
Groove width = 0.65 mm, and
Backwall width = 0.2 mm.

14.4.6.2 Blanking

The second process analyzed is blanking [34]. This is a separation process in which the workpiece is cut out from a sheet because of the movement of a punch. The produced part is stressed in shear until separation occurs. Figure 14.7 shows the plastic and shearing stages that are simulated (Deform2D environment) by continuous remeshing, and once the accumulated damage value reaches the critical value, fracture starts by element removal.

The simulation model is axial symmetrical, the sheet is modeled as plastic, and the punch and the die are rigid. The density of the mesh is very high in the area affected by the shearing phenomenon.

The simulation parameters are as follows [33]:

Sheet material: Low carbon steel (stress strain law $\sigma = K\,\varepsilon^n$, where $K = 123$ ksi and $n = 0.26$),

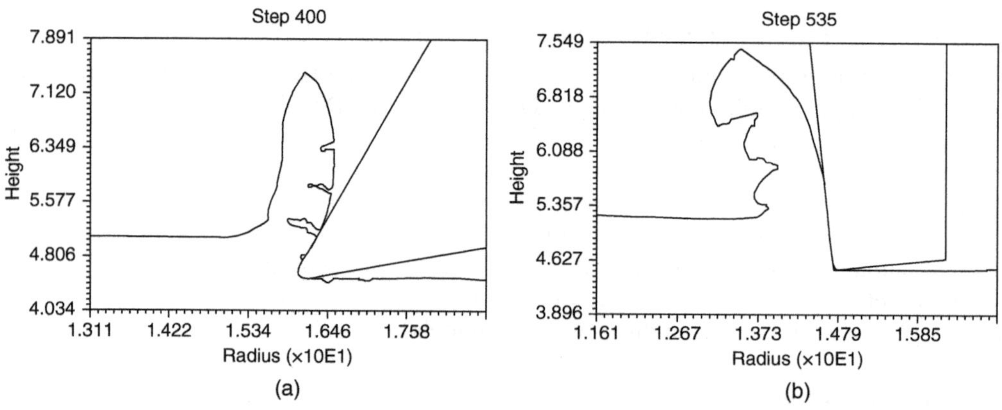

Figure 14.6 Orthogonal cutting simulations result for a tool rake angle of 30° (a) and −6° (b) and a cutting speed of 200 m/min after a cutting length of 6 mm. *Source:* Ceretti et al. 1997 [33]. Reproduced with permission of Elsevier.

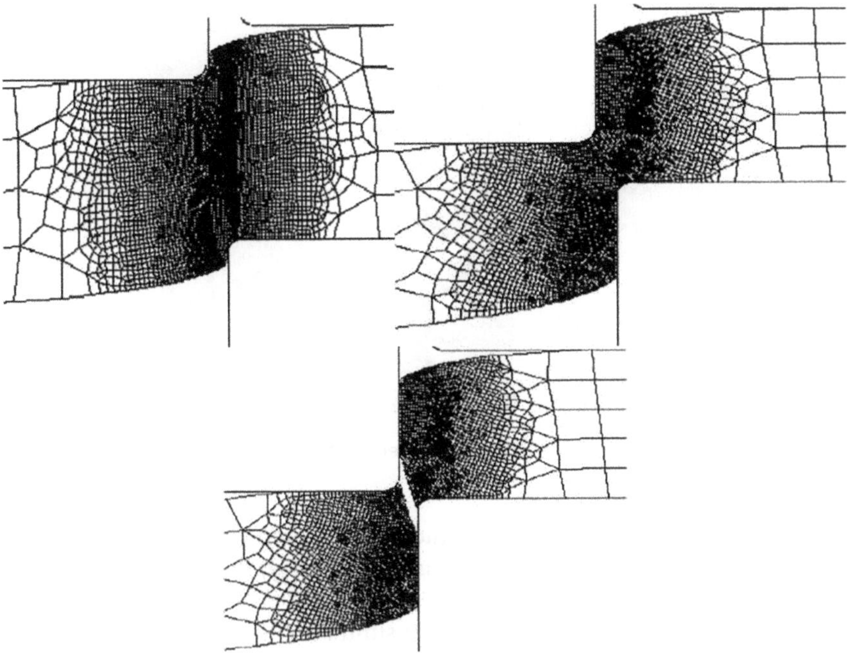

Figure 14.7 Simulated plastic, shearing, and fracture stages during the blanking process. *Source:* Ceretti et al. 1997 [33]. Reproduced with permission of Elsevier.

Sheet thickens: 0.031 in.,
Punch speed: 0.5 in./s,
Punch diameter: 1.492 in, die diameter 1.502 in. punch, and die radii 0.002 in.,
Lubrication: oil $\tau = m\ \tau_{\lim}$ ($m = 0.05$), and
Damage criterion: McClintock ($C_{cr} = 0.8$) [8].

14.4.6.3 Forward Extrusion

The third process investigated is forward extrusion of a cylindrical piece performed in three passes. The geometry of the die at each pass is reported in Table 14.3 [33].

Figure 14.8 shows the results obtained. In particular, there are no cracks to be seen at the end of the first and second passes, while several chevron cracks forms during the third pass. The Cockroft and Latham criterion has been used for internal fracture simulation with a critical damage equal to 0.422.

The results obtained with the simulations agree qualitatively in terms of growth and propagation of chevron defects with the results found in literature [35].

Table 14.3 Die geometry for each extrusion pass.

Pass ($N°$)	Φ Entrance (in.)	Φ Exit (in.)	Half die angle (°)
1	1.051	0.896	22
2	1.061	0.826	22
3	1.071	0.742	22

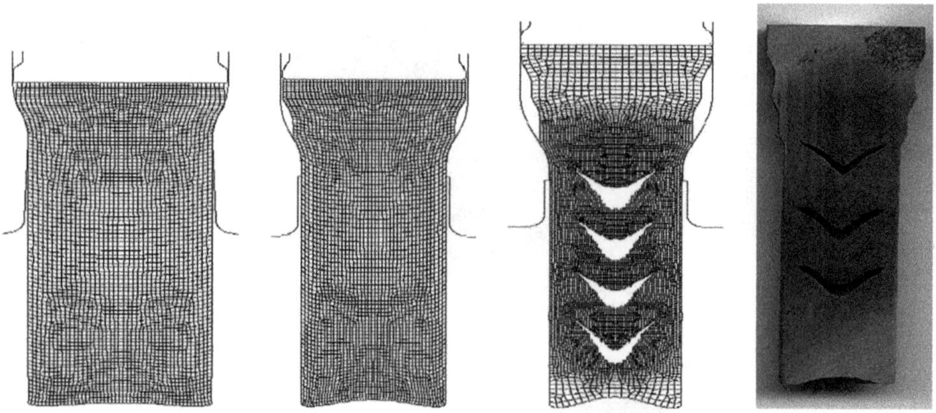

Figure 14.8 Simulation of first, second, and third pass of the produced part and comparison with experiments.

The last example related with modeling of material fracture is related with surface cracking defects in forward extrusion [36]. The worked material is a brass alloy; the ingot has 330 mm diameter, 2000 mm length, and 650 °C initial temperature.

The critical damage value was defined according to the simulation results and the experimental evidence and was set equal to $C_{cr} = 0.09$. As soon as one element reaches the critical damage value, it is deleted from the mesh.

The simulation output is presented in Figure 14.9. The surface cracking defect and the matching with experiments are evident.

14.4.6.4 Die Wear in 2D Stamping FEM Model

Die wear prediction is a fundamental task when designing a new tool or die. In fact, the choice of the appropriate die geometry (rake angles, fillet radii, etc.), die material, and lubrication conditions at the die workpiece interface affects the die wear, which results in bad final quality of the produced part and in die life reduction. The knowledge of the critical points for wear development will allow a better die design, changing the die geometry or thermal treating the

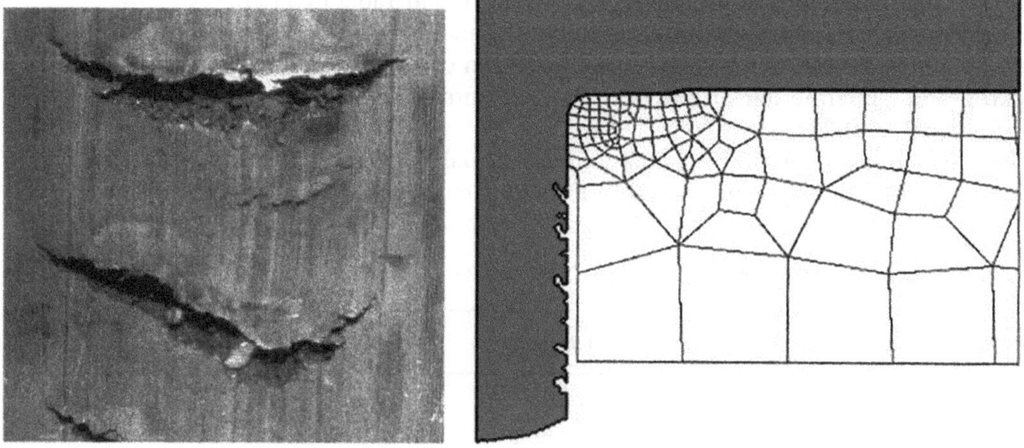

(a) (b)

Figure 14.9 Surface cracking defect: experiment (a) and simulation (b) comparison.

die surface or using harder material for all or part of the dies themselves, and the identification of the optimal lubrication conditions.

The model described in Giardini [37] was developed in Deform 2D and had shown a very good agreement with experimental observations.

The abrasive wear model [27] described in previous sections as

$$Z_{AB} = \int K_1 \frac{p^{a_1} \cdot v^{b_1} \cdot dt}{H_d^{c_1}} \tag{14.19}$$

was modified so to be implemented in the simulation code in the following way:

$$Z_{AB} = K_1 \cdot \sum_{j=1,n} \frac{p_j^{a_1} \cdot v_j^{b_1}}{\left(H_d^{c_1}\right)_j} \cdot \Delta t_j \tag{14.20}$$

for each mesh node of the simulated dies. In this expression, n is the total number of steps, Δt_j is the time interval, p_j is the pressure, and v_j the node sliding velocity. H_d is evaluated once the temperature of the node is known according to Figure 14.10.

The developed program (postprocessor of the FE analysis) is able to extract the worn tool profile from the simulation database and calculates, for each step and for each node of the two dies, the incremental Z_{AB} values from the beginning to the end of the process.

In particular, the considered process is represented in Figure 14.11, and it refers to the industrial hot forging of a flange between closed dies with flash formation. In Figure 14.12, the most worn areas, as the experimental evidence demonstrated, are reported.

The process was simulated, and the results obtained in terms of wear forecast were compared with the actual die profiles showing a very good agreement between them (Figure 14.12).

Since the model was correctly validated, it was also used to study the influence of the lubrication conditions (i.e. the friction factor in the simulation) on the tool wear.

The reported example shows how it is possible to use simulation model to predict not only the workpiece final characteristics but also to predict how the equipment (dies in this case) may wear due to the process conditions.

14.4.6.5 Micro Processes and Simulation

When moving to microprocesses, some aspects of the process that can be neglected in macroprocess show their influence. Some examples are the grain size or the friction surface extension that becomes to be relevant when compared to the dimension of the part.

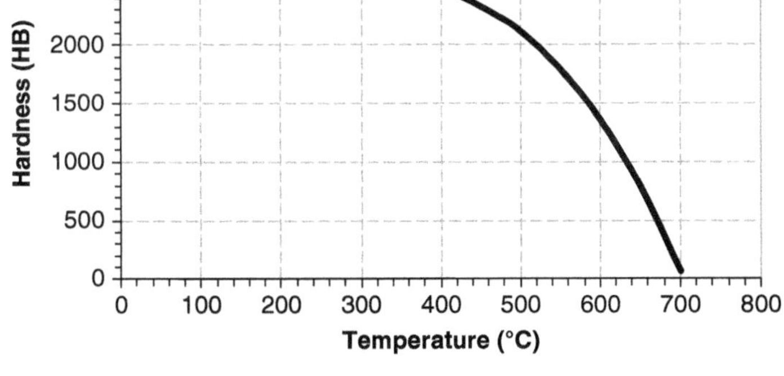

Figure 14.10 Die material hardness as function of the temperature.

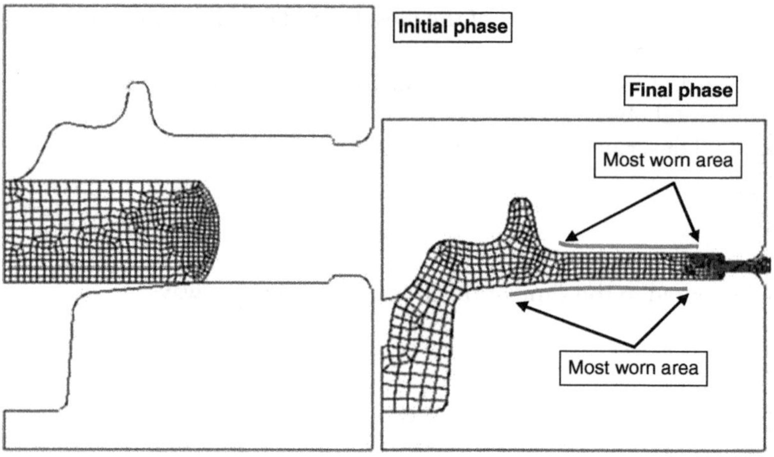

Figure 14.11 Schematic of the hot forming process.

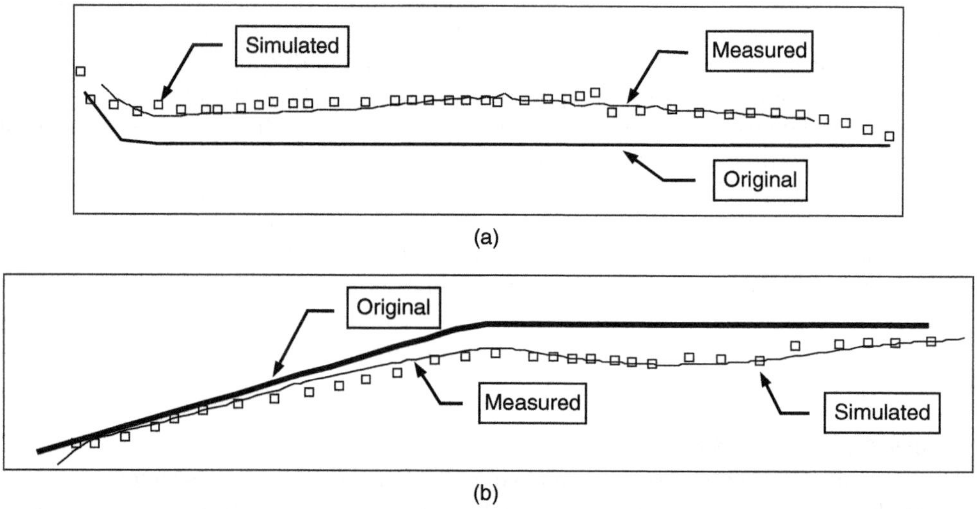

Figure 14.12 Comparison between simulated (squares) and actual (continuous line) profiles of the (a) upper and (b) lower dies.

These phenomena are called *size effects*, that is the deviation from the lineary continuum theory, when the scale is reduced to microsize. When the scale is reduced to such a level, some of the rules change, and a new set of theories have to be applied. In fact, when the size of the component is within orders of magnitude of the physical elements of the process, e.g. material grain size, the linear theory of the macrodomain breaks down. The contributors to size effect can be put into three groups:

- *Density effects*: They are related to the inhomogeneity of materials at small scales. If the grain size is considered to be of the order of 40 μm, a 0.5-mm feature would only contain about 12 grains over the cross-section, resulting in the properties of the individual grains influencing the forming process;
- *Shape effects*: They are closely related to the surface-to-volume ratio of the component. That ratio is an important parameter when considering an ejection or handling situation, where

the friction is a factor of the component surface area, and the component strength is related to the material volume.

- *Structural effects*: This group is made of physical elements which experience a physical length limitation, where it's not practical to scale the micro-geometry or where the secondary scaling effects come into play. Surface roughness is an example of a quantity that is not fully scalable.

Some examples of the main process parameters affected by size effect are as follows:

- *Temperature*: In bulk-forming operations, thermal energy is often supplied to the workpiece to increase its temperature to enhance formability and reduce forming forces. Since the dimensions of the workpiece are under 1 mm or less, it is often preferred to work at room temperature, or higher but under the recrystallization temperature as well.
- *Flow stress*: It is found in many researches that the flow stress increases with the increasing of the length scale. Chan et al. [38] studied the flow stress of a microspecimen in microcompression test and found that the flow stress is lower on the surface of the specimen since the outer grains can move more freely with respect to the inner ones (Figure 14.13). The scale effect on friction is therefore more evident when we can find just few grains along the workpiece thickness.

Lai et al. [39] proposed a model in which the behavior of the material changes depending on the positions of the grain related to the position in the cross-section of the workpiece, this model is called mixed material model. By this issue, the model can be divided into two different terms: the size-dependent model (σ_{dep}) and the size-independent model (σ_{ind}). This theory can represent the behavior of the single crystal model and even the polycrystal model.

$$\begin{cases} \sigma(\varepsilon) = \sigma_{\text{ind}} + \sigma_{\text{dep}} \\ \sigma_{\text{ind}} = M\tau_{\text{R}}(\varepsilon) + \dfrac{k(\varepsilon)}{\sqrt{d}} \\ \sigma_{\text{dep}} = \eta\left(m\tau_{\text{R}}(\varepsilon) - M\tau_{\text{R}}(\varepsilon) - \dfrac{k(\varepsilon)}{\sqrt{d}} \right) \end{cases} \tag{14.21}$$

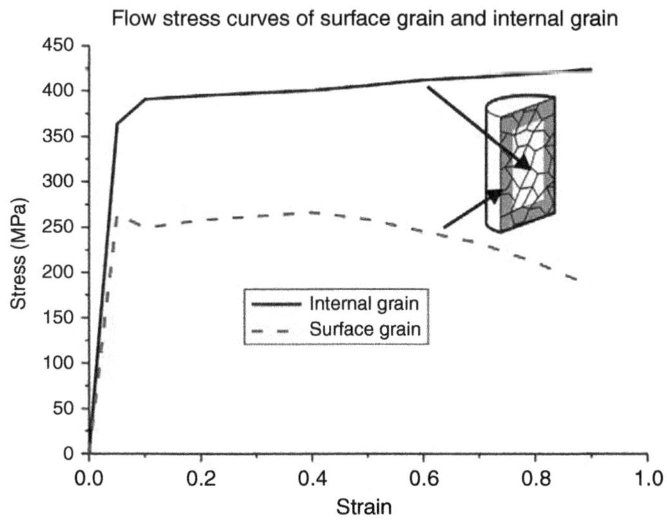

Figure 14.13 The flow stress curves of the surface and internal grains. *Source:* Chan et al. 2011 [38]. Reproduced with permission of Elsevier.

where

$k(\varepsilon)$ is the locally intensified stress needed to propagate general yield across the polycrystal grain boundaries,
M is the orientation factor for the polycrystal model,
m is the orientation factor for the single crystal model,
τ_R is the critical shear stress resolved of the single crystal, and
d is the grain size.

This model has been well confirmed by the experimental and numerical results.

- *Friction*: Since surface roughness is a microfeature in all cases, it is greater in microcomponent when compared to the part dimensions with respect to the macrosize. Test conducted using double cup extrusion test [40] showed that the friction coefficient will increase by a factor of 20 when the experiment is scaled by a factor of 8. This is due to the fact that for microscale surface, more surface asperities reside close to the boundaries of the workpiece, where they are less likely to form lubricant pockets under hydrostatic pressure. This influences the surface contact area, leading to an increase in friction force (Figure 14.14). It is therefore evident that friction is a key challenge to be overcome in microbulk forming. To reduce friction, forming load shall be reduced, lubricant changed ($MoSO_2$-based lubricant or commercial silicon paste showed good results) or forming die polished.

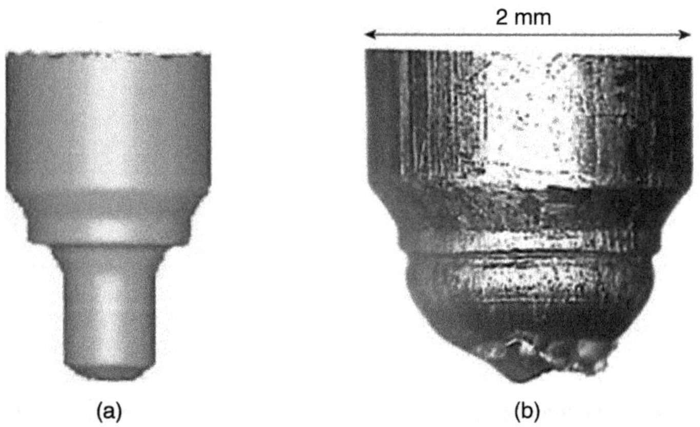

Figure 14.14 (a) 3D simulation of preform for a microaxle, and (b) the experiment result. *Source:* Qin 2000 [40]. Reproduced with permission of Elsevier.

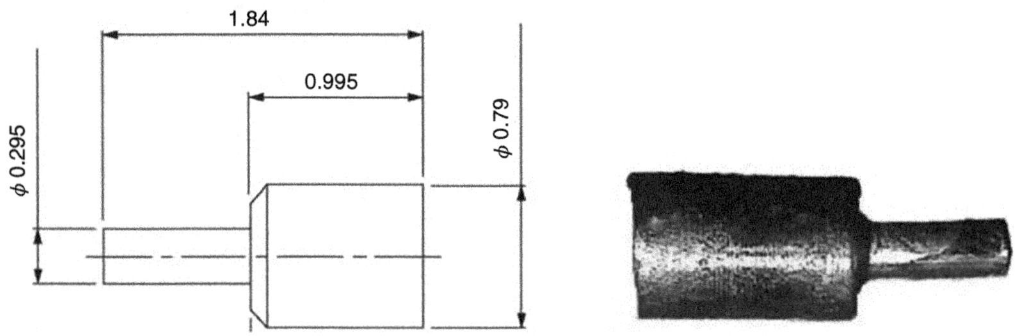

Figure 14.15 Example of the extruded micropin. *Source:* Qin 2000 [40]. Reproduced with permission of Elsevier.

Microextrusion Modeling The microbulk forming process can be simulated in the same way as conventional macrosize bulk process with few important exceptions. Most simulation models hold an underlying assumption of a continuous and isotropic material model. This means that the grain structure of the material is not taken into account, meaning that the model is only valid down to a certain size. As discussed before, a material can be modeled in such a way that if it is indeed isotropic and the number of grain structures across all features is greater than

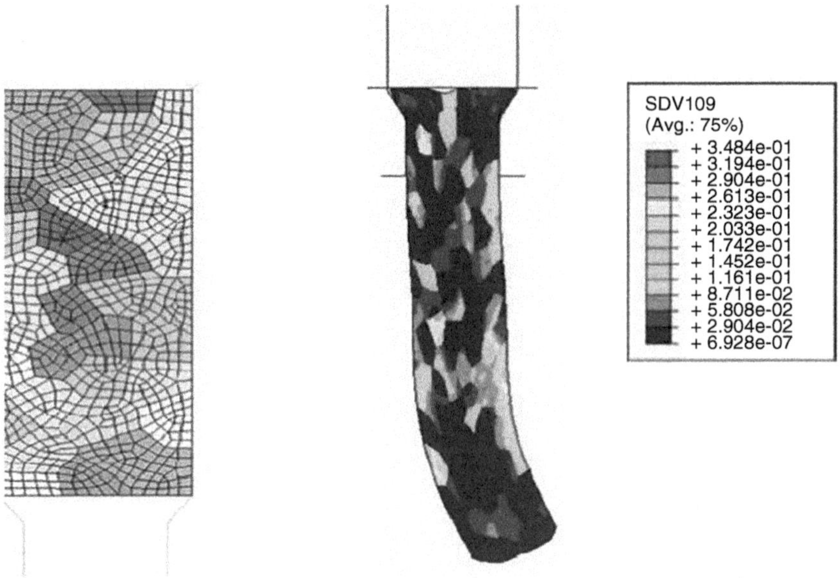

Figure 14.16 (a) Grain distribution in microextrusion and (b) results. *Source:* Cao et al. 2010 [42]. Reproduced with permission of Springer.

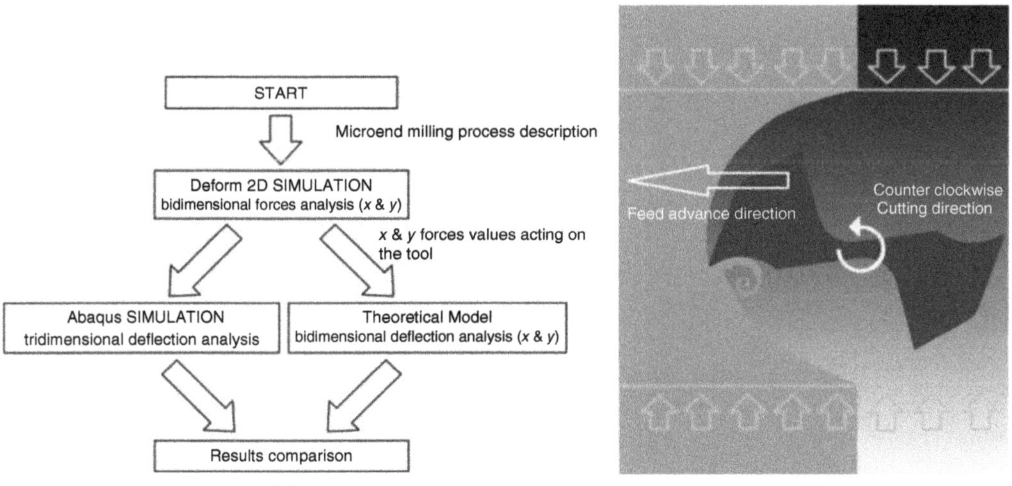

Figure 14.17 (a) Methodology flow chart and (b) deflection direction limits. *Source:* Farina et al. 2011 [48]. Reproduced with permission of AIP Publishing.

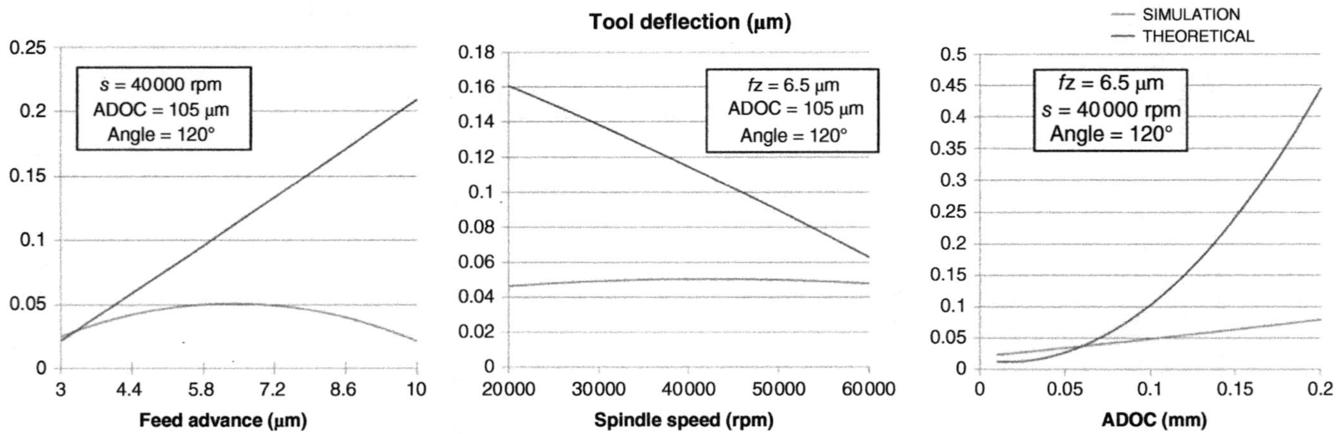

Figure 14.18 Comparison between simulative and theoretical models. *Source:* Farina et al. 2011 [48]. Reproduced with permission of AIP Publishing.

8–10 grains. Figure 14.15 shows a forming experiment in which a small copper pin was extruded where, with an average grain size of about 80 μm, there are less than four grains across the pin, thereby invalidating the assumption of isotropy of the material. If possible, a material with a finer grain structure can be chosen, or the grain size can be changed by heat treatment or mechanical remodeling, so to keep the assumption of homogeneity and isotropy in the material. Otherwise, some authors proposed material constitutive model based on microcompression test and its applicability [41]. Cao and Zhuang [42] has studied the influence of the size effect in the microextrusion process as shown in Figure 14.16.

Micromachining Simulation Microscale milling cannot be considered as the simple downsize from the macroprocess since it has its own characteristics with regard to the cutter edge radius effect [43], microtool stiffness [44], minimum chip thickness [45], microtool failure mechanisms [46], and size effects in general [47].

If the process kinematics does not change, then cutting and surface generation change dramatically in microscale. This is mostly due to the different cutting mechanics induced by the tool-workpiece engagement, cutting geometry relative characteristics between the microend milling tool (flexural strength, tool tip radius, and tool material grain size) and the workpiece material microstructure.

Figure 14.17 shows the methodology flow chart for tool deflection study in micromilling simulations reported in the study by Farina et al. [48]. Results (Figure 14.18) show the difference in tool deflection between microend milling and conventional one. The main difference is represented by the axial depth of cut (ADOC), where the small stiffness of the tool becomes the most affecting parameter of the process.

References

1 Johnson, G.-R. and Cook, W.H. (1983). A constitutive model and data for metals subjected to large strains, high strain rates and high temperatures. In: *Proceedings of the 7th International Symposium on Ballistics*, 541–547. The International Ballistics Society.
2 Oxley, P.L.B. (1989). *Mechanics of Machining*. England: Ellis Horwood Limited.
3 Childs, T.H.C., Maekawa, K., Obikawa, T., and Yamane, Y. (2000). *Metal Machining: Theory and Applications*, 1e. Butterworth-Heinemann.
4 El-Magd, E., Scholles, H., and Weisshaupt, H. (1996). Adiabatic flow curves of metallic materials at high strain rates. *Materialwissenschaft und Werkstofftechnik* 27: 408–413.
5 Cockroft, M.G. and Latham, D.J. (1968). Ductility and workability of metals. *Journals of the Institutes of Metals* 96: 33–39.
6 Gelin, J.C. (1990). Finite element analysis of ductile fracture and defects formation in cold and hot forging. *Annals of CIRP* 39: 215–217.
7 Lemaitre, J. (1985). A continuous damage mechanics model for ductile fracture. *Journal of Engineering Materials Technology* 107: 83–89.
8 McClintock, F.A. (1968). A criterion for ductile fracture by the growth of holes. *Journal of Applied Mechanics* 35 (2): 363–371.
9 Oh, S.I., Chen, C.C., and Kobayashi, S. (1979). Ductile fracture in axisymmetric extrusion and drawing. *Transaction of the ASME, Journal of Engineering for Industry* 101: 36–44.
10 Oyane, M. (1972). Criteria of ductile fracture strain. *JSME* 15: 1507–1513.
11 Iwata, K., Osakada, K., and Terasaka, Y. (1984). Process modeling of orthogonal cutting by the rigid-plastic finite element method. *Transaction of the ASME, Journal of Engineering for Industry* 106: 132–138.

12 Marusich, T.D. and Ortiz, M. (1995). Modeling and simulation of high speed machining. *International Journal of Numerical Methods in Engineering* 38: 3675–3694.
13 Cockroft M.G. and Latham D.J. (1966). A Simple Criterion of Fracture for Ductile Metals. National Engineering Laboratory, Report 216.
14 Bridgeman, P.W. (1964). *Studies in Large Plastic Flow and Fracture*, 51–57. Cambridge: Harvard University Press.
15 Yoshida, F. and Wanheim, T. (1987). The prediction of surface cracking based on the model material technique. *Annals of CIRP* 36: 165–168.
16 Park, J.J. and Kobayashi, S. (1981). Internal cracking in symmetrically opposed extrusion. *Annals of CIRP* 30: 175–177.
17 Clift, S.E., Hartley, P., Strugess, C.E.N., and Rowe, G.W. (1990). Fracture prediction in plastic deformation process. *International Journal of Mechanical Science* 32: 1–17.
18 Dung, N.L. and Mahrenholtz, O. (1987). A criterion for the ductile fracture in cold forging. *2nd ICTP* 2: 1013–1020.
19 Gurson, A.L. (1977). Continuum theory of ductile rupture by void nucleation and growth: yield criteria and flow rules for porous ductile media. *Transaction of ASME* 99 (1): 2–15.
20 Dawson, P., Lee, Y., and Kumar, A. (1992). Void growth in drawing and extrusion of ductile metals. *JMPT* 32: 119–134.
21 Vujovic, V. and Shabaik, A.H. (1986). A new workability criterion for ductile metals. *Transaction of ASME JEMT* 108: 245–249.
22 Bolt, P.J. (1989). Prediction of ductile fracture. PhD dissertation. Einthoven University of Technology.
23 Kudo, H., Sato, K., and Aoi, K. (1967). On cold forgiability test. *Annals of the CIRP* 16: 309–318.
24 Ceretti, E., Filice, L., and Micari, F. (2001). Basic aspects and modeling of friction in cutting. *International Journal of Forming Processes* 4 (1–2): 73–87.
25 Shirakashi, T. and Usui, E. (1973). Friction characteristics on tool face in metal machining. *JSPE* 39 (9): 966–971.
26 Altan, T. (2001). *Short Course on closed die hot forging with flash, Engineering Research Center for Net Shape Manufacturing*. The Ohio State University.
27 Felder, E. and Montagut, J.L. (1980). Friction and wear during hot forging of steels. *Tribology International* 13 (2): 61–68.
28 Painter, B., Shivpuri, R., and Altan T. (1995). Computer-Aided Techniques for the Prediction and Measurement of Die Wear During Hot-Forging of Automotive Exhaust Valves. ERC Report NO. ERC/NSM-B-95-06.
29 Usui, E., Shirakashi, T., and Kitagawa, T. (1978). Analytical prediction of three dimensional cutting process. Part 3. Cutting temperature and crater wear of carbide tool. *Transaction of ASME* 100: 236–243.
30 Takeyama, H. and Murata, T. (1963). Basic investigations on tool wear. *Transaction of the ASME* 85: 33–38.
31 Filice, L., Micari, F., Settineri, L., and Umbrello, D. (2007). Wear modelling in mild steel orthogonal cutting when using uncoated carbide tools. *Wear* 262: 545–554.
32 Attanasio, A., Ceretti, E., and Giardini, C. (2008). Simulation of tool wear in 3D FEM environment considering the tool geometry update: the developed algorithm. In: *11° CIRP Conference on Modelling of Machining Operations*, 199–206. Gaithersburg, MD: National Institute of Standards and Technology (NIST).
33 Ceretti, E., Taupin, E., and Altan, T. (1997). Simulation of metal flow and fracture applications in orthogonal cutting, blanking and cold extrusion. *Annals of CIRP* 46 (1): 187–190.
34 Taupin, E., Breitling, J., and Altan, T. (1996). Material fracture and burr formation in blanking; results of FEM simulations and comparison with experiments. *JMPT* 59: 68–78.

35 Kim, H., Yamanaka, M., and Altan, T. (1994). Prediction of Ductile Fracture in Cold Forging by FE Simulations. ERC Report N°. ERC/NSM-B-94-42.

36 Ceretti, E., Maccarini, G., Giardini, C., and Bugini, A. (1999). Process optimization and defects analysis in reverse extrusion through FEM simulation and experimental tests, 6th ICTP. International Conference on Technology of Plasticity, University Erlangen-Nürnberg, Erlangen, Germany.

37 Giardini, C., Ceretti, E., Maccarini, G., and Bugini, A. (2000). Friction and wear in hot forming. CIRP International Workshop Flow Stress and Friction in Cutting and Forming.

38 Chan, W.L., Fu, M.W., and Lu, J. (2011). The size effect on micro deformation behaviour in micro scale plastic deformation. *Materials and Design* 32 (1): 198–206.

39 Lai, X., Peng, L., Peng, H. et al. (2008). Material behavior modelling in micro/meso-scale forming process with considering size/scale effects. *Computational Materials Science* 43 (4): 1003–1009.

40 Qin, Y. (2010). *Micro-manufacturing Engineering and Technology*. Oxford: Elsevier Inc. ISBN: 13: 978-0-81-551545-6.

41 Chan, W.L., Fu, M.W., and Lu, J. (2011). Experimental and simulation study of deformation behavior in micro-compound extrusion process. *Materials and Design* 32 (2): 525–534.

42 Cao, J., Zhuang, W., Wang, S., and Lin, J. (2010). Development of a VGRAIN system for CPFE analysis in micro-forming applications. *The International Journal of Advanced Manufacturing Technology* 47 (9–12): 981–991.

43 Schmidt, J. and Tritschler, H. (2004). Micro cutting of steel. *Microsystem Technologies* 10: 167–174.

44 Kim, C.J., Mayor, J.R., and Ni, J. (2004). A static model of chip formation in microscale milling. *ASME Journal of Manufacturing Science and Engineering* 126: 710–718.

45 Kim, C.J., Bono, M., and Ni, J. (2002). Experimental analysis of chip formation in micro-milling. *Technical Paper Society of Manufacturing Engineer* (MR): MR2–MR159.

46 Tansel, I., Rodriguez, O., Trujillo, M. et al. (1998). Micro-end-milling: wear and breakage. *International Journal of Machine Tools and Manufacture* 38 (12): 1419–1436.

47 Lucca, D.A., Rhorer, R.L., and Komanduri, R. (1991). Energy dissipation in the ultraprecision machining of copper. *Annals of the CIRP* 40: 69–72.

48 Farina, S., Thepsonti, T., Ceretti, E., and Özel, T. (2011). Determination of specific forces and tool deflections in micro-milling of Ti–6Al–4V alloy using finite element simulations and analysis. *AIP Conference Proceedings* 1353: http://dx.doi.org/10.1063/1.3589588.

15

Advanced Grinding

Taghi Tawakoli and Amir Daneshi

Institute of Grinding and Precision Technology (KSF), Hochschule Furtwangen University, Furtwangen im Schwarzwald, Germany

15.1 Introduction

Grinding is the most important process in manufacturing of many precision components specially those made of ceramics, glass, and tungsten carbides owing to their unique characteristics in terms of surface quality and morphology, part accuracy, and process time. Although this process is classified as a traditional process, demands for finishing new materials, new design of the components, the challenges associated with grinding energy, and environmental concerns, among others, necessitate the progressive development of grinding process. This chapter briefly presents advanced methods of grinding with a short review of the grinding principles. New abrasive grains, new classification of wheel conditioning, approaches in green grinding processes, and employments of ultrasonic vibrations in grinding are involved in this chapter.

15.2 Grinding Wheels

Grinding wheels are composed of the stochastically distributed abrasive grains held together by the bonding material. The abrasive grit materials may be classified into natural and synthetic materials. The natural grits including corundum (with 80–95% Al_2O_3), emery (about 60% Al_2O_3), and natural diamonds have lower strength compared to the synthetic grits. There are also other natural grits usually used for polishing papers and belts, which are not in the scope of this book. Synthetic grits that are widely used in grinding wheels consist of different types of corundum (pure, semipure, pure white, monocrystalline, ceramic, or their mixture), silicon carbide (SiC), synthetic diamonds, and cubic boron nitride (CBN). Additionally, the microcrystalline corundum, i.e. sol–gel (SG), TGX, and CubitronII were introduced in recent years (Figure 15.1). Microfractures in sol–gel grits due to their microcrystalline structure enhance their toughness. The high aspect ratio (about 1 : 8) of the TGX grits in addition to their microcrystalline structure make them an excellent choice for precision grinding. These grits are suitable for grinding with long contact zone, such as creep-feed grinding of turbine blades that are made from nickel-based superalloys because of the enhanced coolant flow into the contact zone. The needle shape of TGX grits favors producing high porous wheels that are suitable for creep-feed applications. The cubitronII grits are the grains with triangular sections

Modern Manufacturing Processes, First Edition. Edited by Muammer Koç and Tuğrul Özel.

Figure 15.1 The microstructure of microcrystalline aluminum oxide grits including SG, TGX, and CubitronII, which benefit the advantages of microfracture during grinding (a); TGX grits with high aspect ratio that are suitable for creep-feed grinding (b; Saint Gobain Abrasive); cubitronII grits with triangular sections that increased the productivity in terms of wheel wear and process time (c; 3M™).

as illustrated in Figure 15.1c, which increased the productivity in terms of the wheel wear and process time owing to the microfractures to occur during grinding operation.

Based on the abrasive type, the grinding wheels are categorized into two separate classes, conventional and superabrasive wheels. Conventional wheels contain aluminum oxide (Al_2O_3) or silicon carbide (SiC) grains, and superabrasive wheels refer to the grinding wheels with diamond or CBN grains. In the following, the properties and applications of these two grinding wheel classes are described. It is followed by a short view of different bonding materials.

15.2.1 Conventional Wheels

Properties of conventional wheels depend on different factors, including grit material, grit size, wheel hardness, and porosity. These factors are specified according to the standard marking systems for aluminum oxide and silicon carbide wheels and are printed by the wheel manufacturer on the wheel. For instance, in the marking "93A46 H9V601W," the first three digits, i.e. 93A indicate the grain material. The letter "A" represents aluminum oxide, and "C" refers to silicon carbide. The number "93" shows the type of the aluminum oxide, which is the private marking of the manufacturer. The number "46" shows the grain size in the form of mesh number. The larger the mesh number, the smaller the grain size and eventually finer wheel. The letter "H" is the wheel grade that indicates its hardness. Hardness of the wheel depends on the bonding material, bond amount, and abrasive grain content. The number "9" shows the wheel structure. The larger structure number represents the more porous wheel and less volume percentage of the abrasive grits in the wheel. It is worth mentioned that the bond amount and therefore wheel hardness is decreased with increasing porosity in constant grain concentration. The letter "V" in the marking system indicates the vitrified bond of the wheel, which may be replaced by "B" for resin-bonded wheels or "R" for the rubber-bonded wheels. There are also other binder materials, such as shellac, oxychloride, and silicate bonds, which are less commonly used and are indicated in the marking system, respectively, by the letters "E", "O", and "S". The wheel bond indicator may optionally be followed by the letters or digits, such as 601W, which identifies the bond structure or properties and are private marking for the manufacturer.

Various chemical compositions and structural characteristics of the conventional abrasive grits introduce wide ranges of aluminum oxide and silicon carbide grits with different hardness and friability. The harder grits benefit the advantages of self-sharpening owing to the fracture or fragment when they dull during operation. Thereof, harder abrasives are widely used in precision grinding. Since the fine grits are usually produced by crushing of the coarser grits, they are less friable compared with the larger grain sizes. Although silicon carbide (SiC) grits are harder than aluminum oxide ones, their application is limited to the nonferrous metals, hard cast irons, and ceramic materials. Silicon carbide wheels cannot be used for ferrous metals because they have chemical reactivity with iron and steel alloys.

15.2.2 Superabrasive Wheels

Superabrasive wheels are composed of CBN or diamond grains set in the vitrified, resin-, or metal-bonding material. Although vitrified CBN wheels are extensively used for grinding of metallic materials, diamond wheels with vitrified bonds are rarely produced. This is due to the higher thermal stability of the CBN grits (up to more than 1400 °C) than that of the diamond (about 800 °C), which enables the CBN grits to be fired in much higher temperature for producing wide range of vitrified bonds [1]. Unlike the conventional wheels, only a thin layer of the abrasive is placed on a metal hub in superabrasive wheels. The hub materials are usually from steel. They may be replaced by alumina, titanium, or carbon fiber–reinforced plastic (CFRP) due to their lower expansion compared to steel. The superabrasive grits are produced in various forms with different strengths, which make them suitable for an individual application. Resin-bonded diamond wheels with relatively low strength of the diamond grains are used in grinding of tungsten carbides, whereas metal-bonded diamond wheels with stronger diamond grains would be a proper choice for grinding of brittle materials such as ceramics, glass, and stone. The use of diamond wheels in grinding of ferrous metals results in excessive diamond wear because of the graphitization and carbon diffusion into the iron.

CBN wheels are extensively applied in grinding of steels and nonferrous high-strength materials. CBN grits with monocrystalline structure have lower toughness compared to the polycrystalline CBN with submicron crystal size.

15.3 Bond Materials

The bond material properties are required to hold the abrasive grains together in the conditions faced during operation in terms of the chemical reaction with the coolant–lubricant, withstanding the forces and temperature, as well as releasing the dulled grains in proper time. Hence, grinding wheels are produced with different bonds depending on their application. For heavy-duty or cutoff grinding processes, which face high forces and shocking loads, resin bond wheels are used. Resinoid bonds are also proper for diamond wheels to grind cemented carbides and CB wheels to grind steels. However, this bond is not suitable when using alkaline fluids because the strength of the resin is reduced, and its bonding to the abrasive is weakened [1]. Rubber bonds are usually used in wet cutoff that the resin bond is avoided. Although vitrified bonds are limited by the wheel speed (about 120 m/s), for precision grinding in which form and geometrical tolerances are crucial, utilizing vitrified bond is recommended.

Metal bonds that are only used in superabrasive wheels can be produced by sintering or electroplating methods. Sintered metal bonds are commonly made from sintered bronze through powder metallurgy process. By alerting bronze composition, the wheel grade is varied.

In electroplated wheels, the binder material is nickel on which a single layer of the superabrasive is placed by electroplating process. These wheels are suitable for precision grinding of metallic materials.

15.4 Grinding Wheel Conditioning

Grinding is mostly the final step in machining of the components with high demands on accuracy and surface quality. The grinding wheel performance has a great influence on the quality of the workpiece and process efficiency. The performance of the wheel is determined by a great variety of influencing factors including, among others, the conditioning of grinding wheels. The purpose of conditioning is to produce the required geometrical accuracy and grinding wheel topography for specified grinding operation. During the grinding process, due to the wear of the grits and the bond, the grinding wheel is subjected to macrowear and microwear. Macrowear may change the shape of the tool profile, which must always match the desired workpiece geometry. The microwear, however, leads to changes in the topography of the wheel, as a result of which the cutting surface of the wheel becomes either too smooth or too rough. Accordingly, the appropriate conditioning of the grinding wheel is a prerequisite for an efficient grinding process.

To date, conditioning of grinding wheels has been classified into profiling, sharpening, and cleaning of grinding wheel, where profiling and sharpening come under the general term of dressing [2, 3]. Regarding the particular attention to the use of structured wheels in recent years, the new class of conditioning so-called structuring is added to this classification. The new classification of grinding wheel conditioning is illustrated in Figure 15.2.

With profiling, the required geometrical and run out accuracy of the grinding wheel is produced. During the profiling, the bond and the grits are removed simultaneously. Thereof, the grinding wheel surface possesses after profiling little or no grit protrusion as illustrated in the left panel of Figure 15.3. This particularly happens after profiling of resin-bonded and metal-bonded superabrasive wheels. In this condition, the wheel does not show a sufficient cutting performance. Hence, the desired grit protrusions and the associated chip pockets are regenerated during a sharpening process after the profiling. During the sharpening, the grits of the outer most abrasive layer are exposed by cutting back the bond, whereby the chip pockets are generated as shown in right panel of Figure 15.3 [4–6].

During grinding, the materials being cut may partially be compressed and adhered into the space between the grains. This phenomenon so-called wheel loading leads to a reduction in

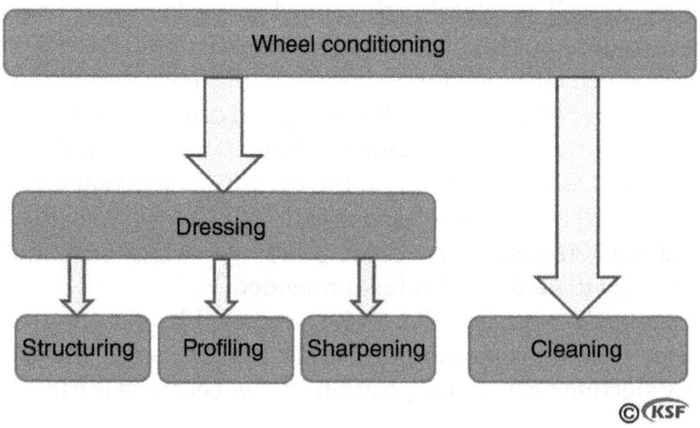

Figure 15.2 The new classification of grinding wheel conditioning.

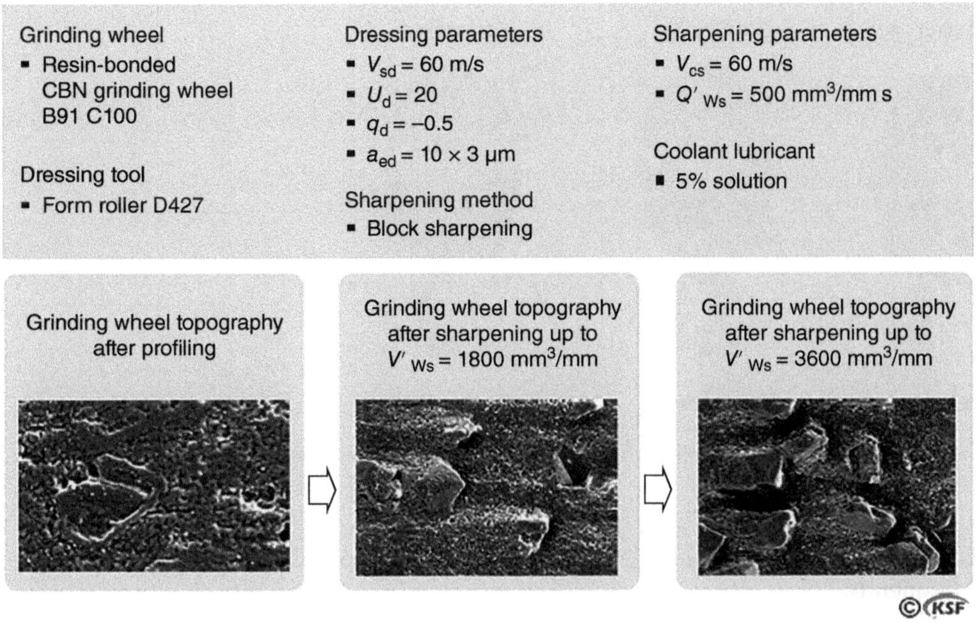

Figure 15.3 With sharpening of the grinding wheels after profiling, the chip pockets and required grain protrusion are regenerated.

chip space, which can decrease the cutting performance of the wheel. With the cleaning process, the loaded chips, loosened grits, and bond residues are removed from the wheel surface [7].

Based on the removal mechanism of grit and/or bond, the conditioning processes may be classified into two groups, mechanical and nonmechanical processes. In the first group, grits and/or the bond are removed by cutting or splitting the grits and/or bond using a diamond or diamond-free dressing tool. In the second group, the dressing process is carried out using nonmechanical forms of energy such as thermal and electrical energy. Laser dressing and electrolytic inprocess dressing (ELID) are two examples of this group.

15.4.1 Profiling, Sharpening, and Cleaning of Grinding Wheels

As previously described, dressing operation involves profiling and sharpening of grinding wheels. In industry, the profiling of a wheel is carried out almost exclusively by a cutting process with a stationary or rotary dressing tool. The tools used for profiling are predominantly diamond dressers, in order to guarantee a high degree of geometric and dimensional accuracy on the wheel to be dressed. Based on the kinematics of the dressing operation, dressers can be classified into two groups, stationary and rotary dressers. Figure 15.4 illustrates different methods of profiling with stationary and rotary dressers. Stationary dressers have no movement in the circumferential direction of the grinding wheel. The wheel is dressed by axial movement of the dresser along the contour of the grinding wheel. In industry, stationary diamond dressers are used in different forms and with different diamond abrasive tips. They may contain only one single diamond grain or multiple diamond grains. Almost all the stationary diamond dressers are only suitable for profiling or dressing rectilinear and simple wheel profiles. Since stationary diamond dressers contain only a small amount of diamond, they are subject to a significant wear when dressing superabrasive CBN and diamond wheels. Thereof, to meet the

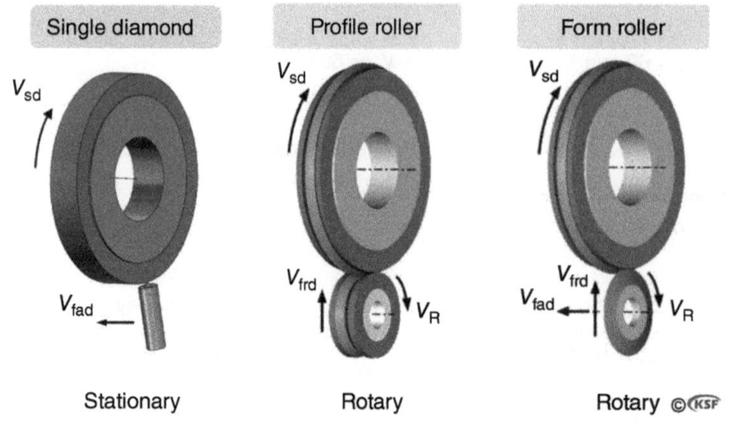

Figure 15.4 Different methods of grinding wheel profiling by stationary and rotary dressers.

demands on geometric and dimensional accuracy of the wheel profile, the dresser needs to be changed frequently.

In addition to the axial movement, rotary dressers have an axial rotation. The relative speed in the contact zone is determined by the peripheral speed of both the dresser and the wheel. There are both diamond-free and diamond-containing rotary dressers. Nowadays, diamond-free rotary dressers are used in few cases. Rotary diamond dressers have a diamond layer on their periphery or front face. With rotary diamond dressers, the desired geometric and dimensional accuracy of the wheel can be obtained over a much longer period of time compared to the stationary diamond dressers. This lies in the fact that the rotary diamond dressers contain much more diamond, as a consequence of which they are subject to much less wear. In addition, when profiling with rotary tools, the topography of the grinding wheel due to the kinematics of the process can be influenced in a broader spectrum. Common rotary diamond dressers are diamond profile rollers and form rollers as shown in Figure 15.4.

For the profiling of superabrasive grinding wheels, SiC wheels can also be used [8]. Currently, diamond grinding wheels are profiled predominantly with SiC wheels [9, 10]. When profiling with a SiC wheel, the bond of the wheel is cut back by the abrasive action of the SiC grits so that the grits on the grinding wheel are loosened and released. The grit size of the SiC wheel is between 80 and 30 US Mesh; a larger grit size is recommended for dressing diamond grinding wheels with a coarser grit [11].

The topography of the wheel and consequently the grinding process are influenced by the sharpening conditions. Grinding forces are reduced with increasing specific sharpening material removal rate Q'_{Ws}, particularly when grinding with resin-bonded diamond wheels. In this case, high grinding forces occur with the unsharpened diamond wheel owing to very small chip pockets and high cutting edge density caused by the low depth of surface smoothness. Furthermore, the greater grit retention force due to the low grit protrusion leads to longer diamond grits retention in the bond. The consequence of this can be grit flattening, which in turn causes a rise in grinding forces. On the other hand, the extended period in which the grit remains in the bond lowers the radial wear. The workpiece roughness is also influenced strongly by specific sharpening material removal rate. It can be stated that the higher the specific sharpening material removal rate, the rougher the ground surface. This is attributed to the flattened grits, greater density of the cutting edges, and consequently more overlapping of the scratch traces of the abrasive grains on the workpiece [12].

Because of the porosity of vitrified bonded superabrasive wheels, they can be efficiently used in many cases without sharpening. A superabrasive grinding wheel can be sharpened with loose or bonded alumina or silicon carbide grits. Predominantly, the sharpening process is carried out with a vitrified bonded alumina or silicon carbide block (so-called sharpening block). The sharpening process can be carried out either by passing the wheel through the sharpening block or by plunging the wheel into the sharpening block.

The cleaning of a grinding wheel serves mainly to remove the loading chips from the pores of a grinding wheel, but primarily without any intention to change the geometry or topography of the wheel. Wheel loading impairs the chip-forming process due to the reduction of the available chip spaces and the restriction of coolant lubricant in the grinding contact zone. Hence, the wheel loading results in higher grinding forces and higher workpiece roughness values as well as higher wheel wear [13, 14]. In addition, the wear on the diamond dressers is higher during dressing of a loaded grinding wheel compared to the unloaded one.

Chip loading occurs particularly when grinding ductile steels, aluminum alloys, or titanium alloys [8]. During the grinding of tough materials such as nickel-chrome alloys, the grinding wheel becomes heavily loaded, as well. In this case, continuous dressing process is employed to keep the abrasive grains sharp and remove the loaded chips. Continuous dressing is a special type of dressing, in which a conventional wheel is dressed continuously during the grinding process with a diamond profile roller. As a result, the wheel always has an accurate profile during grinding and a sharp cutting face.

Sharpening and dressing exhibit a cleaning effect, although in practice, the wheels are cleaned rather with a high-pressure/high-speed jet of coolant during the grinding process. High-pressure/high-speed jetting with coolant is carried out with a special cleaning nozzle. The cleaning effect of jetting with coolant increases with increasing jet pressure, velocity, and flow rate. However, too high jet pressure must be avoided because it can damage the abrasive layer [15, 16].

In the newly developed diamond profile roller, T-Dress, instead of diamond grains which completely cover the metal body in conventional profile rollers (Figure 15.5a), a thin layer of chemical vapor deposition (CVD) diamond is set on the metal body (Figure 15.5b). With this profile roller, the removal mechanism of grinding wheel grits differs from when dressing by the conventional diamond profile rollers. Figure 15.6 illustrates the different shapes of the grits after dressing by T-Dress and conventional diamond roller. The new shape of the abrasive grit tip obtained during dressing by T-Dress influences the chip formation of grinding so that

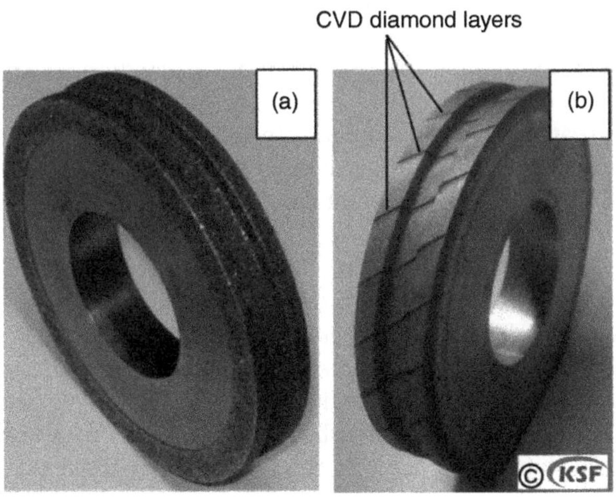

Figure 15.5 The conventional diamond dresser consist of numerous diamond grains which act as the abrasive (a). The newly developed dresser, T-Dress, with point contact concept between the dresser and grinding wheel. This dresser is partially covered by CVD diamond segments which are placed on the metal body (b).

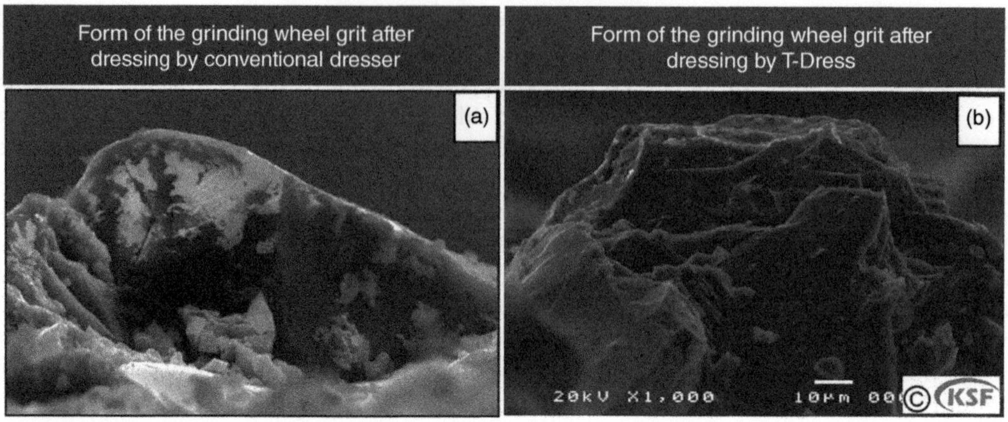

Figure 15.6 After dressing by T-Dress, the forms of the grinding wheel abrasive grains are different from those dressed by conventional diamond dresser consisting of diamond grains.

significantly finer ground surfaces were obtained with no rise in grinding forces. This is due to the fragments of the grit tips, which produce more sharp cutting edges. Unlike the conventional dressers that are fully covered by the diamond grains, the T-Dress roller is covered only partially by the CVD diamond segments. Hence, when dressing by T-Dress, a point contact between the grinding wheel and dresser occurs instead of a linear contact. This lowers the dressing forces and temperature and favors maintaining the profile accuracy during dressing.

In the foresaid methods, the removal mechanisms of abrasive grains are based on the mechanical contact between the diamond grains and grinding wheel grains. The mechanical methods face a number of limitations such as diamond dresser wear and unpredictable changes in diamond form. In the new methods of dressing, the removal process is carried out using nonmechanical forms of energy such as thermal and electrical energy. In this regard, laser dressing and ELID are expressed in this chapter.

Laser conditioning is a method in which the surface of the grinding wheel is heated in controlled conditions by the laser beam. With this process, the components of the abrasive layer and also the chips embedded in the pores of the abrasive layer can reach their melting points. Laser conditioning is influenced both by the specification of the wheel and also by the conditions of the laser irradiation.

Laser conditioning is carried out best with an Nd:YAG laser because not only it is possible to achieve a high degree of focus with the beam of the Nd:YAG laser but also it has the best absorption behavior and, therefore, uses energy to effect on the bond of the grinding wheel. Besides, by means of optical fibers, the beam of the Nd:YAG laser can be directed safely and flexibly to the machining point, which is of great advantage particularly in industrial applications [17]. In the newest developments, the picosecond Yb:YAG laser is used in dressing operation through which the heat generation is minimized, and the thermal damages to the wheel bond or grains are reduced.

The laser beam may be irradiated on the grinding wheel surface with various arrangements shown in Figure 15.7. With a radial beam relative to the irradiated surface of the wheel, the laser beam can produce a uniform irradiated area. This arrangement is therefore best suited for cleaning the wheel and for the controlled cutting back of bond material, namely for sharpening. For profiling with close tolerances on geometry, the radial irradiation would have to be carried out with a depth control and also with a cw-Q-switched laser beam (with pulses of a few nanoseconds) [18]. Unlike with conventional sharpening processes, with radial irradiation, no

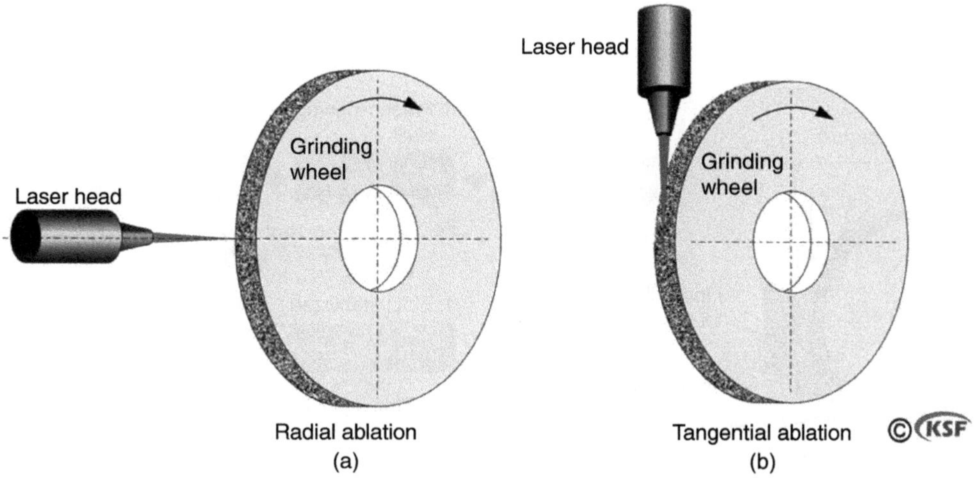

Figure 15.7 The laser beam can be irradiated (a) radially or (b) tangentially to the grinding wheel surface.

bond bridges are produced behind the grits that have a supporting effect on the grit and therefore, have a positive effect on wheel wear [19]. Another principal irradiation arrangement is the beam being directed tangentially to the wheel. With this arrangement, it is possible both to produce bond bridges and to remove roundness errors. However, with this arrangement, the beam power needed to achieve a process speed comparable to that with radial irradiation is greater because the irradiation area is larger, and the focus of the beam is only partially used. In laser conditioning, the influencing factors such as beam power, irradiation time, and defocus distance should be set so that by heating the surface of the wheel, the desired melting and eventually the required removal of the elements of the abrasive layer and chips embedded in the pores of the abrasive layer is achieved.

Laser dressing is mostly applied for superabrasive grinding wheels. Cleaning and dressing operations may be performed by the laser beam. Although, the power needed for dressing is higher than that for cleaning.

With proper laser irradiation, conventional and superabrasive grinding wheels can be both cleaned and dressed. Fusion and evaporation of chips are the main mechanisms in the laser cleaning of a wheel [20, 21]. The wheel can only be cleaned effectively and reliably by laser irradiation if all the process parameters are selected appropriately. With laser dressing, a higher laser power flux input is generated compared to laser cleaning. This leads to melting (followed by resolidification) and/or evaporation of the components of the grinding wheel, as a consequence of which the topography of the wheel changes in terms of its morphology and composition. Depending on the cooling speed, resolidification produces new microstructures [22].

The electro-assisted conditioning methods are used for preparing metal-bonded superabrasive grinding wheels, whereby the bond is removed by an electrical process. These conditioning methods can be controlled readily and are very efficient. ELID is an electro-assisted conditioning method that is in widespread use. With ELID, the metal bond is removed continuously during the grinding process in accordance with the principle of anodic metal dissolution in electrolysis, whereby the blunt grits are continuously dissolved from the grinding wheel, and the sharp grits below are exposed. ELID is used mainly for fine-grained metal-bonded diamond grinding wheels [17]. It can also be used for fine-grained tough metal-bonded CBN wheels in the grinding of glass and ceramics. Cast iron bonds and cast iron fiber bonds are the most suitable application of ELID [23]. The ELID setup and mechanism are shown in Figure 15.8.

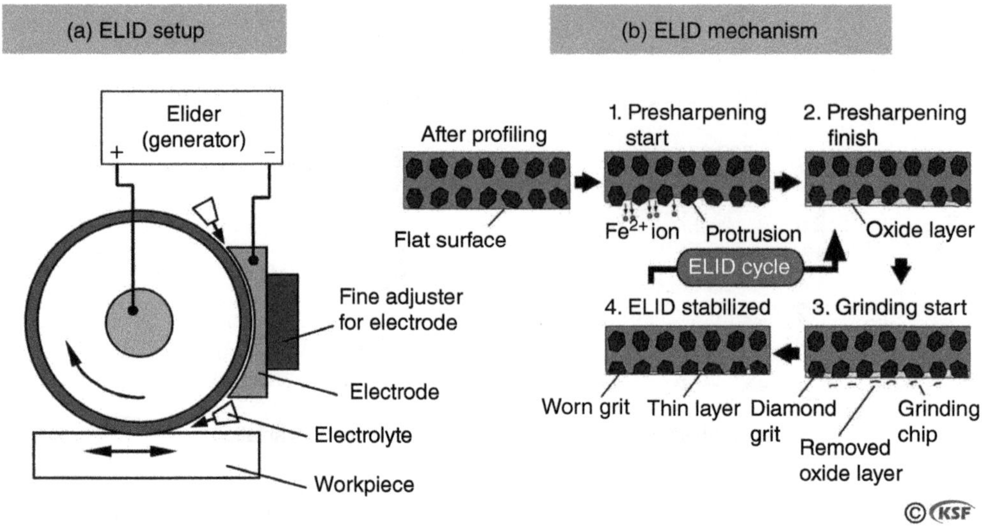

Figure 15.8 Schematic representation of ELID dressing process.

An ELID system essentially consists of a metal-bonded grinding wheel, a counter-electrode, a DC pulse power supply, and an electrolytic coolant with a high pH (Figure 15.8a). The grinding wheel, which acts as the anode in the electrolysis cell, is connected via a brush contact on the wheel spindle to the positive pole of the power generator. The counter-electrode, which is adapted to the profile of the wheel and acts as the cathode, is made of copper and is placed at a distance of 0.1–0.3 mm to the grinding wheel [23, 24]. The negative electrode must cover at least one-sixth of the wheel's active surface and a width that is 2-mm wider than the wheel rim thickness [23]. The electrolytic coolant has low specific conductivity in the range from 1 to 3 mS/cm. In addition to being supplied into the machining area, the coolant is injected directly into the gap between the wheel and the counter-electrode by additional nozzles [24].

The ELID grinding process consists of three stages: the setup stage, the electrolytic presharpening stage, and the ELID cycle stage [25]. The setup stage takes place with a newly fitted grinding wheel, with balancing and precision profiling minimizing the run out of the wheel. This stage should achieve a run out of less than 2–4 μm [23]. After profiling, the grits and the bond are at equal height (Figure 15.8b). In the presharpening stage, the grit protrusion needed at the start of the grinding process is produced by electrolysis (Figure 15.8b-stage1). At this stage, the wheel surface has good conductivity, and hence, the current is as great as that set at the power supply [26]. Presharpening is performed at low speed and takes about 10–30 minutes [23]. Continuing the electrolysis leads to forming of an oxide layer on the surface of the wheel, which acts as an insulating layer and therefore prevents further oxidation of the wheel. The oxide layer is easily removed by the abrasive contact in grinding (Figure 15.8b-stage3). If the oxide layer reaches a certain thickness during the grinding operation, then the electrolysis is started again automatically (Figure 15.8b-stage4), as a result of which the blunt grits are removed from the grinding wheel, and the sharp grits below are exposed. This cycle is called the ELID cycle, which occurs repeatedly throughout the grinding process [26]. The rate at which bond material is removed is set by the ELID parameters, namely voltage, current, and the gap between the electrodes. The rate of bond removal should correspond to the development of wear on the abrasive grits [23].

15.4.2 Grinding Wheel Structuring

Structuring of the grinding wheels is important from two aspects, namely optimization of chip formation in grinding and producing microstructures or macrostructures on the workpiece surface. In the grinding, the ratio of the grains which take part in real chip formation to the total number of grains on the grinding wheel is very low even less than 1% [27, 28]. The rest of the grains either have no contact with the workpiece during the process or rub/plough the material without chip formation. These grains are active as they are in contact with the workpiece and have a great role in the generation of heat, energy, and also grinding forces [29]. This affects workpiece quality and limits the production rates, which can be achieved by grinding. Structuring of grinding wheels with the aim of reducing the number of the grains which only rub or plough has been introduced in recent year. The solution is based on reduction of the number of grains that actually come to contact with workpiece, so called dynamic cutting edges. Using a single diamond dresser or diamond form roller to generate circumferential grooves on the grinding wheel surface is the most common method that is readily executed. The dressing overlap ratio U_d is the quotient from the engagement width (engagement between the dresser and grinding wheel) and the axial dressing feed and shows how often a point on the peripheral face of the wheel is engaged by the dressing tool. As shown in Figure 15.9a, the diamond form roller moves axially with the overlap ratio smaller than 1 ($U_d < 1$). Thereof, the contact between the grinding wheel grains and workpiece is reduced. The percentage of the contact can be controlled by varying the dressing parameters. In one case, a 25% contact resulted in 40% reduction in grinding forces when cylindrical plunge dry grinding with a corundum wheel [30]. With a backward movement of the dressing roller on the grinding wheel, a structure illustrated in Figure 15.9b may be created. In this figure, a schematic representation of the cylindrical plunge grinding by the structured wheel is illustrated, as well. Structuring during the operation is the great advantage of this method compared to other methods of structuring.

Laser beam was used to make CBN wheels with microstructures and macrostructures. In this method, number of holes is generated on the wheel surface as shown in Figure 15.10. In this figure, with the laser, 20% of the contact layer is removed (right). These wheels showed a good potential to reduce the grinding temperature which is a great challenge in dry grinding.

T-Tool is an innovative tool system with discontinuous or intermittent cutting process. This tool is plated with superabrasive grains in a single layer. As it is illustrated in Figure 15.11, various slots and teeth are located on the outer surface of the tool. In contrast to conventional tools, the superabrasive coating is not situated on the outer surface of the tool but rather on the inner flanks. The innovative concept for this tool is its function as a multilayer superabrasive

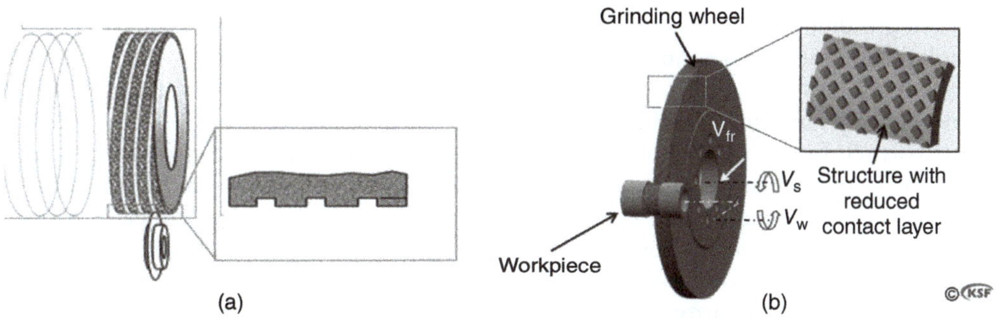

Figure 15.9 Generation of different structures on the grinding wheels to optimize the chip formation in grinding. *Source:* Tawakoli et al. 2012 [30]. Reproduced with permission of Elsevier.

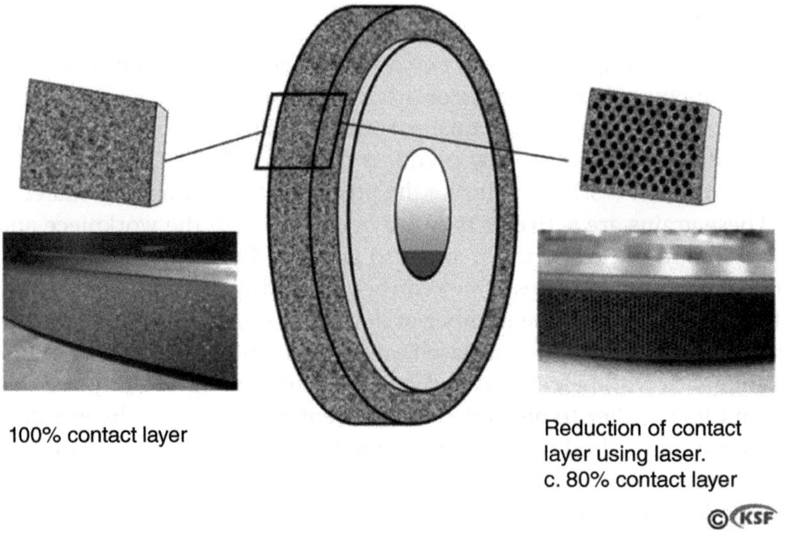

Figure 15.10 Application of laser beam to make microstructures and macrostructures on the grinding wheel surface (right; Tesch GmbH). *Source:* From Rabiey 2010 [31].

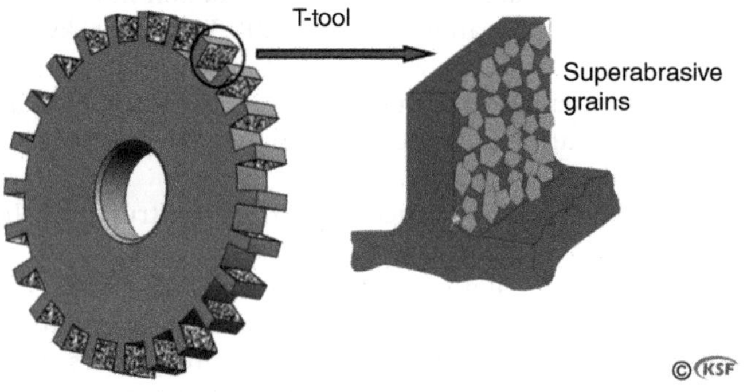

Figure 15.11 Schematic representation of the innovative T-Tool. With this tool, grinding forces and temperature are reduced and coolant reached easier to the grinding contact zone.

tool, although the tool has just one layer of superabrasive. The roughness of ground workpiece, the grinding forces, and grinding power are thus relatively constant throughout the wheel life. T-Tool can guarantee a low grinding temperature, a long tool life, fine and consistent generated surface roughness and simple coolant conduction system. In contrast with conventional electroplated grinding wheels, this tool can be dressed and needs no sharpening [32].

Although the ground surface roughness value increases with using T-Tool, the combination of this tool and ultrasonic-assisted grinding results in a ground surface even finer than that achievable with conventional grinding wheels.

The microstructures on the contact area of the machine parts have many advantages from tribological aspects, such as optimum lubrication and reduction of friction in the contact area of the parts. Various methods such as chemical, thermochemical, and mechanical processes are employed to create the microstructure on the part surfaces. Among them, dressing processes is used to generate microstructures on the part contact surfaces (workpiece). For this purpose, the microstructures are generated on the grinding wheel surface by one of the forgoing inprocess structuring. Afterward, the generated patterns are mapped on the workpiece through grinding process by a definite setup of the operation parameters. The lower production costs compared to other methods of surface structuring is the advantage of surface structuring by dressing. This lies in the fact that with these methods, the desired surfaces are produced during grinding process, which in turn, is a relatively high speed process. Microtwists with depth of 0.8 μm and the twist angle 5° were produced on a round part during grinding operation by this method. The form of the microtwist was generated by a form roller on the wheel surface. Afterward, the twist was mapped on the workpiece surface during the cylindrical plunge grinding process.

15.5 Grinding Force and Energy

In the grinding, the material removal process occurs when the abrasive grains on the grinding wheel penetrate the workpiece material. The mechanism of grain–material interaction, which is a function of grinding wheel topography and grinding parameters, has a decisive influence on the process performance. This may be identified by different methods, among which measurement of grinding forces and temperature are the most common [1]. Grinding forces are the sum of the forces on each single abrasive grain contacting with material and can be separated into two vectors in tangential and normal direction. The grinding power is obtained from multiplying grinding tangential force, F_t, by cutting speed, v_c:

$$P = F_t \cdot v_c \tag{15.1}$$

The effect of table speed, v_w, on the power is neglected in this equation because the workpiece speed is usually much smaller than wheel speed. Therefore, higher grinding force results in higher power that is not desired in grinding. Although the higher wheel speed is directly proportional to the power in Eq. (15.1), it decreases the grinding tangential force in other side. The grinding tangential force increases with higher workpiece speed and depth of cut. The grinding specific energy "u" which is defined as the energy per unit volume of material removal is calculated by the following equation:

$$u = \frac{P}{Q_w} \tag{15.2}$$

where Q_w is the material removal rate and indicates the volume of the removed material per unit time. In grinding, it is usually desired to keep the specific grinding energy as low as possible.

15.6 Thermal Damages in Grinding

Temperatures are generated during grinding as a consequence of the energy expended by the process [1]. In thermal analyses of grinding processes, the heat transferred to the workpiece is determined based on the moving heat source theory. With these analyses, temperature rise in a definite distance of the workpiece surface is calculated. However, due to the complexities with these analyses, only the simple form of the analyses to calculate the maximum temperature rise, θ_m, is expressed in this book. In many practical cases, the maximum temperature rise is of particular interest. If the heat source is moving sufficiently fast so that heat conduction in the direction of motion is much slower than the heat source velocity (like in most of the grinding operations), the maximum temperature rise for a uniform heat source can be approximated as [33, 34]:

$$\theta_{\mathrm{m}} = \frac{1.595 q_{\mathrm{w}} \alpha^{1/2} a_{\mathrm{e}}^{1/4} d_{\mathrm{e}}^{1/4}}{k v_{\mathrm{w}}^{1/2}} \tag{15.3}$$

where a_e is the depth of cut, v_w represents the workpiece speed, d_e the wheel diameter, and the values α and k are thermal diffusivity and heat conductivity of the workpiece material [1]. The q_w in Eq. (15.3) is the heat flux into the workpiece and increases with rising depth of cut and workpiece speed. The parameter q_w depends on the fraction of the generated heat that is transferred to the workpiece. This fraction is known as the energy partition and indicated with ε. The heat flux to the workpiece is obtained as:

$$q_{\mathrm{w}} = \frac{\varepsilon F'_{\mathrm{t}} v_{\mathrm{c}}}{b l_k} \tag{15.4}$$

where l_k is the grinding contact length and "b" the grinding width. In order to calculate the q_w, it is necessary to specify the energy partition to the workpiece, ε. Various techniques have been used to measure the energy partition in grinding. The energy partition can be obtained using temperature matching and inverse heat transfer methods [33].

The workpiece temperature rise induces thermal stresses to the workpiece that results in tensile residual stresses. The workpiece microstructure may deteriorate due to the tensile residual stress. The microstructure degradation due to grinding is a common challenge in grinding of nickel-based superalloys. On the other hand, the grinding mechanical forces induce compressive residual stress to the workpiece that is desirable. In case the compressive residual stress is higher than tensile one (in the same direction or crystalline plane), the resulting residual stress would be compressive. It is obvious that the residual stress must be considered for every crystalline plane of the material structure.

15.7 Environmentally Friendly Grinding

Employment of coolant–lubricants is the most common method to reduce the temperature in grinding process. Due to the high costs associated with the coolant–lubricants, as well as the health and ecological problems rising when working with them, individual attention has been paid to reduction or elimination of coolant–lubricants in machining processes. Dry grinding and grinding with minimum quantity lubricant (MQL) are the two methods developed to approach environmentally friendly grinding process.

The desire to eliminate fluid usage in grinding processes represents a considerable challenge in moving toward green grinding due to the high energy input and high heat flux generated within the grinding zone. In order to perform the efficient grinding process without thermal damages, the chip formation needs to be optimized so that the friction between the grinding wheel and workpiece is minimized. In cylindrical dry grinding of a hardened 100Cr6 material with a CBN structured wheel with 25% contact layer, the grinding forces and specific energy are considerably reduced. Figure 15.12a illustrates that the area-specific grinding energy E''_c values that are accounted as the criteria for thermal damages and are directly proportional to the specific energy progressively fall with lowering the percentage of contact layer between the grinding wheel and workpiece. In this case, with a 75% reduction in contact layer (25% contact layer), dry grinding operation is carried out with the area-specific grinding energy smaller than that in wet grinding condition. This is due to the reduction of number of cutting edges, which

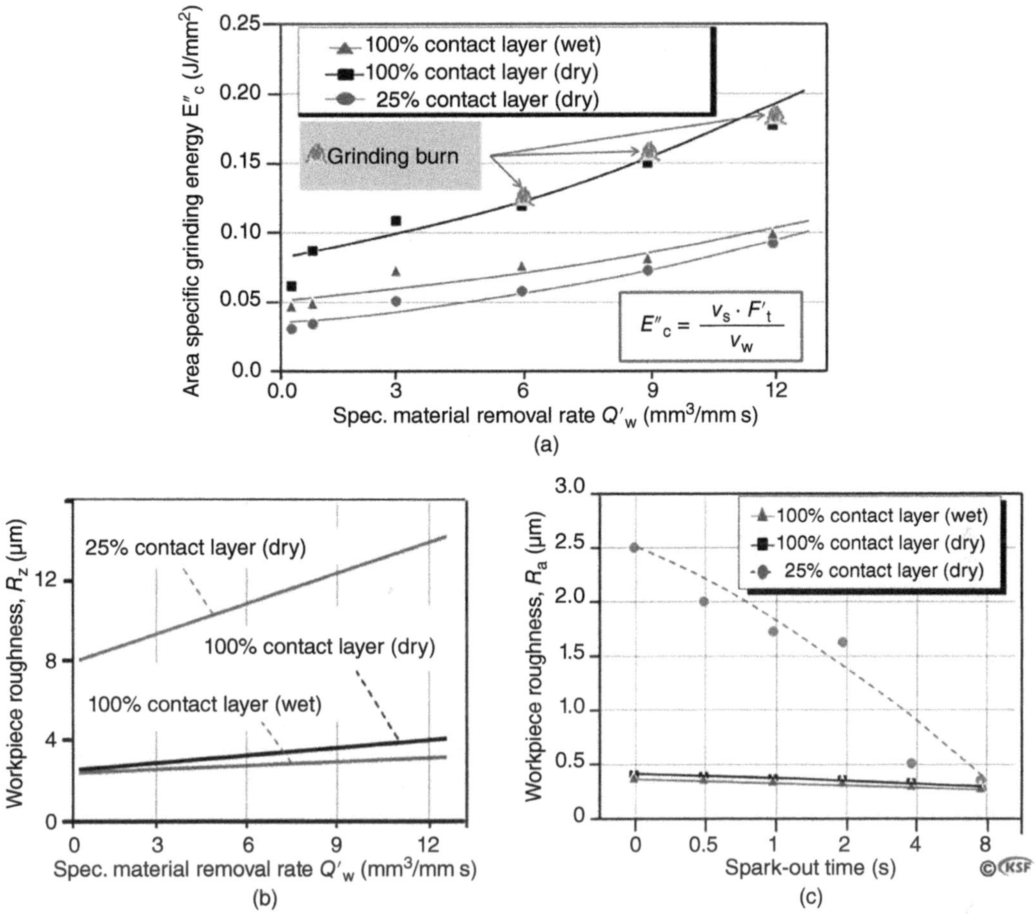

Figure 15.12 The results of plunge cylindrical grinding with structured and nonstructured CBN wheel, B126 C125 vitrified in both dry and wet condition (Workpiece 100Cr6, 62HRC). *Source:* Tawakoli et al. 2012 [30]. Reproduced with permission of Elsevier.

leads to a larger chip thickness and consequently lower friction during the grain-workpiece engagement. In the other side, the ground surface becomes rougher when the contact layer is reduced as shown in Figure 15.12b. This is corrected by additional spark-out time at the end of the operation as shown in Figure 15.12c. On the whole, the dry grinding operation may be performed with the structured wheel with no burn mark and comparable to that of wet grinding.

The other alternative for fluid (wet) grinding is MQL grinding. This process uses a MQL and is referred to as near dry grinding. In this process, aerosols are oil droplets dispersed in a jet of air, oil droplets carried by the air fly directly to the tool working zone, providing the needed cooling and lubricating actions [35, 36]. The physical and chemical properties of the coolant–lubricants determine their effectiveness in the grinding process. By reaction of the coolant–lubricant with the workpiece material in the contact zone, intermediate layers can emerge to separate the surfaces and reduce friction [37]. A successful coolant–lubricant might be thought to form a low shear strength layer between the grain wear flat face and the workpiece surface, which leads to reduce the friction [38].

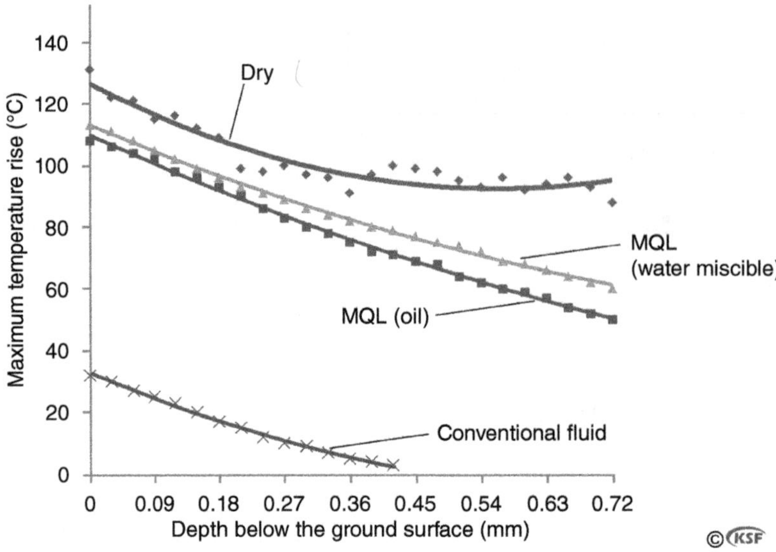

Figure 15.13 The maximum temperature rise vs. the depth "z" in grinding of a 100Cr6 (62HRC) workpiece by the CBN wheel B126 resinoid bond.

The workpiece maximum temperature in MQL grinding is higher than in fluid (wet) grinding but lower than in dry grinding. Figure 15.13 indicates the temperature of a 100Cr6 (62HRC) workpiece surface and subsurface in different depth below the surface when grinding with CBN. It is seen that oil has a better cooling effect on the process compared with the water-miscible in MQL grinding.

Superabrasives and SG corundum wheels with porous vitreous bond and coarse grains are identified as a key technology to make the MQL grinding possible. Under specific conditions, considering coolant–lubricant and wheel type, oil mist parameters, nozzle position, and grinding parameters, MQL is able to compete with conventional fluid cooling delivery. In this case, low grinding forces are evident with MQL, and a reasonable ground surface quality can be achieved. In MQL grinding with Al_2O_3 abrasive wheels, the energy partition varies in the range of 73–77%. However, the energy partition is about of 82% for dry grinding. In fluid grinding with Al_2O_3 wheels, cooling by the fluid at the contact zone reduces the energy partition to less than 36%.

Much lower energy partition to the workpiece (52% for dry grinding, 46% for MQL grinding, and 14% for fluid grinding) is achieved using resin bond CBN wheel owing to the high thermal conductivity of the CBN abrasive. It can be stated that superabrasive wheels can improve the insufficient cooling problem of MQL grinding to certain extent.

Despite the good lubrication of MQL technique in grinding process, it cannot meet the grinding cooling requirements comparing with fluid grinding. This is the main reason limiting the widespread application of MQL in grinding. Nevertheless, using special additives in grinding coolant–lubricants improves the tribological properties of the base fluids, which can enhance lubricating the contact zone, decrease the grinding forces, and consequently lower the heat generation at the contact zone. Although much chip loading occurs in MQL grinding compared to fluid grinding, the ground surface quality obtained by MQL grinding is comparable with that of fluid grinding. Removing the chips from the machining environment, which is performed by the grinding fluid, is another issue confining the application of MQL grinding, to date, to research studies.

15.8 High-efficiency Deep Grinding (HEDG)

From a technical viewpoint, surface grinding can be divided into three distinct techniques, reciprocating grinding, creep-feed grinding, and high-efficiency deep grinding (HEDG). HEDG is understood to mean creep-feed grinding at high wheel speeds and high removal rates. A common feature of creep-feed grinding and HEDG is a large depth of cut (a_e = 0.2–25 mm or more). The essential characteristics distinguishing HEDG from creep-feed grinding are the significantly higher wheel speed and much higher workpiece speed employed in HEDG. HEDG is characterized by the following features:

- High wheel speed ($v_c > 80\,\text{m/s}$)
- Large depth of cut ($a_e = 0.2\text{–}25\,\text{mm}$ or more)
- Medium to high workpiece speed ($v_w = 0.5\text{–}10\,\text{m/min}$ or more)

In addition to the machine settings, intensive application of coolant is also required (oil, high pressure, high throughput, and separate cleaning system).

The specific removal rate that is defined is the arithmetical product of the depth of cut and the workpiece speed

$$Q'_w = a_e \cdot v_w \tag{15.5}$$

It follows that, if the technical conditions are adhered to, the removal rates attainable with HEDG exceed by a factor of 100 or more those which can be achieved by conventional reciprocating and creep-feed grinding.

Raising the depth of cut a_e increases the lengths of contact, which give rise to higher temperature due to the increasing friction. On the other hand, increasing the workpiece speed leads to a lower grinding temperature in special conditions. These considerations show that an increase in the specific removal rate is to be achieved primarily by raising the workpiece speed, as the surface temperature of the workpiece then remains lower. A high workpiece speed is, therefore, one of the fundamental prerequisites for HEDG. Furthermore, a high wheel speed serves not only to reduce the grinding forces and increase the grinding capacity of the wheel but it also promotes the lower workpiece surface temperatures. This relationship is shown in Figure 15.14. This figure shows the temperature curve relative to wheel speed for two different grinding wheels. The wheels were an electroplated CBN wheel (B252) and a bakelite bond aluminum oxide wheel suitable for use at high speeds. Up to the wheel speed of 100 m/s, the temperature increases with both wheels, although that of the aluminum oxide wheel is at a higher level. When the wheel speed is increased further, the temperature falls back again, attaining at 130 m/s approximately the same level as at 60 m/s.

The qualitative effect of workpiece speed and specific stock removal rate on the surface-zone temperature is shown in Figure 15.15. If the process is set up correctly, the surface-zone temperature reached in creep-feed grinding lies in ranges where the workpieces do not suffer any damages. Between the creep-feed grinding range and the HEDG range, there is a zone characterized by higher temperatures. The high temperature in this critical range, which lies just beyond the conventional creep-feed range, is one of the factors responsible for the failure of attempts to achieve higher specific removal rates using the creep-feed technique.

15.9 Ultrasonic-Assisted Grinding (UAG)

In ultrasonic-assisted machining, the kinematics of the conventional machining process are superimposed with an additional vibration in the micrometer range at frequencies in the

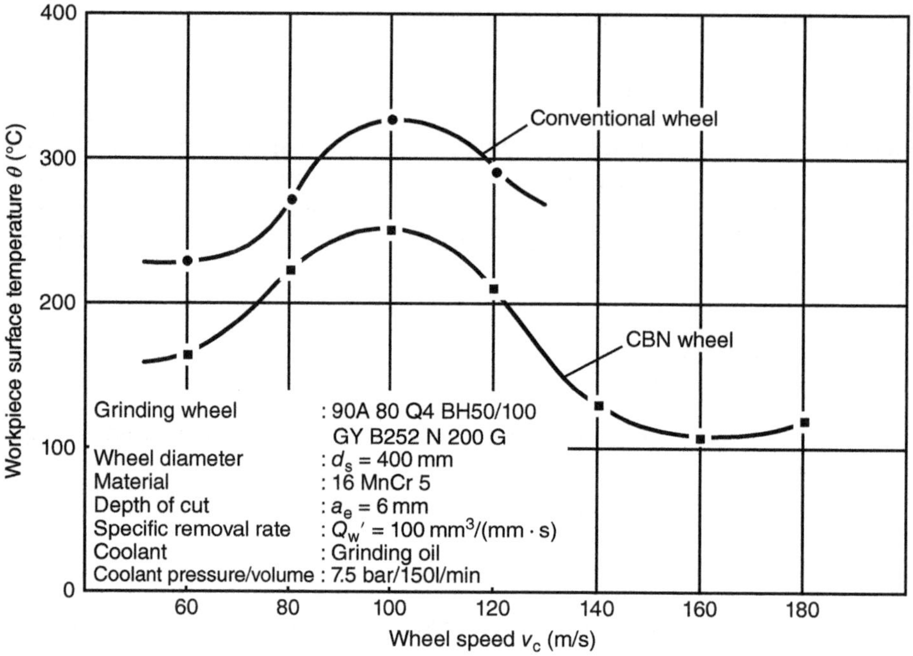

Figure 15.14 The workpiece surface temperature as a function of wheel speed for CBN and aluminum oxide grinding wheels.

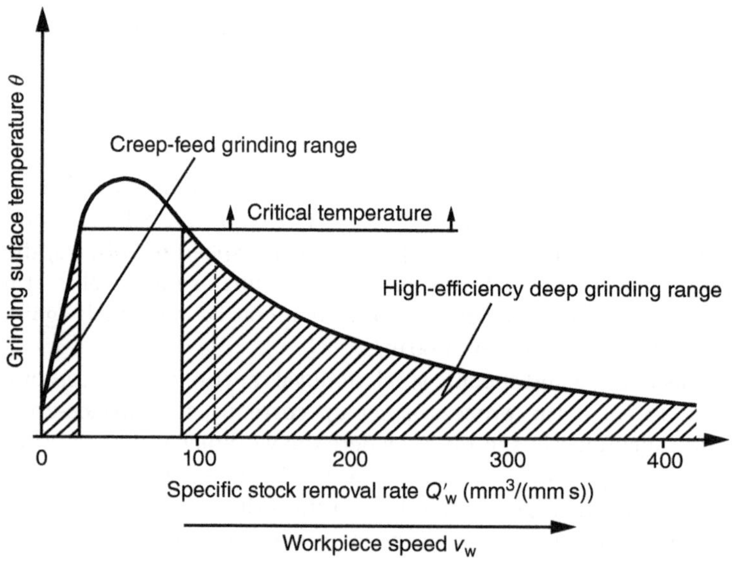

Figure 15.15 Effect of workpiece speed and specific removal rate on surface-zone temperature.

ultrasonic range, as a result of which a considerable improvement in the process, such as reduction in active machining forces, improvement in the supply of coolant into the process zone, reduction in tool wear, etc. can be achieved [39–42]. In the ultrasonic-assisted machining, the ultrasonic vibrations are produced by means of an ultrasonic vibration system that

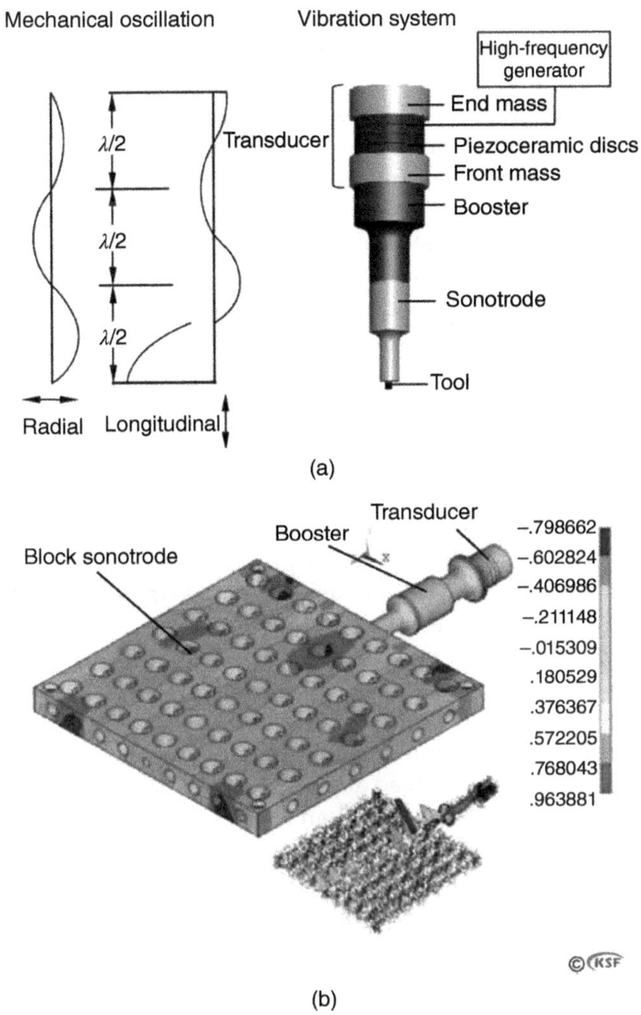

Figure 15.16 Schematic illustration of the ultrasonic systems (a) the conventional system (b) the new developed system.

converts electrical energy to mechanical vibrations and transfers the vibrations through the workpiece or tool into the process zone. This system generally consists of a transducer, booster, and sonotrode as illustrated in Figure 15.16a. Elastomechanical longitudinal ultrasonic vibrations are produced by the conversion of electrical energy to mechanical vibration, for the most part in piezoceramic but also, in some cases, magnetostrictive ultrasonic transducers. The high-frequency electric alternating voltage that is needed is produced by converting low-frequency mains voltage in a voltage generator. The longitudinal vibrations that are produced in the transducer are periodic, elastic deformations in the micrometer range at frequencies in the ultrasonic range. The ultrasonic amplitude produced in the ultrasonic transducer is often very small compared with the amplitudes needed in cutting process. The amplitude of the vibration can be increased considerably by a narrowing of the cross-section of a transition element fixed on the ultrasonic transducer so-called booster. Then, a sonotrode flanged to it transfers the vibration amplitudes through the tool into the process zone.

Decisive for the effectiveness of the ultrasonic vibration for the process is that all the elements, through which the vibration passes, exhibit suitable mode shape. In order to ensure that an increase in amplitude takes place that is largely free of losses, the system must vibrate at one of its natural frequencies (resonance). In terms of the design of the system, this means that the lengths of the individual elements must correspond to the half wavelength of the vibrations or a full-number multiple of it, so that in the entire ultrasonic system a standing wave is generated.

In the conventional ultrasonic vibration systems, the tool or workpiece that is directly attached to the sonotrode must either be very small or operate with a special construction as a part of the vibration chain. However, new developments make it possible to impose the ultrasonic vibrations to industrial parts with different shapes and sizes. In order to vibrate workpieces with different shapes, dimensions, and weights, a special block sonotrode with through holes is used. This block sonotrode is a specially designed plate with a multiresonant frequency response. Figure 15.16b shows this ultrasonic system with its resonance vibration at 20 kHz simulated using finite elements method. The multiresonance behavior of the block sonotrode can only be achieved when the positions and diameters of the through holes are adequately designed. In addition, the block sonotrode serves as the workpiece clamping plate and therefore should possess a certain rigidity to ensure an accurate process.

The UAG process increases the infeed angle (or engagement angle), the maximum scratch depth (penetration depth), and makes the cutting process intermittent. Ultrasonic vibrations change the nature of the cutting process and transform it into a process with a multiple-impact interaction between the abrasive grits and the workpiece. The larger the vibration amplitude, the higher the kinetic energy with which the grits strike the work surface. Regarding the kinematic of the process, increasing vibration amplitude increases the maximum cutting depth (or the maximum uncut chip thickness) and the infeed angle while decreasing the contact length (or chip length) and the contact time, which results in an increased brittle material removal. The material removal in brittle materials mainly takes place by the induced cracks resulting from the contact between the tool and the workpiece. Hence, the ultrasonic impact action and intermittent grinding allow the lateral cracks to propagate and cause more microcracking propagation in the cutting zone and therefore higher material removal rate. The grinding forces and wheel wear rate decrease when imposing ultrasonic vibrations to the brittle materials.

In grinding of tough materials, such as nickel-based superalloys, the chip formation is improved by means of ultrasonic vibration. Due to the high-frequency interaction of active grains on the workpiece, the cutting process in ultrasonic-assisted grinding process becomes discontinuous, and ultrasonic impact action occurs, causing the material to begin to rollover more easily as well as more microcracking propagation in the cutting zone, which both make an effective interaction between grits and the workpiece surface [43]. Therefore, the grinding forces and frictional effects are decreased, so that less plastic deformation occurs. In addition, as the sliding speed in the ultrasonic assisted grinding process caused by ultrasonic vibrations is higher than that in the conventional grinding process, the coefficient of friction reduces [41]. From the diagram in Figure 15.17, the reduction in workpiece temperature when the ultrasonic vibration is imposed to the workpiece in grinding of a nickel-based superalloy is obvious. In addition to the lower grinding forces and friction, improvement in coolant supply into the grinding zone is the other reason for the lower workpiece temperature obtained during ultrasonic-assisted grinding.

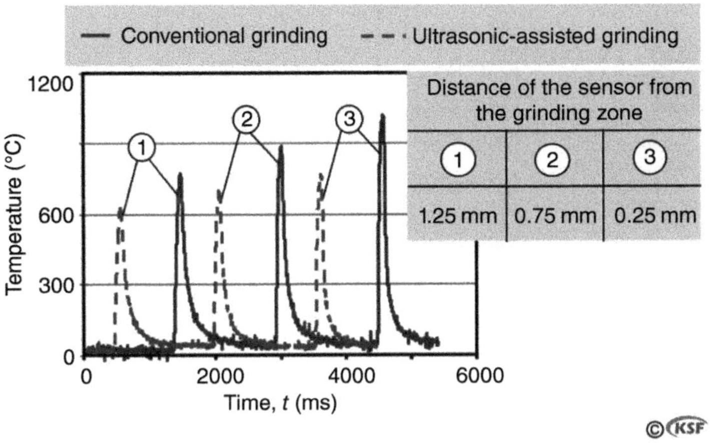

Figure 15.17 Workpiece temperature is reduced when employing ultrasonic vibrations in grinding (workpiece: Nimonic A80, wheel: CBN B126 vitrified bond).

15.10 Ultrasonic-Assisted Dressing

In ultrasonic-assisted dressing, ultrasonic vibrations are superimposed on the conventional acting movement of the process. It is possible to use ultrasonic with both stationary and rotary dressers. Engagement conditions and the resulting splitting and fracture on the abrasive grits and the bond are different in conventional and ultrasonic-assisted dressing. If the stationary dressing tool vibrates in the radial direction of the grinding wheel, the grits and brittle hard bond of the grinding wheel split more intensively [44]. Besides, the sinusoidal active path of the dressing tool in ultrasonic-assisted dressing produces a rougher grinding wheel topography with a smaller number of kinematic cutting edges. The rougher wheel topography with sharper abrasive grits leads to lower grinding forces and greater roughness values of the ground surface. Although, in some cases, lower surface roughness values were reported in grinding with a CBN wheel and a resin bond SiC wheel dressed with ultrasonic assistance. Therefore, the effect of ultrasonic assistance on the generation of the wheel topography during dressing with stationary dressers and the resulting workpiece roughness clearly depends on the wheel specification and the grinding conditions.

Diamond dresser wear can be reduced by ultrasonic assistance. The reduction in wear comes back to the improved coolant supply and the reduction in friction coefficient with ultrasonic assistance, which reduces the heat generation in the dressing contact zone. As a result, the risk of graphitization of the dressing diamond, which is directly related to the forming of wear on the dressing diamond, is reduced [45].

Superabrasive CBN and diamond grinding wheels are dressed almost exclusively with rotary diamond dressers. Thereof, the use of ultrasonic assistance in dressing with diamond rollers is of great importance. However, realization of this subject has much complexities regarding the dimensions and the weight of the roller dressers and dressing spindles. For this purpose, an ultrasonic system shown in Figure 15.18 is applied, in which the form roller is alternately stretched and compressed in the radial direction in micrometer ranges at a frequency in the ultrasonic range [46]. The vibration system consists of a piezoceramic acoustic converter and a sonotrode that functions also as the shaft of the spindle. By resonance, standing longitudinal waves with fixed vibration, antinodes, and nodes are produced in the vibration chain. On the other hand, the standing longitudinal wave leads to a periodic enlargement and reduction of

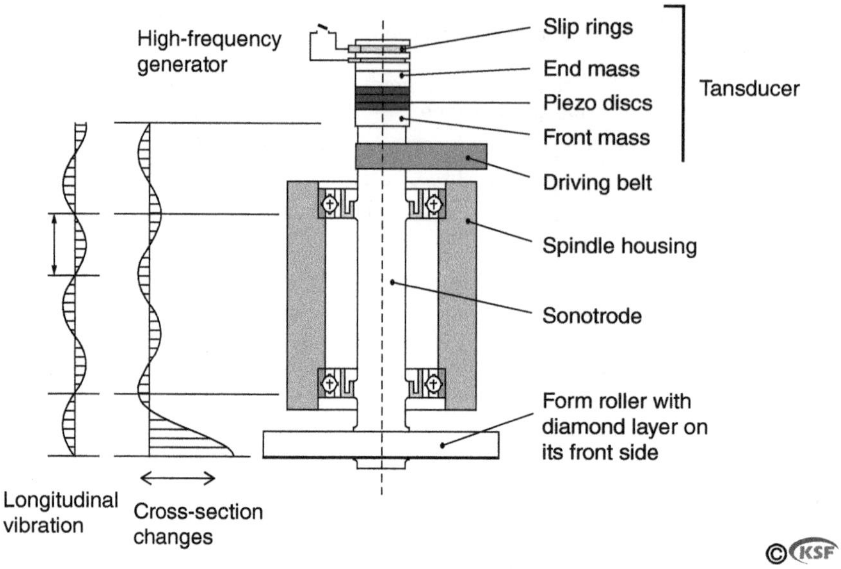

Figure 15.18 Schematic representation of the ultrasonic dressing system for vibrating form rollers.

the waveguide cross-section. The cross-section change is greatest in the nodal points and becomes zero on the longitudinal vibration antinodes, where no local longitudinal extensions occur. The size of the change in cross-section depends on the longitudinal vibration amplitude, cross-section diameter, and Poisson's ratio. As the diameter of the sonotrode increases, the radial amplitude increases. The form roller, coated with diamond on the face is located at a nodal point of the sonotrode. Therefore, the periodic change in cross-section of the vibration node leads to transmission of the radial uniform vibrations into the form roller. This radial vibration is amplified by the special shape of the dressing roller [47]. The engagement paths of the diamonds of the form roller are influenced by the radial vibration of the form roller. This leads to the formation of different wheel topographies and consequently different grinding results with conventional and ultrasonic-assisted dressing [7].

With ultrasonic assistance, the lengths of the diamond engagements are shorter. This leads to a greater theoretical effective peak-to-valley roughness on the grinding wheel. Besides, the penetration angles are greater with ultrasonic-assisted dressing, resulting in greater normal loads and therefore, more intensive splitting of the CBN grits when they impact with the diamond grits. As a result, the CBN wheels dressed with ultrasonic assistance have a sharper abrasive layer with a smaller number of kinematic cutting edges, which causes lower grinding forces and greater workpiece roughness values. This is confirmed by the diagrams in Figure 15.19a. Performing ultrasonic-assisted dressing in up-cutting mode results in higher grinding forces and lower workpiece roughness values compared with down-cutting mode. This is, in turn, attributed to the flatter diamond engagement paths at a higher relative dressing speed.

The greater engagement angle and lower overlapping of the diamond paths in ultrasonic-assisted dressing compared to the conventional dressing causes higher rate of crack formation in both CBN grits and bond. These point act as a weak point in grinding operation and therefore, intensify the grinding wheel wear. This is seen in Figure 15.19b in which the wheel wear rise between 30% and 40%. In addition, ultrasonic-assisted dressing causes higher diamond dresser wear than the conventional dressing method. This is also seen in Figure 15.19c for a dressing of a CBN wheel.

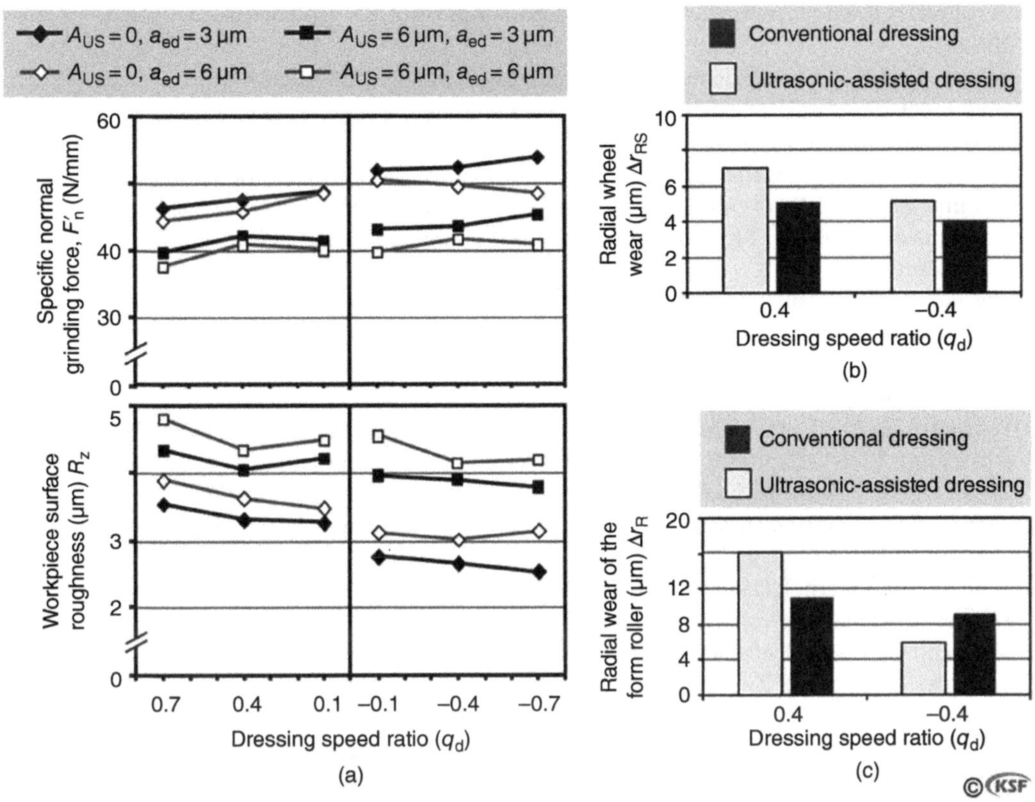

Figure 15.19 Effects of ultrasonic-assisted dressing on the grinding and dressing process when dressing a vitrified CBN wheel B126 and grinding of the workpiece 100Cr6 (60HRC).

The different effects of ultrasonic assistance on the wear behavior of the roller dresser in up dressing and down dressing are due to the different principal wear mechanisms that are dominant in the two dressing methods. Although in down dressing, the microsplitting that results from the high mechanical loading is the principal wear mechanism in the development of wear on the dressing diamonds, in up dressing, the graphitization of the dressing diamonds resulting from the abrasive friction is the principal mechanism of wear on the dressing diamonds [4, 45]. On the other hand, ultrasonic assistance improves the flow of coolant into the dressing contact zone and reduces the coefficient of friction between the active dressing partners, as a result of which the heat generation occurs in the dressing contact zone, and therefore, the intensity of graphitization decreases. Therefore, in up-dressing, ultrasonic assistance leads to a reduction in dressing diamond wear, whereas in down-dressing, it leads to intensity the wear mechanism [7].

References

1 Malkin, S. and Guo, C. (2008). *Ginding Technology: Theory and Applications of Machining with Abrasive*. Industrial Press, Inc. ISBN: 978-0-8311-3247-7.

2 Salje, E. and Harbs, H. (1990). Wirkungsweisen und Anwendungen von Konditionierungsverfahren. *Annals of the CIRP* 39: 337–340.

3 Tonshoff, H.K. and Heuer, W. (1992). Konditionieren von CBN-Schleifscheiben. *IDR* 26: 236–240.
4 Schulz, A. (1996). Das Abrichten von keramisch gebundenen CBNSchleifscheiben mit Formrollen. Doctoral dissertation. RWTH Aachen.
5 Stuff, D. (1996). Einsatzvorbereitung keramisch gebundener CBNSchleifscheiben. Doctoral dissertation. RWTH Aachen.
6 Wimmer, J. (1995). *Konditionieren von hochharten Schleifscheiben zum Schleifen von Hochleistungskeramik*, FBK-Produktionstechnische Berichte, Band 18 (ed. Warnecke). Kaiserslautern.
7 Tawakoli, T. and Rasifard, A. (2011). Dressing of grinding wheels. In: *Machining with Abrasives* (ed. M.J. Jackson and J.P. Davim). New York: Springer-Verlag. ISBN: 978-1-4419-7301-6.
8 Klocke, F. and Konig, W. (2005). *Fertigungsverfahren*, Schleifen, Honen, Lappen. 4. Auflage, Band 2. Springer-Verlag. ISBN: 3-5402-3496-9.
9 Friemuth, T. (2002). Herstellung spanender Werkzeuge. Fortschr.-Ber. VDI Reihe 2, Nr. 615, Dusseldorf, VDI-Verlag, zugl. Habilationsschrift Universitat Hannover.
10 Liebe, I. (1996). *Auswahl und Konditionierung von Werkzeugen fur Ausenrund-Profilschleifen technischer Keramiken*. Dissertation Technische Universitat Berlin. UNZE Verlaggesellschaft Potsdam. ISBN: 3-8167-4509-1.
11 Holz, R. and Sauern, J. (1998). *Schgleifen mit Diamant und CBN. Firmenschrift – Schleiftechnisches Handbuch*. Norderstedt: Ernst Winter & Sohn GmbH & Co.
12 Uhlmann, E. (1993). Tiefschleifen hochfester keramischer Werkstoffe. Dissertation. TU Berlin.
13 Schmaltz, L.H. (1979). Zusetzung von Schleifscheiben. Dissertation. RWTH Aachen.
14 Zhang, L.C., Suto, T., Noguchi, H., and Waida, T. (1995). A study of creep-feed grinding of metallic and ceramic materials. *Journal of Materials Processing Technology* 48: 267–274.
15 Cameron, A., Bauer, R., and Warkentin, A. (2010). An investigation of the effects of wheel-cleaning parameters in creep-feed grinding. *International Journal of Machine Tools and Manufacture* 50 (1): 126–130.
16 Sinot, O., Chevrier, P., and Padilla, P. (2006). Experimental simulation of the efficiency of high speed grinding wheel cleaning. *International Journal of Machine Tools & Manufacture* 46: 170–175.
17 Hoffmeister, H.W., Hlavac, M., and Pekarek, M. (2008). *Verfahren und Trends beim Konditionieren von hochharten und konventionellen Schleifscheiben*. 7. Seminar "Moderne Schleiftechnologie". ISBN: 978-3-00-024493-3.
18 Harbs, U. (1997). *Beitrag zur Einsatzvorbereitung hochharter Schleifscheiben*, Zugl.: Dissertation, Universitat Braunschweig. Vulkan-Verlag. ISBN: 3-8027-8638-6.
19 Hoffmeister, H.W. and Maiz, K. (2002). *Laserkonditionieren hochharter Schleifwerkzeuge* (ed. J. Hesselbach). 10. Internationales Braunschweiger Feinbearbeitungskolloquium (10. FBK), 07-09. Braunschweig.
20 Jackson, M.J., Khangar, A., Chen, X. et al. (2007). Laser cleaning and dressing of vitrified grinding wheels. *Journal of Materials Processing Technology* 185: 17–23.
21 Chen, X. and Pashby, I. (2004). Comparison of laser cleaning of Al_2O_3 and CBN grinding wheels. *Key Engineering Materials* 257 (258): 365–370.
22 Khangar, A. and Dahotre, N.B. (2005). Morphological modification in laser-dressed alumina grinding wheel material for microscale grinding. *Journal of Materials Processing Technology* 170: 1–10.
23 Marinescu, I.D., Rowe, W.B., Dimitrov, B., and Inasaki, I. (2004). *Tribology of Abrasive Machining Processes*. William Andrew, Inc. ISBN: 0-8155-1490-5.
24 Klocke, F., Demmer, A., Klink, A., and Gruntzig, M. (2006). ELID Prazisionsschleifen von Keramikformwerkzeugen. *Werkstatttechnik-Online, Jahrgang* 96: 348–352.

25 Ren, C.Z., Hua, J.H., Wu, Z.Y. et al. (2006). Influence of the state of passivating films in electrolytic in-process dressing (ELID) grinding process. *Key Engineering Materials* 304–305: 24–28.

26 Fathima, K., Schinhaerl, M., Geiss, A. et al. (2010). A knowledge based feed-back control system for precision ELID grinding. *Precision Engineering* 34 (1): 124–132.

27 Tawakoli, T. and Rabiey, M. (2006). Trockenschleifen, Möglichkeiten und Grenzen. *6. Seminar "Moderne Schleiftechnologie und Feinstbearbeitung"*, Stuttgart.

28 Yegenoglu, K. (1986). Berechnung von Topographiekenngrößen. Dissertation. RWTH Aachen.

29 Tönshoff, H.K., Peters, J., Inasaki, T., and Paul, T. (1992). Modelling and simulation of grinding processes. *Annals of the CIRP* 41 (2): 677–688.

30 Tawakoli, T., Heisel, U., Lee, D.H., and Daneshi, A. (2012). An experimental investigation on the characteristics of cylindrical plunge dry grinding with structured cBN wheels. *Procedia CIRP* 1: 399–403.

31 Rabiey, M. (2010). Dry grinding with CBN wheels, the effect of structuring. Dissertation. Stuttgart University.

32 Tawakoli, T. (1994). T-tool neues hochabrasives Werkzeugsystem zum Schleifen und Abrichten, Industrie Diamanten Rundschau. *IDR* 2: 78–81.

33 Malkin, S. and Guo, C. (2007). Thermal analysis of grinding. *Annals of the CIRP* 56 (2).

34 Jaeger, J.C. (1942). Moving sources of heat and the temperature at sliding contacts. *Proceedings of the Royal Society of New South Wales* 76: 203–224.

35 Davim, J.P. (2008). Chapter 7: Machining: fundamentals and recent advances. In: *Ecological Machining: Near-Dry Machining* (ed. V.P. Astakhov). Springer-Verlag London Limited.

36 Astakhov, V.P. (2006). *Tribology of Metal Cutting*. Great Britain: Elsevier Ltd.

37 Klocke, F. (2009). *Manufacturing Process 2: Grinding, Honning, Lapping*, RWTH edition. ISBN: 978-3-540-92258-2.

38 Tawakoli, T., Hadad, M.J., and Sadeghi, M.H. (2010). Investigation on minimum quantity lubricant-MQL grinding of 100Cr6 hardened steel using different abrasive and coolant-lubricant types. *International Journal of Machine Tools & Manufacture* 50: 698–708.

39 Uhlmann, E. (1998). Surface formation in creep feed grinding of advanced ceramics with and without ultrasonic assistance. *Annals of CIRP* 47 (1): 249–252.

40 Daus. N.A. (2004). Ultraschallunterstütztes Quer-Seiten-Schleifen. PhD Dissertation. Technische Universität Berlin.

41 Tawakoli, T. and Azarhoushang, B. (2008). Influence of ultrasonic vibrations on dry grinding of soft steel. *International Journal of Machine Tools and Manufacture* 48 (14): 1585–1589.

42 Brehl, D.E. and Dow, T.A. (2008). Review of vibration-assisted machining. *Precision Engineering* 32: 153–172.

43 Qu, W., Wang, K., Miller, M.H. et al. (2000). Using vibration assisted grinding to reduce subsurface damage. *Precision Engineering Journal* 24: 329–37.

44 Tawakoli, T. and Rasifard, A. (2009). Ultraschallunterstutztes Abrichten von CBNSchleifscheiben, ein neue Alternative zum konventionellen Abrichten, Abschlussbericht zum durch das BMBF im Rahmen des FH3-Programmes geforderten Forschungsprojekt.

45 Linke, B. (2007). Wirkmechanismen beim Abrichten keramisch gebundener Schleifscheiben. Dissertation. RWTH Aachen.

46 Tawakoli, T., Rasifard, A., and Azarhoushang, B. (2011). New developments in ultrasonic-assisted grinding and dressing. In: *4th International Multi-Conference on Engineering and Technological Innovation*, Orlando, FL, USA (19–22 July 2011), 210–214. IMETI. ISBN: 13.978-1-936338-36-8.

47 Rasifard, A. (2011). Ultraschallunterstütztes Abrichten von keramisch gebundenen CBN-Schleifscheiben mit Formrollen. PhD Dissertation. Stuttgart University.

16

Electro-Discharge Machining (EDM)

Muhammad P. Jahan

Department of Mechanical and Manufacturing Engineering, Miami University, Oxford, OH, 45056, USA

16.1 Introduction

Electro-discharge machining (EDM) is one of the earliest nontraditional manufacturing processes, having an inception 50 years ago in a simple die-sinking application. Nowadays, it is the most widely used nontraditional machining process for the production of injection molds and dies, cutting tools and components with complex shapes, and mass production of common objects. EDM is used in a large number of industrial areas: automotive industry, electronics, domestic appliances, machines, packaging, telecommunications, watches, aeronautic, toys, and surgical instruments.

There are many advantages of EDM over traditional methods such as milling, turning, or grinding. The major advantage is that any material that conducts electricity can be machined, regardless of its hardness (hardened steel, tungsten carbide, special alloys for aerospace applications, for example). As a noncontact process, EDM experiences no or very small mechanical stress on the machined piece, compared to the mechanical stress experienced in conventional manufacturing processes due to cutting forces. In addition, no rotation of workpiece or tool is necessary, and the machines have a high autonomy. The complex cutting geometry, sharp angles, and internal corners can be machined with precise accuracy (~1 μm) and acceptable surface finish (<100 nm). On the other hand, the disadvantages are relatively low material removal rate (MRR) (order of 100 mm^3/min), surface modification of the machined workpiece ("white layer" and heat affected zone, typical depth ~50 μm), and limited size of workpiece and tool, for example [1].

The historical roots of EDM date back to the discovery of electric discharges. In 1770, English Physicist Joseph Priestley first studied the erosive effect of electrical discharges [2]. Furthering Priestley's research, the EDM process was invented by two Russian scientists, Dr. B.R. Lazarenko and Dr. N.I. Lazarenko in 1943 [3]. In their efforts to exploit the destructive effects of an electrical discharge, they developed a controlled process for machining of metals. They invented a simple servo controller that helped to maintain the gap width between the tool and the workpiece. This reduced arcing, and new pulse generator made EDM machining more profitable. The discharge generator effect used by this machine, known as the Lazarenko Circuit, was used for many years in the construction of generators for electrical discharge. One of the two principle types of EDM, the die-sinking process was defined as early as in the 1940s [4] with the advent of the pulse generators, planetary and orbital motion techniques, computer numerical control (CNC) and the adaptive control mechanism. The evolution of the wire-EDM in the 1970s [5] was due to powerful generators, new wire tool electrodes, better mechanical concepts,

Modern Manufacturing Processes, First Edition. Edited by Muammer Koç and Tuğrul Özel.

improved machine intelligence, and better flushing. In 1952, the manufacturer Charmilles created the first machine using the spark machining process and presented for the first time at the European Machine Tool Exhibition in 1955. In 1969, Agie launched the world's first numerically controlled wire-cut EDM machine. Seibu developed the first CNC wire EDM machine in 1972, and the first system was manufactured in Japan [6].

Although EDM process has found widespread applications in the industry over last 70 years, true micro-EDM came to existence first in 1968 when Kurafuji and Masuzawa [7] demonstrated the first application of micro-EDM by drilling a minute hole in 50-μm-thick carbide plate. Since then, significant amount of research efforts has been focused on the development of EDM and micro-EDM processes.

16.2 Principle of the EDM Process

16.2.1 Physics

EDM is the process of machining electrically conductive materials by precisely controlled sparks that occur between an electrode and a workpiece in the presence of a dielectric fluid. In this process, a potential is applied between electrically conductive electrode and the workpiece material. A gap is maintained between the tool and the workpiece. A charged electrode is brought toward the oppositely charged work piece by the servo mechanism. The electrical field is strongest (energy density of 10^{11}–10^{14} W/m^2) at the point where the distance between the electrode and work piece is least [1]. As the number of ionic (charged) particles increases, the insulating properties of the dielectric fluid begin to decrease along a narrow channel centered in the strongest part of the field. A current is established, as the dielectric fluid becomes less of an insulator. Heat builds up rapidly as current increases, and the high thermal energy vaporizes some of the fluid, work piece, and electrode. Also, a discharge channel begins to form between the electrode and workpiece. Such intense-localized heat flux leads to extreme instantaneous confined rise in temperature, which would be in excess of 10 000 °C [8]. The high thermal energy vaporizes dielectric fluid, work piece, and electrode. The high thermal energy vaporizes dielectric fluid, work piece, and electrode as the discharge channel becomes broader between the electrode and workpiece. Toward the end of the pulse duration, the current and voltage have stabilized. Heat and pressure within the vapor bubble have reached their maximum, and some metal is being removed. At the beginning of the pulse interval, current and voltage drop to zero. The temperature decreases rapidly, collapsing the vapor bubble and causing the molten metal to be expelled from the workpiece. Finally, the crater is formed by the implosion of the vapor bubble. The expelled metal solidifies into tiny spheres dispersed in the dielectric, which will be flushed away by the dielectric. The volume removed by a single spark is small, in the range of 10^{-6}–10^{-4} mm^3, but this basic process is repeated typically 10 000 times per second [1]. Just after a discharge, the next discharge depends upon the distribution of debris in the gap and the generally random surface irregularities. After a discharge, the gap will have debris, gas bubbles, and tar (arising from disintegrated hydrocarbon dielectric fluid). If sufficient time is allowed for deionization of the dielectric, and flushing is adequate, the occurrence of the next discharge is random in time and space. Otherwise, concentration of debris and ions in the region of preceding discharges is likely to propagate further discharges in the same region. The process mechanism of the EDM and related voltage and current responses over the time is shown in Figure 16.1.

16.2.2 Material Removal Mechanism

Although, the basic principles of the EDM process discussed by many researchers are nearly same, there are several theories on the material removal mechanism of the EDM process.

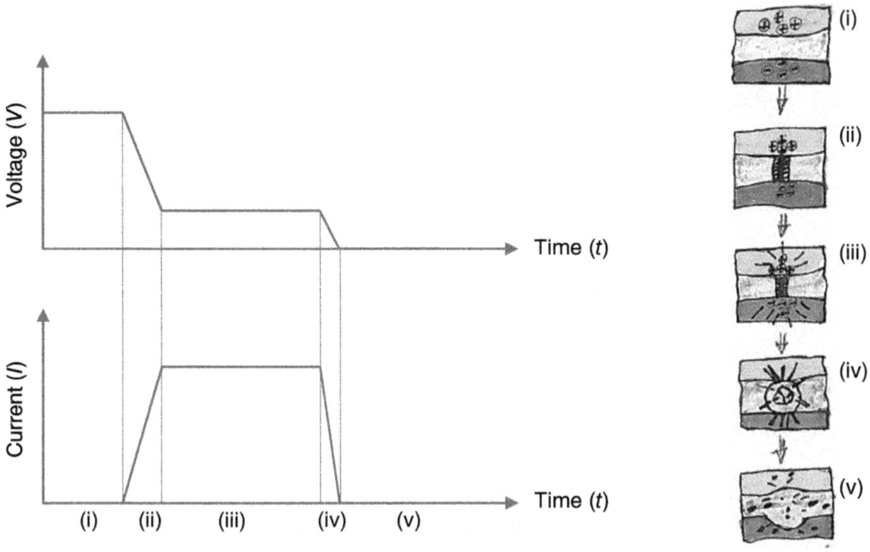

Figure 16.1 Principle of the EDM process showing the gap phenomena and respective voltage and current waveforms: (i) prebreakdown: voltage applied between electrode and the workpiece; (ii) breakdown: dielectric breakdown creates the plasma channel; (iii) discharge: heating, melting, and evaporation of materials; (iv) end of discharge: plasma implosion and removal of molten metal; (v) postdischarge: solidifying and flushing of debris from the gap.

Several theories have been forwarded in attempts to explain the complex phenomenon of "erosive spark" named as the following: electromechanical theory, thermomechanical theory, and thermoelectric theory [9–12]. The electromechanical theory suggests that abrasion of material particles takes place as a result of the concentrated electric field [11]. The theory proposes that the electric field separates the material particles of the workpiece as it exceeds the forces of cohesion in the lattice of the material. This theory neglects any thermal effects. Experimental evidence lacks support for this theory. On the other hand, the thermomechanical theory suggests that material removal in EDM operations is attributed to the melting of material caused by "flame jets"[10, 11]. These so-called flame jets are formed as a result of various electrical effects of the discharge. However, this theory does not agree with experimental data and fails to give a reasonable explanation of the effect of spark erosion. Finally, the thermoelectric theory, best supported by experimental evidence, suggests that metal removal in EDM operations takes place as a result of the generation of extremely high temperature generated by the high intensity of the discharge current [12]. Although well supported, this theory cannot be considered as definite and complete because of difficulties in interpretation.

16.3 EDM System Components

The basic and integral components of an EDM system are as follows:

1) Pulse generator unit for discharging
2) The servo system to feed the tool
3) Dielectric reservoir, pump, and circulation system
4) Working tank with work holding device
5) X–Y table accommodating the working table
6) The tool holder

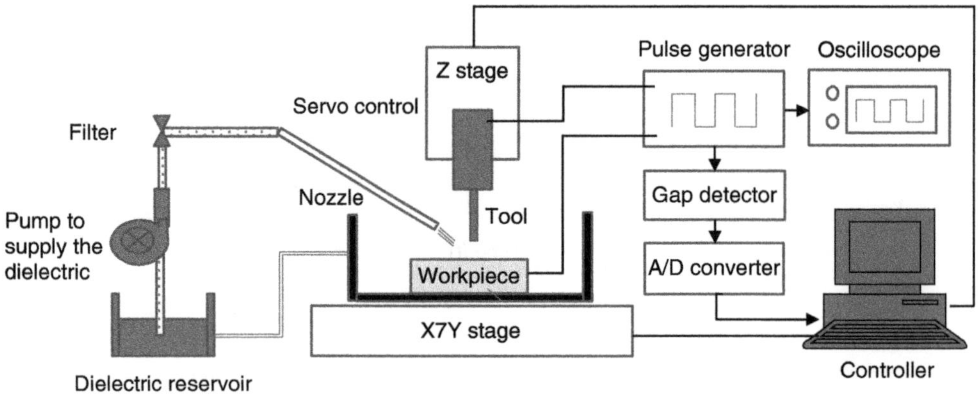

Figure 16.2 Schematic diagram showing different units of the EDM system.

Figure 16.2 presents schematic diagram of the EDM system showing its major components. The brief descriptions of the three major components of the EDM system are provided in the following section.

16.3.1 Pulse Generators or Discharging Unit

16.3.1.1 RC-Type Pulse Generator

The RC-type pulse generator was the first type used for EDM, and it is still being used in finishing and micromachining because the conventional transistor pulse generators do not produce a constant-energy pulse that is sufficiently short [13, 14]. Figure 16.3a shows the schematic representation of basic RC-type pulse generator [13]. In a RC or relaxation type circuit, discharge pulse duration is dominated by the capacitance of the capacitor and the inductance of the wire connecting the capacitor to the workpiece and the tool [15]. The frequency of discharge (or discharge repetition rate) depends upon the charging time that is decided by the resistor (R) used in the circuit. Therefore, "R" should not be made very low because arcing phenomenon can occur instead of sparking, and a critical resistance is desirable, which will prevent arcing [16]. Discharge energy is determined by the used capacitance and by the stray capacitance that exists between the electric feeders, tool electrode holder, and worktable and between the tool electrode and workpiece. This means that the minimum discharge energy per pulse is determined by the stray capacitance. Hence, all the charge stored in the stray capacitance is discharged to the working gap together with the charge stored in the capacitor wired to the circuit. In the final finishing, when minimum discharge energy is necessary, the capacitor

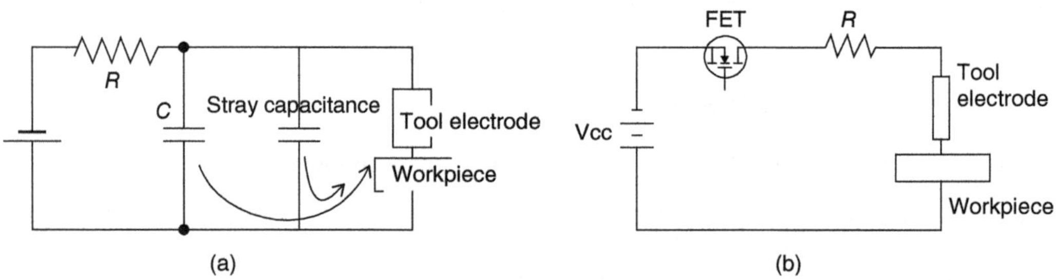

Figure 16.3 Schematic representation of (a) RC type and (b) transistor-type pulse generators. *Source:* Han et al. 2004 [14]. Reproduced with permission of Elsevier.

is not wired, and machining is conducted with the stray capacitance only [15]. It can easily generate pulses with high peak current values and short durations as can be seen from Figure 16.4a [14], allowing efficient and accurate material removal, and meanwhile achieving the required surface quality. Finally, pulse conditions with shorter discharge duration and higher peak current provide better surface roughness due to smaller discharge crater [13].

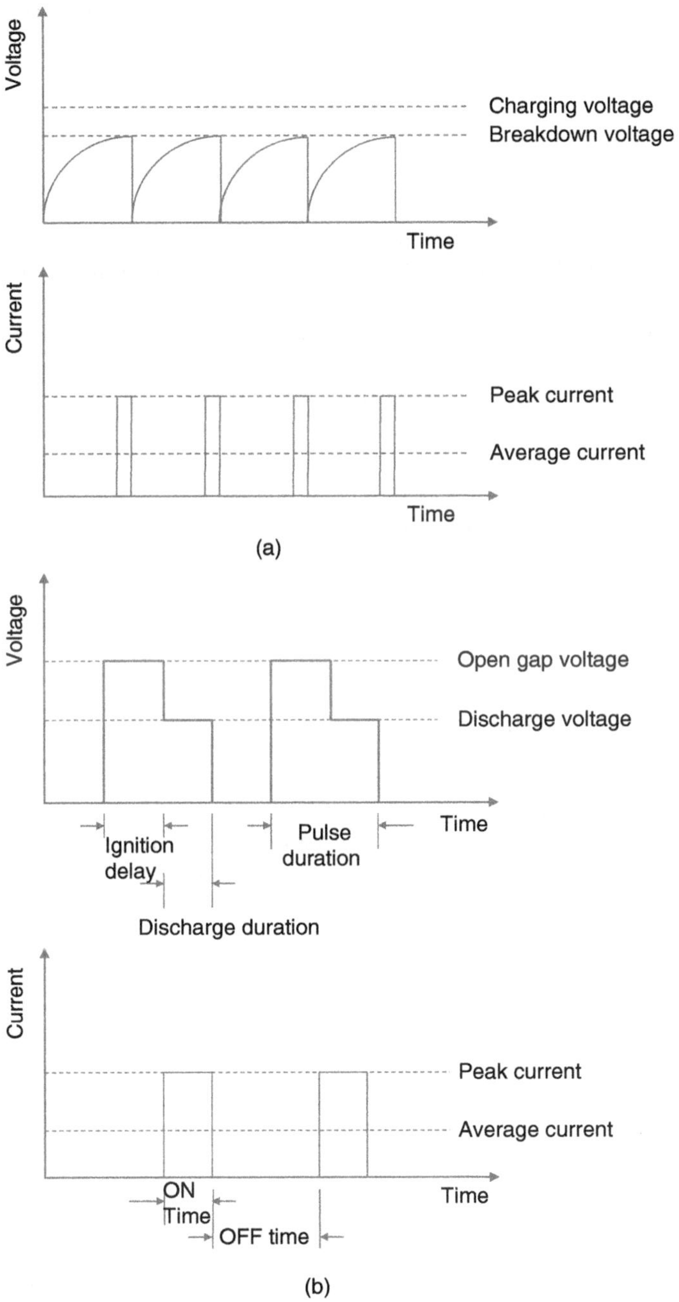

Figure 16.4 The ideal voltage–time and current–time characteristics curve or waveforms generated by (a) RC-type pulse generator and (b) transistor-type pulse generator.

The advantages of RC-type pulse generator are its capability of generating smaller discharge energy using stray capacitance and ability to generate very short pulse duration [17]. As a result, RC-type pulse generator can provide controlled and accurate MRR with very good surface finish. RC-type pulse generator usually has extremely low MRR from its low discharge frequency due to the time needed to charge the capacitor. Moreover, thermal damage can occur easily on the workpiece if the dielectric strength is not recovered after the previous discharge, and the current continues to flow through the same plasma channel in the gap without charging the capacitor [14].

16.3.1.2 Transistor-Type Pulse Generator

The transistor-type pulse generator is widely used in conventional EDM as it provides higher removal rate due to its high discharge frequency. The pulse duration and discharge current can arbitrarily be changed depending on the machining characteristics required. Figure 16.3b shows the schematic representation of basic transistor pulse generator [14]. A series of resistances and transistors are connected in parallel between the direct current power supply and the discharge gap. The discharge current increases proportionally to the number of transistors that are switched on at the same time. The field-effect transistor (FET) operates the switching ON–OFF of gate control circuit. In order to generate a single pulse, gap voltage is monitored to detect the occurrence of discharge, and after preset discharge duration, the FET is switched off. However, there is a delay in signal transmission from the occurrence of discharge to the switching off of the FET due to the time constants in voltage attenuation circuit, pulse control circuit, and insulating circuit and gate drive circuit for the FET [14].

The transistor-type pulse generator can precisely control the ON-time and OFF-time during machining, as can be understood from Figure 16.4b [18]. The transistor-type pulse generator can generate higher discharge energy and thus able to provide higher MRR. However, it is difficult to keep the constant discharge duration shorter than several tens of nanoseconds using the transistor-type pulse generator [14].

16.3.2 Servo Control System or Gap Control Unit

In the EDM process, to ensure stable and efficient machining, the gap condition between the electrode and workpiece must be maintained at the desired state by servo-controlled electrode feed mechanism. A stable gap control system also enables better dimensional accuracy of micromachined features [15]. During the EDM, the discharging gap between the tool electrode and workpiece is of several microns or even less, and therefore requires design of special servo controls compared to conventional EDM. The microcontroller is the most important module of a servo control system, which sends controlled digital signal, translates analog to digital signal by A/D module, and drives the motor by pulse width modulation (PWM) module [19]. To every microfeeding, shift detection result must be feedback to the microcontroller, to adjust the output shift in order for precise control of servo control system. The servo control can be operated based on different algorithms or principles such as predicting the gap distance and offsetting tool position, ignition delay time, average gap voltage, the average delay time, etc. [20].

Ignition delay time (t_d) is an important indicator of the isolation condition of discharge gap. Larger gap width causes longer ignition delays, resulting in a higher average voltage. Tool feed speed increases when the measured average gap voltage is higher than the preset servo reference voltage and vice versa [18]. On the contrary, the feed speed decreases, or the electrode is retracted when the average gap voltage is lower than the servo reference voltage, which is the case for smaller gap widths resulting in a smaller ignition delay. Thus, short circuits caused by debris particles and humps of discharge craters can be avoided. Also, quick changes in the

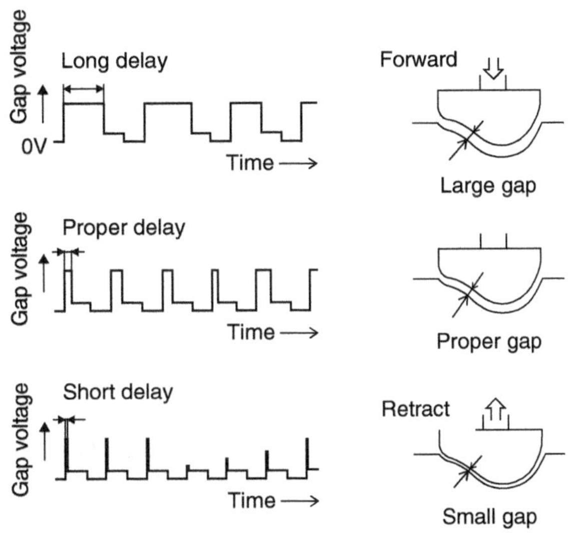

Figure 16.5 Principle of servo control system based on ignition delay. *Source:* Kunieda et al. 2005 [13]. Reproduced with permission of Elsevier.

working surface area, when tool electrode shapes are complicated, do not result in hazardous machining. In some cases, the average ignition delay time is used in place of the average gap voltage to monitor the gap width [21]. In other attempts, gap-monitoring circuits were developed to identify the states and ratios of gap open, normal discharge, transient arcing, harmful arcing, and short circuit [22]. These ratios were used as input parameters for online EDM control based on various control strategies. The servo feed control shown in Figure 16.5 keeps the working gap at a proper width [13].

16.3.3 Dielectric Circulation System or Flushing Unit

The dielectric circulation system is the integral part of the EDM system. The dielectric system consists of dielectric fluid, dielectric reservoir, pump to supply the dielectric fluid into the work tank, filter to remove the debris particles from the dielectric and to ensure recirculation of fresh dielectric to the machining zone, pipe and nozzles to supply the dielectric into the gap between the workpiece and the electrode, and flushing pressure control valve to maintain/control the flushing pressure of the dielectric. Dielectric fluid is the most important component of the flushing unit, provides a known electrical barrier between the electrode and workpiece, and acts as a means for removal of spark debris from the spark gap. The major functions of the dielectric circulation system or flushing unit are as follows:

1) To distribute the dielectric flow through the spark gap to remove gaseous and solid debris generated during EDM
2) To introduce fresh and clean dielectric fluid to the machining region
3) To flush away the chips or metal particles generated in the spark gap
4) To maintain the dielectric temperature well below its flash point
5) To act as a cooler for cooling the electrode and workpiece

16.4 Analysis of the Pulses Used in the EDM Process

Depending upon the situation in the gap which separates electrode and workpiece in EDM, principally four different electrical pulses may be distinguished: open circuit or open voltage,

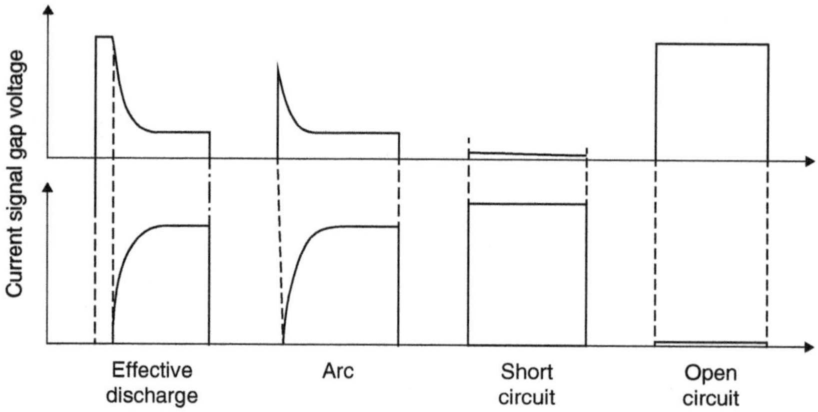

Figure 16.6 Different types of pulses generated during the EDM process. *Source:* Snoeys et al. 1982 [23]. Reproduced with permission of Elsevier.

effective discharges or real sparks, arcs, and short circuits [23]. They are usually defined on the basis of time evolution of discharge voltage and (or) discharge current as shown in the Figure 16.6 [23]. Their effect upon material removal and tool wear may differ quite significantly.

Open voltages, occurring when the distance between electrode and workpiece is too large, obviously do not contribute to any material removal or electrode wear. When contact between tool and workpiece takes place, a short circuit occurs, which also does not contribute to material removal. The range of the electrode distances in between these two extreme cases can be considered to be a practical working gap yielding actual discharges, i.e. sparks and arcs. Both pulse types do show a characteristic voltage drop across the gap during a pulse. The difference between sparks and arcs is quite difficult to establish. It is believed that arcs occur in the same spot or on the electrode surface and may therefore severely damage tool and workpiece. It is assumed that arcs occur when the plasma channel of the previous pulse is not fully deionized; the current during the following pulse will flow by preference along the same current path. Therefore, in such a case, no time is required to form a new gaseous current path. The formation of the gaseous channel is normally considered to be necessary to initiate a new spark breakdown. This peculiarity of EDM arcs is often proposed as a discrimination characteristic with respect to effective discharges or real sparks. It is believed that only "sparks" really contribute to material removal in a desired mode. Until now, it remains an open question how much arcs contribute in terms of material removal and tool wear.

Arc and short-circuit pulses are considered as harmful pulses, and normal pulses are the most desired type of pulses in EDM [24]. The objective of any successful EDM process is to increase the percentage of normal discharge pulses and reduce percentage of ineffective or harmful pulses. Different pulse discrimination system or adaptive control system can be used to calculate and control percentage of pulses, thus controlling the machining stability and performance of EDM [25, 26].

16.5 Brief Overview of the EDM Parameters

The parameters of the EDM process can be broadly categorized into operating parameters or process-control parameters and performance parameters or machining characteristics. The operating parameters are those that have significant effect on the machining performance.

Changing those operating parameters based on the performance requirement and/or applications can change the machining conditions. The operating parameters can be electrical, which the discharging unit and pulse generator maintain. These parameters are polarity, open circuit voltage or gap voltage, peak current, pulse duration, pulse interval, pulse frequency, duty factor, etc. The nonelectrical parameters are mainly the material properties of workpiece, electrode, and dielectric. Another important set of operating parameters is the motion and gap control parameters, which are primarily responsible for the machining stability and accuracy. The motion parameters include electrode and workpiece rotation and vibration, tool electrode geometry, design and shape, wire tension and vibration, etc. Servo feed and EDM speed are the two important gap control parameters. In addition, the flushing types and flushing pressure also have significant effect on the EDM process performance.

The performance parameters or machining characteristics are the outcomes by which the final product is evaluated. The most widely used performance parameters are MRR, electrode or tool wear ratio (TWR), and surface roughness (SR). In addition, the spark gap, gap width, or overcut is used for the measurement of dimensional accuracy of the part.

The performance parameters are greatly dependent on the operating parameters. The MRR greatly depends on the process parameters. A higher value of discharging voltage, peak current, pulse duration, duty cycle, and lower values of pulse interval can result in higher MRR. On the other hand, the pulse condition with shorter discharge duration and lower peak current brings about both lower tool electrode wear ratio and better surface roughness. It is seen that the tool wear characteristics are associated with material properties, especially the boiling point [27]. The volumetric wear ratio of the electrode becomes small for the electrode material with high boiling point, high melting point, and high thermal conductivity, which is independent of the workpiece materials [28]. The increase in gap voltage, peak current, and pulse duration usually results in increased electrode wear ratio. The surface topography and roughness are largely determined by eroded crater size and crater uniformity relating to discharge pulse energy [29, 30]. Low open circuit voltage, shorter pulse duration with higher frequency, and enough higher pulse interval produce small craters and hence, lesser surface roughness [31]. The surface roughness slightly increases with the increase of peak current value up to certain level and then vigorously increases with any increase in peak current [32]. When compared against different dielectric fluid pressure, surface roughness shows slightly decreasing trend with increasing pressure. Again, depending on the electrical and thermal properties of the workpiece and electrode materials, the surface roughness varies [28]. It has been found that the spark gap has a proportional relationship with the gap voltage [33]. However, the spark gap and surface roughness are also influenced by pulse on time. The peak current and the applied energy influence the spark gap. Therefore, the main parameters affecting spark gap were identified as open circuit voltage, peak current, and pulse on time.

16.6 EDM Variants: Working Principles and Application Examples

16.6.1 Die-sinking EDM

Die-sinking EDM, also called sinker EDM, cavity type EDM, or volume EDM is one of the two most common types of EDM. In die-sinking EDM, the electrode is shaped and produces its negative form into the workpiece. The typical part geometry is a complex 3D shape, often with small or odd-shaped angles. The wear has to be very low, in order to keep the electrode's original shape unmodified during the whole machining process. The electrode is generally in copper or graphite, and the dielectric is oil. The asymmetry in the MRR is thus crucial for die-sinking EDM. The servo controller monitors the gap conditions (voltage and current) and synchronously controls the different axes to machine the mirror image of the tool. Figure 16.7 shows the schematic representation of the die-sinking EDM [34].

Die-sinking EDM is mainly used to produce injection molds and complex 3D structures in industry. High wear-resistant composites based on refractory materials such as tungsten–copper or cemented carbide are preferred in die-sinking EDM. Conventional flushing strategies cannot be used in micro die-sinking EDM, since flushing through the electrodes is not possible, because of their small dimensions. Together with the extremely small gap width, the poor flushing conditions put great demands on the feed control in die-sinking micro-EDM [28]. Figure 16.8 presents application examples of die-sinking EDM for machining dies for mobile phone mold [35] and cavities for micromixing device [36].

16.6.2 Wire EDM

In wire-EDM, the electrode is a continuously circulating metallic wire (typical diameter 0.1 mm, generally in steel, brass or copper), which cuts the workpiece along a programmed path. The basic mechanism of wire-EDM is same as EDM, and the material is removed as a result of

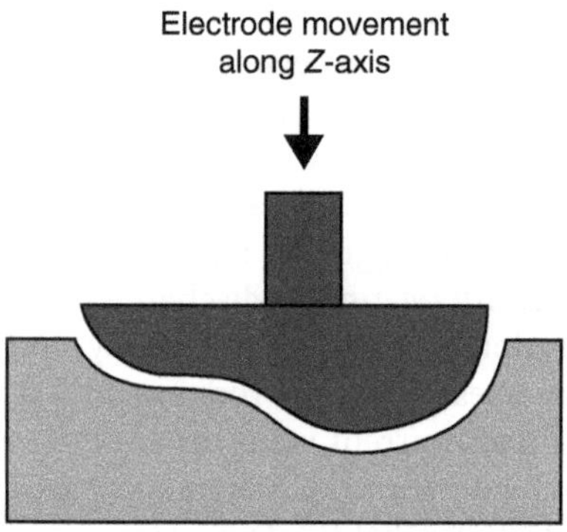

Figure 16.7 Schematic diagram showing the principle of die-sinking EDM. *Source:* Bleys et al. 2004 [34]. Reproduced with permission of Elsevier.

(a)

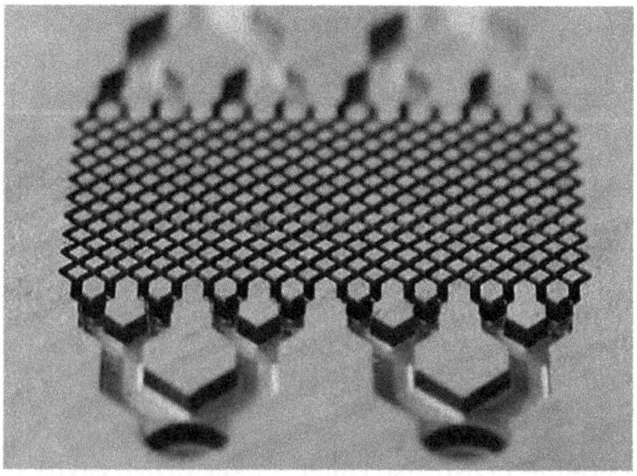

(b)

Figure 16.8 Application examples of die-sinking EDM: (a) die for mobile phone cover making [35]; (b) die-sinking electrode for micromixing device made of fine-grained graphite. *Source:* Uhlmann et al. 2005 [36]. Reproduced with permission of Elsevier.

series of electric sparks between the workpiece and wire electrode. Since controlling the horizontal position of the upper wire guide relative to the lower guide can change wire orientation, all types of ruled surfaces can be cut by wire-EDM process. As discharge currents with a high peak value over a short duration are used, current is supplied through both the upper and lower feeding brushes to obtain a quick rise in the discharge current by reducing the inductance and to avoid wire breakage due to Joule heating. Tension is applied to the wire to reduce vibration and deflection, which deteriorates cutting accuracy. Mostly, the deionized water is used as dielectric rather than dielectric oil, and the most commonly used wires are brass, tungsten, and coated wire. The wire is capable of achieving very small cutting angles. The wear is a lesser problem than in die-sinking because eroded portions of the wire are continuously replaced by new ones due to the wire circulation. Figure 16.9 represents the schematic illustration of the wire-EDM process.

Besides separating workpieces, this technique enables the fabrication of highly complex recesses and complex components and is used in prototyping, in tool and mold manufacturing, and in series production. Wire-EDM is the most commonly used process for the direct production of micromechanical devices and for microstamping tools. Axisymmetric products can be machined by micro wire EDM method with a rotating workpiece system [37]. Figure 16.10

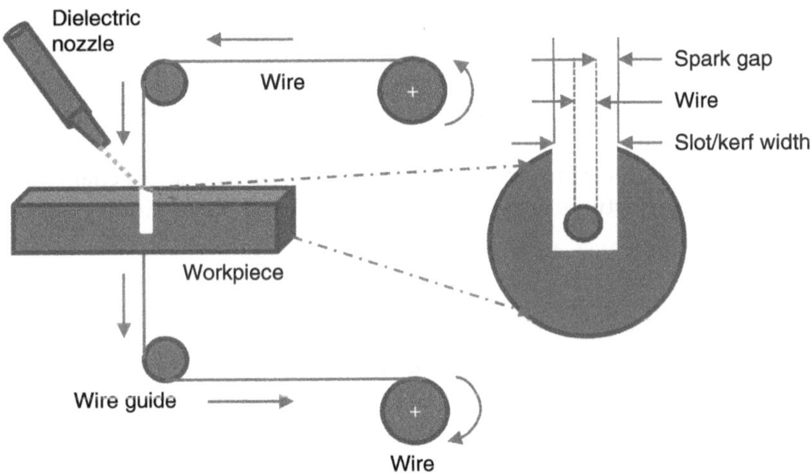

Figure 16.9 Schematic diagram showing the working principle of the wire-EDM.

(a) (b) (c)

Figure 16.10 Application examples of wire EDM: (a) complex molds for injection molding [35], (b) Gears machined by wire EDM [38], and (c) Chinese pagoda (1.25 mm × 1.75 mm) machined by the wire EDM. *Source:* Liao et al. 2005 [39]. Reproduced with permission of IOP Publishing.

shows the features machined by EDM and micro-EDM. The images of complex mold [35] and gears [38] machined by wire-EDM are shown in Figure 16.10a,b, respectively. The capability of micro wire EDM is fully exploited in the machining of a complex Chinese pagoda (1.25 mm × 1.75 mm) shown in Figure 16.10c [39].

16.6.3 EDM Milling

EDM milling is an improvement on EDM. Similar to numerically controlled milling, it uses a universal tool electrode of simple shape (e.g. spherical or cylindrical) that moves along a complex path to machine hard materials. In this process, usually tubular or cylindrical microelectrodes are employed to produce the desired complex shape by scanning. A cylindrical electrode rotates around its axis (Z-axis) with the scanning movements in X and Y directions. The contour of a particular layer is specified in the part program of CNC. However, electrode compensation is an important factor to consider as electrode length is reduced after scanning every layer. Figure 16.11 shows the schematic representation of milling EDM [34]. The EDM milling process solves the problem of manufacturing accurate and complex-shaped electrodes for die sinking of three-dimensional cavities. In addition, EDM milling enhances dielectric flushing due to high-speed electrode rotation. However, the main limitation of the EDM milling is that complex shapes with sharp corners cannot be machined because of the rotating tool electrode.

EDM milling is now replacing the die-sinking EDM in many applications of producing dies and 3D cavities, as die-sinking EDM requires fabrication of complex 3D electrode using other machining processes. Both blind 3D cavities and through structures can be machined using the EDM milling process. The application of EDM milling is more pronounced at microscale, as fabrication of microelectrodes with 3D features for die-sinking micro-EDM is a real challenge. Figure 16.12 shows example of 3D structures fabricated by the EDM and micro-EDM process [40–42].

16.6.4 EDM Drilling

In EDM drilling, electrodes are used to “drill” holes in the workpiece. Both tubular and solid electrodes are used for the drilling purpose. For tubular electrodes, jet flushing inside the tube is used, whereas for solid electrodes, dielectric is fed to the machining zone by either suction or injection through predrilled holes. The problem with deep small-hole drilling by EDM is that forming and clamping the long electrode is difficult. Therefore, the microelectrodes are used to fabricate on-machine with high accuracy using different micro-efficiency deep grinding (EDG) processes. This also solves the clamping problem because the electrode is clamped

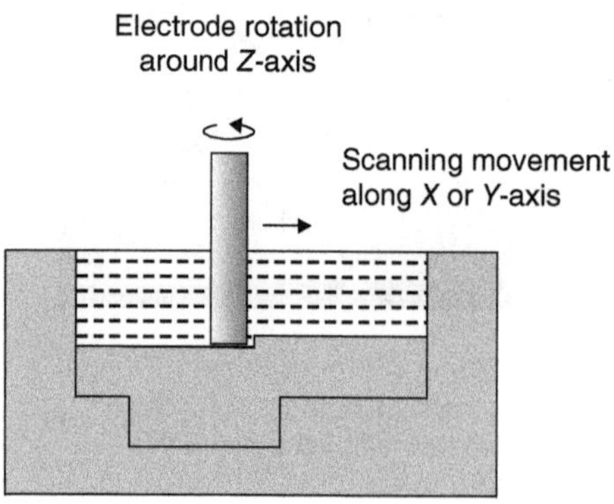

Figure 16.11 Schematic showing the principle of milling EDM [34].

Figure 16.12 Application examples of EDM milling: (a) micropyramids produced by 3D EDM milling [40]; (b) 1/8 ball in a square cavity [41], and (c) a complex 3D cavity by the CLU method (combination of the linear compensation method [LCM] and the uniform wear method [UWM]). *Source:* Yu et al. 2010 [42]. Reproduced with permission of IOP Publishing.

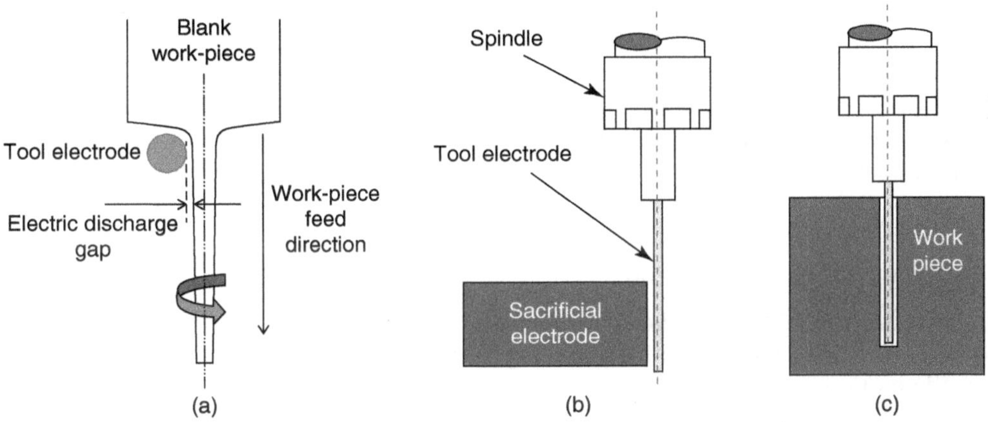

Figure 16.13 EDM drilling process: (a) fabrication of electrode using wire electro-discharge grinding (WEDG) process; (b) fabrication of electrode using block-EDG process; and (c) EDM drilling using the fabricated electrodes. *Source:* Jahan et al. 2011 [44]. Reproduced with permission of Elsevier.

before the forming process and never reclamped until the hole is machined [43]. Figure 16.13 shows the consecutive processes of on-machine electrode fabrication by wire electro-discharge grinding (WEDG) and Block-EDG, and EDM drilling with fabricated electrode [44].

All kinds of microholes; e.g. irregular, tapered, curved, as well as inclined holes can be produced by EDM drilling process. Machining of high-aspect ratio holes for fuel injection nozzle and cooling channels in turbine blades made of hard alloys are few typical applications of EDM drilling. Figure 16.14 shows the fabrication of high-aspect ratio microholes with different diameters [45] and application of micro-EDM drilling in automobile and aerospace industries [46].

16.6.5 Rotary Disc Electrode Electrical Discharge Machining (RDE-EDM)

The rotary disc electrode electrical discharge machining (RDE-EDM) is one of the variant process of EDM process based on removing material by means of a series of recurring electrical discharges between a rotary tool called disc (thickness ranging from micron level to 1 mm) and the work material in the presence of a dielectric fluid (kerosene, distilled water). The disc is commonly made of conductive hard materials such as tool steel or carbide and the thickness

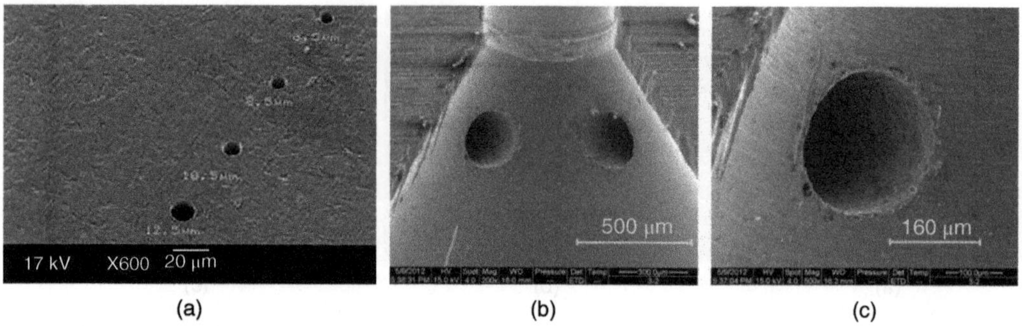

Figure 16.14 Application examples of EDM and micro-EDM drilling: (a) 6.5-μm hole machined on 50 μm stainless steel plate [45]; (b) fuel injector for diesel engine in automobile [46]; and (c) magnified image of microhole shown in Panel (b). *Source:* Hao et al. 2013 [46]. Reproduced with permission of Elsevier.

ranging from several microns to few millimeters. The commonly used dielectric fluids for this process are kerosene and deionized water [47]. The cutting speed of the RDE-EDM process is reported to be twice of that of the conventional abrasive sawing method [48, 49].

The process is used for cutting of billets and bars with smaller kerf and burr-free surface. Figure 16.15 shows the schematic representation of the working principle of RDE-EDM process [48]. The fabricated microstructures using the RDE-EDM process for different applications are presented in Figure 16.16 [48, 50, 51].

16.6.6 Dry or Near-Dry EDM Milling

The dry or near-dry EDM can be die-sinking type or milling type. However, the application of dry EDM is found to be more effective as milling EDM due to the improved flushing of debris by electrode motion. The dry EDM uses oxygen or air as the dielectric medium [52]. On the other hand, the near-dry EDM uses liquid–gas mixture or mist as the dielectric [53]. The mist dielectric can be mixture of gas medium (air/nitrogen) and liquid (water/kerosene) [54]. In this process, usually a thin-walled tubular electrode is used through which high-pressure gas or air is supplied to the machining zone, as can be seen from Figure 16.17 [52]. The role of the gas is to act as dielectric, removing the debris from the gap, and cooling of the inter electrode gap. Although the dry and near-dry micro-EDM has been found to be a variant of die-sinking or milling micro-EDM with gas medium of dielectric, there are significant differences in plasma characterization, gap control, or even the material removal mechanism [54].

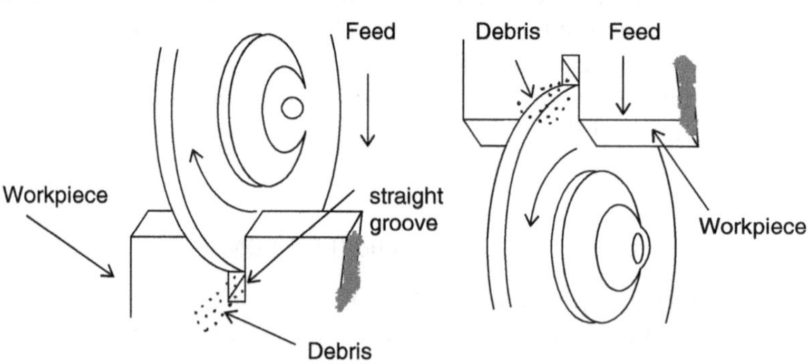

Figure 16.15 Schematic diagrams showing the working principle of rotary disc electrode electrical discharge machining (RDE-EDM). *Source:* Chow et al. 1999 [48]. Reproduced with permission of Elsevier.

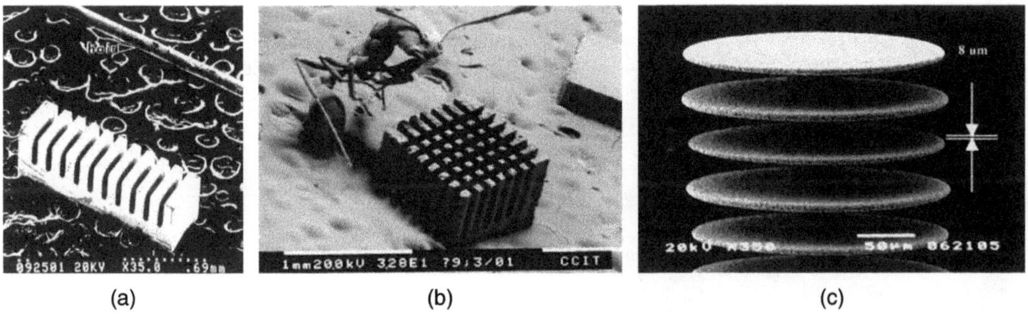

Figure 16.16 Application examples of RDE-EDM; (a) scanning electron microscopy (SEM) image of a multiple-slit microstructure in Ti–6Al–4V fabricated by rotating disc electrode (RDE). *Source:* Chow et al. 1999 [48]. Reproduced with permission of Elsevier. (b) SEM image of a multiple-slit micro-structure in Ti–6Al–4V by RDE electrode. *Source:* Chow et al. 2008 [50]. Reproduced with permission of Elsevier. (c) Series pattern microdisc fabricated by multiple rotating disk electrode (MRDE). *Source:* Kuo et al. 2004 [51]. Reproduced with permission of Elsevier.

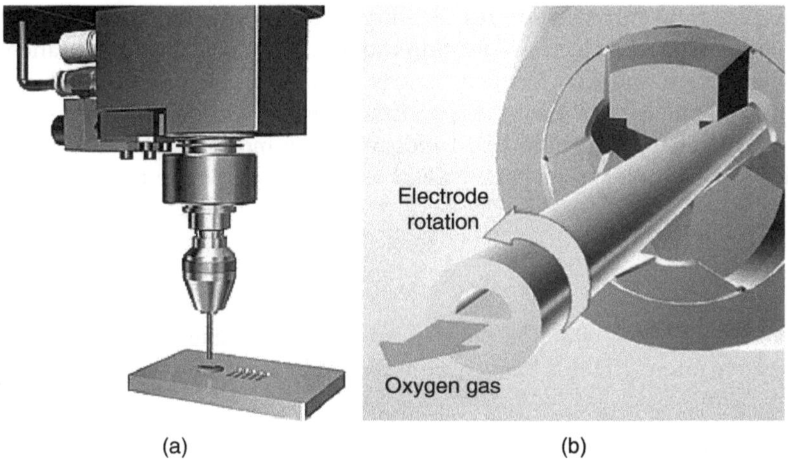

Figure 16.17 Schematic diagram showing the working principle of the dry EDM process using gas as dielectric. *Source:* Yu et al. 2004 [52]. Reproduced with permission of Elsevier.

So far, the dry EDM has been used for replacing conventional die-sinking and EDM milling. As the removal of debris becomes more difficult in die-sinking EDM without the help of dielectric flushing, the dry EDM milling is found to be used more extensively than die-sinking. Dry EDM milling is capable of machining spherical surface [52], fine-curved slots [55], and mirror-like machined surface [53] as reported in literature. Figure 16.18 shows example of machining spherical surface using dry EDM milling [52].

16.6.7 Planetary EDM

A significant enlargement of the field of application of die-sinking EDM or EDM drilling is the so-called planetary EDM. In the planetary or orbital EDM, a spatial translation movement of the tool electrode overlaps the conventional rectilinear sinking movement. The planetary or orbital EDM using the movement of the electrode was found to be very useful particularly in drilling of microholes as flushing is more difficult for a thin electrode [56]. A common problem

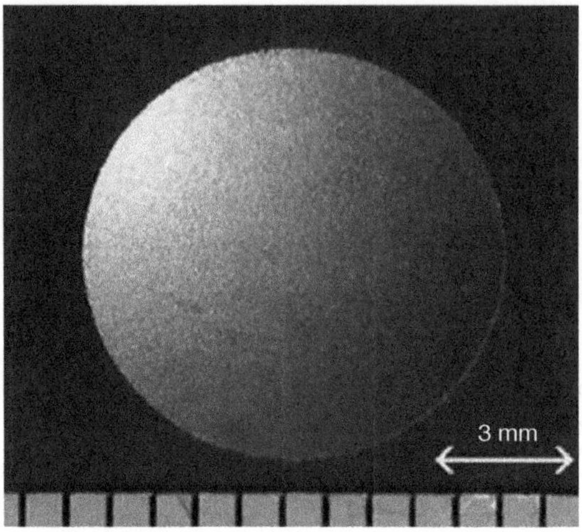

Figure 16.18 Application examples of dry EDM milling in machining of spherical surface. *Source:* Yu et al. 2004 [52]. Reproduced with permission of Elsevier.

during the die-sinking EDM or EDM drilling is the debris accumulation, which becomes worst during machining high-aspect ratio microstructures. Adding a relative motion between the electrode and workpiece, other than the electrode feeding motion, produces a wide clearance between them for fluid circulation and then reduces debris concentration, resulting a high MRR, low electrode wear ratio, and higher machining accuracy. Figure 16.19 shows the schematic representation of the planetary or orbital EDM indicating the motion and path of the electrode [57]. The application examples of planetary EDM and micro-EDM are presented in Figure 16.20 [58, 59].

16.6.8 Reverse EDM

The reverse EDM is a variant of EDM, which is being used more extensively at microscale nowadays [60–62]. The reverse micro-EDM is being accomplished by reversing the polarity so

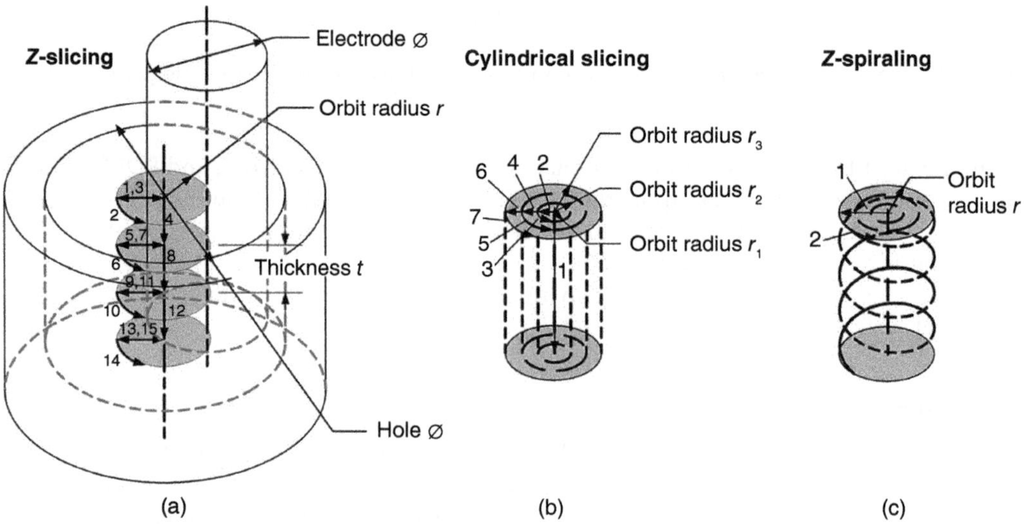

Figure 16.19 Schematic diagram showing the working principle of planetary EDM. *Source:* Bamberg and Heamawatanachai 2009 [57]. Reproduced with permission of Elsevier.

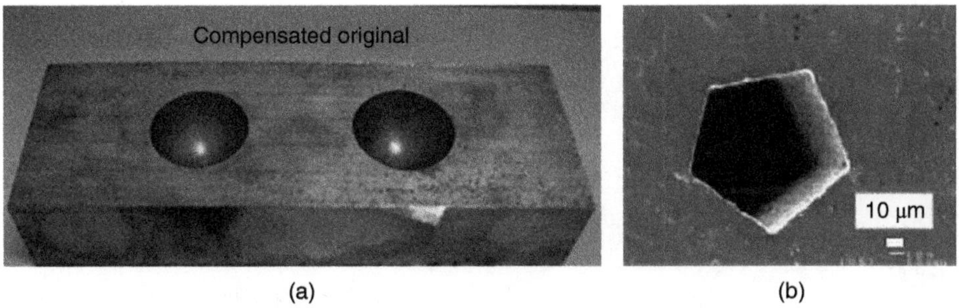

Figure 16.20 Application examples of planetary EDM; (a) blind cavities fabricated by the original and orbital electrode movement. *Source:* Wang and Shan 2005 [58]. Reproduced with permission of Elsevier. (b) Pentagonal microhole using orbital micro-EDM with electrode feed of 100 μm and offset of tool paths by 15–20 μm. *Source:* Yu et al. 2002 [59]. Reproduced with permission of Elsevier.

that the material erodes from the larger diameter rod. The polarity of the electrode and sacrificial workpiece is interchanged during reverse micro-EDM, so that the electrode can be extruded among the arrays of microholes. In this process, the electrode is considered as workpiece and is fed down to the holes of the metal plate to be machined by electrical discharge, which occurs within machining gap. However, the regions that correspond to the holes are not machined. Finally, microelectrodes are machined as many as the holes. The principle of reverse micro-EDM is presented schematically in Figure 16.21. The fabricated arrays of microelectrodes with similar or different diameters using the reverse EDM/micro-EDM are shown in Figure 16.22 [61, 62].

16.7 Examples of Research Advances in EDM and Micro-EDM

Although EDM process is found to be capable of machining difficult-to-cut materials irrespective of their hardness, it suffers from several shortcomings such as lower machining speed, heat-affected zone, higher tool wear, poor surface finish, etc. Therefore, in recent years, the research focus has been moved toward the development of hybrid machining processes combining EDM with conventional machining processes. The method of two constituent processes may be applied

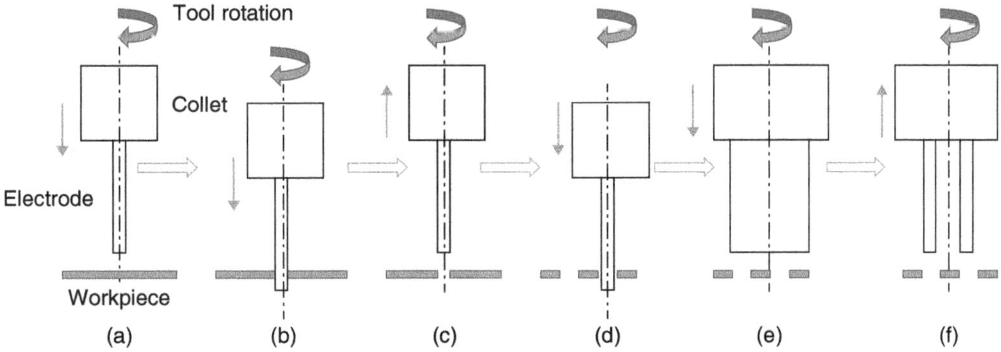

Figure 16.21 Schematic representation of fabricating microelectrode using reverse micro-EDM, (a–d) fabricating microhole using conventional EDM and (e, f) using the fabricated microhole in reverse micro-EDM to make microelectrode.

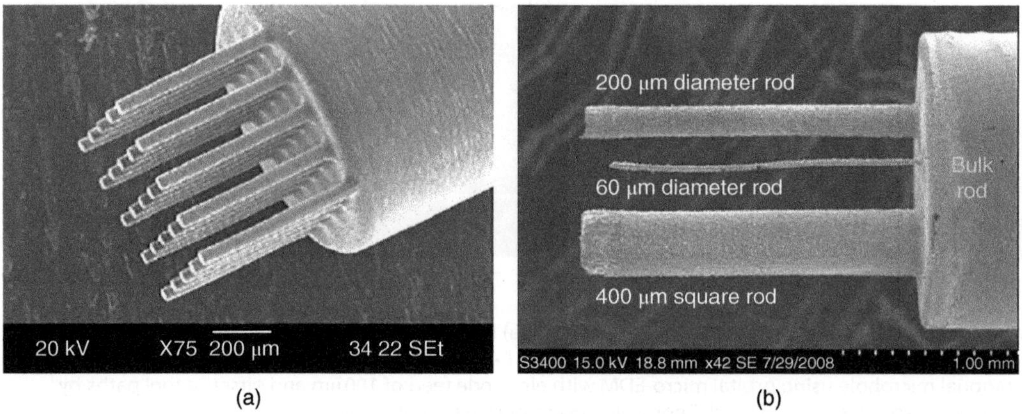

Figure 16.22 Application examples of reverse EDM/micro-EDM: (a) arrays of microelectrodes fabricated by the reverse micro-EDM. *Source:* Kim et al. 2006 [61]. Reproduced with permission of IOP Publishing. (b) Microelectrodes of three different diameters in one step using reverse micro-EDM. *Source:* Mujumdar et al. 2010 [62]. Reproduced with permission of SAGE Journals.

sequentially and simultaneously to constitute compound and hybrid machining processes, respectively. Compound machining is defined as the combination of two different machining processes in a single setup applied one after another. On the other hand, the hybrid machining process is defined as the integrated application or combination of different physical active principles in a single process. In the following sections, some recent developments in the areas of EDM-based compound and hybrid machining processes along with application examples will be presented.

16.7.1 Abrasive Electro-Discharge Grinding (AEDG)

Abrasive electro-discharge grinding (AEDG) is a hybrid process, which combines EDM and grinding. In AEDG process, mechanical abrasion of a metal bonded diamond wheel is combined with the electro-erosion of EDM, as can be seen from Figure 16.23a [63]. The removal of conductive or partially conductive material is by a combination of rapid, repetitive spark discharges between workpiece and rotating tool, separated by a flowing dielectric fluid and also by a mechanical action of irregularly shaped abrasive particles on the periphery of the wheel [63]. Simultaneous synergic mechanical energy and energy of electric spark discharges increase the machining productivity and improve surface quality [64, 65]. Therefore, material is removed by the combined effect of electro-erosion and microcutting process (mechanical effect of abrasives), as shown in Figure 16.23b [66]. An increase in performance measures of the AEDG process becomes evident when machining superhard materials, engineering ceramics, sintered carbides, and metal composites. It can be seen from Figure 16.24a that the MRR increases significantly in AEDG process compared to conventional grinding [63]. However, the surface finish should be optimized based on the requirement of the product performance. If the grinding process is allowed to dominate in material removal, then smoother surface finish can be obtained [66]. On the other hand, if the primary objective is faster machining speed, then EDM will play the major role in material removal mechanism, and hence, comparatively poorer surface finish will be obtained (Figure 16.24b) [66].

16.7.2 Abrasive Wire EDM (AWEDM)

The working principle of the abrasive wire electro-discharge machining (AWEDM) process is nearly same as the AEDG process. In this process, the abrasive grains are embedded within the

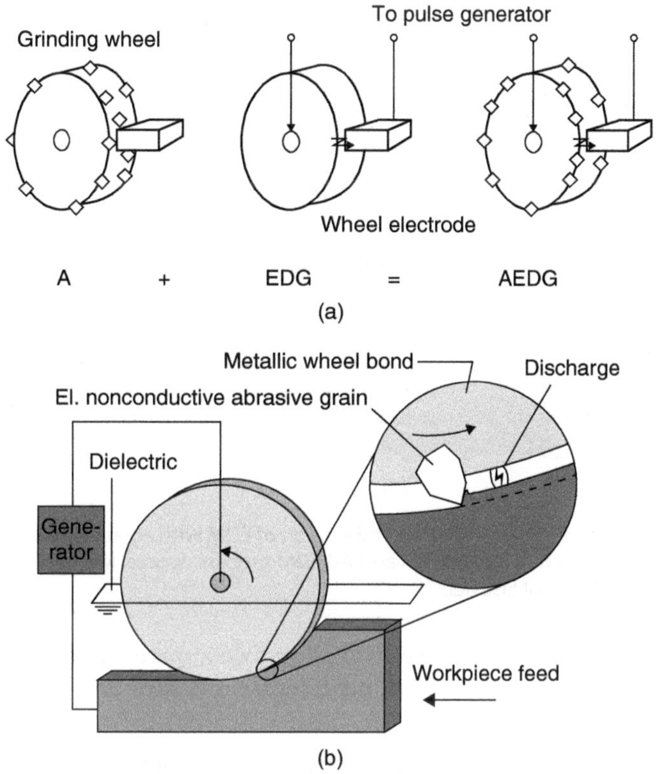

Figure 16.23 (a) Principle of abrasive electrical discharge grinding (AEDG) process. *Source:* Kozaka et al. 2016 [63]. Reproduced with permission of Elsevier. (b) Schematic diagram showing the process mechanism. *Source:* Koshy et al. 1997 [66]. Reproduced with permission of Elsevier.

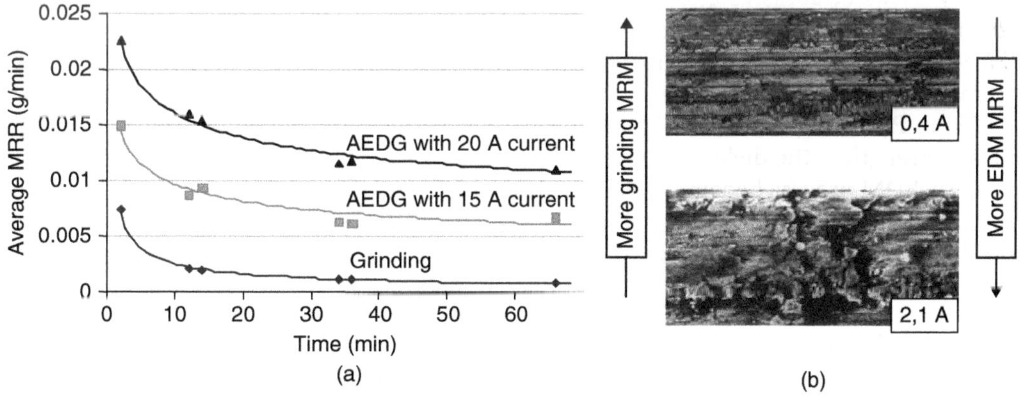

Figure 16.24 (a) Comparison of the MRR between grinding as a single process and AEDG. *Source:* Kozaka et al. 2016 [63]. Reproduced with permission of Elsevier. (b) Controlling surface finish by changing the intensity of either EDM or grinding process. *Source:* Koshy et al. 1997 [66]. Reproduced with permission of Elsevier.

wire used to perform the wire electro discharge machining (WEDM). Therefore, the material removal process is the combination of spark erosion and the mechanical abrasion [67, 68]. The combined mechanical and electro-discharge action helps to minimize the recast layer and at the same time, enhances the MRR [26]. The working principle of the AWEDM process is presented in Figure 16.25a [68]. The reduction or elimination of recast layer in AWEDM compared to conventional WEDM can be seen from Figure 16.25b [68]. The experiments were

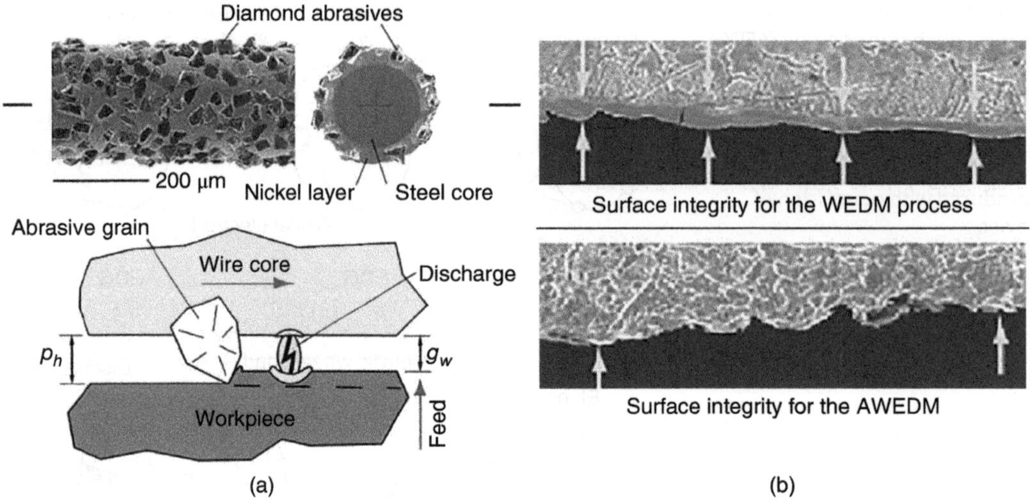

Figure 16.25 (a) AWEDM: combination of EDM with an abrasive process; (b) comparison of recast layer formation between WEDM and AWEDM process. *Source:* Menzies and Koshy 2008 [68]. Reproduced with permission of Elsevier.

conducted on a nickel 600 alloy to compare the extent of the recast layer in WEDM and AWEDM, as the machining of recast-free surfaces is of interest.

16.7.3 Powder-Mixed EDM and Micro-EDM

The major difference of the powder-mixed electro-discharge machining (PMEDM) with the previously discussed abrasive-assisted EDM processes is that in PMEDM, conductive and semiconductive powder materials are mixed with the dielectric instead of with the tool electrode [69]. Another difference lies in the material removal mechanism of both the processes. Unlike the abrasive-assisted EDM processes, the abrasive action of the powder is very minimal or negligible in PMEDM [70]. In PMEDM process, electrically conductive powder reduces the insulating strength of the dielectric fluid and increases the spark gap between the tool and the work piece. EDM process becomes more stable as the flushing becomes easier at higher spark gap [71]. As a result, the overall machining efficiency, MRR, and SQ improve. The sparking is uniformly distributed among the powder particles in the spark gap and thus, reducing the intensity of a single spark, which results in uniform shallow craters instead of a single broader crater. Thus, the surface finish improves. The characteristics of the powder such as the size, type, and concentration influence the dielectric performance [72]. Figure 16.26 shows the differences in the spark gap, power density, and gap explosive pressure between conventional and PMEDM [71]. Figure 16.27 shows the surface finish at different regions of a mold cavity machined by conventional EDM and PMEDM [73]. A significant reduction in the surface roughness can be observed at every region of the mold cavity after machining by PMEDM.

16.7.4 Vibration-Assisted EDM and Micro-EDM

In recent years, the assistance of vibration during the EDM and micro-EDM process has been found to be very effective in enhancing machining stability and overall EDM performance. The higher efficiency gained by the employment of vibration assistance is mainly attributed to the improvement in dielectric circulation that facilitates the debris removal and the creation

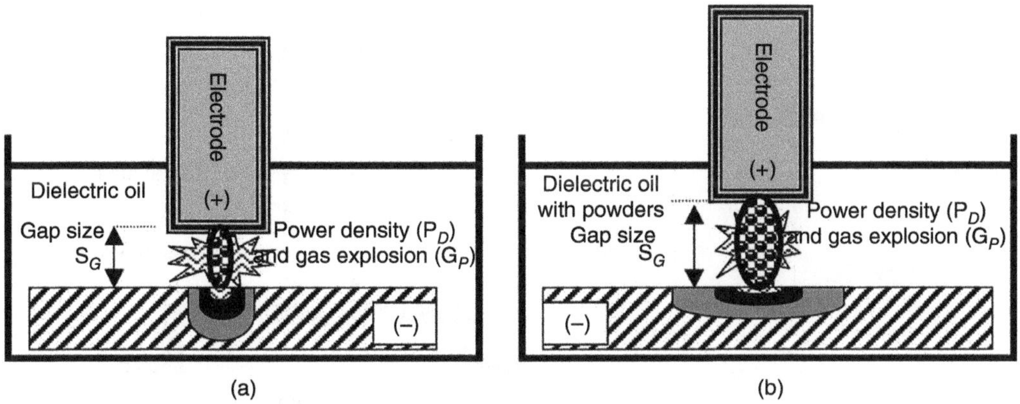

Figure 16.26 Schematic diagram showing the comparison of working principle between (a) conventional EDM and (b) powder-mixed EDM. Powder-mixed EDM has larger spark gap, lower gas explosive pressure, and lower single-crater size. *Source:* From Tzeng and Chen 2005 [71].

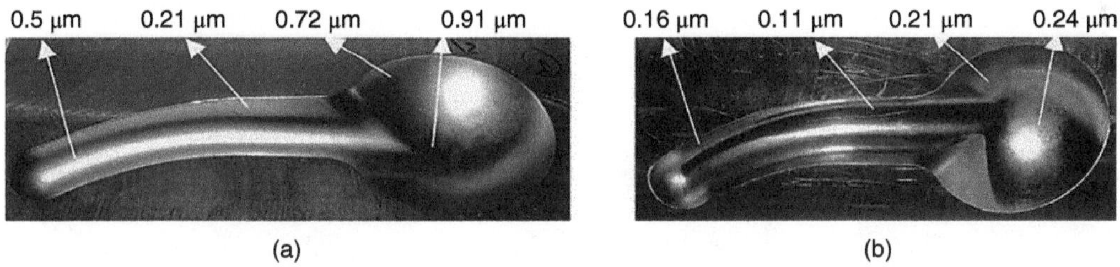

Figure 16.27 Fabrication of mold cavity (a) without and (b) with the powder-mixed dielectric. Significant reduction of roughness has been observed after applying silicon powder. *Source:* Pecas and Henriques 2003 [73]. Reproduced with permission of Elsevier.

of a large pressure change between the electrode and the work piece, as an enhancement of molten metal ejection from the surface of the work piece [74]. Research has been conducted on the feasibility of applying both ultrasonic [75] and low frequency vibration [76] during EDM. In addition, the vibration has been applied in both the tool electrode [77] and the workpiece [76]. It is found that tool electrode vibration can be applied more effectively in conventional EDM, as applying vibration to several micron tool electrodes will affect the micro-EDM performance negatively [76]. On the other hand, the workpiece vibration assistance has been found to be suitable for both EDM and the micro-EDM process [75, 76]. Although the vibration assistance is found to be more effective on micro-EDM drilling and fabrication of microfeatures, it has been practiced in conventional die-sinking and wire EDM also [78]. Figure 16.28 shows the schematic representation of the developed vibrating device and vibration mechanism for generating tool vibration [77] and workpiece vibration [75] during EDM. Figure 16.29 shows different types of microfeatures machined by vibration-assisted EDM and micro-EDM [75, 76, 79].

16.7.5 EDM/Micro-EDM and ECM-Combined Process

The objective of incorporating electrochemical machining process (ECM) with EDM or micro-EDM is to improve the surface finish generated by EDM/micro-EDM. The surface machined by EDM is relatively rough due to microcraters and cracks produced by the microdischarges.

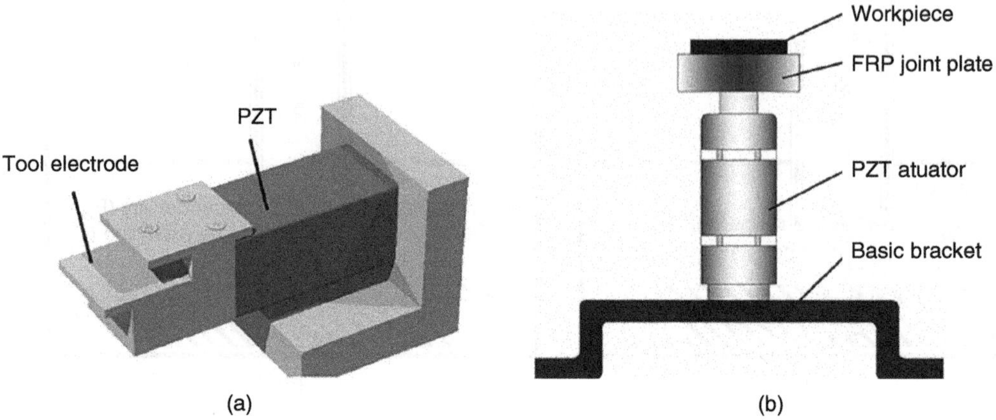

Figure 16.28 (a) Mechanism of applying vibration to the tool electrode. *Source:* Endo et al. 2008 [77]. Reproduced with permission of Elsevier. (b) To the workpiece [75] during vibration-assisted EDM. *Source:* Tong et al. 2008 [75]. Reproduced with permission of Elsevier.

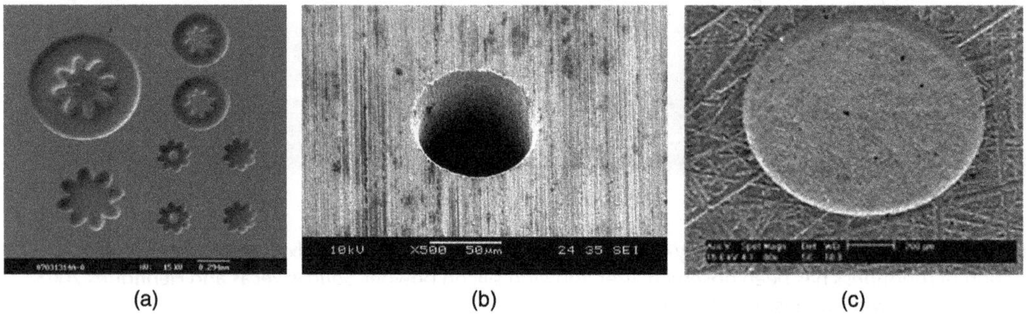

Figure 16.29 Microfeatures machined by vibration-assisted micro-EDM; (a) microstructures with noncircular cross-section by workpiece vibration-assisted micro-EDM. *Source:* Tong et al. 2008 [75]. Reproduced with permission of Elsevier. (b) High-aspect ratio (16.7) microhole machined by workpiece vibration-assisted micro-EDM. *Source:* From Jahan 2010 [76]. (c) Profile on Cu–W surface using vibration-assisted powder-mixed EDM. *Source:* Prihandana et al. 2009 [79]. Reproduced with permission of Elsevier.

The deionized water is used in the combined process as it can serve as the dielectric in the EDM process, as well as can work as an electrolyte medium for ECM at certain machining conditions [80]. The surface after applying ECM becomes much smoother compared to that of EDM, as well as the peak-to-valley distances of craters (R_{max}) reduce significantly.

In addition to sequential EDM and ECM processes, concurrent EDM and ECM processes have also been reported [81, 82], in which the discharging or dissolution takes place in the same cycle during machining. Therefore, the material removal is a combination of electro-discharge and anodic dissolution. The concurrent EDM and ECM processes have advantages of both higher MRR and improved surface finish due to combined action. Figure 16.30 shows the discharging and anodic dissolution steps and the corresponding change in gap during the micro-EDM and ECM process [80]. The improvements of surface finish both for compound and hybrid processes are presented in Figure 16.31 [82]. In addition to application of the combined process in die-sinking, the ECM-assisted wire-EDM is also investigated for the machining of low-resistance silicon wafers [83]. High efficiency of slicing ingots with low wire consumption is achieved by the use of a specific detergent capability of the electrolyte [83].

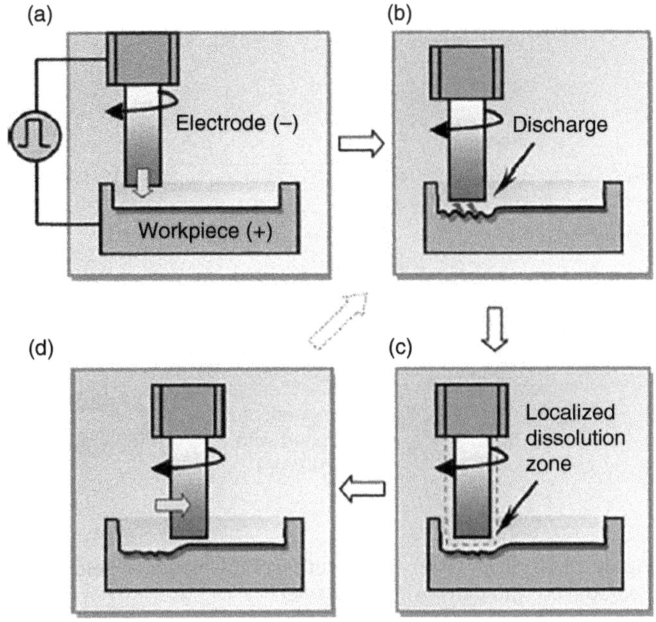

Figure 16.30 Mechanism of EDM–ECM combined process showing the discharging and dissolution cycles of EDM and ECM: (a) vertical feeding; (b) discharge; (c) dissolution; and (d) horizontal feeding. *Source:* Nguyen et al. 2013 [80]. Reproduced with permission of Elsevier.

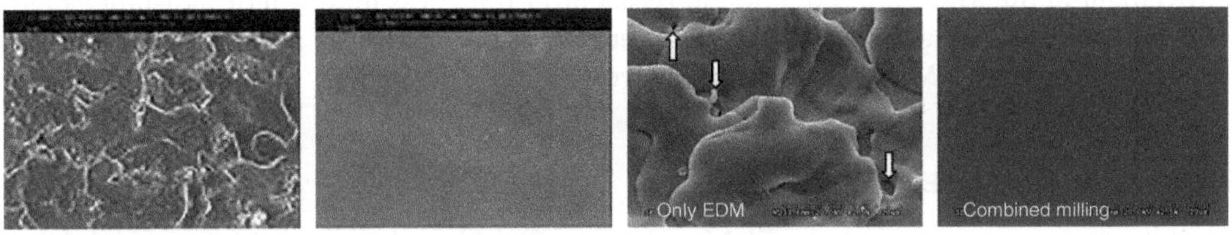

(a) (b) (c) (d)

Figure 16.31 (a) Surface generated by EDM; (b) surface generated by ECM followed by EDM (sequential process); (c) surface generated by EDM milling process; (d) surface generated by EDM and ECM simultaneous milling process. *Source:* Zeng et al. 2012 [82]. Reproduced with permission of Elsevier.

16.7.6 Magnetic Field–Assisted EDM/Micro-EDM

The magnetic field–assisted electro-discharge machining (MFEDM) is a process, where magnets are attached to EDM system, near the machining zone. The strong magnetic force developed by the magnet, as shown in Figure 16.32 [84], accelerates the removal of debris from the machining zone and thus, assists in efficient removal of debris. The gap conditions of EDM would significantly affect the stability of machining progress. The machining performance would be improved by expelling debris from the machining gap fast and easily. Implementing magnetic force perpendicular to the electrode's rotational force produces a resultant force that is efficient in transporting debris out of the hole during machining. A debris particle in a magnetic field–assisted micro EDM is subjected to two kinds of forces, namely the magnetic force and the centrifugal force. The resultant force on the debris particle is given by the vector addition of the magnetic force and centrifugal force, which helps to flush out the debris particles from the machining zone, thus improving the machining stability, MRR, reducing tool wear, and overall improving EDM performance [85]. It has been reported that for a magnetic material, the MRR increased nearly three times that of a hole cut without the magnetic field [85]. Figure 16.32 shows the schematic representation of MFEDM [84]. Figure 16.33a,b shows the

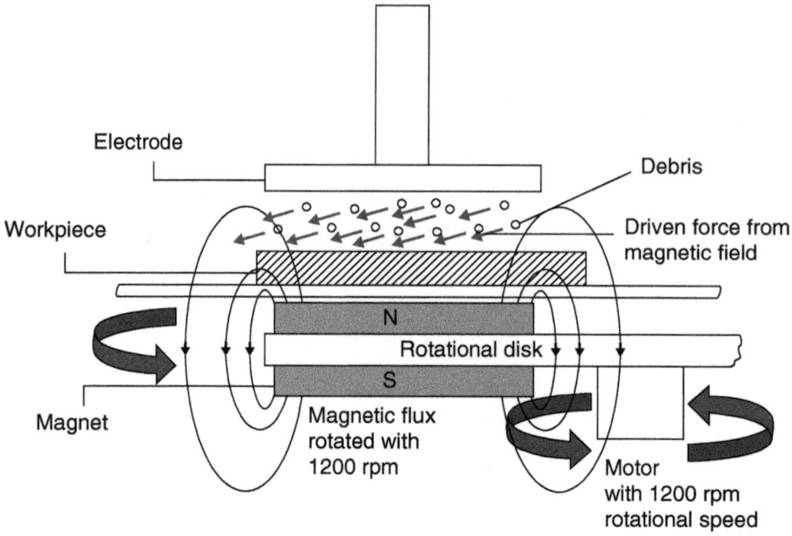

Figure 16.32 Schematic representation of the magnetic field–assisted EDM. *Source:* Lin and Lee 2008 [84]. Reproduced with permission of Elsevier.

improvement in the MRR and reduction in recast layer thickness in magnetic assistance EDM compared to conventional EDM [84]. The increment in aspect ratio of microhole in magnetic field–assisted micro-EDM can be understood from Figure 16.33c [86].

16.7.7 Laser-Assisted EDM and Micro-EDM

One of the major problems with the EDM and micro-EDM is its lower machining speed. Laser machining has comparatively higher machining speed although it suffers from the formation of heat-affected zone and recast layer. The accuracy is also an issue in laser machining. Therefore, combined laser and EDM/micro-EDM process can reduce the negative effects of each single process. The combined process, that is laser-assisted electro-discharge machining (LAEDM), has been found to be useful for micro assembly and for drilling high-quality holes.

A novel process has been developed to solve the issues by combining the micro-EDM and Nd-YAG laser welding workstation in one single setup [87, 88]. The micro-EDM process fabricates the parts, and the Nd-YAG laser performs microjoining to produce the assembly, so the whole processes from microfabrication to microassembly can be completed on the same system. The Nd-YAG laser possesses the advantageous features of heating concentration and rapid cooling and is well suited to microwelding. This can overcome the problem of small dimensional assembly. Figure 16.34a shows an example of pin-plate assembly by the combined micro-EDM and laser welding process [87]. The pins are fabricated by micro-EDM and welded to the plate by laser spot welding.

In addition, sequential laser and micro-EDM process has been invented in order to machine high-aspect ratio inclined hole with less burr and recast layer [89]. A laser beam first drills a pilot hole, then an end mill is used to machine the diffuser portion of the hole and provide a flat surface for the drill entrance side. Micromechanical drilling then finishes the holes.

Moreover, a combined micro-EDM milling and laser ablation process was reported [90]. A hybrid machine tool was developed to combine ultra-short-pulsed laser ablation and micro-EDM milling in one setup. No reclamping is necessary in this process. The limitations and challenges of the single processes have been reduced in the proposed combined process.

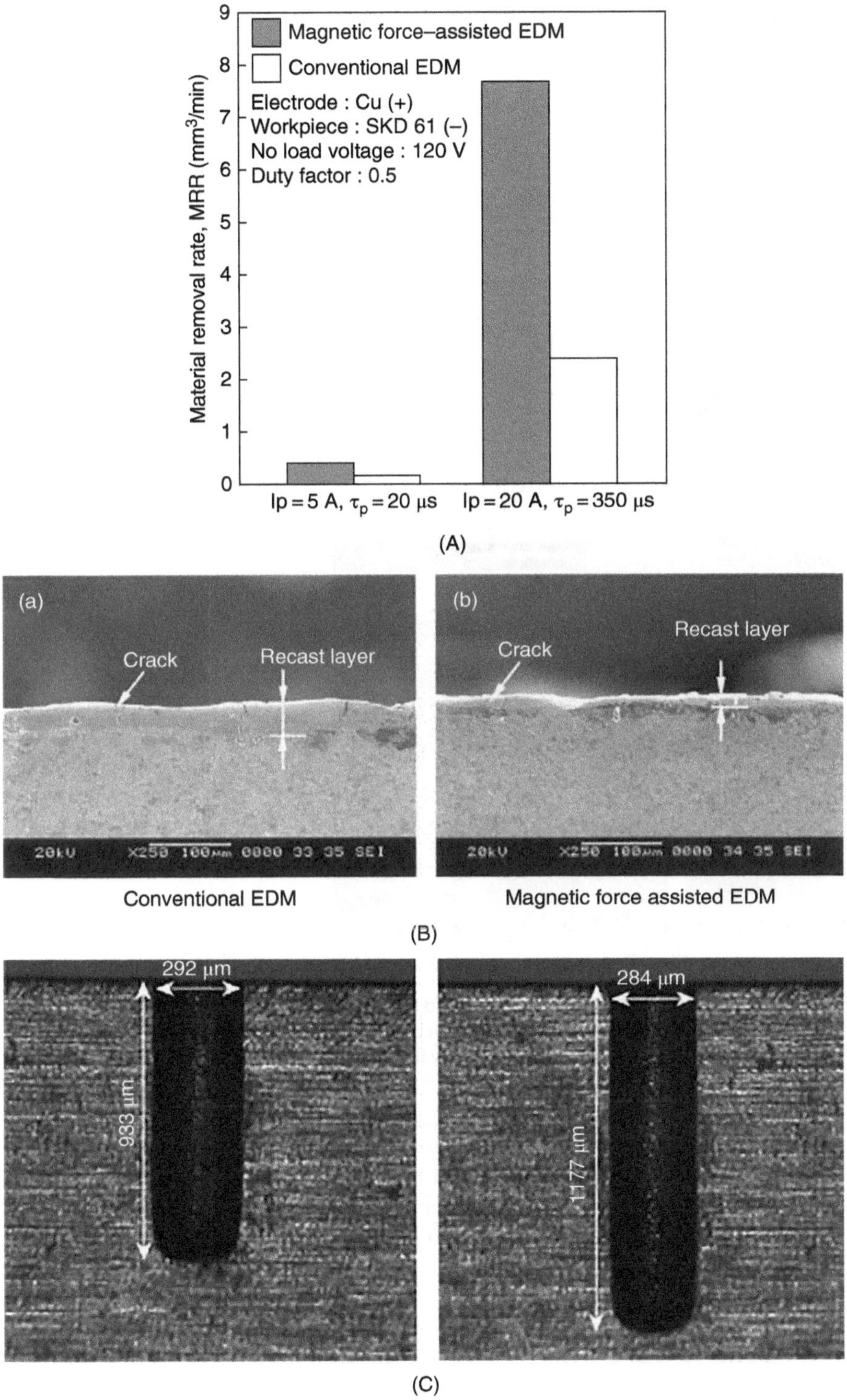

Figure 16.33 (A) Comparison of MRR between conventional and magnetic force–assisted EDM. *Source:* Lin and Lee 2008 [84]. Reproduced with permission of Elsevier., (B) recast layer thickness obtained by conventional and magnetic force–assisted EDM [84], and (C) cross-section of microhole machined in conventional and magnetic force–assisted micro-EDM. *Source:* Yeo et al. 2004 [86]. Reproduced with permission of IOP Publishing.

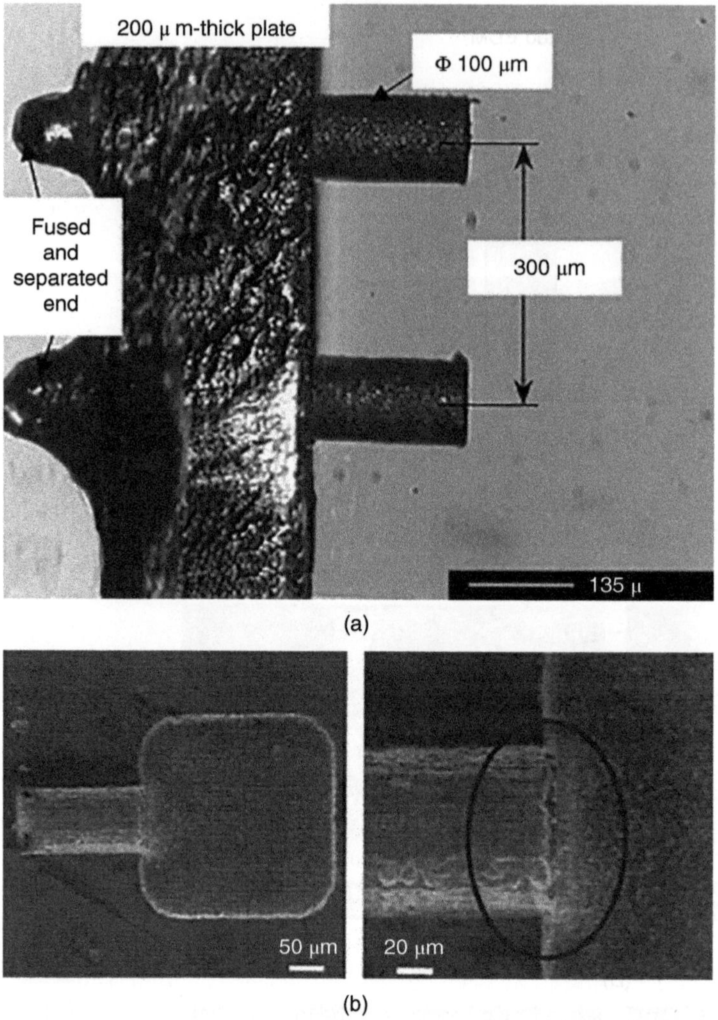

Figure 16.34 (a) Pin and plate microassembly by micro-EDM and laser welding. *Source:* Kuo et al. 2002 [87]. Reproduced with permission of Springer. (b) Micro-structure machined by combined micro-EDM and laser ablation (left part is machined by micro-EDM and right part is by laser ablation). *Source:* Weber et al. 2009 [90]. Reproduced with permission of American Vacuum Society.

A given structure is divided into different parts that can only be machined by a particular process, for example if the structure contains any part that is smaller than 100 μm, then it can only be machined by μ-EDM or it has to be machined by laser beam machining (LBM). Figure 16.34b shows an example showing a microstructure with two different pockets machined by micro-EDM and laser ablation, respectively [90].

16.8 Research Focus Toward Micro- and Nano-EDM

In recent years, due to the growing trend of miniaturization, continuous research has been carried out for applying the EDM process at micro and nano scale. Micro-EDM has become a powerful bulk micromachining technique and hence a useful tool for fabricating

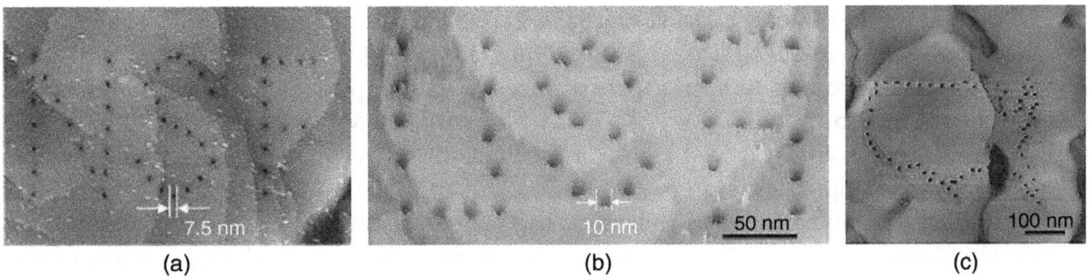

Figure 16.35 Machining examples of dry nano-EM: (a) machining of "NSF" using 50 nanofeatures with average feature size of 7.5 nm by dry nano-EM; (b) machining of letters "USA" using 10 nm diameter holes by dry nano-EM process; and (c) image of machined "Map of USA" by dry nano-EM. *Source:* Jahan et al. 2012 [96]. Reproduced with permission of Elsevier.

microstructures for microelectromechanical systems (MEMS). The micro-EDM, as a MEMS tool, can be applicable to any type of electrical conductor, including all kinds of metals and alloys as well as doped semiconductors [91]. In addition, as a noncontact machining technique, micro-EDM can be easily applied to thin, fragile, and/or soft materials regardless of their mechanical properties. Due to extensive research, several micro-EDM–based compound and hybrid processes have been developed, which can fabricate very high aspect ratio microfeatures [92]. In addition, due to the development of reverse micro-EDM and micro-EDM – LIGA (lithographie, galvanoformung, abformung) hybrid processes, mass production of the high-aspect ratio microfeatures becomes possible. Micro-EDM has been reported in various MEMS applications, such as microhole array (20 × 20 array, 20-μm diameter, 60-μm pitch) machined by LIGA-fabricated arrays of microelectrodes [93], graphite honeycomb microstructures (hexagonal pitch of 70-μm and wall thickness of 16-μm) formed by batch-mode micro-EDM [93], superhard alloy gears batch produced by micro-EDM, and 7-mm-long planar stent sample as cut by micro-EDM from 50-μm-thick stainless-steel foil [94].

In recent years, the researchers have been successful to apply EDM at nanoscale [95]. A novel nanomachining process termed as "nano-electro-machining (nano-EM)" has been developed by modifying a scanning tunneling microscopy (STM) platform. The nano-EM setup can be compared to the conventional die-sinking EDM. In the nano-EM process, Platinum–Iridium (Pt–Ir [80 : 20]) or Tungsten (99.9%W) is used as a tool electrode, atomically flat gold surface as the workpiece, and *n*-decane as the dielectric, similar to conventional EDM. It has been demonstrated that the process is capable of fabricating sub 10 nm features with consistency and accuracy under both the liquid and air dielectric media [96]. The material removal mechanism of the nano-EM is associated with the breakdown of dielectric, which generates intense heat at the gap between the nano-EM tool and the workpiece causing localized ionization and evaporation of materials. The average dimensions of the 50 nanofeatures machined using dry nano-EM are found to be 7.528, 8.077, and 0.6390 nm in X, Y, and Z directions with standard deviation of ±1.314, ±1.410, and ±0.1015 nm in X, Y, and Z directions [96]. Figure 16.35 shows the machining examples of dry nano-EM, where arrays of nanoholes have been fabricated with Pt-Ir tool in atomically flat gold surface using a single-step machining [96].

16.9 Summary

The EDM is becoming an inevitable process in the manufacturing industries to meet the increasing demands of aerospace parts with difficult-to-cut materials, tools and dies, and miniaturized components with intricate shape and strict accuracy for MEMS applications. Many of the variants of EDM and micro-EDM are already in use by the industry extensively, and some of the processes are still in development phase. The present chapter provides a comprehensive overview of the EDM process, its working principle, system components, variants, parameters, and research advances on the EDM, micro-EDM, and nano-EDM processes. First, a brief discussion on the physics and material removal mechanism of the EDM process is presented. A discussion on the important system components of the EDM process has been provided including discharging unit or pulse generators, servo control system or gap control unit, and dielectric circulation system or flushing unit. The chapter also provided an overview of the major variants of EDM along with the application examples from both industry and academic research. Finally, the chapter focused on the research advances on the development of EDM-based compound and hybrid processes and their applications at macro, micro, and nano scale.

Although EDM has established itself as an independent process in the manufacturing industries, there are a number of issues remain to be solved before EDM process can become one of the top manufacturing processes. The major challenges of EDM process and its variants are slow production rate, postprocessing requirement to improve the surface finish, and low throughput. Extensive research is being carried out nowadays in order to solve those shortcomings of EDM by integrating either other conventional processes or introducing assistance of outside physical sources. In addition to the process development, continuous research has been carried out on improvement of the process stability and gap phenomena, design of new pulse generator and servo controller, and development of new machine tool with hybrid machining platform. Finally, some of the areas for future research on EDM and micro-EDM are improving the positioning accuracy, machine intelligence, online monitoring and developing knowledge-based system for hybrid machine tools.

References

1 Descoeudres, A. (2006). Characterization of electrical discharge machining plasmas. PhD thesis. ÉCOLE POLYTECHNIQUE FÉDÉRALE DE LAUSANNE, THÈSE No. 3542

2 Priestley, J. (1775). Experiments on the circular spots made on pieces of metal by large electrical explosions. In: *The History and Present State of Electricity with Original Experiments*, 3e, vol. II. London: Franklin Classics.

3 Lazarenko, B.R. (1943). *About the Inversion of Metal Erosion and Methods to Fight Ravage of Electric Contacts.* Moscow (in Russian): WEI-Institute.

4 Ho, K.H. and Newman, S.T. (2003). State of the art electrical discharge machining (EDM). *International Journal of Machine Tools & Manufacture* 43: 1287–1300.

5 Ho, K.H., Newman, S.T., Rahimifard, S., and Allen, R.D. (2004). State of the art wire electrical discharge machining. *International Journal of Machine Tools and Manufacture* 44: 1247–1259.

6 Webzell, S. (2001). That first step into EDM. In: *Machinery*, 159, (4040), 41. Kent, UK: Findlay Publications Ltd.

7 Kurafuji, H. and Masuzawa, T. (1968). Micro-EDM of cemented carbide alloys. *Journal of The Japan Society of Electrical Machining Engineers* 2 (3): 1–16.

8 Schumacher, B.M. (2004). After 60 years of EDM the discharge process remains still disputed. *Journal of Materials Processing Technology* 149: 376–381.

9 Eubank, P.T., Patel, M.R., Burrufet, M.A., and Bozkurt, B. (1993). Theoretical models of the electrical discharge machining process: III. The variable mass, cylindrical plasma model. *Journal of Applied Physics* 73: 7900–7909.

10 Snoeys, R. and Van Dijck, F. (1971). Investigation of electro discharge machining operations by means of thermo-mathematical model. *CIRP Annals* 20: 35–37.

11 Van Dijck, F.S. and Dutre, W.L. (1974). Heat conduction model for the calculation of the volume of molten metal in electric discharges. *Journal of Physics D: Applied Physics* 7: 899–910.

12 Yeo, S.H., Kurnia, W., and Tan, P.C. (2007). Electro-thermal modelling of anode and cathode in micro-EDM. *Journal of Physics D: Applied Physics* 40: 2513–2521.

13 Kunieda, M., Lauwers, B., Rajurkar, K.P., and Schumacher, B.M. (2005). Advancing EDM through fundamental insight into the process. *Annals of CIRP* 54 (2): 599–622.

14 Han, F., Yamada, Y., Kawakami, T., and Kunieda, M. (2004). Improvement of machining characteristics of micro-EDM using transistor type isopulse generator and servo feed control. *Precision Engineering* 28: 378–385.

15 Rajurkar, K.P., Levy, G., Malshe, A. et al. (2006). Micro and nano machining by electro-physical and chemical processes. *Annals of CIRP* 55 (2): 643–666.

16 Wong, Y.S., Rahman, M., Lim, H.S. et al. (2003). Investigation of micro-EDM material removal characteristics using single RC-pulse discharges. *Journal of Materials Processing Technology* 140 (1–3): 303–307.

17 Jahan, M.P., Wong, Y.S., and Rahman, M. (2008). A comparative study of transistor and RC pulse generators for micro-EDM of tungsten carbide. *International Journal of Precision Engineering and Manufacturing* 9 (4): 3–10.

18 Rajurkar, K.P. and Wang, W.M. (1991). On-line monitor and control for wire breakage in WEDM. *Annals of the CIRP* 40 (1): 219–222.

19 Yang, G.H., Liu, F., and Lin, H.B. (2011). Research on an embedded servo control system of micro-EDM. *Applied Mechanics and Materials* 120: 573–577.

20 Zhang, L., Jia, Z., Liu, W., and Li, A. (2012). A two-stage servo feed controller of micro-EDM based on interval type-2 fuzzy logic. *The International Journal of Advanced Manufacturing Technology* 59 (5–8): 633–645.

21 Altpeter, F. and Perez, R. (2004). Relevant topics in wire electrical discharge machining control. *Journal of Materials Processing Technology* 149 (1–3): 147–151.

22 Snoeys, R., Dauw, D., and Kruth, J.P. (1980). Improved adaptive control system for EDM processes. *Annals of the CIRP* 29 (1): 97–101.

23 Snoeys, R., Dauw, D., and Jennes, M. (1982). Survey of adaptive control and detection systems. *Annals of the CIRP* 31 (2): 483–489.

24 Mahdavinejad, R.A. and Mahdavinejad, A. (2005). ED machining of WC–Co. *Journal of Materials Processing Technology* 162–163: 637–643.

25 Yeo, S.H., Aligiri, E., Tan, P.C., and Zarepour, H. (2009). *An Adaptive Speed Control System for Micro Electro Discharge Machining*, CPl 181, Third Manufacturing Engineering Society International Conference (ed. V.J. Segui and M.J. Reig), 61–72. American Institute of Physics.

26 Liao, Y.S., Chang, T.Y., and Chuang, T.J. (2008). An on-line monitoring system for a micro electrical discharge machining (micro-EDM) process. *Journal of Micromechanics and Microengineering* 18: 035009 (8pp).

27 Tsai, Y.-Y. and Masuzawa, T. (2004). An index to evaluate the wear resistance of the electrode in micro-EDM. *Journal of Material Processing Technology* 149: 304–309.

28 Jahan, M.P., Wong, Y.S., and Rahman, M. (2009). A study on the fine-finish die-sinking micro-EDM of tungsten carbide using different electrode materials. *Journal of Materials Processing Technology* 209: 3956–3967.

29 Jahan, M.P., Anwar, M.M., Wong, Y.S., and Rahman, M. (2009). Nanofinishing of hard materials using micro-EDM. *Proceedings of the Institution of Mechanical Engineers, Part B: Journal of Engineering Manufacture* 223: 1127–1142.

30 Jahan, M.P., Rahman, M., and Wong, Y.S. (2011). Study on the nano-powder mixed sinking and milling micro-EDM of WC–Co. *International Journal of Advanced Manufacturing Technology* 53: 167–180.

31 Masuzawa, T., Yamaguchi, M., and Fujino, M. (2005). Surface finishing of micropins produced by WEDG. *Annals of the CIRP* 54 (1): 171–174.

32 Hewidy, M.S., El-Taweel, T.A., and El-Safty, M.F. (2005). Modelling the machining parameters of wire electrical discharge machining of Inconel 601 using RSM. *Journal of Materials Processing Technology* 169: 328–336.

33 Amorim, F.L. and Weingaertner, W.L. (2005). The influence of generator actuation mode and process parameters on the performance of finish EDM of a tool steel. *Journal of Materials Processing Technology* 166 (3): 411–416.

34 Bleys, P., Kruth, J.P., and Lauwers, B. (2004). Sensing and compensation of tool wear in milling EDM. *Journal of Materials Processing Technology* 149: 139–146.

35 Haas, P. Hydrodynamics of the electrical discharge machining process (EDM). http://cmefe.ch/Publication/2012_hepia_cmefe_EDM_e_bd.pdf (accessed on 07 April 2018).

36 Uhlmann, E., Piltz, S., and Doll, U. (2005). Machining of micro/miniature dies and moulds by electrical discharge machining – recent development. *Journal of Materials Processing Technology* 167: 488–493.

37 Klocke, F. and Lung, D. (2001). Micro contouring by EDM with fine wires. In: *Proceedings of the 13th International Symposium for Electromachining*, 767–782. Bilbao, Spain.

38 Intricate Grinding and Machine Specialties, Inc. Wire EDM. http://www.intricategrinding.com/wire-edm.html (accessed on 07 April 2018).

39 Liao, Y.-S., Chen, S.-T., and Lin, C.-S. (2005). Development of a high precision tabletop versatile CNC wire-EDM for making intricate micro parts. *Journal of Micromechanics and Microengineering* 15 (2): 245–253.

40 Sarix 3D EDM milling. http://www.sarix.com/en/3d-milling-en/ (accessed 07 April 2018).

41 Narasimhan, J., Yu, Z.Y., and Rajurkar, K.P. (2004). Tool wear compensation and path generation in micro and macro EDM. *Transaction of NAMRI/SME* 32: 151–158.

42 Yu, H.-L., Luan, J.-J., Li, J.-Z. et al. (2010). A new electrode wear compensation method for improving performance in 3D micro EDM milling. *Journal of Micromechanics and Microengineering* 20 (5): 055011.

43 Masuzawa, T., Tsukamoto, J., and Fujino, M. (1989). Drilling of deep micro-holes by EDM. *Annals of the CIRP* 38 (1): 195–198.

44 Jahan, M.P., Rahman, M., and Wong, Y.S. (2011). A review on the conventional and micro-electrodischarge machining of tungsten carbide. *International Journal of Machine Tools and Manufacture* 51 (12): 837–858.

45 Asad, A.B.M.A., Masaki, T., Rahman, M. et al. (2007). Tool-based micro- machining. *Journal of Materials Processing Technology* 192–193: 204–211.

46 Tong, H., Li, Y., Zhang, L., and Li, B. (2013). Mechanism design and process control of micro EDM for drilling spray holes of diesel injector nozzles. *Precision Engineering* 37: 213–221.

47 Soni, J.S. and Chakraverti, G. (1997). Performance evaluation of rotary EDM by experimental design technique. *Defense Science Journal* 47 (1): 65–73.

48 Chow, H.M., Yan, B.H., and Huang, F.Y. (1999). Micro slit machining using electro-discharge machining with a modified rotary disk electrode (RDE). *Journal of Materials Processing Technology* 91: 161–166.

49 Yan, B.H., Wang, C.C., Liu, W.D., and Huang, F.Y. (2000). Machining characteristics of Al_2O_3/6061Al composite using rotary EDM with a disk like electrode. *International Journal of Advance Manufacturing Technology* 16: 322–333.

50 Chow, H.-M., Yang, L.-D., Lin, C.-T., and Chen, Y.-F. (2008). The use of SiC powder in water as dielectric for micro-slit EDM machining. *Journal of Materials Processing Technology* 195 (1–3): 160–170.

51 Kuo, C.-L. and Huang, J.-D. (2004). Fabrication of series-pattern micro-disk electrode and its application in machining micro-slit of less than 10 μm. *International Journal of Machine Tools and Manufacture* 44: 545–553.

52 Yu, Z., Jun, T., and Kunieda, M. (2004). Dry EDM of cemented carbide. *Journal of Material Processing Technology* 149: 353–357.

53 Tao, J., Shih, A.J., and Ni, J. (2008). Near-dry EDM milling of mirror-like surface finish. *International Journal of Electrical Machining* 13: 29–33.

54 Subbu, S.K., Karthikeyan, G., Ramkumar, J., and Dhamodaran, S. (2011). Plasma characterization of dry μ-EDM. *International Journal of Advanced Manufacturing Technology* 56 (1–4): 187–195.

55 Yu, Z., Takahashi, J., Nakajima, N. et al. (2005). Feasibility of 3-D surface machining by dry EDM. *International Journal of Electrical Machining* 10: 15–20.

56 Egashira, K., Taniguchi, T., and Hanajima, S. (2006). Planetary EDM of micro holes. *International Journal of Electrical Machining* 11: 15–18.

57 Bamberg, E. and Heamawatanachai, S. (2009). Orbital electrode actuation to improve efficiency of drilling micro-holes by micro-EDM. *Journal of Materials Processing Technology* 209: 1826–1834.

58 Wang, G. and Shan, Y. (2005). Compensation of electrode orbiting in electrical discharge machining based on non-uniform offsetting. *International Journal of Machine Tools and Manufacture* 45 (14): 1628–1634.

59 Yu, Z.Y., Rajurkar, K.P., and Shen, H. (2002). High aspect ratio and complex shaped blind micro holes by micro EDM. *Annals of the CIRP* 51 (1): 359–362.

60 Weiliang, Z., Zhenlong, W., and Desheng, D. (2006). A new micro-EDM reverse copying technology for microelectrode array fabrication. In: *Proceedings of International Technology and Innovation Conference 2006 (ITIC 2006) (CP524)*, 1633–1636. Hangzhou, China.

61 Kim, B.H., Park, B.J., and Chu, C.N. (2006). Fabrication of multiple electrodes by reverse EDM and their application in micro ECM. *Journal of Micromechanics and Microengineering* 16 (4): 843–850.

62 Mujumdar, S.S., Mastud, S.A., Singh, R.K., and Joshi, S.S. (2010). Experimental characterization of the reverse micro-electrodischarge machining process for fabrication of high-aspect-ratio micro-rod arrays. *Proceedings of the Institution of Mechanical Engineers, Part B: Journal of Engineering Manufacture* 224: 777.

63 Kozaka, J., Skrabalak, M.Z., and Skrabalak, G. (2016). Development of advanced abrasive electrical discharge grinding (AEDG) system for machining difficult-to-cut materials. *Procedia CIRP* 42: 872–877.

64 Zhang, J.H., Ai, X., Lee, K.W., and Wong, P.K. (1996). Study on electro-discharge diamond wheel grinding (EDGM) of ceramic materials. *Materials and Manufacturing Processes* 11 (5): 763.

65 Koshy, P., Jain, V.K., and Lal, G.K. (1996). Mechanism of material removal in electrical discharge diamond grinding. *International Journal of Machine Tools and Manufacture* 36 (10): 1173.

66 Koshy, P., Jain, V.K., and Lal, G.K. (1997). Grinding of cemented carbide with electrical spark assistance. *Journal of Materials Processing Technology* 72 (1): 61–68.

67 Lauwers, B. (2011). Surface integrity in hybrid machining processes. *Procedia Engineering* 19: 241–251.

68 Menzies, I. and Koshy, P. (2008). Assessment of abrasion-assisted material removal in wire EDM. *CIRP Annals – Manufacturing Technology* 57 (1): 195–198.

69 Kansal, H.K., Singh, S., and Kumar, P. (2007). Technology and research developments in powder mixed electric discharge machining (PMEDM). *Journal of Materials Processing Technology* 184: 32–41.

70 Jahan, M.P., Wong, Y.S., Rahman, M., and Asad, A.B.M.A. (2009). An experimental investigation on the surface characteristics of tungsten carbide for the fine-finish die-sinking and scanning micro-EDM. *International Journal of Abrasive Technology* 2 (3): 223–244.

71 Tzeng, Y.-F. and Chen, F.-C. (2005). Investigation into some surface characteristics of electrical discharge machined SKD-11 using powder-suspension dielectric oil. *Journal of Materials Processing Technology* 170: 385–391.

72 Jahan, M.P., Rahman, M., and Wong, Y.S. (2010). Modeling and experimental investigation on the effect of nanopowder mixed dielectrics in fine-finish micro-EDM of tungsten carbide. *Proceedings of the Institution of Mechanical Engineers, Part B: Journal of Engineering Manufacture* 224 (11).

73 Pecas, P. and Henriques, E.A. (2003). Influence of silicon powder mixed dielectric on conventional electrical discharge machining. *International Journals of Machine Tools and Manufacture* 43: 1465–1471.

74 Abbas, N.M., Solomon, D.G., and Bahari, M.F. (2007). A review on current research trends in electrical discharge machining (EDM). *International Journal of Machine Tools & Manufacture* 47: 1214–1228.

75 Tong, H., Li, Y., and Wang, Y. (2008). Experimental research on vibration assisted EDM of micro-structures with non-circular cross-section. *Journal of Materials Processing Technology* 208 (1–3): 289–298.

76 Jahan, M.P., Saleh, T., Rahman, M., and Wong, Y.S. (2010). Development, modeling, and experimental investigation of low frequency workpiece vibration-assisted micro-EDM of tungsten carbide. *Journal of Manufacturing Science and Engineering* 132 (5): 054503 (8 pp).

77 Endo, T., Tsujimoto, T., and Mitsui, K. (2008). Study of vibration-assisted micro-EDM-the effect of vibration on machining time and stability of discharge. *Precision Engineering* 32 (4): 269–277.

78 Guo, Z.N., Lee, T.C., Yue, T.M., and Lau, W.S. (1997). A study of ultrasonic-aided wire electrical discharge machining. *Journal of Materials Processing Technology* 63 (1–3): 823–828.

79 Prihandana, G.S., Mahardika, M., Hamdi, M. et al. (2009). Effect of micro-powder suspension and ultrasonic vibration of dielectric fluid in micro-EDM processes – Taguchi approach. *International Journal of Machine Tools and Manufacture* 49 (12–13): 1035–1041.

80 Nguyen, M.D., Rahman, M., and Wong, Y.S. (2013). Transitions of micro-EDM/SEDCM/micro-ECM milling in low-resistivity deionized water. *International Journal of Machine Tools and Manufacture* 69: 48–56.

81 Nguyen, M.D., Rahman, M., and Wong, Y.S. (2012). Simultaneous micro-EDM and micro-ECM in low-resistivity deionized water. *International Journal of Machine Tools and Manufacture* 54–55: 55–65.

82 Zeng, Z., Wang, Y., Wang, Z. et al. (2012). A study of micro-EDM and micro-ECM combined milling for 3D metallic micro-structures. *Precision Engineering* 36 (3): 500–509.

83 Wang, W., Liu, Z.D., Tian, Z.J. et al. (2009). High efficiency slicing of low resistance silicon ingot by wire electrolytic-spark hybrid machining. *Journal of Materials Processing Technology* 209: 3149–3155.

84 Lin, Y.C. and Lee, H.S. (2008). Machining characteristics of magnetic force-assisted EDM. *International Journal of Machine Tools and Manufacture* 48: 1179–1186.

85 Heinz, K., Kapoor, S.G., DeVor, R.E., and Surla, V. (2011). An investigation of magnetic-field-assisted material removal in micro-EDM for nonmagnetic materials. *Journal of Manufacturing Science and Engineering* 133: 021002 (9 pages).

86 Yeo, S.H., Murali, M., and Cheah, H.T. (2004). Magnetic field assisted micro electro-discharge machining. *Journal of Micromechanics and Microengineering* 14: 1526–1529.

87 Kuo, C.-L., Huang, J.-D., and Liang, H.-Y. (2002). Precise micro-assembly through an integration of micro-EDM and Nd-YAG. *International Journal of Advanced Manufacturing Technology* 20: 454–458.

88 Kuo, C.-L., Huang, J.-D., and Liang, H.-Y. (2003). Fabrication of 3D metal microstructures using a hybrid process of micro-EDM and laser assembly. *International Journal of Advanced Manufacturing Technology* 21: 796–800.

89 Lin, L., Diver, C., Atkinson, J. et al. (2006). Sequential laser and EDM micro-drilling for next generation fuel injection nozzle manufacture. *Annals of the CIRP* 55 (1): 179–182.

90 Weber, P., Haupt, S., and Schulze, V. (2009). Hybrid machining of microstructures using a combination of electrical discharge machining milling and laser ablation. *Journal of Vacuum Science and Technology B* 27 (3): 1327–1329.

91 Takahata, K. (2009). *Micro-Electro-Discharge Machining Technologies for MEMS*, Source: Micro Electronic and Mechanical Systems (ed. K. Takahata), 143–164. Croatia: INTECH, Chapter 10.

92 Rahman, M., Asad, A.B.M.A., Masaki, T. et al. Integrated hybrid micro/nano-machining. ASME 2007 International Manufacturing Science and Engineering Conference (MSEC2007), Atlanta, Georgia, USA (15–18 October 2007). Paper no. MSEC2007-31009, pp. 197–209.

93 Takahata, K. and Gianchandani, Y.B. (2002). Batch mode micro-electro-discharge machining. *IEEE/ASME Journal of Microelectromechanical Systems* 11 (2): 102–110.

94 Takahata, K. and Gianchandani, Y.B. (2004). A planar approach for manufacturing cardiac stents: design, fabrication, and mechanical evaluation. *IEEE/ASME Journal of Microelectromechanical Systems* 13 (6): 933–939.

95 Malshe, A.P., Virwani, K.R., Rajurkar, K.P., and Deshpande, D. (2005). Investigation of nanoscale electro machining (nano-EM) in dielectric oil. *Annals of the CIRP – Manufacturing Technology* 54: 175–178.

96 Jahan, M.P., Malshe, A.P., and Rajurkar, K.P. (2012). Experimental investigation and characterization of nano-scale dry electro-machining. *Journal of Manufacturing Processes* 14 (4): 443–451.

17

MicroMilling Operations

Simon S. Park[1], Martin B.G. Jun[2], and Gerardo García[3]

[1] Department of Mechanical and Manufacturing Engineering, University of Calgary, Calgary, Alberta, Canada
[2] School of Mechanical Engineering, Purdue University, 585 Purdue Mall, West Lafayette, IN, 47907, USA
[3] Center for Innovation in Design and Technology, Tecnológico de Monterrey, Monterrey, México

17.1 Introduction

The miniaturization of high-accuracy components is becoming increasingly important for various modern industries. Miniature systems can provide portability, large surface area per volume ratio, lower power consumption, lower sample requirements, and improved laminar flow characteristics. The application of micro parts in different fields, such as microfluidic devices (lab-on-chips), electronics, micromolds, microsensors and actuators, and biomedical tools (catheters and surgical tools) and implants, resulted in the development of various micro-manufacturing methods. Unlike in photolithographic processes, ultraprecision micromachining is a viable technology for the fabrication of small devices in the micro/meso range. The cost-effectiveness of microcomponents is increased when using manufacturing processes that do not require expensive photolithographic masks and setup. Especially, fabrication of 3D structures of high aspect ratio and complex geometry, utilizing a variety of materials is very important. The principles of the mechanical material removal done in micromachining are similar to those of conventional machining. Therefore, the use of existing infrastructure and software would provide the flexibility and efficiency in micromilling that permits an economical fabrication of smaller batch sizes [1].

There are several critical issues associated with micromechanical machining and machine tools: these come mainly from the miniaturization of the components, tools, and processes. A comparison between conventional macromachining and micromachining operations is depicted in Table 17.1. Very small vibrations and excessive forces can be detrimental to the longevity of tools and part tolerances due to the size of micro-end mills; it is also very difficult to detect damage, wear, or fracture in the cutting tools. Furthermore, conventional macro cutting force models, such as Merchant and Oxley's sharp edge theorems, cannot be directly applied in the prediction of micro cutting forces because the chip thickness in micromachining operations is in the same size range as the edge radius of the tool, and this results in large negative rake angles and elastic–plastic deformations of the workpiece material. In the microscale, the specific cutting energy required for material removal is higher and when the depth of cut is less than the minimum uncut chip thickness (MUCT), and the chips may not form [2–4]. In addition, the accurate measurement of very small cutting forces is challenging because even a small amount of sensor noise can give a false cutting force signal. The high rotational speeds

Modern Manufacturing Processes, First Edition. Edited by Muammer Koç and Tuğrul Özel.

Table 17.1 Comparison and challenges associated with macromachining vs. micromachining operations.

	Macro	Micro	Challenges associated with micro machining
Machine tool accuracy	10 μm	<1 μm	Requirement of highly accurate stages and control
Rotational spindle speeds	0–10 000 rpm (high torque)	20 000–500 000 rpm (low torque)	Changing dynamics due to centrifugal, gyroscopic, and thermal effects
Tool material	Large grain tungsten carbide or HSS	Small grain tungsten carbide (usually two fluted cutters)	May require special coatings and small grain WC; edge radius sharpness is important for micromills
Diameter of tools	Above 2 mm	25 μm to 2 mm	Require precision ground flutes. Eccentricity plays an important role
Rigidity of machine	Rigid	Not very rigid	Natural frequency of micromachines are higher than macromachines
Cutting force prediction	Sharp edge theorem	Shearing, ploughing, and elastic recovery models	Need to consider minimum uncut chip thickness (MUCT) and ploughing effects
Cutting force measurement	Table dynamometer	Sensor fusion and compensation (accelerometers, noncontact displacement sensors, AE and force sensors)	Requires very high signal bandwidth for measurement due to high spindle speeds
Tool deflection	Minor	Major	Requires a compensation scheme
Burr removal	Manual/semi	Difficult	Removal of burrs without damage is challenging

used in micromilling processes produce a high cutting force frequency regime that requires of a higher frequency bandwidth than those of commercial force sensors.

The micro cutting of steel has recently received a strong research interest with the advent of miniaturized systems using a variety of materials, especially for molding of polymeric components. Several research laboratories and universities [5–12] have active research programs in micromachining. The consensus that can be derived from micromachining research is that micro cutting technology is important for the fabrication of 3D structures (i.e. high aspect ratio and complex geometries) using a variety of materials, especially steel. According to Sandia Lab [6], the need to fill the gap between nano/micro- and macro-domains is becoming an increasingly important research topic.

Several survey papers on micromachining have addressed the importance of microfabrication techniques. Hesselbach [13] examined the state-of-the-art in microproduction technology and analyzed the position of German industry within the limitations and economic potential of this industry. Although outlining different fabrication processes, they conclude that in micro cutting, there is a demand for the fabrication of steel micromolds for injection molding or embossing processes. Masuzawa [14] examined nonphoto etching technologies for micromachining and classified the basic machining phenomena of the processes required to fabricate microstructures as force, vaporization, ablation, dissolution, deformation, solidification, lamination, and recomposition. Masuzawa concluded that the size of microfeatures currently being requested is of approximately 100 μm or smaller, with micromachining technology in the research stage being able to create less than 5 μm features. Alting [15] broadly examined the field of microengineering, from the design and development stages to modeling and fabrication, with a focus on how microproducts and their production systems can be designed. In their discussion of micromechanical cutting processes, the authors found major challenges to be that excessive cutting forces limit the accuracy due to the deflection of tools and workpieces and that tool wear on the edge radius also influences cutting forces. They also described microfabrication systems such as the microfactory [12] and concluded that "many benefits will come from applying a systematic approach to microengineering," with the achievement of complete process integration as a fundamental issue.

Dornfeld [16] reviewed the mechanical micromachining field by covering the process physics, including materials, microstructural effects, chip formation, and surface and edge finish; software and simulation tools used for modeling; workpiece and design issues that arise during micromolding and in the creation of micropatterns; machine tools, cutting tools and metrology; and concluded with an outlook for future developments. The review stated that further development is required in the area of design and fabrication of cutting tools in order to provide tools with adequate geometry tolerances, precise edge definition, high stiffness, and better coatings. Liu [2] presented a survey of the mechanics of micromachining processes, focused on chip formation as influenced by MUCT, elastic–plastic deformation, and the nonhomogeneity in microworkpieces. They suggested that research is necessary for the modeling of the micromachining process addressing thermal effects, dynamics, improved methods for estimating MUCT, and the nano-micro-meso continuum. Koc [17] presented the fundamentals of micromanufacturing that explains the use of microforming and micromachining technologies for the design and manufacture of microproducts. They have displayed the fundamental principles, capabilities, performance specifications, and the size effects of micromanufacturing processes. Gandarias [18] compiled a global review of micromilling technology in which he included micromilling machines, micro cutting tools, and workpiece materials.

Despite several challenges still encountered in micromechanical milling, such as low microtool rigidity, material removal rates, and spindle rotational speeds as well as the inability to use diamond cutting tools on ferrous materials, micromilling can provide the flexibility required to fabricate complex 3D geometries out of metallic, composite, polymeric, and ceramic materials. This chapter examines tools, sensors, and machines required for micromilling operations. The MUCT that determines the formation of chips is formulated with respect to the edge radius of the tool. In addition, the modeling of micromilling forces with micro cutting phenomena considerations is examined. The cutting force model considers the shearing and ploughing cutting regime both of which depend on the MUCT. In addition, the measurement of tool tip dynamics using the receptance coupling method is discussed for compensating tool deflections and minimizing microchatter.

17.2 Machine Tools for Micromilling

The size and quality of the microproducts that can be manufactured is limited by the characteristics of the machine tools used to produce them, including their overall accuracy and their dynamic performance. The capabilities of the machine tool are vital to such product requirements as size, accuracy, surface roughness, and dimensional repeatability. In this section, the details of micro cutting tools, machine tools, and auxiliary processes are examined for micromilling operations.

The three main systems of precision machine tools are spindle, precision stage, and controller. The spindle rotates the cutting tool in order to remove material as it is moved across the workpiece. The precision stage moves the workpiece relative to its base, while the cutting tool moves vertically as the spindle moves on the column it is mounted on. The relative motion between the workpiece and the cutting tool is generated as the controller converts position and velocity commands into signal voltages, which are sent to the precision stage or spindle motors [19]. The resulting movements in the machine tool components are then measured with sensors that feed the data back to the controller.

Micromachine tools are equipped with several features that increase machined part accuracy by limiting the amount of vibration, spindle run out, and thermal expansion of the machine tool components. Machine tools for micromilling generally have a minimum quantity lubricant (MQL) system, since conventional cooling methods are not suitable for micromilling operations mainly on account of the large impact forces of the fluids on the cutting tools. MQL concepts apply spray with fine droplets and sufficiently high velocity to cool, lubricate, and flush away fine chips generated during cutting processes. MQL provide the low surface tension, viscosities, and particle sizes required to enter the cutting zone of micromilling processes.

17.2.1 Spindles

In micromachining applications, the rotational speeds of the spindle are extremely high in order to maintain acceptable productivity. To achieve these high rotational speeds, a direct motor spindle, or motorized spindle, is utilized. These spindles do not have gears or couplings and consequently provide a more precise positioning, since they eliminate the mechanical backlash caused by the gearbox. Motorized spindles generally have two sets of bearings, and these are most commonly ball bearings. There are other types of spindle bearings used in micromilling spindles, each with their own strength and trade-offs.

When the torque requirements are high and speeds are below 160 000 rev/min, electric motors with hybrid-angular contact ceramic bearings are commonly used. These bearing systems have ceramic balls on a high angle tandem arrangement and provide the spindle with high axial stiffness and low friction properties. When a higher spindle speed is required, air bearing spindles with air turbines are typically used, but they produce relatively low torque. Often to achieve higher speeds, ultraprecision machine tools are retrofitted with high-speed spindles compatible with conventional tool holder interfaces. Air-bearing spindles that exceed 200 000 rev/min are commercially available and are commonly found in micromilling machines. However, air spindles exhibit a nonlinear behavior when subjected to intermittent loads.

The large "power to volume" ratio of high-speed spindles generate large amounts of heat that need to be dissipated in order to ensure part accuracy [20]. The heat generated is mitigated by a series of heat sinks and heat transfer systems. These include a convection system that uses fluid to cool the motor and bearings; a convection and radiation system that drives away heat to surrounding lubricating medium; and systems that conduct heat through the housing flange to the spindle head. Thermal deformations induced by the heat generated from the spindle may cause errors on the axial depth as large as 30 μm [21]. These vertical inaccuracies can be partially compensated by running a warmup sequence to lengthen the spindle prior to cutting operations.

When the spindle reaches very high speeds, centrifugal forces and gyroscopic moments need to be considered as they may affect the overall spindle dynamics [20]. Another challenge associated with high-speed spindles is tool eccentricity [22, 23]. To monitor this issue, dynamic tool run out is often measured utilizing multiple noncontact displacement sensors with a high-frequency bandwidth. Some spindles have embedded sensors (e.g. piezoelectric force, capacitance, inductance sensors, etc.) that measure chatter, acceleration, and cutting forces. Furthermore, balanced high-precision collets and tool holders are necessary to minimize the tool run out.

17.2.2 Drives and Guides

An ultraprecision stage (i.e. XY table) is essential to achieve high accuracy when fabricating microstructures. The precision stage is a working table resting on a set of guides that are actuated by a ball screw or a linear motor feed drive. There are several types of machine tool guides, each with different characteristics. Roller-based guides have high performance and modular integration with machine tools; hydrostatic guides have high accuracy, stiffness, and damping; aerostatic, magnetic, and vacuum guides have the high precision but can only handle small loads [24]. Precision machines commonly use linear guides with cylindrical noncirculating roller bearings under a constant preload. Also, for high-precision applications, it is important to avoid "axis stacking" by having dynamically independent axis in order to eliminate alignment and positioning errors caused by "cross talk" effects.

Micromilling machine tool drives generally operate with a ball screw or linear direct system. Ball screw drives are often used to deliver relatively high torque, especially since they generate low amounts of heat and exhibit low wear [25]. However, the main drawback of a ball screw drive is that it moves nonlinearly and with backlash [11]. To decrease the amount of backlash, the nuts in ball screw drives are preloaded with a spacer, an offset between the leads, or by using oversized ball bearings [26].

Linear drive motors and control systems are also commonly used in ultraprecision machine tools. Compared to conventional drive mechanisms such as ball screws, linear motors have no accumulated errors from friction and motor-coupling, no loss of accuracy due to wear, and no backlash. The typical accuracy for precision machine tools using linear drive systems is ±1 μm. Linear drives are capable of delivering high accelerations for low-weight workpiece applications since their acceleration capacity is inversely proportional to the mass they require to move. Linear drives are more costly than ball screw drives due to the elaborate guide systems such as aerostatic, hydrostatic, or magnetic guides. Also, the linear system does not exert high torque, and it is sensitive to changes in workpiece size.

Micromilling operations would not be feasible without the technological advances in drives and guides. Available drive systems provide accurate positioning with low amounts of friction and geometric deformation with the high frequency control bandwidth. However, more development is required to increase real-time adaptive compensations and control and to minimize vibrations.

17.2.3 Ultraprecision Machine Tools

There have been two schools of thoughts for micromilling machine tools. One is pushing to minimize the footprint of overall machine tools by promoting the "micro factory" concept. The other promotes a temperature–controlled large-scale ultraprecision micromachining center. The micro factory concept is predominantly in a research phase, but the advantages include reduction in energy, space, materials, and cost. The greatest advantage of micro factories is their mobility (portability) and their ease on which they can be installed in any building. This can be particularly advantageous if a micro factory is installed inside an operating room to machine sculpted geometries on implants dependent on individual requirements. A micro factory could also be used in a remote location.

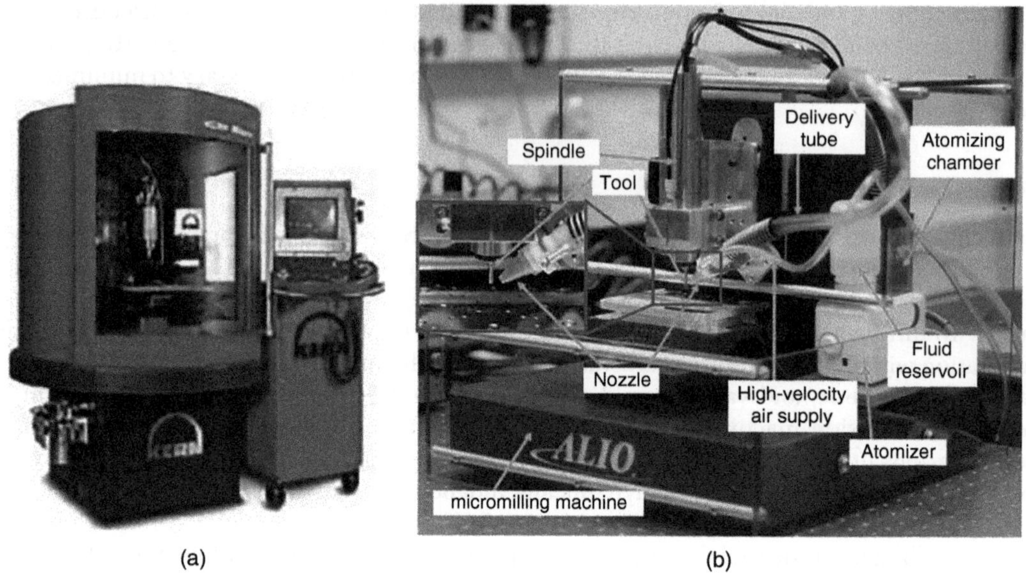

Figure 17.1 Ultraprecision machine tools for micromilling: (a) Kern micro and (b) customized micromilling machine.

Whereas ultraprecision machine tools have several advantages over micro factories, they are more robust and can provide better static and dynamic stiffness and make it easier to integrate complementing equipment such as monitoring and tool handling systems. Both of these types of machines need to be isolated from the floor's vibrations due to precise nature of micromilling operations. To achieve this vibration isolation, machine tools use active or passive systems as well as a polymer concrete or granite base to improve the rigidity and damping characteristics [27].

Ultraprecision macromachine tools have high rigidity, good damping, and move precisely with the aid of precision sensors and actuators. However, the large scale and precisely controlled machining environment adds considerable costs to the fabrication of miniature components. Examples of precision micromachining centers are depicted in Figure 17.1.

17.2.4 Coolant and Lubricant Delivery

In micromilling, the limitations of microtooling technologies in the form of large cutting edge radii, poor geometry control, and the resulting ploughing mechanism of the MUCT effect lead to increased levels of tool wear and tool failure [28]. Thus, short tool-life stands as one of the main limitations in achieving higher productivity. The application of cutting fluids benefits the machining operation through promoting longevity in cutting tools [29]. However, there is limited understanding of both the requirements and the application effectiveness of cutting fluids as the size-scale of the process is reduced.

Many researchers have used different cooling methods for micromachining. Nitrogen cooling was used by Friedrich [30] for cooling the workpiece to its near-brittle state to lower the specific cutting energy and consequently the required feed in order to machine at low cutting speeds. Yan et al. [31] looked at surface roughness of a single crystal silicon workpiece under different coolant conditions (dry, kerosene, and water) in ultraprecision cutting with a diamond cutting tool to study ductile mode machining. However, it is likely that traditional macromachining process cutting fluid systems such as flood, high-pressure coolant application, and liquid-nitrogen–based cooling are not viable approaches in micromilling. The impact force of the cutting fluids may be greater than the cutting forces, which are in the order of a few Newtons [2], resulting in tool deflection or even damage. Also, reduced levels of surface tension and viscosity and very small particle sizes will likely be required to penetrate into the small cutting zone through the boundary layer of air around the tool periphery, which is formed at high rotational speeds [32]. Furthermore, due to increased ploughing in micromachining, there is increased friction at the tool–workpiece interface, and consequently, good lubricating properties of the cutting fluid are essential and important for extended tool life.

The MQL concept (Figure 17.2a) involving low-volume sprays of oil delivered in compressed air has been gaining more attention. This lubrication technique decreases the amount of cutting fluids used in cutting operations while maintaining a similar performance. Efforts related to MQL were initiated to reduce the cost and adverse environmental impacts of the metalworking fluids (MWFs). It is reported that over 7.5 billion liters of cutting fluids were used by North American manufacturers in 2002, and costs associated with the use of MWFs often constitute between 7% and 17% of total production costs, as compared to 4% for tooling cost [33]. Expenses associated with MWFs include the cost of installing a supply system, fluid purchase and system maintenance, and discarded fluid (waste) treatment. The cost of waste fluid treatment is substantial because at least two-thirds of MWFs used need to be disposed every year and treated before disposal [34]. Because most MWFs are petroleum-based oils, responsible handling of waste fluid is needed to avoid contamination of lakes, rivers, and groundwater. Such handling may include the pretreatment and treatment of cutting fluid wastes. The cost of fluid pretreatment/treatment can be higher than the purchase price of the fluid itself.

The MQL technology is the most feasible method of applying cutting fluids to micromilling operations because it uses small amounts, and fluids are applied in the form of droplets with diameters smaller than a few hundred micrometers. These droplets applied as a spray do not interfere with the operation so much. Recently, many researchers have used MQL approach in micromilling operations [35–38] and shown the effectiveness of the MQL technology in micromilling. Jun et al. [32, 39] recently developed atomization-based cutting fluid application technology similar to the MQL technology, which uses a focused spray containing one order of magnitude smaller droplets, as shown in Figure 17.2b. It was shown that significant

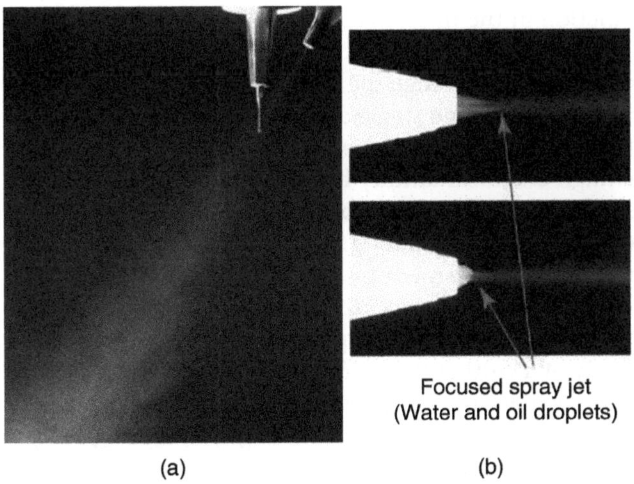

Figure 17.2 MQL with low-volume sprays of oil delivered in compressed air: (a) lubricant with oil sprayed on cutting tool and (b) focused spray jet.

temperature decrease and improvement in tool life have been achieved with this technology. The results show that for micromilling operations, because the cutting zone is small and the tool is rotating at high rotational speed, focused spray with fine droplets and sufficiently high velocity is needed to effectively cool, lubricate, and flush away fine chips generated.

17.2.5 Micro-end Mills

The size of precision micro-cutting tools (henceforth referred to as "micro-tools") determines the limit of the size and accuracy of microstructure features. Smaller tools have decreased thermal expansion relative to their size, increased static stiffness from their short structure, and increased dynamic stability from their higher natural frequency [40]. Commercially available micro-end mills can be as small as 50 μm in diameter, with their helix angle fabricated by precision grinding.

Carbide micro-tools are commonly used in micromilling operations because they provide optimum hardness, heat resistance, strength, toughness, and high abrasion resistance [41]. Carbide cutting tools are made from very fine grain tungsten carbide (WC) powders sintered with cobalt-binding agents. The properties of the cutting tool can be changed by modifying WC powder grain size as well as the cobalt contents. As the Hall-Petch relation states that the yield strength of the cutting tool increases as the grain size is decreased since the number of grain boundaries are increased and these boundaries interrupt dislocation propagation. Reducing the cobalt content makes the carbide harder, but at the expense of higher brittleness.

WC tools are coated in order to improve their wear-resistance characteristics. Tungsten carbide tools that are coated with titanium nitride (TiN) or variations have shown to minimize excessive flank wear, edge radius wear, chipping, and burrs [42]. Current coatings adapt a process conceived for macroscale cutting tools and show a presence of droplet-like defects of micrometric dimensions. Monolayer nanostructures with low silicon content coatings exhibit a good trade-off between tool wear reduction and productivity, as well as a reliability increase. Other types of coatings include cubic boron nitride (CBN) and diamond-like carbon (DLC), which provide high hardness and low friction coefficients; can be deposited as nanometric layers through a vapor deposition process. These thinner coatings allow the production of sharper cutting edges and have shown no droplet-like defects [43]. The effective cutting edge radius is the sum of the cutting tool radius and the coating thickness. The thicker the tool coating results in, the larger the effective cutting edge [44].

Tool edge radius has an important function in the micromachining since it determines the MUCT. A smaller cutting edge radius results in the smaller MUCT, which in turn lowers the cutting forces by removing the chips predominantly through the shearing instead of ploughing.

Recently, the demand for diamond (polycrystalline and single crystal)–based micro-tools has increased. These tools are used when tight tolerances are required when machining nonferrous alloys such as nickel-plated alloys. The high chemical affinity between diamond and ferrous materials causes severe wear [40, 45], limiting its use on nonferrous workpieces. Polycrystalline diamond (PCD) tools are also extensively used in micromilling applications as cutting inserts mounted on carbide cutting tool bodies. PCD tools are made by sintering diamond grains with cobalt to create a thin layer of diamond grain on a tungsten carbide board [46]. These bonded layers are shaped into cutting inserts using wire electric discharge erosion or grinding process and are then brazed onto the metallic body of a cutting tool. Single-crystal diamond tools are particularly advantageous for micromilling operation since they are the material on which the sharpest cutting edge radius can be made [47]. However, PCD can be tougher than single-crystal diamond on account of its random crystallographic orientation [48]. PCD cutting tools have a higher wear resistance, give better surface finishes, and can be used at higher cutting speeds than carbides. But the higher hardness of PCD also makes them more brittle and less suitable for interrupted cutting operations. PCD tools can be used to remove material of difficult to cut hard material, and PCD tools are limited to two-dimensional geometries; therefore, they lack the complex edge preparation commonly found on carbide cutting tools.

Single-crystal diamonds can have a cutting edge radius smaller than 100 nm, whereas typically coated WC can range from 1 to 5 μm [49–51]. Straight edge PCD end mills can simultaneously mill both the side and bottom surfaces with submicron surface roughness, and when these tools have smaller rake angles, they have better stiffness and lower tool wear rate [52]. Significantly smaller structures can be machined when using single-edged micromilling cutters made of diamond than with conventional milling cutters made of tungsten carbide [53].

17.2.6 Measurements and Monitoring

Real-time feedback of measurements made by different sensors allows machine tools to monitor and take corrective actions which in turn increase precision and help select appropriate machining parameters. Feed drives in machine tools integrate position, velocity, acceleration, and load measurements to improve milling stability [19]. Machine tools generally use optical encoders to perform position measurements, which can also provide speed through digital differentiation. Sensors are also used to measure the axial thermal expansion of the spindle and cutting tool. Chatter detection is typically done with microphones, acoustic emission sensors, lasers, or accelerometers, while cutting forces are measured with miniature table dynamometers placed under the workpiece or with rotating dynamometers placed between the tool clamping and the milling spindle based on piezoelectric load cells. Cutting force measurement is the most effective method for monitoring tool conditions since it provides high signal-to-noise ratios and best represents the state of machine tools and machining operations [54]. Malekian et al. developed a model that monitors the tool wear of micro-end mills by fusing force, acceleration, and acoustic emission signals using the adaptive neural fuzzy inference system (ANFIS) algorithm [55].

When the spindle rotates at high speeds, the bandwidth of the force sensor bandwidth is limited, and this distorts the cutting forces measurements. For example, when a spindle rotates a two flute endmill above 150 000 rpm, the frequency bandwidth of the force sensors needs to be higher than 5 kHz in order to accurately capture forces. This requires of at least two or three harmonics which in turn requires the frequency bandwidth of sensors to be at 10 kHz or greater. Park et al. developed a signal-processing method that uses a Kalman filter to increase the bandwidth of sensors by removing the noise from the measurements through a cutting force reconstruction. The Kalman filter is a dynamically compensating algorithm that minimizes the influence of structural modes in x, y, and z directions to improve the sensing of the piezo-electric forces. The Kalman filter modifies the data obtained from the sensors by integrating it with a prediction based on the system's dynamic model in order to reduce the amount on noise in the measurements. The weights in the algorithm are calculated continuously based on the estimated uncertainty of the systems dynamic model, and by increasing the gain on the algorithm the filter increases the dependence of the output on the data from the sensors and becomes more responsive and "noisy" [56].

Process measurements and monitoring in micromilling operations require different equipment and techniques than those used in the macroscale. The fragility of the tools, high frequency requirements, the lack of conventional feedback, and the need for an increase in process efficiencies make measuring and monitoring a critical area for future research development.

With an increase in real-time feedback reliability, better process parameters can be obtained and a higher part precision can be achieved.

17.2.7 Burr Removal

Macroscale deburring processes often lack the precision and finesse needed to finish off a micro part. Conventional macro deburring processes can damage a micrometric surface and would not be able to polish the hard to reach areas with high aspect ratio geometries. There are several burr removal techniques created for micro applications. Abrasive flow machining, chemical etching, diamond particle polishing, and laser polishing are some of effective methods.

Abrasive flow machining has several variants; most are composed of two cylinders with abrasive media that force material flow through the workpiece to cause material removal [57]. The burrs are removed by a shearing motion or flattened by elastic deformation. The abrasive fluid moves through the workpiece with a one directional flow, a two-way reciprocating motion, which tends to cause sealing complications, or with an orbital movement, which provides the most uniform abrasion.

Electrochemical-aided abrasive flow machining (ECAFM), magnetorheological-assisted abrasive flow finishing (MRAFF), and ultrasonic-assisted abrasive flow machining (UAAFM) are the types of hybrid abrasive flow machining that have the best results with hard metallic materials. ECAFM, a polymeric electrolyte carries abrasive particles of alumina oxide or silicon carbide and has also been used on stainless steel. MRAFF causes abrasive particles to form chains with the application of an electric field, and these are used to shear peaks on workpiece [58]. This method has been proven effective for the deburring of stainless steel with silicon carbide particles. UAAFM removes material by applying ultrasonic vibrations to the workpiece, which causes material removal with surface friction caused by organic polymers with alkenes, alkanes, sulfoxides, and esters [59]. UAAFM produces a glazed finish on the surface and can remove materials on hard surfaces made of Inconel 718. Machined components can also be polished or remove burrs with a nylon brush and abrasive silicon carbide or diamond particles. This process has resulted in 100 nm improvements in roughness with a minimal groove depth change of 8 mm [60]. The requirements to effectively remove burrs and polish microcomponents without damaging the machined components are pressing.

Micromilling machine tools are made with components specialized for high rotational speed with the highest positional accuracy. They are tailored to deal with the scale reduction that increases demands on cutting tools as well as on heat removal systems. In microscale and nanoscale, the machine tool operators cannot aid themselves with the audible and visual feedback used in the conventional macroscale. For this reason, measurements and monitoring of the systems are crucial as they interpret the data sent by sensors and provide the dependable process feedback. Micromilling machine tools must deal with the inefficient cutting process inherent due to fragility of tools and components as well as with the large amounts of heat generated on account of the high rotational speeds. In addition to the aforementioned complications, micromilling operations cannot rely on conventional macro polishing and deburring processes. Further studies are needed to address some of these challenges in micromachine tools and auxiliary processes.

17.3 Micromilling Forces

Since micro-tools have finite edge radii, the conventional theory of machining is not sufficient to describe the material removal mechanism at the microscale. Unlike the conventional sharp-edge cutting model, chip formation in micromachining occurs along the rounded edge of micro-tools, which results in a negative rake angle and affects the ploughing and shearing forces. Furthermore, the chips are not formed when the chip thickness is smaller than a minimum or critical value [61]. The accurate prediction of the critical chip thickness in micromachining is important for accuracy, productivity, and process optimization. In addition, due to the finite edge radius of the micro-tools, an elastic recovery of the material occurs beneath the flank face of the tools [62], which affects cutting forces and surface generation.

A mechanistic force model is used to predict micro-end milling forces for both the shearing and ploughing dominant cutting regimes. The model assumes that there is a MUCT that determines whether the cutting is predominantly in the shearing or in the ploughing regime. The cutting process is dominated by the ploughing mechanism when the depth of cut is below the MUCT.

17.3.1 Minimum Uncut Chip Thickness (MUCT)

A schematic of material flow around an orthogonal tool with a finite edge radius is shown in Figure 17.3. A stagnation point is assumed on the tool, below which the material flows downward without any chip formation. This phenomenon is called ploughing, which is plastic deformation without material removal. Above this stagnation point, the material flows up and forms chips. The stagnation point, which happens at a critical or minimum angle, θ_m, with an effective negative rake angle (i.e. $\alpha_e = \pi/2 - \theta_m$), determines the value of the MUCT, namely

$$h_c = r_e\left(1-\cos\theta_m\right) \tag{17.1}$$

where r_e is the edge radius of the tool and h_{er} is the elastic recovery.

In order to find the critical angle, finding the stagnation point (A) is necessary, and shear strain at the stagnation point should be infinite since the material is stagnant. Based on this premise, we employed the shear strain in orthogonal machining, expressed as the following:

$$\varepsilon = \cot\varphi + \tan\left(\varphi-\alpha_e\right) = \frac{\cos\alpha_e}{\sin\varphi\cos\left(\varphi-\alpha_e\right)} \tag{17.2}$$

By substituting the Merchant's shear angle formulation, the shear strain can be rewritten as the following:

$$\varepsilon = \frac{\cos\alpha_e}{\sin\left(\frac{\pi}{4}-\frac{1}{2}\left(\beta_s-\alpha_e\right)\right)\cos\left(\frac{\pi}{4}-\frac{1}{2}\left(\beta_s+\alpha_e\right)\right)} \tag{17.3}$$

Figure 17.3 Machining using an edge radius tool.

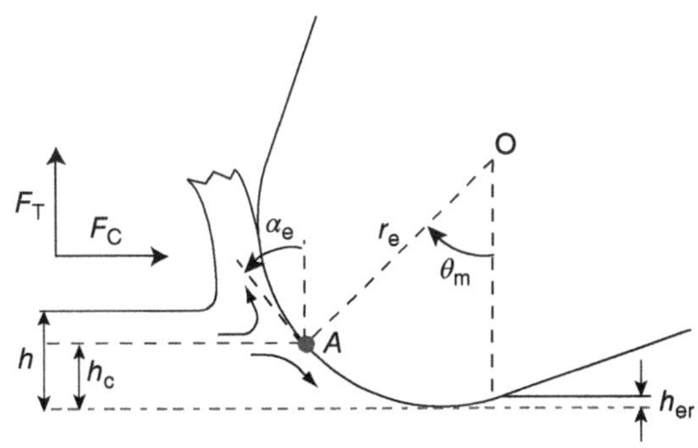

By equaling the denominator of the equation to zero and solving the equation, the critical rake angle (α_e) at the MUCT (h_c) becomes the following:

$$\alpha_e = \beta_s - \frac{\pi}{2} \tag{17.4}$$

Therefore, the stagnant point occurs when the critical angle is at the friction angle as shown in Eq. (17.5).

$$\theta_m = \frac{\pi}{2} - \alpha_e \approx \beta_s \tag{17.5}$$

Increasing the friction between the material and tool rake face increases the critical angle, which results in a decrease of the MUCT based on Eq. (17.1). According to Kita [63], a higher friction between the material and the tool rake face causes a downward flow of material or ploughing; whereas, at low frictions, the material flows up and forms a chip. This is in agreement with our experimental findings, which demonstrate that increasing friction causes more material to be ploughed under the tool and more ploughing occurs.

The MUCT model identifies the critical angle based on the friction angle with respect to the edge radius. The model is essential to providing a trend in which the critical angle changes with respect to the friction. In the formulation of the critical angle, several assumptions have been made. Material shearing is assumed to happen at a shear angle; whereas, in reality, shearing occurs in a large shearing zone, especially for negative rake angles. The effects of material hardening, cutting velocity, and change of temperature during the machining operations, which can change the friction coefficient, have not been directly considered.

Based on the force ratio measured from orthogonal cutting tests for aluminum, the critical angle (θ_m) was identified as a range from 31° to 42°. The average value for the force ratio was found to be approximately 0.8 for different depths and feeds, which corresponds to a critical angle (θ_m) of approximately 39°, representing the MUCT of $0.23r_e$. The result was in agreement with Basuray's work, where he found the MUCT to be $0.21r_e$.

In order to experimentally investigate the presence of a MUCT, the depth of cut was incrementally increased; and the chip formation was explored using the digital microscope. It was observed that, for chip thicknesses below 3 μm, chips were not formed continuously but rather intermittently. This discontinuous chip formation was primarily due to the formation of material piled up in front of the tool edge during ploughing; and, when the height of the piled-up material reaches a critical value, a chip is formed. When the depth of cut was greater than 3 μm ($0.23r_e$), chips started to form continuously, which showed that the MUCT has been passed. Figure 17.4a,b show the formation of the discontinuous and continuous chips at two different depths of cut (2 and 5 μm, respectively). It was also observed that this critical value was not dependent on the different feed rates used in the experimental tests. Based on the MUCT formulations, we used the value obtained in the formulation of micromilling forces.

For orthogonal cutting tests, we have retrofitted the spindle with a fixture to hold an orthogonal WC orthogonal tool where the edge radius of the orthogonal tool is approximately 13 μm utilizing measurement software (Digimizer•).

17.3.2 Elastic Recovery

When machining is performed in the microscale, the uncut chip thickness is comparable in size to the edge radius of the cutting tools. When the uncut chip thickness is smaller than a certain amount, an elastic recovery occurs under the tool flank face and this significantly affects

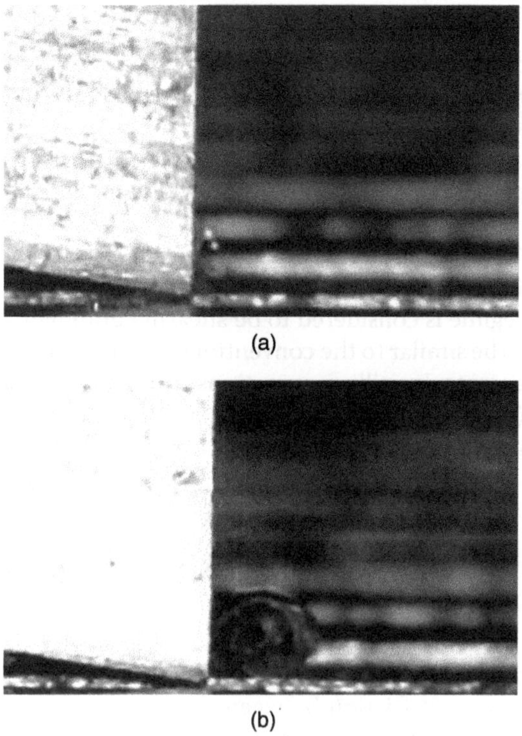

Figure 17.4 Formation of the chip above and below the MUCT: (a) depth of cut of 2 μm and (b) depth of cut of 5 μm at 100 μm/s.

the cutting forces. This elastic recovery rate needs to be identified, in order to account for the surface generation and cutting forces during the cutting operations.

The elastic recovery rate of the material can be determined directly by inspecting the remaining grooves from the scratch tests. In this technique, with the aid of a profilometer, the surface profile of the groove is recognized as shown in Figure 17.5. The area of the groove under the original surface ("abc" in Figure 17.5) is then calculated. By knowing this area and the projected area of the tool in the vertical direction ("ABC" in Figure 17.5), the elastic recovery rate of the material is acquired from the following:

$$p_e = \frac{S_{ABC} - S_{abc}}{S_{ABC}} \tag{17.6}$$

It was found that the average elastic recoveries, p_e, of the brass 360 and aluminum 6061 are approximately 24% and 10%, respectively, which are comparable to Jardret's [62] work. The elastic recovery rate can slightly change with respect to depths of cut.

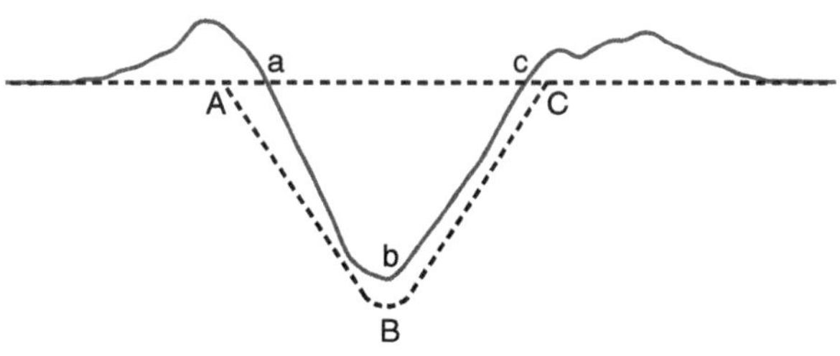

Figure 17.5 Surface profile and projected area of the conical tool.

17.3.3 Mechanistic Micromilling Force Modeling

Since micromachining processes are susceptible to excessive cutting forces, which are detrimental to the parts and tools, the real mechanisms of material removal are needed to be understood and correctly modeled. This process includes the identification of critical chip thickness and elastic recovery for the correct prediction of surface generation and the modeling of ploughing forces for the case where material is not sheared from the workpiece.

17.3.3.1 Shearing Dominant Regime

Beyond this MUCT, the elastoplastic effects are assumed to be negligible, and the cutting regime is considered to be shearing dominant. In this case, the cutting mechanism is assumed to be similar to the conventional cutting mechanism that considers the shearing and edge coefficients. In milling operations, when the uncut chip thickness is greater than the MUCT value ($h > h_c$), the tangential (dF_{ts}) and radial (dF_{rs}) shearing cutting forces acting on a differential flute element with height dz can be modeled as follows:

$$
\begin{aligned}
dF_{rs} &= \left[K_{rc} h(\theta_i(z)) + K_{re}\right] dz \\
dF_{ts} &= \left[K_{tc} h(\theta_i(z)) + K_{te}\right] dz
\end{aligned}
\tag{17.7}
$$

where K_{rc}, K_{tc}, K_{re}, and K_{te} are the radial and tangential cutting and edge coefficients, respectively. The cutting coefficients represent shearing of the workpiece, and the edge components represent friction between the tool and the workpiece. In order to come up with the micro cutting force model for the shearing dominant regime, the identification of cutting constants, which can be obtained from experiments, is imperative.

17.3.3.2 Ploughing Dominant Regime

The micromilling undergoes both ploughing and shearing when chip formation does not occur because when the uncut chip thickness is less than the critical thickness; instead, there is ploughing and partial elastic recovery of the material. When the chip thickness increases due to the rotation, shearing occurs. The ploughing forces are modeled as proportional to the volume of interference between the tool and the workpiece. The elastic recovery of the workpiece is experimentally obtained using scratch testing. The following equation is formulated in the ploughing dominant cutting regime:

$$
\begin{aligned}
dF_t &= \begin{cases} (K_{tc} h + K_{te}) dz & \text{when} \quad h \geq h_c \text{ (shearing)} \\ (K_{tp} A_p + K_{te}) dz & \text{when} \quad h < h_c \text{ (ploughing)} \end{cases} \\
dF_r &= \begin{cases} (K_{rc} h + K_{re}) dz & \text{when} \quad h \geq h_c \text{ (shearing)} \\ (K_{rp} A_p + K_{re}) dz & \text{when} \quad h < h_c \text{ (ploughing)} \end{cases}
\end{aligned}
\tag{17.8}
$$

where K_{rp}, K_{tp} are ploughing constants; h is the chip thickness, which is a function of elastic recovery, run out, and dynamics; and A_p is the interference volume. The friction forces are considered constant for different conditions and modeled with the edge coefficients, which are obtained from the shearing dominant cutting regime.

A comprehensive chip thickness model developed by Jun et al. is used to compute the correct chip thickness, including the effects of the trochoidal tool path, MUCT, elastic recovery, and tool vibrations. Figure 17.6 shows the surface generation and chip thickness computation in the presence of elastic recovery, which is represented as the shaded region, for an arbitrary axial

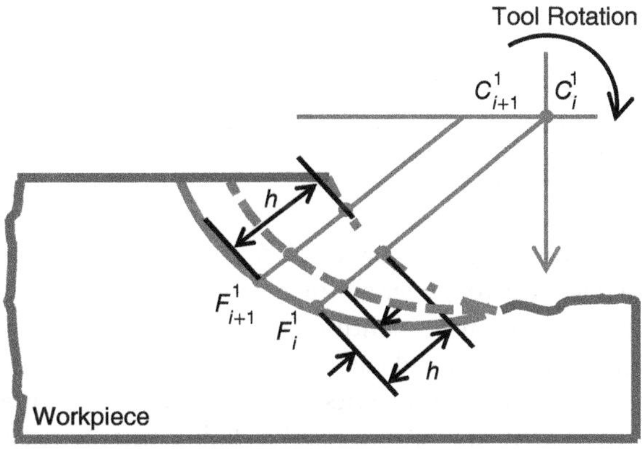

Figure 17.6 Chip thickness model considering elastic recovery.

slice. Points C and F represent the tool center and cutting edge locations, respectively. The superscript denotes the tooth path number, and the subscript represents the rotational angle. Point I is found at the intersection between the previously generated surface from the previous tooth path and the line connecting C and F for the current tooth path. The chip thickness can be formulated as follows:

$$h = \max\left(0, \left\|C_i^j F_i^j\right\| - \left\|C_i^j I_i^{j-1}\right\|\right) \tag{17.9}$$

The interference area, A_p, can be obtained by examining the height of the elastic recovery $h_{er} = p_e x h$, where p_e is the elastic recovery rate; and the effective rake angle $\alpha_p = \cos^{-1}(1 - h/r_e)$ is obtained as shown in Figure 17.7.

The plowed area, A_p (shaded area of Figure 17.7), can be obtained as:

$$A_p \approx \frac{1}{2} r_e^{\,2} \left(\alpha_p + \gamma\right) + \frac{1}{2} r_e \left(l_1 - l_2\right) \tag{17.10}$$

where

$$\begin{aligned} l_1 &\approx \left(\frac{h_{er} - r_e\left(1 - \cos\gamma\right)}{\sin\gamma}\right) \\ l_2 &\approx \sqrt{r_e^2 + l_1^2}\,\sin\left(\alpha_p + \gamma + \psi\right) \\ \psi &= \tan^{-1}\left(l_1 / r_e\right) \end{aligned} \tag{17.11}$$

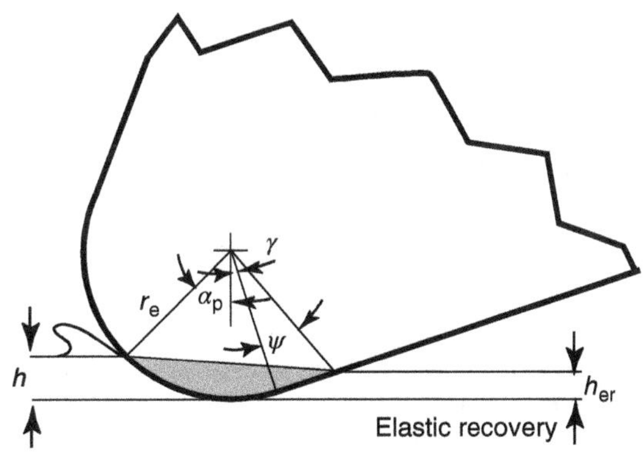

Figure 17.7 Ploughing area and elastic recovery.

Table 17.2 Cutting constants for Al 7075 at 60 000 rev/min, 50 μm depth of cut, and full immersion.

K_{tc} (N/mm^2)	K_{rc} (N/mm^2)	K_{te} (N/mm)	K_{re} (N/mm)	K_{tp} (kN/mm^3)	K_{rp} (kN/mm^3)
3440	2350	1.9	1.3	760	1150

Depending on the MUCT, h_c, the computed forces can be summed up among all the engaged axial slices (dz) over all the cutting flutes to obtain the total forces using Eq. (17.8). The cutting constants can be identified by performing the experimental tests with varying chip thicknesses (i.e. feed rates) and can compensate for the cutting forces using the expanded KF method [64]. The cutting coefficients are found via a nonlinear curve fitting by minimizing the error of instantaneous cutting results. Full immersion cutting data were used for this purpose, since they showed better agreement with the theoretical data. Table 17.2 depicts the identified cutting constants for aluminum 7075. The tangential and radial cutting constants for micromilling operations is considerably higher than that of macro cutting constants mainly due to size effects.

When the uncut chip thickness is smaller than the critical value, an elastic recovery occurs that can affect the chip formation, cutting forces, and surface generation during machining operations. The elastic recovery can be experimentally obtained by using a conical indenter to perform scratch tests. The acquired elastic recovery and MUCT values were used as the parameters of the developed model for micromilling.

Several cutting tests at various conditions with different feed rates and spindle speeds have been performed. The MUCT was found to be approximately 0.7 μm based on Eqs. (17.1)–(17.5). The simulations were performed based on the developed algorithm with the obtained mechanistic cutting coefficients and considerations of dynamics and run out. Figure 17.8a shows the comparison between the forces at half immersion at a spindle speed

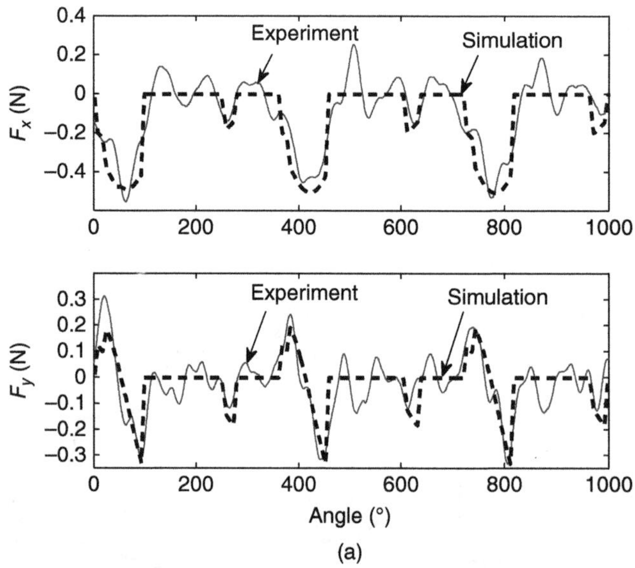

Figure 17.8 Comparison of forces between the simulation and KF compensated experimental data. 500-μm diameter end mill, depth of cut: 50 μm, feed rate: 0.6 μm/flute; (a) half immersion and spindle speed: 60 krpm and (b) full immersion and spindle speed: 90 krpm

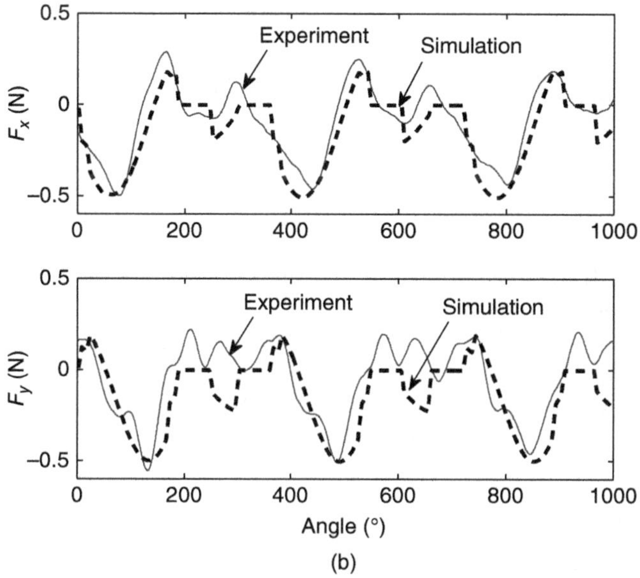

Figure 17.8 (*Continued*)

of 60 000 rev/min at 0.6 μm/flute, which is in the ploughing dominant cutting regime. There is a good agreement between the forces with an average deviation of approximately 10%. Similarly, the comparison at full immersion at a spindle speed of 90 000 rev/min and feed rate of 0.6 μm/flute is depicted in Figure 17.8b. The deviations mainly come from small vibration caused by the table movement, material nonuniformity, rubbing of the tool against the workpiece, material hardening, and nonlinearities, which have not been considered in the model. When we also compared the shearing dominant forces at the feed rate of 8 μm/flute, the simulation results depict a good agreement with the experimentally measured forces. Consequently, the proposed mechanistic model can effectively predict micro-end milling forces at high rotational speeds. The accurate prediction of micromilling forces is important in order to optimize the process.

17.4 Tool Tip Dynamics

Due to the fragility of micro-end mills, the tool deflection cannot be ignored in order to achieve tight tolerance requirements. Furthermore, instability known as regenerative chatter occurs. Chatter is an unstable, self-excited vibration that occurs as a result of an interaction between the dynamics of the machine tool and the workpiece [54, 65–67]. To investigate both the tool deflection and micromilling chatter, the accurate measurement of tool tip dynamics is vital.

Unlike macroscale machine tools, performing the impact hammer test at the tool tip of micromills is not feasible, due to the fragility of the miniature tools. Often, micro-tools are modeled as cantilever beams. However, this may not correctly represent the dynamics at the tool tip. To take advantage of both analytical and experimental modal analyses (EMA), the receptance coupling (RC) method can be used to obtain the dynamics at the tool tip, as depicted in Figure 17.9. This method mathematically combines the results of EMA of the spindle and machine tool (i.e. Substructure B) with the result of finite element analysis of the

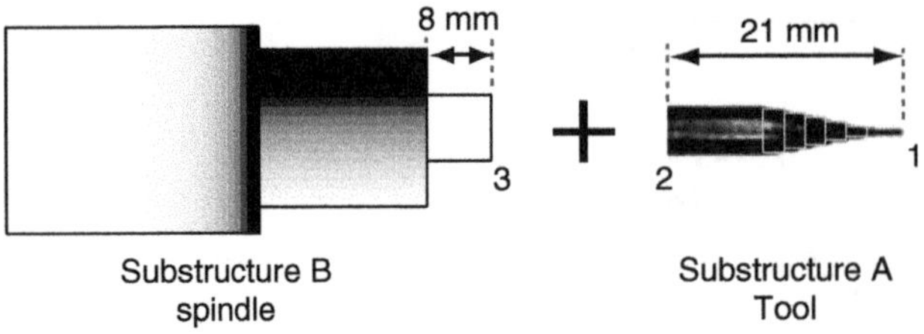

Figure 17.9 Receptance coupling of a spindle and an arbitrary end mill. *Source:* Rahnama et al. 2009 [68]. Reproduced with permission of Elsevier.

tool (i.e. Substructure A) to come up with the overall dynamics at the tool tip. The rotational dynamics is obtained through gauge tools outlined in [69].

A miniature impact hammer (Dytran 5800) and a capacitance sensor (Lion Precision C3-D) are used for EMA tests at the 3 mm shank (location 3) in Figure 17.9. The FE analysis of tungsten carbide tool is performed with density = 14 300 kg/m^3, Young's modulus = 580 GPa, Poisson's ratio = 0.28, and damping ratio ≈ 0.01. The tool tip dynamics can be obtained through the following equation [70]:

$$\mathbf{G}_{11} = \mathbf{X}_1 / \mathbf{F}_1 = \underbrace{\mathbf{H}_{11}}_{\text{FE}} - \underbrace{\mathbf{H}_{12}}_{\text{FE}} (\underbrace{\mathbf{H}_{22}}_{\text{FE}} + \underbrace{\mathbf{H}_{33}}_{\text{EMA}})^{-1} \underbrace{\mathbf{H}_{21}}_{\text{FE}} \quad (17.12)$$

where G and H denote the assembled and substructure dynamics, respectively. Figure 17.10 depicts the curve fitted tool tip dynamics using the receptance coupling method. The modes occurring between 6 and 7 kHz are mainly due to the shank and taper section of the tool. Once the tool dynamics can be identified, the static tool deflection compensation and regenerative chatter stability can be acquired.

The static deflection of the tool cannot be discounted in micro-cutting force analysis. Tool deflections of about 100 μm are commonly observed in micromilling, and these can be compensated by manipulating NC paths using a free-form deformation technique [71]. Ni investigated the effect of static tool deflection by assuming the tool as a simple cantilever beam. Their model corrected the actual depth of cut by subtracting the deflection of the cutting tool from the commanded depth of the cut. Dow [72] compensated for the deflection errors in microtools by predicting the cutting and thrust forces. Since at micro-scales, cutting force influences tool deflection, and tool deflection influences the cutting force, cutting force models should include this coupling.

Chaotic dynamic cutting processes, such as chatter, pose a significant problem in micromachining because they result in excessive vibration that can lead to catastrophic failure.

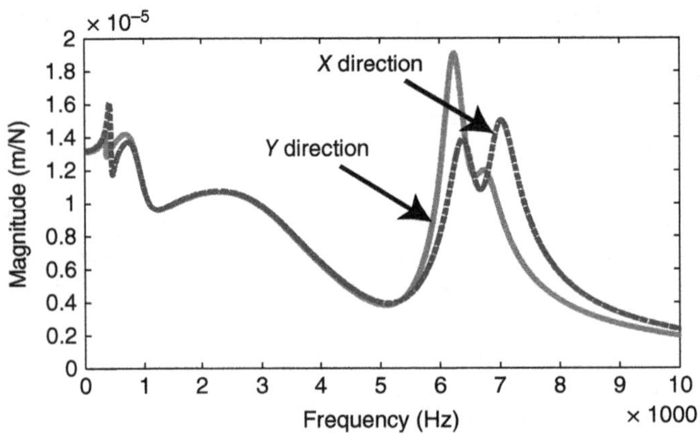

Figure 17.10 Tool tip dynamics using the RC Method. *Source:* Rahnama et al. 2009 [68]. Reproduced with permission of Elsevier.

Figure 17.11 Wavy surface and phase shift in machining.

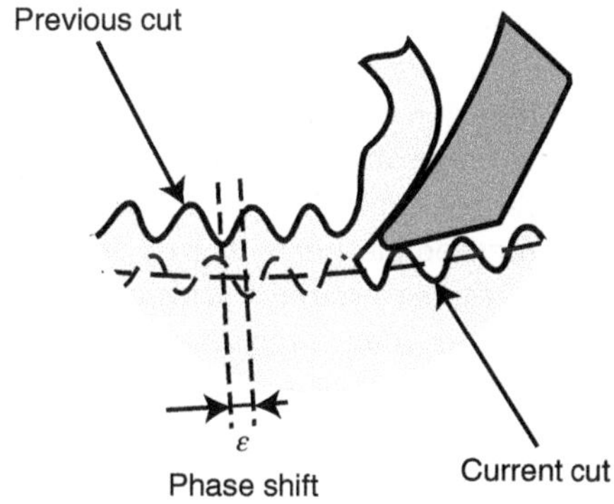

In traditional regenerative chatter stability, the occurrence of chatter is dependent on three factors: cutting conditions, workpiece material properties, and the dynamics of the machine tool spindle system. There are several differences between conventional macro-chatter stability vs. micromilling chatter stability. In micromachining, chatter stability needs to consider both the elasticity of the workpiece due to low feed rates, as well as the conventional macro-regenerative waves that result in a dynamic change in chip thickness. Figure 17.11 shows the phase shift caused by the dynamic change in chip thickness. Regenerative chatter occurs when the chip thickness becomes oscillatory due to vibrations between the previous and subsequent cuts.

High-speed machines also present new challenges for micromachining. At higher speeds, the dynamics of machine tools change, with an unbalanced spindle producing centrifugal and gyroscopic effects [73]. Modeling and analysis of high-speed spindle dynamics with various bearing configurations and thermal expansions are important. The majority of spindle research has dealt with static stiffness optimization rather than dynamic performance. The few papers addressing high-speed spindle dynamics were far below the operation range of micromachining [8, 73]. In the microscale, material property homogeneity cannot be assumed;

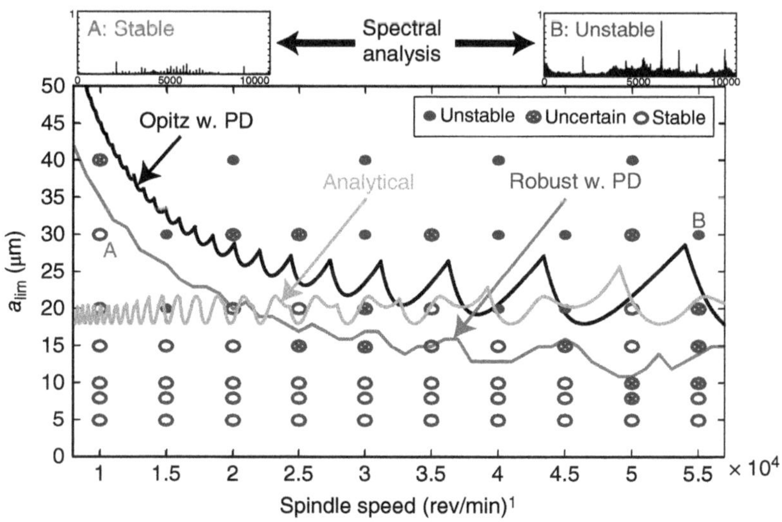

Figure 17.12 Comparison of micromilling stability lobes with conventional chatter, robust, analytical models and experiments at slotting of Al7075. *Source:* Rahnama et al. 2009 [68]. Reproduced with permission of Elsevier.

thus, the cutting coefficients used in chatter control models will vary from macromachining processes due to anisotropy of materials. Instability in micromachining is a very challenging problem due to various nonlinear effects and time-variant dynamics. In particular, the feed rates of micromilling operations can affect the stability due to elastic deformation [74]. To address these challenges, robust chatter stability based on the edge theorem and zero exclusion method has been utilized as shown in Figure 17.12 [68]. The robust chatter stability finds the guaranteed stability within bounds of changing parameters such as dynamics and cutting coefficients. The results show that similar chatter phenomena occur even at the micro level, and chatter can cause excessive tool wears and premature tool failures.

17.5 Summary

The ability to accurately fabricate miniature components is becoming very important in numerous applications. Translating the macromachining knowledge and shifting its paradigm where appropriate are essential for making micromilling processes more flexible, accurate, and productive. Rapid development of micromechanical machining in recent years is based on transferring macromachining knowledge to micromachining by shrinking the size of tools and increasing the rotational speeds of the spindle.

In micromilling, the tool is in direct contact with the workpiece, and the tool follows the desired tool path, fabricating complex 3D geometries using various materials. Whereas macro and micro milling operations are two similar processes, as they both remove the workpiece material by milling tools, there are some differences between cutting mechanisms in micro and macro milling processes. The chip thickness or the linear feed rate in micromilling is considerably less than macromilling process to prevent excessive tool deflections or breakage. In order to compensate the negative effect of low feed rate on the productivity, micromilling operations are employing very high rotational spindle speed that can exceed 100 000 rev/min. When the feed rate is comparable to the edge radius of the tool, the MUCT phenomena, which determine the ploughing and shearing cutting regimes, can occur. The MUCT can be identified by finding the stagnant point on the edge of the orthogonal tool through finding the friction angle in the orthogonal cutting operations. Based on this information, a mechanistic force model can be formulated to predict micro-end milling forces for both the shearing and ploughing dominant cutting regimes.

The components in micro machine tools are different than conventional macromachining centers. Furthermore, macroscale auxiliary processes (such as deburring and polishing) are not applicable in the microscale. The fragility of tools used in micromilling operations requires tool deflection compensation to minimize regenerative chatter. To compensate and predict, the accurate acquisition of tool tip dynamics is done through receptance coupling, which mathematically couples experimental FRFs and analytical FRFs to predict the overall tool tip dynamics. High-quality microcomponents with a good surface finish can only be delivered by using the optimal processing parameters that consider tool deflections and chatter.

Despite the progresses in micromilling research over the last decade to overcome micromilling challenges, more research is needed in order to prolong the longevity of micro-tools and to improve the machining efficiency. Developments are needed to obtain accurate measurements of finished components as well as the edge radius of tools. In-situ micro metrology systems are needed to correct errors during the machining operations. A better handling of micromachined components is required to overcome the fragility of miniature components, as well as the releasing of microcomponents to overcome the Van der Waals force between miniature components and grippers. Further investigations are needed on the utilization of hybrid machining methods such as elliptical vibration–assisted or laser-assisted machining to minimize cutting forces and improve the surface finish while improving the productivity.

References

1 Weck, M., Huemmler, J., and Petersen, B. (1997). Assembly of hybrid microsystems in a large-chamber scanning electron microscope by use of mechanical grippers. In: *Micromachining and Microfabrication*, 223–229. International Society for Optics and Photonics.

2 Liu, X., DeVor, R.E., Kapoor, S.G., and Ehmann, K.F. (2004). The mechanics of machining at the microscale: assessment of the current state of the science. *Journal of Manufacturing Science and Engineering* 126 (4): 666–678.

3 Vogler, M.P., DeVor, R.E., and Kapoor, S.G. (2003). Microstructure-level force prediction model for micro-milling of multi-phase materials. *Journal of Manufacturing Science and Engineering* 125 (2): 202–209.

4 Weule, H., Hüntrup, V., and Tritschler, H. (2001). Micro-cutting of steel to meet new requirements in miniaturization. *CIRP Annals-Manufacturing Technology* 50 (1): 61–64.

5 Bang, Y.B., Lee, K.M., and Oh, S. (2005). 5-Axis micro milling machine for machining micro parts. *The International Journal of Advanced Manufacturing Technology* 25 (9): 888–894.

6 Benavides, G.L., Adams, D.P., and Yang, P. (2001). Meso-machining Capabilities. Sandia Report.

7 El-Fatatry, A. (2003). EUSPEN Vision Online: Transferring Microsystems Technology. NEXUS report.

8 Howe, R.T. (2003). *Micro Systems Research in Japan*. World Technology Evaluation Center (WTEC).

9 Kussul, E.M., Rachkovskij, D.A., Baidyk, T.N., and Talayev, S.A. (1996). Micromechanical engineering: a basis for the low-cost manufacturing of mechanical microdevices using microequipment. *Journal of Micromechanics and Microengineering* 6 (4): 410.

10 Kussul, E., Baidyk, T., Ruiz-Huerta, L. et al. (2002). Development of micromachine tool prototypes for microfactories. *Journal of Micromechanics and Microengineering* 12 (6): 795.

11 Okazaki, Y., Mishima, N., and Ashida, K. (2004). Microfactory: concept, history, and developments. *Journal of Manufacturing Science and Engineering* 126 (4): 837–844.

12 Tanaka, M. (2001). Development of desktop machining microfactory. *Riken Review* (34).

13 Hesselbach, J., Raatz, A., Wreg, J. et al. (2004). International state of the art of micro production technology. *Production Engineering* 11 (1): 29–36.

14 Masuzawa, T. (2000). State of the art of micromachining. *CIRP Annals-Manufacturing Technology* 49 (2): 473–488.

15 Alting, L., Kimura, F., Hansen, H.N., and Bissacco, G. (2003). Micro engineering. *CIRP Annals-Manufacturing Technology* 52 (2): 635–657.

16 Dornfeld, D., Min, S., and Takeuchi, Y. (2006). Recent advances in mechanical micromachining. *CIRP Annals-Manufacturing Technology* 55 (2): 745–768.

17 Koç, M. and Ozel, T.G. (2011). Fundamentals of micro-manufacturing. In: *Micro-Manufacturing: Design and Manufacturing of Micro-Products*, 1–23. Wiley.

18 Gandarias, E. (2007). MICROM: a revolutionary monitoring system to detect tool breakages & collisions, enhance machine cycles and introduce a new probing concept in micromilling. Thesis. Mondragon Unibertsitatea.

19 Altintas, Y. (2012). *Manufacturing Automation: Metal Cutting Mechanics, Machine Tool Vibrations, and CNC Design*. Cambridge University Press.

20 Abele, E., Altintas, Y., and Brecher, C. (2010). Machine tool spindle units. *CIRP Annals-Manufacturing Technology* 59 (2): 781–802.

21 Bissacco, G., Hansen, H.N., and De Chiffre, L. (2005). Micromilling of hardened tool steel for mould making applications. *Journal of Materials Processing Technology* 167 (2): 201–207.

22 Lu, X., Jamalian, A., and Graetz, R. (2011). A new method for characterizing axis of rotation radial error motion: Part 2. Experimental results. *Precision Engineering* 35 (1): 95–107.

23 Ozdoganlar, O.B., Epp, D.S., Reu, P.L., and Sumali, H. (2004). *Development of a Testing Facility for Experimental Investigation of MEMS Dynamics*. ASME.

24 Altintas, Y., Verl, A., Brecher, C. et al. (2011). Machine tool feed drives. *CIRP Annals-Manufacturing Technology* 60 (2): 779–796.

25 Byrne, G., Dornfeld, D., and Denkena, B. (2003). Advancing cutting technology. *CIRP Annals-Manufacturing Technology* 52 (2): 483–507.

26 Verl, A. and Frey, S. (2010). Correlation between feed velocity and preloading in ball screw drives. *CIRP Annals* 59 (1): 429–432.

27 Huo, D. and Cheng, K. (2010). Experimental investigation on micromilling of oxygen-free, high-conductivity copper using tungsten carbide, chemistry vapour deposition, and single-crystal diamond micro tools. *Proceedings of the Institution of Mechanical Engineers, Part B: Journal of Engineering Manufacture* 224 (6): 995–1003.

28 Jun, M.B.G., Liu, X., DeVor, R.E., and Kapoor, S.G. (2006). Investigation of the dynamics of micro-end milling, Part 1: model development. *Journal of Manufacturing Science and Engineering, Transactions of the ASME* 128: 893–900.

29 Shaw, M.C., Pigott, J.D., and Richardson, L.P. (1951). Effect of cutting fluid upon chip-tool interface temperature. *The American Society of Mechanical Engineers – Transactions* 73 (1): 45–52.

30 Friedrich, C.R. (2000). Near-cryogenic machining of polymethyl methacrylate for micromilling tool development. *Materials and Manufacturing Processes* 15 (5): 667–678.

31 Yan, J., Syoji, K., and Tamaki, J. (2003). Some observations on the wear of diamond tools in ultra-precision cutting of single-crystal silicon. *Wear* 255 (7–12): 1380–1387.

32 Jun, M.B.G., Joshi, S.S., DeVor, R.E., and Kapoor, S.G. (2008). An experimental evaluation of an atomization-based cutting fluid application system for micromachining. *Journal of Manufacturing Science and Engineering* 130 (3): 031118–031111.

33 Adler, D.P., Hii, W.W.-S., Michalek, D.J., and Sutherland, J.W. (2006). Examining the role of cutting fluids in machining and efforts to address associated environmental/health concerns. *Machining Science and Technology* 10 (1): 23–58.

34 Abdalla, H.S., Baines, W., McLntyre, G., and Slade, C. (2007). Development of novel sustainable neat-oil metal working fluids for stainless steel and titanium alloy machining. Part 1. Formulation development. *International Journal of Advanced Manufacturing Technology* 34 (1–2): 21–33.

35 Kajaria, S., Chittipolu, S., Adera, S., and Hung, W.N. (2012). Micromilling in minimum quantity lubrication. *Machining Science and Technology* 16 (4): 524–546.

36 Lee, P.H., Nam, T.S., Li, C., and Lee, S.W. (2010). Experimental study on meso-scale milling process using nanofluid minimum quantity lubrication. *Transactions of the Korean Society of Mechanical Engineers, A* 34 (10): 1493–1498.

37 Li, K.-M. and Chou, S.-Y. (2010). Experimental evaluation of minimum quantity lubrication in near micro-milling. *Journal of Materials Processing Technology* 210 (15): 2163–2170.

38 Li, K.-M. (2012). Experimental study on minimum quantity lubrication in mechanical micromachining. *Advanced Materials Research* 579: 193–200.

39 Rukosuyev, M., Goo, C.S., Jun, M.B.G., and Park, S.S. (2010). Design and development of cutting fluid system based on ultrasonic atomization for micro-machining. In: *Transactions of NAMRI/SME, XXXVIII*, 97–104. Society of Manufacturing Engineers.
40 Cox, D., Newby, G., Park, H.W., and Liang, S.Y. (2004). *Performance Evaluation of a Miniaturized Machining Center for Precision Manufacturing*. ASME.
41 Prakash, L.J. (1995). Application of fine grained tungsten carbide based cemented carbides. *International Journal of Refractory Metals and Hard Materials* 13 (5): 257–264.
42 Aramcharoen, A., Mativenga, P.T., Yang, S. et al. (2008). Evaluation and selection of hard coatings for micro milling of hardened tool steel. *International Journal of Machine Tools and Manufacture* 48 (14): 1578–1584.
43 De Cristofaro, S., Funaro, N., Feriti, G.C. et al. (2012). High-speed micro-milling: novel coatings for tool wear reduction. *International Journal of Machine Tools and Manufacture* 63: 16–20.
44 Bouzakis, K.D., Michailidis, N., Skordaris, G. et al. (2003). Optimisation of the cutting edge roundness and its manufacturing procedures of cemented carbide inserts, to improve their milling performance after a PVD coating deposition. *Surface and Coatings Technology* 163: 625–630.
45 Shabouk, S. and Nakamoto, T. (2002). Micro machining of single crystal diamond by utilization of tool wear during cutting process of ferrous material. *Journal of Micromechatronics* 2 (1): 13–26.
46 Paggett, J.W., Drake, E.F., Krawitz, A.D. et al. (2002). Residual stress and stress gradients in polycrystalline diamond compacts. *International Journal of Refractory Metals and Hard Materials* 20 (3): 187–194.
47 Ng, C.K., Melkote, S.N., Rahman, M., and Senthil Kumar, A. (2006). Experimental study of micro-and nano-scale cutting of aluminum 7075-T6. *International Journal of Machine Tools and Manufacture* 46 (9): 929–936.
48 Bachmann, P.K. and van Enckevort, W. (1992). Diamond deposition technologies. *Diamond and Related Materials* 1 (10): 1021–1034.
49 Brinksmeier, E., Gläbe, R., and Osmer, J. (2006). Ultra-precision diamond cutting of steel molds. *CIRP Annals-Manufacturing Technology* 55 (1): 551–554.
50 Huo, D., Cheng, K., and Wardle, F. (2010). Design of a five-axis ultra-precision micro-milling machine – UltraMill. Part 1: holistic design approach, design considerations and specifications. *The International Journal of Advanced Manufacturing Technology* 47 (9): 867–877.
51 Liu, K., Li, X.P., Rahman, M. et al. (2007). A study of the effect of tool cutting edge radius on ductile cutting of silicon wafers. *The International Journal of Advanced Manufacturing Technology* 32 (7): 631–637.
52 Cheng, X., Wang, Z., Nakamoto, K., and Yamazaki, K. (2010). Design and development of PCD micro straight edge end mills for micro/nano machining of hard and brittle materials. *Journal of Mechanical Science and Technology* 24 (11): 2261–2268.
53 Denkena, B., Hoffmeister, H.W., Reichstein, M. et al. (2006). Micro-machining processes for microsystem technology. *Microsystem technologies* 12 (7): 659–664.
54 Tlusty, J. and Andrews, G.C. (1983). A critical review of sensors for unmanned machining. *CIRP Annals-Manufacturing Technology* 32 (2): 563–572.
55 Malekian, M., Park, S.S., and Jun, M.B. (2009). Tool wear monitoring of micro-milling operations. *Journal of Materials Processing Technology* 209 (10): 4903–4914.
56 Welch, G. and Bishop, G. (1995). An Introduction to the Kalman Filter. Technical Report 95-041 (5 April 2004). University of North Carolina-Chapel Hill.
57 Cheema, M.S., Venkatesh, G., Dvivedi, A., and Sharma, A.K. (2012). Developments in abrasive flow machining: a review on experimental investigations using abrasive flow machining

variants and media. *Proceedings of the Institution of Mechanical Engineers, Part B: Journal of Engineering Manufacture* 226 (12): 1951–1962.
58 Jha, S. and Jain, V.K. (2004). Design and development of the magnetorheological abrasive flow finishing (MRAFF) process. *International Journal of Machine Tools and Manufacture* 44 (10): 1019–1029.
59 Rajesha S. (2011). Some studies to enhance the capabilities of abrasive flow machining process. PhD dissertation. IIT Roorkee.
60 Mathai, G. and Melkote, S. (2012). Effect of process parameters on the rate of abrasive assisted brush deburring of microgrooves. *International Journal of Machine Tools and Manufacture* 57: 46–54.
61 Vogler, M.P., Liu, X., Kapoor, S.G. et al. (2002). Development of meso-Scale Machine Tool (mMT) Systems. Technical papers-society of manufacturing engineers-all series-.
62 Jardret, V., Zahouani, H., Loubet, J.L., and Mathia, T.G. (1998). Understanding and quantification of elastic and plastic deformation during a scratch test. *Wear* 218 (1): 8–14.
63 Kita, Y., Ido, M., and Hata, S. (1978). The mechanism of metal removal by an abrasive tool. *Wear* 47: 185–193.
64 Altintas, Y. (2004). Dynamic compensation of spindle integrated force sensors with Kalman filter. *Journal of Dynamic Systems, Measurement, and Control* 126: 443.
65 Budak, E. and Kops, L. (2000). Improving productivity and part quality in milling of titanium based impellers by chatter suppression and force control. *CIRP Annals-Manufacturing Technology* 49 (1): 31–36.
66 Tobias, S.A. and Fishwick, W. (1958). Theory of regenerative machine tool chatter. *The Engineer* 205: 199–203.
67 Tobias, S.A. (1965). *Machine-Tool Vibration*. Wiley.
68 Rahnama, R., Sajjadi, M., and Park, S.S. (2009). Chatter suppression in micro end milling with process damping. *Journal of Materials Processing Technology* 209 (17): 5766–5776.
69 Mascardelli, B.A., Park, S.S., and Freiheit, T. (2008). Substructure coupling of microend mills to aid in the suppression of chatter. *Journal of Manufacturing Science and Engineering* 130 (1): 011010.
70 Park, S.S., Altintas, Y., and Movahhedy, M. (2003). Receptance coupling for end mills. *International Journal of Machine Tools and Manufacture* 43 (9): 889–896.
71 Biermann, D., Krebs, E., Sacharow, A., and Kersting, P. (2012). Using NC-path deformation for compensating tool deflections in micromilling of hardened steel. *Procedia CIRP* 1: 132–137.
72 Dow, T.A., Miller, E.L., and Garrard, K. (2004). Tool force and deflection compensation for small milling tools. *Precision Engineering* 28 (1): 31–45.
73 Jorgensen, B.R. and Shin, Y.C. (1997). Dynamics of machine tool spindle/bearing systems under thermal growth. *Journal of Tribology* 119 (4): 875–882.
74 Kim, B., Schmittdiel, M.C., Degertekin, F.L., and Kurfess, T.R. (2004). Scanning grating micro interferometer for MEMS metrology. *Journal of Manufacturing Science and Engineering* 126 (2004): 807–812.

18

Laser Machining

Dani Teixidor[1], Inés Ferrer[1], Luis Criales[2], and Tuğrul Özel[2]

[1] Department of Mechanical Engineering and Industrial Construction, University of Girona, Girona, Spain
[2] Manufacturing and Automation Research Laboratory, Department of Industrial and Systems Engineering, School of Engineering, Rutgers University, Piscataway, NJ 08854, USA

18.1 Introduction

The use of industrial lasers in material processing, material removal, and micromanufacturing is a rapidly growing area of research. Laser-based material processing has been used in thermal processing, shock peening, surface treatment, cleaning of surfaces, welding, melting and polishing, and scribing as well as a material removal mechanism as in micromachining of basic geometric features on a wide variety of materials. Lasers can provide unique solutions in material removal, offer the ability to manufacture otherwise unattainable devices, and yield cost-effective solutions to complex manufacturing processes that demand high levels of flexibility.

Further developments in pulsed laser techniques and systems have increased the applicability of laser machining technology in the production systems. Hence, laser machining technology has become a viable alternative to conventional methods for producing complex and microsized features on difficult-to-process materials and is being increasingly employed in industry because of the advantages described previously.

Laser machining can be defined as material processing and removal by means of ablation that is applied to a wide range of materials: metals and nonmetals, soft and hard, difficult-to-machine, etc. and enables the production of parts with complex geometries in high precision without expensive tooling. Due to its nature as a noncontact material removal process, laser machining removes much less material, involves highly localized heat input to the workpiece, minimizes distortion, and does not require tooling. These are significant advantages when compared with other conventional material removal processes. In particular, the process is not limited by common constraints such as maximum tool force, buildup edge formation, or tool chatter, which have to be taken into account when considering other more traditional machining methods such as conventional turning, milling, and drilling.

Lasers are can be categorized into two groups based on their mode of operation: continuous wave (CW) and noncontinuous (pulsed) lasers generated typically using q-switching, cavity dumped or mode-locking methods. In a CW laser, the power output is nearly constant over time, while in a pulsed laser, the transmitted energy can be observed in pulses of a given duration at a given repetition rate. Conventional CW and pulsed laser ablation are currently

Modern Manufacturing Processes, First Edition. Edited by Muammer Koç and Tuğrul Özel.

used in many fields, such as shape cutting and hole drilling in sheet materials, marking, scribing, etching, and patterning on surfaces of medical devices, and directly fabricating microchannels for microfluidics. Pulsed lasers achieve much higher intensities (and peak power levels) than CW lasers and are the preferred solution for the fabrication of microsized structures, where ablation is required. Furthermore, pulsed laser can be classified with respect to the pulse duration. Long-pulsed lasers typically have pulse duration of a few nanoseconds, while short-pulsed laser duration is in the order of picoseconds. Nanosecond pulse lasers have applications in remote hole drilling on aluminum and silicon, edge deletion of thin film solar cells. Short-pulsed lasers operate at picosecond pulse duration and are used in thin-film ablation, dicing of silicon wafers, cutting of stents, and deep hole drilling in ceramics. Finally, ultrashort-pulsed lasers can generate pulses with as short as a few femtoseconds, resulting in vast peak power, ideal for chemical ablation. These types of laser are commonly used for repairing, trimming, marking, scribing, texturing, cutting, and drilling, as presented in Table 18.1.

The process of material removal with laser machining is influenced by the physical characteristics of both the laser beam and the workpiece, but it is intrinsically determined by the way that these characteristics interact with each other. The main laser parameters that can be chosen and controlled are wavelength, laser power or pulse energy, intensity or fluence, pulse duration, pulse repetition rates, and peak power. Rarely can these parameters be modified without changing the laser itself, although a few exceptions exist. For example, harmonic generation units can be used to modify the incident beam's frequency and obtain wavelengths that vary between ultraviolet (UV) and infrared (IR) ranges. Since laser ablation occurs only when the substrate material strongly absorbs the wavelength of the transmitted radiation, it is necessary to study whether a specific material will absorb the energy from the incident beam. Hence, material properties such as surface finishing, surface coating, reflectivity, absorptivity, thermal diffusivity, and thermal conductivity are parameters that can be varied and will determine the efficiency of the laser machining process. Additionally, there are process parameters that can be controlled and modified in order to obtain optimal machining results. The selection of the pulse repetition rate, the scanning speed, and the pulse intensity significantly affects the quality of the machined features, as well as the material removal rate of the overall process.

Table 18.1 Commonly used lasers in industry.

Laser type	Wave-length (nm)	Power (W)	Pulse energy	Pulse duration	Repetition rate (kHz)
Q-switched Nd:YAG	1064	1–35	8 mJ	5–100 ns	1–400
	532	0.5–20	5 mJ	5–70 ns	1–300
	355	0.2–10	3 mJ	5–50 ns	15–300
	266	0.5–3	<1 mJ	5–30 ns	15–300
Nanopulsed fiber lasers	1070	10–50	0.5–1 mJ	1.5–10 ns	20–200
	532	18	0.3 mJ	4–200 ns	60–300
		5–10	14 μJ	1.3 ns	300–600
Disk-type	1030	750	<80 mJ	30 ns	5–100
	1030	50	<250 μJ	<10 ps	200–800
	515	30	150 μJ	<10 ps	200–800
	343	10	50 μJ	<10 ps	200–800
Ti:Sapphire	800 (central) (700–980)	0.5–2	0.25–0.9 mJ	100–150 fs	1–5

18.2 Laser–Material Interaction

The word "laser" has its origin as an acronym for "light amplification by stimulated emission of radiation." In this case, "light" denotes electromagnetic radiation of any frequency, ranging from ultraviolet to infrared waves. Therefore, laser radiation can be thought of as an electromagnetic wave with good coherence. When the electromagnetic radiation is incident on the surface of a specific material, various phenomena occur, including reflection, refraction, absorption, scattering, and transmission. The most important phenomenon in laser processing of materials is the absorption of radiation.

Absorption of light consists of the interaction of the electromagnetic radiation waves with the electrons of the material; therefore, absorption factors include the wavelength of the laser and the spectral absorptivity characteristics of the material being machined. The absorption of laser radiation in the material is generally expressed by the Beer–Lambert law:

$$I(z) = I_0^{-\alpha z} \tag{18.1}$$

where I_0 is the initial laser intensity (or power), α is the absorption coefficient, and z is the distance from the incidence region. Thus, once inside the material, absorption causes the intensity of the light to decay with depth at a rate determined by the material's absorption coefficient α. In general, α is a function of wavelength and temperature, but for constant α, intensity decays exponentially with depth.

The laser energy absorbed by the material during laser–material interaction is converted into heat by degradation of the ordered and localized primary excitation energy. The conversion of light energy into heat and its subsequent conduction into the material establishes the temperature distributions in the material. Depending on the magnitude of the temperature rise, it can cause various physical effects in the material, including heating, melting, and vaporization of the material. Furthermore, the ionization of vapor during laser irradiation may lead to generation of plasma. These effects of laser material interaction are schematically presented in Figure 18.1.

18.2.1 Laser Ablation

Laser ablation is the removal of material from a substrate by direct absorption of laser energy. In ablation, the chemical bonds between atoms are torn apart by the excess amount of energy absorbed by the material's electrons. Therefore, the onset of ablation occurs above a threshold fluence, which will in turn depend on the absorption mechanism, the properties of the material, its microstructure, morphology characteristics, and the presence of defects. Additionally, laser parameters such as wavelength and pulse duration also have an effect of laser ablation. The laser–material interaction during ablation is complex and involves interplay between the photothermal (vibration heating) and photochemical (bond breaking) processes. Regardless, the dominant process involved is the heating of the material and subsequent phase changes from its solid phase to its vapor phase, passing through the liquid phase.

At low fluence levels, photothermal mechanisms for ablation include material evaporation and sublimation. The absorbed laser energy is converted into thermal energy in the material. The subsequent temperature rise at the surface may facilitate the material removal due to generation of thermal stresses. However, photothermal ablation is usually accompanied by heating and collateral damage to the surrounding areas, which is an undesirable, often detrimental factor in the manufacturing process. When the incident laser energy is sufficiently large, the temperature at the surface exceeds the boiling point, causing rapid vaporization.

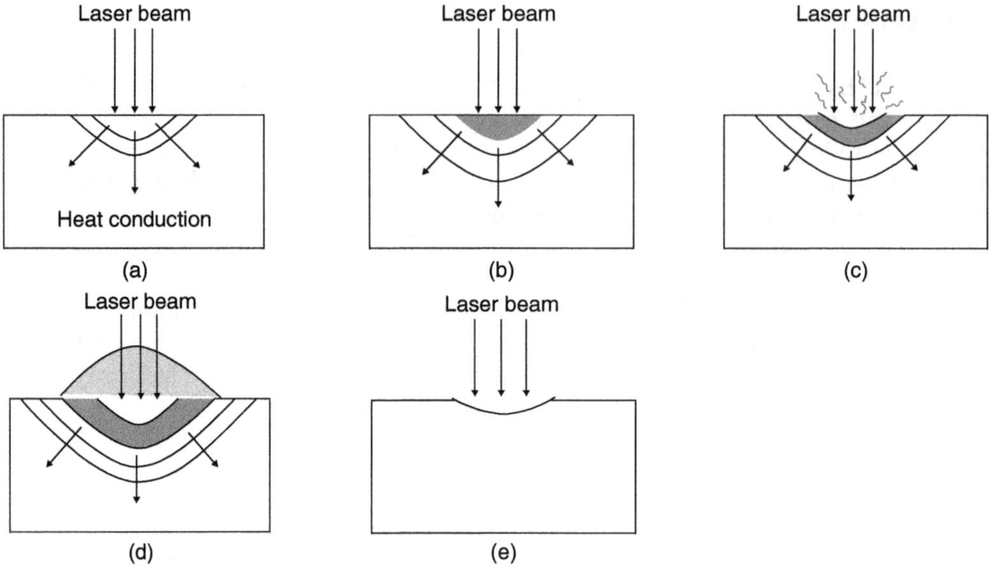

Figure 18.1 Laser material interaction. (a) Heating, (b) surface melting, (c) surface vaporization, (d) Plasma formation, and (e) ablation.

These thermal mechanisms can be understood as thermodynamic phase changes in response to the high temperatures.

When the excitation time is shorter than the time required to heat the material so that the temperature reaches the boiling point, nonthermal, photochemical ablation mechanisms can occur. In photochemical ablation, the energy of the incident photon causes the direct bond breaking of the molecular chains in the organic materials resulting in material removal by molecular fragmentation without significant thermal damage. The photon energy must be greater than the bond energy for photochemical ablation to occur.

The temporal pulse length of the laser has a significant effect on the dynamics of the ablation process. In general, as the pulse length is shortened, energy is more rapidly deposited into the material, leading to faster material ejection. The volume of material that is directly excited by the laser has less time to transfer energy to the surrounding material before being ejected. Therefore, the ablated volume becomes more precisely defined by the laser's spatial profile and optical penetration depth, and the remaining material has less residual energy, which reduces the heat-affected zone. The advent of short-pulsed and ultrashort-pulsed lasers has opened up a wide array of possibilities, by making extremely localized laser heating a possibility.

18.3 Laser Processing of Materials

As explained previously, laser-absorption mechanisms depend on the characteristics of the material. Absorption, reflectivity, and thermal diffusion and conductivity of different kinds of materials affect the laser machining process. Hence, the use of a laser system with the adequate wavelength and the selection of the optimum set of process parameters will depend on the optical/thermal properties of the material. The following sections further develop the interaction between various laser systems and different materials.

18.3.1 Laser Processing of Metals And Alloys

When electromagnetic waves interact with metals, the energy absorbed raises the temperature of the incident material. The laser beam heats, melts, and vaporizes the metal in a process called metal sublimation. However, due the high ablation threshold of metals and their high reflectivity at most common laser wavelengths, the energy absorbed may not be sufficient to achieve the softening of the material necessary to substantially induce the process of removal of the material. The use of short-pulsed lasers with proper choice of laser parameters may still achieve thermal softening in highly reflective metals.

Several authors used lasers processing systems to machine steel [1–4], copper [5], aluminum [6, 7], and other metals. Heyl et al. [8] and Dumitru et al. [9] machined hard metals with a UV laser and a femtosecond laser, respectively, presenting favorable experimental results. Other authors reported favorable results while machining different metals using a femtosecond Ti:sapphire laser ([10–12], and [13]).

18.3.2 Laser Processing of Polymers

Polymers are the materials of choice for disposable lab-on-chip devices, such as those utilized in microfluidics systems, because of their favorable, highly flexible thermal and mechanical properties, biocompatibility, and optical transparency. Polymers exhibit strong absorption in the ultraviolet (UV) and infrared (IR) wavelengths but weak absorption at visible wavelengths, between 390 and 700 nm. However, the way in which polymers react to laser irradiation is somewhat different compared to the reaction observed in metals. It is believed that UV lasers produce a cooler excitation in polymers, meaning a lower rise of temperature in the heat-affected zone. On the other hand, IR laser irradiation causes the most molecular vibration through the thermal process. Regardless, the material properties of most polymers vary strongly as a function of temperature. This implies that even slight changes in temperature can have a strong effect on machining results, and localized laser heating can be used effectively for increasing the productivity of the process and the resulting product's characteristics. At UV wavelengths (200–400 nm), the material removal mechanism in polymers is usually thermal evaporation. Below 400 nm, the polymeric material is removed most often than not by chemical ablation. Various research investigations have been performed utilizing both UV wavelength lasers [4, 14–17] and IR wavelength CO_2 lasers ([18, 19]; and [20]) to machine microfluidic channels of polymer materials.

Polymers can be also be machined by ultrashort laser pulses, as shown in Figure 18.2. Femtosecond laser pulses provide a unique micromachining tool as they can modify transparent materials at wavelengths at which they are normally transparent, by depositing energy

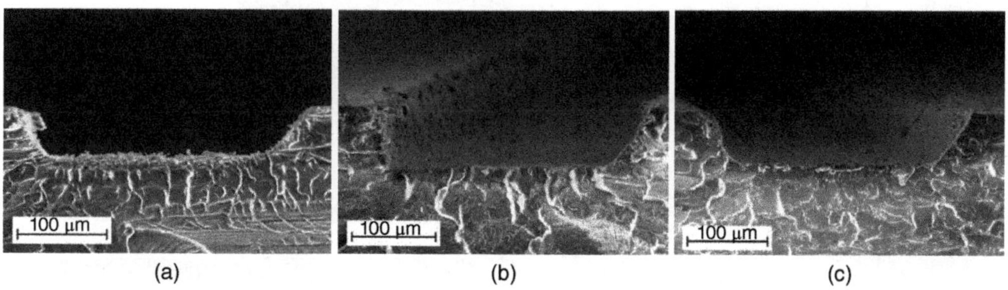

Figure 18.2 Multiscan microchannels ablated with a femtosecond laser in (a) PMMA, (b) PS, and (c) COP. *Source:* Suriano et al. 2011 [21]. Reproduced with permission of Elsevier.

through high-order nonlinear absorption processes. One of such processes is optical breakdown [22]. Several authors [21, 23] used femtosecond Ti:sapphire lasers to machine polymer materials as well.

18.3.2.1 Laser Processing of Composites

The use of carbon-fiber–reinforced plastic carbon fiber-reinforced plastic (CFRP) materials in aerospace, automotive, and marine industries is rapidly increasing due to their lighter weight and superior performance when compared to more traditional materials such as aluminum or titanium alloys. Therefore, CFRP composites have become major structural materials and are considered as substitutes for metals in many weight-critical components. However, certain challenges exist in laser processing of CFRP when the goal is to minimize or eliminate the dimensions of the heat-affected zone (HAZ) in the polymer matrix or when the objective is to maintain high processing speeds. The properties of conventional materials such as metals, ceramics, and polymers are isotropic, meaning that the machining quality is the same in all possible directions. However, CFRP composite materials are laminated with a different fiber orientation bound together in a polymer matrix according to its application. Each constituent retains its own chemical, physical, and mechanical properties, thus posing a challenge in laser processing due to the possibility of large disparities for a given material property of the two constituents at elevated temperatures. Anisotropic heat conduction at different fiber orientation directions is another problem that negatively affects the HAZ. Although the use of lasers for machining of composite materials has not been studied extensively, some authors researched the drilling of holes in CFRP materials with UV lasers [24–26] and femtosecond lasers [27]. Riveiro [28] studied laser cutting of CFRP material using a CO_2 laser and reported the experimental results.

18.3.3 Laser Processing of Ceramics

Laser machining of ceramics is challenging mainly due to their brittleness and large scattering exhibited at common laser wavelengths, which restricts energy absorption. A combination of short-pulse frequencies and short wavelengths usually presents best results [29]. The elevated hardness and thermal stability, low electrical conductivity, and high wear resistance have made structural ceramics such as alumina (Al_2O_3), silicon nitride (Si_3N_4), silicon carbide (SiC), and magnesia (MgO) useful for various applications in aerospace, electronics, automotive, medical, and semiconductor industries. Figure 18.3 presents results for the machining of aluminum

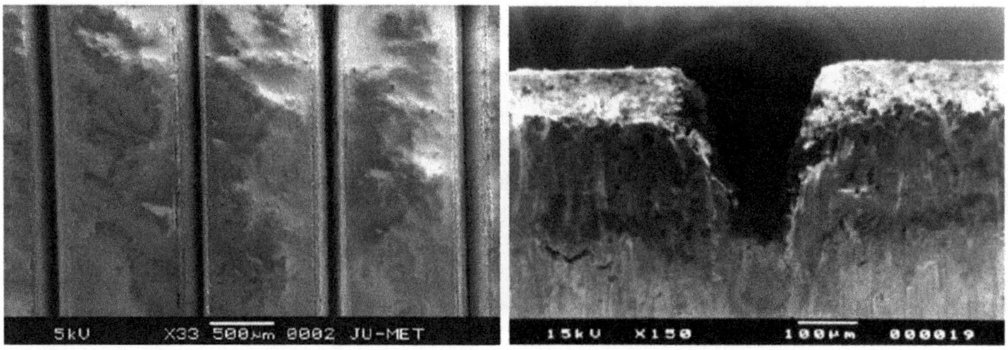

Figure 18.3 Microgroove of aluminum titanate machined with an Nd:YAG laser. *Source:* Dhupal et al. 2007 [30]. Reproduced with permission of Springer.

titanate. Samant et al ([31, 32], and [33]) presented results for machining alumina, silicon nitride, silicon carbide, and magnesia via one-dimensional machining (drilling), two-dimensional machining (cutting), and three-dimensional machining (milling) with a millisecond-pulsed Nd:YAG laser (1064 nm near-IR wavelength).

Gilbert et al. [34] generated a series of lines and pockets on the surface of the AlN using Nd:YAG lasers of UV and near-IR wavelengths. The near-IR laser showcased better material removal rates but less machining accuracy. If smaller, accurate features are desired, and the use of an UV laser would be preferred. Other authors used Nd:YAG lasers to machine grooves and cavities in alumina (Al_2O_3) and aluminum titanate (Al_2TiO_5) ([30, 35]; and [36]).

18.3.3.1 CVD Diamond

Synthetic Chemical vapor deposition (CVD) diamond, an allotropic form of carbon which may be thought of as a type of ceramic, is an attractive material for laser processing since it could be utilized effectively in IR optical applications, detectors, sensors, and thermal management systems. Among all existing materials, diamond is the most difficult one to machine because of its hardness and inertness; additionally, similar to glass, it is highly transparent over a broad range of the optical spectrum. Diamond can be ablated in two separate ways: the first requires that the laser radiation is highly absorbed by diamond, and the second method involves the mechanism of laser-induced graphitization. Various authors have used femtosecond Ti:sapphire lasers [37, 38] and nanosecond Nd:YAG lasers [39, 40] in their research to achieve laser processing of diamond.

18.3.3.2 Silicon

Due to its nature as a semiconductor, silicon is widely utilized in electronic devices, microelectromechanical systems (MEMS), and thin-film applications [41]. Laser etching of silicon permits the formation of a wide variety of structures, since, unlike wet etching, it is independent of the crystal plane orientation. Although the laser machining of silicon has not been investigated in depth, several authors used femtosecond lasers [41, 42] and UV lasers [29, 43] for machining silicon.

18.3.3.3 Glass

Glass is an amorphous (noncrystalline) ceramic material. Micromachining of hard and brittle glass finds applications in biochemistry, biomedicine, lab-on-chip devices, sensors, and Bio-MEMS devices. Some of the difficulties encountered in machining are due to the brittleness and poor thermal properties of most glasses, making the fabrication of finely machined features a challenging task. There is risk of laser-induced microcracking and other laser-induced collateral damage such as debris and poor surface quality. The absorption of light by some glass types can be characterized by very nonlinear behavior. However, since glass is monopaque, absorption occurs largely within the volume of the material, rather than on the surface.

Glass exhibits strong optical absorption at deep UV and IR wavelengths with much weaker absorption at visible and near-infrared wavelengths. Thus, researchers have utilized a nanosecond pulse excimer laser at 355 nm to machine-fused silica [44, 45]. An alternative consists of utilizing lasers with ultrashort pulse duration, which can modify transparent materials at wavelengths at which they are normally transparent, by means of optical breakdown. Many authors used femtosecond lasers (Ti:sapphire) in this manner to machine fused silica [46–49], borosilicate glass [50–52], among other glass materials [53, 54]. Other authors [29, 55] conducted comparative studies machining glass bulk materials using different types of pulsed lasers, UV, and femtosecond.

18.4 Laser-Processing Parameters

There are several key laser parameters directly influencing the ablation mechanism and modifying the energy within a material. A large reduction in laser power or an increase of scanning speed will result in incomplete penetration of the cutting zone and/or poor ablation quality. Abundant research work has been done with respect to how laser process parameters affect the quality of the resultant surfaces or geometrical features using experimental analysis tools. Several authors [1, 2, 56] have studied the influence of pulse intensity, scanning speed, and pulse frequency on the quality of the resulting sample and on the material removal rate of the process. The objective of this research is to establish relationships between process parameters and the identification of the optimum set of process parameters, which provide the best results.

18.4.1 Pulse Duration

During laser ablation, pulse duration (also called pulse width) has a significant effect on feature quality. Chichkov et al. [10] investigated laser ablation of different materials with femtosecond-, picosecond-, and nanosecond-pulsed lasers. Figure 18.4 clearly portraits the effect of different pulse durations for laser machining. Several authors [29, 57, 58] developed similar studies evaluating the effect of the pulse duration on laser processing. In conclusion, there are different mechanisms for the interaction of light and matter when applying laser pulses of different lengths.

As mentioned previously, continuous wave (CW) lasers emit radiation at a laser power that continuously depends on the pump power. CW lasers are used for applications that require a thermal impact, i.e. welding or melt cutting. Pulsed systems with long laser pulses (~1 ms) are used when higher power densities are required on short timescales, such as microwelding. Short laser pulses (pulse duration 1 ns to 1 μs) reduce thermal impact on the material. For material removal to be possible, an energy density above the ablation threshold is required. A short laser pulse with an energy density above the ablation threshold is absorbed by the material, and a part of this material is transformed into an expanding plasma plume within several picoseconds. Short laser pulses are generated through resonator Q-switching. Due to the high peak power, Q-switched solid state lasers allow for efficient frequency conversion and, therefore, are able to deliver visible or UV wavelengths.

Ultrashort laser pulses are defined as pulses with pulse duration below 10 ps. In opposition to short laser pulses, the energy is deposited in a time period significantly shorter than the relaxation time between the electron system and the lattice. The vaporization and

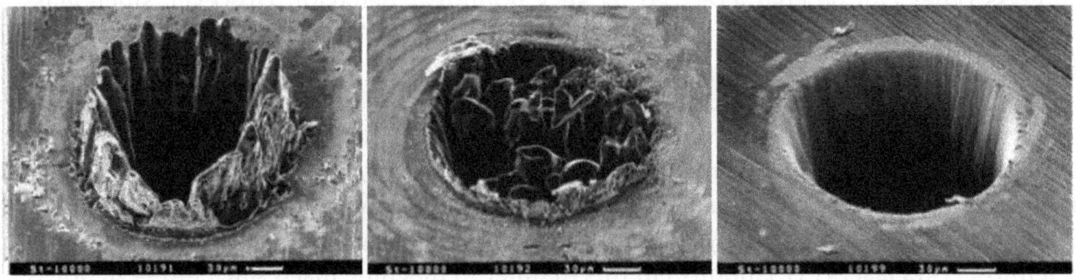

(a) (b) (c)

Figure 18.4 Laser ablation with (a) nanosecond pulse, (b) picosecond pulse, and (c) femtosecond pulse. *Source:* Chichkov et al. 1996 [10]. Reproduced with permission of Springer.

plasma formation take place much faster than the heat conduction throughout the material. This phenomenon leads to a decrease in thermal impact and allows material processing without thermally affecting the surrounding material when processing at moderate fluence levels.

18.4.2 Pulse Repetition Rate

The repetition rate (or pulse frequency) defines the number of pulses per second used for machining. The scanning speed can be scaled by the repetition rate. Thermal impact occurs by increasing the repetition rate, even for ultrashort laser pulses.

When the energy is sufficient, every pulse affects the workpiece. If the pulse rates were low, the energy would leave the thermal zone and would be of no use to the machining process. If the residual heat were retained by a rapid repetition rate, thereby limiting the time for thermal conduction, the thermal effect on the work material would be more efficient. On the other hand, a pulsed laser has an upper limit in pulse repetition, which restricts its usage.

The thermal impact is caused by two different factors. The first is heat accumulation, where increasing the repetition rate leads to a reduction of time for heat diffusion into the workpiece. With higher repetition rates, the heat injected into the material cannot be transferred out of the interaction zone. Hence, the temperature of the workpiece rises. This effect leads to a formation of molten material when applying ultrashort laser pulses. The second cause of thermal impact is particle shielding. Due to the short interval between two subsequent pulses, ablated airborne particles are located in the region of the laser radiation. The subsequent laser pulse interacts with these particles and leads to plasma ignition above the workpiece, which in turn creates an additional heat source close to the workpiece surface. The impact of these two effects strongly depends on the laser fluence, the thermal conductivity of the processed material, as well as the size and geometry of the ablated structures. Both effects appear at repetition rates above 100 kHz for metals.

18.4.3 Wavelength

The selection of the optimum laser wavelength for a particular machining process is influenced by the minimum feature size and the optical properties of the work material. As presented in Figure 18.5, the characteristics of absorption, reflectivity, and thermal diffusion of a given material are different for each laser wavelength. Therefore, the same result can be achieved with significantly less laser power when using green visible light radiation instead of IR radiation. Even after taking the conversion efficiency into account, the energy balance is still superior for converted laser systems. However, due to the complexity of the conversion technology, cost efficiency plays an important role in determining the required wavelength for a particular process.

18.4.4 Beam Quality

The beam quality (M^2) is a property of the laser system. It is essentially a measure of how tightly a laser beam can be focused under certain conditions. The spatial intensity distribution of laser radiation mainly depends on the laser source. Depending on the design of the resonator, different transverse modes can be emitted. The best possible beam quality is achieved for a diffraction-limited Gaussian distribution, which corresponds to $M^2 = 1$. This value is closely approached ($M^2 \leq 1.1$) by many lasers, in particular by solid-state bulk lasers operating on a single transverse mode, and by fiber lasers based on single-mode fibers.

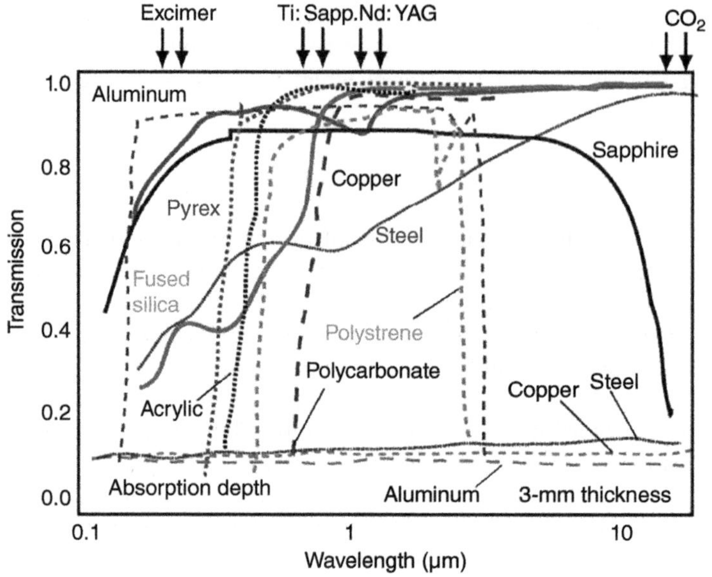

Figure 18.5 Relationship between wavelength and transmission for some materials. *Source:* Lee & Özel 2007 [59]. Reproduced with permission of ASME.

High beam quality is desired to decrease the spot size as well as to increase the working distance, characteristics which are required in many laser material processes such as cutting, drilling, and marking. An additional motivation for high beam quality is its well-defined intensity distribution in the focal plane for the use of beam-shaping elements.

18.4.5 Laser Power

A major processing control parameter is the on-target laser power (W). The controlled variation of this parameter could enable processing advantages, where different functional properties of the exposed base material are realized as a result of the irradiation conditions.

In laser material processing, the power can be controlled either internally to the laser using inherent excitation and light amplification characteristics or externally by utilizing a light valve and modulator.

With the advent of the solid-state laser and with particular care in the design of thermal management, it is now possible for lasers to vary the output power without incurring significant loss in pulse-to-pulse stability. In fact, the current generation of pulsed lasers has the capability to create any pulse amplitude profile and alter it on a pulse-to-pulse level. Depending on the application, laser power can be represented in terms of fluence (J/m^2) or pulse energy (J).

18.4.5.1 Pulse Energy

The pulse energy (J) is the total optical energy content of a pulse, i.e. the integral of its optical power over time. For single pulses, e.g. from a Q-switched laser, the pulse energy may be measured with a pyroelectric device. For regular pulse trains, the pulse energy is often calculated by dividing the average power, measured with a powermeter, by the pulse repetition rate.

Typical pulse energies from Q-switched lasers range typically from microjoules to millijoules, and in the case of large systems, to multiple joules or even kilojoules. Mode-locked lasers achieve much lower pulse energy levels (picojoules, nanojoules, or sometimes several microjoules) due to their high-pulse repetition rates and occasionally due to limiting nonlinear effects in the laser resonator (see Table 18.1). Much higher energies of ultrashort pulses can be achieved by amplifying pulses at a lower repetition rate, as obtained with a pulse picker or a regenerative amplifier. Some authors focused their research on the effects of the pulse energy on the quality of the machined cavities ([60, 61], and [41]).

18.4.5.2 Fluence

The fluence or the energy density (J/m^2) is the most important laser parameter affecting the ablation result, using short and ultrashort laser pulses since the peak power can be high even with low pulse energy. The fluence is defined as the energy density on the workpiece. The fluence is calculated dividing the pulse energy by the spot size. The fluence determines the ablation diameter and depth, as well as the thermal impact. The ablation depth and the square of the ablation diameter are proportional to the logarithm of the fluence.

18.4.5.3 Peak Power

The peak power of a pulse is the maximum occurring power. Due to the short pulse durations, peak powers can become very high even for moderately energetic pulses. The peak power is calculated by dividing the pulse energy by the pulse duration. The peak power must be able to soften the workpiece but must not be strong enough to cause direction ablation. Optimal values of laser beam intensity exist such that extremely localized material softening occurs.

18.4.6 Pulse Overlap

The pulse overlap describes the spatial overlap between two subsequent laser pulses. It is not a parameter of the laser system itself but depends on parameters of the process:

$$PO = 1 - \frac{V_{rel}}{\varnothing f_{pulse}} \tag{18.2}$$

where the V_{rel} is the relative speed between laser spot and workpiece (scanning speed), f_{pulse} is the pulse repetition rate, and Ø is the focal diameter. Samant and Dahotre [32] developed the calculation of this parameter and its incidence in the process. In experimental studies [4, 14, 62], a pulse overlap of around 75% has been determined to be appropriate. Higher pulse overlap often shows a significant decrease of quality due to thermal effects.

18.5 Laser Drilling

Drilling has been one of the first applications in laser machining. The common industrial applications of laser drilling include cooling holes in aircraft turbine blades, optical apertures, flow orifices, and apertures for electron beam instruments [63]. Laser drilling is an established and well-known practice in the field of micromanufacturing. Application examples are broad, varying in quality, processing time, cost, and several other conditions.

Laser drilling is a noncontact, precise, and reproducible technique that can be used to form small diameter and high-aspect ratio holes in a wide variety of materials. The advantages of laser drilling include the ability to drill holes in difficult-to-machine materials such as super alloys, ceramics, and composites, without the high tool wear rate normally associated with conventional machining of these materials. In laser drilling, the high intensity, stationary laser beam is focused onto the surface at power densities sufficient to heat, melt, and subsequently eject the material in both liquid and vapor phases. There are four approaches to laser drilling, as described in Figure 18.6: single-pulse, pulsating, trepanning, and helical drilling [64].

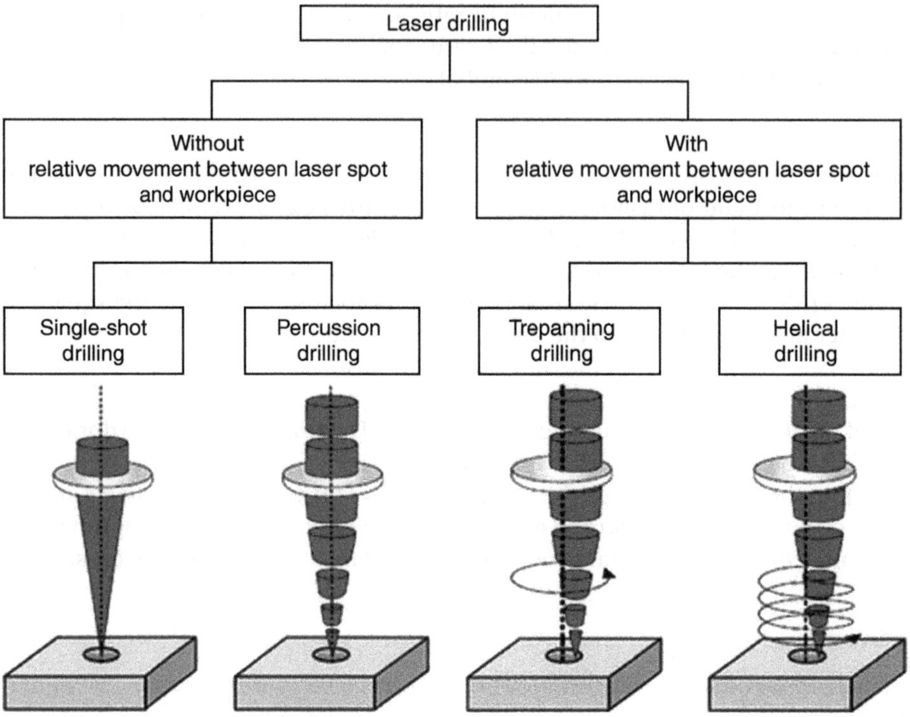

Figure 18.6 Laser drilling approaches.

18.5.1 Laser Drilling Without Relative Motion Between Laser Spot and Workpiece

This process is characterized by the little to no presence of relative motion between the laser spot and the workpiece during the drilling process. Therefore, the hole diameter is very close in magnitude to the focal diameter. This simplifies the process but also decreases its flexibility when a modification of the drilled hole diameter is required.

18.5.1.1 Single-Pulse Drilling

In single-pulse drilling, only one laser pulse hits the target and constitutes the complete drilling. Therefore, in this method, very high pulse energies are required to be able to ablate the material. The pulse duration is usually in the range of several hundred microseconds. The effectiveness of this approach is limited by the material thickness. A typical material thickness for single-pulse drilling is ~1 mm for steel materials. The achievable aspect ratio is approximately 1 : 10. Holes drilled with single-pulse laser drilling often show a decrease in diameter on the backside of the workpiece, as laser energy is dissipated as the depth from the surface increases.

18.5.1.2 Percussion Drilling

Percussion drilling consists of multiple laser pulses directed to the same spot to ablate the material and create a hole. Percussion drilling can be used to increase the accuracy of the machining process, by ablating smaller volumes with each pulse, and to increase the dimensions of the resulting hole, up to several micrometers in depth.

Percussion drilling is the most common approach for laser drilling without relative motion between the laser spot and the workpiece. More efficient laser drilling and better quality holes are obtained through the use of multiple low-energy laser pulses compared to those obtained

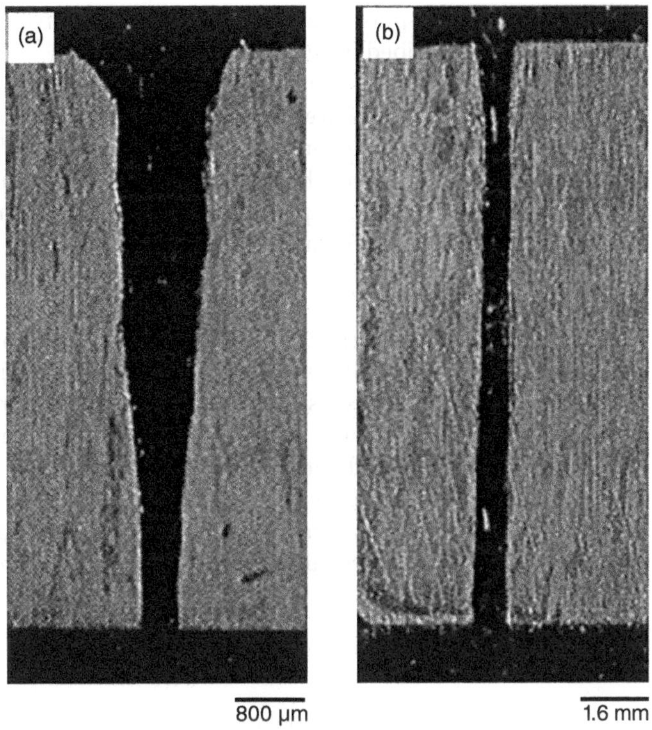

Figure 18.7 Longitudinal cross-sections of laser-drilled hole in (a) 4-mm and (b) 8-mm thick Ti–6Al–4V (pulse width: 0.7 ms, pulse energy: 7 mJ, and pulse frequency: 4 kHz). *Source:* Bandyopadhyay et al. 2005 [65]. Reproduced with permission of Elsevier.

utilizing a single high-energy laser pulse approach [65]. Figure 18.7 portrays the longitudinal cross-section images of holes created using laser drilling on a Ti–6Al–4V sample.

A wide array of authors has investigated this particular drilling approach for different materials. Wang et al. [27] investigated the number of pulses necessary to achieve complete drilling through the sample, as well as the average drilling rates with different laser fluence levels, sample thickness, repetition rates, and ambient gas pressures, utilizing a nanosecond-pulsed laser on stainless steel. Bandyopadhyay et al. [65] found the focal position, the pulse energy, and the pulse duration to be the most significant process parameters influencing the hole quality in the laser drilling of thick sections of Inconel 718 and Ti–6Al–4V. Biswas et al. [66] studied the influence of several parameters such as lamp current, pulse frequency, air pressure, and thickness of the material on the hole circularity/geometry at the exit level, as well as the hole tapering effect on a gamma-titanium-aluminide sample. Mishra et al. modeled and optimized laser beam percussion drilling of aluminum sheet and Inconel 718.

18.5.2 Laser Drilling With Relative Motion Between Laser Spot and Workpiece

This particular laser drilling process is characterized by relative motion between the laser spot and the workpiece. Although the approaches that are described in this section are classified as drilling processes, they may also be considered as laser cutting and laser milling processes. They permit the creation of holes with different diameters independent of the laser spot size.

18.5.2.1 Trepanning Drilling

In trepanning drilling, a hole is produced by drilling a series of overlapping holes around the circumference of a circle of given diameter so as to cut out a contour of the plate. It is usually performed with long-pulsed lasers. This approach consists of the laser cutting of holes, so the process itself is described in the laser cutting section.

Few authors have studied this drilling approach. Choudhury et al. [67] studied the effect of laser power, assist gas pressure, and cutting speed and stand-off distance on hole tapering and hole circularity, for laser trepanning of polymeric materials. Yilbas et al. [68] investigated laser trepanning of small diameter holes into a Ti–6Al–4V sample. Ashkenasi et al. [69] presented laser trepanning of different tapered through holes with an entrance diameter of 90–150 μm, in a 1-mm-thick sample of several metals and ceramics.

18.5.2.2 Helical Drilling

Helical drilling is characterized by drilling a hole while employing more than one revolution of the laser spot around the circumference of a circle. It can also be described as multipass trepanning drilling, as the process is the same as trepanning drilling but with multiple repetitions. With each pass, the ablation front moves downward on a helical path, ergo its name. The amount of material ablated with each pulse is smaller compared to trepanning, which leads to an increase in contour accuracy and a decrease in the thermal load on the workpiece. It is generally performed with short and ultrashort-pulsed laser sources. The general approach can be better understood as a laser milling process that is described in a subsequent section.

18.6 Laser Cutting

Laser cutting is a two-dimensional machining process in which material removal is obtained by focusing a high intensity laser beam on the workpiece. It is a high-speed, repeatable, and reliable method for a wide variety of material types of varying thicknesses, producing very narrow and clean-cut dimensional features. The advantages of laser cutting have to do with the high speed of the process, as well as the fine and precise cut dimensions that can be obtained. In particular, very narrow kerf width is necessary so that this process can be utilized for fine and profile cutting [70].

As explained previously, lasers can now be used to machine a wide variety of materials, including metals, plastics, ceramics, and composites. Due to its high machining precision and processing speed, laser cutting offers tremendous economic benefits when applied to production lines. One of the most popular applications is the micromachining of coronary stents, which requires high-quality cuts with very small kerf widths. Many authors investigated the cutting of stents in stainless steel [71–73] and nitinol, among other materials [74]. Figure 18.8 shows the highly accurate channels that can be obtained after applying a laser cutting process to four different materials, using a short-pulsed laser.

Laser cutting techniques can be subdivided into two groups according to the mechanisms responsible for material ejection at the cutting front.

18.6.1 Melt Cutting

Melt cutting is a process in which the laser provides the energy necessary to heat and melt the material but does not deliver the energy to remove the material from the processing zone. Therefore, a "melting pool" is created. In melt cutting, the laser induced melt is removed through an external source: a coaxial cutting gas jet that is applied by a gas nozzle close to the laser material interaction zone.

Therefore, the laser optical equipment and the gas nozzle can be thought of as a single unit, which is moved over the workpiece surface along the cutting path. In a steady state cutting process, a fraction of the laser radiation is absorbed by the solid workpiece surface resulting in

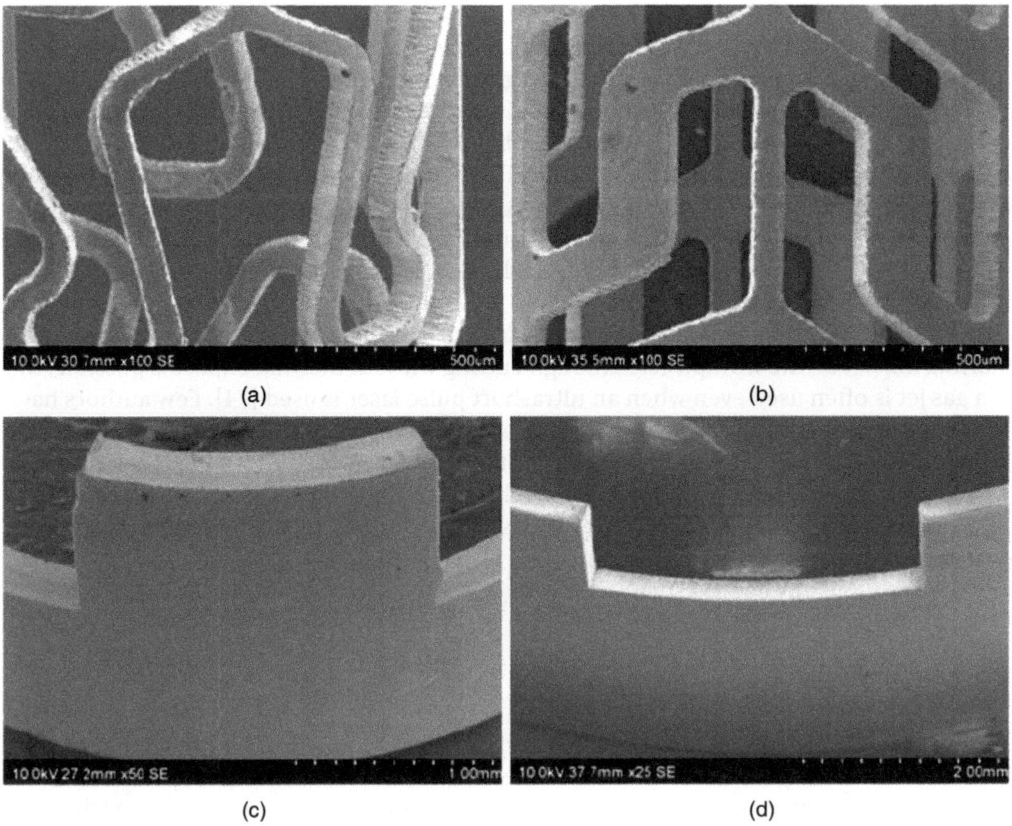

Figure 18.8 SEM images of picoseconds laser cutting of stents (a) nitinol, (b) cobalt chromium, (c) silver, and (d) titanium. *Source:* Muhammad et al. 2011 [74]. Reproduced with permission of Springer.

preheating of the material. Another fraction of the laser radiation is absorbed at the liquid cutting front, where heat diffusion and advection transport the heat to the solid liquid interface, i.e. the melting front.

This mechanism of laser cutting is the most common. Choudhury et al. evaluated the effect of laser power, cutting speed, and compressed air pressure on the material HAZ, the surface roughness, and the dimensional accuracy. They used a CW CO_2 laser in conjunction with air-compressed gas for laser cutting of polypropylene (PP), polycarbonate (PC), and polymethylmethacrylate (PMMA) polymers. Pfeifer et al. [75] used a nanosecond Nd:YAG laser to cut nitinol sheets, in order to properly investigate the influence of different cutting parameters on the cut quality and the resulting mechanical properties of the material. Yan et al. [76] studied the effects of different parameters on striation characteristics of laser cutting of alumina sheets with ~1-mm thickness using a nanosecond Nd:YAG laser. Yilbas et al. [77] analyzed the effect of cutting parameters on the kerf size variations of the CO_2 laser cutting of thick steel sheets with nitrogen as an assisted gas.

18.6.2 Laser Ablation Cutting

The use of pulsed laser radiation or high brilliance CW lasers leads to vaporization and, eventually, to plasma formation at the laser material interaction zone. The resulting pressure gradient at the gas melt interface is responsible for melt ejection. Hence, the processes that use ablation as the main mechanism do not require an external gas jet to remove the material, which is melted or vaporized. Pulsed lasers with high peak intensity are used for laser ablation cutting, resulting in melting, vaporization, and plasma ignition of the material. Although the plasma plume expands, pressure gradients at the gas liquid interface lead to the ejection of molten and vaporized material from the cutting zone. Laser ablation cutting is mostly performed with high pulse energy (>10 μJ) and repetition rates in the kHz regime. Laser ablation cutting can be classified according to the pulse duration into short-pulsed and ultrashort-pulsed laser cutting. However, the most commonly utilized method is ultrashort pulse laser cutting.

Due to the very high peak intensity of ultrashort pulses, nonlinear effects dominate the absorption of laser radiation. Therefore, almost every material can be cut regardless of the wavelength used. Furthermore, ultrashort-pulsed laser cutting is considered to not generate any thermal impact on the workpiece. Although cutting without thermal impact is possible, the use of a gas jet is often used even when an ultrashort pulse laser is used [74]. Few authors have researched laser cutting ablation without the use of an assisting gas. Yung et al. [78] investigated the micromachining of nitinol using a 355 nm (UV) nanosecond Nd:YAG laser, in particular the influence of process parameters on the kerf geometry and cutting quality. Wang et al. [79] studied the effects of various conditions on the cutting quality of alumina substrates with a Ti:sapphire femtosecond laser. Li et al. [80] also investigated femtosecond laser processing of nitinol without any gas assistant.

18.7 Laser Milling

Laser milling is based on the generation of overlapping multiple shallow grooves by scanning a single laser beam to systematically remove the surface layer of the workpiece material. Vaporization of the material during laser scanning is a primary material removal mechanism for the formation of each groove. Overlapping of the grooves is realized by applying a continuous motion feed to the workpiece, perpendicular to the laser scanning direction. The process is repeated to remove the additional layers of material, applying intermittent feed motion perpendicular to the previously machined surface to either the workpiece or the focusing lens.

The laser beam is directed along the z-axis, while the workpiece is moved in x- and y-directions by linear translational stages (see Figure 18.9). During machining, the workpiece is

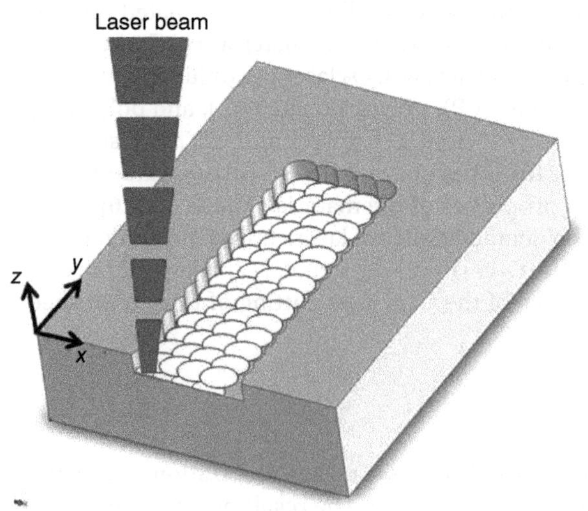

Figure 18.9 Schematic illustration of the laser milling process. *Source:* Teixidor et al. [4]. Reproduced with permission of Elsevier.

translated relative to the focal spot of the beam in the $+x$-direction. The formation of a continuous groove is facilitated by adjusting the workpiece translation speed and the repetition rate of the laser.

Laser milling is a very complex process because it is affected by the characteristics of both the laser beam and the workpiece and is mainly determined by how the laser beam and the workpiece interact. Moreover, the structure and the material properties change after each pass of the laser. Therefore, material removal will vary with each laser pass. To properly understand laser milling, authors have focused their research in three distinct levels of the process.

18.7.1 Single-Shot Laser Milling

The first level consists of the study of the grooves machined utilizing a single shot of the laser. The characteristics of a crater/hole produced with a single shot can be considered a good approximation and be used to predict the machining process of a superposition of several grooves in the different directions of laser motion.

Dobrev et al. [6] developed a model of the crater formation to investigate the factors affecting the resulting surface finish in laser milling. Kumar and Gupta [81] investigated the dependence of groove depth on laser power, repetition rate, number of scans, and gas pressure in the generation of micronotches with a nanosecond-pulsed fiber laser in stainless steel and aluminum. Bordatchev and Nikumb [60] investigated experimentally the effect of pulse energy on the accuracy, precision, and surface quality of copper machined parts. The results showed that the crater diameter and depth increase in accordance with an increase in the laser pulse energy. Yousef et al. [61] used a nanosecond Nd:YAG laser in order to machine craters on brass, copper, and stainless steel as bulk material. They investigated, analyzed, and modeled the effect of the pulse energy on the geometry of the crater (diameter, depth, and volume of material removed) and final surface profile.

18.7.2 Single-Pass Laser Milling (Laser Scribing)

Laser scribing creates only superficial grooves and does not generate a complete through cut. The process is performed in a single pass of the laser, which moves along the x- and y- axes, as described by Figure 18.9. Thus, the machining is completed in just one layer. One of the most common applications of laser scribing is selective ablation of layer systems such as in thin-film solar cells [82–84]. The film-solar cells consist of several layers of materials. Some of these layers have to be patterned without delamination and without causing damage to other layers. Another application of laser scribing can be found in the manufacturing of microfluidic devices, which are used to manipulate liquids and gases in diagnostics and biomedical research. The objective is to generate channels in transparent materials, either a polymer or a glass-based material, having cross-sectional dimensions in the order of 10–500 μm [19, 62].

18.7.3 MultiPass Laser Milling (3D Laser Milling)

The multipass approach is the culmination of the laser milling process. Researchers have studied the effects of process parameters on the final geometry of the machined parts. Investigations into multipass laser milling vary significantly, from simple geometries like channels in a macro level [1, 32] to the machining of completed parts [85, 86]. Other authors have studied the process of machining complex geometries with round shapes, angles, or taper walls [2, 30]. Teixidor et al. [4] investigated the influence of laser process parameters on dimensional precision and surface quality in laser milling of hardened AISI H13 tool steel. This research group machined

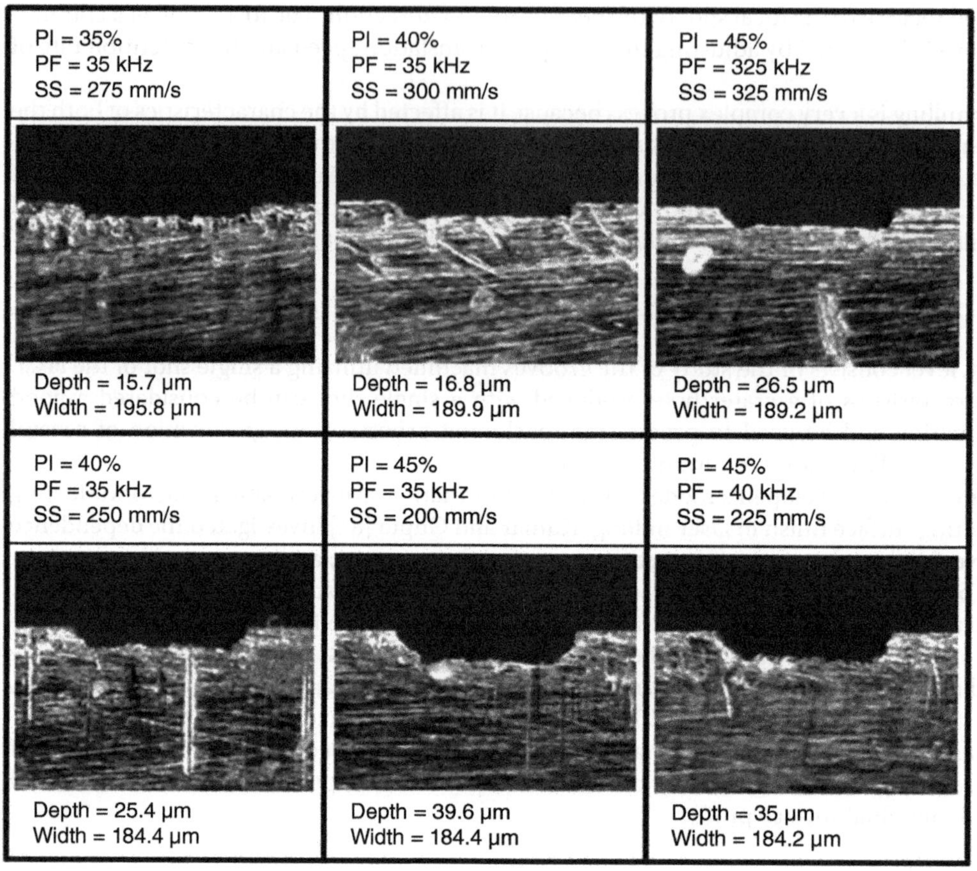

Figure 18.10 Images of microchannels. *Source:* Teixidor et al. [4]. Reproduced with permission of Elsevier.

microchannels of 200 μm in width and 50 μm in depth. Figure 18.10 shows the images obtained from six of the microchannels manufactured by varying scanning speed (SS), pulse frequency (PF), and pulse intensity (PI).

18.8 Concluding Remarks

In this chapter, the basics of several laser machining processes have been introduced. Currently, laser machining is used in many industrial processes such as drilling, cutting, welding, marking, and milling. These processes can be found in a wide variety of industrial sectors ranging from semiconductor fabrications to medical device manufacturers, the solar cell industry, and many other fields. A variety of different material responses can be achieved depending on the material selection and the laser parameters, allowing processes to be designed and optimized to permanently alter the material's surface chemistry, crystal structure, and morphology.

As laser systems continue to be developed with an ever broadening range of capabilities, research will continue to increase the overall knowledge of the field; laser machining will continue to improve performance of finalized parts in existing applications and will broaden the scope of action to novel applications and materials, whose use in industry was unthinkable in the past.

References

1 Bartolo, P., Vasco, J., Silva, B., and Galo, C. (2006). Laser micromachining for mould manufacturing: I. The influence of operating parameters. *Assembly Automation* 26 (3): 227–234.

2 Ciurana, J., Arias, G., and Ozel, T. (2009). Neural network modeling and particle swarm optimization (PSO) of process parameters in pulsed laser micromachining of hardened AISI H13 steel. *Materials and Manufacturing Processes* 24 (3): 358–368.

3 Dhara, S.K., Kuar, A.S., and Mitra, S. (2007). An artificial neural network approach on parametric optimization of laser micro-machining of die-steel. *The International Journal of Advanced Manufacturing Technology* 39 (1–2): 39–46.

4 Teixidor, D., Ferrer, I., Ciurana, J., and Özel, T. (2013). Optimization of process parameters for pulsed laser milling of micro-channels on AISI H13 tool steel. *Robotics and Computer-Integrated Manufacturing* 29 (1): 209–218.

5 Bustillo, A., Sedano, J., Villar, J.R. et al. (2008). AI for modelling the laser milling of copper components. In: *IDEAL 2008. LNCS*, vol. 5326 (ed. C. Fyfe, D. Kim, S.-Y. Lee and H. Yin), 498–507. Heidelberg: Springer.

6 Dobrev, T., Dimov, S.S., and Thomas, a J. (2006). Laser milling: modelling crater and surface formation. *Proceedings of the Institution of Mechanical Engineers, Part C: Journal of Mechanical Engineering Science* 220 (11): 1685–1696.

7 Dubey, A.K. and Yadava, V. (2008). Optimization of kerf quality during pulsed laser cutting of aluminium alloy sheet. *Journal of Materials Processing Technology* 204 (1–3): 412–418.

8 Heyl, P., Olschewski, T., and Wijnaendts, R.W. (2001). Manufacturing of 3D structures for micro-tools using laser ablation. *Microelectronic Engineering* 57–58: 775–780.

9 Dumitru, G., Lüscher, B., Krack, M. et al. (2005). Laser processing of hardmetals: physical basics and applications. *International Journal of Refractory Metals and Hard Materials* 23 (4-6): 278–286.

10 Chichkov, B.N., Momma, C., Nolte, S. et al. (1996). Femtosecond, picosecond and nanosecond laser ablation of solids. *Applied Physics A: Materials Science & Processing* 63 (2): 109–115.

11 Nedialkov, N.N., Imamova, S.E., Atanasov, P.a. et al. (2005). Mechanism of ultrashort laser ablation of metals: molecular dynamics simulation. *Applied Surface Science* 247 (1-4): 243–248.

12 Zhao, J., Huettner, B., and Menschig, A. (2001). Microablation with ultrashort laser pulses. *Optics and Laser Technology* 33: 487–491.

13 Cheng, J., Perrie, W., Edwardson, S.P. et al. (2009). Effects of laser operating parameters on metals micromachining with ultrafast lasers. *Applied Surface Science* 256 (5): 1514–1520.

14 Criales, L.E., Orozco, P.F., Medrano, A. et al. (2014). Effect of fluence and pulse overlapping on fabrication of microchannels in PMMA/PDMS via UV laser micromachining: modeling and experimentation. *Materials and Manufacturing Processes* 30: 890–901.

15 Roberts, M.A., Rossier, J.S., Bercier, P., and Girault, H. (1997). UV laser machined polymer substrates for the development of microdiagnostic systems. *Analytical Chemistry* 69 (11): 2035–2042.

16 Suriyage, N.U., Ghantasala, M.K., Iovenitti, P., and Harvey, E.C. (2004). Fabrication, measurement, and modeling of electro-osmotic flow in micromachined polymer microchannels. *Proc. SPIE* 5275. *BioMEMS and Nanotechnology* 5275: 149–160.

17 Waddell, E.A. (2006). Laser ablation as a fabrication technique for microfluidic devices. *Methods in Molecular Biology* 321 (1): 27–38.

18 Klank, H., Kutter, J.P., and Geschke, O. (2002). CO(2)-laser micromachining and back-end processing for rapid production of PMMA-based microfluidic systems. *Lab on a Chip* 2 (4): 242–246.

19 Romoli, L., Tantussi, G., and Dini, G. (2011). Experimental approach to the laser machining of PMMA substrates for the fabrication of microfluidic devices. *Optics and Lasers in Engineering* 49 (3): 419–427.
20 Snakenborg, D., Klank, H., and Kutter, J.P. (2004). Microstructure fabrication with a CO_2 laser system. *Journal of Micromechanics and Microengineering* 14 (2): 182–189.
21 Suriano, R., Kuznetsov, A., Eaton, S.M. et al. (2011). Femtosecond laser ablation of polymeric substrates for the fabrication of microfluidic channels. *Applied Surface Science* 257 (14): 6243–6250.
22 Khan Malek, C.G. (2006). Laser processing for bio-microfluidics applications (part II). *Analytical and Bioanalytical Chemistry* 385 (8): 1362–1369.
23 De Marco, C., Eaton, S.M., Suriano, R. et al. (2010). Surface properties of femtosecond laser ablated PMMA. *ACS Applied Materials and Interfaces* 2 (8): 2377–2384.
24 Li, Z.L., Chu, P.L., Zheng, H.Y. et al. (2009). Process development of laser machining of carbon fibre reinforced plastic composites. *SIMtech Technical Reports* 10: 1.
25 Li, Z.L., Zheng, H.Y., Lim, G.C. et al. (2010). Study on UV laser machining quality of carbon fibre reinforced composites. *Composites Part A: Applied Science and Manufacturing* 41 (10): 1403–1408.
26 Yung, K.C., Mei, S.M., and Yue, T.M. (2002). A study of the heat-affected zone in the UV YAG laser drilling of GFRP materials. *Journal of Materials Processing Technology* 122 (2–3): 278–285.
27 Wang, C., Zhang, L., Liu, Y. et al. (2012). Ultra-short pulse laser deep drilling of C/SiC composites in air. *Applied Physics A* 111 (4): 1213–1219.
28 Riveiro, A., Quintero, F., Lusquiños, F. et al. (2012). Experimental study on the CO_2 laser cutting of carbon fiber reinforced plastic composite. *Composites Part A: Applied Science and Manufacturing* 43 (8): 1400–1409.
29 Karnakis, D., Rutterford, G., Knowles, M. et al. (2006). High quality laser milling of ceramics, dielectrics and metals using nanosecond and picosecond lasers. Society of Photo-Optical Instrumentation Engineers. *Proc. SPIE 6106*. In: *Photon Processing in Microelectronics and Photonics*, 610604–610604–11. Society of Photo-Optical Instrumentation Engineers.
30 Dhupal, D., Doloi, B., and Bhattacharyya, B. (2007). Parametric analysis and optimization of Nd:YAG laser micro-grooving of aluminum titanate (Al_2TiO_5) ceramics. *The International Journal of Advanced Manufacturing Technology* 36 (9–10): 883–893.
31 Samant, A.N. and Dahotre, N.B. (2009). Physical effects of multipass two-dimensional laser machining of structural ceramics. *Advanced Engineering Materials* 11 (7): 579–585.
32 Samant, A.N. and Dahotre, N.B. (2010). Three-dimensional laser machining of structural ceramics. *Journal of Manufacturing Processes* 12 (1): 1–7.
33 Samant, A.N., Du, B., and Dahotre, N.B. (2009). In-situ surface absorptivity prediction during 1.06 μm wavelength laser low aspect ratio machining of structural ceramics. *Physica Status Solidi (a)* 206 (7): 1433–1439.
34 Gilbert, T., Krstic, V.D., and Zak, G. (2007). Machining of aluminium nitride with ultra-violet and near-infrared Nd:YAG lasers. *Journal of Materials Processing Technology* 189 (1–3): 409–417.
35 Wang, C. and Zeng, X. (2007). Study of laser carving three-dimensional structures on ceramics: quality controlling and mechanisms. *Optics and Laser Technology* 39 (7): 1400–1405.
36 Tsai, C.-H. and Chen, H.-W. (2003). Laser milling of cavity in ceramic substrate by fracture-machining element technique. *Journal of Materials Processing Technology* 136 (1–3): 158–165.
37 Komlenok, M.S., Kononenko, V.V., Ralchenko, V.G. et al. (2011). Laser induced nanoablation of diamond materials. *Physics Procedia* 12: 37–45.

38 Kononenko, T.V., Komlenok, M.S., Pashinin, V.P. et al. (2009). Femtosecond laser microstructuring in the bulk of diamond. *Diamond and Related Materials* 18 (2-3): 196–199.

39 Butler-smith, P.W., Axinte, D.A., Pacella, M., and Fay, M.W. (2013). Micro/nanometric investigations of the effects of laser ablation in the generation of micro-tools from solid CVD diamond structures. *Journal of Materials Processing Technology* 213: 194–200.

40 Kononenko, T.V., Kononenko, V.V., Pimenov, S.M. et al. (2005). Effects of pulse duration in laser processing of diamond-like carbon films. *Diamond and Related Materials* 14 (8): 1368–1376.

41 Ngoi, B.K., Venkatakrishnan, K., Lim, E.N. et al. (2001). Effect of energy above laser-induced damage thresholds in the micromachining of silicon by femtosecond pulse laser. *Optics and Lasers in Engineering* 35 (6): 361–369.

42 Amer, M.S., El-Ashry, M.A., Dosser, L.R. et al. (2005). Femtosecond versus nanosecond laser machining: comparison of induced stresses and structural changes in silicon wafers. *Applied Surface Science* 242 (1–2): 162–167.

43 Greuters, J. and Rizvi, N. (2003). UV laser micromachining of silicon, indium phosphide and lithium niobate for telecommunications applications. In: *Proceedings of the SPIE*, vol. 4876, Opto-Ireland 2002, 479–486.

44 Böhme, R., Hirsch, D., and Zimmer, K. (2006). Laser etching of transparent materials at a backside surface adsorbed layer. *Applied Surface Science* 252 (13): 4763–4767.

45 Niino, H., Kawaguchi, Y., Sato, T. et al. (2004). Surface microfabrication of fused silica glass by UV laser irradiation. *Photon Processing in Microelectronics and Photonics III* 5339: 112–117.

46 Ben-Yakar, A. and Byerh, R.L. (2002). Femtosecond laser machining of fluidic microchannels for miniaturized bioanalytical systems. *Proceedings of SPIE* 4637: 212–217.

47 Chaitanya Vishnubhatla, K., Clark, J., Lanzani, G. et al. (2009). Femtosecond laser fabrication of microfluidic channels for organic photonic devices. *Applied Optics* 48 (31): G114–G118.

48 Streltsov, A.M. and Borrelli, N.F. (2002). Study of femtosecond-laser-written waveguides in glasses. *Journal of the Optical Society of America B* 19 (10): 2496.

49 Will, M., Nolte, S., Chichkov, B.N., and Tünnermann, A. (2002). Optical properties of waveguides fabricated in fused silica by femtosecond laser pulses. *Applied Optics* 41 (21): 4360–4364.

50 Cai, W. and Piestun, R. (2007). Low pulse-energy micromachining in bulk glass with a short-cavity femtosecond oscillator. *Optical Engineering* 46 (12): 124301.

51 Eaton, S., Zhang, H., Herman, P. et al. (2005). Heat accumulation effects in femtosecond laser-written waveguides with variable repetition rate. *Optics express* 13 (12): 4708–4716.

52 Giridhar, M.S., Seong, K., Schülzgen, A. et al. (2004). Femtosecond pulsed laser micromachining of glass substrates with application to microfluidic devices. *Applied Optics* 43 (23): 4584–4589.

53 Davis, K.M., Miura, K., Sugimoto, N., and Hirao, K. (1996). Writing waveguides in glass with a femtosecond laser. *Optics Letters* 21 (21): 1729–1731.

54 Florea, C., Winick, K.A., and Member, S. (2003). Fabrication and characterization of photonic devices directly written in glass using femtosecond laser pulses. *Journal of Lightwave Technology* 21 (1): 246–253.

55 Nikumb, S., Chen, Q., Li, C. et al. (2005). Precision glass machining, drilling and profile cutting by short pulse lasers. *Thin Solid Films* 477 (1–2): 216–221.

56 Cicală, E., Soveja, A., Sallamand, P. et al. (2008). The application of the random balance method in laser machining of metals. *Journal of Materials Processing Technology* 196 (1-3): 393–401.

57 Jandeleit, J., Horn, A., Weichenhain, R. et al. (1998). Fundamental investigations of micromachining by nano- and picosecond laser radiation. *Applied Surface Science* 127–129: 885–891.

58 Petkov, P.V., Dimov, S.S., Minev, R.M., and Pham, D.T. (2008). Laser milling: pulse duration effects on surface integrity. *Proceedings of the Institution of Mechanical Engineers, Part B: Journal of Engineering Manufacture* 222 (1): 35–45.
59 Lee, W.-H. and Özel, T. (2009). An experimental method for laser micro-machining of spherical and elliptical 3-D objects. *International Journal of Nanomanufacturing* 3 (3): 264–278.
60 Bordatchev, E.V. and Nikumb, S.K. (2003). An experimental study and statistical analysis of the effect of laser pulse energy on the geometric quality during laser precision machining. *Machining Science and Technology* 7 (1): 83–104.
61 Yousef, B.F., Knopf, G.K., Bordatchev, E.V., and Nikumb, S.K. (2003). Neural network modeling and analysis of the material removal process during laser machining. *The International Journal of Advanced Manufacturing Technology* 22 (1–2): 41–53.
62 Teixidor, D., Orozco, F., Thepsonthi, T. et al. (2012). Effect of process parameters in nanosecond pulsed laser micromachining of PMMA-based microchannels at near-infrared and ultraviolet wavelengths. *The International Journal of Advanced Manufacturing Technology* 67: 1651–1664.
63 Knowles, M.R.H., Rutterford, G., Karnakis, D., and Ferguson, a. (2007). Micro-machining of metals, ceramics and polymers using nanosecond lasers. *The International Journal of Advanced Manufacturing Technology* 33 (1–2): 95–102.
64 Sugioka, K., Meunier, M., and Piqué, A. (2010). *Laser Precision Microfabrication*. Springer.
65 Bandyopadhyay, S., Gokhale, H., Sarin Sundar, J.K. et al. (2005). A statistical approach to determine process parameter impact in Nd:YAG laser drilling of IN718 and Ti–6Al–4V sheets. *Optics and Lasers in Engineering* 43 (2): 163–182.
66 Biswas, R., Kuar, A.S., Sarkar, S., and Mitra, S. (2010). A parametric study of pulsed Nd:YAG laser micro-drilling of gamma-titanium aluminide. *Optics and Laser Technology* 42 (1): 23–31.
67 Choudhury, I.a., Chong, W.C., and Vahid, G. (2012). Hole qualities in laser trepanning of polymeric materials. *Optics and Lasers in Engineering* 50 (9): 1297–1305.
68 Yilbas, B.S., Akhtar, S.S., and Karatas, C. (2011). Laser trepanning of a small diameter hole in titanium alloy: temperature and stress fields. *Journal of Materials Processing Technology* 211 (7): 1296–1304.
69 Ashkenasi, D., Kaszemeikat, T., Mueller, N. et al. (2011). Laser trepanning for industrial applications. *Physics Procedia* 12: 323–331.
70 Dahotre, N.B. and Harimkar, S.P. (2008). *Laser Fabrication and Machining Of Materials*. Springer.
71 Kathuria, Y.P. (2005). Laser microprocessing of metallic stent for medical therapy. *Journal of Materials Processing Technology* 170 (3): 545–550.
72 Meng, H., Liao, J., Zhou, Y., and Zhang, Q. (2009). Laser micro-processing of cardiovascular stent with fiber laser cutting system. *Optics and Laser Technology* 41 (3): 300–302.
73 Muhammad, N., Whitehead, D., Boor, A., and Li, L. (2010). Comparison of dry and wet fibre laser profile cutting of thin 316L stainless steel tubes for medical device applications. *Journal of Materials Processing Technology* 210 (15): 2261–2267.
74 Muhammad, N., Whitehead, D., Boor, A. et al. (2011). Picosecond laser micromachining of nitinol and platinum–iridium alloy for coronary stent applications. *Applied Physics A* 106 (3): 607–617.
75 Pfeifer, R., Herzog, D., Hustedt, M., and Barcikowski, S. (2010). Pulsed Nd:YAG laser cutting of NiTi shape memory alloys – influence of process parameters. *Journal of Materials Processing Technology* 210 (14): 1918–1925.

76 Yan, Y., Li, L., Sezer, K. et al. (2012). Nano-second pulsed DPSS Nd:YAG laser striation-free cutting of alumina sheets. *International Journal of Machine Tools and Manufacture* 53 (1): 15–26.
77 Yilbas, B.S., Arif, A.F.M., and Abdul Aleen, B.J. (2009). Laser cutting of large-aspect-ratio rectangular blanks in thick sheet metal: thermal stress analysis. *Proceedings of the Institution of Mechanical Engineers, Part B: Journal of Engineering Manufacture* 223 (1): 063–071.
78 Yung, K.C., Zhu, H.H., and Yue, T.M. (2005). Theoretical and experimental study on the kerf profile of the laser micro-cutting NiTi shape memory alloy using 355 nm Nd:YAG. *Smart Materials and Structures* 14 (2): 337–342.
79 Wang, X.C., Zheng, H.Y., Chu, P.L. et al. (2010). High quality femtosecond laser cutting of alumina substrates. *Optics and Lasers in Engineering* 48 (6): 657–663.
80 Li, C., Nikumb, S., and Wong, F. (2006). An optimal process of femtosecond laser cutting of NiTi shape memory alloy for fabrication of miniature devices. *Optics and Lasers in Engineering* 44 (10): 1078–1087.
81 Kumar, A. and Gupta, M.C. (2010). Laser machining of micro-notches for fatigue life. *Optics and Lasers in Engineering* 48 (6): 690–697.
82 García-Ballesteros, J.J., Torres, I., Lauzurica, S. et al. (2011). Influence of laser scribing in the electrical properties of a-Si:H thin film photovoltaic modules. *Solar Energy Materials and Solar Cells* 95 (3): 986–991.
83 Gečys, P., Račiukaitis, G., Miltenis, E. et al. (2011). Scribing of thin-film solar cells with picosecond laser pulses. *Physics Procedia* 12: 141–148.
84 Lauzurica, S., García-Ballesteros, J.J., Colina, M. et al. (2011). Selective ablation with UV lasers of a-Si:H thin film solar cells in direct scribing configuration. *Applied Surface Science* 257 (12): 5230–5236.
85 Kaldos, A., Pieper, H.J., Wolf, E., and Krause, M. (2004). Laser machining in die making – a modern rapid tooling process. *Journal of Materials Processing Technology* 155–156: 1815–1820.
86 Pham, D.T., Dimov, S.S., and Petkov, P.V. (2007). Laser milling of ceramic components. *International Journal of Machine Tools and Manufacture* 47 (3–4): 618–626.

19

Laser-assisted Machining Operations

Eneko Ukar, Ivan Tabernero, Silvia Martínez, Aitzol Lamikiz, and Asier Fernández

Department of Mechanical Engineering, University of Basque Country UPV/EHU, Bilbao, Spain

19.1 Introduction

There is a growing interest in the industry in improving productivity in manufacturing processes of materials with low machinability properties. Some high added value applications need manufacturing of parts with very high standards that cannot be satisfied with regular materials. Typically, these materials present high mechanical and thermal resistance combined with low thermal conductivity. The high shear strength makes difficult the plastic deformation during machining, and the low thermal conductivity affects directly tool life increasing the tool wear.

Examples of these materials are aeronautical heat-resistant alloys such as nickel-based alloy Inconel 718 and cobalt-based alloy Haynes 25 or titanium alloys, which are commonly used in aircraft engine manufacturing. On the other hand, ceramic materials are also difficult to machine because of their high strength combined with very low thermal conductivity and high brittleness. Ceramics are becoming of higher interest in fields such as aerospace or medicine due to the high temperature and wear resistance but are almost impossible to machine using conventional technologies.

Inconel 718 is widely used in heat resistant part manufacturing because of its high yield and ultimate strength, high fatigue strength, and high working temperature range, up to 700 °C. However, Inconel 718 is difficult to machine because of its high shear strength, high work-hardening capacity, the content of abrasive carbides in the microstructure, the low thermal conductivity becoming in high temperature in the cutting edge, and the strong tendency to form built-up edge and plastic deformation. The heat generated in the cutting process comes from the plastic deformation, the friction between tool and the chip, and the friction between tool and workpiece. After the first cutting pass, once deformed, the material is hardened, and the shear strength increases dramatically. Thus, tool with negative rake angle is necessary to avoid tool break. Main portion of the generated heat is removed with the chip, but the remaining part is still enough to heat the tool reducing its life. Other materials present low machinability, such as ceramics. Ceramic materials present a very hard and brittle structure that leads to tool wear and high surface roughness. If the machining process is carried out at high temperatures with the optimum tool geometry, it is possible to reduce brittleness and obtain good results. Therefore, some materials can be processed using heat-assisted processes.

Modern Manufacturing Processes, First Edition. Edited by Muammer Koç and Tuğrul Özel.

19.2 Heat-assisted Processes

Local heating is an alternative to improve the machinability of materials with high strength at high temperatures. In general, as the temperature increases, the resistance of the material decreases, and higher productivity rates can be achieved with lower tool wear. In the literature, several examples of heat-assisted processes can be found for different materials such as Inconel 718, Haynes 25, and others such as high-strength steels [1–7], aluminum 6061-T6 [8], iron [9, 10], or metal matrix composites (MMC) [11].

There are several ways to heat the material, such as plasma, laser beam, gas torch, induction heating, furnace preheating, electric current heating or electric arc, etc. But in order to get an optimum result, the heating source must be local, fast, and controllable. The heating of material must be local in order to avoid metallurgical changes on the nonmachined workpiece surface. Thus, best results are obtained using high energy density heat sources such as laser and plasma sources. Depending on the heating intensity and duration, it is possible to soften, melt, or vaporize the material.

19.2.1 Plasma-assisted Machining

Plasma arc welding (PAW) has the same principle as tungsten inert gas (TIG) welding but reaching higher temperatures. The plasma state is achieved when a gas is ionized separating ions and electrons. The energy required for gas ionization is achieved by an electric arc between a tungsten electrode and the base metal. The most extended use combines pure argon or helium with small proportions of hydrogen. The same plasma, modifying the arc intensity, is used as heating source in plasma-enhanced machining (PEM) following the scheme depicted in Figure 19.1.

The maximum temperature in the plasma can reach up to 28.000 °C and enables local heating of areas of 4–5 mm diameter. The source integration depends on the process. The orientation of the plasma is a critical point that affects directly the result. In milling process, the plasma torch must be placed ahead of the cutting tool in the feed direction, but since this direction varies during milling operation, the position of the torch must be reoriented for each direction, complicating the integration. In most cases, several torches are used, operating the corresponding one depending on the feed direction. In Figure 19.2, it is shown the typical PEM configuration.

On the other hand, for turning operations, the cutting tool remains stationary, and the plasma torch can be placed more easily in just one position. Thus, main industrial applications of PEM are developed for turning operations.

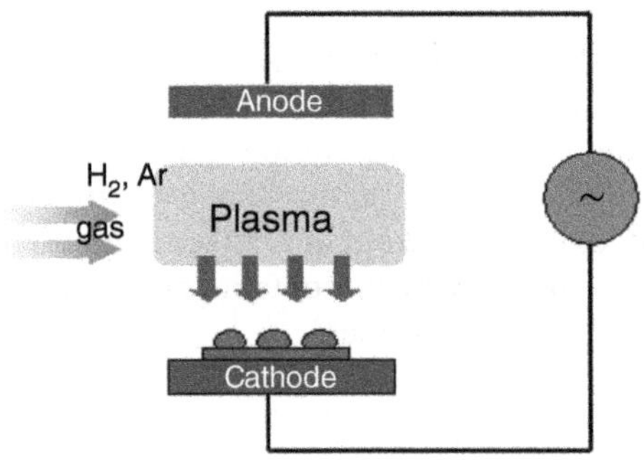

Figure 19.1 Plasma generating scheme.

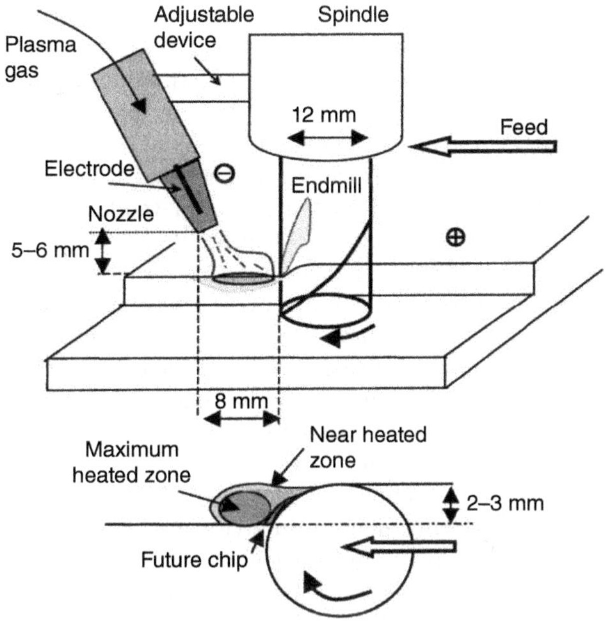

Figure 19.2 PEM setup configuration.

PEM is particularly suitable for low-machinability alloys and especially those in which the mechanical properties decreased only from certain temperature. In most cases, high mechanical strength is related to high shear strength, and therefore, the machinability is very low [12].

The cutting tool is critical in the machining process. The most extensively used tools are sintered tungsten carbide (K5–K10 grade) with TiAlN or TiCN coatings [13]. Despite abrasion wear is reduced using PEM, high temperatures reached have direct effect and shorten tool life. The plasma heats and softens the material, but it also heats the tool and workpiece. This means that the tool life is not increased in the same proportion; hence, the shear strength of the material is reduced.

One possible solution for applications where the tool heating becomes a problem is to use polycrystalline cubic boron nitride (PCBN) or whisker-reinforced ceramic ($Al_2O_3 + CSi_W$) tools. In Figure 19.3, tool wear comparison between conventional and PEM using ceramic tools is shown. Another solution to avoid excessive heating of tool is the hybrid machining with combination of PEM and cryogenic tool cooling using nitrogen [14]. This solution provides

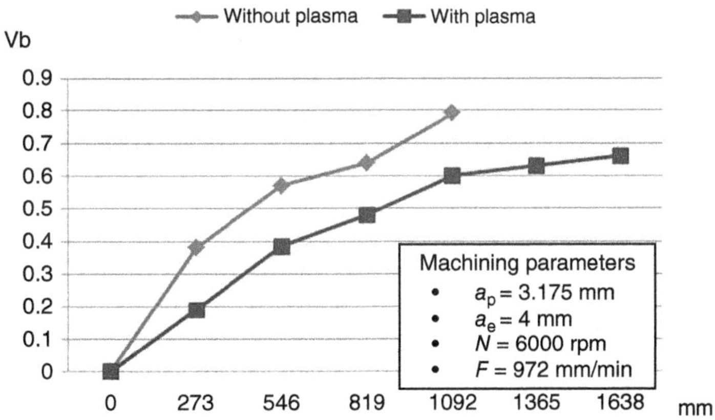

Figure 19.3 Flank wear in conventional vs. PEM ($a_p = 3.175$ mm, $a_e = 4$ mm, $N = 6000$ rpm, and $f = 972$ mm/min).

good results in materials such as Inconel 718, where tool life is improved over 150%, cutting forces are reduced between 30% and 50%, and surface roughness is reduced with improvements about 250%. On the other hand, the integration of a cryogenic cooling nozzle complicates the machining setup.

Regarding processable materials, PEM is not suitable for Ti alloys due its reactivity with most cutting tool materials when temperatures are over 500 °C [15]. On the other hand, due to its low thermal conductivity, there is a tendency of heat accumulation in the surface and Widmanstatten structure formation, that is brittle and with tendency to crack formation and propagation. This becomes a problem for fatigue life considerations.

Among aeronautical alloys, PEM gives best results for Inconel 718. This material is widely used in gas turbine components, cryogenic storage tanks, jet engines, pump bodies and parts, rocket motors and thrust reversers, nuclear fuel element spacers, hot extrusion tooling, high strength bolting, or downhole shafting. In the bibliography, there are several studies that show productivity improvements up to 350% using whisker-reinforced ceramic tools [12, 14, 16–18]. PEM also gives cutting force reduction and high material removal rates. The cutting force reduction allows cutting speeds over 950 m/min and material removal rates over 12 cm^3/min. Thus, considering high productivity and heat affected area on the workpiece, the PEM should be applied on only during roughing operations. However, using conventional techniques on finishing operations, any heat affected area can be removed.

PEM also gives satisfactory results on Co base materials such as Haynes 25 alloy [12, 19, 20]. The process has a positive effect on tool wear reduction. Using carbide tools, PEM provides tangential cutting force reduction of 20–30% and maximum cutting speed of 650 mm/min. The main limitation for Haynes 25 milling is the strain hardening of the work surface. The large amount of plastic deformation during and the allotropic structural transformation from fcc to hcp during cutting process gives a result of surface hardening. The hardness of the material can be increased from 22–25 to 58 HRC during cutting process. The effect of PEM on the structure is very limited, as it can be noticed in Figure 19.4. There is no negative effect on the structure, and the only effect observed is to delay the aforementioned allotropic transformation.

19.2.2 Laser-assisted Machining (LAM)

When a laser beam is focused on the surface, an energy fraction is reflected, and some is absorbed. Depending on the material, the wavelength of the beam, etc. the reflected and absorbed ratio is different. In Figure 19.5, energy absorption ratios for different materials and wavelengths are represented. The energy absorbed by the material is converted into heat. The absorption coefficient varies for different materials and depends on the temperature, surface roughness, or purity.

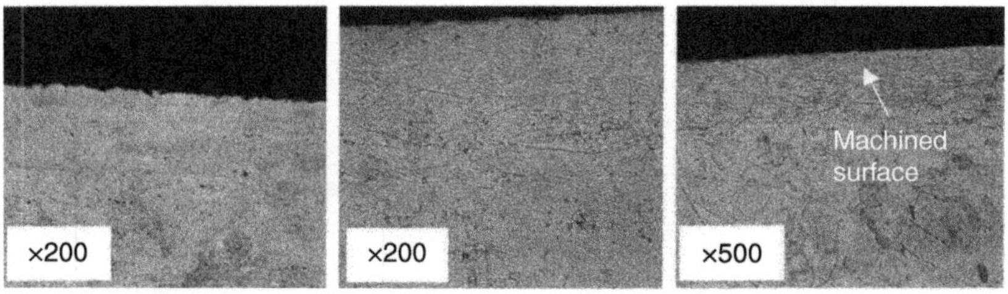

Figure 19.4 Strain hardening in Haynes 25 after plasma assisted milling (PAM) (a) surface previous to machining (b) machined surface ×200 (c) machined surface ×500.

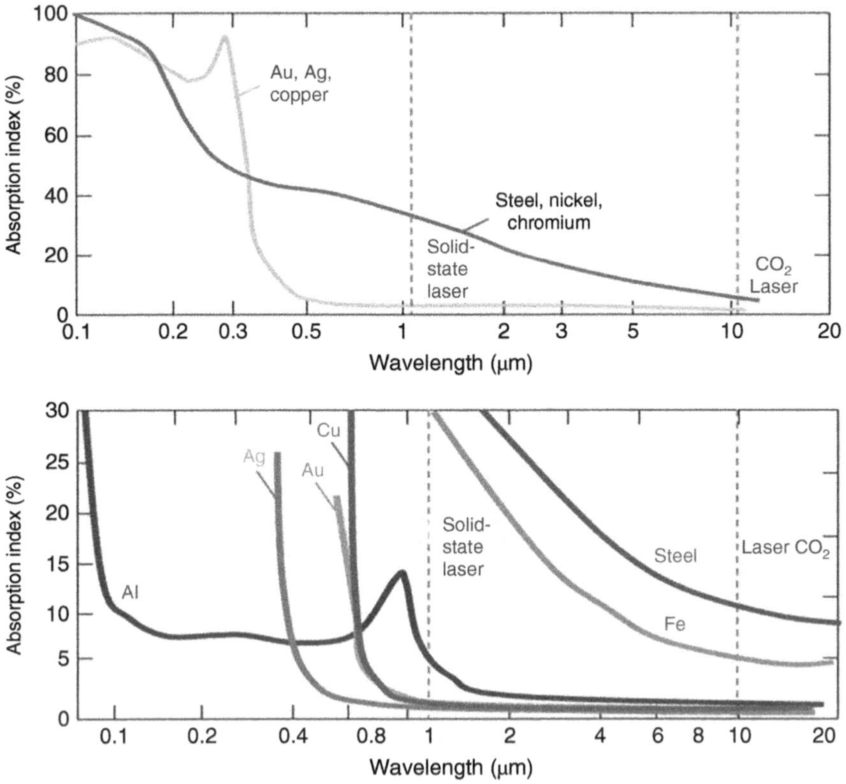

Figure 19.5 Energy absorption for different materials.

In general, laser technology allows higher energy concentration than plasma technology. The spot size of laser beam depends on the laser type and used optics. Last generation industrial lasers are able to focus up to several kilowatt in spots of few microns due to the specific characteristics of the electromagnetic radiation obtained. Compared with other electromagnetic radiation such as sunlight, the resulting radiation is different because it is collimated: the wavefronts are planar in normal to the direction to the propagation and very similar to the input signal in terms of wavelength and polarization. This gives the radiation its characteristic coherence.

Lasers contain four primary components, the active medium, the excitation mechanism, a high reflectance mirror, and a partially transmissive mirror. The excitation mechanism pumps energy into the active medium, and the electrons reach higher energy levels. When those electrons return the fundamental energy status, a photon is emitted. The photons travel inside active medium reflected between mirrors, and when these photons collide with excited electrons, another photons are created in phase with the first ones. This phenomenon enables the amplification of the radiation and creates the laser beam coming across the partially transmissive mirror.

The laser beam is an electromagnetic radiation that presents three different characteristics from other electromagnetic radiations such us sunlight. The radiation is coherent, all waves are in phase; it is monochromatic, all waves present the same wavelength; and it is collimated, all waves travel in parallel directions.

Depending on the active medium, there are three main types of lasers, solid state lasers, gas lasers, and diode lasers. In Figure 19.6, beam parameter product (BPP) quality parameter is

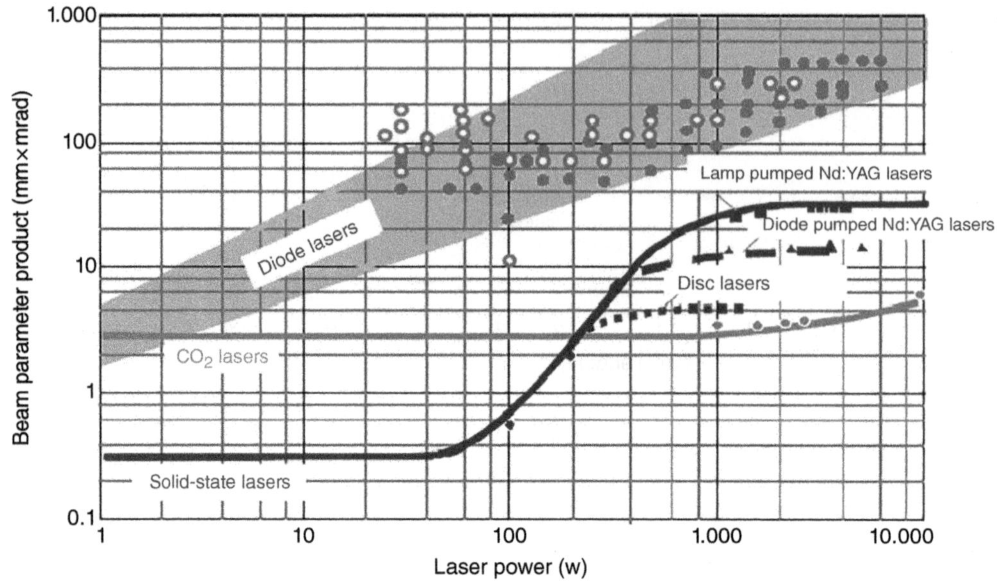

Figure 19.6 Beam quality for different laser sources.

represented for different laser types and power levels. Depending on the laser type, the resulting wavelength varies for different laser types, and some wavelengths can be guided using optical fiber, which simplifies noticeably the implementation in manufacturing processes. Between industrial lasers, diode lasers and Nd:YAG fiber and disk lasers present wavelengths between 800 and 1060 nm and can be guided using optical fiber.

When compared with plasma technology, the laser technology is significantly more expensive, and its use as assistant device for machining operations is only justified in very high added value applications. In the same way as PEM, laser-assisted machining (LAM), is more extended for turning operations. In turning operations, there are two main relative positions for the laser beam as it is shown in Figure 19.7, normal to workpiece or normal to the machining chamfer surface.

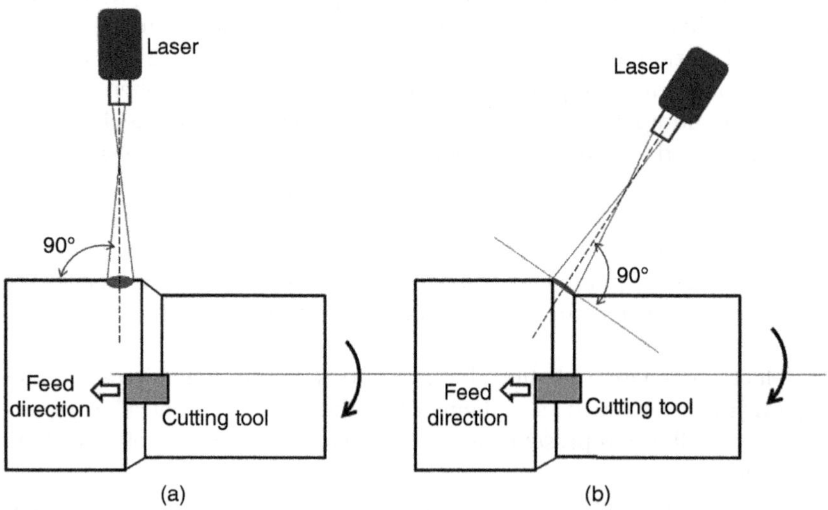

Figure 19.7 Laser orientation in LAM. (a) Normal to workpiece and (b) normal to chamfer.

Using a relative position normal to chamfer surface, higher and uniform reduction of cutting forces is achieved [18, 21, 22]. When laser spot fully covers the chamfer surface, the cutting forces are minimized, but even with partial coverage, the cutting force and tool wear are reduced. In literature, there are studies using multiple laser units heating simultaneously both the unmachined surface and the chamfer surface to create optimal temperature distribution through depth of cut that is reported to result in longer tool life when compared with other solutions [2, 23, 24]. The drawback of this solution is the increased complexity and cost, only justified in very high added value applications.

In general, the closer is laser beam to the cutting tool, the higher the cutting force reduction is and longer tool life is, up to a limit. If laser beam is too close to the cutting tool, the tool may be damaged by overheating, and the chips may fly into the laser beam interfering with the process.

In surface milling, it is important to heat the area above the cutting tool to minimize vibrations and reduce tool wear. In general, the beam can be arranged to be separated from the tool or integrated with the spindle. The easiest way is to set the beam in front of the tool in the feed direction. In LAM, the spot size of the laser beam limits the ability to cover the width of cut by a single spot in most applications. Considering the middle area and the entry and exit area of the cutting tool, it is more interesting to heat the tool entry area in order to avoid tool vibration and rapid wear development. Nevertheless, best results are obtained heating completely the width of cut area by using specific optics or more than one source at the same time [25, 26]. On the other hand, heating the material in front of the tool when feed direction is variable is not easy and requires more than a single beam guiding system.

Among LAM applications, laser-assisted milling of ceramics such as silicon nitride had been reported. PCBN tools in combination with LAM show longer tool life as a result of cutting force reduction [27, 28]. During this process, ceramic material is softened, and the ratio between feed and cutting force is reduced, as well as the cutting force itself [29, 30]. In Figure 19.8, the total cost comparison for conventional and LAM in turning operations of Inconel 718 using same process parameters at different cutting speeds of 2.2 and 3 m/s is represented [23].

Considering life of the cutting tool in LAM applications, dominant wear mechanisms are abrasion, adhesion, and diffusion. Tool wear depends strongly on the maximum temperature reached during the process. The average temperature of the material as it enters the plastic deformation zone is a key factor to make the tool life maximum. This needs to be controlled in

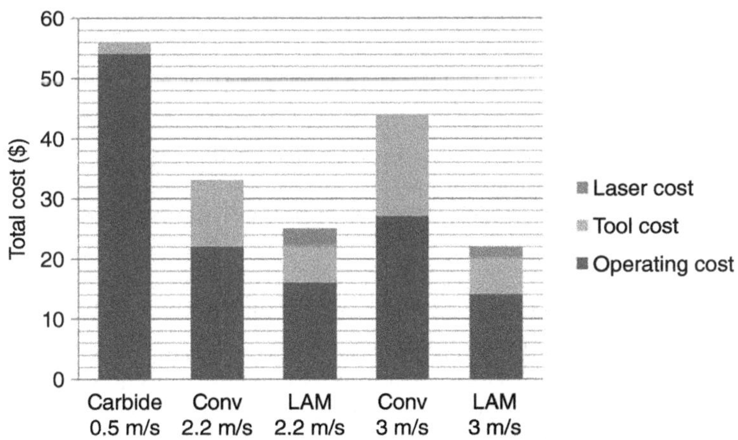

Figure 19.8 Total cost comparison for LAM and conventional machining.

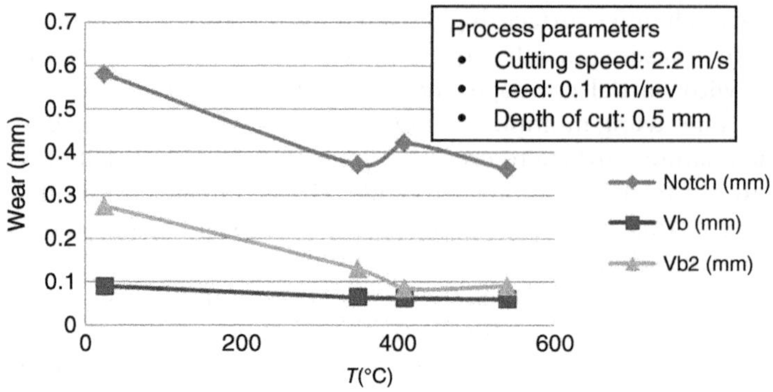

Figure 19.9 Wear evolution with temperature in Inconel 718 machining.

order to get satisfactory results without melting or leaving undesirable microstructures. Lower temperatures make cutting process more difficult due to the high shear strength of the material, and very high temperatures result on high tool wear due to diffusion. This point is confirmed in tests carried out by Anderson et al. [23]. After machining 5000 mm^3 in Inconel 718, Figure 19.9 shows decreasing notch wear and decreasing wear in both, primary flank wear (Vb) and secondary face (Vb2).

On the other hand, by increasing workpiece temperature, tool breakage due to chipping is also avoided, and tool flank wear is more gradual. Thus, it is important to know the temperature distribution through material up to the cutting edge. Milling process allows higher temperatures than turning because of the interrupted cutting and the cooling of the tool teeth when not immersed [31].

One of the most relevant applications of LAM is ceramic machining. In silicon nitride ceramic machining, high temperatures can lead to negative effects on the processed surface due to β-Si_3N_4 silicon nitride grain pullout [27, 28]. On the other hand, in materials such as magnesia partially stabilized zirconia (PSZ), the stress due to thermal cycles can lead to surface microcracking [29].

19.2.3 Direct Laser Machining

If the energy transmitted using a laser is enough, the material can be heated, melted, and vaporized depending on the selected energy level. Direct laser machining is based on the selective elimination of material by melting and vaporizing. A highly focused laser beam allows detail engraving and complex geometries that are not possible with other technologies.

Light moves through space as a wave, but when it encounters matter, it behaves like a particle of energy, a photon. The amount of energy in each photon is calculated by the expression proposed by Planck, $E = h \cdot \nu$, where h is the Planck constant with value of $6.626\,068\,96 \times 10^{-34}$ J s, and ν is the frequency of the wave, which is related with wavelength by the light-speed constant c by following expression $\nu \cdot h = c$.

The spectrum of the visible light is between 400 and 750 nm wavelength. For a wavelength of 500 nm, which is within the visible spectrum, the energy of each photon is 4×10^{-19} J or 2.5 eV being 3 – 10 eV the energy necessary to break the material bonds. This limitation can be solved focusing the energy using a laser (pyrolytic process) or using a more energetic laser (UV) with photons presenting energy levels of 3 – 7 eV, which is the type of laser used to break the chemical bonds in plastics (photolytic process). For metal processing, values of five times sublimation energy are necessary, and for most metal materials, the sublimation energy is close to 4 eV.

Another option to vaporize the material is to liberate a large amount of photons at the same time and hit the electrons with more than one photon at the same time. Thus, the absorption of multiple photons is equivalent to the absorption of a single high-energy photon. In this case, the energy level of each photon is not critical, and different sources can be used, operating in the visible spectrum or even in the infrared (IR) range.

A key parameter determined by the type of laser and the power level is the beam intensity that can be defined as the power per unit area. The intensity evolution through material can be obtained by the following expression:

$$I_x = I_0 \cdot \mathrm{e}^{-\alpha x} \tag{19.1}$$

where I_0 is the source intensity, α is the optical absorptivity, and x is the depth measured from surface. Another relevant parameter is the laser penetration depth. Thus, the upper layer thickness where the beam is absorbed, δ is obtained through $\delta = 2/\alpha$. In metal materials, this layer thickness is about 10 nm. On the other hand, an electron needs 1 ps to convert the energy got from the laser beam into heat, which is called the relaxation time [32]. Therefore, the laser heats a layer of 10 nm in 1 ps.

Once the upper layer of 10 nm is heated, the heat is diffused into the material following the $d = \sqrt{4at}$ expression, where d is the diffusion depth, a is the thermal diffusivity, and t time. Thus, for 1 ps (1×10^{-12} s), the diffusion depth is also 10 nm. In this sense, an ultra-short pulse is considered when the diffusion depth is of same order as the absorption upper layer. In general, the diffusion depth depends on material properties, and pulses close to 1 ps are considered ultra-short. Based on physical considerations, a removal rate of 10 nm/pulse at 1 J/cm^2 is expected on steel, but experiments carried out at Lawrence Livermore show that it is possible to vaporize material with different rates, from 50 nm/pulse at 4 J/cm^2 up to 350 nm/pulse at 14 J/cm^2, which is the saturation limit [33].

In general, for metal vaporization, a pulsed laser gives better result than a laser operating in a continuous wave mode (CW). Pulsed laser allows higher energy concentration in the surface, and the accumulated heat is less. A short interaction time gives better heat diffusion without heat accumulation. There are several industrial solutions for laser pulsation that can be resumed in three technologies: gain switching, Q-switching, and mode-locking [34].

In gain switching, the pulsation is got by the controlled pump of the active medium. Once the critical population inversion is reached, the laser starts to oscillate, and the gain becomes larger than the losses. This situation continues until pumping process is switched off. The pulse duration depends on the excitation mechanism, flash lamp pumping, or diode pumping and can vary from 10 μs to 10 ms. Main drawback of this technology is the limited stability and reproducibility of the pulses. This, added to pulse duration of ms, makes difficult to obtain good results for direct machining with gain switching technology. In Figure 19.10, a texture obtained in stainless steel 316L is shown.

Q-switching technology works with lasers pumped in CW mode and provides regular pulses. The energy is stored in the active medium as excited atoms, and then, the energy is released in a short burst. The energy level of electrons in the active medium can be increased avoiding the stimulated emission by canceling the reflection of the emitted photons. Once the maximum energy level is reached, the energy is released in one short pulse aligning the mirrors of the resonator cavity, and the material is eliminated creating a crater on the surface similar to the one shown in Figure 19.11.

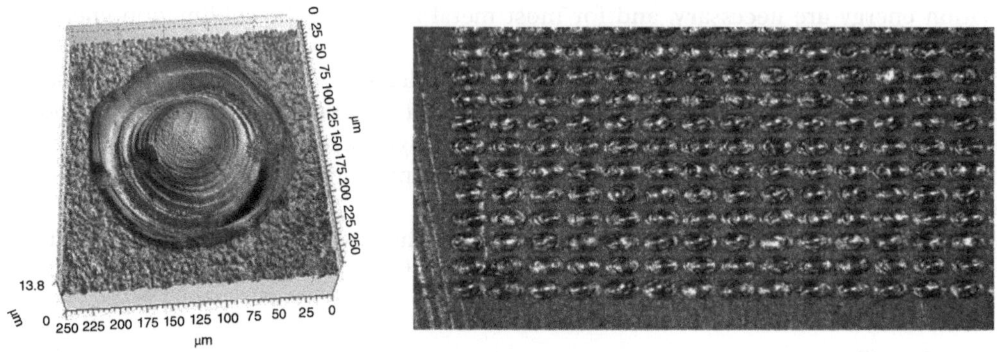

Figure 19.10 Gain switching–based surface texturing.

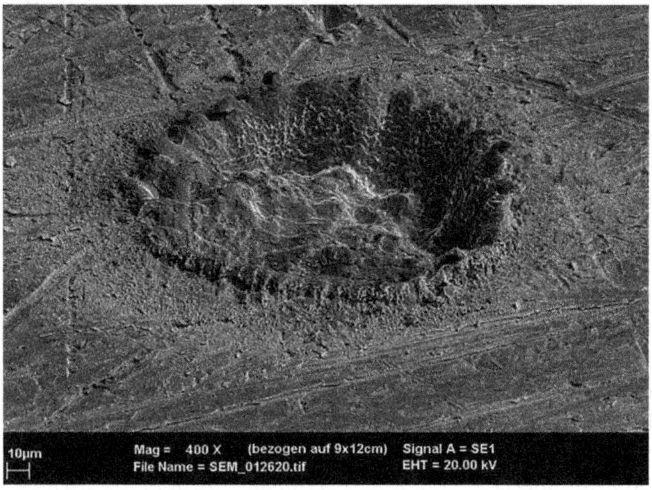

Figure 19.11 Q-switching–based surface modification. *Source:* Courtesy Dr. D. Bleiner.

Physically, there are several configurations to get a Q-switch pulse. The most relevant ones are the following:

- *Spinning reflectors*: The high-reflecting mirror is rotated so that the mirror tilted out of alignment. When the mirror rotates back to alignment, the system is Q-switched, and the pulse is released. On pulse is released in each revolution, and the switching time is typically of few nanoseconds.
- *Electro optic Q-switching*: A polarizing filter is placed inside the resonator cavity, and each rotation of the filter releases one short pulse. With this technology, switching time is fast, less than a nanosecond.
- *Acusto-optic switches*: A transparent element is placed in the resonator cavity. This element exhibits a diffraction effect and diffracts part of the beam out of the cavity alignment when it is excited with intense standing acoustic waves. When the acoustic wave is over, the diffraction disappears and the system emits a pulse. The switching time is typically nearly 100 ns or greater.
- *Saturable absorbers*: The absorber can be a thin film on glass substrate or liquid in glass cells. The absorber acts like a dye with a high cavity loss for the laser wavelength and preventing lasing. When the irradiance from the active medium becomes intense enough, the energy absorbed by the dye optically pumps the dye material causing it to be transparent at laser wavelength. With this technology, switching time is fast, less than a nanosecond.

With mode-locking technology, it is possible to get faster switching times than with Q-switching. Mode locking is used to generate ultra-short pulses with pulse duration in the picosecond to femtosecond range. In this type of laser, the resonator contains an active element, an optical modulator, or a nonlinear passive element, a saturable absorber, which causes the formation of an ultra-short pulse circulating in the resonator. Once reached the steady state, the various effects influencing the circulating pulse are in balance so that the pulse parameters are unchanged after each complete round trip.

- *Active mode-locking*: An acoustic-optic modulator is placed into the cavity. Driven with an electrical signal, this modulator produces sinusoidal amplitude modulation of the light in the cavity. The modulator creates sound waves that interfere with the modes and actuates like a shutter to the light. The scheme is represented in Figure 19.12.
- *Passive mode-locking*: A saturable absorber is used to interfere with the light. The saturable absorber is an optical device that exhibits intensity-dependent transmission and very fast response. In steady state, a pulse is circulating in the resonator. Each time the pulse hits the absorber, it saturates the absorption, and pulse is reflected. The gain can be adjusted to a level that compensates the losses for circulating the pulse. Thus, any light of lower intensity which hits the absorber at other times will experience losses that are higher than the gain, since the absorber cannot be saturated by this light. The scheme is represented in Figure 19.13.

The pulse repetition corresponds to the resonator round trip, and the pulse duration is much lower, typically between 30 fs (30×10^{-15} s) and 30 ps (30×10^{-12} s), in extreme cases down to ≈5 fs. For that reason, the peak power of a mode-locked laser can be orders of magnitude higher than the average power.

Femtosecond (fs) range lasers are suitable for material removing without melting material. This type of laser is used for surface microstructuring, but a low material removal rate and the high cost of the source limit its application to very high added value parts.

On the other hand, in the literature, different solutions can be found that combine direct laser machining and conventional machining. Thus, the laser is used not only to soften the

Figure 19.12 Active mode locking.

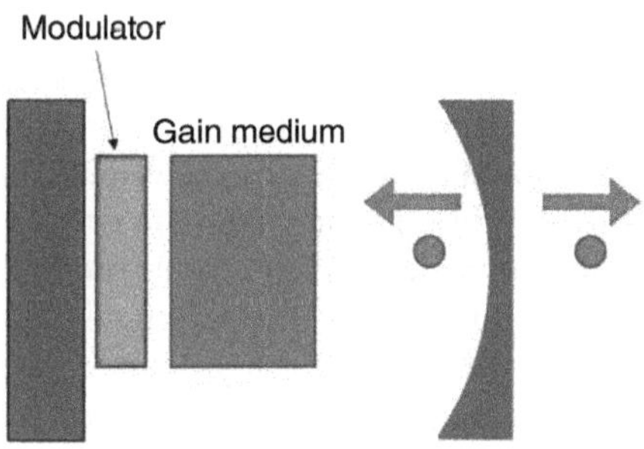

Figure 19.13 Passive mode locking.

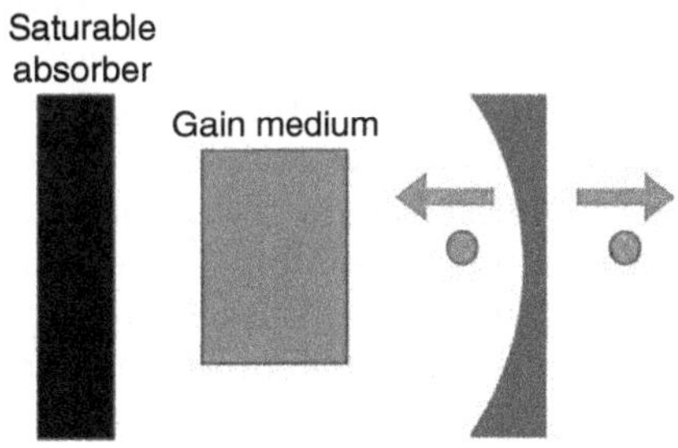

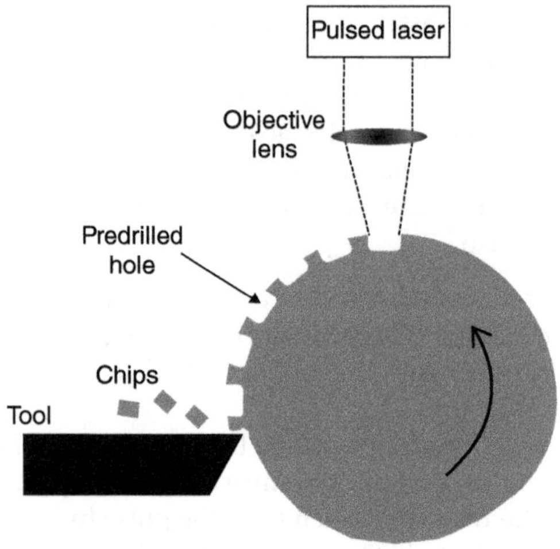

Figure 19.14 Direct laser and conventional machining combination.

material but also to remove part of the material by laser drilling or ablation [35, 36]. In order to reduce the cutting forces, the laser-drilled holes should have a depth not greater than the depth of cut and diameter greater than the chip width. The drilling operation can be done at the same time with machining operation or before cutting. In this case, the operation is carried out in two steps, but drill distribution and shape can be optimized to get maximum reduction of cutting forces and therefore, less tool wear. The experiments conducted show a 50% reduction of cutting forces in Ti–6Al–4V alloy machining. The scheme of the process is represented in Figure 19.14.

19.3 Analysis of LAM Processes

To get best process parameters is necessary a comprehensive analysis of temperature distribution in the cutting area, so thermal field simulation becomes key point to achieve good results.

During the LAM process, the main variable to control is the temperature field. The laser is used to heat locally different areas of the part and thus improve the mechanical properties of the material during the machining process. This saves total machining time besides increasing the tool life by reducing its wear. In Figure 19.15, the decrement of tensile strength of different metal- and ceramic-based materials at high temperature can be observed.

The objective in a thermal study of a manufacturing process as the LAM is to try to predict the final results by minimizing experimentation. Thus, the study of the temperature field and the final results of the process can be approached through three types of methodologies or models that can be combined: an experimental model, a numerical model, or an analytical model. Thus, deciding the methodology to be followed during the thermal study is crucial. A variable that has not been taken into account or an oversized model can be modified with more or less work, but the wrong choice of the methodology to follow means starting over the study.

An experimental model addresses the problem from the beginning using statistical variable techniques such as analysis of variance techniques (ANOVA) or neural networks. In this case, it is not necessary to have initial data regarding the constitutive equations that govern the process or the relationship between the main parameters that come into play. The problem is treated like a "black box" where one can enter variables and get data results. This methodology is especially useful in processes where there is a lack of process knowledge or, due to the complexity of the process, is not possible to take in consideration its constitutive equations. Input

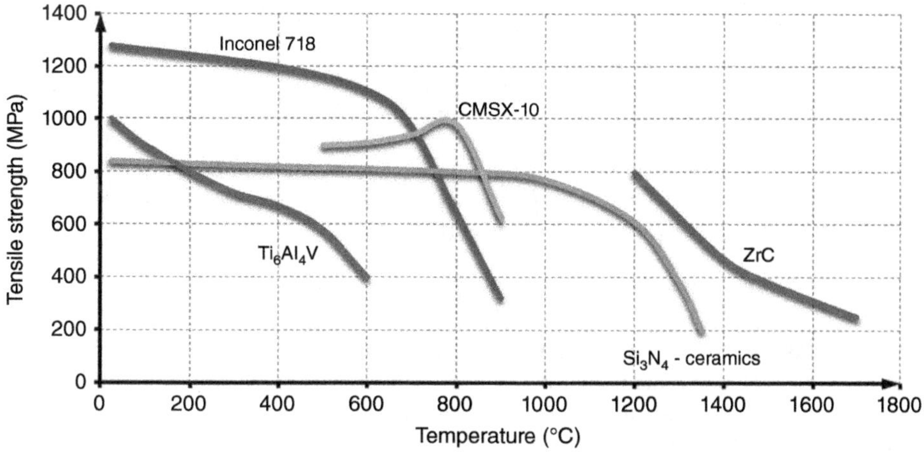

Figure 19.15 Temperature and tensile strength relation for different metal- and ceramics-based materials.

and output parameters' tendencies in the process work space are studied taking the relationship between the outputs and the inputs. For example, in laser-hardening process, there are studies that only using experimental techniques take relationships between input variables, such as laser power and process feed rate, with the transversal geometry of the heat-affected zone (HAZ).

By contrast, on an analytical model, it is necessary to know exactly the process constitutive equations. Moreover, the resolution of these equations is usually done through a very complicated mathematics and using several number of simplifications. Typically, these models suggest exact solution, in 2D or 3D models, for an infinite part under quasistationary conditions. The main disadvantages of analytical models are the simplifications that must be assumed. It is not easy to obtain a solution for complex geometries and beam trajectories, or modeling a pulsed laser. Additionally, in analytical models, it is usual to neglect the variability of physical properties with temperature, such as material conductivity, density, or specific heat.

Finally, there are the numerical models based on the temporal and spatial discretization of the variables involved in the constitutive equations of the process. These equations are solved by numerical methods via software. Actually these models, based on numerical methods, are the most widely used due to its versatility when working with complex part geometries, customized energy distributions in laser beam, or machine parameters which are variable over time, such us beam trajectory, feed rate, or laser power. In general, a physical change in these parameters is relatively simple. With the numerical models, the dynamical thermomechanical field in a multiscale analysis can be studied. But it has to be kept in mind that the limit on the accuracy and complexity of these models is closely linked to the development of computers and specific software.

Thus, among the most widely used numerical methods to resolve the equations involved in the thermal analysis of LAM are the finite element method (FEM) and the finite difference–volume method (FDM–FVM). These equations are, compulsorily, an energy balance on the elements of the part and, optionally, other equations as phase transformations suffered in the material during heating and cooling or heat stolen during the phase transitions.

19.3.1 Energy Source

The modeling of the energy source is a key aspect for the correct estimation of the thermal field and the evaluation of the new material properties enhanced with the temperature in the

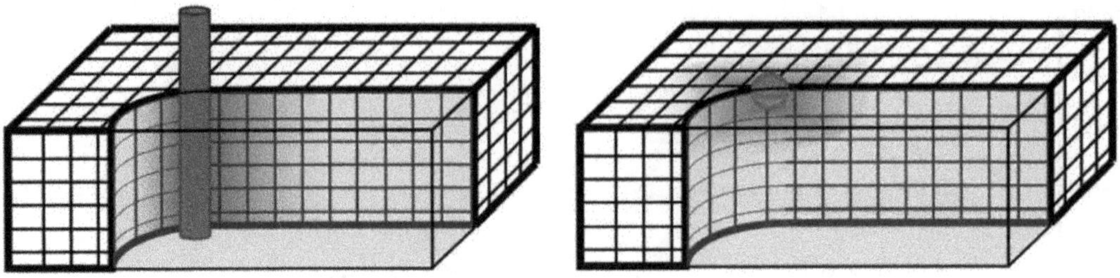

Figure 19.16 Difference between linear source and point source.

HAZ. In the case of lasers, the energy radiated on the workpiece is highly concentrated. Typically, the processes that use a laser beam as energy source are classified into high-penetration processes and low-penetration processes. The most common high-penetration processes are laser welding and laser cutting. Thus, in this type of processes, either total or partial penetration, the laser source is considered and modeled as a line source. On the other hand, in laser surface treatment processes, such as hardening, alloying, texturing, polishing, or LAM processes, laser beam penetration is low, and the source is modeled as a point source at the part surface. The difference between a linear source and a point source is shown in Figure 19.16.

Although the distinction between high and low penetration process is valid for most cases, sometimes are necessary specific models, such as for the simulation of the initial stages of cutting and welding processes (formation of keyhole). In these cases, the combination of linear and punctual source is used to simulate the effects of energy absorption below the keyhole and the plume formation on the surface.

Basic thermal models consider the laser beam as point energy source, but in general, thermal models present a varying energy distribution inside the area of incidence of the laser. This variation depends on the type of laser considered. In the figure are shown different energy distributions, Gaussian, Top-Hat, cylindrical or multimode, depending on the type of laser considered.

For example, a CO_2 laser, commonly used for welding or cutting process, presents Gaussian or a doughnut-shape energy distribution. In processes such as laser hardening or LAM, direct diode or fiber-guided diode lasers are commonly used. Diode lasers have higher energetic efficiency with lower cost than CO_2 lasers and larger spot size. The energy distribution in the spot depends on the optic and laser configuration. Thus, in direct diode lasers, the energy distribution follows rectangular complex Top-Hat shape, and in fiber-guided diode lasers, the resulting energy distribution presents cylindrical uniform shape.

The energy distribution is a key point in thermal modeling of laser-based processes and has to be correctly introduced. In Table 19.1, different laser beam shapes and their corresponding equations are presented.

Laser beam presents a convergent–divergent spatial shape with a maximum power density in the waist. Depending on beam defocusing ratio, the absorbed power and its thermal effects may change considerably. In order to reach accurate results, it is important to consider the laser beam spatial shape. The laser beam quality can be defined by its diameter at the narrowest part (focal point or waist) following BPP factor, which gives an idea of the beam quality or its capacity to converge at a minimum area. As it is shown in Figure 19.17, BPP is the product of a laser beam divergence angle (half-angle) and the radius of the beam at its narrowest point in units of (mm mrad).

Table 19.1 Shapes and equations of energy sources for different types of lasers.

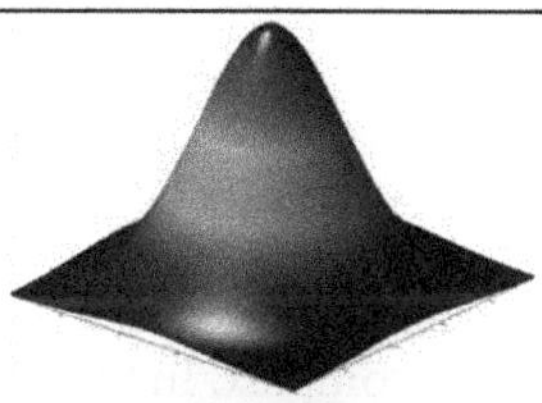	Gaussian source $$I = \frac{2P}{\pi w^2(z)} \exp\left[\frac{-2\left(x^2 + y^2\right)}{w^2(z)}\right]$$
	Cylindrical (uniform) source $$I = \begin{cases} \dfrac{2P}{\pi w^2(z)} & \text{if } \left(x^2 + y^2\right) \le w^2(z) \\ 0 & \text{if } \left(x^2 + y^2\right) > w^2(z) \end{cases}$$
	Top-Hat source $$I = \frac{P}{4AB}\left[\operatorname{erf}\left(\frac{A-2x}{2\sqrt{2}\cdot a}\right) + \operatorname{erf}\left(\frac{A+2x}{2\sqrt{2}\cdot a}\right)\right]\cdot \left[\operatorname{erf}\left(\frac{B-2y}{2\sqrt{2}\cdot b}\right) + \operatorname{erf}\left(\frac{B+2y}{2\sqrt{2}\cdot b}\right)\right];$$ $A(z)$, $B(z)$ – axis dimentions a, b – beam dispersion
	Multimode source (TEMmn) $$I_{\text{mn}} = \frac{2P}{\pi w^2(z)}\left[H_{\text{m}}\left(\frac{\sqrt{2}x}{w(z)}\right) \exp\left(\frac{-x^2}{w^2(z)}\right)\right]^2 \cdot \left[H_{\text{n}}\left(\frac{\sqrt{2}y}{w(z)}\right) \exp\left(\frac{-y^2}{w^2(z)}\right)\right]^2$$
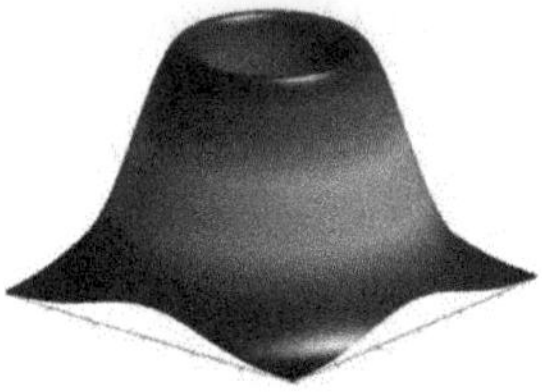	"Doughnut" or "Center cold" mode (TEM01*) $$I_{01^*} = \frac{4P}{\pi w^4(z)}\cdot\left(x^2 + y^2\right)\cdot \exp\left[\frac{-2}{w^2(z)}\cdot\left(x^2 + y^2\right)\right]$$

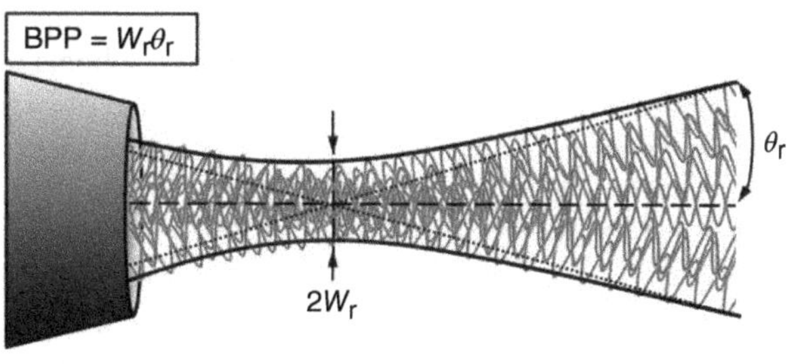

Figure 19.17 Energy distribution in laser beam considering the beam parameter product.

Thus, in a laser beam, spatial model laser beam diameter in different parts of the work space is related with the following equation where z is the distance to beam waist:

$$w(z) = w_r + |z| \tan\left(\frac{\text{BPP}}{w_r}\right) \tag{19.2}$$

BPP gives, as a result, the energy caustic, which is important to consider if the incident angle of the laser beam is not normal to the surface, or if the processed shape is complex. In this situation, it is necessary to consider the real energy distribution on the surface coming from the intersection of the energy caustic and part surface.

19.4 Laser-assisted Applications

The demands of the technologically advanced sectors require the use of new materials such ceramics, titanium alloys, or superalloys. These new materials are difficult to cut and impose the development of new machining processes. In this sense, thermal-assisted processes are becoming more relevant in order to accomplish the machining of these materials. There are different heat sources used in these processes, but laser sources have the suitable characteristics for this purpose. These characteristics include high energy density, high control of the heated zone, and easy inclusion in conventional machines. Thus, laser beam is commonly used to soften the material of the zone where the tool is cutting the workpiece. This local softening of the material improves surface finish, reduces tool wear, and generally results in a better material machinability. These facts are long proved for turning and milling of difficult to cut materials, and nowadays, the research efforts are focused on the development of new laser-assisted processes and new tools for this purpose. For example, Fraunhofer IPT has developed a milling tool that guides laser beam into the tool in order to reach the interaction zone between tool and workpiece. Another use of LAM currently being studied is the micromachining of difficult to cut materials where tool is highly exposed to break because of high forces during machining and low resistance of the tools.

Grinding is another process where the improvements obtained by combination with laser sources are well known. In laser-assisted grinding, the surface of workpiece could be heated just before the grinding tool, softening the material and making easier machining it. Moreover, thermal gradient between workpiece and grinding zone is lower, reducing surface layer damage by cracking. There are three applications of laser-assisted grinding. First, laser spot could be used for hardening workpiece surfaces directly after grinding. Other application is the use of laser beam to heat the workpiece in order to increase the machinability of hard materials and reduce the grinding wheel wear. Finally, the most extended application of this hybrid process is grinding wheel recovery with the objective of replacing conventional truing and dressing.

Truing is the process by which grinding wheels obtain or recover its initial shape. Generally, the realization of convex contours becomes of high difficulty due to the necessary tool wear compensation during the process. On the other hand, sharp- and concave-edged profiles are not possible to achieve by conventional truing. Thus, Walter et al. [37] combine laser and conventional truing processes in order to manufacture grinding wheel with complex shapes. In this application, grinding tool is placed on five axis machine with a high-speed spindle. Laser beam is placed tangentially to grinding wheel in order to generate slots with very low aspect ratios with a minimum tool corner radius below 20 μm.

Dressing of grinding wheels is another process which is possible to combine with laser sources. Laser-assisted dressing is focused on recovering of high-price grinding wheels, such as cubic boron nitride (CBN) and diamond metal bonded wheels. There are two ways to combine laser and dressing processes. It is possible to remove the bond by vaporization and/or melting via heat input without damaging abrasive grains and avoiding wheel deformation by dressing force. When direct removing of bond is not possible, laser beam softens it and allows recovering the grinding wheel minimizing dressing forces and tool wear (Figure 19.18).

Ball burnishing is another finishing process where the addition of heat sources is very useful. It is proved that laser heating improves the plastic deformation of conventional burnishing of hard materials. Tian and Shin [38] proposes a novel laser-assisted burnishing as finishing process to efficiently produce excellent surface finish, cold work hardening, and compressive residual stresses in the workpiece (Figure 19.19). The hybrid process is applied on aging hardened alloy (MP35N) and a quenched steel (AISI 4140), which need a strict control of the tem-

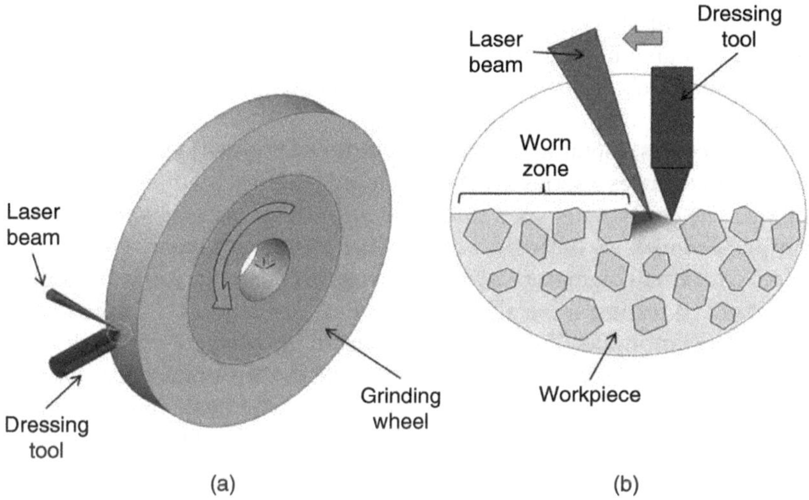

Figure 19.18 Laser-assisted dressing: Process configuration (a) and dressing tool–laser beam interaction zone (b).

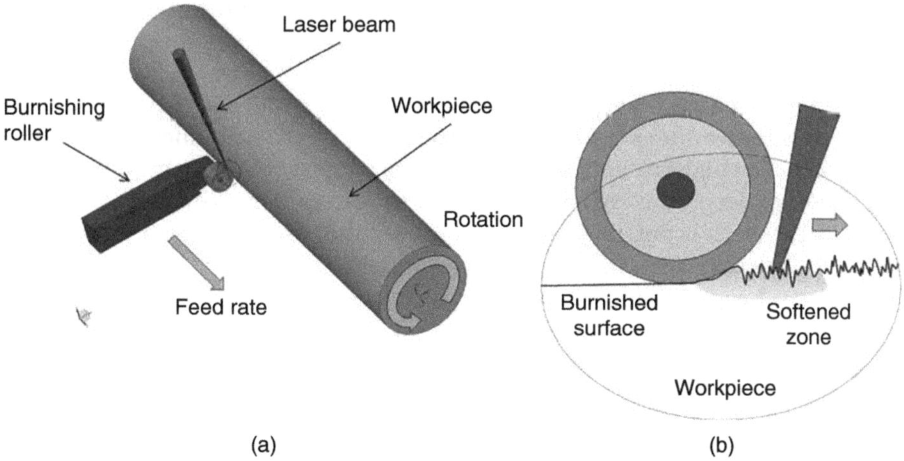

Figure 19.19 Laser-assisted ball burnishing: Process configuration (a) and tool–laser beam interaction zone (b).

perature during the process in order to keep the initial mechanical properties. Thus, compared to conventional burnishing, laser-assisted process improves workpiece finish by 30–40% of initial roughness value. Moreover, forces during the process are reduced, obtaining longer tool life or faster burnishing speed due to the softening of workpiece. Finally, the compressive residual stresses are similar to those obtained by conventional burnishing. Therefore, the combination of both technologies improves the benefits of conventional burnishing during finishing of hard materials.

Coating processes are also susceptible to combine with laser sources. In the laser-assisted cold spraying, the laser beam heats the coating layer just after the spraying spot deposits it on the substrate. The additional heat source melts the coating giving as a result dense and strong bonded depositions. Results show that assisted process obtains more uniform coatings minimizing the porosity and improving the corrosion behavior. Moreover, layer thickness of laser-assisted cold spraying is higher than that obtained by conventional cold spraying, increasing the deposition efficiency of the process. Thermal spraying techniques (plasma, HVOF, flame, etc.) can also be combined with laser sources such as Nd:YAG and diode lasers in order to get bonded dense structures with corrosion and wear resistant coatings on metallic surfaces.

Regarding metal forming processes, metal wires are manufactured by plastic deformation by conventional drawing. This process uses a series of dies which reduce the wire dimensions until it reaches the final diameter. The contact between these dies with the processed wire can produce an uncontrolled surface chemistry where oxides and other elements appear. Thus, applications which require wires with small diameter and high corrosion resistance, such as medical and jewelry applications, are requesting for a drawing process that eliminates the contact between the wire and the deformation die. This way, heat-assisted drawing uses an additional source which increases the wire temperature in order to increase the wire plasticity allowing the maximum section reduction only making use of traction forces. The heat sources initially used were induction and resistance heating, but the control of temperature of these techniques is very difficult for very fine wire diameters. Therefore, the capability of laser sources to concentrate energy in small spots improves temperature control with a minimal energy loss.

The drawing limit of laser-assisted wire drawing process depends on the mechanical and metallurgical properties of the wire material, the traction forces and velocities, the laser source and the temperature field on the wire. Thus, an average section reduction of 32% is achieved for 500 μm in Ni-200 wires, decreasing to 25% for wire of 125 μm initial diameter wire of the same material. For copper and steel wire, it was possible to reduce the cross-section of the wire by more than 20%.

Sheet forming is another process which can be assisted by laser sources. Nowadays, some industries, such as automotive industry, demand formed parts with a thinner sheets using high-strength materials. The high strength and low ductility of these materials make them difficult in cutting, bending, deep drawing and embossing processes with the specified quality and complexity. In this sense, it is possible to heat the zone of the part being formed with laser in order to increase the ductility and improve the process limits.

This hybrid process has been studied and developed by Fraunhofer IPT, which uses a scanner to preheat the sheet for different forming processes. This way, it is possible to shear ninety holes of 1-mm diameter with a sheet thickness of 1 mm for three high-strength materials without considerable tool wear. These process improvements can be extended for bending where it is possible to obtain ratios of one-fourth of the sheet thickness, or for deep drawing where the forming ratio is doubled in comparison with the conventional processing. Finally, the application of laser-assisted forming increases the embossing depth up to 50% of high-strength sheet thickness.

Incremental forming is another process, which could be improved by combining with a laser source. In this process, although low-yield strength and a low-hardening coefficient are desirable for limiting process forces and spring-back, they also result in excessive, unwanted plastic deformation in zones near the forming zone. Thus, such localized heating by laser beam assistance allows a change in material properties at the forming zone. For this purpose, an incremental forming setup by incorporating a laser source that provides the heat input in the unprocessed side the workpiece has been developed at Catholic University of Leuven. Moreover, in order to assure the temperature gradient around the heated spot, a cooling system with combined function was incorporated. Forming tool is mounted on a six-axis robot, and laser beam movement is provided by a three-axis table that allows defocusing it in order to change the beam diameter. Test results show the improvement achieved in terms of process forces and accuracy when using local and dynamic heating by a laser during incremental forming.

Finally, Kalyanasundaram in Iowa State University combines a CO_2 laser and abrasive-free water jet process in order to cut ceramics with high strength and fracture toughness. In this process, the hybrid system uses the low thermal shock resistance of ceramics to control the crack propagation during the machining. Laser spot heats a localized zone, and water jet causes a rapid quenching initiating a crack growth. This method decreases the machining cost and thermal damage of ceramics caused by grinding.

Another example of combining laser sources and water jet is the Laser Microjet® Systems of Synova S.A. in Switzerland. In this case, the water jet is used in order to guide the laser beam into itself obtaining a cylindrical shape. This fact makes possible long working distances without divergence of the beam. Furthermore, the use of the water jet in the system has another two great advantages, firstly, it expels the molten material providing a cleaner cutting surface, and on the other hand, it cools the cutting zone minimizing the HAZ. The use of this system extends from thicknesses of microns to tens of millimeters with a cutting accuracy of $\pm$ 3 μm.

Therefore, it can be concluded that laser-assisted processes are becoming more and more relevant due to their distinct advantages compared to conventional processes. Thus, these processes are continuously developing in order to provide manufacturing solutions to most advanced industrial sectors.

19.5 Conclusions

The following summary statements can be made to conclude the advantages of LAM processes and operations:

- Local heating typically allows faster cutting, lower cutting forces, and longer tool life in several high-strength materials that are complicated to process using conventional techniques. The key on heat-assisted processes is the heating ratio of the material, high enough to soften the material but without affecting the structure of the material below and without damaging the cutting tool.
- The plasma and laser energy sources are best suited to the process. These sources allow high energy concentration in small areas giving more control on the process to avoid overheating. A laser beam can be higher focused than plasma. Thus, plasma source gives better results in roughing operations, with higher productivity ratios, and laser is more suitable for finishing operations, where it is critical to control the heated area.
- LAM process has been successfully applied on different materials, and the most extended application is in Inconel 718. This material is of high interest for the industry due to the numerous applications and the problems associated to conventional machining.

- Considering turning and milling operations, the milling process is more complicated because it is necessary to use more than one source or change the orientation of the laser as a function of the feed direction.
- LAM integration in turning operations is simpler because the tool is statically placed and the cutting process is continuous. However, diffusion wear on the tool can be a problem due to the temperatures reached during the process. So, high temperature–resistant tool materials such as ceramics are strongly recommended. Another point to consider is the chemical affinity between workpiece and tool materials that is normally temperature dependent.
- In general, the results are better as the energy source gets closer to the tool. The limit is the chip interference with the beam and the maximum temperature admissible on the tool.
- For process control, it is necessary to calculate the thermal field to avoid overheating problems. Thermal modeling is an essential tool. In order to achieve good results, it is necessary to consider both the energy distribution of the beam and the laser processing effects on the workpiece.

References

1 Amin, A.K.M.N., Dolah, S.B., Mahmud, M.B., and Lajis, M.A. (2008). Effects of workpiece preheating on surface roughness, chatter and tool performance during end milling of hardened steel D2. *Journal of Materials Processing Technology* 201: 466–470.

2 Ding, H. and Shin, Y.C. (2010). Laser-assisted machining of hardened steel parts with surface integrity analysis. *International Journal of Machine Tools and Manufacture* 50: 106–114.

3 Dumitrescu, P., Koshy, P., Stenekes, J., and Elbestawi, M.A. (2006). High-power diode laser assisted hard turning AISI D2 tool steel. *International Journal of Machine Tools and Manufacture* 46: 2009–2016.

4 Germain, G., Morel, F., Lebrun, J.-L., and Morel, A. (2007). Machinability and surface integrity for a bearing steel and a titanium alloy in laser assisted machining (optimisation on LAM on two materials). *Lasers in Engineering* 17: 329–344.

5 Lajis, M.A., Amin, A.K.M.N., Karim, A.N.M. et al. (2009). Hot machining of hardened steels with coated carbide inserts. *American Journal of Engineering and Applied Sciences* 2: 421–427.

6 Maitya, K.P. and Swain, P.K. (2008). An experimental investigation of hot-machining to predict tool life. *Journal of Materials Processing Technology* 198: 344–349.

7 Özler, L., Inan, A., and Özel, C. (2001). Theoretical and experimental determination of tool life in hot machining of austenitic manganese steel. *International Journal of Machine Tools and Manufacture* 41: 163–172.

8 Jeon, Y. and Pfefferkorn, F. (2008). Effect of laser preheating the workpiece on micro end milling of metals. *Journal of Manufacturing Science and Engineering, Transactions of the ASME* 130: 011004-1–011004-9.

9 Armitage, K. (2005). Laser assisted machining of high chromium white cast iron. Thesis for Master of Engineering. Swinburne University of Technology.

10 Skvarenina, S. and Shin, Y.C. (2006). Laser-assisted machining of compacted graphite iron. *International Journal of Machine Tools and Manufacture* 46: 7–17.

11 Wang, Y., Yang, L.J., and Wang, N.J. (2002). An investigation of laser-assisted machining of Al_2O_3 particle reinforced aluminum matrix composite. *Journal of Materials Processing Technology* 129: 268–272.

12 López de Lacalle, L.N., Sánchez, J.A., Lamikiz, A., and Celaya, A. (2004). Plasma assisted milling heat-resistant superalloys. *Journal of Manufacturing Science and Engineering, Transactions of the ASME* 126: 274–285.

13 Arunachalam, R. and Mannan, M.A. (2000). Machinability of nickel-based high temperature alloys. *Machining Science and Technology: An International Journal* 4: 127–169.

14 Wang, Z.Y., Rajurkar, K.P., Fan, J. et al. (2003). Hybrid machining of Inconel 718. *International Journal of Machine Tools and Manufacture* 43: 1391–1396.

15 Yang, X. and Liu, C.R. (1999). Machining titanium and its alloys. *Machining Science and Technology: An International Journal* 3: 107–139.

16 Leshock, C.E., Kim, J.-N., and C. Shin, Y. (2001). Plasma enhanced machining of Inconel 718: modeling of workpiece temperature with plasma heating and experimental results. *International Journal of Machine Tools and Manufacture* 41: 877–897.

17 Novak, J.W., Shin, Y.C., and Incropera, F.P. (1997). Assessment of plasma enhanced machining for improved machinability of Inconel 718. *Journal of Manufacturing Science and Engineering, Transactions of the ASME* 119: 125–129.

18 Rajagopal, S., Plankenhorn, D.J., and Hill, V.L. (1982). Machining aerospace alloys with the aid of a 15 kW laser. *Journal of Applied Metalworking* 2: 170–184.

19 König, K.W. and Zaboklicki, A.K. (1994). Laser-assisted hot machining processes: technological potentials. In: *Laser Assisted Net Shape Engineering, Proceedings of the LANE'94*, vol. 1 (ed. M. Geiger and F. Vollertsen), 389–404. Bamberg: Meisenbach.

20 Weinert, T. (1994). Relation between process energy and tool wear when turning hardfacing alloys. *CIRP Annals* 43: 97–100.

21 Salem, W.B., Marot, G., Moisan, A., and Longuemard, J.P. (1994). Laser assisted turning during finishing operation applied to hardened steels and Inconel 718. In: *Laser Assisted Net Shape Engineering, Proceedings of the LANE'94*, vol. 1 (ed. M. Geiger and F. Vollertsen), 455–464. Bamberg: Meisenbach.

22 Sun, S., Harris, J., and Brandt, M. (2008). Parametric investigation of laser-assisted machining of commercially pure titanium. *Advanced Engineering Materials* 10: 565–572.

23 Anderson, M. and Shin, Y.C. (2006). Laser-assisted machining of an austenitic stainless steel: P550. *Proceedings of the Institution of Mechanical Engineers – Part B: Journal of Engineering Manufacture* 220: 2055–2067.

24 Shin, Y.C. (2007). Laser assisted machining process with distributed lasers. US Patent 0062920 A1.

25 Thomas, T. and Vigneau, J.O. (1999). Laser-assisted milling process. US Patent 5,906,459.

26 Tian, Y., Wu, B., Anderson, M., and Shin, Y.C. (2008). Laser-assisted milling of silicon nitride and Inconel 718. *Journal of Manufacturing Science and Engineering, Transactions of the ASME* 130: 031013-1–031013-9.

27 Lei, S., Shin, Y.C., and Incropera, F.P. (2001). Experimental investigation of thermo- mechanical characteristics in laser assisted machining of silicon nitride ceramics. *Journal of Manufacturing Science and Engineering, Transactions of the ASME* 123: 639–646.

28 Rozzi, J.C., Pfefferkorn, F.E., Shin, Y.C., and Incropera, F.P. (2000). Experimental evaluation of the laser assisted machining of silicon nitride ceramics. *Journal of Manufacturing Science and Engineering, Transactions of the ASME* 122: 666–670.

29 Pfefferkorn, F.E., Shin, Y.C., Incropera, F.P., and Tian, Y. (2004). Laser-assisted machining of magnesia-partially-stabilized zirconia. *Journal of Manufacturing Science and Engineering, Transactions of the ASME* 126: 42–51.

30 Rebro, P.A., Shin, Y.C., and Incropera, F.P. (2002). Laser-assisted machining of reaction sintered mullite ceramics. *Journal of Manufacturing Science and Engineering, Transactions of the ASME* 124: 875–885.

31 Yang, B. and Lei, S. (2008). Laser-assisted milling of silicon nitride ceramic: a machinability study. *International Journal of Mechatronics and Manufacturing Systems* 1: 116–130.

32 Meijer, J. (2004). Laser beam machining (LBM), state of the art and new opportunities. *Journal of Materials Processing Technology* 149: 2–17.

33 Semak, V. (2001). Laser drilling: from milli to femto, laser solutions course. In: *Proceedings of the ICALEO 2001*, Jacksonville (18 October), 1–9.

34 Meijer, J., Du, K., Gillner, A. et al. (2002). Laser machining by short and ultrashort pulses, state of the art and new opportunities in the age of the photons. *CIRP Annals* 51 (2): 531–550.

35 Kaiser, M. (2001). Thermal–mechanical modeling of laser ablation hybrid machining. Ph.D. thesis. Texas Tech University.

36 Komanduri, R., Lee, M., Flom, D.G. et al. (1982) Pulsed laser pretreated machining. US Patent 4,356,376.

37 Walter, C., Rabiey, M., Warhanek, M. et al. (2012). Dressing and truing of hybrid bonded CBN grinding tool using a short-pulsed fibre laser. *Annals of the CIRP* 61: 279–282.

38 Tian, Y. and Shin, Y.C. (2007). Laser-assisted burnishing of metals. *International Journal of Machine Tools and Manufacture* 47: 14–22.

20

Selective Laser Sintering

Jordi Delgado, Lidia Serenó, Karla Monroy, and Joaquim Ciurana

Department of Mechanical Engineering and Industrial Construction, Product Process and Production Engineering Research Group, School of Engineering, University of Girona, Girona, Spain

20.1 General Overview

Selective laser sintering (SLS), developed and patented at the University of Texas at Austin in the mid-1980s, is an additive manufacturing (AM) technology used to produce three-dimensional (3D) physical objects, layer by layer, from computer-aided design (CAD) models [1]. The technology is based on a laser beam that selectively irradiates and sinters consecutive layers of material, in powder form, to obtain a specific 3D product. Therefore, any powder material capable of absorbing electromagnetic radiation could be used in the SLS process to manufacture final parts or products. SLS, considered one of the most important AM manufacturing processes, may shorten the manufacturing cycle time, reducing production costs and promoting competitiveness, compared to traditional technologies [2–4].

SLS was firstly commercialized by DTM Corporation in 1992 [5]. Nowadays, 3D Systems, who bought DTM in 2001, and Electro Optical Systems (EOS GmbH) are leading the SLS market worldwide commercializing SLS machines and direct metal laser sintering (DMLS) machines, respectively. The greatest potential of these machines lies in their ability to produce specialized and functional products.

20.1.1 Selective Laser Sintering Process

The SLS process uses the energy of a laser to selectively sinter powder into a predefined shape, layer by layer. Figure 20.1 shows the different parts of an SLS machine, including a computer to control the system, a laser with a scanning mechanism, a system to spread the material, the feeding platform, and the building platform. Due to the high tendency to oxidation, an inert atmosphere is required during the entire process.

The process, common for all AM technologies, begins with the generation of a CAD model of the desired object. This 3D CAD model can be obtained by the following: (i) 3D modeling (CAD systems); (ii) importing medical data acquired through scanners such as computer tomography (CT) or magnetic resonance imaging (MRI); or (iii) using reverse engineering. The 3D CAD model is saved as STL (STereoLithography) format, the standard format for AM technologies, based on triangulation that defines the contours of the entire object. More efficient extended STL formats are being implemented to allow printing complex geometries or

Modern Manufacturing Processes, First Edition. Edited by Muammer Koç and Tuğrul Özel.

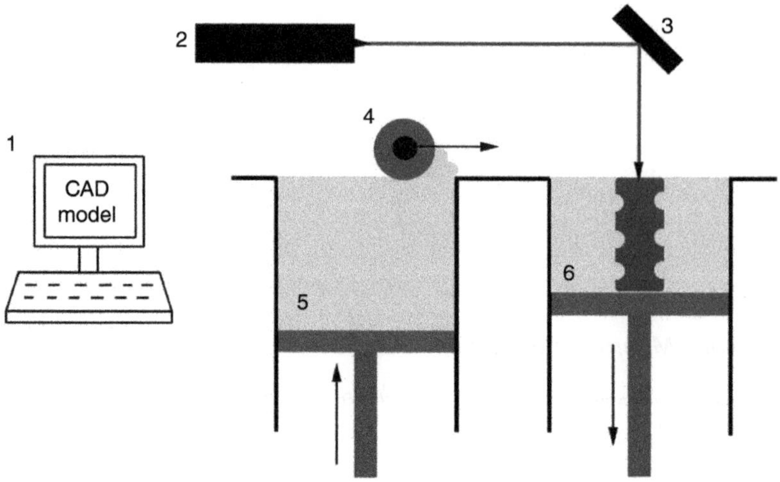

Figure 20.1 Basic components of an SLS machine: (1) computer; (2) laser; (3) scanning mirrors; (4) roller/sweeper; (5) material platform; and (6) building platform.

adding to the file some extra features such as densities, colors, gradients, etc. [6]. Afterward, the file is processed to select the best growth direction, which influences the dimensional accuracy and production time, and is mathematically sliced into thin orthogonal layers to the direction of growth, which determines the path of the scanning beam laser and the thickness of the layers. At this point, the data are sent to the AM machine, which transforms it into internal language.

The SLS process then begins with the spreading of a controlled amount of powder from the material feeding platform into the building platform by a roller o sweeper tool. This area is leveled by the same spreading mechanism to compact the powder and homogenize the surface (powder bed). The powder is typically preheated by a heating mechanism just below its melting temperature to reduce the shrinkage, thermal distortion, porosity, and power laser requirements and also to improve wettability and absorption of the laser. The most common heating mechanisms are infrared heaters, feed cartridges, and resistive heaters [7].

Following, a high-power laser generates a beam that scans a given area, which corresponds to a cross-sectional surface of the sliced CAD model. The laser beam, controlled by a computer and integrated with a scanning system, heats the irradiated zone enough to melt partially or totally the particles of the material and to obtain a solid two-dimensional part. Once the process is finished for the specific layer, the material platform moves up while the building platform moves down, and a new layer of powder is deposited in the building platform by the spreading mechanism. By carefully controlling the energy of the laser, each layer of sintered powder is fused to the underlying one. The process is repeated until the final three-dimensional product is completely sintered. Afterward, the unsintered loose powder may be removed, using brushes, and recycled; whereas the final product may be drawn out of the container. The powder not sintered acts as a support structure during the sintering. Thus, SLS process does not need extra support to produce overhanging parts, which facilitates the postprocessing and enhances the productivity of the process [8]. Several postprocess such as machining, polishing, grinding, coating, painting, thermal processes, etc. can be applied to improve the quality of the final product sintered [9, 10]. Several materials, such as polymers, metals, or ceramic powders, can be used to build products with SLS technology.

20.1.2 Lasers

The most important element in the SLS technology is the laser. Lasers, the power source used by SLS processes, are characterized by their coherence, monochromatism, and high directionality. Up to four components are needed to produce a laser beam: (i) an active medium, where the stimulated emission occurs, confined within a cavity. It determines the wavelength of the laser; (ii) an excitation mechanism, which provides the energy necessary to stimulate the active medium; (iii) a feedback mechanism, which consists of two reflective mirrors at each end of the optical cavity. This mechanism is responsible for the amplification of the light intensity and the direction of the beam; and (iv) optical coupler, which transforms the amplified light into a beam.

The underlying concept of laser operation is based on a stimulated emission. However, three mechanisms (population inversion, stimulated emission, and amplification) are required to produce a high-energy laser beam. During population inversion, electrons (from any molecule, atom, or ion) are optically or electrically pumped from their lowest energy level to a higher one. Lifetime of electrons in their higher level is generally very short, and they rapidly decay into lower energy levels producing emission of laser radiation (photons). Stimulated emission results when incoming photons (stimulating photons) interact with the excited molecule, atom, or ion with population inversion. These photons bring the electrons back to their lower energy state, resulting in emission of radiation. Amplification of laser light is accomplished by a set of well-aligned highly reflecting mirrors at the cavity. The photons move back and forth within the cavity stimulating more and more emissions. Usually, one of the mirrors is not fully reflective allowing the laser beam escape from the cavity. If the laser beam is maintained over time, then it is considered a continuous laser. On the other hand, if the laser beam is emitted discontinuously over time, then it is considered a pulsed laser.

Lasers are generally classified into four different types based on the physical nature of the active medium used: solid-state lasers (such as the Nd:YAG laser), gas lasers (such as the CO_2 laser), semiconductor lasers, and liquid dye lasers.

20.2 Mechanisms

SLS process can be further divided into two different mechanisms or approaches termed indirect and direct SLS.

20.2.1 Indirect Selective Laser Sintering

This sinterization mechanism consists of a mixture of metallic or ceramic materials with a polymer. The metallic or ceramic powder is coated with the polymeric one, named polymer binder. Then, the polymeric powder is sintered by a low-power laser beam, provoking the agglutination of the metallic or ceramic powder within the polymeric mass. Actually, the metallic or ceramic powder particles are bonded together by the polymer. The resulting structure (green part) is highly porous and has poor mechanical strength. To improve its mechanical behavior, several postprocesses need to be applied including postsintering and infiltration [11]. For the postsintering process, the green part is introduced into a controlled furnace to increase its compaction and density and remove the polymer binder. For the infiltration postprocess, a metal with low melting point is spread into the porosities by capillary forces. Once the infiltrating metal reaches its melting point, it is absorbed by the open pores producing a high-density metallic part with metal–metal bonds. Eventually, final parts can be subjected to machining process for dimensional adjustments.

The indirect SLS has some drawbacks associated such as long postprocessing steps and the quality and accuracy of the resulting part manufactured.

20.2.2 Direct Selective Laser Sintering

In direct SLS mechanism, a metallic or ceramic powder is sintered without adding any polymeric binder. Two direct SLS approaches can be found:

20.2.2.1 Solid-State Sintering (SSS)

During Solid-State Sintering (SSS), material particles soften and form solid bonds without reaching their melting temperature, specifically between $T_m/2$ and T_m [12]. However, the temperature needed to create these solid bonds depends on the material and its particle size. Before SSS, the particles are compacted to collapse pores and enlarge particle contact. After laser beam irradiation, the densification process starts with the approximation of two particles [13]. By reducing the surface energy between particles, an interparticle neck is formed and starts growing. The average particle size increases, whereas the number of particles decreases. In SSS, the thermal energy provided by the laser beam is enough to overcome energy barriers and provoke atomic motion. Therefore, the diffusion of vacancies is the driven force governing this SSS process. Higher vacancy concentration is found in the interparticle neck producing a flow of atoms from the surface of the particles to the neck resulting in an increased neck size. Finally, the two particles become a single particle, twice its initial size. Additionally, it is common to observe sintering with simultaneous densification. SSS is considered a slow process and generally associated with high porosity [13]. Besides being a very slow process, one of the main advantages of SSS is the wide variety of materials that can be processed this way. In fact, if the temperature provided by the laser is enough to induce diffusion of vacancies across particle boundaries, all powder materials are able to be processed by SSS. However, to obtain the required final properties, postsintering operations are required (Figure 20.2).

20.2.2.2 Liquid-phase Sintering (LPS)

Liquid-phase sintering (LPS) is a sintering mechanism based on bonding particles by heating them near their melting temperature. The material used consists of a mixture of two metal powders, one with high melting point (structural) and another with low melting point

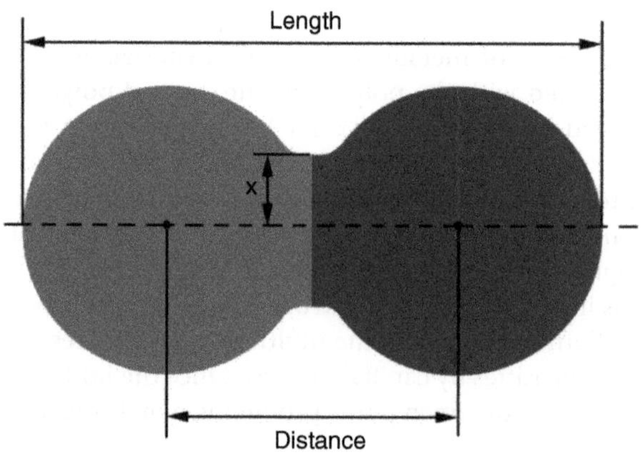

Figure 20.2 Solid-state sintering neck formation.

(binder). When the low melting point material approaches to its melting range, the sintering process begins. First, the application of a laser beam causes the melting of the binder material with lower melting point, which in turn, starts solubilizing the high melting point material. Usually, this liquid phase represents the 5–15% in volume [14]. At the sintering temperature, the solid is soluble in the liquid. During this phenomenon, the liquid wets the solid particles, providing a capillary force that attracts the particles together dissolving the sinter bonds and inducing grain rearrangement [15]. Simultaneously, the high temperature softens the solid grains while pores collect between grains and get wetted by the liquid, promoting densification of the structure and increasing its rigidity [16]. An appropriate amount of flow is critical to eliminate porosity, to further produce a highly dense part and maintain dimensional accuracies. The appropriate amount of flow is controlled by factors such as atmosphere, powder bed temperature, and laser energy density. However, when there is no solubility between the liquid phase and the solid grains, the liquid does not interact with the solid grains but acts only as pore filler [17, 18]. The final product obtained by LPS is a composite with customized properties and a microstructure composed of solid grains with a solidified liquid network and residual pores. This process is mainly driven by capillary forces; however, the velocity of the reaction is determined by the solid–liquid transformation [19].

LPS is an ideal process capable to densify hard materials such as cemented carbides (WC–Co) that cannot be processed using other manufacturing approaches. Other processable materials are as follows: porous bronze (Cu–Sn), tungsten heavy alloys (W–Ni–Cu), copper steels (Fe–Cu–C), and cermets (TiC–Fe). LPS is also suitable to process single powders with two different grain sizes: slight and coarse grains.

20.2.2.3 Selective Laser Melting (SLM)

All aforementioned mechanisms are characterized by producing final parts with poor mechanical properties. Therefore, to obtain high density functional products, selective laser melting (SLM) was developed [20]. SLM is considered a variation of the classical SLS because (i) it entirely brings to fusion the powder material, and (ii) it does not need low melting point elements. During SLM, near-full density products can be produced without special post-treatments; however, a high-power laser is required. Besides, polymers can be completely fused by a laser beam, the appellation SLM is reserved for metallic materials [12].

In SLM, all metallic particles convert from a solid to a liquid state in a very short period of time. The process is normally accompanied by the appearance of residual thermal stresses and the balling effect [21]. The thermal stress can cause distortion of the product, cracks, or even delaminations [22, 23]. The balling effect is the term used to define the formation of isolated spheres of molten material, which decrease the density of the product resulting in unacceptable surface finishes. It occurs when the fused material is unable to fully wet the substrate because of surface tension [7]. In fact, an excessive amount of thermal energy reduces the viscosity of the melted powder [24]. In order to control the process, several parameters (such as laser absorption, surface tension, and viscosity) need to be evaluated for each material. Therefore, the range of commercial SLM materials available is still limited.

At any event, SLM offers the possibility to produce highly complex products with different types of materials (such as stainless steel, titanium alloy, and cobalt-chromium alloys) and with good accuracy for nominal dimensions over 0.4 mm [7]. The final products obtained have almost full density and mechanical properties similar to those assured by traditional technologies. Moreover, SLM can provide customized final products without requiring special surface finishes. On the other hand, this process results not competitive for large batches due to it is quite slow and very expensive.

20.3 Process Parameters

Although SLS is conceptually easy to understand, a great multitude of parameters are involved and influence the process making it very complex to study. The SLS process is mainly based on complex phenomena such as absorption and scattering of laser radiation, heat transfer, phase transformation, exchange between solid and liquid particles, surface tension gradients, evaporation and material emissions, and chemical reactions [25, 26]. Therefore, SLS can be considered a multifaceted process that requires a thorough research and analysis in order to control several parameters including the melting or sintering mechanism, the physics of the melt pools generated, and the main quality attributes (surface roughness, strength, accuracy, hardness, density, and residual stresses). The process parameters are different variables involved in the sintering or melting of a material, and they are used to measure the process.

SLS can be defined by a large number of parameters that may affect the process [25, 27]. However, only a small number of these parameters can critically influence the final properties of the part. Up to 13 are considered crucial to obtain an optimal quality and accuracy [28, 29]. This influence can be studied by evaluating output parameters such as mechanical, physical, metallurgical, and dimensional properties, which in a whole contributes to the building of knowledge that would finally help to control the final properties of parts produced by SLM [30].

The basic principle of the laser beam interaction with the powder bed during SLS is influenced by a number of factors that affect the amount of energy delivered into the surface of the powder, which in turn affects the stability of the molten metal and the grip of the powder to the substrate. Therefore, an adequate combination of the process parameters is critical to ensure the consolidation of all the layers into a single piece [26, 31]. SLS parameters can be classified according to geometric parameters, laser parameters, or material properties (Figure 20.3)

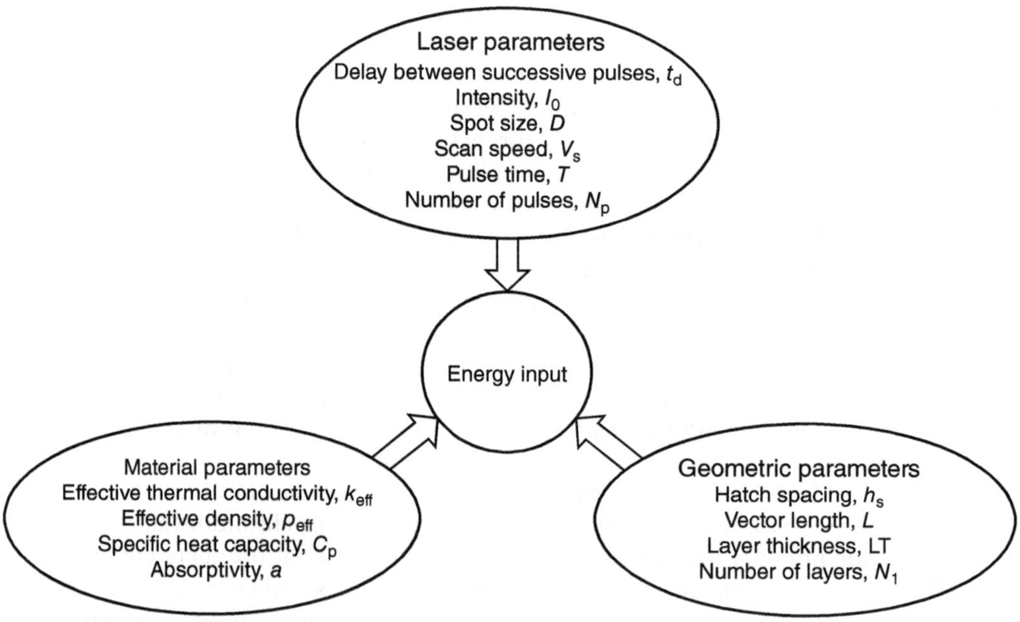

Figure 20.3 Factors that affect the laser sintering process. *Source:* Dahotre and Harimkar 2008 [29]. Reproduced with permission of Springer.

20.3.1 Geometric Parameters

Hatch spacing, scanning strategy, and layer thickness are considered typical geometric parameters.

20.3.1.1 Hatch Spacing

A typical scan pattern is composed by parallel scanning tracks, and the distance between two adjacent tracks is defined as hatch spacing. Depending on the distance between tracks and the track width size, superposition of them (overlapping) may occur. If the separation of two tracks is very large, no intersection between them occurs, producing two distinct strands. In order to obtain a layer of material sintered and not single tracks, hatch spacing is usually less than the laser beam diameter [29] and is associated with the energy density received by the powder material [32].

The overlap between two tracks can be evaluated in two ways: (i) taking into account the laser beam diameter and the separation between the two parallel laser beams and (ii) taking into account the track profile, knowing that the geometry of the track varies depending on process parameters [33]. Results have shown that when hatch spacing is increased, the mechanical properties and density of the parts produced are decreased due to empty spaces [32]. On the other hand, with overlapping of tracks, a wavy surface is obtained, and therefore, mechanical properties are affected due to generation of porosity in the part manufactured [34]. When combining low hatch spacing with low scan speed, thicker parts can be obtained (99.98%). In opposition, with higher hatch spacing and scan speed less dense parts are obtained (95.11%) due to porosity.

20.3.1.2 Scanning Strategy

The scanning strategy is a set of actions based on the definition of laser trajectories and sequences to follow up in order to sinter an object. Usually, each SLS machine has specific software that performs these actions and that is optimized for each material used, reducing scanning time or influencing in some properties of the final part produced. The scanning strategy determines the scanning speed and direction, meaning that mechanical properties, precision, stress, and scanning time will be influenced by it [35].

Figure 20.4 shows three different scanning strategies or patterns used in laser sintering process. Scanning patterns may also affect the quality of the surface leading to surface roughness [37].

Scanning strategy describes the orientation of the manufactured parts. Therefore, whether to use supports or not, the time needed to manufacture the part, and the final mechanical properties will be determined by the orientation. Thus, in order to increase the tensile strength,

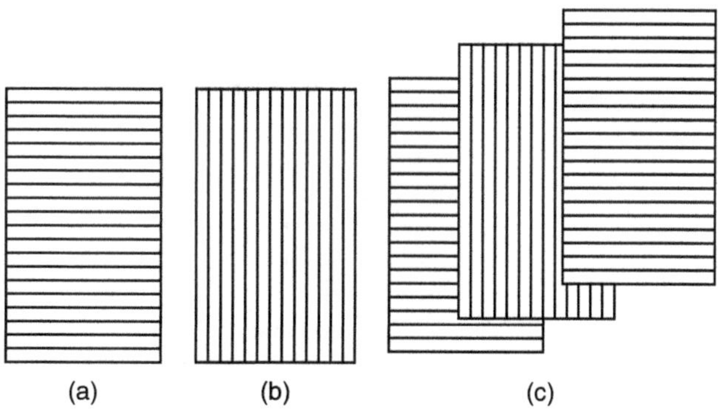

Figure 20.4 Examples of scan patterns, (a) sorted in *X*-direction; (b) sorted in *Y*-direction; (c) sorted alternating *X* and *Y* directions. *Source:* Simchi and Pohl 2003 [36]. Reproduced with permission of Elsevier.

the parts must be oriented with minimal sectional area in order to avoid distortion and the curling effect [32]. To reduce internal stresses, the best strategy involves dividing the area in different parts: scanning the workpiece along its width, scanning the workpiece along its length, scanning the workpiece randomly, or scanning the workpiece trying to minimize the thermal effect across the sectors [38].

Scanning strategy may be defined using vectors. Large vectors provoke a reduction of the temperature reducing its wettability [20]. On the other hand, short vectors show more stable and uniform temperatures, resulting in better mechanical properties [32]. Moreover, the orientation has great influence on the fabrication time, impacting on the part final price.

20.3.1.3 Layer Thickness

The layer thickness is the height of a layer of material that has been deposited in the construction platform. Most technologies operate by a constant and uniform layer height [39], normally around 100 μm for polymer and from 20 to 40 μm for metallic materials. However, layer thickness can be also variable according to the design or features of the 3D model [40, 41]. For instance, for micrometric scale pieces, it is necessary to work with nanometers [11].

In order to obtain an optimal, constant, and uniform layer throughout the entire platform, the layer thickness should be larger than the powder particle size [12]. The maximum layer height is limited by the penetration of the laser into the powder, which is based on laser power, energy density, particle size, power density, heat capacity, and powder thermal conductivity.

As shown in Figure 20.5, the layer thickness parameter influences the generation of a staircase effect in objects without vertical or horizontal walls [32]. As the layer height increases, the staircase effect augments, worsening the quality of the object but lowering the production time and cost [43, 44]. Therefore, layer thickness is an important parameter to consider in seeking a balance between quality and costs.

Surface finish and roughness are the major problems in AM technologies [45]. These quality characteristics can be improved by using smaller layer thicknesses, which due to their high thermal conductivity and the presence of less powder produce more stable molten pools and reduced staircase effect. Although some authors differentiate the surface roughness as the roughness being caused by the grains and the roughness of the layers texture, most authors do not use the two terms isolated but instead they evaluate roughness as the surface texture stated in the ISO [45].

20.3.2 Laser Parameters

Laser parameters can be divided into scanning speed, laser power, and beam size. Laser parameters can also include pulse parameters such as pulse length and the number of pulses per time unit.

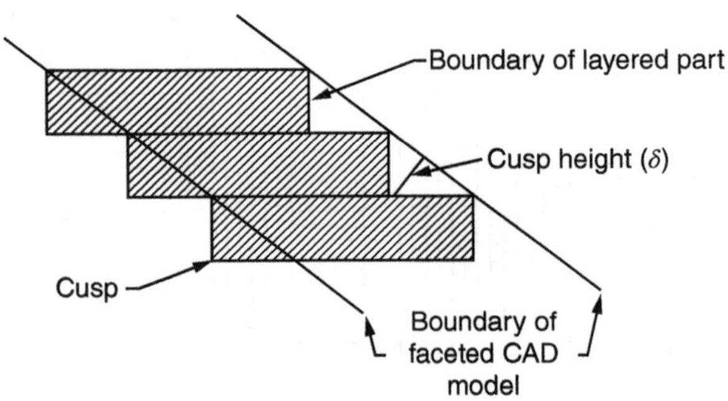

Figure 20.5 Staircase effect. *Source:* Dolenc and Mäkelä 1994 [42]. Reproduced with permission of Elsevier.

20.3.2.1 Scanning Speed

Scanning speed is described by the speed of the laser beam moving within the powder bed. Usually, the movement of the laser beam is performed using galvanometric mirrors of high accuracy, which are properly positioned to cover the entire surface.

Several studies have been developed to study the influence of the scanning speed in the SLS processes [25, 46]. Results showed that when scanning speed increased, the material fused (molten pool) decreased, due to the laser bean irradiated the material more superficially (less deeply). Moreover, it was found that the contact angle of the laser beam decreases when scanning speed increases, producing a lower wetting ability, which can promote pore generation. Finally, by controlling the scanning speed, different porosities and tensile strengths can be obtained [46]. When the scanning speed decreases, high density parts are obtained due to the increased time spent by the laser beam on the material. This effect lowers the viscosity of the material allowing the air to escape, augmenting the adherence between layers, and increasing resistance. Thus, porosity and tension can be controlled by modifying scanning speed.

20.3.2.2 Laser Power

The laser power is determined as the energy of the laser beam per unit of time. It is defined as the distribution of the power density of the laser beam (for continuous wave lasers) or the energy distribution (for pulsed lasers). In both cases, it is referred to the energy supplied by the laser. The power density is physically equivalent to the irradiance. Both irradiance and power density are measured in watts per unit of area. The power density is generally used to describe the distribution of the radiation within a beam, while the irradiance is generally used to describe the distribution of radiation incident on a surface.

The energy density is based on the laser power and irradiated area. When the energy density is low, the samples have a typical large pore structure. When the energy density increases, a combination between sintering and melting processes produce two types of structures: a dense structure (SLM) and another with porosity (SLS) [47]. For each material, there is an optimal and appropriate laser power [48]. This power depends on the wavelength and duration. If the laser exceeds the optimal power, the high temperatures may provoke the evaporation of the sample, losing material and consuming unnecessary energy. Therefore, laser power should be set to ensure the heating of the material up to temperatures below its melting point because laser energy brought by the beam has important effect on sintering result [32].

20.3.2.3 Beam Size

Beam size is defined as the diameter of the laser spot. Usually, the diameter of the sintered area is larger than diameter of the laser beam. Beam size parameter is associated with accuracy of the object manufactured, and it is an important value to determine energy input [49].

20.3.3 Material Properties

In the laser sintering process, there are wide range of materials used, for example polymers, metals, ceramics, and cermets can be used. The interaction of these materials and their properties with the laser have an obvious effect on the working conditions, energy input, part throughput, and postprocessing during the laser sintering process. Some material properties and their influence to laser sintering processes are presented on Table 20.1.

Other phenomena such as the absorption and reflection during the laser material processing are in an extremely important role. As, laser sintering is based on absorption of laser energy, when laser bean incidences powder, it is partly reflected from surface and partly absorbed. Each time laser beam hits particle, part of laser energy absorbs and part reflects. With thin

Table 20.1 Material properties and effect on the laser sintering process.

Property	Significance to SLS process
Particle size distribution (mean and width)	Sinterability, packing efficiency, surface roughness
Particle shape	Packing efficiency
Apparent and tap densities	Packing efficiency
Flowability	Uniform spreading of powder layer
Melting point	Indicator of energy requirements
Strength of green part	Facilitates component handling prior to thermal processing (debinding, sintering, infiltration)
Burnout	Decomposition temperature and ash content of binders
Specific heat, thermal conductivity	Heat transfer in powder bed
Void fraction	Residual porosity in prototypes

Source: From Atre 2001 [50]

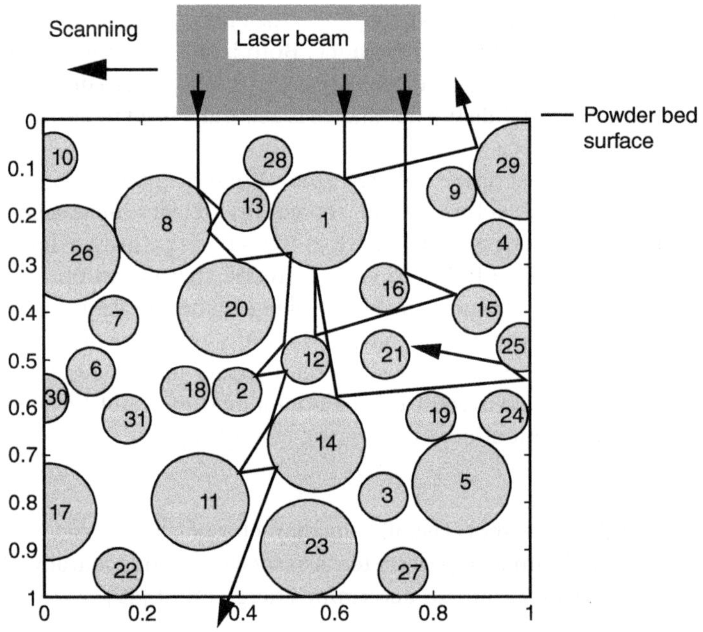

Figure 20.6 Reflection from powder particles. *Source:* From Atre 2001 [50].

enough layer of metallic materials, part of laser energy can be transmitted through material [51, 52]. There are few properties that have to be considered as they have an effect on the absorption and reflection: wavelength, temperature, surface films, angle of incidence, and the material and surface roughness (Figure 20.6).

20.4 Materials

The success of SLS as an additive manufacturing technology results mainly to the possibility to process almost any type of materials to accommodate multiple applications throughout the manufacturing process. Perhaps any powder material that is susceptible to be sintered by thermal energy can be used [52]. The type of material used depends on the final application.

The most important material used in SLS processes are polymers, metals, and ceramics. The interaction of these materials and their properties with the laser affects the working conditions, the energy input, the throughput, and postprocessing.

20.4.1 SLS of Polymers

Polymeric materials were the first and are still the most widely used materials in SLS. Among the various types of polymeric materials, thermoplastics are the most suitable. Different types of thermoplastic polymeric materials include amorphous, semicrystalline, and reinforced or filled polymers [29, 52].

Amorphous polymers consist of long-chain molecules that are arranged in a random manner. The most important characteristic of amorphous polymers is their glass transition temperature, T_g. During SLS, the amorphous polymers are preheated to the temperature slightly less than T_g such that subsequent irradiation with the laser transforms the polymer from rigid glassy state to highly viscous liquid state by increasing the temperature above the glass transition point [29–32]. Initially, amorphous polymers such as polycarbonate powders (PC) were used as starting materials for both experimentation and modeling in the SLS process. Polycarbonates are able to produce parts with very good dimensional accuracy, resolution, and surface finish. However, they can also be related with incomplete sintering problems limiting their use to obtain products for low strength and durability applications [52]. Typical applications for amorphous polymers are the manufacturing silicone rubber and cast epoxy molds [53]. Nevertheless, with recent technology advances, crystalline and semicrystalline materials have been introduced to the SLS process.

Crystalline polymers consist of long-chain molecules arranged in an orderly structure. Perfect or full crystalline polymers are not generally encountered in nature, and therefore, they usually exist in semicrystalline state [29]. During SLS, these polymers are preheated to the temperature slightly below to their melting point. In fact, they can be sintered or melted using a relatively low thermal energy input. SLS of semicrystalline powders is associated with high dense parts with very good mechanical properties; even very close to those obtained using injection processes, making them suitable for high-strength functional applications. However, when approaching to their melting point, semicrystalline polymers also exhibit large specific volume changes resulting in significant shrinkage and distortion, approximately in the order of 3–4% [32]. The most common crystalline polymers used in SLS are nylon (neat, glass-filled, or with other fillers) and polystyrene. They are used to obtain durable and functional parts including snap fits and living hinges and are a good choice for high heat and chemically resistant applications. New grades of nylon can produce very good resolutions and roughness, making them suitable for casting of silicone rubber and epoxy molds. Other polymers available are acrylic styrene (polymethyl methacrylate/polystyrene [PMMA/PS]) for investment casting and elastomer for rubber-like applications [52].

20.4.2 SLS of Reinforced Polymers

Recently, significant interest has been directed toward the SLS of reinforced polymers. Polymer powder can be mixed with other materials to rapidly fabricate composite parts with improved thermal and mechanical properties. Most of the work developed so far has been focused on glass-reinforced polyamide and metal-reinforced polyamide. Compared to plain polyamide, reinforced polyamide (with copper) parts are three and a half times heavier and four times more thermally conductive. They also exhibit similar tensile strength. The copper-reinforced polyamide is used as high-strength material for inserts for injection molds [29, 52]. The utilization of sintered reinforced polymers for advanced medical applications is being studied. SLS can allow the direct fabrication of customized implants for solid and porous bone replacement. Moreover, it can be used to obtain implants of controlled porosity (scaffolds), such as the fabrication of hydroxyapatite–reinforced high-density polyethylene composite (HA-HDPE), which is a bioactive ceramic polymer composite material [29].

20.4.3 SLS of Metals and Ceramics

When high mechanical properties are required, metals and ceramics are considered the most adequate materials to use [29]. Different metals and ceramics are used in SLS processes such as SLM, LPS, or indirect SLS mechanisms.

In the case of metals, the powders can be composed by two different materials or only one material. For the latter, the selection of the working conditions is more critical because the metal must be fully molten, being much more complicated to control in comparison with combined materials where one of the metals is always in solid state, being structurally more stable. The condition where single-metal powders can be processed by SLS must be such that full melting is achieved however excessive heating could lead to irregularities on the final part [30, 54]. The direct SLS process based on SSS uses commonly single component powders and is generally a slow process and almost always associated with high porosity in the final built part. However, if strong and dense metallic parts are desired, high density parts can be fabricated by irradiating an integral metal powder without adding low melting point elements with high-power laser such that a fully selective melting (SLM) takes place [33]. For SLM mechanism, where the material is fully melted, many metals are compatible. The most important are the following: stainless steel, aluminum, cobalt chrome, nickel chrome, steel, titanium, and gold alloys.

20.4.3.1 Stainless Steel

Stainless steel GP1 alloy is characterized by having good corrosion resistance and mechanical properties, especially excellent ductility in laser processed state. It is widely used in a variety of engineering applications. Stainless steel PH1 alloy is characterized by good corrosion resistance and excellent mechanical properties, especially in the precipitation hardened state. This PH1 alloy is widely used in variety of medical, aerospace, and other engineering applications requiring high hardness, strength, and corrosion resistance. Both alloys are ideal for many part-building applications such as functional metal prototypes, small series products, individualized products, or spare parts. Stainless steel alloys are usually processed by SLM using a layer thickness of 20 μm. However, it is also possible to use 40-μm layer thickness to increase the manufacturing speed.

20.4.3.2 Aluminum

AlSi10Mg is a typical casting silicon/magnesium alloy with good casting properties used for manufacturing parts with thin walls and complex geometries. The alloy has very good strength and hardness. It also has good dynamic properties and good thermal properties with low weight. Therefore, it is used to manufacture parts subject to high loads. After sintering, parts can be machined, wire eroded and electrical discharge machined, welded, microblasted, polished, and coated.

20.4.3.3 Cobalt Chrome

The fine powder mixture of cobalt-chrome-molybdenum–based superalloy is characterized by having excellent mechanical properties (strength and hardness), corrosion resistance, and temperature resistance. It is characterized by a fine and uniform crystal grain structure. This kind of alloys are sterilizable, and therefore, perfectly suitable for biomedical prototype applications (dental and medical implants). Moreover, they are also used to manufacture parts with very complex geometries and parts used for high-temperature engineering applications, such as in aeroengines. This material is also ideal for many part-building applications such as functional metal prototypes, small series products, individualized products, or spare parts. Standard processing parameters include SLM with 20-μm layer thickness.

20.4.3.4 Nickel Chrome

Inco 718 is a precipitated-hardened nickel-chromium alloy characterized by having good tensile, fatigue, creep, and rupture strength at temperatures up to 700 °C. This alloy has outstanding corrosion resistance in various corrosive environments. Therefore, it is ideal for many high temperature applications such as gas turbine parts, instrumentation parts, and process industry parts. It also possesses excellent cryogenic properties. Among all metallic materials listed, they exhibit the best densified condition after melting. Standard processing parameters use SLM typically with 20-μm layer thickness. Parts built using EOS Nickel Alloy IN718 can be easily posthardened to 40–47 HRC (HRC, hardness Rockwell C) (370–450 HB) by precipitation-hardening heat treatments.

20.4.3.5 Steel

Maraging steel MS1 is characterized by having very good mechanical properties and being easily heat-treatable using a simple thermal age-hardening process in order to obtain excellent hardness and strength. It is ideal for many tooling applications such as tools for injection molding, die casting of light metal alloys, punching, extrusion, and also for high-performance industrial and engineering parts, for example in aerospace and motor-racing applications. Standard processing parameters include SLM typically with 40-μm layer thickness. Parts built using Maraging steel MS1 can be easily posthardened to more than 50 HRC by age-hardening at 490 °C for six hours.

20.4.3.6 Titanium

Ti6Al4V ELI is a light alloy characterized by having excellent mechanical properties and corrosion resistance combined with low specific weight and biocompatibility. It is ideal for many high-performance engineering applications, such as in aerospace and motor racing, and also for the production of biomedical implants. Standard processing parameters use SLM mechanism. Using standard parameters, the mechanical properties obtained are fairly uniform in all directions. Parts made using this material can be machined, spark-eroded, welded, micro shot-peened, polished, and coated if required, and the unexposed powder can be reused.

Alternatively, for LPS mechanism, where two different components are found, the following materials are the most commonly used: alloys (Fe–Cu, Fe–Sn, Cu–Sn); metals (Al, Cr, Ti, Fe, Cu), ceramics (Al_2O_3, FeO, NiO, ZrO2, SiO_2, CuO), and other alloys (bronze, nickel, Inconel 625) [6, 55–57]. The mechanical properties and characteristics obtained using LPS are based on the weakest composite phase [10]. Therefore, research has been extended for the direct fabrication of hard metal and heavy duty parts, which are used as inserts for stamping dies, deep drawing dies, and cutting tools. Various combinations of powders, such as WC–Co, TiC–Ni/Co/Mo, TiCN–Ni, TiB_2–Ni, ZrB_2–Cu, and Fe_3C–Fe, are now used by LPS mechanism [29].

Indirect SLS mechanism is based on metals or ceramics sintered using a binder material. Organic materials, such as polymers or inorganic materials can be used as binder materials. However, thermoplastic polymers (nylon, polyesters, waxes, nylon, and polycarbonate mixtures) are the most common binding materials used. In indirect SLS, structural metallic or ceramic powders are mixed or coated with binding materials prior to laser sintering. Ammonium dihydrogen phosphate and aluminum can be used as binders, while copper can be used to infiltrate indirect SLS steel parts in order to obtain products with very good tensile strength and conductivity. These products can be easily finished by surface grinding, milling, drilling, polishing, and surface plating postprocesses. Recently, high interest is directed toward the infiltration of water-suspended colloids such as colloids of alumina or silica in the polymer-bound ceramic parts. After drying the colloids, the ceramic parts are burned to eliminate the polymer binder. Infiltration of alumina colloid into alumina ceramic parts tends to substantially improve the part strength during debinding and sintering. Some ceramic parts can also be fabricated directly by SLS without the use of binders, for example, SiC ceramics with chemically induced binding [29].

20.5 Capabilities and Limitations

Market is constantly changing, inducing companies to innovate in new technologies and respond, as quickly as possible, to the increased competition. Market trends are based on product complexity and customization due to customer preferences, along with the demand for lower costs and production times. This situation has motivated the development of manufacturing technologies to make fast, accurate, and cost-effective high-quality products with high added value. An innovative technology that allows accomplishing a wide range of manufacturing requirements is the so-called additive manufacturing (AM). Among the technologies that use AM, SLS is becoming the most demanded. SLS has the ability to produce functional parts with comparable mechanical properties of parts made by traditional processes [11, 58].

Since its introduction to the market, the process seems promising given that its main advantage is the ability to manufacture parts with a significantly higher complexity than traditional processes, without molds and tools, and with a very low wastage of material. As no tools are required for production, the design restrictions normally associated with the subtractive and forming methods are significantly reduced, providing SLS advantages in the design, implementation, and performance of the products. Geometries that were difficult or impossible to manufacture using conventional techniques can be now produced without complications. While in traditional manufacturing process, there is a removal of up to 95% of the material to reach a finished component, SLS machines only use the exact material needed to build the part. Such characteristics are the reason why SLS have lower manufacturing costs and increased performance over conventional processes. Furthermore, no tooling costs are present, and therefore, the cost per part is no longer dependent on the production volume, reducing the development costs by 40–70% and the time to market by up to 90%. Similarly, the use of standard materials has provided new possibilities in creating prototypes and direct manufacturing parts. In consequence, this technology promise to revolutionize many traditional manufacturing processes by providing a new way of designing and developing final pieces shortening design cycles and production due to its additive nature.

SLS is not only faster and cheaper than conventional processes in many areas but it also allows the development of new geometries as internal geometries such as network or mesh structures, porous, channels, or irregular patterns. It permits the creation of cellular structures that offer network structures with a porosity geometrically ordered to create lighter parts. As a result of its potential and capacities, developers of all industries not only see the technical possibilities of the SLS technology in building pieces of metal or plastic but also to include it in their plans mass manufacturing.

Given this potential, many companies are experimenting with SLS technology. However, despite its advantages, several recurring problems have been found, including porosity, balling effect, low surface quality, poor mechanical properties, oxidation, and shrinkage [58–60]. Furthermore, as the technology has been developed, new applications have been proposed at a faster rate than the range of materials that can be processed with this technology. Therefore, SLS materials can be considered a limitation. Likewise, the converting process of SLS into a robust manufacturing is now held back by inherent detriments such as precision, surface quality, limiting materials, lack of repeatability, and industrial regulations [61]. Therefore, despite many applications and advantages of this process, there has been found a knowledge gap in the SLS process, which needs to be solved. In this context, high accuracy can only be achieved in this process when the nominal dimensions are above 0.4 mm. Linked to the accuracy, sometimes it is required the design of supporting structures to anchor the parts and to prevent the movement during the sintering process. Therefore, avoiding surface distortion and obtaining a correct geometry. The use of these structures has been reported as a major constraint associated with the geometry of the parts. Moreover, surface finish and roughness are two main drawbacks of this technology. The additive component and the staircase effect generate roughness problems. Evidences have also shown that the environment during the laser sintering process requires a strict control as it is reflected in the surface quality of the part produced. During SLS processes, high temperature irradiating the sintered powder is need to reach the powder melting or sintering temperature, thereby, a natural tendency for material oxidation and degradation is present. This effect can be minimized by injecting an inert gas that displaces or decreases the oxygen content in the fabrication chamber [53, 55]. Moreover, in some SLS processes, an inefficient control of the atmosphere during the process can result in poor wettability between the liquid phase and solid phase, otherwise, when an efficient control is achieved, smoother surfaces and a decrease of balling effect occurs [12].

Currently, the technology can only fabricate fully dense parts without the use of post-processes, in a very limited range of materials. In theory, any material that is available in powder form can be processed by the laser sintering. In practice however, this is not the case [11, 28]. Most metals are not produced directly in powder form, and therefore, they must be converted to an appropriate particle size prior to sintering. Obtaining a correct particle size and morphology is vital because the properties of the powder affect processing capacity, deposition, sinterability, surface roughness, and porosity. Thus, the availability of material in powder form is the first major obstacle in the acquisition of suitable materials. In fact, it is highly complex to develop new materials and especially to find their optimal process parameters due to this kind of machines are barely flexible and open.

In SLM, to achieve parts with high density, the powder particles are completely molten. The laser melting process is accompanied by a development of residual stresses, resulting from the high heat gradient in the material [38]. The excess of energy can cause undesirable phenomenon as the vaporization of the material during melting process, since the temperature can rise beyond the melting temperature of the material. An additional problem is the "balling" effect. It is the formation of isolated spheres with a diameter equal to the laser beam; this occurrence inhibits the deposition density and gives the powder a very low melt viscosity [47]. Furthermore, laser sintering is a process where all particles sinter in a very short period of time. The process becomes more difficult to control due to the high heat transfers [28]. Consequently, the quality certification of SLS processes is very complicated and is considered an obstacle to solve to further implement efficiently and extensively SLS processes in the industry.

References

1 Beaman, J.J. and Deckard, C.R. (1989). Selective laser sintering with assisted powder handling. US Patent 4, 938, 816.
2 Edson, C.S., Masanari, S., Kozo, O., and Tahar, L. (2006). Rapid manufacturing of metal components by laser forming. *International Journal of Machine Tools & Manufacture* 46: 1459–1468.
3 Liu, J.H., Shi, Y.S., Lu, Z.L. et al. (2007). Manufacturing metal parts via indirect SLS of composite elemental powders. *Materials Science and Engineering A* 444: 146–152.
4 Asgharzadeh, H. and Simchi, A. (2005). Effect of sintering atmosphere and carbon content on the densification and microstructure of laser-sintered M2 high-speed steel powder. *Materials Science and Engineering A* 403: 290–298.
5 Cooper, K.G. (2001). *Rapid Prototyping Technology: Selection and Application*. New York: Marcel Dekker. ISBN: 978-0-8247-0261-8.
6 Yin, Z.W. (2011). Direct generation of extended STL file from unorganized point data. *Computer-Aided Design* 43: 699–706.
7 Campanelli, S.L., Contuzzi, N., Angelastro, A., and Ludovico, A.D. (2010). Capabilities and performances of the selective laser melting process. In: *New Trends in Technologies: Devices, Computer, Communication and Industrial Systems* (ed. M.J. Er). InTech. ISBN: 978-953-307-212-8.
8 Xu, F., Loh, H.T., and Wong, Y.S. (1999). Considerations and selection of optimal orientation for different rapid prototyping systems. *Rapid Prototyping Journal* 5: 54–60.
9 Rosochowski, A. and Matuszak, A. (2000). Rapid tooling: the state of the art. *Journal of Materials Processing Technology* 106: 191–198.
10 Panjan, P., Dolinsek, S., Dolincek, M. et al. (2005). Improvement of laser sintered tools with PVD coatings. *Surface and Coatings Technology* 200: 712–716. 1–4 Special Issue.
11 Dewidar, M.M., Lim, J.-K., and Dalgarno, K.W. (2008). A comparison between direct and indirect laser sintering of metals. *Journal of Material Science Technology* 24: 227–232.
12 Kruth, F.P., Mercelis, P., and Van Vaerenbergh, F. (2005). Binding mechanisms in selective laser sintering and selective laser melting. *Rapid Prototyping Journal* 11 (1): 26–36.
13 Agarwala, M., Bourell, D., Beaman, J. et al. (1995). Direct selective laser sintering of metals. *Rapid Prototyping Journal* 1 (1): 26–36.
14 Kang, C.H. and Yoon, D.N. (1981). Coarsening of cobalt grains dispersed in liquid copper matrix. *Metallurgical and Materials Transactions A* 12 (1): 61–69.
15 Huppmann, W.J., Riegger, H., Kaysser, W.A. et al. (1979). The elementary mechanisms of liquid phase sintering, I rearrangement. *International Journal of Materials Research* (Z. Metallkd) 70: 707–7013.
16 Shaw, T.M. (1993). Model for the effect of powder packing on the driving force for liquid-phase sintering. *Journal of the American Ceramic Society* 76 (3): 664–670.
17 Johnson, J.L. and German, R.M. (1996). Solid-state contributions to densification during liquid-phase sintering. *Metallurgical and Materials Transactions B* 27 (6): 901–909.
18 Johnson, J.L. and German, R.M. (2001). Role of solid-state skeletal sintering during processing of Mo-Cu composites. *Metallurgical and Materials Transactions A* 32 (3): 605–613.
19 German, R.M. (1996). *Sintering Theory and Practice*. New York: Wiley. ISBN: 978-0-471-05786-4.

20 Yang, J., Ouyang, H., and Wang, Y. (2010). Direct metal laser fabrication: machine development and experimental work. *International Journal of Advanced Manufacturing Technology* 46: 1133–1143.

21 Kruth, J.P., Froyen, L., Van Vaerenbergh, J. et al. (2004). Selective laser melting of iron-based powder. *Journal of Materials Processing Technology* 149: 616–622.

22 Fischer, P., Romano, V., Weber, H.P. et al. (2003). Sintering of commercially pure titanium powder with a Nd:YAG laser source. *Acta Materialia* 51 (6): 1651–1662.

23 Nickel, A.H., Barnett, D.M., and Prinz, F.B. (2001). Thermal stresses and deposition patterns in layered manufacturing. *Materials Science and Engineering A* 317: 59–64.

24 Kannatey-Asibu, E. Jr. (2009). *Principles of Laser Materials Processing*. New Jersey: Wiley. ISBN: 978-0-470-17798-3.

25 Van Elsen, M., Al-Bender, F., and Kruth, J. (2008). Application of dimensional analysis to selective laser melting. *Rapid Prototyping Journal* 14 (1): 15–22.

26 Yadroitsev, I. and Smurov, I. (2010). Selective laser melting technology: From the single laser melted track stability to 3D parts of complex shape. *6th International Conference on Laser Assisted Net Shape Engineering, LANE*, Erlangen, Germany (24 September 2010).

27 Childs, T.H.C., Hauser, C., and Badrossamay, M. (2004). Mapping and modelling single scan track formation in direct metal selective laser melting. *CIRP Annals Manufacturing Technology.* 53 (1): 191–194.

28 Rombouts, M. (2006). Selective laser sintering/melting of iron-based powders. Tesis U.D.C. 621.762. Katholieke Universiteit Leuven.

29 Dahotre, N.B. and Harimkar, S.P. (2008). *Laser Fabrication and Machining of Materials*, 558. New York: Springer. [e-book] Springer Science + Business Media, LLC. From: http://www.metapress.com/content/v01441.

30 Delgado, J., Sereno, L., Ciurana, J., and Hernandez, L. (2012). *Methodology for analyzing the depth of sintering in the building platform*. 5th International Conference on Advanced Research in Virtual and Physical Prototyping, VR@P 2011, Leiria, Portugal.

31 Delgado, J., Sereno, L., Ciurana, J., and Hernandez, L. (2012). Influence of process parameters in the first melting layer of a building platform in a SLM machine. *5th International Conference on Advanced Research in Virtual and Physical Prototyping*, VR@P 2011, Leiria, Portugal.

32 Gibson, I. and Shi, D. (1997). Material properties and fabrication parameters in selective laser sintering process. *Rapid Prototyping Journal* 3 (4): 129–136.

33 Wang, D., Yang, Y., Su, X., and Chen, Y. (2012). Study on energy input and its influences on single track, multi-track, and multi-layer in SLM. *International Journal of Advanced Manufacturing Technology* 58 (9–12): 1189–1199.

34 Wang, Y., Bergstrom, J., and Burman, C. (2006). Characterization of an iron-based laser sintered material. *Journal of Materials Processing Technology* 172 (1): 77–87.

35 Khaing, M.W., Fuh, J.Y.H., and Lu, L. (2001). Direct metal laser sintering for rapid tooling: processing and characterisation of EOS parts. *Journal of Materials Processing Technology* 113 (1–3): 269–272.

36 Simchi, A. and Pohl, H. (2003). Effects of laser sintering processing parameters on the microstructure and densification of iron powder. *Materials Science and Engineering A.* 359 (1–2): 119–128.

37 Teknologian kehittamiskeskus TEKES (1997). *Rapid Prototyping*. Mallien, prototyyppien ja tyokalujen pikavalmistus, 75. 52/973 Helsinki: Tekes-julkaisu.

38 Mercelis, P. and Kruth, J. (2006). Residual stresses in selective laser sintering and selective laser melting. *Rapid Prototyping Journal* 12 (5): 254–265.

39 Liao, Y. and Chiu, Y. (2001). A new slicing procedure for rapid prototyping systems. *International Journal of Advanced Manufacturing Technology* 18 (8): 579–585.
40 Kulkarni, P. and Dutta, D. (1996). An accurate slicing procedure for layered manufacturing. *Computer Aided Design* 28 (9): 683–697.
41 Zhou, M.Y. (2004). Adaptive slicing of functionally graded material objects for rapid prototyping. *International Journal of Advanced Manufacturing Technology* 24 (5–6): 345–352.
42 Dolenc, A. and Mäkelä, I. (1994). Slicing procedures for layered manufacturing techniques. *Computer-Aided Design* 26 (2): 119–126.
43 Hope, R.L., Roth, R.N., and Jacobs, P.A. (1997). Adaptive slicing with sloping layer surfaces. *Rapid Prototyping Journal* 3 (3): 89–98.
44 Ma, W. and He, P. (1999). Adaptive slicing and selective hatching strategy for layered manufacturing. *Journal of Materials Processing Technology* 89–90: 191–197.
45 Paul, B.K. and Baskaran, S. (1996). Issues in fabricating manufacturing tooling using powder based additive freeform fabrication. *Journal of Materials Processing Technology* 61 (1–2): 168–172.
46 Li, R., Liu, J., Shi, Y. et al. (2009). 316L stainless steel with gradient porosity fabricated by selective laser melting. *Journal of Materials Engineering and Performance* 19 (5): 666–671.
47 Tolochko, N.K., Arshinov, M.K., Gusarov, A.V. et al. (2003). Mechanisms of selective laser sintering and heat transfer in Ti powder. *Rapid Prototyping Journal* 9 (5): 314–326.
48 Shiomi, M., Yoshidome, A., Abe, F., and Osakada, K. (1999). Finite element analysis of melting and solidifying processes in laser rapid prototyping of metallic powders. *International Journal of Machine Tools and Manufacture* 39 (2): 237–252.
49 Senthilkumaran, K., Pander, P.M., and Rao, P.V.M. (2009). Influence of building strategies on the accuracy of parts in selective laser sintering. *Materials and Design* 30 (8): 2946–2954. [e-journal] From: http://www.sciencedirect.com/.
50 Atre, S.V. and German, R.M. (2001). *Materials for SLS. LIA Handbook of Laser Material Processing*, 556–557. Laser Institute of America: Magnolia Publishing, Inc.
51 Steen, W.M. (2003). *Laser Material Processing*, 3e, 48. London: Springer-Verlag.
52 Kruth, J.P., Wang, X., Laoui, T., and Froyen, L. (2003). Lasers and materials in selective laser sintering. *Assembly Automation* 23 (4): 357–371. [e-journal] From: http://www.emeraldnight.com/0144-5154.htm.
53 Seitz, S., van de Crommert, S., Esser, K.K., and McAlea, K. (1997). Materials and applications for the selective laser sintering process. In: *Proceedings of the 7th International Conference on Rapid Prototyping*, 23–33. SPIE.
54 Tolochko, N.K., Mozzharov, S.E., Yadroitsev, I.A. et al. (2004). Balling processes during selective laser treatment of powders. *Rapid Prototyping Journal* 10 (2): 78–87.
55 Kruth, J.-P., Froyen, L., Morren, B., and Bonse, J.E. (1997). Selective laser sintering of WC-Co 'hard metal' parts. In: *Proceedings of the 8th International Conference on Production Engineering (ICPE-8), Rapid Product Development*, 149–156. London: Chapman and Hall. ISBN: 0-412-81160X.
56 Kruth, J.-P. et al. (1996). Basic powder metallurgical aspects in selective metal powder sintering. *Annals of the CIRP* 45 (1): 183–186.
57 Laoui, T., Bonse, J., Froyen, L., and Kruth, J.-P. (1998). Influence of powder parameters on selective laser sintering of tungsten carbide-cobalt. In: *Proceedings of the 7th European Conference on Rapid Prototyping and Manufacturing*, Aachen, Germany (7–9 July 1998), 271–279.
58 Gu, D., Hagedon, Y.C., Meiners, W. et al. (2012). Densification behavior, microstructure evolution, and wear performance of selective laser melting processed commercially pure titanium. *Acta Materialia* 60: 3849–3860.

59 Kruth, J.P., Badrossamay, M., Yasa, E. et al. (2000). Part and material properties in selective laser melting of metals. *Proceedings of the 16th International Symposium on Electromachining (ISEM – XVI)*, Shanghai, China (19–23 April 2010).

60 Yadroitsev, I., Pavlov, M., Bertand, Ph., and Smurov, I. (2009). Mechanical properties of samples fabricated by selective laser melting. *14th European Meeting Prototyping & Rapid Manufacturing*, Paris (France).

61 Hague, R., Mansourb, S., and Saleha, N. (2004). Material and design considerations for rapid manufacturing. *International Journal of Production Research* 42 (22): 4691–4708.

Index

a

abrasive electro-discharge grinding (AEDG) 386, 387
abrasive flow machining 412
abrasive wire electro-discharge machining (AWEDM) process 386–388
activation temperature 329
active drawbeads
  control techniques
    closed-loop control system 8, 9
    open loop control system 8
  optimal path 9
  use of 7–8
active mode-locking 461
acusto-optic switches 460
additive manufacturing technology 473, 474
adiabatic heating 211
ADOC. *see* axial depth of cut (ADOC)
advanced and ultrahigh strength steels (A/UHSSs) 239
AEDG. *see* abrasive electro-discharge grinding (AEDG)
"93A46 H9V601W," 344
AlSi10Mg 484
aluminum 484
amorphous polymers 483
AMOTEC (Asahi molding technology with CO_2) 125
and methods 245–254
ankle prosthesis 58
arc melting drop/suction casting method 142–143
arcs pulses 376
asymmetric incremental sheet forming (AISF) process 52
atomization 72
A/UHSSs. *see* advanced and ultrahigh strength steels (A/UHSSs)
AWEDM. *see* abrasive wire electro-discharge machining (AWEDM) process
μ-AWM. *see* micro abrasive waterjet machining (μ-AWM)
axial depth of cut (ADOC) 339

b

back-propagation neural network (BPNN) 123, 124
ball burnishing 467–468
beam parameter product (BPP) 464, 466
beam quality (M^2), laser machining parameters 435–436
beam size parameters 481
Beer–Lambert law 429
binder hold down force (BHDF) 4
biopsy micro forceps (BMF) jaw 59, 60
blankholder 3–6
  BHF or BHDF 4
  deformation sequences 4, 5
  tools 3–4
blankholder force (BHF) 4
blanking 330–331
BMF. *see* biopsy micro forceps (BMF) jaw
BMGs. *see* bulk metallic glasses (BMGs)
boundary element method (BEM) 129
BPNN. *see* back-propagation neural network (BPNN)
Brinksmeier's classification 152
bulk glass formers 137

Modern Manufacturing Processes, First Edition. Edited by Muammer Koç and Tuğrul Özel.

bulk metallic glasses (BMGs) 70
arc melting drop/suction casting method 142–143
cap casting method 144–145
centrifugal casting method 145
copper mold casting method 144, 145
high-pressure die casting method 143–144
mechanical properties and usage 139–140
metal foaming method 146, 147
water-quenching method 141–142

c
Cadmould 84
cap casting method 144–145
carbide micro-tools 410
carbon-fiber–reinforced plastic carbon fiber-reinforced plastic (CFRP) 432
cavity type EDM. *see* die-sinking EDM
centrifugal atomization 71
centrifugal casting method 145, 146
ceramic matrix composite (CMC) parts 84
ceramic microsystems 85
ceramics 432–433, 451
chaotic dynamic cutting processes 420–421
chemical fabrication techniques 72
chemical vapor deposition (CVD) technique 38, 294, 307, 433
chip formation, in grinding 353
μ-CIM. *see* micro ceramic injection molding (μ-CIM)
"closed-loop control" system 8, 9
CMC. *see* ceramic matrix composite (CMC) parts
CO_2-assisted plasticization 125
coatings 294
cobalt chrome 485
Cockroft and Latham's criterion 320–321
cold compaction 73
cold die powder compaction manufacturing process 65, 66
cold forming
advantages 188
Oxley model 317
for semifinished products 35
sliding sheet drawing 324
V-shaped cold and hot-formed tubes 244
cold isostatic compaction 83
cold powder compaction
compacted copper powder 77
compacting pressures 75, 76
density dependence of properties 75, 76
die compaction 74–75, 77–78
double action compaction 78
double-action pressing 77
floating die 77
green powder compacts 78
post-compaction operations 75
pressure and density, mathematical relation between 76
single-action pressing 77
uniaxial compaction 75
compacted parts 75
compacted powder 75
compacting pressures 75, 76
composites 432
compound machining 386
continuous wave (CW) lasers 434
Continuum Theory of Plasticity 24
controllable process variables 223
control-volume-based finite-element-method (CVFEM) 129
conventional mechanical machining technique 90–91
conventional wheels, in grinding 344–345
copper mold casting method 144, 145
Coulomb friction model 324
cranial implant 58
critical damage value 323
critical-to-quality factors (CTQ) 108
crystalline polymers 483
cubic boron nitride (CBN) 293–294, 346, 410
"cutting stream," 159
CVD. *see* chemical vapor deposition (CVD) technique

d
DBRF. *see* drawbead restraining force (DBRF)
3D CAD model 473, 474
debinding 82
deep drawing 47, 204
deep drawing or casting 15, 32
deep reactive-ion etching (DRIE) 119–120
"define, measure, analyze, improve, and control" (DMAIC) procedure 108
degree of freedom (DOF) 12
densification 67, 73
design of experiments (DOE) 122, 123

diamond–based micro-tools 410
diamond grinding wheels 363
diamond-like carbon (DLC) 410
diamond turning. *see* microturning
die and investment casting 83
die compaction 74–75, 77
die holder 187
dielectric circulation system 375
die-sinking EDM 377–378
diffusion tool wear model 329
direct laser machining 458–462
direct selective laser sintering
  liquid-phase sintering 476–477
  selective laser melting 477
  solid-state sintering 476
disc 381
discharging unit. *see* pulse generator
3D laser milling. *see* multipass laser milling
DMAIC. *see* "define, measure, analyze, improve, and control" (DMAIC) procedure
3D model 128–129
DOE. *see* design of experiments (DOE)
DOF. *see* degree of freedom (DOF)
double-action pressing 77
drawbead restraining force (DBRF) 5, 6–7
drawbeads 3–6
  active 7–9
  bending and re-bending deformations 5
  clamping (or locking) phase 4
  deformation damages 5, 6, 8
  drawing (or pulling) phase 4
  and EMBR, comparison 11
  sidewall curl occurrences 8
  types 7
DRIE. *see* deep reactive-ion etching (DRIE)
dry EDM 382–383
dual phase (DP) steels 241, 243
ductile fracture
  criteria
    Cockroft and Latham 320–321
    effective strain 320
    McClintock 321–322
    other criteria 322
  critical damage value 323
  prediction of 319
ductile fracture criteria 320–322
ductile-regime finish machining. *see* microgrinding
ductile-regime grinding. *see* microgrinding

e

EBL. *see* electron beam lithography (EBL)
ECAFM. *see* electrochemical-aided abrasive flow machining (ECAFM)
ECM. *see* electro-chemical micromachining (ECM)
eddy current 253
EDM. *see* electro-discharge machining (EDM)
elastic recovery 414–415
elastomechanical longitudinal ultrasonic vibrations 361
electrical/chemical micromanufacturing processes
  ECM 169–170, 171
  micro WEDM, μ-WEDM 168–169
electro-assisted conditioning methods, of grinding wheel 351
electrochemical-aided abrasive flow machining (ECAFM) 412
electrochemical machining process (ECM) 389–391
electro-chemical micromachining (ECM) 169–170, 171
electro-deposition 71
electro-discharge machining (EDM) 295
  advantages 369
  components 371–372
    dielectric circulation system/flushing unit 375
    pulse generators/discharging unit 372–374
    servo control system/gap control unit 374–375
  compound and hybrid machining processes
    AEDG 386, 387
    AWEDM 386–388
    deionized water 390
    ECM-combined process 389–391
    LAEDM 392, 394
    MFEDM 391–393
    micro-EDM 388–389
    PMEDM 388, 389
    vibration-assisted EDM 388–389
  die-sinking 377–378
  drilling 380–381
  electrical pulses 375–376
  historical roots 369–370
  material removal mechanism 370–371

electro-discharge machining (EDM) (*Contd.*)
  micro-EDM 394–395
  milling 380
  near-dry 382–383
  parameters 376–377
  physics 370
  planetary 383–384
  principle 370–371
  RDE-EDM 381–382
  reverse 384–385
  WEDM 378–380
electroforming 118–119
electroforming method 118–119
electrohydraulic forming
  methodology/basics 273
  pressure chamber 275
  pressure pulse and exploding wire technique 273–275
  typical applications and equipment 275
  wave shaper 275
electrolytic deposition 70
electrolytic inprocess dressing (ELID) 351–352
electromagnetic blank restraining (EMBR) 11
electromagnetic forming and magnetic pulsed welding
  coil (and field shaper) 270–272
  generation of current pulse 269–270
  methodology/basics 267–269
  optional female forming die 272
  other applications 272–273
  typical applications and equipment 272
electromechanical theory 371
electron beam-based micromanufacturing processes 170–172
  electron beam drilling 172
  electron beam welding 172–173
electron beam drilling 172
electron beam lithography (EBL) 119
electron beam welding 172–173
electro optic Q-switching 460
El-Magd's model 316, 318–319
EMBR. *see* electromagnetic blank restraining (EMBR)
energy density (J/m^2) 437
expansion injection molding 126
experimental modal analyses (EMA) 419–420
explosive forming
  methodology/basics 275–278
  setup and equipment 278
  typical applications 278

f
FEA. *see* finite element analysis (FEA)
femtosecond (fs) range lasers 461
finite element analysis (FEA) 24
  in hot stamping 257–258
finite element method (FEM) 51, 129
finite volume method (FVM) 129
FLC. *see* forming limit curves (FLC)
FLD. *see* forming limit diagram (FLD)
floating die 77
fluence 437
"fluidity," 128
fluid usage, in grinding processes 356
flushing unit. *see* dielectric circulation system
focused energy beam machining 91
forming limit curves (FLC) 30
forming limit diagram (FLD) 30, 52–53, 191, 204–205

g
gas atomization 74
gas phase 70
geometric parameters
  hatch spacing 479
  layer thickness 480
  scanning strategy 479–480
GFA. *see* glass forming ability (GFA)
glass 433
glass forming ability (GFA) 137
green body. *see* compacted powder
green compact 166
green parts. *see* compacted parts
grinding 343
  bond materials 345–346
  environmentally friendly 356–358
  force and energy 355
  high-efficiency deep grinding 359, 360
  laser-assisted 466
  thermal damages 355–356
  ultrasonic-assisted dressing 363–365
  ultrasonic-assisted machining 359–363
  wheels
    abrasive grit materials 343
    conditioning processes 346–347
    conventional 344–345
    dressing 466–467

profiling, sharpening and cleaning 347–352
structuring 353–355
superabrasive 345

h

hard boring 302
hard broaching process 302
hardening phenomena 194
hard machining
characteristics of 302–304
cutting tools
coating application and materials 307
materials 305–306
tool microgeometry 307–308
tool wear and failure 308
mechanics of 304–305
operations
hard boring 302
hard broaching process 302
in hard milling 302
in hard turning 301–302
saw-tooth shape chip formation 304
surface quality and integrity
microhardness 310
microstructural alterations 309–310
residual stresses 311
surface defects 309
surface roughness 310–311
hard milling 302
hard turning 301–302
hatch spacing 479
heat affected zone (HAZ) 261
heat-assisted processes
direct laser machining 458–462
laser-assisted machining (*see* laser-assisted machining (LAM) operations)
plasma arc welding 452–454
helical drilling 440
high-efficiency deep grinding (HEDG) 359, 360
high-pressure die casting method 143–144
high-speed forming (HSF)
advantages 266
disadvantages 266–267
electrohydraulic forming 273–275
electromagnetic forming 267–273
explosive forming 275–278
magnetic pulsed welding 267–273
material characterization 279
variations 267
velocity/metrology measurements 279–283
high-speed machining (HSM)
applications and future trends 297–298
machine tools 290–292
mold cavity 297
motor block by milling 297
overview 289–290
processes and capabilities 290
sculptured surfaces 291
tools for
coating application and materials 294
cutting tool materials 292–294
tool microgeometry 294–296
tool wear and failure 296–297
vertical machine tool 291
high-speed milling 296, 298
hot consolidation processes 74
hot pressing process 79
hot runner technology 97
hot spot 251
hot stamping
advantages and disadvantages 243–245
automotive parts and potentials 241–243
coating and lubrication 248–251
conduction heating 252–253
conventional chamber furnace (radiation) 252
cooling for 254–255
design and optimization 256
FEA 257–258
heating 251–252
induction heating 253–254
materials 247–248
modeling and analysis 255–256, 257
process
types 245–247
process control 255, 256
process description
and methods 245–254
and motivation 240
research and development trends and needs 258–262
steel grades 241
tools 251
HRC (hardness Rockwell C) 301
HSF. *see* high-speed forming (HSF)
HSM. *see* high-speed machining (HSM)

hybrid-constrained multiobjective genetic algorithm 24
hybrid isothermal/thermomechanical approach 225
hybrid machining process 386
hybrid manufacturing 151, 152, 178
hydroforming
  classification 16
    integrated operations 19–23
    sheet hydroforming 17–19
    tube hydroforming 17
  forming limits 29–30
    buckling and wrinkling 30
    necking and bursting 29–30
  forming tools 37–38
  machine systems 38–39
  process development and design
    part design 33–35
    semifinished products 35–37
  process parameters 23–29
    sheet hydroforming 26–29
    tube hydroforming 24–26
  tribology 31–32, 40
hydrometallurgical processing 72
"hysteresis loss," 253

i

Inco 718 485
Inconel 718 451
incremental sheet forming (ISF)
  accuracy 55, 57
  case study 59
  deformation mechanisms 53–54
  formability
    FLD 52–53
    forces 54
    ISF deformation mechanisms 53–54
  functioning principle 47
  future trends
    application fields 58
    materials 58–59
    sustainability 59
  materials 52
  process cycle 51
  process parameters 55, 56
    bidirectional 55
    spiral 55
    unidirectional 55
  simulations 57–58
    historical timeline 58
  stretch forming 50–51
  variants
    roboforming 51
    SPIF 49
    stretch forming with ISF 50–51
    TPIF 49–50
    warm ISF 50
  warm ISF 50
incremental sheet metal forming (ISMF) 52
indirect selective laser sintering 475–476
inert gas atomization 74
injection compression molding
  advantages 106
  cycle of 106
  mold base 107
injection molding
  developments of 93
  equipment 94–95
  injection compression molding 106–107
  machine 91
    and cycle 91–92
    piston type 95
    screw plunger hybrid 95
    screw type 95
  materials 93–94
  microparts 124–127, 128
  microstructured parts 116–124, 127–128
  mold manufacturing technology 103–105
  polymer materials 94
  precision parts 105, 127
  processing 95–103
  processing parameters 107–109
  simulation of 127–131
  size scales 91
  structure of 92–93
  thin-wall parts 109–115, 127
injection molding, process study 126–127
in-mold assembly technology 99–101, 102
in-process monitoring of process parameters 122
ion beam machining (IBM) 295
ISF. *see* incremental sheet forming (ISF)
isothermal heating 189

j

jar mill 71
Johnson–Cook (JC) models 207
Johnson–Cook's model 316, 317

k

Kalman filter 411
Keeler–Goodwin diagram 52
knee implant 58

l

LAEDM. *see* laser-assisted electro-discharge machining (LAEDM)
Lagrangian or Eulerian algorithms 129
laser ablation 175
laser-assisted electro-discharge machining (LAEDM) 392, 394
laser-assisted machining (LAM) operations
  active medium 455–456
  aeronautical heat-resistant alloys 451
  analysis
    analytical models 463
    energy source 463–466
    experimental model 462
    numerical models 463
    thermal study 462
    variable to control 462
  applications
    abrasive-free water jet process 469
    ball burnishing 467–468
    coating processes 468
    dressing 466–467
    grinding 466
    hybrid process 468
    incremental forming 469
    metal forming processes 468
    sheet forming 468
    truing 466
  ceramic 451
  components 455
  direct laser machining 458–462
  energy absorption ratios 454–455
  excitation mechanism 455
  heat-assisted processes 452
  Inconel 718 451
  orientation 456
  plasma arc welding 452–454
  surface milling 457
laser-assisted microforming
  laser ablation 175
  laser cutting 175–176
  laser drilling 178, 179
  laser etching 178
  laser micro welding 176–177
    of polymers 177–178
  other laser-assisted microforming processes 178
laser-assisted microforming processes 178
laser beam 353, 354, 464, 466
  selective laser sintering process 474
laser beam machining 178
laser conditioning, of grinding wheel 350
laser cutting 175–176
laser drilling 178, 179
laser etching 178
laser machining technology
  continuous wave 427–428
  cutting
    ablation 441–442
    advantages 440
    melt 440–441
  definition 427
  drilling
    applications 437
    approaches 438
    with relative motion between laser spot and workpiece 439–440
    without relative motion between laser spot and workpiece 438–439
  laser-based material processing 427
  laser–material interaction 429
    ablation 429–430
    Beer–Lambert law 429
  materials
    ceramics 432–433
    composites 432
    metals and alloys 431
    polymers 431–432
    properties 428
    removal process 428
  milling
    beam 442
    multipass approach 443–444
    single-pass 443
    single-shot 443
    vaporization 442
  parameters 428
    beam quality 435–436
    power 436–437
    pulse duration 434–435
    pulse overlap 437
    repetition rate 435
    wavelength 435
  pulsed 428

laser microwelding 176–177
  of polymers 177–178
laser parameters 480
  beam size 481
  laser power 481
  scanning speed 481
lasers 475
  power 481
laser scribing. *see* single-pass laser milling
layer thickness 480
Lazarenko Circuit 369
LCS. *see* low carbon steel (LCS)
Levi and von Mises correlations 24
limiting drawing ratio 191
linear drive motors and control systems 407
liquid atomization 70
liquid phase (chemical precipitation) 70
liquid-phase sintering (LPS) 476–477
lithography technique involvement 151–152
long-pulsed lasers 428
low carbon steel (LCS) 219
lubricant 326–327

m

machining 83
"macro-scale" ceramic components 85
macroscale deburring processes 412
Maekawa model 316, 317
magnetic field–assisted electro-discharge machining (MFEDM) 391–393
magnetorheological-assisted abrasive flow finishing (MRAFF) 412
manufacturing approach (top-down *vs.* bottom-up) 150–151
maraging steel MS1 485
market projections 149–150
martensitic steels (MSs) 241
material forming processes 65
material modeling
  and applications
    El-Magd's model 318–319
    Johnson–Cook's model 317
    Maekawa model 317
    Oxley model 317
  and characterization 315
  of contact, friction, and wear 323–329
  in 2D and 3D simulations
    blanking 330–331
    die wear in 2D stamping FEM model 332–333, 334
    forward extrusion 331–332
    micro processes and simulation 333–339
    orthogonal cutting 329–330
  failure models
    critical damage value 323
    ductile fracture criteria 320–322
    prediction of ductile fracture 319
  parameters influencing friction
    lubricants 326
    process parameters 326
    workpiece and tool materials 326
McClintock's criteria 321–322
mechanical alloying 71
mechanical comminution
  attrition 70
  compression 70
  impaction 70
  method of powder production 71
  shearing fractures 70
  solid phase dispersion strengthening (mechanical alloying) 70–71
mechanical disintegration, principle of 71
mechanical forming operations
  bending 20–21
  collar forming 21, 22
  deep drawing 19–20
  integrated flattening (crushing) 21
  joining processes 21–22, 23
  piercing and cutting 23
    inward–outward piercing 23
  stamping operations 21
mechanical micromachining 117–118, 120
  and LIGA, combination of 120
mechanical processes
  micro abrasive waterjet machining 159–161
  microdrilling 156
  microgrinding 157–158
  micromachining processes 155
  micromilling 155
  micropolising 158
  microturning 156–157
mechanical production techniques 71–72
melt atomization 71
melt cutting 440–441
melt spinning method 141
Membrane Theory 24

MEMS. *see* micro-electro mechanical system (MEMS)
MET. *see* micro engineering technologies (MET)
metal foaming method 146, 147
metal matrix composite (MMC) 84
metal powder production techniques 69–70
metals and alloys 431
metals and ceramics
  selective laser sintering materials 484–486
metal working fluids (MWFs) 409
micro abrasive waterjet machining (μ-AWM) 159–161
micro ceramic injection molding (μ-CIM) 167
micro deep-drawing 163, 164
microdrilling 156
micro-EDM process 91, 132
  assistance of vibration 388–389
  and electrochemical machining process 389–391
  and LAEDM 392, 394
  and MFEDM 391–393
  PMEDM 388, 389
micro-efficiency deep grinding (EDG) processes 380
micro-electro mechanical system (MEMS) 89, 149, 395
micro embossing 165
micro engineering technologies (MET) 149
micro etching. *see* laser etching
micro extrusion 162–163
microfabrication systems 405
microforging 161
microforming processes
  micro deep-drawing 163, 164
  micro extrusion 162–163
  microforging 161
  micro hydroforming 165–166
  micro injection molding 166–167
  micro stamping 164–165
  powder compaction 167–168
  progressive microforming 163, 164
microgrinding 157–158
microhardness 310
micro hydroforming 165–166
micro injection molding 166–167
micromachining processes 155
micromanufacturing
  categorization 154
  classification
    Brinksmeier's classification 152
    effect of process 152–154
    lithography technique involvement 151–152
    manufacturing approach (top-down *vs.* bottom-up) 150–151
  electrical/chemical micromanufacturing processes 168–170
  electron beam-based micromanufacturing processes 170–172
  laser-assisted microforming 173–179
  market projections 149–150
  mechanical processes 154–161
  microforming processes 161–168
  trends in 150, 151
micromilling 91, 155
micromilling operations
  challenges 405
  force model 412–413
    elastic recovery 414–415
    minimum uncut chip thickness 413–414
    ploughing dominant regime 416–419
    shearing dominant regime 416
  macromachining *vs.* 403, 404
  Merchant and Oxley's sharp edge theorems 403
  microcomponents 403
  micro cutting of steel 404
  microfabrication techniques 404, 405
  micromechanical cutting processes 405
  miniaturization 403
  tools
    burr removal 412
    capabilities 405
    coolant and lubricant delivery 408–410
    drives and guides 407
    measurements and monitoring 411–412
    micro-end mills 410–411
    microproducts 405
    minimum quantity lubricant (MQL) 406
    spindles 406–407
    ultraprecision 407–408
  tool tip dynamics 419–422
"micro nano manufacturing technologies (MNMT)" 149
microparts 128
  CO_2-assisted plasticization 125
  injection molding, process study 126–127
  X-melt process 126

micropolising 158
micro processes and simulation
  friction 336
  microextrusion modeling 337, 339
  micromachining simulation 337, 338, 339
  size effects
    density effects 334
    process parameters 335
    shape effects 334–335
    structural effects 335
micro stamping 164–165
micro stripping. *see* laser etching
microstructured parts 127–128
  materials 116, 117
  molding machine and mold design 121
  mold manufacturing technologies
    deep reactive-ion etching 119–120
    electroforming 118–119
    electron beam lithography 119
    mechanical micromachining 117–118, 120
  process parameters
    design of experiments approach 122, 123
    effect of 121–122
    in-process monitoring 122
    multiple quality optimization 122–124
microstructures or micro systems technology (MST) 149
microturning 156–157
micro wire electro discharge machining (micro WEDM, ì-WEDM) 168–169
"midplane" model 128
milling. *see* mechanical comminution
miniature injection molding processes 129–130
miniature systems 403
minimum quantity lubrication (MQL) 305, 356–358, 406, 409
minimum uncut chip thickness (MUCT) 405, 413–414
mixed material model 335
MMC. *see* metal matrix composite (MMC)
mode-locking technology 461
Moldflow 84, 132
mold manufacturing technology
  etching 104–105
  laser micromachining 104
  LIGA/UV-LIGA 104
  mechanical micromachining 104
  micro EDM 104
mold visualization technology 101–103, 103–105
monocrystalline diamond (MCD) 294
MPF. *see* multipoint forming (MPF) technology
MRAFF. *see* magnetorheological-assisted abrasive flow finishing (MRAFF)
MUCT. *see* minimum uncut chip thickness (MUCT)
multipass laser milling 443–444
multiple quality optimization of process parameters 122–124
multipoint forming (MPF) technology 12, 13

n
nano-electro-machining (nano-EM) 395
nano-electro mechanical systems (NEMS) 149
nanosecond pulse lasers 428
nanotechnology-based coatings 260
natural grits, grinding wheels 343
near-dry EDM 382–383
near dry grinding 357
NEMS. *see* nano-electro mechanical systems (NEMS)
nickel chrome 485
nitrogen cooling 408
nonisothermal heating 189

o
"open loop control" system 8
open voltages, EDM process 376
orthogonal cutting 329–330
orthogonal experimental method 113–114
Oxley model 316, 317

p
palate implant 58
parameters influencing friction 326
passive mode-locking 461
peak power 437
pendant-drop melt extraction techniques 141
percussion drilling 438–439
phase-breakdown or top-down approach 70
phase buildup or bottom up approach 70
phase transformation methods 70
physical vapor deposition (PVD) technique 38, 294, 307
PID. *see* proportional-integral-derivative (PID) type control
PIM. *see* powder injection molding (PIM)
piston and anvil technique 141

planetary EDM 383–384
plasma arc welding (PAW) 452–454
plastic injection molding 83
ploughing dominant regime 416–419
PM. *see* powder metallurgy (PM) process
PMEDM. *see* powder-mixed electro-discharge machining (PMEDM)
polycrystalline diamond (PCD) 293–294, 410–411
polymers 431–432
  selective laser sintering materials 483
powder compaction 167–168
powder forging 79, 80
powder forming
  consolidation techniques
    cold powder compaction 74–79
    hot pressing process 79
    powder forging 79
  powder injection molding 82–84
  powder production 69–73
  reasons for using PM route 67–69
  sintering 79–82
  technologies 74
powder injection molding (PIM)
  advantage of shape complexity 82
  injection stage 84
  MIM 83–84
  MMC or CMC 84
  powder/binder mix 82–83, 84
  Ti-MIM field 84
powder metallurgy (PM) process 65
  industrial sectors 66
  porous bearings 68
  product cost effectiveness 67
  rapid solidification processing 68–69
  reasons for using 67, 68
  T15 tool 67–68
powder-mixed electro-discharge machining (PMEDM) 388, 389
powder production
  atomization 72
  chemical fabrication techniques 72
  mechanical production techniques 71–72
  metal powder production techniques 69–70
  microstructure and morphology 70
  phase-breakdown or top-down approach 70
  phase buildup or bottom up approach 70
  precipitation from solution 72–73
power, laser machining parameters 436–437
"power to volume" ratio, of spindles 406
precision injection molding machines 96
precision parts 105, 127
precision/ultraprecision grinding. *see* microgrinding
presses 187
press hardening (die quenching) 245
press tools 11
pressure pulse and exploding wire technique 273–275
process description
  and methods 245–254
  and motivation 240
progressive microforming 163, 164
proportional-integral-derivative (PID) type control 255
pulsed laser ablation in liquid (PLAL) 175
pulse duration, laser machining parameters 434–435
pulse energy (J) 436–437
pulse generator
  RC-type 372–374
  transistor-type 374
pulse overlap, laser machining parameters 437
punch-and-die or stamping die 187
PVD. *see* physical vapor deposition (PVD) technique
pyrolysis. *see* thermal decomposition process

q

Q-switching technology 459, 460
quenching 256

r

rapid burn-off zones 81
rapid solidification processing (RSP) 68–69
rapid thermal cycling technology 96–97, 98
rapid thermal response (RTR) molding
  Bayer polycarbonate CD2000 110–111
  experimental flow patterns 111, 112, 113, 114
  experimental mold surface temperature 110, 112
  setup for 110, 111
  side insulation elements 110, 111
RC-type pulse generator 372–374
real sparks, EDM process 376
receptance coupling (RC) method 419

refractive index variation 108
reinforced polymers 483–484
repetition rate, laser machining parameters 435
residual stresses 311
resinoid bonds 345
reverse EDM 384–385
roboforming 51
roller-based guides 407
rotary disc electrode electrical discharge machining (RDE-EDM) 381–382
RSP. *see* rapid solidification processing (RSP)
RTR. *see* rapid thermal response (RTR) molding
"*r/t* ratio," 242

s

saturable absorbers 460
scalable filling 109–110
scanning speed 481
scanning strategy 479–480
Schley friction model or shear model 324–325
selective laser melting (SLM) 477
selective laser sintering (SLS) process 473
  additive manufacturing technology 473, 474
  basic components 474
  capabilities and limitations 486–488
  3D CAD model 473, 474
  laser beam 474
  lasers 475
  materials 482–483
    metals and ceramics 484–486
    polymers 483
    reinforced polymers 483–484
  mechanisms
    direct 476–477
    indirect 475–476
  parameters
    classification 478
    geometric 479–480
    laser 480–481
    material properties 481–482
    multitude of 478
self-lubricating bearings 68
semifinished products
  anisotropic forming behavior 35
  customized blank and tubular product 36–37
  high-pressure sheet forming 36
  tube hydroforming 35–36
servo control system/gap control unit 374–375
Shack–Hartmann wavefront sensor-based metrology system 108
shear fractures 242
shearing dominant regime 416
sheet hydroforming
  friction zones 31
  high-pressure sheet forming
    body component 19
    single blanks and double blanks 17–18
  hydromechanical deep drawing 17, 19
  lubricants for 31–32
  process parameters
    bursting pressure 27–28
    in hydromechanical deep drawing 28–29
    loads in high-pressure sheet forming 26, 27
    range of process controls 27–28
sheet metal forming. *see* stamping
short-circuit pulses 376
short-pulsed lasers 428
silicon 433
simple suction casting method 138
single-action pressing 77
single-crystal diamonds 411
single-pass laser milling 443
single point incremental forming (SPIF) 49, 57
single-pulse drilling 438
single-shot laser milling 443
sinker EDM. *see* die-sinking EDM
sintering 166
  high temperature sintering 81
  inadequate sintering 81
  ISO definition 79–80
  mixtures of hydrogen and nitrogen 82
  porosity or density 81
  temperatures for powder metal alloys and ceramics 80–81
sintering or hot extrusion 78
Six Sigma approach 108
slides, or side-actions 93
slip casting 83
slow machining process 169
sol–gel grits, microfractures in 343, 344
solid-state reduction process 72

solid-state sintering (SSS) 476
SPIF. *see* single point incremental forming (SPIF)
spin forming 193
spinning reflectors 460
splat cooling 137
splat quenching or piston 141
splitting (tearing) 241, 249
stainless steel GP1 alloy 484
stamping 187
  active drawbeads
    control techniques of 8–9
    use of 7–8
  blankholder 3–6
  DBRF 6–7
  drawbeads 3–6
  flexibility 11–12
  future challenges 12
  VBHF 10–11
stamping press 187
standard injection molding simulation
  3D model 128–129
  simulation example 129
stop 187
strain 316
strain rate 316
stress 316
stretch forming 227
  with ISF 50–51
stretching 204
superabrasive wheels 345–346
synthetic grits, grinding wheels 343

t

Taguchi method. *see* orthogonal experimental method
Taguchi screening procedure 108
tailor-welded blanks (TWBs) 12, 260, 262
temperature 316
Theory of Shells 24
thermal coupling 258
thermal damages, in grinding 355–356
thermal decomposition process 70, 72
thermoelectric theory 371
thermolysis. *see* thermal decomposition process
thermomechanical theory 371
thin-wall parts 127
  effect of processing parameters 112–115
  experimental design and results 114, 115
  levels of factors 114, 115
  rapid thermal response molding 110–112
  scalable filling 109–110
Ti6Al4V ELI 485
time-temperature transformation curves (TTT) 254
tin oxide (SnO_2) 73
titanium 485–486
tool edge radius 410
tool wear model
  abrasive and adhesive 328–329
  diffusion 329
TPIF. *see* two point incremental forming (TPIF)
transistor-type pulse generator 374
trepanning drilling 439–440
truing 466
T-Tool 353, 354
"tube bulge test," 233
tube hydroforming
  components, T or Y-shaped 17, 18
  external pressurization 17, 18
  finite element analysis 24
  friction zones 31
  inner pressurization 25–26
  with internal pressurization 17
  loads in 24
  lubricants for 31–32
  moment of bursting 25
  principle of 15, 16, 17
  process parameters 24–26
  production machine 39
  range of process controls 24, 25
tungsten carbide (WC) tools 410
TWBs. *see* tailor-welded blanks (TWBs)
twin fluid melt atomization 71
twin roller quenching technique 141
two point incremental forming (TPIF) 49–50, 57

u

UAAFM. *see* ultrasonic-assisted abrasive flow machining (UAAFM)
ultimate tensile strength (UTS) 241
ultraprecision machine tools 407–408
ultrashort laser pulses 434–435
ultrashort-pulsed lasers 428
ultrasonic-assisted abrasive flow machining (UAAFM) 412
ultrasonic-assisted dressing 363–365

ultrasonic-assisted grinding (UAG) 359–363
Ultrasonic vibration-assisted polishing method 158
uniform wall angle (UWA) geometries 54
UV-LIGA 91, 132

V

vacuum venting technology 97–99, 100, 101
vapor condensation 70
variable blankholder force (VBHF) 10–11
variotherm process 121
VBHF. *see* variable blankholder force (VBHF)
vibration-assisted EDM 388–389
volume EDM. *see* die-sinking EDM

W

warm forming 220
  advantages 188
  formability of materials 189, 191, 215
  nonisothermal warm forming setup 189
  recrystallization temperature 190
warm hydroforming
  warm hydromechanical deep drawing 230–231
  warm sheet hydroforming 220–230
  warm tube hydroforming 231–237
warm hydromechanical deep drawing (Warm HMD) 220, 221, 230–231
warm ISF 50
warm sheet hydroforming
  current/future trends 230–231
  with die 229–230
  implementation of 220–223
  material characterization 229
  optimization of parameters 229
  parameters in 223–228
  studies on 228–229
warm sheet hydroforming with die (Warm SHF-D) 220, 221, 227
warm stamping
  analyze techniques on formability 205–215
  benefits and usage areas 187–188
  effect of temperature
    and deformation on elasticity modulus 198–201
    on forming limit diagrams 204–205
    interrelation and strain rate 196–198
    on springback 201–204
    on strain hardening 194–196
  future directions 215–216
  lubrication, effects of 215
  and recent developments 188–194
warm tube hydroforming
  implementation of 231–232
  parameters in 232
  studies on 232–237
water atomization 74
water-quenching method 141–142
wavelength, laser machining parameters 435
wave shaper 275
WEDM. *see* wire electro discharge machining (WEDM) technology
windshield beam member 243
wire electro discharge machining (WEDM) technology 295, 378–380
work/strain hardening 194
wrinkling 241

X

X-melt process 126

Z

Zerilli–Armstrong models 207